Volume 84

ASPHALT PAVING TECHNOLOGY 2015

JOURNAL OF THE ASSOCIATION OF ASPHALT PAVING TECHNOLOGISTS

Portland, Oregon
March 8–11, 2015

AAPT
Association of Asphalt Paving Technologists

DES*tech* Publications, Inc.

Asphalt Paving Technology 2015—Volume 84

Produced by:
DEStech Publications, Inc.
439 North Duke Street
Lancaster, Pennsylvania 17602 U.S.A.

Printed in the United States of America
10 9 8 7 6 5 4 3 2 1

Main entry under title:
Asphalt Paving Technology 2015—Volume 84

ISSN No. 0270-2932

Editor: Rebecca S. McDaniel

Managing Editor: Eileen P. Soler

Executive Editor: R. Michael Anderson

Guest Editors: Brian Prowell
Geoffrey Rowe
Randy West
Gaylon Baumgardner
Jo Daniel
Ray Bonaquist
Y. Richard Kim (RMPD)
Reynaldo Roque (RMPD)

Journals of the Association were printed for the meetings listed below and may be obtained from the AAPT Office.

1974	Williamsburg, Vol. 43	15.00	1999	Chicago, Vol. 68	75.00
1975	Phoenix, Vol. 44	20.00	1999	75th Ann, Volume 68A	25.00
1977	San Antonio, Vol. 46	25.00	2000	Reno, Vol. 69	75.00
1982	Kansas City, Vol. 51	40.00	2001	Clearwater, Vol. 70	80.00
1983	Atlanta, Vol. 52	45.00	2002	Colorado Springs, Vol. 71	80.00
1984	Scottsdale, Vol. 53	50.00	2003	Lexington, Vol. 72	80.00
1986	Clearwater, Vol. 55	50.00	2004	Baton Rouge, Vol. 73	90.00
1988	Williamsburg, Vol. 57	55.00	2005	Long Beach, Vol.74	90.00
1989	Nashville, Vol. 58	55.00	2006	Savannah, Vol.75	100.00
1990	Albuquerque, Vol. 59	55.00	2007	San Antonio, Vol. 76	100.00
1991	Seattle, Vol. 60	60.00	2008	Philadelphia, Vol. 77	125.00
1992	Charleston, Vol. 61	60.00	2009	Minneapolis, Vol. 78	150.00
1993	Austin, Vol. 62	60.00	2010	Sacramento, Vol. 79	150.00
1994	St. Louis, Vol. 63	60.00	2011	Tampa, Vol. 80	180.00
1995	Portland, Vol. 64	70.00	2012	Austin, Vol. 81	180.00
1996	Baltimore, Vol. 65	70.00	2013	Denver, Vol. 82	200.00
1997	Salt Lake City, Vol. 66	75.00	2014	Atlanta, Vol. 83	200.00
1998	Boston, Vol. 67	75.00	2015	Portland, Vol. 84	200.00

Reprints of Articles included in Volumes which are out-of-print may be obtained from the AAPT Office.

ISSN 0270-2932

Printed in the United States of America by DEStech, Lancaster, PA, US

ASSOCIATION OF ASPHALT PAVING TECHNOLOGISTS 2015

W.J. Emmons Annual Award

The Board of Directors announces the recipients of the W.J. Emmons Annual Award of the Association of Asphalt Paving Technologists for the best paper presented at the Annual Meeting in Atlanta, Georgia held March 16-19, 2014, to be:

Winner:

Balanced RAP/RAS Mix Design System for Project-Specific Service Conditions

by Fujie Zhou[a], Sheng Hu[a], Tom Scullion[a], and Robert Lee[b]

[a] Texas A&M Transportation Institute
[b] Texas DOT

Runner Up:

Development of a Binder Fracture Test to Determine Fracture Energy

Tianying Niu[a], Reynaldo Roque[b], and George Loop[b]

[a] Mayo Clinic
[b] University of Florida

AAPT Scholarship Fund

The Board of Directors announces the 2015 winners of the AAPT Scholarship to be:

Rostyslav Shamborovskyy, Rutgers University
Michael Vrtis, Auburn University

AAPT Board of Directors 2014-2015

From left:
1st Row: Jo Sias Daniel, Geoffrey Rowe, Brian Prowell, Randy West, Gaylon Baumgardner

2nd Row: Rebecca McDaniel, Jeyna Rowe, Marcia Prowell, Ronda West, Ramon Bonaquist, Maria Bonaquist

2015-2016 Board of Directors

President: Geoffrey Rowe

First Vice-President: Randy West

Second Vice-President: Gaylon Baumgardner

Director-at-Large: Audrey Copeland

Director-at-Large: Matthew Corrigan

Past President: Ramon Bonaquist

Past President: Brian Prowell

Moreland Herrin

President of the Association in 1978

Professor Moreland Herrin died on July 6, 2015. Professor Herrin had been a member of AAPT since 1955 and was a Life and Honorary Member of the Association.

He received his bachelor and master's degree from Oklahoma State University and later received his PhD from Purdue University.

He was a registered professional civil engineer who worked as a consultant and researcher. He taught civil engineering first at Oklahoma State and then for more than 30 years at the University of Illinois Civil and Environmental Engineering Department. He retired from University of Illinois in 1987 and continued on as Professor Emeritus.

In addition to AAPT, Moreland was a member of ASCE, ASTM, American Society for Engineering Education, Phi Kappa Phi and Sigma Xi.

He became a life member of AAPT in 1985, served on the Nominating Committee, Award Committee, Paper Solicitation and on the Board of Directors from 1975-1981 and President in 1978.

Moreland's expertise and friendly nature will be missed by the members of AAPT.

Introduction of Jim Scherocman

2015 Honorary Member of AAPT

by E. Ray Brown

It is my pleasure to introduce Mr. James (Jim) Scherocman as a newly elected Honorary Member of the Association of Asphalt Paving Technologists (AAPT). Jim has been actively involved in materials, mix design, and construction of asphalt pavements for almost 50 years and is clearly deserving of this award. Jim has been very active in all activities of AAPT since he joined the organization in 1967. Since joining, Jim has not missed one AAPT Annual Conference. He has held all of the AAPT offices including serving as President in 1991-92.

Jim received his education at Ohio State University where he received a BS and MCE degrees in Civil Engineering. He also received an MBA from Ohio State University after completing the requirements for his Civil Engineering degrees. Jim is a Professional Engineer in the states of Ohio, Indiana, and Arizona.

He has been very active in a number of other organizations including: TRB, ASTM, ASCE, and the Canadian Technical Asphalt Association (CTAA). He has received recognition for his service to TRB by being selected as an emeritus member and has been recognized for his service to the CTAA by being selected as an honorary member.

Jim has worked for a number of organizations that were involved in hot mix asphalt technology. Jim worked as Assistant Flexible Pavement Engineer for the Ohio DOT, Area Engineer for the Asphalt Institute, Executive Director of the Indiana Asphalt Pavement Association, Chief Engineer for Shumaker Bros Industries, Chief Paving Engineer for the Barber-Greene Company, and Director of Marketing—Asphalt Additives for the Carstab Division of Morton, Thiokol, Inc. These diverse responsibilities have provided Jim with a very broad perspective on production and placement of asphalt pavements. Since 1986, Jim has been a self-employed consulting engineer.

Jim has conducted many pavement workshops on mix production, placement, compaction, longitudinal joints, and QC/QA of asphalt pavement construction. He is a frequent speaker at state and national meetings related to construction of HMA.

Jim has been an author or co-author of more than 10 AAPT papers and has offered discussion on many papers as well as served as an active participant in AAPT business meetings. He has also authored or co-authored a number of additional papers related to asphalt pavement construction that have been published by TRB, NCHRP, ASTM, and ASCE. Probably his most notable publication has been the Hot Mix Asphalt Paving Handbook that was originally published in 1991. The second edition was published in 2000. This handbook, which was sponsored by the American Association of State Highway and Transportation Officials, the National Asphalt Pavement Association, the Federal Aviation Administration, the Federal Highway Administration, and the U.S. Army Corps of Engineers, has been widely used for a number of training courses and has been invaluable for those involved in the production and placement of hot mix asphalt. Jim was also one of the engineers involved in the forensic study of the WesTrack asphalt pavement sections. A report of the findings was published by FHWA that included the analysis and recommendations of Jim and the other members of the forensic team.

Jim has a wide range of experience in hot mix asphalt construction. Due to this vast experience in pavement construction, Jim is often asked to provide training, participate in forensic studies, serve as a conference speaker, and serve as an expert witness on legal disputes related to construction of hot mix asphalt. Over the years, Jim has shared many of his experiences through his presentations and discussions of AAPT papers.

Jim has made many contributions to the HMA Industry and to AAPT and is very deserving of this recognition as an AAPT Honorary Member. I am happy to be able to introduce him as the newest Honorary Member of AAPT.

Jim Scherocman Remarks

My daughter is here with me and my son was supposed to be here also but his wife is ill and he was not able to make it. I did not know that they were coming to Portland. My daughter, Jenny, also called Jen, has two sons, a four year old and a one year old. Although she was at the AAPT meeting in South Carolina when I was President in 1992 and although she has been at several AAPT meetings since then, I never expected her to come to Portland because of her two kids.

What is also interesting is that my son, Matt, and I flew from Cincinnati to Minneapolis together on Sunday morning. He was headed to Phoenix and I was coming here. Little did I know that he was planning to come to Portland after the trip to Arizona. If his wife had not become ill, he would have been here with Jen and me also, but he needed to return to Cincinnati to take care of her and his two children.

This morning about 11:00 AM, I am talking to several people at the meeting when Jenny walks in. You talk about being surprised!!! It is noted that none of my "friends" on the Board of Directors, as well as Eileen and Charlena and all the other people who knew about my kids being invited to this dinner, said anything to me about it. So I was very surprised and, of course, very pleased that Jen made it. Both of my kids have been very supportive of me over the years.

It is indeed an extreme privilege to be elected an Honorary Member of AAPT. What is interesting to me, however, is that AAPT has been in existence for 90 years and there have only been 34 honorary members in all that time. And, very frankly, we need to do a much better job, in my opinion, of recognizing people who have been very active in this group, not just me, but many, many other individuals. I hope that the Board will look into this situation and get more people nominated and voted on for honorary membership because only 34 people so recognized over a time period of 90 years is not nearly enough. There are many individuals who have been very active in AAPT who also should be so recognized. I hope that the Board will make an effort to get more people recognized and rewarded for all the efforts they have put into this group.

As was stated, I have attended every AAPT meeting since 1968 in Atlanta, Georgia. I keep coming back to these meetings for several reasons. One is to interact with the people who are here. As a young engineer back in the late 60's, I came to talk to people, to pick their brains, to get ideas from them, and to ask them a whole bunch of dumb questions that I did not know the answers to. This group was excellent for that purpose. And for the younger members here, I suggest to you to find people who have been around for a while and ask them questions. Find out what you can learn from them. It is amazing what you can learn if you just ask. I am very, very pleased that I was able to take advantage of those people who had been around much longer than I had been at that time, who had better knowledge than I did, and who took the time to answer my dumb questions and basically try to help me. I very much appreciate that group of people who helped me gain a much better understanding of the asphalt paving industry.

Getting to know people at the AAPT meeting gave me the opportunity to call someone if I had a problem. Thus, if I did not know the answer to something or did not think that I had all the information that I needed, I could call someone and say "Hey, have you seen this particular problem?" And people went out of their way to provide that information to me, and relationships developed that I just cannot say enough great things about. Those relationships are worth everything. This is what

AAPT is all about--the ability to interact with other people in this particular business. It made a big difference to me at that time and still makes a big difference to me to this day.

For those people in the room, I would suggest a couple of things. Number one, stay current and keep reading. A lot of people after they get out of school, die professionally. They stop reading and they stop finding out what is going on in the industry. About 46 years ago, I started reading one technical paper every single day of the year. Still do. If I miss a day, I read two technical papers the next day. I devote thirty minutes daily to keeping current in the asphalt paving business. I strongly suggest that everyone here in the audience start doing that. Find out what other people are doing. You will discover that we reinvent the wheel in this business a lot of times. If you do the reading, you can save yourself a lot of time and effort without building a lot of flat tires when you reinvent the wheel.

Attend meetings like AAPT. Everyone you need to talk to is here. You have ample opportunity to interact with them, ask them questions, and find out what is going on. Whether it is here at AAPT, or CTAA, or TRB, or whatever meeting, get involved. Volunteer to help, get on committees, and actively participate in the group. That is how you learn. That is how you find out who can help you when you have a question or problem that you need an answer to.

In addition, present papers. Write papers for AAPT. Write papers for CTAA and TRB. It is amazing how quickly you can find out just how ignorant you are of certain "facts" when you write a technical paper and everyone says: "Did you consider this issue, or that issue, or whatever"? But that is how you learn. Writing and publishing papers is very, very beneficial in increasing your knowledge and understanding of the asphalt paving industry.

And finally, again, ask questions. If you don't know, ask, because somebody else has had the same problem that you've got. It is amazing how fast you can find the answers to your questions and problems if you just ask people: "Have you seen this problem? What is the cause? What is the solution?" It saves a lot of time and effort.

I will make one more comment. My late wife, June, is, of course, not with us tonight. I wish she was because I am sure that she would be very pleased with my selection as an Honorary Member of AAPT. I could let her make the acceptance speech for me. She was very special and very, very beneficial to me in the asphalt paving business.

I accept this honor from AAPT with the greatest gratitude. Thank you.

Table of Contents

TECHNICAL SESSIONS

CD TABLE OF CONTENTS

In addition to the papers listed above, the following items are also included in the CD

Aging and Rejuvenators: Evaluating Their Impact on High RAP Mixtures Fatigue Cracking Characteristics Using Advanced Mechanistic Models and Testing Methods

Walaa S. Mogawer[a], Alexander Austerman[a], Reynaldo Roque[b], Shane Underwood[c], Louay Mohammad[d], and Jian Zou[b]

[a]*University of Massachusetts Highway Sustainability Research Center (HSRC), Fall River, MA, 02723*

[b]*University of Florida, Gainesville, FL 32611*

[c]*Arizona State University, Tempe, AZ 85287*

[d]*Louisiana State University, Dept. of Civil and Environmental Eng., Louisiana Transportation Research Center, Baton Rouge, LA 70803*

ABSTRACT. *Fatigue cracking is a major distress in asphalt mixtures. It is highly dependent on many factors including aging. During mixing and construction this is referred to as short-term aging, and aging during the service life of the pavement is referred to as long-term aging. Using larger amounts of Reclaimed Asphalt Pavement (RAP) in new paving mixtures presents a concern that the resultant mixtures may be prone to more fatigue cracking during the service life of the pavement. This is due to the asphalt binder in the RAP being significantly aged. In new paving mixtures, this already aged binder will be exposed to additional short and long-term aging. To alleviate the effect of the aged RAP binder on the cracking susceptibilities of new asphalt mixtures, generally a softer binder is used. However, several studies have indicated that asphalt rejuvenators can allow more aged binder to be incorporated into mixtures than a softer binder alone.*

The objectives of this study were to evaluate the effect of long-term aging on the fatigue characteristics of high-RAP mixtures modified with rejuvenators using the conventional fatigue test and recently developed mechanistic models and tests. Also the results from these fatigue tests were compared to see if they provided similar performance trends in regards to each other. Four tests/models were used: Four Point Flexural Beam Fatigue Test, HMA Fracture Mechanics Model, Simplified Visco-Elastic Continuum Damage Model (SVECD), and the Semi-Circular Bending (SCB) Test.

The oral presentation was made by Dr. Mogawer.

This is a reproduction of an article whose final and definitive form has been published in *Road Materials and Pavement Design*© 2015 Taylor & Francis. The article is available online at: http://dx.doi.org/10.1080/14680629.2015.1076996

The results indicated that the long-term aging in accordance with the AASHTO R 30 specification (five days aging at 85°C) did not have a significant effect on the fatigue characteristics of the high RAP mixture with and without rejuvenators.

Comparison of the fatigue test rankings suggested that they did not show universal agreement in the rankings of the mixtures. Also, the repeated load cyclic tests (beam fatigue and SVECD) exhibited a wide range of performance compared to the constant rate fatigue tests and analysis (HMA-FM models and SCB). The reason for these variations is not known at this time and requires further study. Field trials are needed to determine the actual ranking trend that should be obtained in the laboratory tests. This may be the only way to separate and evaluate the quality of the results obtained by these methods.

KEYWORDS: aging, reclaimed asphalt pavement, asphalt rejuvenator, fatigue cracking performance

1.0 Introduction

Cracking is a major distress in asphalt mixtures that manifests in the field in the form of fatigue, low temperature, longitudinal, block, reflection, and slippage cracking. For this study one of the most typically encountered forms of cracking, fatigue cracking, was the focus of investigation.

Fatigue cracking is highly dependent on many factors including aging. Asphalt mixtures encounter both short- and long-term aging. Aging during mixing and construction is referred to as short-term aging. Aging during the service life of the pavement is referred to as long-term aging. Short-term aging is mainly a result of the production process (mixing and compaction) including the elevated temperatures and presence of moisture. Long-term aging is a result of exposure to a combination of environmental conditioning factors after placement (ultraviolet radiation, temperature, rainfall, and in-place service time) and also in-service loading (Abbas et al., 2002). It continues to occur throughout an asphalt mixture's service life. A prominent component of long-term aging is oxidation which results from the interaction of asphalt binder components with both oxygen and ultraviolet radiation. During the first Strategic Highway Research Program, laboratory procedures were developed to simulate the field hardening of asphalt binders and mixtures including oxidative aging (Bell et al., 1994). These procedures are American Association of State Highway and Transportation Officials (AASHTO) R 28 "Standard Practice for Accelerated Aging of Asphalt Binder Using a Pressurized Aging Vessel (PAV)" and AASHTO R 30 "Standard Practice for Mixture Conditioning of Hot-Mix Asphalt (HMA)" (AASHTO, 2011).

Mixture aging is affected by both internal and external variables. Internal variables include asphalt and aggregate properties, mixture asphalt content, asphalt film thickness and air void content (Kandhal and Chakraborty, 1996). External variables are the short- and long-term variables noted previously. These variables interact throughout the service life of the pavement causing an asphalt mixture to undergo rheological (elastic, viscous, and cracking behaviors) and chemical changes including an increase in the stiffness of the asphalt binder. The result of aging is that it increases an asphalt mixture's susceptibility to cracking (Abbas et al., 2002; Glover et al., 2005). Utilization of larger amounts of previously aged materials like Reclaimed Asphalt Pavement (RAP) in new paving mixtures further complicates the efforts to quantify and understand the effects of aging and its impact on the fatigue cracking potential of mixtures.

Due to the increased cost of asphalt binder and Federal Highway Administration's policy to increase environmental stewardship, the industry has been looking for ways to increase the amount of RAP in asphalt mixtures. RAP is comprised of an asphalt binder that has been hardened due to long-term aging along with re-usable aggregates. Using larger amounts of RAP in new paving mixtures presents a concern that the resultant mixtures may be prone to more fatigue cracking during the service life of the pavement. This is due to the asphalt binder in the RAP being significantly aged. In new paving mixtures, this already aged binder will be exposed to additional short- and long-term aging. Studies have confirmed that increased aging will reduce the stress relaxation capacity of the binder which means it decreases the cracking resistance of the mixture (Molenaar et al., 2010; Daniel et al., 2013). To alleviate the effect of the hardened RAP binder on the cracking susceptibility of new asphalt mixtures generally a softer binder is used. However, several studies have indicated that asphalt rejuvenators can allow more aged binder to be incorporated into mixtures than a softer binder alone. Even so, for recycling purposes, rejuvenators are not encouraged by some state agencies because of potential rutting-related concerns.

For mixtures that incorporate materials with high RAP contents (with or without using a softer binder and/or a rejuvenator) fatigue cracking models and tests that rely on the mechanistic and volumetric properties of mixtures are needed. The classic method for evaluating asphalt mixture fatigue in the United States is the standardized flexural beam fatigue test outlined in AASHTO T 321 "Standard Method of Test for Determining the Fatigue Life of Compacted Hot-Mix Asphalt (HMA) Subjected to Repeated Flexural Bending" (AASHTO, 2011). The primary outcome of this method, when carried out in a series of experiments with mixtures at different temperatures and/or strain amplitudes, is a relationship between strain amplitude, modulus, and fatigue life. This relationship can then be incorporated in some empirical or mechanistic-empirical algorithm of varying complexity to extract the fatigue performance of asphalt pavements. There are certain limitations in this approach with respect to the mathematical rigor of the pavement prediction algorithm since it must rely on two separate analyses (one to extract the pavement response and one to predict the damage). Likewise, the material model does not directly consider any specific cracking mechanism and in the end it is not known what individual factors are responsible for the empirically calibrated coefficients.

For the methodology to yield reasonably accurate results the calibration data set must be sufficiently large so as to properly cover the entire range of conditions expected even though this calibration set may in fact include some redundancy in characterizing the contribution from individual mechanisms.

Recently, researchers have used the theories of fracture mechanics and continuum damage to develop models and tests that can evaluate the fatigue cracking initiation and cracking propagation phenomenon of asphalt mixtures from a less empirical and more mechanism-centric perspective (Roque et al., 2004; Underwood and Kim, 2010; TRB, 2012; Wu et al., 2005). These models and tests provide another means to ascertain the effects of long-term aging on the performance of high RAP content mixtures. Also, since these models and tests were derived based on different theories, it is important to determine if they provide similar performance in regards to each other and to the flexural bending beam fatigue test.

2.0 Objectives

The objectives of this study were as follows: (1) study the effect of long-term aging on the fatigue characteristics of high RAP mixtures modified with rejuvenators using both conventional and recently developed mechanistic models and tests; and (2) to determine if these models/tests provide similar performance trends in regards to each other and to the flexural bending beam fatigue test.

3.0 Description of Models and Tests Used in the Study

The following outlines the tests and models utilized to evaluate the effects of long-term aging on the fatigue cracking characteristics of the mixtures included in this study. Four tests/models were used:

- Four Point Flexural Beam Fatigue Test
- HMA Fracture Mechanics Model
- Simplified Visco-Elastic Continuum Damage Model (SVECD)
- Semi-Circular Bending (SCB) Test

Each of these tests/models follows different testing protocols such as sample size, specimen shape, testing temperatures, etc. Nevertheless, each of them has shown good correlation to field performance. Accordingly, it is expected that, although these tests/models were developed based on different theories, they would provide similar fatigue cracking performances for the mixtures evaluated in this study.

3.1 Four Point Flexural Beam Fatigue Test

One of the most common and historically used laboratory test procedures chosen to evaluate the fatigue cracking resistance of asphalt mixtures is the four point flexural beam fatigue test. The typical test protocols are AASHTO T 321, "Determining Fatigue Failure of Compacted Asphalt Concrete Subjected to Repeated Flexural Bending" (AASHTO, 2011), and ASTM D7460, "Determining Fatigue Failure of Compacted Asphalt Concrete Subjected to Repeated Flexural Bending" (ASTM, 2009). In order to investigate the effects of aging and rejuvenators on the fatigue cracking characteristics of high RAP content asphalt mixtures, beam fatigue testing was conducted in accordance with AASHTO T 321. This test was conducted in strain control mode at a load frequency of 10Hz applied using a sinusoidal waveform. Specimens were tested at strain levels of 500 $\mu\varepsilon$ and 750 $\mu\varepsilon$. The number of cycles to failure was determined by fitting an exponential function to the flexural stiffness versus number of cycles and then evaluating the number of cycles it took to decrease the initial stiffness by 50% of the initial stiffness measured at the 50th cycle.

3.2 HMA Fracture Mechanics Model

The HMA fracture mechanics model was developed based on the fact that an asphalt mixture has a fundamental dissipated creep strain energy limit ($DCSE_f$) (also called the fracture threshold), which has been determined to be independent of loading mode (controlled stress or controlled strain) and history. Cracking will initiate or propagate in any region of an asphalt mixture where induced damage exceeds the threshold. The following asphalt mixture properties are used in the model: (i) resilient modulus (M_R). (ii) creep rate (dependent on creep compliance Power Law parameters D_1 and m-value), (iii) tensile strength (S_t), (iv) fracture energy limit (FE_f), and (v) dissipated creep strain energy limit ($DCSE_f$). These properties can be obtained using the Superpave indirect tensile (IDT) test system. Detailed procedures regarding testing, data reduction and interpretation are described elsewhere (Roque et al., 1997; Zhang et al., 2001; Roque et al., 2002).

Based on a detailed analysis and evaluation of 22 field test sections (in service for over 10 years) throughout the state of Florida, including conducting Superpave IDT tests on aged field cores and comparing predicted cracking performance by the HMA fracture mechanics model with the observed field performance, an energy-based fracture criterion was developed at the University of Florida (UF), consisting of an Energy Ratio (ER) and associated minimum values of ER required for adequate cracking performance (Roque et al., 2004). The ER parameter, defined as the ratio of $DCSE_f$ over the minimum dissipated creep strain energy required, is expressed in Equation 1:

$$ER = a \times DCSE_f / \left(m^{2.98} \times D_1\right) \quad [1]$$

Where: $a = 0.0299\sigma_t^{-3.1} \times (6.36 - S_t) + 2.46 \times 10^{-8}$

And where, σ_t (in psi) is the tensile stress in the asphalt layer; S_t is tensile strength (in MPa), $DCSE_f$ is dissipated creep strain energy limit (in kJ/m^3), D_1 (in 1/psi) and m-value are creep compliance Power Law parameters. Previously established minimum values of ER (ER_{min}) required are (i) 1.0 for no greater than 5 million ESALs, (ii) 1.3 for no greater than 10 million ESALs, and (iii) 1.95 for no greater than 20 million ESALs, which were determined by taking into account the change in pavement structure with varying traffic levels.

The ER accounts for the effects of pavement structural characteristics (σ_t) and material properties ($DCSE_f$, S_t, m, D_1) on cracking performance. The higher the value of the ER, the better the expected cracking performance of the section. Therefore, ER can be used to integrate asphalt mixture properties in the pavement design process as well as to predict the performance of in-service pavement sections.

3.3 Simplified Visco-Elastic Continuum Damage Model (SVECD)

Researchers (Underwood and Kim, 2010) have shown that models developed based on continuum damage theory (CDT) can be used to explain and predict the fatigue life of an asphalt mixture. By theory, these methods are generally considered to be most applicable up to the point of macrocrack development, many times referred to as crack initiation. In the laboratory, this point is clearly defined by a localization of deformation at the macrocrack site. In the field, initiation is less clearly defined since it is often taken as the time when a crack becomes visible on the pavement surface. For top-down cracking such a definition may correspond to the localization based definition. However, in the case of bottom-up cracking, this definition actually represents both the initiation of cracking and the propagation of the crack through the pavement layers. While their direct theoretical application is limited to initiation, CDT approaches may be applied to model post-localization behaviors of pavements (Kim et al., 2008; Underwood et al., 2009; Park and Kim, 2013). Justification for this extension lies in implicit correlations between the initiation and propagation mechanics of asphalt mixtures (Roque et al., 2010). These correlations may be related to the observation that macrocrack propagations in asphalt mixtures, at least at moderate to high temperatures where fatigue is a primary concern, are associated with a microcrack dominated process zone ahead of the macrocrack tip. While CDT can be used for the prediction of highly developed pavement cracking, it is the more commonly held belief that after the point of crack initiation, other techniques including fracture mechanics and approximate cohesive zone methods are needed for the further evolution of cracking. Even without explicitly integrating these techniques with CDT, pavement predictions based on CDT can be beneficial as the aim for pavement design is to produce pavements that do not exhibit larger cracks.

CDT methods have been evolving for asphalt concrete since at least the late 1980's, and their characterization and analysis protocols involve more complex calculus than is used in traditional beam fatigue analysis. The method used in this paper is the Simplified Visco-Elastic Continuum Damage (SVECD) model developed by Underwood and Kim (2010). This technique allows the fatigue life of an asphalt mixture at various strain-stress amplitudes under different temperatures to be predicted from its dynamic modulus and cyclic fatigue data. For the purposes of this paper, cyclic fatigue tests were performed at strain levels between approximately 250με and 550με at 15°C. These conditions were determined following AASHTO TP 107 "Determining the Damage Characteristic Curve of Asphalt Concrete from Direct Tension Cyclic Fatigue Tests." The experiments themselves do not preclude any particular analysis and could be interpreted using the same protocols used in beam fatigue testing. However, for modeling purposes the primary material function determined from these tests is the relationship between the material integrity, denoted by the variable C, and the amount of damage, S. This function is referred to as the damage characteristic function and has been found to be a unique function irrespective of temperature and mode of loading (Underwood and Kim, 2010).

3.4 Semi-Circular Bending (SCB)

The semi-circular bend (SCB) test has been favored by many researchers due to the ease of sample preparation including cores removed from the field and the quick and simple testing procedure (Adamson et al., 1996; Li and Marasteanu, 2010). It is a based on fracture mechanics principles (Anderson, 2005) that offers the potential of assessing the cracking resistance of asphalt mixtures in the laboratory in the design phase as well as in QA (quality assurance) testing activities. Cracking mechanisms in asphalt mixtures have been studied since the 1960s. Various test methods were proposed to predict the fracture behavior of asphalt mixtures, including the single edged notched beam (SENB) test, the SCB test, the indirect tensile test (IDT) fracture parameter, and the disk-shaped compact tension test (DCT) test. Each of these test procedures has unique advantages and disadvantages (Li and Marasteanu, 2010). The SCB test is based on fracture mechanics and the J-integral, which considers the elasto-plastic/visco-plastic relationships of asphalt mixtures rather than the stress intensity factor (K) commonly used in linear elastic fracture mechanics (Mull et al., 2006).

Mull et al. (2006) evaluated the use of the semi-circular bend (SCB) configuration to characterize the fatigue crack propagation of asphalt mixtures. Three mixtures were evaluated for fatigue susceptibility using SCB fracture parameters and scanning electron microscopy to identify fatigue damage species associated with each mixture. The research found that the SCB specimen is suitable for both static and fatigue fracture characterization (Mull et al., 2006).

Elseifi et al. (2012) utilized the SCB test at 25°C in the evaluation of a number of asphalt mixtures, including mixtures with a high content of RAP. Results of the experimental program were used to validate a three-dimensional finite element model, which was used to interpret and to analyze the failure mechanisms in the SCB test. The experimental program showed that the SCB test results successfully predicted the fracture performance of the mixtures and was able to differentiate between them in terms of cracking resistance (Elseifi et al., 2012).

4.0 Experimental Plan

In order to address the objectives of this study, an experimental plan was developed as shown in Figure 1.

5.0 Materials

All materials for this study (binders, aggregates, and RAP) were obtained from sources in the state of Massachusetts. Rejuvenators were obtained from multiple sources in the United States.

5.1 Rejuvenators

A total of four asphalt rejuvenators were selected for evaluation in this study. Rejuvenators were selected so that different types were represented as well as those already commonly used and those based on emerging green technologies. Details about each rejuvenator are outlined in Table 1.

Table 1. *Rejuvenator descriptions and details.*

Rejuvenator ID	Details
Aromatic Oil	Commonly used in pavement preservation activities
Paraffinic Oil	Commonly used in pavement preservation activities
Organic Blend #1	Green chemistry product
Organic Blend #2	Organic oils based

The proposed dose of each asphalt rejuvenator was based on a previous study (Mogawer et al., 2013) which suggested a dosage of 0.5% by weight of total RAP in the mixture. Based on the design binder content and percent asphalt binder in the RAP, this dosage equaled 9.0% rejuvenator by weight of recycled binder. This dosage was utilized throughout this study. Each rejuvenator was added directly to the pool of heated binder immediately prior to mixing for each specimen fabricated.

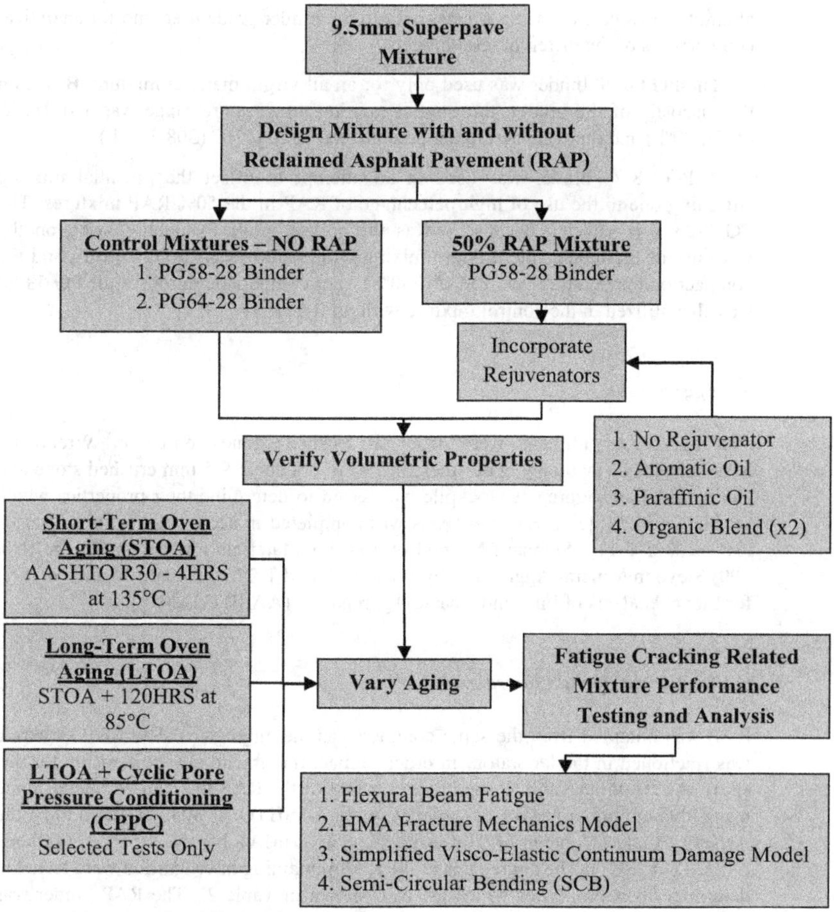

Figure 1. *Experimental plan.*

5.2 Asphalt Binder

For this study, two different asphalt binders were utilized. The binders consisted of a PG 64-28 which is typically specified in the Northeast and the typically available PG 58-28 softer grade binder utilized for RAP mixtures. These binders were used in an attempt to evaluate each test's sensitivity to the binder grade used and for analytical comparisons of the different tests/models.

The PG 64-28 binder was used only for an all virgin material mixture. Based on the viscosity of the binder, the mixture mixing temperature range was 161-165°C (322-329°F) and the compaction temperature was 153-157°C (308-315°F).

A PG 58-28 binder was used in an attempt to offset the potential mixture stiffening due to the use of high percentage of RAP in the 50% RAP mixtures. The PG58-28 was selected because it was the softest grade available. Based on the viscosity of the binder, the mixture mixing temperature was 150°C (300°F) and the compaction temperature was 138°C (280°F). For comparison purposes the PG 58-28 was also utilized in the control mixture without RAP.

5.3 Aggregates

The aggregates utilized were from a crushed stone source in Wrentham, Massachusetts. Two aggregate stockpiles were obtained: 9.5 mm crushed stone and stone dust. Each aggregate stockpile was tested to determine their properties which are shown in Table 2. Sieve analysis was completed in accordance with AASHTO test method T 11, "Standard Method of Test for Materials Finer Than 75-µm (No. 200) Sieve in Mineral Aggregates by Washing," and T 27, "Standard Method of Test for Sieve Analysis of Fine and Coarse Aggregates" (AASHTO, 2011).

5.4 Reclaimed Asphalt Pavement (RAP)

RAP was obtained from the same contractor as the aggregates. The RAP stockpile was fractioned in the laboratory in order to meet the gradation requirements for this study. As shown in Table 2, the binder content of the RAP material was determined using the ignition oven in accordance with AASHTO T 308, "Determining the Asphalt Binder Content of Hot Mix Asphalt (HMA) by the Ignition Method" (AASHTO, 2011). The aggregates in the RAP remaining after ignition were tested to determine their properties which are also shown in Table 2. The RAP binder was extracted and recovered in accordance with AASHTO T 164, "Standard Method of Test for Quantitative Extraction of Asphalt Binder from Hot Mix Asphalt (HMA)," and AASHTO T 170, "Standard Method of Test for Recovery of Asphalt Binder from Solution by Abson Method" (AASHTO, 2011). As shown in Table 2, the continuous and performance grade of the recovered RAP binder was determined in accordance with AASHTO R 29, "Grading or Verifying the Performance Grade of an Asphalt Binder," and AASHTO M 320, "Standard Specification for Performance-Graded Asphalt Binder" (AASHTO, 2011).

Table 2. *Virgin aggregate, RAP, and recovered RAP binder properties.*

Sieve Size	9.5 mm	Stone Dust	RAP Aggregates Post Ignition
19.0 mm	100	100	100
12.5 mm	99.4	100	100
9.5 mm	93.8	100	100
4.75 mm	29.7	99.7	76.8
2.36 mm	5.2	83.7	57.6
1.18 mm	2.8	57.1	43.3
0.600 mm	2.3	38.6	31.1
0.300 mm	2.1	24.9	19.8
0.150 mm	1.8	15.9	12.1
0.075 mm	1.5	10.9	8.3
RAP Binder Content, % =			5.6%
RAP Binder Continuous Grade =			82.0-21.8
RAP Binder Performance Grade =			82-16

6.0 Description of Mixture Aging Procedures

For this study, four different aging schemes were utilized as described in the following. For mixture fatigue cracking performance evaluations, testing and analysis was conducted using the short-term, long-term, and long-term plus cyclic pore pressure conditioning (CPPC) aging schemes presented.

For long-term aging, it should be noted that all Superpave gyratory compacted (SGC) specimens were aged in the compacted state prior to any cutting/trimming for a specific test. Only the long-term aged flexural beam fatigue specimens were cut prior to long-term aging. This was due to the large weight (25kg (55lb)) and dimensions (150 mm wide, 150 mm tall, and 450 mm long (6 in. wide, 6 in. tall, 17.5 in. long)) of the slabs from which the beams were cut. Aging of the smaller cut beams was done in an attempt to ensure thorough and consistent aging between all beams tested in this study.

6.1 Mixture Design and Volumetric Verification Aging

For mixture design and volumetric verifications, specimens of each mixture were batched, mixed and aged at the respective compaction temperature previously noted for 2 hours ± 5 minutes in a loose state in accordance with AASHTO R 30, "Standard Practice for Mixture Conditioning of Hot Mix Asphalt (HMA)" Section 7.1, Mixture Conditioning for Volumetric Mixture Design (AASHTO, 2011). Specimens were stirred after 60 ± 5 minutes to maintain uniform conditioning. After aging, specimens were immediately compacted.

6.2 Short-Term Oven Aging (STOA)

STOA was conducted in accordance with AASHTO R 30, Section 7.2, Short-Term Conditioning for Mixture Mechanical Property Testing (AASHTO, 2011). For STOA aging, specimens of each mixture were batched, mixed and aged at 135 ± 3°C (275± 5°F) for 4 hours ± 5 minutes in a loose state. Specimens were stirred after 60 ± 5 minutes to maintain uniform conditioning. After aging, specimens were returned to the compaction temperature and immediately compacted.

6.3 Long-Term Oven Aging (LTOA)

LTOA was conducted in accordance with AASHTO R 30, Section 7.3, Long-Term Conditioning for Mixture Mechanical Property Testing (AASHTO, 2011). For LTOA aging, specimens of each mixture were first subjected to STOA aging as described previously. Next, the compacted STOA specimens were cooled for 16 ± 1 hours. In accordance with AASHTO R 30, Subsection 7.3.4, Long-Term Conditioning of Prepared Test Specimens, the compacted specimens were aged in an 85 ± 3°C (185 ± 5°F) oven for 120 ± 0.5 hours (five days). After aging, the specimens were allowed to cool at room temperature which took approximately 16 hours. The specimens were not handled or disturbed until after they had completely cooled to room temperature.

6.4 LTOA + Cyclic Pore Pressure Conditioning (CPPC)

The CPPC system was developed at the University of Florida to induce damage in specimens due to combined effects of moisture and load (Birgisson et al., 2007; Isola et al., 2014). It was determined that CPPC can induce internal pressure (stress) in a tensile mode within the air voids that is similar to the effect of repeated load induced by traffic on mixtures in the field. Specifically, pore water under pressure in mixtures can cause premature failure of the mixture, through loss of adhesion between asphalt binder and aggregates (i.e. stripping of the asphalt film) or through loss of cohesion within the asphalt binder or mastic. In this study, the CPPC system was used for additional conditioning of asphalt mixtures after LTOA to simulate the effects of moisture and cyclic internal pressure on changes in mixture fracture properties after oxidative aging.

The structural core of the system is a triaxial cell modified for cyclic pore pressure conditioning of asphalt specimens, which consists of two round stainless steel plates separated by posts or struts and encased with a Plexiglass cylinder. Once sealed, the entire package creates an enclosed cavity capable of being pressurized. Figure 2 shows the tabletop triaxial chamber containing cut specimens for Superpave Indirect Tensile (IDT) tests. As can be seen, spacers were used in between specimens to facilitate water infiltration and to protect the gauge points from being damaged during the conditioning process.

The conditioning of the asphalt specimens takes place by exerting cyclic pore pressure on all water accessible voids of the specimens. This is accomplished by vacuum saturating specimens and then placing them into the airtight, water-filled triaxial chamber. Deaerated water is forced into the chamber to build up pressure. This pressure is transferred to every surface with which the water is in contact. When this pressure is cycled in the chamber at a constant frequency in the form of a sine wave, it can simulate the pumping action that occurs in the field when vehicles travel over wet pavement.

Figure 2. *Tabletop triaxial chamber.*

Prior to insertion in the chamber, the specimens were first subjected to a two-cycle saturation process. Each cycle included a 15-minute vacuum saturation period at 85 kPa (± 7.0 kPa) followed by a 20-minute submergence period at atmospheric pressure. No specific saturation levels were targeted since each mixture has a unique void structure that may enhance or reduce its saturation capacity. The specimens were then placed into the tabletop triaxial chamber, carefully filled with water, and subjected to a combination of pore pressure cycles and temperature determined during previous research conducted at the University of Florida (Birgisson et al., 2007). Specifically, water pressure in a sine waveform at a frequency of 0.33 Hz and an amplitude of 69 kPa, ranging from 34.5 to 172.5 kPa, was applied for 5800 cycles at 25°C. Immediately after CPPC, specimens were kept in a water bath for two days at 10°C, the temperature used for Superpave IDT tests.

7.0 Mixture Design

The target gradation for the mixtures utilized in this study is shown in Table 3. This target gradation was developed to meet the requirements for a 9.5 mm Superpave mixture in accordance with AASHTO M 323, "Superpave Volumetric Mix Design," and AASHTO R 35, "Superpave Volumetric Design for Hot Mix Asphalt" (AASHTO, 2011).

Table 3. *Target mixture gradation and specification.*

Sieve Size	Sieve Size (mm)	Target Gradation for All Mixtures	Superpave 9.5mm Specification
1/2"	12.5 mm	100	100 min
3/8"	9.5 mm	98	90-100
No. 4	4.75 mm	85	90 max
No. 8	2.36 mm	58	32-67
No. 16	1.18 mm	42	-
No. 30	0.600 mm	27	-
No. 50	0.300 mm	15	-
No. 100	0.150 mm	9	-
No. 200	0.075 mm	6.0	2-10
Binder Content		6.5%	-

Control mixtures utilizing all virgin materials were designed utilizing the control PG 64-28 binder and the softer PG 58-28 binder for comparison purposes. Then RAP was incorporated to replace a portion of the virgin materials in the control mixtures. RAP was added to replace 50% of the mixture aggregates with RAP aggregates. The aggregate gradation for the control and 50% RAP mixtures was identical. The 50% RAP mixture was then designed utilizing only the softer PG 58-28 binder. Because the gradations were identical for the mixtures developed, mixture design verifications were performed for the 50% RAP mixtures incorporating each rejuvenator at the design binder content determined for the control mixture. Verifications were completed assuming 100% contribution of the RAP binder.

To incorporate the RAP into the mixtures, a procedure that was used in a prior study utilizing similar materials was followed (Mogawer et al., 2013). This procedure was utilized to eliminate moisture in the RAP stockpile material and to optimize the blending between the aged and virgin binders in the mixture. The procedure steps were:

1. The RAP was air dried until a constant mass was achieved, which typically took three to five days.

2. The RAP was further dried for two days at 60°C (140°F).

3. The RAP was added to heated aggregate two hours prior to adding the binder during the mixing process.

The design Equivalent Single Axle Loads (ESALs) for this project was selected as 0.3 to < 3 million which is consistent with surface course mixtures in New England. The design Superpave gyratory compactive effort for this ESALs level was N_{design} =

75 gyrations. The results of the mixture design indicated that all mixtures met the volumetric requirements within typical production tolerances.

8.0 Dynamic Modulus Results

The baseline for understanding fatigue performance was establishing the stiffness behavior of each mixture. The dynamic modulus, |E*|, was measured according to the method outlined in AASHTO TP 79, "Standard Method of Test for Determining the Dynamic Modulus and Flow Number for Hot Mix Asphalt (HMA) Using the Asphalt Mixture Performance Tester (AMPT)," with the exception that the test temperatures were limited to 4°, 20° and 40°C for the multiple loading frequencies. The results are summarized in the mastercurves shown in Figures 3a and 3b. The amount of stiffening that occurs due to aging is summarized by temperature in Figure 4. The aging ratio shown on the y-axis of this plot is calculated by taking the ratio of $|E^*|_{LTOA}$ to $|E^*|_{STOA}$. From this figure it can be seen that the relative amount of stiffening caused by LTOA varies by mixture and rejuvenator type. It is also seen in Figures 3a and 3b that the modulus from the rejuvenated mixtures shows a larger spread in the STOA case relative to the LTOA case. Since there was no reduction in dynamic modulus from the STOA to LTOA condition for mixtures with rejuvenators, this suggested that for these rejuvenators there is no continued softening of the mixtures. Therefore, increased rutting susceptibility of these mixtures with rejuvenators may not be a concern.

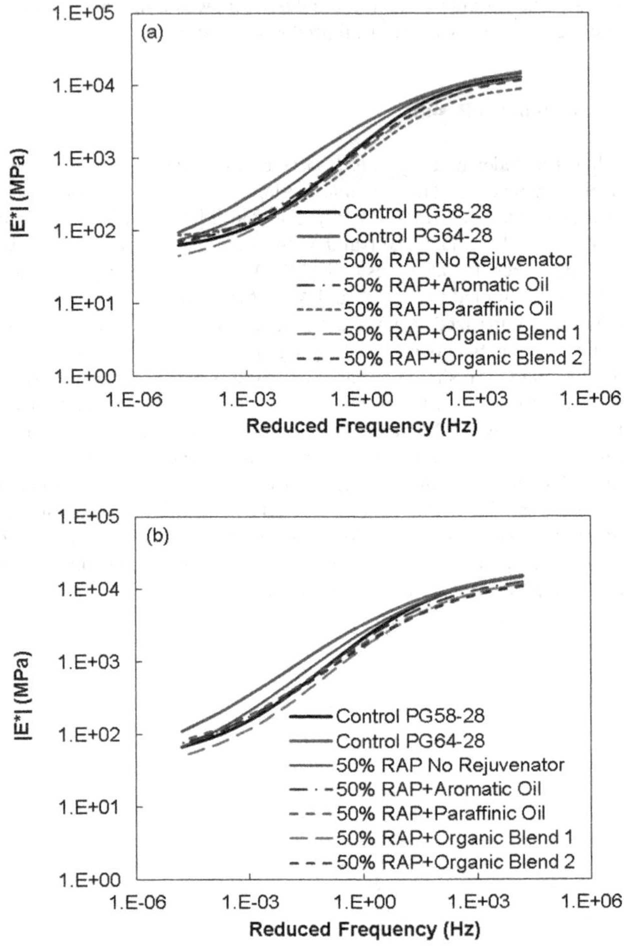

Figure 3. *Results from |E*| characterization; (a) STOA (b) LTOA.*

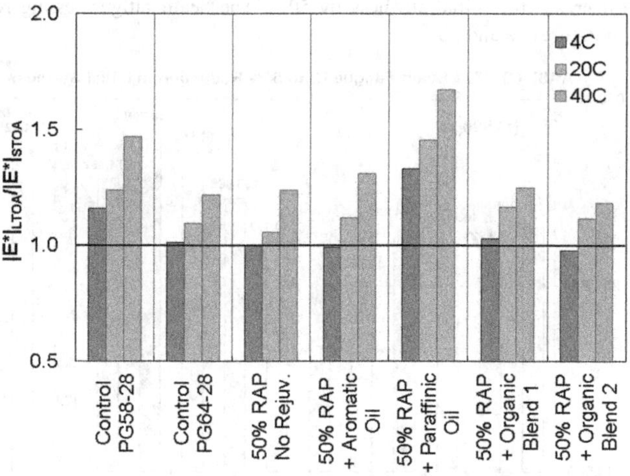

Figure 4. *Results from |E*| characterization: (c) aging ratio at 10 Hz.*

9.0 Fatigue Cracking Tests and Results

This section outlines the testing details and results of each fatigue cracking test. Unless otherwise specified, these tests were conducted at a 15°C (59°F) test temperature which represents the intermediate temperature for the Northeast.

9.1 Four Point Flexural Beam Fatigue Test

For this test, fatigue specimens were cut from slabs. Slabs with dimensions of 150 mm wide, 150 mm tall, and 450 mm long (6 in. wide, 6 in. tall, and 17.5 in. long) were fabricated for each mixture using the IPC Global Pressbox slab compactor. From each slab, beams with dimensions of 63 mm wide, 50 mm tall, and 380 mm long (2.5 in. wide, 2 in. tall, 15 in. long) were cut such that the sides had smooth faces. The air voids of the final cut specimens were 7±1%. Beam specimens were conditioned at the test temperature of 15°C (59°F) for at least two hours prior to testing.

Each beam fatigue test was conducted in strain control mode at a loading frequency of 10Hz applied using a sinusoidal waveform. Specimens were tested at strain levels of 500με and 750με. The 500με and 750με were selected because all mixtures lost 50% of their initial stiffness after at least 10,000 cycles. The number of cycles to failure was determined by fitting an exponential function to the flexural stiffness versus number of cycles and then evaluating the number of cycles that it took to decrease the initial stiffness by 50%. The beam fatigue testing results are shown in Figures 5a and 5b.

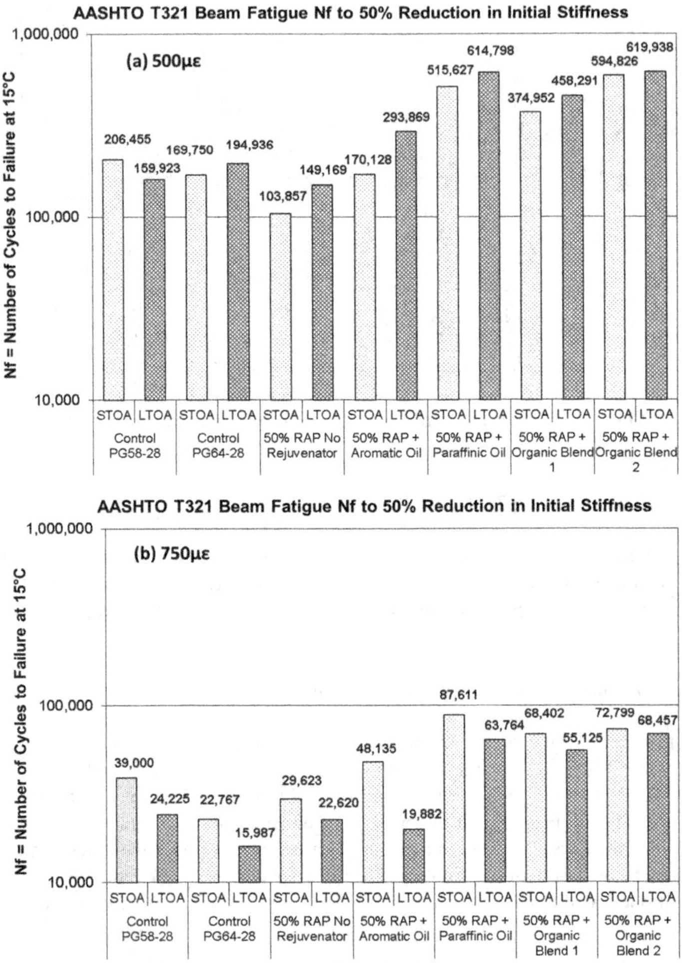

Figure 5. *Fatigue cracking results from the flexural beam fatigue—(a) 500με (b) 750με.*

Comparing the properties of the 50% RAP mixture with no rejuvenator and those of the control PG 58-28 mixture it appeared that the introduction of high RAP content resulted in a reduction in the number of cycles to failure. All mixtures with rejuvenators generally exhibited higher numbers of cycles to failure. Based on the aging condition only, the beam fatigue data showed opposite trends. At 500με the 50% RAP LTOA aged specimens had better fatigue performance than the STOA specimens. This trend was reversed for the 750με level.

9.2 HMA Fracture Mechanics Model

9.2.1 Mixture Properties Determined by the Superpave IDT

Results obtained using the Superpave IDT tests are presented in Figures 6 and 7 showing the effect of the aging levels and the effects of the high RAP content as well as rejuvenators on mixture properties.

9.2.1.1 Effect of aging levels. FE limits (FE_f), representing the fracture tolerance of the mixtures, are shown in Figure 6a. It is clear that the FE limits of all mixtures were reduced after LTOA. The FE limits of the three mixtures with no rejuvenator (the control PG 58-28, the PG 64-28, and the 50% RAP mixtures) further decreased after CPPC, while the FE limits of all four mixtures with rejuvenators increased. The results of creep rate in Figure 6b showed that the damage-related property of all mixtures greatly decreased due to LTOA, indicating a significant reduction in the rate of damage accumulation of the mixtures.

The creep rates, however, remained about the same after CPPC. The results of resilient modulus testing, representing stiffness of the mixtures, are presented in Figure 7a. As can be seen, the resilient modulus of all mixtures increased due to LTOA. This property was affected by CPPC to a lesser extent, but varied in both directions. Similarly, the tensile strength of all mixtures increased due to LTOA, and in general, slightly increased after CPPC (Figure 7b). Overall, the conditioning levels had greater effects on FE limit and creep rate than on the other two properties. It is important to note that the results of a separate study on laboratory conditioning procedures (Roque et al., 2010) showed that CPPC was able to induce changes in fracture properties consistent with those observed in field cores, while oxidation alone (STOA or LTOA) was not able to induce relevant changes. Therefore, it was determined that the predicted cracking performance based on the properties obtained after CPPC should be used in the final ranking of all the mixtures.

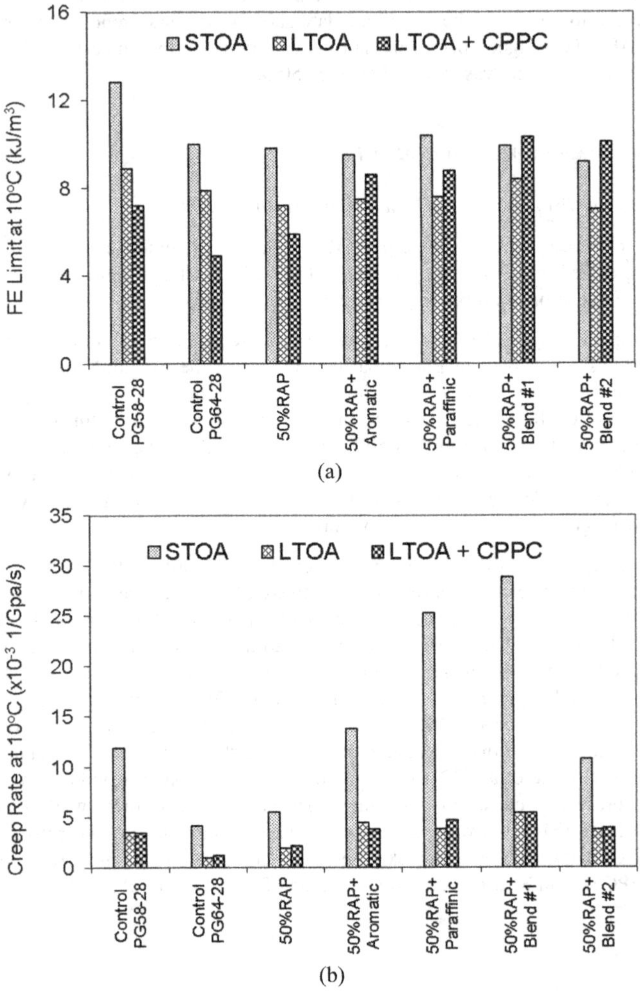

Figure 6. *Mixture properties: (a) FE limit, (b) creep rate determined using Superpave IDT at three aging levels.*

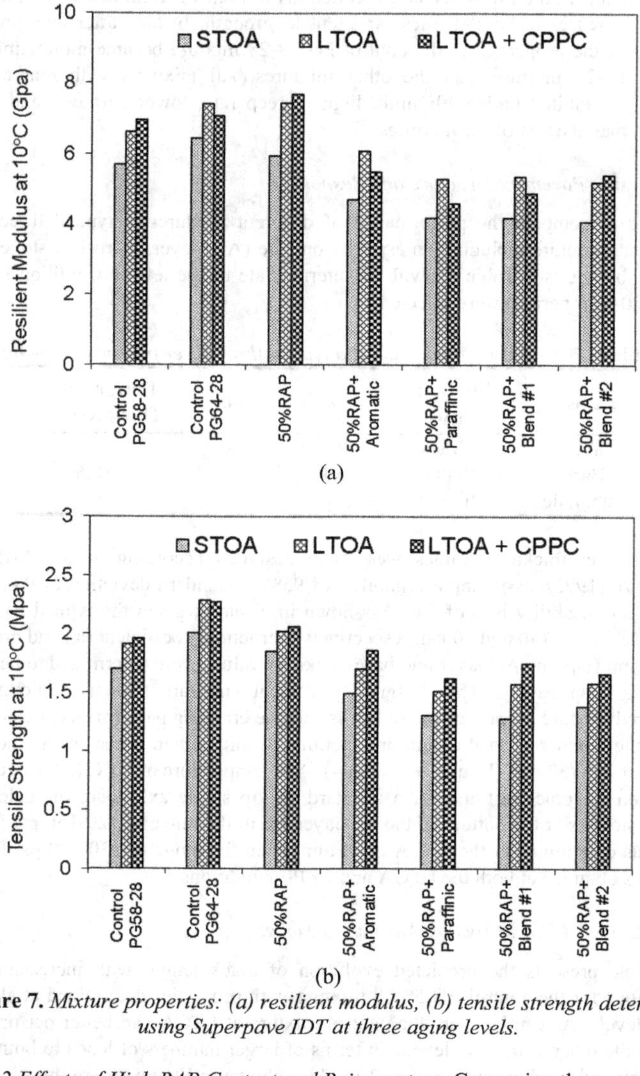

Figure 7. *Mixture properties: (a) resilient modulus, (b) tensile strength determined using Superpave IDT at three aging levels.*

9.2.1.2 Effects of High RAP Content and Rejuvenators. Comparing the properties of the 50% RAP mixture with no rejuvenator and those of the control PG 58-28 mixture (see Figures 6a, 6b, 7a, and 7b) it appeared that the introduction of high

RAP content resulted in fairly large reductions in both FE limit and creep rate, but smaller increases in both stiffness and tensile strength. In fact, after incorporating 50% RAP, the properties of the control PG 58-28 mixture became more similar to the PG 64-28 mixture than the other mixtures. All mixtures with rejuvenators generally exhibited higher FE limit, higher creep rate, lower stiffness, and lower strength than the rest of the mixtures.

9.2.2 Design Pavement Structure and Responses

In order to compare the performance of different mixtures, a typical three-layer pavement structure including an asphalt concrete (AC) layer, a crushed stone base, and a subgrade (see Table 4), with an intermediate traffic level (10 million ESALs over a 20-year period) were selected.

Table 4. *Design layer thickness for a typical three-layer pavement structure.*

Layers	Modulus (psi)	a_i	m_i	Design Layer Thickness (in.)
AC	M_R	0.40	–	7
Base	10,000	0.14	1.2	10.5
Subgrade	40,000	–	–	–

The layer thickness values were then designed according to the AASHTO procedure (1993), assuming a reliability of 95%, a standard deviation of 0.4, and a design serviceability loss of 2.0. As shown in Table 4, given the typical base and subgrade moduli (40 and 10 ksi, respectively), structural coefficient (a_i) and drainage coefficient (m_i), the AC layer and base thickness values were determined to be 7 and 10.5 in., respectively. The designed pavement structure was used along with measured mixture properties to predict its relative cracking performance, determined as N_f, the number of load repetitions required to initiate and propagate a crack to a length of 2 in./50 mm (Roque et al., 2004). At a temperature of 10°C (a fairly critical condition for cracking) and for a standard 18-kip single axle load, the calculated tensile stresses at the bottom of the AC layer are in the range of 120-150 psi for AC modulus determined at the STOA condition, or in the range of 130-160 psi for AC modulus obtained at both the LTOA and CPPC conditions.

9.2.3 Results of HMA Fracture Mechanics Model

Figure 8a presents the predicted evolution of crack length with increasing load repetitions for the control PG 58-28 mixture with properties determined at all three aging levels. As can be seen, the control mixture at LTOA had better performance than at the other two aging levels, in terms of larger numbers of loads to both crack initiation and crack propagation to 2 in. Also shown in Figure 8a are the calculated ER values using Equation 1, which support that the control mixture had the best performance at the LTOA aging level. However, it should be noted that both the HMA fracture mechanics model and associated ER criterion are recommended for use in relative comparison of performance at a similar aging level, because the healing potential is not considered in the current form of the model, which can substantially affect the cracking performance (Zou et al., 2012). Therefore, the predicted performance of all mixtures was compared at individual aging levels subsequently. For simplicity, the predicted performance indicators for all mixtures (both ER and N_f) were normalized with respect to the predicted values for the control PG 58-28 mixture. The final results are presented in Figures 8b, 9a, and 9b for the STOA, LTOA, and CPPC aging levels, respectively.

In general, the results showed that for all three aging levels and in terms of both ER and N_f predictions, the PG64-28 mixture had the best performance, and the 50% RAP mixture with no rejuvenator outperformed the control PG58-28 mixture. The performance (both ER and N_f) of high RAP mixtures with rejuvenators were generally worse than the control PG 58-28 mixture except for the CPPC aging level, at which better performance was observed.

The ER and N_f predictions generally provided consistent rankings among high RAP mixtures with rejuvenators for both LTOA and CPPC aging levels. At LTOA, the mixture with Paraffinic Oil and the one with Organic Blend #2 appeared to outperform the other two mixtures modified with different rejuvenators. At CPPC, Organic Blend #2 appeared to be the best rejuvenator that improved the cracking performance of the high RAP mixture. Surprisingly, the ER and N_f ranked differently the performance of mixtures with rejuvenators at the STOA aging level. It should be noted that the coefficients of the ER equation were calibrated based on measured properties and observed performance of 22 field test sections (Roque et al., 2004). When mixture properties beyond the range of properties used in the calibration are encountered, the ER equation may lead to erroneous results. The N_f results are directly predicted by the HMA fracture mechanics model, which are not affected by the data sets used in the calibration of the ER equation. Therefore, the N_f results are more accurate than the ER results. In this specific case, it was found that the creep rates of all mixtures modified with rejuvenators are far beyond the range used in the calibration, i.e., 0.2-6.2 ($\times$ 10^{-3} 1/Gpa/s), which appeared to desensitize the ER results to the creep rates. According to the N_f results, the mixture with Organic Blend #2 outperformed the mixtures modified with other rejuvenators at STOA.

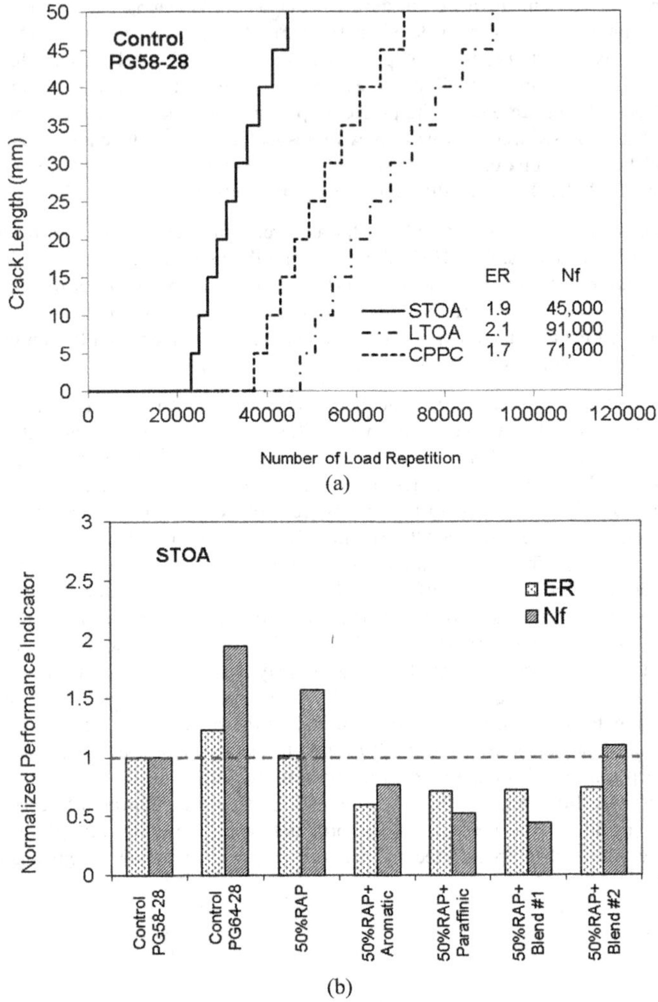

Figure 8. *Model prediction: (a) number of loads resulting in a 50 mm (2 in.) long crack (N_f) for the PG 58-28 control mixture at three aging levels; (b) normalized N_f with respect to that of the control mixture at STOA.*

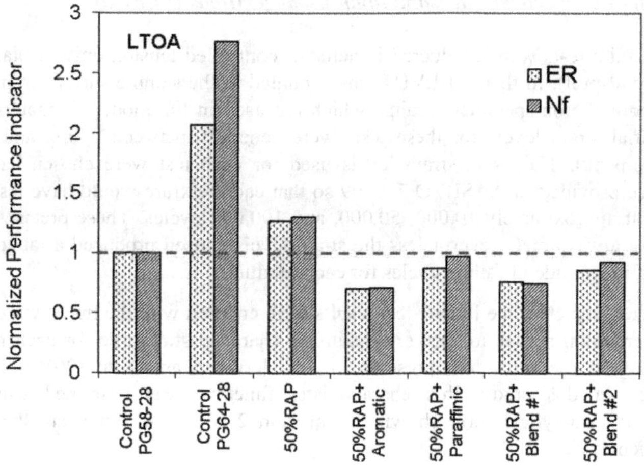

(a)

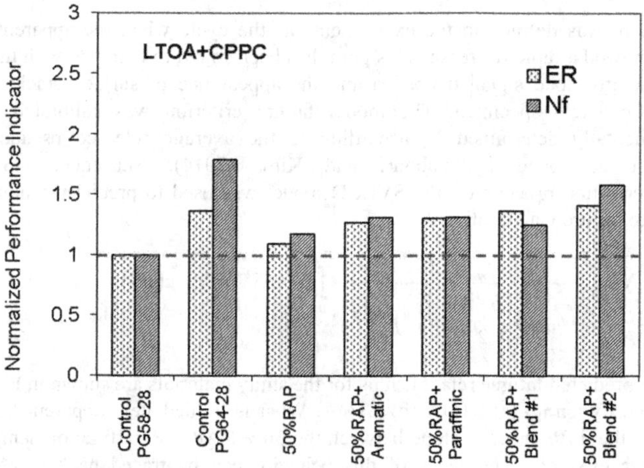

(b)

Figure 9. *Model prediction: (a) comparing normalized N_f at LTOA; (b) comparing normalized N_f at LTOA+CPPC.*

9.3 Simplified Visco-Elastic Continuum Damage Model (SVECD)

The uniaxial tests were conducted in actuator controlled tension-only displacement mode (pull-pull). In this test LVDTs are mounted to the sample surface to measure and record the on-specimen strains, which are used in the model characterization. The initial strain levels for these tests were generally between 250με and 550με (peak-to-peak). The exact strain levels used for each test were chosen based on guidance provided in AASHTO TP 109 so that each mixture would have tests with failure at approximately 10,000, 50,000, and 100,000 cycles. These precise targets were not always met; nevertheless the strain levels chosen produced a range in the order of magnitude of failure cycles for each mixture.

Once completed, the fatigue test results were coupled with the linear viscoelastic characterization results to derive the damage characteristic curve for each mixture according to the procedure presented by Underwood and Kim (2010). For the mixtures in this study, this characteristic function was described using the exponential analytical form shown in Equation 2, where a and b are the fitting coefficients.

$$C = \exp\left(aS^{b}\right) \qquad [2]$$

Failure was defined in the experiments as the cycle when the apparent phase angle showed a rapid decrease. This point has been found to coincide with the onset of substantial load signal distortion and the appearance of surface cracks on the asphalt concrete specimen. The model failure criterion was calibrated to this experimentally determined N_f according to the average released pseudo strain criterion developed by Sabouri and Kim, (2014). Subsequent to these characterization procedures, the SVECD model was used to predict the number of cycles to failure via Equation 3.

$$N_f = \frac{f_{red} \times 2^{\alpha}}{\left[\left(\varepsilon_{0,pp}\right)\dfrac{(\beta+1)}{2}\left(|E^*|\right)\right]^{2\alpha} K_1} \int_0^{S_{Nf}} \left(-abS^{b-1}e^{aS^{b}}\right)^{-\alpha} (dS) \qquad [3]$$

The predicted fatigue relationships for the study materials are shown in Figure 10 separated by aging, (a) STOA (b) LTOA. What is immediately apparent from this graph is the differences in slope between the 50% RAP + Paraffinic oil sample and the other mixtures. The cause of this behavior can be traced back to a greater sensitivity of the failure criteria to the rate of change in the average pseudo strain energy, G_R, than the other mixtures. It should be noted that the experimental data for this rejuvenator, although gathered over a more narrow range of strain levels, also showed a greater sensitivity to strain than the other cases.

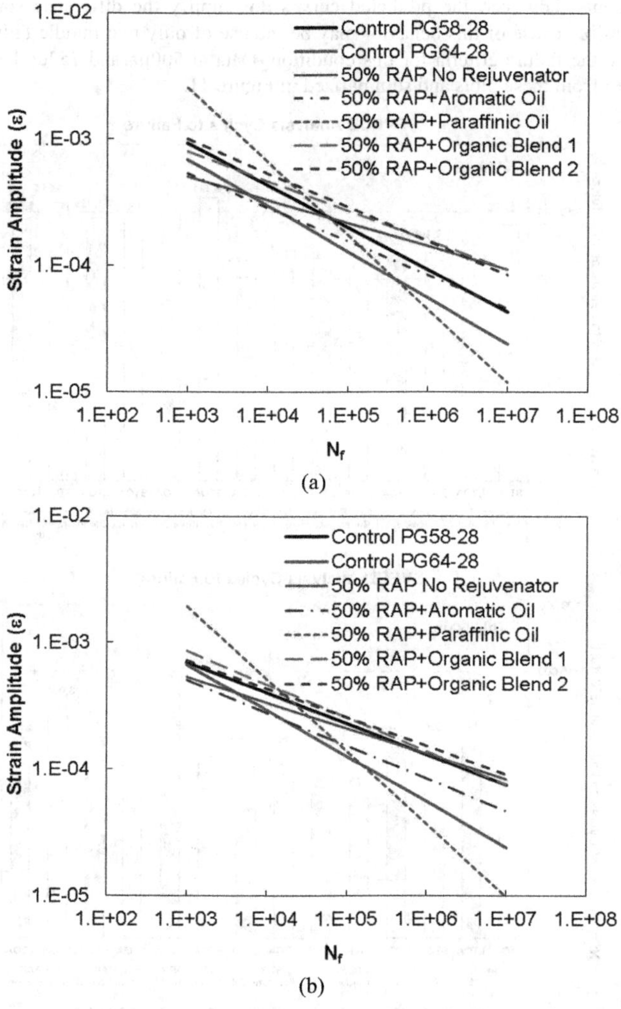

Figure 10. *Predicted fatigue life from SVECD model (a) STOA (b) LTOA.*

It is also interesting to observe that the failure curve for the 50% RAP with no rejuvenator case is actually expected to outperform the control mixtures at lower strain amplitudes. These behaviors are supported with the trends in the axial experiments; however, the predicted curves do amplify the difference somewhat. One possible cause of this behavior may be the use of only two middle failure tests to define the failure criteria for this condition. Data at 500με and 750με have been extracted from these plots and summarized in Figure 11.

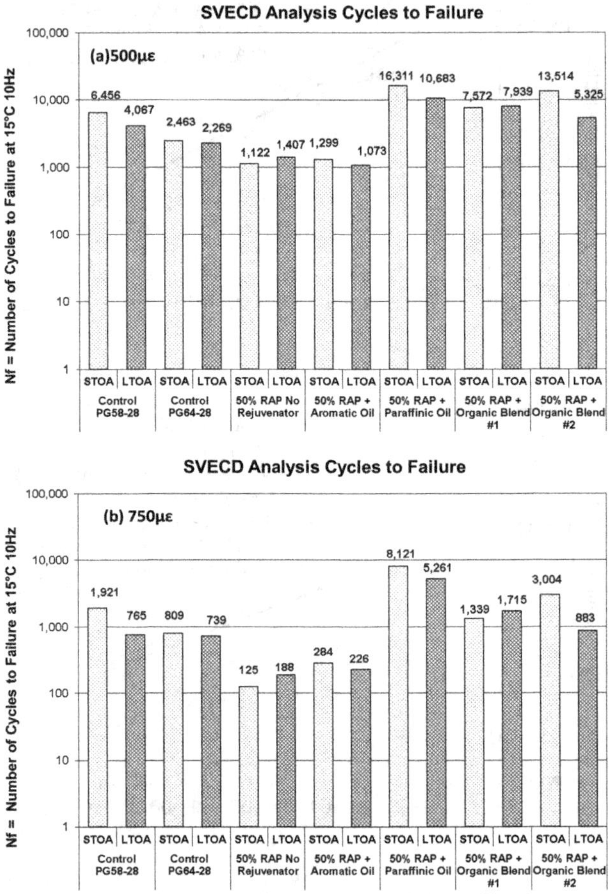

Figure 11. *SVECD analysis results - (a)500με (b)750με.*

The data in Figure 11 indicated that generally the mixture performed better after STOA aging than LTOA aging. This trend matched the beam fatigue results at the 750με level, but not the 500με level. Overall when comparing the SVECD predicted results to those from the beam fatigue test, it is found that the rankings were similar. The largest exception to this case is the Control PG 64-28 case. These results are truly unexpected. The beam fatigue tests suggest relatively little difference between the two control cases whereas the SVECD model suggests a reduction by nearly 60-70% in the fatigue life. Reasons for the differences in behavior are not immediately clear and require further investigation. It is noted, however, that this difference cannot be associated with the SVECD model itself since the experimental data used in model characterization also demonstrated noticeably inferior performance from the PG 64-28 asphalt mixture.

9.4 Semi-Circular Bending (SCB) Test

Cracking potential was evaluated using the SCB test procedure, based on fracture mechanics (FM) principles (Anderson, 2005) and suggested by Mohammad and co-workers (Wu et al., 2005), Figure 12. The critical strain energy release rate, also called the critical value of J-integral (J_c) was used to describe the mixture's resistance to fracture:

$$J_c = \left(\frac{U_1}{b_1} - \frac{U_2}{b_2} \right) \frac{1}{a_2 - a_1} \qquad [4]$$

Where: J_c= critical strain energy release rate (kJ/mm^2);
b = sample thickness (mm);
a = notch depth (mm);
U = strain energy to failure (N.mm); and
dU/da = change of strain energy with notch depth

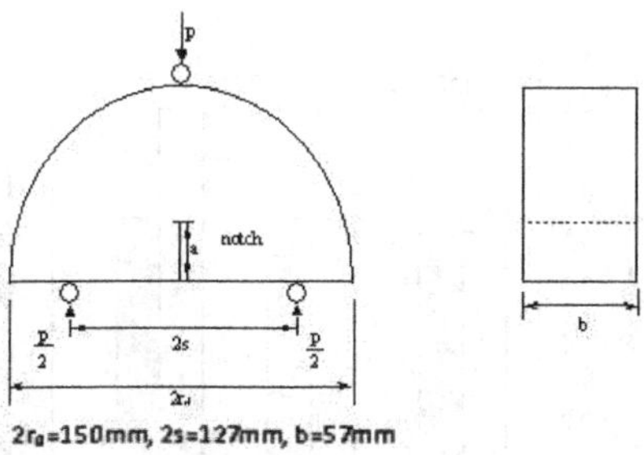

Figure 12. *Semi-circular bending test setup.*

To determine the critical value of J-integral (J_c) using Equation 4, semi-circular specimens with at least two different notch depths should be tested to determine the change of strain energy with notch depth (dU/da). In this study, three notch depths of 25.4 mm, 31.8 mm, and 38 mm were tested to increase the accuracy of slope calculation (dU/da) by fitting a regression line to the change of strain energy with notch depth.

The semi-circular specimen was loaded monotonically until fracture failure under a constant cross-head deformation rate of 0.5 mm/min. in a three-point bending load configuration. The load and deformation were continuously recorded. This test was performed at a temperature of 15°C. Specimens were compacted in the SGC to an air void level of 7 ± 0.5%. Triplicate specimens were utilized for this test. In general, the coefficient of variation was within 15% for the samples tested. High J_c values are desirable for fracture-resistant mixtures

Figure 13 presents the calculated critical fracture resistance (J_c) values for the asphalt mixtures evaluated in the study. The control mixture containing PG 64-28 binder exhibited better J_c values than the control mixture containing PG 58-28 binder at STOA condition; however at the LTOA condition, both mixtures possessed similar J_c values. This indicates that the LTOA condition affected control mixture PG 64-28 more than the one with PG 58-28. Mixtures containing 50% RAP and no rejuvenator; 50% RAP and aromatic oil; and 50% RAP and organic blend 1 showed similar fracture resistance performances to control mixture containing PG 64-28. On the other hand, mixtures containing 50% RAP and paraffinic oil, and 50% RAP and organic blend 2 presented similar fracture resistance performances to the control mixture containing PG 58-28. It is worth noting that the addition of aromatic oil rejuvenator did improve the fracture resistance at the LTOA condition when compared to control mixture PG 64-28. On the other hand, paraffinic oil and organic blend 2 rejuvenators did not improve the mixture resistance to fracture at both STOA and LTOA conditions.

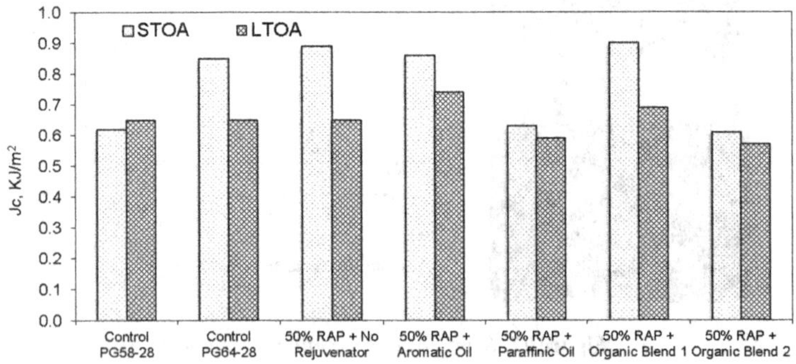

Figure 13. *Semi-circular bending test results.*

10.0 Comparative Analysis of the Different Fatigue Tests

An objective of this study was to evaluate whether different fatigue tests show the same trends. In order to accomplish this objective, the results from each test for the PG64-28 control mixture were normalized. A performance indicator ratio was calculated for each test result by dividing it by the value for the control PG 64-28 mixture. Therefore, if a specific test indicated for a particular mixture the same performance as the PG 64-28 mixture, the performance indicator ratio was a value of 1.0. The control PG 64-28 mixture was selected because it utilized the typically specified binder in the state of Massachusetts. The asphalt binders, aggregates, and RAP for the mixtures were obtained from a location in Massachusetts. The aim was to develop a 50% RAP mixture that performed similar to a typical mixture, which in this study was the PG 64-28 control mixture.

The performance indicator ratios are shown in Figures 14 and 15 for STOA and LTOA respectively. Figure 14a and 15a show specific performance indicator ratios for beam fatigue and SVECD at a 500με while all other indicator ratios remain constant based on the STOA aging condition. Similarly, Figures 14b and 15b show performance indicator ratios for the beam fatigue and SVECD at 750με while all other indicator ratios remain constant based on the LTOA aging condition.

If two tests were comparable, their performance indicator ratios would have the same trends for the same mixtures tested, but not necessarily the same magnitude. Overall, the performance indicator ratios did not agree among the various methodologies. Furthermore, the performance indicator ratios did not show universal agreement in the trends of the mixtures, meaning whether the ratio went up or down compared to a ratio of 1.0 for the PG 64-28 control mixture. Also, the repeated load cyclic tests (beam fatigue and SVECD) exhibited a wide range of performance compared to the constant rate fatigue tests and analysis (HMA-FM models and SCB). Furthermore, as shown by the performance indicator ratios, the two repeated cyclic load tests (beam fatigue and SVECD) often provided large increases in the performance of the 50% RAP mixture when a rejuvenator was added. The ratios for these mixtures were very high compared to the ratios for PG 58-28 and PG 64-28 mixtures. It is hypothesized that this means blending in the binders was incomplete, causing these mixtures to have areas of binder with relatively low stiffness. These two tests are sensitive to mixture stiffness. However, the very high performance indicator ratios provided by some of the SVECD tests strongly suggest that this test might be overly sensitive to stiffness. If this hypothesis concerning blending is correct, in general the constant rate fatigue tests and analyses were not sensitive to these areas of low stiffness. Repeated loads must be used to measure this effect. The reason why some of the constant rate fatigue tests and analyses provided no change, or even a decrease, in the performance of the 50% RAP mixture when a rejuvenator was added is not known.

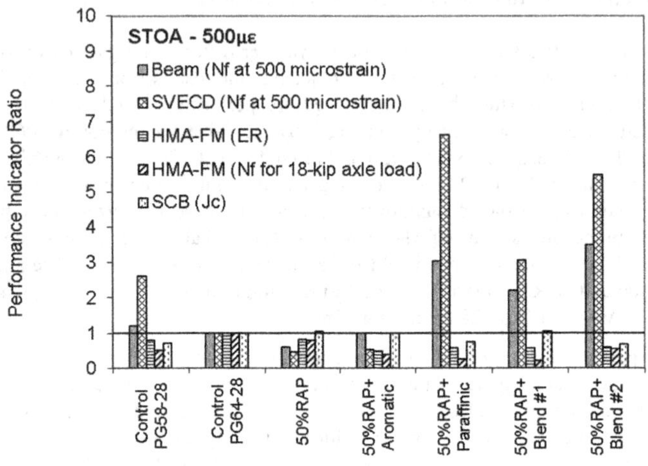

(a)

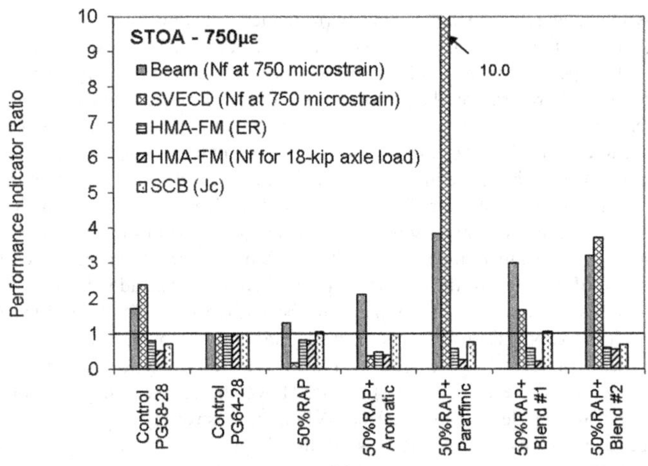

(b)

Figure 14. *Comparative analysis of normalized test results- STOA aging (a) 500με (b) 750με.*

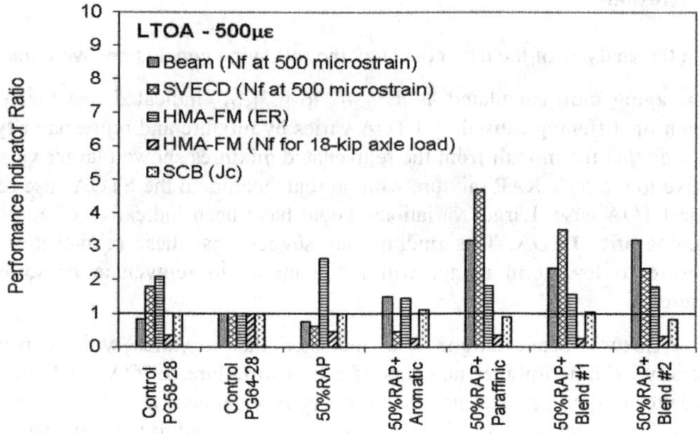

(a)

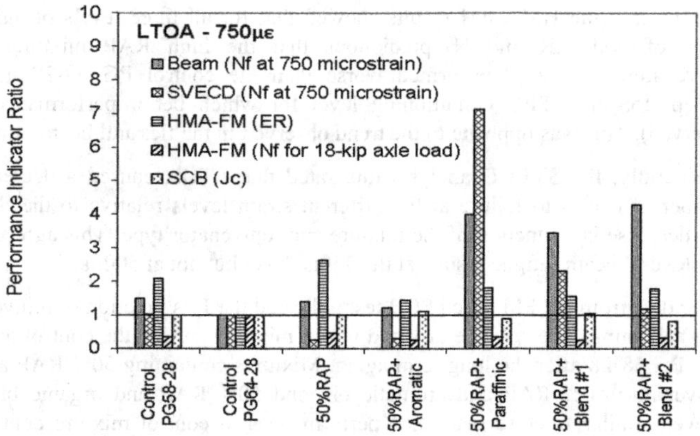

(b)

Figure 15. *Comparative analysis of normalized test results- LTOA aging (a) 500με (b) 750με.*

Field trials are needed to determine the actual ranking trend that should be obtained in the laboratory tests. This may be the only way to separate and evaluate the quality of the results obtained by these methods.

11.0 Conclusions

Based on the analysis of the data collected, the following conclusions were made:

- The aging ratio calculated as $|E^*|_{LTOA}$ to $|E^*|_{STOA}$ indicated that the relative amount of stiffening caused by LTOA varies by mixture and rejuvenator type. It was seen that the moduli from the rejuvenated mixtures showed larger variations relative to the 50% RAP mixture with no rejuvenators in the STOA case relative to the LTOA case. Larger variations would have been indicative of continuous softening after LTOA. This finding may suggest that these rejuvenators, after exposure to long-term aging, will not continue to rejuvenate or soften the mixtures.

- The flexural beam fatigue tests indicated that mixtures with rejuvenators generally exhibited higher numbers of cycles to failure. LTOA aged specimens had better fatigue performance than the STOA specimens at a 500 $\mu\varepsilon$ test level. This trend was reversed for the 750$\mu\varepsilon$ test level. This led to the conclusion that a strain level should be used that is representative of the actual loading conditions where these mixtures will be placed.

- In general, the HMA-FM results showed that for all three levels of aging in terms of both ER and N_f predictions that the high RAP mixtures with rejuvenators generally performed worse than the control PG 58-28 mixture (except for the CPPC conditioning level for which better performance was observed). This was opposite to the trend observed in the flexural beam fatigue.

- Generally, the SVECD analysis illustrated that LTOA caused a decrease in number of cycles to failure at the different strain levels relative to the STOA. The decrease is a function of the mixture and rejuvenator type. This agreed with the flexural beam fatigue results at the 750$\mu\varepsilon$ level but not at 500$\mu\varepsilon$.

- The data from the FM based SCB test indicated the J_c is function of rejuvenator type and aging. The J_c value dropped for all mixtures except the control mixture with PG 58-28 after the long term aging. Mixtures containing 50% RAP and no rejuvenator; 50% RAP and aromatic oil; and 50% RAP and organic blend 1 showed similar fracture resistance performances to control mixture containing PG 64-28 after STOA and LTOA. On the other hand, mixtures containing 50% RAP and paraffinic oil, and 50% RAP and organic blend 2 presented similar fracture resistance performances to control mixture containing PG 58-28. It is worth noting that the addition of aromatic oil rejuvenator did improve the fracture resistance at LTOA condition when compared to control mixture PG 64-28. On the other hand, paraffinic oil and organic blend 2 rejuvenators did not improve the mixture resistance to fracture at both STOA and LTOA conditions.

- Both FM based J_c values and HMA-FM results showed that the PG 64-28 mixture had the best performance, and the 50% RAP mixture with no rejuvenator outperformed the control PG 58-28 mixture after STOA.

• Comparison of the performance indicator ratios showed that the fatigue tests did not provide universal agreement. Also, the two repeated load cyclic tests (beam fatigue and SVECD) exhibited a wide range of performance compared to the constant rate fatigue tests and analysis. The reason for these variations is not known at this time and requires further study. The repeated load cyclic tests indicated the mixtures with RAP and rejuvenators had the best fatigue performance, except when using aromatic oil, whereas the constant rate fatigue tests and analysis generally provided no constant trend in fatigue performance among the mixtures. The reason for this trend is not known and requires further study.

• Field trials are needed to determine the actual ranking trend that should be obtained in the laboratory tests. In addition to field trials, chemical and rheological analyses of the binders and rejuvenators are recommended to provide an insight into the mixtures' performance. This may help to separate and evaluate the quality of the results obtained by these methods.

• Future work is needed to establish the causes of the differences in trends amongst the fatigue tests. Because of these differences no particular test can be recommended at this time.

12.0 References

Abbas, A., Chai, B.C., Masad, E., and Papagiannakis, T., "The Influence of Laboratory Aging Method on the Reheological Properties of Asphalt Binders," *American Society for Testing and Materials (ASTM) Journal of Testing and Evaluation*, Vol. 30, Issue 2, March 2002.

Adamson, R.M., Dempsey, J.P., and Mulmule, S.V., "Fracture Analysis of Semi-Circular and Semi-Circular-Bend Geometries," *International Journal of Fracture*, Vol. 77, 1996, pp. 213–222.

American Association of State Highway and Transportation Officials (AASHTO), *Standard Specifications for Transportation Materials and Methods of Sampling and Testing*, 31st Edition, AASHTO, Washington, D.C., 2011.

American Association of State Highway and Transportation Officials, *AASHTO Guide for Design of Pavement Structures*, AASHTO, Washington, D.C., 1993.

American Society for Testing and Materials (ASTM), *Annual Book of ASTM Standards*, ASTM International, Section 4, Volume 04.03, 2009.

Anderson, T. J., *Fracture Mechanics—Fundamentals and Application*, Third Edition, CRC Press, Taylor and Francis Group, Florida, 2005.

Bell, C.A., Wieder, A.J., and Fellin, M. J., *Laboratory Aging of Asphalt-Aggregate Mixtures: Field Validation*, Report SHRP-A-390, Strategic Highway Research

Program, Transportation Research Board, National Research Council, Washington, D.C., 1994.

Birgisson, B., Roque, R., Page, G.C., Wang, J., "Development of New Moisture Conditioning Procedure for Hot-Mix Asphalt," *Transportation Research Record: Journal of the Transportation Research Board,* No. 2001, Transportation Research Board of the National Academies, Washington, D.C., 2007, pp. 46–55.

Daniel, J.D., Gibson, N., Tarbox, S., Copeland, A., and Andriescu, A., "Effect of Long Term Aging on RAP Mixtures: Laboratory Evaluation of Plant Produced Mixtures," *Journal of the Association of Asphalt Paving Technologists*, Vol. 82, 2013, pp. 327–365.

Elseifi, M., Mohammad, L. N., Ying, H., Cooper, S., "Modeling and Evaluation of the Cracking Resistance of Asphalt Mixtures Using the Semi-Circular Bending Test at Intermediate Temperature," *Journal of Road Materials and Pavement Design*, Vol. 13, No. S1, 2012, pp. 124–139.

Glover, C.J., Davison, R.R., Domke, C.H., Ruan, Y., Juristyarini, P., Knorr, D.B., and Jung, S.H., *Development of New Method for Assessing Asphalt Binder Durability with Field Validation,* Report FHWA/TX-05/1872-2, Texas Department of Transportation, Austin, TX, August 2005.

Isola, M., Chun, S., Roque, R., Zou, J., Koh, C., Lopp, G., "Development and Evaluation of Laboratory Conditioning Procedures to Properly Simulate Mixture Property Changes Effectively in the Field," *Transportation Research Record: Journal of Transportation Research Board,* No. 2447, Transportation Research Board of the National Academies, Washington, D.C., 2014, pp. 74–82.

Kandhal, P.S., and Chakraborty, S., "Effect of Asphalt Film Thickness on Short-and Long-Term Aging of Asphalt Paving Mixtures," Presented at the 75th Annual Meeting of the Transportation Research Board, 1996 and National Center for Asphalt Technology (NCAT), Report 96-1, January 1996.

Kim, Y.R., Baek, C., Underwood, B.S., Subramanian, V., Guddati, M.N., Lee, K., "Application of Viscoelastic Continuum Damage Model Based Finite Element Analysis to Predict the Fatigue Performance of Asphalt Pavements," *KSCE Journal of Civil Engineering*, Vol. 12, No. 2, 2008, pp. 109–120.

Li, X.J., and Marasteanu, M.O., "Using Semi Circular Bending Test to Evaluate Low Temperature Fracture Resistance for Asphalt Concrete," *Journal of Experimental Mechanics*, Vol. 50, 2010, pp. 867–876.

Mogawer, W.S., Booshehrian, A., Vahidi, S. and Austerman, A.J., "Evaluating the Effect of Rejuvenators on the Degree of Blending and Performance of High RAP, RAS, RAP/RAS Mixtures," *Journal of the Association of Asphalt Paving Technologist*, Vol. 82, 2013, pp. 403–436.

Molenaar, A., Hagos, E., and Van de Ven, M., "Effects of Aging on the Mechanical Characteristics of Bituminous Binders in PAC," *Journal of Materials in Civil Engineering*, Vol. 22, No. 8, American Society of Civil Engineers, 2010, pp. 779–787.

Mull, M.A., Othman, A., and Mohammad, L., "Fatigue Crack Growth Analysis of HMA Employing the Semi-Circular Notched Bend Specimen," *Transportation Research Board 85th Annual Meeting Compendium of Papers,* CD-ROM. 2006.

Park, H.J., and Kim, Y.R., "Investigation into Top-Down Cracking of Asphalt Pavements in North Carolina," *Transportation Research Record: Journal of the Transportation Research Board,* No. 2368, Transportation Research Board of the National Academies, Washington, D.C., 2013, pp 45–55.

Roque, R., Birgisson, B., Drakos, B., and Dietrich, B., "Development and Field Evaluation of Energy-Based Criteria for Top-down Cracking Performance of Hot Mix Asphalts," *Journal of the Association of Asphalt Paving Technologists*, Vol. 73, 2004, pp. 229–260.

Roque, R, Birgisson, B, Sangpetngam, B, Zhang, Z., "Hot Mix Asphalt Fracture Mechanics: A Fundamental Crack Growth Law for Asphalt Mixtures," *Journal of the Association of Asphalt Paving Technologists*, Vol. 71, 2002, pp. 816–27.

Roque, R., Buttlar, W.G., Ruth, B.E., Tia, M., Dickson, S.W., Reid, B., *Evaluation of SHRP Indirect Tension Tester to Mitigate Cracking in Asphalt Pavements and Overlays*, Final Report of the Florida Department of Transportation, University of Florida, Gainesville, FL, 1997.

Roque, R., Zou, J., Kim, Y.R., Baek, C.M., Thirunavukkarasu, S., Underwood, B.S., Guddati, M.N., *Top-Down Cracking of Hot Mix Asphalt Layers: Models for Initiation and Propagation*, Project No. 1-42A, Web Document No. 162, National Cooperative Highway Research Program, Transportation Research Board, National Research Council, Washington, D.C., 2010.

Sabouri, M.R., and Kim, Y.R., "Development of a Failure Criterion for Asphalt Mixtures under Different Modes of Fatigue Loading," *Transportation Research Record: Journal of Transportation Research Board,* No. 2447, Transportation Research Board of the National Academies, Washington, D.C., 2014, pp. 117–125.

Transportation Research Board (TRB), *Applications of Advanced Models to Understand Behavior and Performance of Asphalt Mixtures*, Transportation Research Circular E-C 161, Transportation Research Board, National Research Council, Washington, D.C., January 2012.

Underwood, B.S., and Kim, Y.R., "Improved Calculation Method of Damage Parameter in Viscoelastic Continuum Damage Model," *International Journal of Pavement Engineering*, Vol. 11, No. 6, 2010, pp 459–476.

Underwood, B.S, Kim, Y.R., Savadatti, S., Thirunavukkarasu, S., Guddati, M.N., "Simplified Fatigue Performance Modeling of ALF Pavements using VECD+3-D Finite Element Modeling," 7th *International RILEM Symposium on Advanced Testing and Characterization of Bituminous Materials*, Rhodes, Greece, May 2009.

Wu, Z., Mohammed, L.N., Wang, L.B., Mull, M.A., "Fracture Resistance Characterization of Superpave Mixtures Using the Semi-Circular Bending Test," American Society for Testing and Materials (ASTM), *Journal of ASTM International*, Vol. 2, No.3, March 2005.

Zhang, Z, Roque, R, Birgisson, B, Sangpetngam, B., "Identification and Verification of a Suitable Crack Growth Law," *Journal of the Association of Asphalt Paving Technologists*, Vol. 70, 2001, pp. 206–41.

Zou, J., Roque, R., and Byron, T., "Effect of HMA Aging and Potential Healing on Top-Down Cracking Using HVS," *International Journal of Road Materials and Pavement Design*, 13(3), 2012, pp. 518–533.

13.0 Discussion

JOHN D'ANGELO: Walaa, nice work. Great presentation too. I wanted to ask you a little about, you've two different types of tests. You've got the cyclic test and then the fracture test. But you also have to look at… They are actually looking at two different mechanisms for the failure, and instead of trying to say that they tell you the same thing, I think we have to go back and say how do we combine the two different aspects? Again, one is, you know, at certain strain levels is the material able to handle that strain tolerance? The other one is, once you're going to fail it, because in the fracture test you are going to fail those materials, how large a stress does it take to propagate that crack? So it's a combination of tying the two together as opposed to saying that they both tell you the same type of thing. And I think that's the key part that we haven't put together yet, but that's really more in line with what we need to do.

WALAA MOGAWER: Thank you John. I agree with your comment completely. It would be good to look at combining both mechanisms to predict fatigue cracking. How, I don't know.

ELIE HAJJ: Thank you, Walaa, for the work and the presentation. We've got to be careful with how to look into the data from the traditional beam fatigue test. Looking at your strain levels, you are getting a better fatigue life at 500 microstrain and a lower fatigue life at 750 microstrain. So if I'm trying to visualize the relationship here, you are getting a flatter curve overall. And then when comparing fatigue data, you shouldn't be only just looking at the fatigue relationship, but you should combine the mixture relationship with its stiffness into a simple mechanistic analysis to assess fatigue life. Depending on the pavement structure and mixture stiffness, different strain levels are generated at the bottom of the asphalt layer hence leading to different fatigue lives. The comparison should not be done at a constant strain level but based on the calculated fatigue life from the strain level at the bottom of the asphalt layer which is mixture dependent.

WALAA MOGAWER: Agree.

ELIE HAJJ: Does that make sense?

WALAA MOGAWER: Thanks for the comment. Yes, however, our goal was to see if the beam fatigue would capture the effect of aging on the fatigue performance of RAP mixtures designed using rejuvenators. Also, we realize that the strain levels we used were too high—the 500 and 750.

ELIE HAJJ: Correct. Lower strain levels are typically observed in typical pavement structures.

WALAA MOGAWER: Correct. In reality, the strain levels are expected to be lower.

ELIE HAJJ: The point is you shouldn't be basing your conclusions on just looking at the number of repetitions to failure at 500 and 750 microstrain rather than have your final conclusions based on the analysis of combining pavement structure, mixture stiffness, and fatigue curve.

WALAA MOGAWER: Right. However, just for relative comparison and not predicting fatigue life, I believe the number of cycles to failure is a good indicator. Thank you.

NIMA ROOHI: Thank you for the presentation. As always, it was a nice presentation.

WALAA MOGAWER: Thanks.

NIMA ROOHI: Walaa, one question and a comment. The continuation of the discussion, just gathering a comparison between the number of cycles versus energy without considering the material characteristics and the number of cycles approach, may cause some of the differentiations here. So I think it may be really the time to move to more material characteristics compared to just the number of cycles like we were doing at different ranges of temperatures. That's one discussion point. The second discussion point – I missed a couple slides, I guess – I don't know how many mix designs you used in the study.

WALAA MOGAWER: We did the mix design for the 64-28, which is all virgin. And then we did the mix design with 58-28. Then we verified all the mix designs with the rejuvenators and the RAP.

NIMA ROOHI: All right. So my assumption is that you are using the same gradation for all of them.

WALAA MOGAWER: Yes.

NIMA ROOHI: That's one.

WALAA MOGAWER: We used individual sieving to make sure that we are getting the same gradation for all the mixtures.

NIMA ROOHI: That totally makes sense. I just wanted to give probably a thinking direction here that adding more mix design, first of all, my understanding is that different mix designs have different sensitivities to the effect of rejuvenation oils. So if you change the gradation, change the mix design, a different type of rejuvenating oil, whether it's more efficient or less efficient in terms of dosage you need to soften the material, make it more flexible, may be better in terms of durability and cracking that you're taking a look at. So my concern is that taking into conclusion using one mix design may be appropriate. You may want to use like a couple of mix designs and then try to take a look at the trend between the two types of results you're getting, the cyclic test, for example, and fracture test. And, of course, taking into conclusion for each specific type of rejuvenator whether they are good in performance or not, using one mix design, again, may be a little concerning. But thank you for the presentation.

WALAA MOGAWER: You're welcome. Yeah, we stayed with one mix design, but, again, I don't want anyone to be misled. We're not recommending any test or saying this test is better than this. Yes, we looked at just these tests because based on the literature these tests were correlated to the field, so our expectation was they give the same trend.

FRANK FEE: Walaa, my question was going to be, how did you handle the RAP? This was all lab mix designs, right?

WALAA MOGAWER: Yes.

FRANK FEE: How did you handle the RAP and how did you prepare the samples? Did you preheat the RAP? Did you use a superheated aggregate? How did you prepare the mixes?

WALAA MOGAWER: We air dried the RAP for few days and we kept measuring the weight to make sure it's a constant weight - that took few days. And then before mixing by two hours, we put the RAP on top of the hot aggregate.

FRANK FEE: Okay. I think the issue we have to deal with is understanding how that relates to what happens in the field, all right? It's a time-temperature relationship, and I think you're going to see a completely different action, depending on how you handle that. If you're going to do it in the laboratory, you probably want to start in the field and come back to find out what that relationship is, all right? So we're talking about doing five different mix designs. You need to do one that's related to the field, all right? We need to work in that direction. As you're going to go forward, you've got to understand how everybody handles this differently in a laboratory right now. That's a problem, all right?

WALAA MOGAWER: Right, right.

FRANK FEE: It wasn't so bad for 10% or 15%. It's a big deal for 50%.

WALAA MOGAWER: Yes.

FRANK FEE: Because the interaction is going to depend on what you do in that laboratory.

WALAA MOGAWER: I'm a big believer, Frank, for any study that I do that plant-produced mix should be used. However, to get a 50% RAP plant-produced mix is not easy. But I agree with you—the lab is different than the field.

GRANT WOLLENHAUPT: I'm going to exercise a moderator's prerogative and jump in, and then we'll take one more question. But in light of our members' meeting discussion on how to get new members and thinking from an industry perspective, if we want to bring value and show value in the testing of what you guys are doing, maybe a suggestion is instead of normalizing your data to a mix design that's a virgin mix and expensive and probably rare in your area, definitely in ours, maybe normalizing the data for a more realistic or common mix design so that when we have membership hearing this, they can think in terms of, "This is what I do back home. I can improve my mix design if I do X," as opposed to, "Here's a perfect scenario. I don't know how that relates to what I'm doing right now." So maybe a suggestion is normalize your data on something that's being used in that area.

WALAA MOGAWER: That's actually a very good comment. I mean, we learned our lesson so we're doing another study now, and we're normalizing to the typical mix with 15% RAP.

GRANT WOLLENHAUPT: Last question.

ZIA ALAVI: Such a nice work.

WALAA MOGAWER: Thank you.

ZIA ALAVI: I always enjoy such collaborative efforts. I believe with more research on this topic, actually we can solve the problem better. My question is about your experimental plan. I wonder how you select the doses of the rejuvenating agent. I believe the best way to select appropriate dosages is to perform some binder tests and find the equivalent doses for different rejuvenating types instead of selecting arbitrary ones, even based on the industry experience.

WALAA MOGAWER: That's a good point. The dosage selection was based on a previous study that we did using the same binders and rejuvenators. So what we did in the previous study was to find the proper dosage based on binder PG grading. Our ultimate goal was that the resultant binder which combines 50% virgin binder, 50% RAP binder, and a rejuvenator would be a PG 64-28. We did this by incrementally adding the rejuvenators at different dosages until we reached a continuous grade similar to the continuous grade of a virgin PG 64-28.

ZIA ALAVI: Okay.

WALAA MOGAWER: So it's fortunate in this case that for all rejuvenators, we were able to keep the dosage the same.

ZIA ALAVI: Thank you. It is clear now!

WALAA MOGAWER: You're welcome.

GRANT WOLLENHAUPT: Sorry, I have to cut you off. We're running behind, but continue this discussion later over lunch.

A Comprehensive Evaluation of the Fatigue Behavior of Plant-Produced RAP Mixtures

Mohammadreza Sabouri[a], Thomas Bennert[b], Jo Sias Daniel[c], Y. Richard Kim[d]

[a]*Braun Intertec Corporation, Minneapolis, MN*

[b]*Civil and Environmental Engineering, Rutgers, The State University of New Jersey, Piscataway, NJ*

[c]*Civil Engineering, University of New Hampshire, Durham, NH*

[d]*Department of Civil, Construction and Environmental Engineering, North Carolina State University, NC*

ABSTRACT. In this study, the fatigue performance of twelve plant-produced mixtures from New Hampshire and Vermont that contain reclaimed asphalt pavement (RAP) contents of 0% to 40% by total weight of mixture was evaluated. The mixture tests included dynamic modulus, uniaxial fatigue, beam fatigue, and overlay tests. Also, the simplified viscoelastic continuum damage (S-VECD) model failure criterion, called the G^R method, was applied and input to the linear viscoelastic pavement analysis for critical distresses (LVECD) program to predict the fatigue behavior of the tested mixtures in thin and thick asphalt pavement structures. In order to explain the observed fatigue behavior, the performance grades (PGs) of the binders that were extracted and recovered from the mixtures were determined. In general, the addition of RAP resulted in an increase in the stiffness of the materials. The magnitude of the impact of higher RAP percentages varied with each set of mixtures. The S-VECD model and beam fatigue test data showed a loss of fatigue resistance for high percentage RAP mixtures in most of the cases. The overlay tester results showed clear drops in performance at higher RAP contents. The impact of lowering the PG of the virgin binder to compensate for higher levels of RAP also was studied. Lowering the PG led to improvement in the fatigue properties and was found to be a convenient practice. The changes in the measured properties also appeared to be a function of mix design variables that included the stiffness of the RAP, asphalt content, and production parameters such as silo storage times. In some cases, the effects of these factors outweighed the impact of the RAP level or PG of the virgin binder in the mixtures.

KEYWORDS: reclaimed asphalt pavement, fatigue, asphalt, failure criterion, damage

---

The oral presentation was made by Dr. Sabouri.

This is a reproduction of an article whose final and definitive form has been published in *Road Materials and Pavement Design*© 2015 Taylor & Francis. The article is available online at: http://dx.doi.org/10.1080/14680629.2015.1076997

1.0 Introduction

Each year millions of tons of asphalt concrete debris is produced from aging infrastructure in the United States. The disposal of this waste in landfills has been a traditional solution to disposing of it, but the scarcity of landfill areas, environmental regulations, and costs have hindered the safe disposal of these waste products. As an alternative solution, the use of reclaimed asphalt pavement (RAP) has been proven to provide technical, economical, and environmental benefits. Over recent decades, agencies have been seeking ways to reap the benefits of RAP since noticing a significant reduction in production costs when RAP is used. Using RAP reduces the cost of purchasing and transporting new aggregate and binder for asphalt mixtures and reduces the energy associated with extracting and processing nonrenewable natural resources for pavement construction, rehabilitation, and maintenance. Therefore, state highway agencies have been moving toward increasing the percentages of RAP in hot mix asphalt (HMA) pavements (Copeland, 2011). Incorporating 15% to 20% RAP into mixtures has been accomplished comfortably; however, concerns about fatigue performance and the need to use softer performance grades (PG) binders has limited the amount of HMA that has more than 15% to 20% RAP that can be produced in many areas of the northeast United States (Copeland, 2011).

The main concern about using higher percentages of RAP is that RAP contains asphalt binder that has been aged, so incorporating higher RAP contents into HMA can produce mixtures that are stiffer than the same mixtures without RAP. Consequently, such mixtures would be less workable and more susceptible to field failure, such as fatigue cracking (Bonaquist, 2005; McDaniel et al., 2007). Also, certain production factors, such as the RAP source, virgin binder PG, production temperature, and plant type, can affect the degree of blending between RAP and virgin asphalt and, consequently, affect the performance of the resultant asphalt mixture (Mogawer et al., 2012; Mensching et al., 2014; Sabouri et al., 2015).

To mitigate the increase in stiffness, some specifications have recommended the use of a softer binder when RAP is incorporated into the mixture. However, the use of high RAP contents and a softer binder may still result in a mixture that is very stiff. Mixtures that are very stiff may experience low-temperature cracking and may crack prematurely in pavements that experience high levels of deflection (Copeland, 2011). Also, if poor blending occurs between the soft binder and the RAP binder, the mixture may still be susceptible to rutting, moisture damage, and even cracking.

Despite these problems, departments of transportation (DOTs) in the United States have been encouraged to increase the amount of RAP in flexible pavements because of economic considerations and environmental concerns. The primary objective is to encourage the use of recycled materials in the construction of highways to the maximum economical and practical extent possible with equal or improved performance (Al-Qadi et al., 2012).

In this study, a new method that allows mechanistic evaluation of the effect of RAP on the performance of asphalt mixtures was used. The simplified viscoelastic continuum damage (S-VECD) model is a continuum damage mechanics-based model that has been applied effectively to predict the performance of asphalt concrete mixtures under different loading conditions (Hou et al., 2010; Daniel and Kim, 2002; Chehab et al., 2003; Underwood and Kim, 2009). Zhang et al. (2013) developed an energy measure that represents the rate of damage growth using the S-VECD model and can predict fatigue failure. However, Zhang et al. developed this approach based on a controlled crosshead (CX) mode of loading test data only. Later, Sabouri and Kim (2014) applied Zhang et al.'s failure criterion to different modes of loading and found it to be mode of loading-dependent. They proposed a new failure criterion (the G^R method) that remedies the problems associated with Zhang et al.'s failure criterion. This new failure criterion is considered to be independent of mode of loading, strain amplitude, and temperature. Later, Norouzi et al. (2014) successfully applied the G^R method to investigate the effect of incorporating high percentages of RAP on the performance of asphalt mixtures in terms of fatigue cracking.

2.0 Objectives

The purpose of this study is to evaluate the fatigue behavior of plant-produced asphalt mixtures that contain up to 40% RAP. This study used available mixture fatigue testing methods and procedures: the uniaxial fatigue test with the S-VECD model, the beam fatigue test, and the overlay test. Using plant-produced mixtures is a better representation of what would be expected in terms of field fatigue performance.

3.0 Materials and Test Methods

In this study, in order to investigate the fatigue behavior of RAP mixtures, experiments were performed using state agency-approved plant-produced mixtures from New Hampshire and Vermont. These mixtures were produced during the 2010 construction season. The primary test variables were the percentage of RAP in the mixture and the virgin binder PG. Table 1 presents a summary of the twelve plant-produced mixtures. These mixtures are designated as XXLC in this study, with XX being the percentage of incorporated RAP and LC designating that the test samples were reheated and compacted at the research laboratory.

Table 1. *Plant-produced mixtures.*

Mixture Name	Plant Type	NMAS (mm)	Virgin PG	RAP Content (%) by total wt. of mix			
				0	20	30	40
VTeXXLC	batch	9.5	64-28	x	x	x	x
VTaXXLC	batch	9.5	52-34	x	x	x	x
NHeXXLC	drum	12.5	64-28	x	x	x	x

The Vermont (VT) mixtures were produced in an H&B five-ton batch plant located in Williston, Vermont. The mixing time was approximately 60 seconds. After production, the mixtures were immediately discharged into trucks for sampling; they were not stored in a silo. The New Hampshire (NH) mixtures were produced in a 2008 Gencor Ultra drum plant with 400-ton/hour capacity located in Portsmouth, New Hampshire. The mixing time was estimated to be approximately 40 seconds. Table 2 shows the general mixture design information for all the VT and NH mixtures.

The VT mixtures had a nominal maximum aggregate size (NMAS) of 9.5 mm with optimum asphalt contents ranging from 6.5% to 6.8%. The VT virgin binders had continuous grades of PG 56.3-32.5 (m-controlled by 4°C) and PG 64.4-30.2 (m-controlled by 1.5°C), respectively. The RAP used in the VT mixtures had a continuous grade of PG 73.8-25.2. The extracted and recovered asphalt contents from the plant-produced mixtures were lower in terms of percentage than the design optimum, with the VTe mixtures having significantly lower percentages than the design optimum. The NH mixtures had a NMAS of 12.5 mm with an optimum asphalt content of 5.7%. The virgin binder had a continuous grade of PG 66.3-29.5 (slightly m-controlled). The RAP used in the NH mixtures had a continuous grade of PG 85.5-13.2. The asphalt binder contents for the NHe00LC and NHe40LC mixtures were slightly above optimum while the asphalt binder contents for NHe20LC and NHe30LC were below optimum.

As Table 2 shows, the extracted and recovered asphalt content is usually lower and sometimes quite different from the design asphalt content, which could be due to several factors, including: (1) not knowing the 'true' asphalt content of the RAP material in the asphalt plants, (2) the variability in the RAP stockpile due to the RAP being brought from various projects/locations, and (3) plant production control issues.

Table 2. *Summary of study mixture properties*

Mix	Virgin PG	NMAS (mm)	Design AC (%)	E&R* AC (%)	E&R* PG			VMA	VFA	RAP AC (%)	RAP E&R* PG	%RAP by total wt.	% RAP by binder wt.**
					High	S	m						
VTa00LC	52-34	9.5	6.7	6.58	65.4	-36.9	-28.3	20.2	76.3	--	--	0	0.00
VTa20LC	52-34	9.5	6.8	6.27	68.3	-35.3	-28.1	18.8	81.9	5.41	70-22	20	17.3
VTa30LC	52-34	9.5	6.6	6.13	71.4	-34.8	-26.3	17.7	82.5	5.41	70-22	30	26.4
VTa40LC	52-34	9.5	6.6	6.12	68.6	-33.4	-21.0	18	77.8	5.41	70-22	40	35.4
VTe00LC	64-28	9.5	6.5	5.84	67.4	-30.8	-28.1	20.3	71.5	--	--	0	0.00
VTe20LC	64-28	9.5	6.7	5.46	69.6	-30.4	-27.0	18.7	79.7	5.41	70-22	20	19.8
VTe30LC	64-28	9.5	6.6	5.31	74.7	-30.0	-23.0	19.1	75.9	5.41	70-22	30	30.6
VTe40LC	64-28	9.5	6.6	6.02	78.0	-30.4	-24.9	18.2	76.4	5.41	70-22	40	36.0
NHe00LC	64-28	12.5	5.7	5.84	71.8	-29.5	-28.4	14.9	74.8	--	--	0	0.00
NHe20LC	64-28	12.5	5.7	5.46	76.7	-28.9	-24.1	14.4	79.9	4.79	82-10	20	15.5
NHe30LC	64-28	12.5	5.7	5.31	78.1	-36.9	-28.3	14.5	81.3	4.79	82-10	30	23.3
NHe40LC	64-28	12.5	5.7	6.02	79.8	-35.3	-28.1	14.5	82.1	4.79	82-10	40	30.3

*E&R: Extraction and recovery

**%RAP by recovered binder weight

For the VT mixtures, the aggregate gradations were consistent, with the VTe40LC mixture being slightly coarser and the VTe20LC mixture being slightly finer than the others. For the NH mixtures, the gradations for the NHe20LC and NHe30LC mixtures were similar to that of the virgin mixture. The percentages passing the #100 and #200 sieves were lower in the NHe40LC mixture than in the other mixtures. Figure 1 presents a graph of the mixture gradations for all the control mixtures (i.e., the mixtures without RAP). RAP mixture gradations were very similar to their corresponding control mixture.

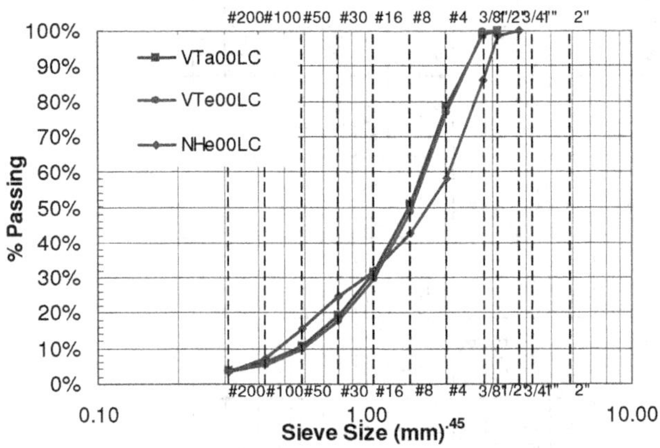

Figure 1. *Non-RAP mixture gradations.*

Table 3 shows the plant production information for all the study mixtures. The VT mixtures were produced between 146°C and 171°C (295°F and 340°F) and immediately discharged for sampling. The ambient temperatures during production were 4°C to 10°C (40°F to 50°F) during cloudy, wet weather. Moisture was observed in the 30% and 40% RAP mixtures. The NH mixtures were produced between 157°C and 168°C (315°F and 335°F). The virgin mixture was produced at 166°C (330°F) and the RAP mixtures were produced between 154°C and 157°C (310°F and 315°F). All the NH mixtures were stored in a silo prior to being discharged into delivery trucks. However, the NHe00LC mixture was stored in the silo for a much longer time (6 hours) than the NHe20LC and NHe30LC mixtures (1.25 and 1.0 hours, respectively).

Table 3. *Plant production information*

Mix	PG	Mixing Times Dry/Wet (s)	Discharge Temp. (°C/°F)	Silo Storage time (hr)
VTa00LC	52-34	6/36	171/340	0
VTa20LC	52-34	10/36	162/324	0
VTa30LC	52-34	14/36	160/320	0
VTa40LC	52-34	13/36	149/300	0
VTe00LC	64-28	6/36	166/330	0
VTe20LC	64-28	10/36	149/300	0
VTe30LC	64-28	13/36	161/322	0
VTe40LC	64-28	13/36	146/295	0
NHe00LC	64-28	n/a	166/330	6
NHe20LC	64-28	n/a	157/315	1.25
NHe30LC	64-28	n/a	168/335	1.0
NHe40LC	64-28	n/a	168/335	n/a

3.1 Specimen Preparation

The study materials were loose mixtures that were collected and cooled to room temperature and then stored in a stable, environmentally controlled room at air temperatures lower than 25°C (77°F) in sealed, five-gallon metal buckets.

Reheating the loose mix for sample fabrication was conducted at temperatures that were 10°C lower than those measured from the discharge temperatures (Table 3). All specimens were produced using loose mix that had been reheated only once, as it was feared that additional reheating cycles might affect the general stiffness properties of the HMA as well as potentially increased the possible blending of the RAP with the virgin asphalt binders.

Each bucket of loose mix was placed in the oven at the appropriate splitting temperature with the bucket lid on. After about an hour, the lid was removed, and the bucket was kept in the oven for one more hour. Then, the temperature of the mixture was checked to confirm that the center of the mixture was at least 75°C. The material was then placed into a splitting mold that divided the loose mix into three portions. The loose mix was scooped from each of the three portions and weighed to achieve the required testing sample weight. The individual samples were then heated to an appropriate mixing temperature and compacted (Table 3). The whole splitting process took less than ten minutes to complete. No loose mix was reheated for more than three hours. The compacted specimens were wrapped and sealed in black plastic and stored in a stable, environmentally controlled room at air temperatures lower than 25°C (77°F).

3.2 Test Protocols

The binder testing was conducted according to the test protocols described below. Additionally, four main mixture tests were carried out in this study: (1) the dynamic modulus test to determine the linear viscoelastic characteristics, (2) the uniaxial fatigue test using the S-VECD model to describe the viscoelastic damage characteristics, (3) the flexural beam fatigue test, and (4) the overlay test. The uniaxial and beam fatigue tests are considered to be crack initiation fatigue tests whereas the overlay test is a crack propagation fatigue test. The binder and mixture laboratory test protocols are described in the following subsections.

3.2.1 Binder Testing

The asphalt binders were extracted and recovered from sampled loose mix obtained from the asphalt plant. The asphalt binder from the loose mix was extracted and recovered in accordance with AASHTO T 164, "Standard Method of Test for Quantitative Extraction of Asphalt Binder from Hot-Mix Asphalt (Method A)," and ASTM D5404, "Recovery of Asphalt from Solution Using the Rotatory Evaporator," using trichloroethylene (TCE) as the extracting solvent. After the recovery process, the asphalt binders were tested for the respective high temperature performance grades in accordance with AASHTO M 320, "Standard Specification for Performance-Graded Asphalt Binder." The low temperature cracking properties of the asphalt binders were also evaluated in accordance with AASHTO R 49, "Determination of Low Temperature Performance Grade (PG) of Asphalt Binders." The recovered asphalt binders were treated as rolling thin-film oven (RTFO) -aged asphalt binders, which assumes that the aging that occurred at the asphalt plant is equivalent to the aging that occurs during RTFO aging. In addition, during the extraction and recovery process, the asphalt binder content was determined in accordance with AASHTO T 164.

3.2.2 Dynamic Modulus Testing

Dynamic modulus testing was performed in load-controlled mode in axial compression following the protocol given in AASHTO TP 79, except for the target strain level. Tests were completed for all the mixtures at 4°C, 20°C, 40°C, and 54°C and at frequencies of 25, 10, 5, 1, 0.5, and 0.1 Hz. Load levels were determined by a trial and error process so that the resulting strain amplitudes were between 50 and 75 microstrain and the testing order was from low to high temperatures and from high to low frequencies in order to minimize the damage to the specimens. Complex modulus values were obtained from the final six cycles of each loading series, i.e., when the material reached the steady state. The dynamic modulus values were fitted for the coefficients of the sigmoidal function and time-temperature shift factors by optimizing the dynamic modulus mastercurves. After determining the shift factors, the dynamic modulus was converted to the relaxation modulus, E(t), of the Prony series form to obtain a constitutive relationship between the strain and stress in the time domain. Finally, a power term, α, used in S-VECD theory, was calculated from the maximum log-log slope, m, of the relaxation modulus and time using the relationship, $\alpha = 1 + \frac{1}{m}$.

3.2.3 Uniaxial Fatigue Testing using the Simplified Viscoelastic Continuum Damage Model

Cyclic testing was performed in crosshead-controlled (CX) mode of loading following the protocol given in AASHTO TP 107. In this study, all the cyclic tests were performed at four to six different amplitudes to cover a range of numbers of cycles to failure (from 1,000 to 100,000).

The fatigue failure for each of the specimens tested in the CX cyclic tests was determined using Reese's approach, which is based on the change in phase angle behavior (Reese, 1997). The phase angle increases until strain localization occurs, and then it drops suddenly. This sharp sudden decrease in the phase angle occurs around the failure point, which makes the determination of the number of cycles to failure accurate and consistent in laboratory testing.

Test temperatures were determined using Equation 1. This equation can be used to determine the maximum test temperature based on the PG of the base binder. Also, a test temperature higher than 19°C is not recommended.

$$T(^{\circ}C) \leq \frac{\textit{High temperature binder PG grade} + \textit{Low temperature binder PG grade}}{2} - 3 \leq 19^{\circ}C \qquad [1]$$

The S-VECD failure criterion (the G^R method) also was applied to the data. According to this criterion, a characteristic relationship exists between the rate of change of the averaged released pseudo strain energy (G^R) during fatigue testing and the final fatigue life (N_f). This relationship has been proven to be present in both RAP and non-RAP mixtures (Sabouri and Kim, 2014). The G^R can be calculated using Equation 2. This failure criterion combines the advantages of the S-VECD model and this characteristic relationship, which both originate from fundamental mixture properties.

$$G^R = \frac{\frac{1}{2}\int_0^{N_f} \left(\varepsilon^R{}_{0,ta}\right)_i^2 (1 - F_i)}{N_f{}^2} \qquad [2]$$

Where: $\left(\varepsilon^R_{0,ta}\right)_i$ = pseudo strain amplitude at cycle i, and

F_i = pseudo stiffness at cycle i.

All the uniaxial fatigue specimens were compacted at a height of 178 mm and a diameter of 150 mm using the Superpave gyratory compactor (SGC). To obtain specimens of uniform air void distribution, these samples were cored to a diameter of 100 mm and cut to a height of 130 mm (Lee et al., 2014).

Before proceeding to testing, the air void ratios were measured using the CoreLok method for each specimen for quality control. All of the uniaxial fatigue specimens had an air void content within the range of 6.0±0.5 percent. To minimize the aging effect, the specimens were sealed in plastic bags and stored carefully in a cabinet if they were not tested immediately after fabrication. No specimens were tested later than two weeks after they were cored and cut. All specimens were glued to metal plates at both ends using epoxy before they were placed in the machine for testing.

3.2.4 Beam Fatigue Testing

Flexural fatigue testing was conducted using the flexural beam fatigue test procedure outlined in AASHTO T 321. The applied tensile strain levels used for the fatigue evaluation were 300, 500, 600, 700 and 900 micro-strains. However, the number of tensile strain levels used during the testing was sometimes reduced when the amount of loose mix available for testing was limited. Also, due to limitations in material quantities, only one replicate per strain level was conducted.

AASHTO T 321 is a test procedure to evaluate the crack initiation properties of an asphalt mixture. Therefore, 'fatigue life' for this test is defined as the time at which crack initiation begins.

Samples used for the flexural beam fatigue tests were compacted using a vibratory compactor designed to compact brick samples that are 400 mm in length, 150 mm in width, and 100 mm in height. After the specimen compaction was complete, the samples were trimmed to within the recommended dimensions and tolerances specified under AASHTO T 321. The test conditions utilized were those recommended by AASHTO T 321, including the test temperature of 15°C.

3.2.5 Overlay Testing

The overlay test, suggested by Zhou and Scullion (2005), has been shown to provide good correlations to field cracking for both composite pavements (Zhou and Scullion, 2005; Bennert et al., 2009) and flexible pavements (Zhou et al., 2007; Bennert and Maher, 2013). The overlay tester evaluates the asphalt mixture's ability to resist or retard crack propagation. The sample preparation and test parameters used in this study follow that of TxDOT Tex-248-F (2009) testing specifications. These conditions include 25°C (77°F) test temperature, opening width of 0.635 mm (0.025 in.), cycle time of 10 s (5 s loading, 5 s unloading), and specimen failure defined as 93% reduction in initial load.

Five replicate specimens were tested for each mixture. The low and high values were discarded, and the remaining three values were used to calculate the average value and standard deviation.

4.0 Results and Discussion

4.1 Performance-Related Mixture Testing and Analysis

4.1.1 Stiffness: Dynamic Modulus

Figure 2 and Figure 3 show the averaged dynamic modulus values of the replicates for the VTe (PG 64-28) and VTa (PG 52-34) mixtures, respectively, on the same graph. As Figure 2 shows, the dynamic modulus values of the VTe mixtures with different percentages of RAP do not show significant differences. In other words, it seems that incorporating RAP into the VTe mixtures does not significantly affect the mixture properties in terms of dynamic modulus values. In contrast, for the VTa mixtures, incorporating RAP increased the stiffness of the mixtures. Also, as the RAP percentage increased, the stiffness values increased as well (Figure 3).

One possible reason for the different behavior between the VTe and VTa mixes could be the greater difference in the asphalt binder grades between the RAP binder and the virgin binder in the VTa mixtures than in the VTe mixtures. The VT RAP binder has a PG of 70-22, which is not very different from the PG of the virgin binder used in the VTe mixtures (PG 64-28). Therefore, adding a high percentage of RAP did not affect the stiffness significantly, whereas in the case of the VTa mixture with a base PG of 52-34, the difference between the virgin binder PG and the RAP binder PG is more pronounced and, as a result, adding more RAP caused a significant increase in mixture stiffness. Moreover, the VTa mixtures have higher asphalt contents (more than 0.5%, except for the 40% RAP mixtures, which are only 0.1% apart) than the VTe mixtures. Also, the 30% VTe has about 5% more RAP by weight of binder than the 30% VTa. The difference in asphalt binder grades may have contributed to the different stiffness behavior observed between the VTe and VTa mixes. In Figure 4, the stiffness values of the VTe and VTa mixtures are compared at the same level of RAP percentage. As these graphs suggest, as the RAP content increased, the impact of the binder PG decreased, and the stiffness values of the mixtures with the different binders became increasingly similar, likely due to the increased amount of recycled material.

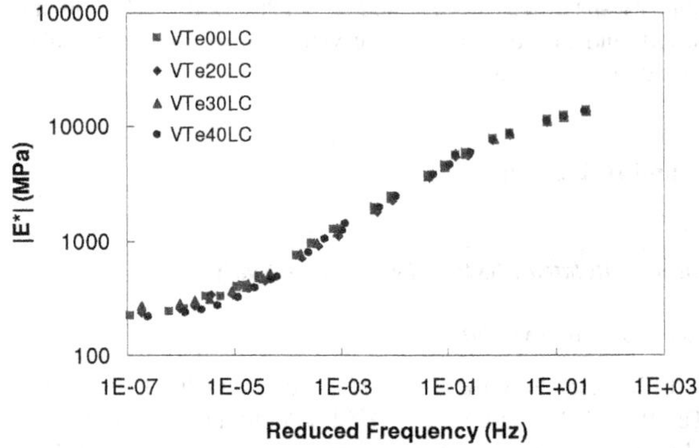

Figure 2. *Dynamic modulus test results for VTe mixtures.*

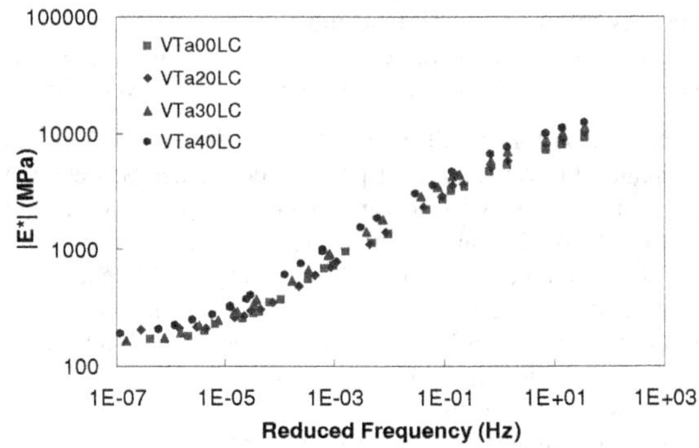

Figure 3. *Dynamic modulus test results for VTa mixtures.*

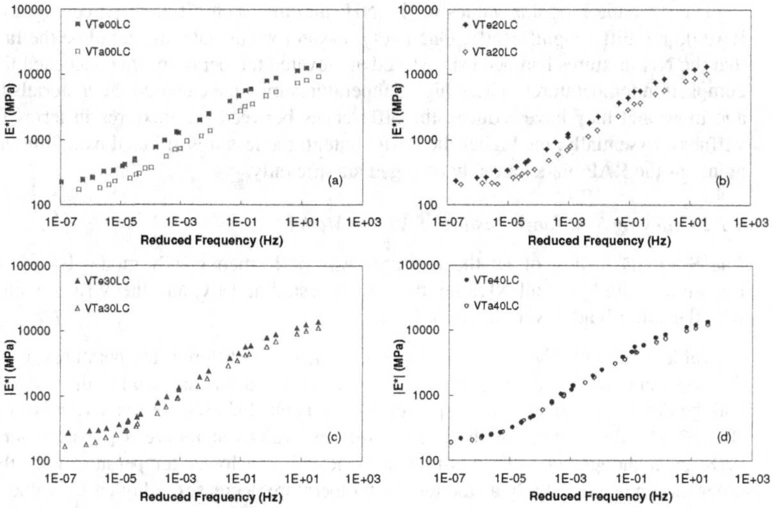

Figure 4. *Comparision of dynamic modulus values between VTe and VTa mixtures for (a) 0%, (b) 20%, (c) 30%, and (d) 40% RAP.*

Figure 5 shows the dynamic modulus values of the averaged replicates for all the NH mixtures on the same graph. As this graph shows, the non-RAP NHe00LC mix has the lowest value, and the NHe30LC has the highest value. It should be noted that the 30% RAP mix was 0.7% lower in total asphalt content than the 40% RAP mix, resulting in similar binder replacement values.

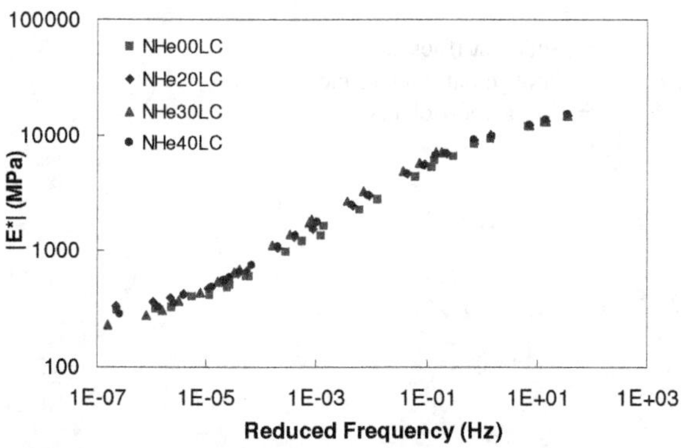

Figure 5. *Dynamic modulus test results for NHe mixtures.*

The dynamic modulus values of the NH mixtures with different percentages of RAP do not differ significantly. One likely reason for this outcome could be the fact that the NH mixtures had been silo-stored at elevated temperatures that exceeded the compaction temperatures. These high temperatures may have caused the materials to age more and may have reduced the differences between the mixtures in terms of stiffness. Essentially, the higher the RAP content, the less it is affected by additional aging, as the RAP binder already has aged significantly.

4.1.2 Cracking Resistance Testing: S-VECD Model

The S-VECD testing of all the mixtures was performed in CX mode. Based on Equation 1, the VTe and NHe mixtures were tested at 13°C and the VTa mixtures with the softer binder were tested at 7°C.

Table 4 presents the S-VECD model parameters at different temperatures using the exponential form for describing the relationship between pseudo stiffness (C) and the damage parameter (S) (Equation 3). As Table 4 shows, for the VTa mixtures (PG 52-34), the average of the pseudo stiffness values at failure, C_F, was around 0.18, even though the VTa mixtures were tested at a lower temperature than the other mixtures and usually as the testing temperature decreases, a higher C_F value is expected. This value increased to 0.24 for the VTe mixtures (PG 64-28). The value increased to 0.30 for the NH mixes. This trend may suggest that the NH mixes were the most brittle materials among the three sets of mixtures, with the VTe mixtures next, followed by the VTa mixtures with the softer binder and higher asphalt content that were more ductile than the other mixtures. This same general trend also was found for the average asphalt contents. That is, the VTa mixtures have the highest average asphalt content and the NH mixtures have the lowest average asphalt content of the three mixture types.

$$C = e^{aS^b} \quad [3]$$

Where: C = pseudo stiffness,
S = damage parameter, and
a, b = fitting coefficients.

Table 4. *Exponential fit parameters for VECD model for mixtures.*

Mix Type	Alpha	Temp (°C)	a	b	C_F
VTa00LC	4.64	7	-0.00040878	0.6374	0.16
VTa20LC	4.05	7	-0.00057296	0.6152	0.17
VTa30LC	4.19	7	-0.00036720	0.6437	0.21
VTa40LC	4.25	7	-0.00076187	0.5809	0.20
VTe00LC	4.31	13	-0.00021314	0.6768	0.24
VTe20LC	4.06	13	-0.00017425	0.6921	0.23
VTe30LC	4.08	13	-0.00054682	0.5951	0.25
VTe40LC	4.00	13	-0.00022793	0.6691	0.24
NHe00LC	4.13	13	-0.00006653	0.7906	0.31
NHe20LC	4.14	13	-0.00008771	0.7563	0.31
NHe30LC	4.19	13	-0.00013500	0.7113	0.27
NHe40LC	4.22	13	-0.00009147	0.7470	0.30

Figure 6 shows the fitted curves for all the VT mixtures on the same graph. The mixtures with the same virgin binder grade are clustered together with similar damage characteristic curves. The curves for the VTe mixtures are above those for the VTa mixtures. The performance of all the VTe mixtures is shown to be very similar, with the 30% RAP mixture showing a slightly higher curve than the other mixtures. For the VTa mixtures, it seems that the addition of RAP moves the position of the C vs. S curves higher where the 40% RAP mixture shows the highest curve among the VTa mixes. This trend is observed as well in the dynamic modulus test data for these mixtures.

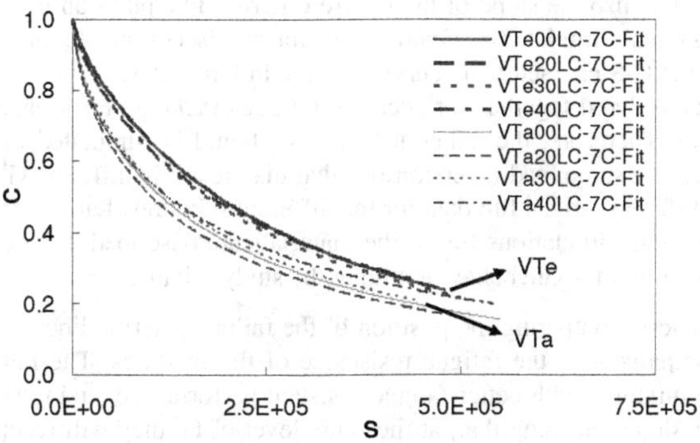

Figure 6. *Fitted results for CX testing for VT mixtures.*

Figure 7 shows the fitted curves for all the NH mixtures on the same graph. As this graph shows, the 0% RAP mix has the lowest position, and it seems that by adding more RAP, the C vs. S position moves to a higher position in general.

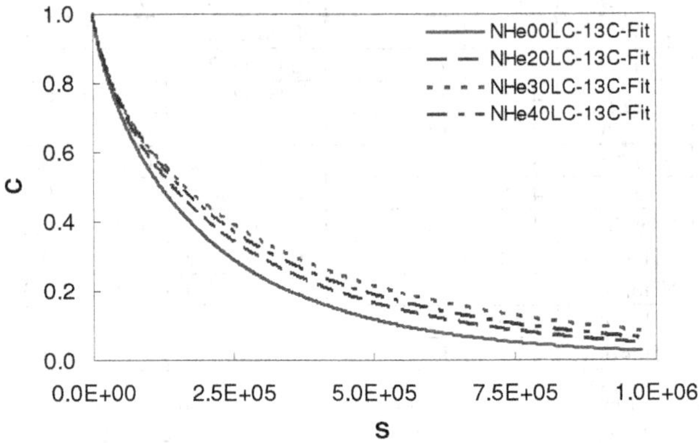

Figure 7. *Fitted results for CX testing for NH mixtures.*

The S-VECD failure criterion was applied to all the study mixtures. It should be noted that comparing fatigue life data at the same level of G^R may not be fair, because under the same loading history, different materials could respond differently causing different strains. That is, different materials could experience different levels of G^R under the same loading history due to their different stiffnesses. Also, not only the position but also the slope of the failure criterion line plays an important role in pavement system simulations. Overall, the linear viscoelastic (complex modulus) properties, the C-S characteristic curve, and the failure criterion (position and slope) should be considered together for accurate fatigue cracking performance prediction of a mixture. Therefore, the material behavior should be simulated in a pavement system under the same loading conditions that may result in different G^R values and, therefore, different fatigue life data for the different mixtures. Later in this paper, the pavement system simulations for all the materials are described in order to obtain a true comparison of fatigue behavior among the study mixtures.

Nevertheless, comparing the position of the failure criterion lines can be used as a rough comparison of the fatigue resistance of the mixtures. The failure criterion lines of the mixtures with better fatigue-resistant performance tend to stay high with a shallower slope, meaning that, at the same level of G^R they will result in a higher N_f value.

Figure 8 shows all the VTe mixture failure criterion data on the same plot. As this graph shows, the relationship between G^R and N_f is significant for all the mixtures. Also, the characteristic relationships for the four VTe mixtures seem to be at about the same position, which suggests similar fatigue behavior for these mixtures. Based on this, it is concluded that high quality HMA with up to 40% RAP can be designed and can meet the desired fatigue resistance performance.

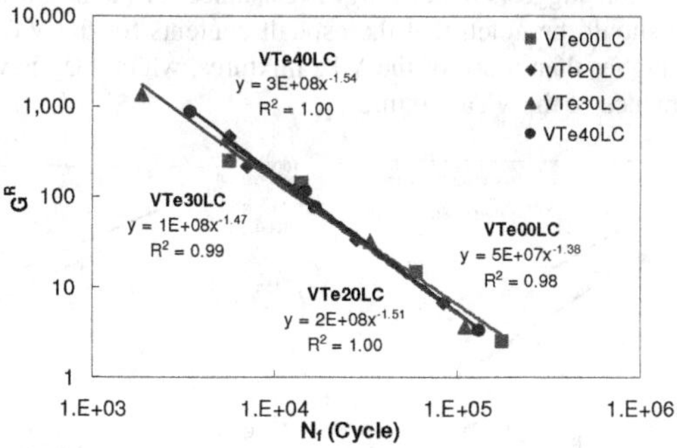

Figure 8. *Fatigue failure criterion for VTe mixtures at 13°C.*

Figure 9 shows all the VTa mixture failure criterion data on the same plot. Again, the relationship between G^R and N_f is significant for all the mixtures. The VTa00LC mixture shows the best performance, followed by the VTa20LC mixture. The VTa30LC and VTa40LC mixtures are very similar.

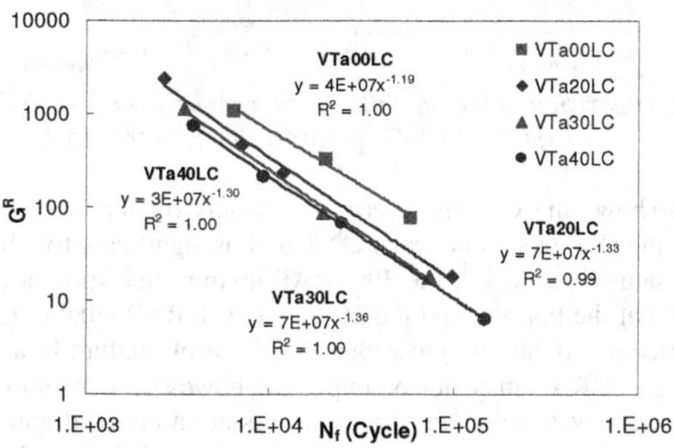

Figure 9. *Fatigue failure criterion for VTa mixtures at 7°C.*

Like before, one possible reason for this different behavior between the VTe and VTa mixes could be the greater difference in the asphalt binder grades of the RAP binder and virgin binder in the VTa mixtures, which impose a greater effect on the VTa mixes that have a softer base binder.

Figure 10 shows the impact of the different binder grades for the different RAP contents. It seems that as the RAP content increases, the impact of the binder grade decreases and the performance of the mixtures with the different binders becomes increasingly similar, likely due to the increased amount of recycled material. Also, for all the RAP percentages, the VTa failure criterion line lies above the VTe failure criterion line, which suggests better fatigue resistance for the mixture with the softer base binder. It should be noted that the asphalt contents for the VTa mixtures were 0.1% to 0.8% higher than those of the VTe mixtures, which may have improved the fatigue performance of the VTa mixtures.

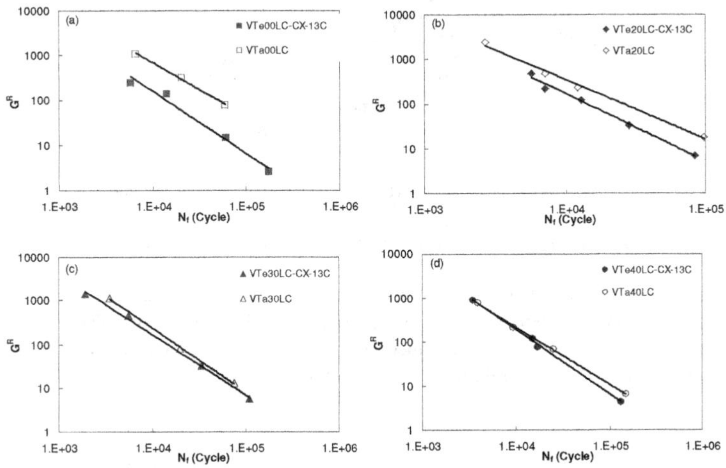

Figure 10. *Comparision of fatigue failure criterion between VTe and VTa mixtures for (a) 0%, (b) 20%, (c) 30%, and (d) 40% RAP.*

Figure 11 shows the G^R failure criterion results for the NH mixtures. As this graph shows, the relationship between G^R and N_f is significant for all four mixtures. Also, for the same level of G^R, the 30% RAP mixture line stays at the top, the 0% RAP mix stays at the bottom, and the 20% and 40% RAP mixture lines are located in between. Better fatigue resistance than the Control mixture in a mixture with a high percentage of RAP may not be expected; however, as mentioned before, the NHe00LC mixture was silo-stored for six hours at an elevated temperature, which could have affected its fatigue performance. Also, it should be noted that the slope of the NHe00LC line is less steep than the other mixtures, which would affect its fatigue life prediction in pavement system simulations. More definitive conclusions about the NH mixes can be made once the pavement system simulations are performed using the LVECD model.

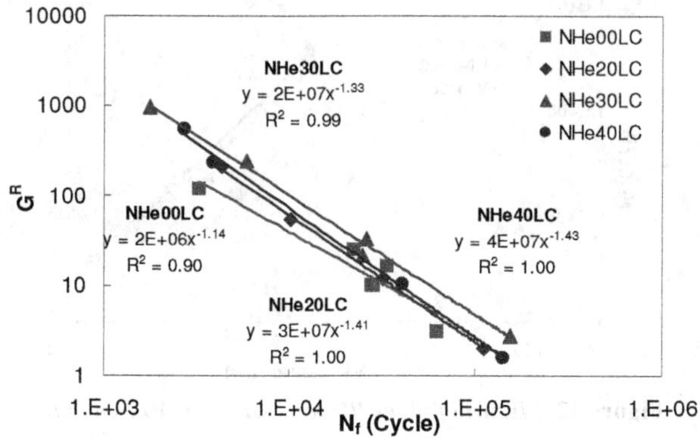

Figure 11. *Failure criterion for NH mixtures at 13°C.*

4.1.3 Cracking Resistance Testing: Beam Fatigue

Beam fatigue testing was performed according to AASHTO T 321, but due to material limitations, only one replicate per strain level was performed. Figure 12 shows the test results for the VTe mixtures. As this graph indicates, all four of the VTe mixtures resulted in almost identical flexural fatigue performance, and no conclusions could be drawn as to which mixture performed the best. These results are in line with the results of S-VECD testing where no significant difference in the fatigue properties of the VTe mixtures was seen.

Figure 13 presents the flexural fatigue test results of the VTa mixtures. This graph indicates that the VTa30LC mixture performed the worst out of the three mixtures evaluated for a specific strain level, whereas the VTa00LC and VTa20LC mixtures performed similarly. There was not enough material to evaluate the VTa40LC mixture. It was difficult to draw conclusions regarding the ranking of the three mixtures because of the limited test data (only one replicate was performed), and also because of the scatter around the fit lines (especially for the 0% RAP mixes).

Figure 14 shows the flexural fatigue test results for the NH mixtures. These results seem to be more scattered than the data from the two VT mixtures. The NHe00LC and NHe40LC mixtures performed similarly to one another and achieved the highest flexural fatigue life among the NH mixtures. Also, the NHe20LC mixture performed the worst. The NHe30LC mixture results were somewhat inconclusive, as the mix performed poorly at low tensile strain levels, but its performance improved significantly at higher tensile strain levels. The differences in the beam fatigue rankings may be affected by the difference in asphalt content where the 0% and 40% RAP mixtures had higher actual asphalt contents than the 20% and 30% RAP mixtures.

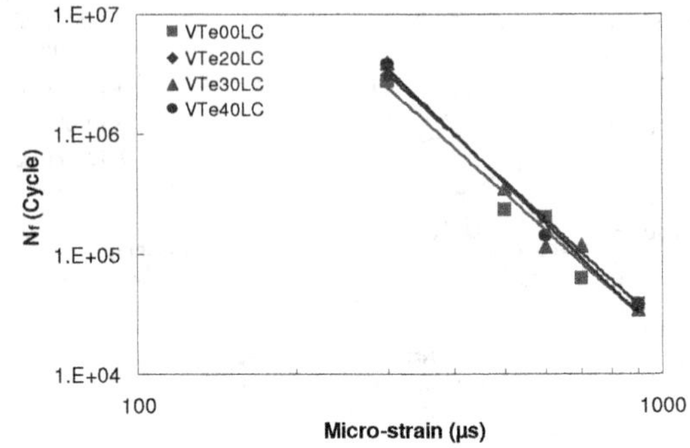

Figure 12. *Flexural fatigue life test results for VTe mixtures.*

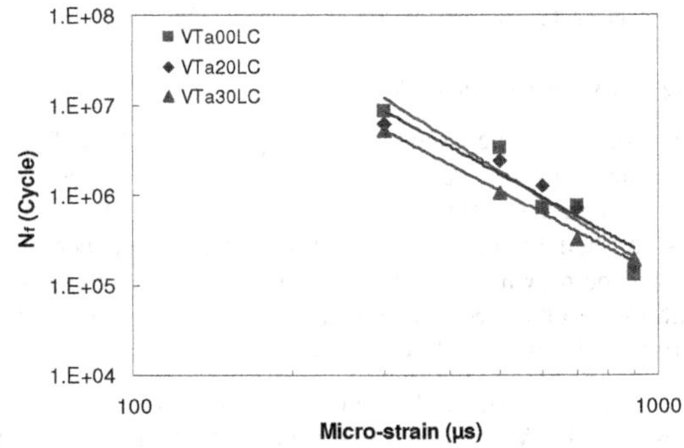

Figure 13. *Flexural fatigue life test results for VTa mixtures.*

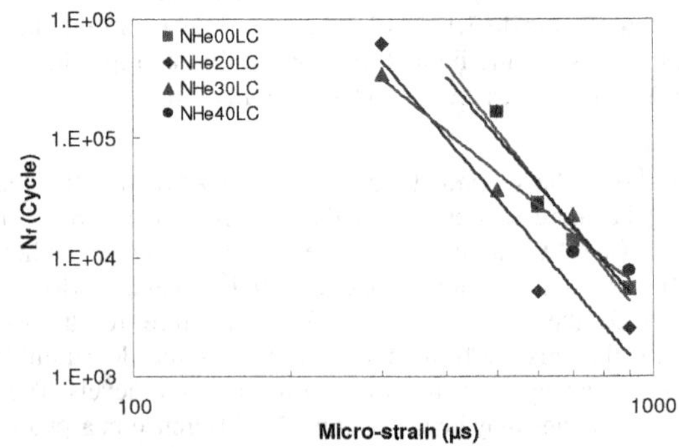

Figure 14. *Flexural fatigue life test results for NH mixtures.*

Comparisons were made for each of the different RAP contents to determine if the use of a softer asphalt binder would improve the fatigue properties of the asphalt mixtures, as presented in Figure 15. These results confirm the S-VECD results. The use of a softer binder with 0.1% to 0.8% more asphalt binder resulted in an improvement in flexural fatigue resistance at the identical RAP content, but as the RAP content increased, the impact of the binder PG decreased, and the performance of the mixtures with the different binders became similar, likely due to the increased amount of recycled material.

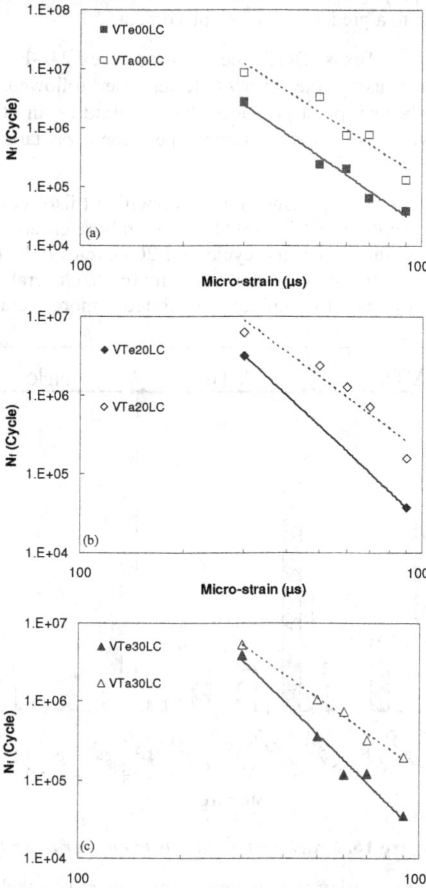

Figure 15. *Comparision of flexural fatigue life test results between VTe and VTa mixtures for (a) 0%, (b) 20%, and (c) 30% RAP.*

4.1.4 Cracking Resistance Testing: Overlay Tester

The plant-produced mixtures were also evaluated for their cracking resistance using the overlay tester. The overlay tester applies tension loading (in displacement control) to test specimens while recording load, displacement, temperature, and time measurements. The main difference between this testing method and the other two cracking resistance test methods discussed in this paper is that this test is a crack propagation test, not a crack initiation test. That is, the test measures a mixture's ability to resist the propagation of cracks from the bottom of the specimen to the top of the specimen due to a predetermined displacement.

For this study, a Texas DOT specification (Tex-248-F, 2009) for testing bituminous mixtures using the overlay tester was followed. Specimens were fabricated from reheated plant-produced loose mixtures in the SGC and then trimmed. The air void level of the trimmed specimens was targeted for 6.5% ±0.5 percent.

The parameters for testing included a joint opening (displacement) of 0.635 mm (0.025 in.), test temperature of 15°C (59°F), and a failure criterion of 93% reduction in the load measured during the first cycle or 1,200 cycles (whichever occurs first). The averaged results of the tests are shown in Figure 16. Generally, the mixtures that exhibited more cycles to failure exhibited more cracking resistance.

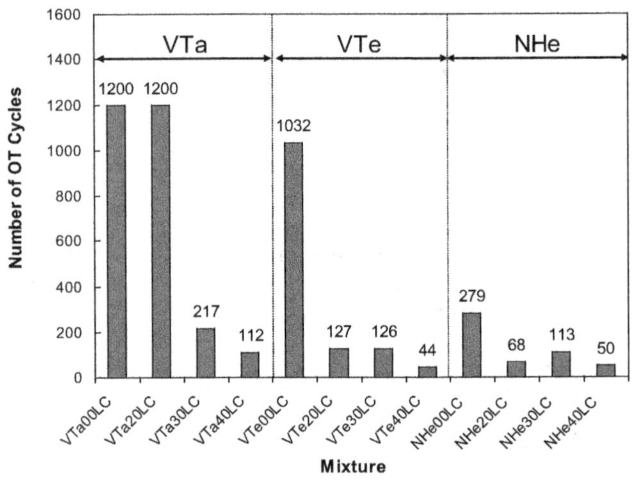

Figure 16. *Cracking test results from overlay tester.*

The data from the overlay tests indicate that, generally, cracking resistance was reduced as the percentage of RAP in the mixture increased. These data agree with the results from the stiffness tests that indicated that the addition of RAP stiffened the resultant mixture. These data also agree with the S-VECD and beam fatigue test data. Stiffer mixes generally are more susceptible to cracking at moderate to high levels of deflection. Also, the data indicate that the softer binder improved the cracking characteristics of the VT mixtures.

Like the S-VECD and beam fatigue test results, the NH mixture data trend showed a slightly unexpected variation, i.e., where the number of overlay tester cycles increased at 30% RAP. It is interesting to note that the same trend was observed for the low temperature PG of the extracted binder, which showed an improvement from 20% to 30% RAP.

4.2 Pavement Analysis

In this study, LVECD simulations were used to evaluate long-term pavement system performance. Eslaminia et al. (2012) developed this layered viscoelastic structural program with the material level continuum damage model to calculate the stresses and strains necessary for predicting fatigue behavior using the S-VECD model. This program is based on an efficient framework developed by combining time scale separation and layered viscoelastic analysis and uses fast-Fourier transforms to perform the three-dimensional viscoelastic calculations under moving loads quickly without the need to account for the previous loading history at a given point. This model provides simulation times that are orders of magnitude smaller than those found in conventional three-dimensional finite element methods. The assumption behind the transforms is that loading occurs in the same repeated manner each time, so the behavior is cyclic and has a steady-state response. This assumption is reasonable and is used to characterize 20 years of pavement performance within an hour.

4.2.1 Inputs

The inputs required for the LVECD simulations are the standard components needed for most pavement simulations: design time, structural layout, traffic, and climate. The typical design time for asphalt pavements is 20 years, so the simulations are run to 20 years at the same traffic rate to predict future performance.

Figure 17 presents the pavement structure that was used in the LVECD simulations. Simulation was done on thin (strain-controlled) and thick (stress-controlled) pavement structures. In the thin pavement case, the asphalt concrete layer is 100 mm (4 in.) thick with an aggregate base of 200 mm (8 in.), while for the thick pavement a full depth asphalt layer of 300 mm (12 in.) is assumed. The asphalt layer is described as viscoelastic with damage. Therefore, these layers need the Prony series representation of the dynamic modulus values, shift factors, and S-VECD model values. The aggregate base and subgrade were modeled using linear elastic properties. The layer properties are listed in Table 5.

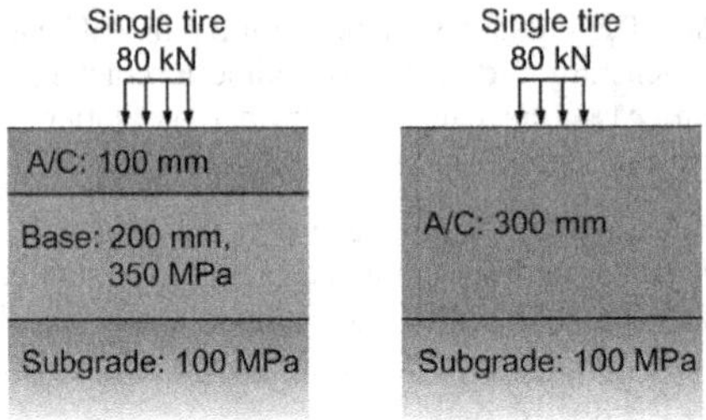

(a) Thin Pavement (b) Thick Pavement

Figure 17. *Pavement structure used in LVECD simulations for (a) thin pavement, and (b) thick pavement.*

Table 5. *Design structure.*

Layer	Modulus (MPa)	Poisson's Ratio	Material Type
AC	$\|E^*\|$	0.30	Viscoelastic with Damage
Aggregate Base	350	0.35	Linear Elastic
Subgrade	100	0.35	Linear Elastic

The traffic information is listed in Table 6. For convenience, a single tire from the standard axle loading of 18 kips on one axle is applied at the center of the pavement. The LVECD model simulates 20 years of asphalt concrete pavement performance using climate and traffic data. The average annual daily truck traffic (AADTT) count is 2000, so the total number of wheel applications is 14.4 million for 20-year simulations.

For the climate data, the pavement was assumed to be placed in Boston, Massachusetts. The LVECD program uses pavement temperatures obtained from the Enhanced Integrated Climatic Model (EICM). The EICM program provides hourly temperatures of asphalt pavements in terms of pavement depth.

Table 6. *Traffic inputs for LVECD.*

Vehicle	ESAL
Design Velocity (kph)	100
AADTT	2000
Pressure Distribution	Constant
Contact Area	Rectangular
Aspect Ratio (length/width)	11/7
Tire Pressure (MPa)	0.21
Growth Type	No Growth
Lane Distribution Factor	1

4.2.2 Fatigue Performance

In the LVECD program analysis, the results (i.e., for damage evolution, stress, and strain) can be evaluated in terms of spatial distribution and time history distribution. For fatigue resistance evaluation, the LVECD program calculates the damage growth (i.e., reduction of the secant modulus) and the damage factor that is defined in Equation 4 based on Miner's law. If the damage factor is equal to zero, the element does not have any damage. A damage factor of one indicates failure of the element. Examples of the damage factor contour output as well as the secant modulus contour output are shown in Figure 18. The plot shows the results for different time periods to show the progression of the damage. The plots show the cross-section of the pavement with the direction of traffic heading into the page.

$$Damage\ Factor = \sum_{i=1}^{T} \frac{N_i}{N_{fi}} \qquad [4]$$

Where: T = total number of periods,
N_i = traffic for period i, and
N_{fi} = allowable failure repetitions under the conditions that prevail in period i.

4.2.3 Fatigue Performance Prediction

The LVECD model fatigue predictions were performed for all the study mixtures. In order to compare the mixtures in terms of fatigue resistance, the numbers of failure points (the points with the damage factor of '1') during the design period were compared for the thin pavement structures. For the thick pavements, because the structure was very strong, no failure occurred even after the 20-year simulations; so, in order to make the comparison, the numbers of points with the damage factor of '0.1' were compared. It is noted that the fatigue performance predicted from the LVECD program has not been fully calibrated against the field performance data. A preliminary comparison of the LVECD predicted damage and the percent cracking area measured from in-service pavements is presented in Norouzi and Kim (2014). However, transfer functions that convert the damage predicted from the LVECD to the percent cracking area have not been developed yet. Therefore, the LVECD predictions presented in the remaining portion of this paper use the number of failure points to evaluate the effects of different RAP mixtures on the pavement performance.

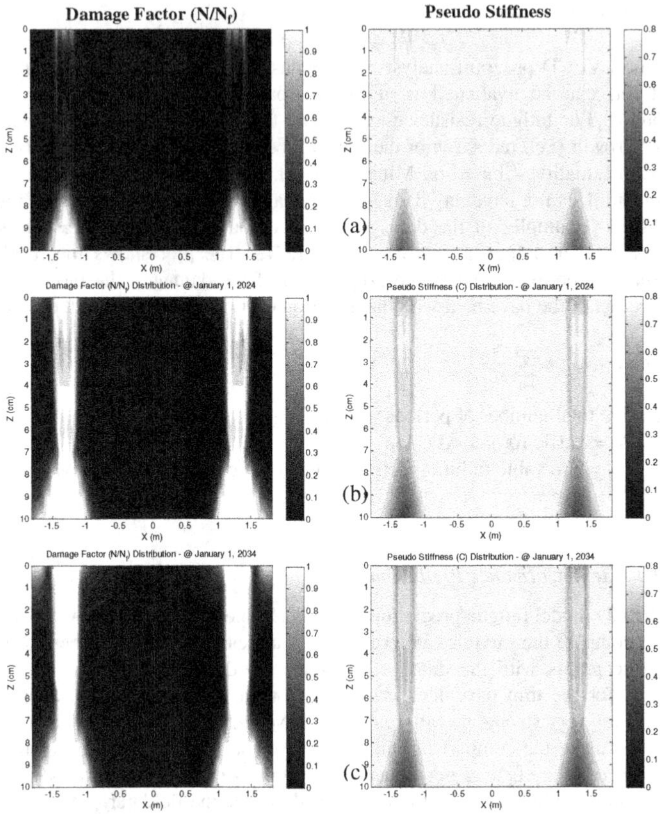

Figure 18. *Example of damage factor and secant modulus contours after (a) 1 year, (b) 10 years, and (c) 20 years.*

Figure 19 shows the results for the VT mixtures for both the thin and thick pavement structures. In general, the thick pavements performed better than the thin pavements; this outcome suggests that higher RAP contents can be used comfortably for thicker pavements. Only one simulation was run for the VTe mixtures because the dynamic modulus curves (i.e., the C versus S damage characteristics curves) and the failure criterion lines were effectively the same lines, and thus, no difference in performance within the LVECD system would be expected.

As Figure 19 shows, the VTa (PG 52-34 and higher asphalt content) mixtures, even with a high percentage of RAP, performed better than the VTe (PG 64-28) mixtures. Figure 19 shows the comparison for the VTa mixtures for the thin pavement structure. The VTa00LC and VTa20LC mixes show no failure, suggesting that adding up to 20% RAP did not affect the fatigue behavior of the VTa mixtures. By increasing the RAP level to 30% and 40%, the pavement system shows some failure points. The behavior of the VTa30 and VTa40 mixtures is very similar. These data agree well with the overlay test data that showed no drop in number of overlay test cycles with the addition of up to 20% RAP. VTa mixtures do not show any points with the damage ratio exceeding 0.1 in the thick pavement simulation (Figure 19 (d)).

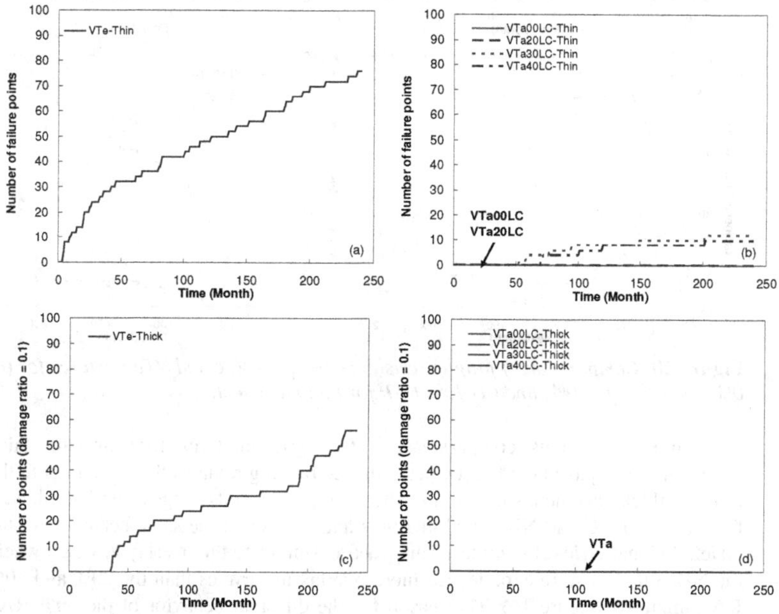

Figure 19. *Fatigue cracking simulation results for thin pavements for (a) VTe and (b) VTa mixtures, and for thick pavements for (c) VTe and (d) VTa mixtures.*

Figure 20 shows the impact of the different binder PGs at the different RAP contents. For all the RAP percentages, the VTa mix shows fewer failure points than the VTe mixture, which suggests better fatigue resistance of the mixture with the softer base binder. This impact seemed to lessen once the incorporated RAP exceeded 20%.

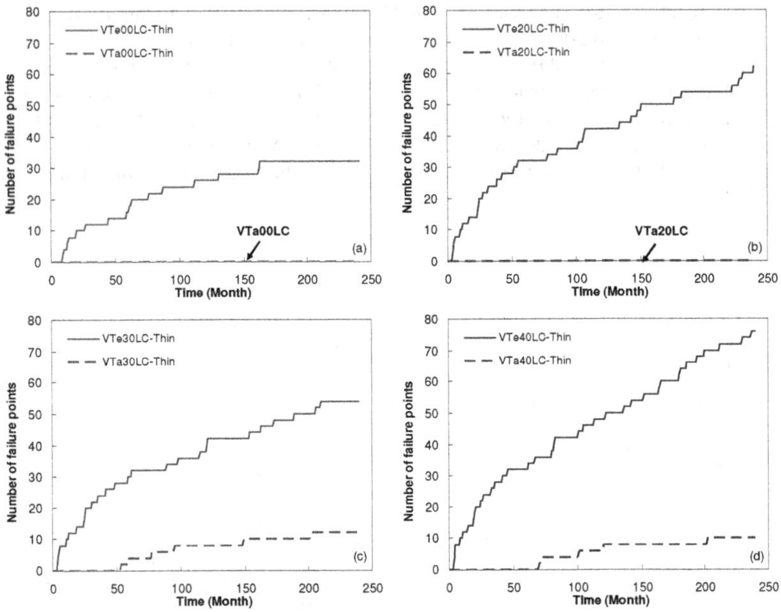

Figure 20. *Comparision of fatigue resistance between VTe and VTa mixtures for (a) 0%, (b) 20%, (c) 30%, and (d) 40% RAP for thin pavement.*

Figure 21 presents comparisons of the NHe mixtures for thin and thick pavements. The pavement fatigue performance ranking remains the same in both the thin and thick pavement simulations. The simulation results suggest similar behavior for the NHe00LC and NHe30LC mixtures and also about the same behavior for the NHe20LC and NHe40LC mixes. This trend is similar to the overlay test data where the NHe30LC mixture experienced more overlay test cycles than the 20% and 40% RAP mixtures (Figure 16). The reason for the different behavior of the 30% RAP mixture is not clear, but this difference was captured in much of the test data presented throughout this study. All the mixtures are compared in terms of fatigue resistances on the same graph in Figure 22. As this graph shows, the VTa mixtures exhibited the best fatigue resistance among all the mixtures in both thin and thick cases. The NHe mix performed better than the VTe mix at 0% and 30% RAP, but it performed worse than the VTe mix at 20% and 40% RAP.

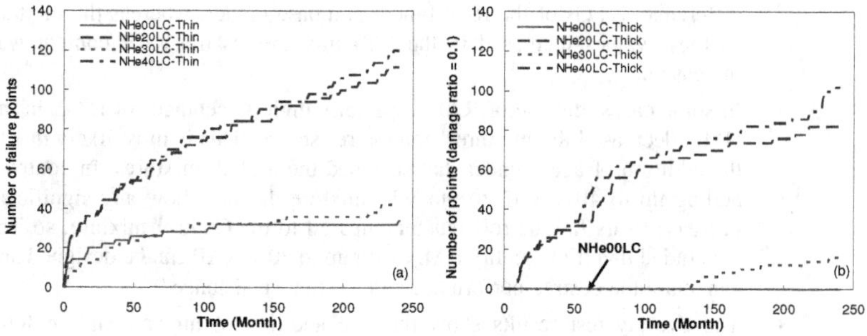

Figure 21. *Fatigue cracking simulations for NHe mixtures for (a) thin pavement and (b) thick pavement.*

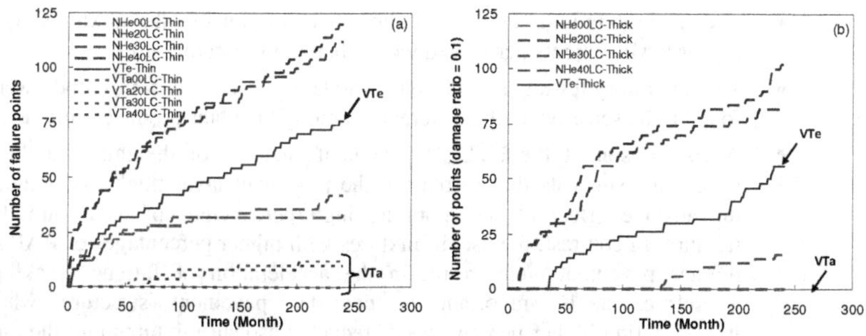

Figure 22. *Fatigue cracking simulation comparison for all the mixtures for (a) thin pavement and (b) thick pavement.*

5.0 Conclusions

In this study, the fatigue performance of twelve plant-produced mixtures from New Hampshire and Vermont that contained RAP contents of 0% to 40% by total weight of mixture was evaluated. Production data, such as plant type, mixing and discharge temperatures, and silo storage time, were collected. A summary of the findings and the conclusions that can be drawn from this study are as follows.

- Plant production parameters, such as silo storage time, can affect the mixture stiffness values and cracking properties. Therefore, in order to document the research findings properly, it was important to document how the mixtures were produced and handled prior to testing.
- The data indicate no significant change in dynamic modulus values as the RAP content was increased in the VTe mixtures, likely because of the small

difference in PGs of the RAP binder and base binder, whereas the mixture stiffness values increased in the VTa mixtures as the RAP content was increased.

- In some cases, the use of RAP, especially high percentages of RAP, in the HMA decreased the mixtures' fatigue resistance. This is most likely due to the addition of aged binder that stiffened the asphalt mixtures. In contrast, adding up to 40% RAP to the VTe mixture did not show any significant decrease in its fatigue resistance compared to the Control mixture, so it is concluded that high quality HMA with up to 40% RAP can be designed and can attain the desired performance, i.e., fatigue resistance.
- The overlay test results show that the addition of any amount of RAP significantly decreased the resistance to crack propagation for both the NH and VTe mixtures. The VTa mixtures did not show a decrease in the resistance to crack propagation until the RAP content was above 20%.
- The flexural fatigue test results show that, in general, the mixtures with higher RAP contents performed worse than the virgin mixtures.
- The laboratory testing and pavement modeling discussed in this study were found to be sensitive to the differences among the plant-produced mixtures.
- A comparison of the LVECD pavement analyses of the thin and thick pavements suggests that increasing the pavement layer thickness reduces the adverse effect of incorporating high percentages of RAP into the mixture. In contrast, the use of mixtures with higher percentages of RAP in thinner pavements may result in the acceleration of fatigue cracking. Therefore, it is important to consider pavement structure when implementing higher percentages of recycled asphalt mixtures into the mix design.
- The S-VECD, beam fatigue, and overlay test results, and the LVECD pavement analysis all showed that the use of a softer grade binder for the VT mixtures had the desired effect of improving the cracking resistance of the mixtures.

6.0 Acknowledgement

The authors would like to acknowledge the financial support from the Transportation Pooled Fund 5(230) study, *Evaluation of Plant-Produced High-Percentage RAP Mixtures in Northeast.*

7.0 References

AASHTO, M 320, "Standard Specification for Performance-Graded Asphalt Binder."

AASHTO, R 49, "Determination of Low-Temperature Performance Grade (PG) of Asphalt Binder".

AASHTO, T 164, "Standard Method of Test for Quantitative Extraction of Asphalt Binder from Hot-Mix Asphalt (HMA)."

AASHTO, T 321, "Determining the Fatigue Life of Compacted Hot-Mix Asphalt (HMA) Subjected to Repeated Flexural Bending."

AASHTO, TP 79-12, "Developing Dynamic Modulus Master Curves for Hot Mix Asphalt (HMA) Using the Asphalt Mixture Performance Tester (AMPT)."

AASHTO, TP 107, "Determining the Damage Characteristic Curve of Asphalt Concrete from Direct Tension Cyclic Fatigue Tests."

ASTM, D5404, "Recovery of Asphalt from Solution from Solution Using the Rotatory Evaporator."

Al-Qadi, I. L., Q. Aurangzeb, and S. H. Carpenter, *Impact of High RAP Content on Structural and Performance Properties of Asphalt Mixtures*, Research Report FHWA-ICT-12-002, Illinois Center for Transportation, Rantoul, IL, 2012.

Bennert, T., and A. Maher, "Forensic Study on the Cracking of New Jersey's LTPP Specific Pavement Study (SPS-5) Sections," *Transportation Research Record: Journal of the Transportation Research Board*, No. 2371, Transportation Research Board of the National Academies, Washington, D.C., pp. 74–86, 2013.

Bennert, T., W. Worden, and M. Turo, "Field and Laboratory Forensic Analysis of Reflective Cracking on Massachusetts Interstate 495," *Transportation Research Record: Journal of the Transportation Research Board*, No. 2126, Transportation Research Board of the National Academies, Washington, D.C., pp. 27–38, 2009.

Bonaquist, R., *Laboratory Evaluation of Hot Mix Asphalt (HMA) Mixtures Containing Recycled or Waste Product Materials Using Performance Testing,* Publication FHWA-PA-2005-006-98-32(19), Pennsylvania Department of Transportation, Office of Planning and Research, 2005.

Chehab, G. R., Y. R. Kim, R. A. Schapery, M. Witczack, and R. Bonaquist, "Characterization of Asphalt Concrete in Uniaxial Tension Using a Viscoelastoplastic Model," *Asphalt Paving Technology: Journal of the Association of Asphalt Paving Technologists*, Vol. 72, pp. 315–355, 2003.

Copeland, A., *Reclaimed Asphalt Pavement in Asphalt Mixtures: State of the Practice,* Publication FHWA-HRT-11-021, Turner-Fairbank Highway Research Center, Federal Highway Administration, McLean, VA, 2011.

Daniel, J. S., and Y. R. Kim, "Development of a Simplified Fatigue Test and Analysis Procedure Using a Viscoelastic Continuum Damage Model," *Journal of the Association of Asphalt Paving Technologists*, Vol. 71, pp. 619–650, 2002.

Eslaminia, M., S. Thirunavukkarasu, M. N. Guddati, and Y. R. Kim, "Accelerated Pavement Performance Modeling Using Layered Viscoelastic Analysis," *Proceedings of the 7th International RILEM Conference on Cracking in Pavements*, Delft, The Netherlands, 2012.

Hou, T., B. S. Underwood, and Y. R. Kim, "Fatigue Performance Prediction of North Carolina Mixtures Using Simplified Viscoelastic Continuum Damage Model," *Journal of the Association of Asphalt Paving Technologists*, Vol. 79, pp. 35–80, 2010.

Lee, J. S., A. Norouzi, and Y. R. Kim, "Determining Specimen Geometry of Cylindrical Specimens for Direct Tension Fatigue Testing of Asphalt Concrete, Submitted to *ASTM Journal of Testing and Evaluation*, 2014.

Mensching, D., J.S. Daniel, T. Bennert, M. Medeiros, M. Elwardany, W. Mogawer, E. Hajj, and M. Alavi, "Low Temperature Properties of Plant-Produced RAP Mixtures in the Northeast," *Road Materials and Pavement Design*, Vol 15, Supplement 1, pp. 1–27, 2014.

McDaniel, R., A. Shah, G. Huber, and V. Gallivan, "Investigation of Properties of Plant-Produced RAP Mixtures," *Transportation Research Record: Journal of the Transportation Research Board*, No. 1998, Transportation Research Board of the National Academies, Washington, D.C., pp.103–111, 2007.

Mogawer, W. S., T. Bennert, J. Daniel, R. Bonaquist, A. Austerman, and A. Booshehrian, "Performance Characteristics of Plant Produced High RAP Mixtures," *Road Materials and Pavement Design*, Vol. 13, Supplement 1, pp. 183–208, 2012.

Norouzi, A. and Y. R. Kim, "Mechanistic Evaluation of the Fatigue Cracking in Asphalt Pavements," Submitted to *International Journal of Pavement Engineering*, 2014.

Norouzi, A., M. Sabouri, and Y. R. Kim, "Evaluation of the Fatigue Performance of High RAP Asphalt Mixtures," *Asphalt Pavements*, CRC Press, Journal of Taylor and Francis Group, pp. 1069–1077, 2014.

Reese, R., "Properties of Aged Asphalt Binder Related to Asphalt Concrete Fatigue Life," *Journal of the Association of Asphalt Paving Technologists*, Vol. 66, pp. 604–632, 1997.

Sabouri, M., and Y. R. Kim, "Development of Failure Criterion for Asphalt Mixtures under Different Modes of Fatigue Loading," *Transportation Research Record: Journal of Transportation Research Board*. No 2447, pp. 117–125, 2014.

Sabouri, M, T. Bennert, J. Daniel, and Y. R. Kim, "Evaluating Laboratory-Produced RAP Mixtures in Terms of Rutting, Fatigue, Predictive Capabilities, and High RAP Content Potential," *TRB 94th Annual Meeting Compendium of Papers*, 2015.

Texas Department of Transportation (TxDOT), *Test Procedure for Overlay Test,* TxDOT Designation Tex-248-F, January 2009, ftp://ftp.dot.state.tx.us/pub/txdot-info/cst/TMS/200-F_series/pdfs/bit248.pdf, Accessed June 15th, 2011.

Underwood, B. S., and Y. R. Kim, "Analytical Techniques for Determining the Endurance Limit of Hot Mix Asphalt Concrete," *Proceedings of International Conference on Perpetual Pavement,* 2009.

Zhang, J., M. Sabouri, Y. R. Kim, and M. N. Guddati, "Development of a Failure Criterion for Asphalt Mixtures under Fatigue Loading," *Road Materials and Pavement Design,* Vol. 14, Supplement 2, pp. 1–15, 2013.

Zhou, F., S. Hu, and T. Scullion, *Development and Verification of the Overlay Tester Based Fatigue Cracking Prediction Approach,* Report FHWA/TX-07/9-1502-01-8, 2007.

Zhou, F. and T. Scullion, "Overlay Tester: A Simple Performance Test for Thermal Reflective Cracking," *Journal of the Association of Asphalt Paving Technologists,* Vol. 74, pp. 443–484, 2005.

8.0 Discussion

PAVEL KRIZ: I see I have to be fast at AAPT to ask my question, so I'm doing it now. Thank you for your presentation. I think the previous talk showed us that testing method selected for fatigue is really the key here, right? I mean, you can get totally opposite ranking of your mixtures depending on what method you choose. So I say you should be a little bit careful with your conclusion that using a soft grade, and this is by two grade softer, is a better approach than using the current grade used in Vermont. Also I read your paper and I think there are a few important messages you haven't mentioned in your talk, so just to remind the audience. The Vermont mixes with the harder grade had about a percent less binder in them, so they're dry mixes. Those mixes would be fatigue prone even though you just don't have that much RAP in it, okay? Then on the silo time, there were two New Hampshire RAP mixes in silo for 1 hour and 1.25 h. Only the control sample with no RAP was in silo for 6 h but there was no blending between the binders to be observed. Generally, 1–1.25 hour silo storage is a very short time for RAP and virgin binder blending. It would have to be about 12 hours. We had the paper last year here on the RAP—virgin binder diffusion. We also did a paper on binder diffusion in mix at CTAA in fall last year, and we saw that in mixes you need about 10 to 12 hours at 150°C to achieve complete blending between the two binders, the RAP and the virgin. So the silo times had they been longer would have been interesting. And I have one question. I notice that your New Hampshire mixes were sampled in 2008, which is seven years ago. Do you know how they perform in the field?

MOHAMMADREZA SABOURI: As far as I know, at this point we don't have any information about the field performance, but I agree that it would be very good if we can compare these results with the field data. In another study at NCSU which involved some mixtures that we had field performance data for, we performed the S-VECD failure criterion, and it did a good job in predicting the field fatigue performance. Thanks for your other good comments.

JO DANIEL: Just a clarification: the mixtures that were used for this project were actually not placed in the field. They were produced specifically for this research project at the end of the season and were not placed, so we do not have field performance on them.

ESHAN DAVE: Really good presentation, also the paper. I really enjoyed having these three types of test all put together on same mixes, same with the previous presentation.

MOHAMMADREZA SABOURI: Thank you.

ESHAN DAVE: Going back to the silo storage question, so I think partly Jo's answer clarified what I was going to ask. Because I was going to ask, after silo storage, how long before the mix is actually placed, because there is also transportation time, placement etc.? So I don't think that's an issue here in this study because it's essentially just plant-produced mixes used for lab testing. One question on your LVECD simulations, did that primarily only account for traffic loading or was there also temperature loading, especially low temperature side of things? Was there any evaluation there?

MOHAMMADREZA SABOURI: You mean the thermal cracking? It is not included here because at the time of this study it had not been incorporated into the LVECD, but as of right now, it's been added and LVECD does thermal stress predictions as well.

ESHAN DAVE: And I really liked your G^R plots because that really helps you distinguish between mixtures and look at things from, in my opinion, actual cracking mechanisms. Did that help you answer some of those unexpected trends in one of the mix? In case of one mix, you have trends that are somewhat unexpected when you did the LVECD analysis, but here they do fall in a correct order. So any ideas why that this could have been happening?

MOHAMMADREZA SABOURI: Thank you. We had the most difficulties with the New Hampshire mixtures. They were showing different behaviors when we did S-VECD or beam fatigue. So I believe something is going on with the New Hampshire mixtures, but it was not the case for the other two mixes from Vermont.

ESHAN DAVE: And if my understanding is correct, the damage-characteristic curve is what drives the LVECD damage.

MOHAMMADREZA SABOURI: Yes. The damage-characteristic curve is one of the inputs though. We also need to input dynamic modulus mastercurve and the G^R failure criterion for the LVECD simulations.

ESHAN DAVE: Thank you.

MOHAMMADREZA SABOURI: Thank you.

FUJIE ZHOU: Great presentation.

MOHAMMADREZA SABOURI: Thanks.

FUJIE ZHOU: I saw you used three types of fatigue tests to evaluate different RAP mixes from Vermont and New Hampshire. It seems to me, based on your S-VECD analysis, in terms of evaluating those three types of mixes, it always looks like the Vermont A, mix A, is better than Vermont E and better than the New Hampshire mix.

MOHAMMADREZA SABOURI: Yes, that's true.

FUJIE ZHOU: You also have the three types of fatigue tests. Are all these three types of fatigue tests showing the same trend or not?

MOHAMMADREZA SABOURI: As I showed in my presentation, except for the New Hampshire mixes, for Vermont mixes the S-VECD, the beam fatigue, and also the overlay results showed the same trends; VTa, with the softer binder, is performing better than VTe in all of them. This was the case for all the RAP percentages. Did I answer your question?

FUJIE ZHOU: Yes, but if I quickly look at one of the chart by the Texas overlay test versus S-VECD analysis, it looks to me, you know, as a clear trend from the overlay test, the Vermont A is better than Vermont E and then better than New Hampshire.

MOHAMMADREZA SABOURI: VTa is outperforming the other two mixtures in the overlay tester data because incorporating up to 20% RAP didn't show any decrease in the number of OT cycles. This is the case for S-VECD results too. In another words, VTa is doing better than the other mixtures from both crack propagation and crack initiation points of view.

FUJIE ZHOU: All right. Thank you.

GRANT WOLLENHAUPT: Just a comment off of Fujie's too. Going back to the mixture properties slide, as you keep saying, the VTe mixtures did worse. We have to keep in mind that they're, you know, a percent lower on their design AC.

MOHAMMADREZA SABOURI: Good point, and I mentioned it in my presentation. In the case of Vermont, we are looking at the effect of two factors at the same time, one is the softer base binder and the other is the higher binder content. These two factors happened to help VTa mixtures in term of fatigue properties.

BAOSHAN HUANG: I'm glad you brought these two slides. Obviously, the VTa and the VTe initially start with very similar or same mix design. However, plant mixture turns out VTe is way low in the actual asphalt content. I'm not sure this is even an acceptable mixture or not. Do you think this is an acceptable mixture?

MOHAMMADREZA SABOURI: Yes but this is what we got from the plant, so at least it can tell us that this kind of thing could happen when we're dealing with plant-produced mixtures. We would like to keep every factor at its design value, but unfortunately, it is not always the case in the plant.

BAOSHAN HUANG: Okay. And the second question is a quick observation from your data from your own lab testing. Usually we know that you have raw mixtures, A and B blended as C. The C should have the value between A and the B. Looking from the blending from VTa, for example, your RAP is 70-22 and your asphalt is -34?

MOHAMMADREZA SABOURI: Yes.

BAOSHAN HUANG: And your blending, somehow your lab results gives -16. Seems to be an interpolation range.

MOHAMMADREZA SABOURI: So you are asking about the extraction and recovery PG grades?

BAOSHAN HUANG: Yeah. Do you feel comfortable with your lab results?

MOHAMMADREZA SABOURI: Not really. That's why I didn't talk about the extraction and recovered PG grades in my conclusion. I believe that when we extract and recover asphalt binder from RAP mixtures, RAP binder gets completely mixed up with the base binder as a result of the extraction and recovery process. This is not the case in reality. It is why I didn't draw my conclusions based on the extraction and recovery PG grades, but more on the mechanical testing results.

BAOSHAN HUANG: My simple question is: do you feel comfortable with this kind of data?

MOHAMMADREZA SABOURI: No. It's why I did not draw my conclusion based on those results.

BAOSHAN HUANG: Yeah. You have -34 blended with -22, you get -16. Do you feel comfortable?

MOHAMMADREZA SABOURI: That's a valid point. No. It's why I'm not mentioning that in my presentation that much. Thank you.

Effect of Mineral Filler on Changes in Molecular Size Distribution of Asphalts during Oxidative Aging

Raquel Moraes[a], Hussain U. Bahia[a]

[a]*University of Wisconsin-Madison, 3350 Engineering Hall, 1415 Engineering Drive, 53706, Madison, WI, USA*

ABSTRACT. *Asphalt binder is always present in contact with fine and coarse mineral particles in the pavement and roofing shingles. Therefore, the mechanisms of oxidative aging of binders are expected to be influenced not only by the characteristics of the binder but also by the properties of the mineral surface due to the molecular interaction between these two components. The purpose of this study is to develop a fundamental understanding of the mechanisms of interaction between binder and filler and the resulting changes in the molecular size distribution after oxidative aging. In this study, a method to characterize the behavior of mastics (binder mixed with mineral dust) with aging was developed by using the Pressure Aging Vessel (PAV). The changes due to this aging are measured by monitoring the mastic's |G*| aging index (ratio of complex modulus before and after aging). The results clearly indicate that fillers can significantly change the effect of oxidative aging of binders. Gel Permeation Chromatography (GPC) testing results support the mentioned findings regarding |G*| changes, as the presence of mineral filler appears to decelerate the rate of production of larger molecular size oxidation products in the binder phase of mastics. An implication of the findings is that change in molecular size distribution of asphalts during aging—a controlling factor in mechanical and thermo-volumetric behavior—can be engineered by proper selection of fillers in the mastic phase. It is thus concluded that by selecting a proper type and concentration of filler, aging of pavement or roofing shingles and, consequently, durability and performance, can be controlled.*

KEYWORDS: *oxidative aging, asphalt mastics, molecular size distribution.*

The oral presentation was made by Dr. Moraes.

This is a reproduction of an article whose final and definitive form has been published in *Road Materials and Pavement Design*© 2015 Taylor & Francis. The article is available online at: http://dx.doi.org/10.1080/14680629.2015.1076998

1.0 Introduction

Asphalt ages through oxidation, changing its chemical composition and molecular structure, with these changes affecting its mechanical properties. Excessive molecular structuring leads to a brittle material which tends to crack (Robertson, 1991). As a result, chemical aging leads to a global stiffening and embrittlement of the asphalt binder, which in turn increases the cracking susceptibility under either traffic load or thermal stress (Lesueur, 2009).

Aging of pavements increases the elastic behavior and decreases the time-dependent viscoelastic behavior of the material, thus stress relaxes relatively slower when compared to fresh pavements, critically affecting thermal cracking susceptibility. Furthermore, aged pavements accumulate damage more quickly. Therefore, through various mechanisms, the durability and service lives of asphalt pavements are greatly affected by aging.

In a Hot Mix Asphalt (HMA) mixture, the majority of the asphalt-aggregate interface area is within the fine fraction (particles smaller than 0.075 mm), which is embedded in the asphalt binder due to the higher surface area when compared with the coarse aggregate (Anderson et al., 1992). In asphalt shingles, the roofing material is a true composite composed of asphalt binder and more than 80% by weight of mineral and rock.

The current state of knowledge is limited in terms of fundamental understanding of the mechanisms of interaction between asphalt binder and mineral filler with respect to aging characteristics of the mastic. The aging performance implications of filler properties in HMA pavements and asphalt shingles is unclear due to the various interaction effects of filler with the asphaltic phase, which are known to depend on the filler content, mineralogy, and its surface area.

Fundamental understanding of the asphalt-filler interaction during the aging of the effective binder is, therefore, needed to determine an appropriate method for characterization of aging effects on asphaltic materials' properties, and to improve current analysis methods used to select durable and aging-resistant materials.

2.0 Literature Review

2.1 Changes in Asphalt Structure during Oxidative Aging

Asphalt is a colloidal dispersion of asphaltenes into an oily matrix constituted by saturates, naphthene aromatics, and resins (which compose the maltene fraction) (Figure 1). The asphalt physico-chemical behavior depends on the concentration of its different fractions. Thus, a variation in its composition strongly affects its mechanical properties and chemical reactivity.

In general, the mechanical or structural properties of asphalt binders are related to the intermolecular structuring among polar components. Asphalt binders are considered as a continuum of molecules with a gradual transition in polarity, molecular weight, and functionality (Kriz et al., 2008). The ranking of the asphalt fractions regarding increase in molecular polarity is: saturates, naphthene aromatics, polar aromatics (resins), and asphaltenes (Corbett, 1969). In an asphalt molecule, the polar matrix is responsible for the elastic component of the material, while the continuous non-polar phase gives the viscous component (Robertson, 1991; Read and Whiteoak, 2003). This is due to association of the polar molecules that leads to large structures, in some cases even to three-dimensional networks, i.e., 'GEL' type asphalt. The degree to which this association takes place depends on the temperature, the molecular weight distribution, the concentration of the polar aromatics and the solvency power of saturates and aromatics in the maltene phase. If the concentration and molecular weight of the asphaltenes are relatively low, the result will be a 'SOL' type asphalt. These colloidal states can largely reflect the rheological properties of asphalt.

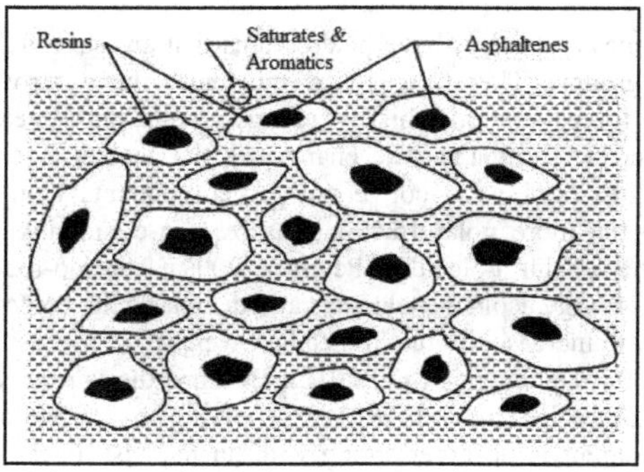

Figure 1. *Schematic of asphalt components (Moraes, 2014).*

Oxidation causes aged asphalts to contain more polar molecules. Because polar molecules associate into a matrix, which is dispersed in less polar and non-polar materials, polar oxidation products greatly alter the state of dispersion of asphalt components. This results in a change of the mobility and of the chemical reactivity of asphalt molecules susceptible to oxidation (Robertson, 1991; Petersen et al., 1996). Therefore, during the oxidative aging process of asphalt binders, the concentration of polar functional groups increases, resulting in an immobilization of molecules through intermolecular association (Petersen, 2009). Hence, the molecules or molecular agglomerates lose sufficient mobility to flow past one another under thermal or mechanical stress, resulting in embrittlement of the asphalt, making it more susceptible to fracturing or cracking and resistant to healing (Tarefder and Arisa, 2011). Figure 2 illustrates schematically the effects of temperature, time, and oxidation on asphalt performance.

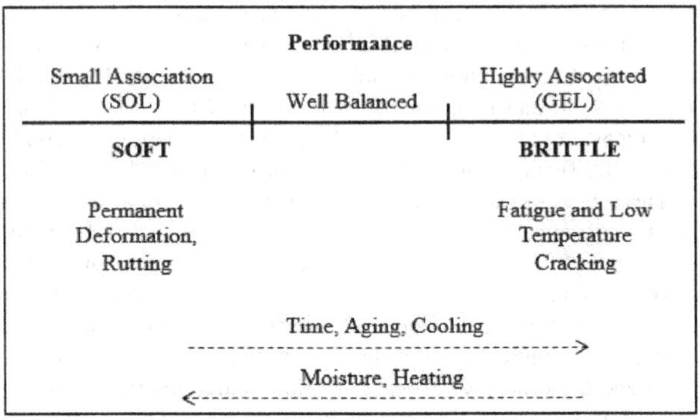

Figure 2. *Schematic of effects of temperature, time, and oxidation on asphalt performance ("association" refers to the asphaltenes-resins association) (Moraes, 2014).*

Asphaltenes and resins have heteroatoms that are strongly associated with polar functional groups. Therefore, these molecules have strong interaction and/or association forces and are highly reactive with highly electronegative oxygen (Tarefder and Arisa, 2011). The changes in the asphalt fractions during oxidation have been observed as a conversion of components from the more non-polar fractions to the more polar fractions, as oxygen-containing functional groups are formed in the asphalt molecules (Petersen, 2009). Thermo-oxidative aging increases the content of large asphalt molecules and decreases the content of small molecules, leading to an increase in the molecular weight of the asphalt binders (Lu and Isacsson, 2000; Dehouche et al., 2012). Because the various asphalt fractions have different reactivities toward oxidation, during oxidative aging a reduction of aromatics content is observed along with an increase in the content of resins and asphaltenes (Dehouche et al., 2012). Oxidation causes the oils (i.e., naphthene aromatic and saturates) to convert to resins, and the resins to convert to asphaltenes (Rostler and White, 1962). The change in asphaltene-maltene ratio that occurs during oxidation affects the stiffness properties of the asphalt (O'Sullivan, 2011). Changes in the content of saturates are almost negligible due to their inert nature (Liu et al., 1998).

According to Tarefder and Arisa (2011), the association, agglomeration, and interaction of asphaltenes and resins mainly depend upon the temperature and the amount of oxygen available. Petersen et al. (1974) showed that at high temperatures 87% of the total oxidation occurred in the polar aromatic fractions (i.e., asphaltenes and resins) of the asphalt. Because of high molecular mobility at high temperatures, resins and asphaltenes molecules become unassociated and susceptible to oxidation (Tarefder and Arisa, 2011).

2.2 Asphalt-Aggregate Molecular Interaction during Oxidative Aging

Since asphalts binders are used in conjunction with mineral aggregates in pavements and roofing, the molecular interaction between binders and mineral fillers is of concern. Barbour et al. (1974) showed that when the asphalt fractions were oxidized separately, the mineral aggregate surfaces were found to catalyze the oxidation of the lower polarity fractions. Plancher and Petersen (1976) showed that the beneficial effect of hydrated lime addition to the asphalt binder in reducing oxidative hardening is due to two effects: (*i*) lime reduces the formation of oxidation products by the removal of oxidation catalysts; and (*ii*) lime removes reactive polar molecules that would interact with oxidation products.

Therefore, mineral filler particles adsorbing oxidative functional groups as well as the polar fractions of the asphalt binder dictates the changes in the mechanical properties (e.g.' stiffness, viscosity (Curtis et al., 1993)) and thermo-volumetric properties (e.g.' coefficients of thermal contraction and glass transition temperature (Clopotel, 2012)) of the asphalt mastics.

It is important to mention that change in molecular size distribution of asphalt binders during aging is a controlling factor in mechanical and thermo-volumetric behavior of binders (Shen et al., 2006; Anderson, 1987; Garrick, 1994; Al-Abdul Wahhab, 1999; Delaporte, 2007; Delaporte, 2008). Additionally, the classification based on molecular size was found to be effective in analyzing the aging process that affects asphalt binders (Jennings et al., 1980; Churchill et al., 1995; Kim et al., 2006; Bowers, 2013).

2.3 Oxidative Aging Evaluation of Asphalts Using Gel Permeation Chromatography

Gel Permeation Chromatography (GPC) is used to determine the molecular size distribution (MSD) of asphalt binders (aged, unaged, modified and unmodified), providing a distinct and reproducible molecular-size distribution curve (chromatogram) of the asphalt sample in solution (Altgelt, 1965; Gehrke and Ruyle, 1968; Jacobson et al., 1968; Snyder, 1969; Bynum and Traxler, 1970; Haley, 1971; Jennings et al., 1980; Kim and Burati, 1993; Churchill et al., 1995).

In this method, the asphalt binder is dissolved in a solvent and is injected into the GPC system. The injected sample travels through a series of columns which separates the sample based on molecular size. The larger molecular size particles exit the columns first and are detected by the system's detectors. The smaller molecular size particles travel into the pores of the columns and, therefore, have longer retention times. As a result, a molecular size distribution (which can be thought of as analogous to a type of sieve analysis of the sample) is obtained (Figure 3).

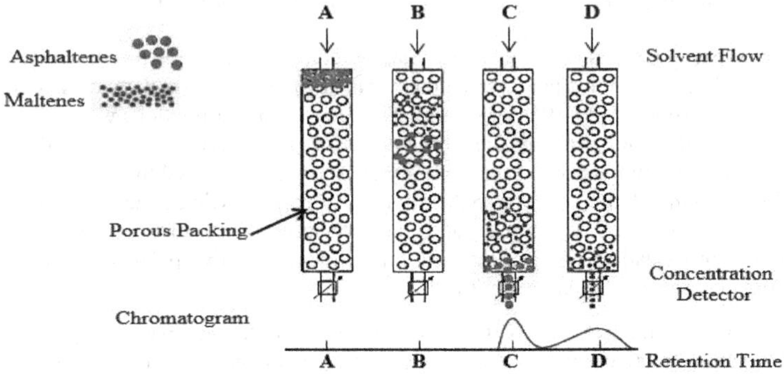

Figure 3. *Example of gel permeation chromatography (Moraes, 2014).*

One of the great advantages of Gel Permeation Chromatography is its ability to separate by molecular size rather than by solubility or adsorptivity. The GPC is a simple separation technique available that responds to molecular weight alone and not to chemical structure. This feature makes GPC especially suited for fractionating complex mixtures, like asphalt binders. For complex materials, ordinary fractionation methods fail because they are usually based on solubility, which in turn differentiates by both molecular weight and chemical structure. Regular chromatographic methods are based on either solubility or on adsorptivity, which also depends on both molecular properties (Altgelt, 1965).

The Gel Permeation Chromatography test is adopted from the polymer industry and includes the classification of the chemical composition of the asphalt binder into three groups based on molecular size by analyzing the GPC chromatographic profiles (chromatograms) of the binders (Al-Abdul Wahhab et al., 1999; Xiao et al., 2009). These groups are the large molecular size (LMS), medium molecular size (MMS), and small molecular size (SMS) (Figure 4).

The molecular size distribution can have a significant effect on the physical properties of the asphalt binders such as viscosity, penetration, temperature susceptibility, and viscosity ratio (Garrick, 1994; Al-Abdul Wahhab et al., 1999). Additionally, the classification based on molecular size was found to be effective in analyzing the aging process that affects asphalt binders (Jennings et al., 1980; Churchill et al., 1995; Kim et al., 2006; Bowers, 2013).

The chromatograms provide some insight about what fractions of the asphalt binder were affected after oxidative aging. Upon aging, the LMS fraction tends to increase (Glover et al., 1988). The increment of LMS due to aging was known to create increasing viscosity (Jennings et al., 1980; Churchill et al., 1995) and loss of consistency (Noureldin and Wood, 1989). The LMS has been related to low temperature cracking of pavements (Jennings et al., 1980; Plummer and Zimmerman, 1984). Other studies (Zenewitz and Tran, 1987; Glover et al., 1988; Enustum et al., 1990) have also found that the lower the LMS and higher the SMS contents, the higher is the asphalt "tenderness".

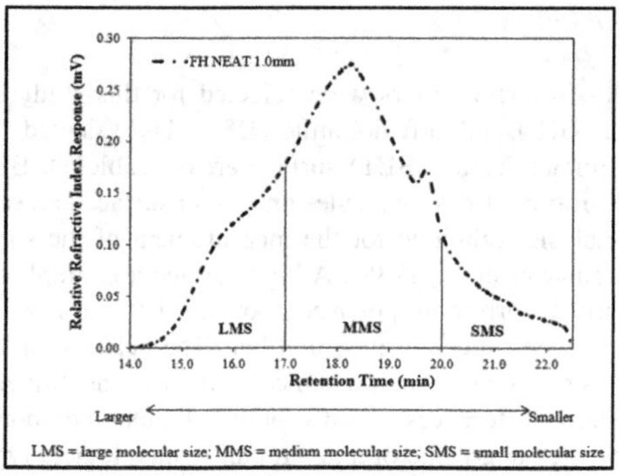

Figure 4. *Typical GPC chromatogram of an asphalt binder showing three elution times fractions (Moraes, 2014).*

The GPC chromatogram defines the molecular size distribution (MSD) of asphalt binders. The changes in the molecular size distribution due to aging can significantly influence the asphalt binder consistency and consequently its physical properties (Shen et al., 2006). Since asphalts binders are used in conjunction with mineral aggregates in pavements and roofing, the interactions between binders and these minerals and the resulting changes in molecular size distribution after oxidative aging are of concern and are thus an important focus of the present study.

3.0 Research Hypothesis and Objective

Considering asphalt mastics, it is hypothesized that if an interaction between the mineral filler and the asphalt matrix exists, the molecular properties of the asphalt matrix will be affected. The nature of the molecular change will therefore depend specifically on the type of interaction involved. It follows that an asphalt mastic, in which a significant fraction of the binder is in contact with a mineral filler surface and which has interaction between filler and binder, should exhibit the degree of interaction (regardless of mechanism) by changes in the viscoelastic properties of the asphalt matrix. The present paper aims to show that changes in molecular size distribution of asphalts binders during aging can be engineered by proper selection of fillers in the mastic phase.

4. 0 Materials and Methods

4.1 Materials

4.1.1 Mineral Fillers

Three types of mineral fillers were selected for this study: hard limestone (LH_1), hard granite (GH_1) and soft dolomite (DS_2). The selected fillers present different Braunauer-Emmett-Teller (BET) surface areas (Table 1). BET theory explains the physical adsorption of gas molecules on a solid surface and serves as the basis for an important analysis technique for the measurement of the specific surface area of a material (Braunauer et al., 1938). A high surface area implies a single filler particle can potentially adsorb more polar components as there is a higher interface surface between the binder and the mineral filler. The selected fillers have also different mineralogy as revealed by their chemical compositions. Mineral fillers with different chemistry show differences in adsorption of polar components from the asphalt binder, since each aggregate of a given bulk mineralogical type has a unique surface chemistry. All fillers used for production of mastics in this study passed through the #200 sieve (0.075 mm).

Table 1. *Filler characteristics and chemical composition (Faheem et al., 2010).*

Filler Characteristics								
Filler	Rigden Voids (%)	Fine Modulus	Methylene Blue Volume	Specific Gravity	Loss on Ignition	Solubility (%)	Surface Area (m^2/g)	Average Pore Dimension (Å)
LH_1	32.2	5.63	0.62	2.65	0.1	1	1.07	148.6
GH_1	42.6	4.06	2.35	2.66	0.36	2.3	2.17	224.5
DS_2	29.4	4.73	1.82	2.7	0.35	0.7	1.98	186.7

Filler Chemical Composition						
Filler	CaO	SiO_2	Al_2O_3	FeO	MgO	K_2O
LH_1	43.1	5.0	1.9	0.1	5.4	0.3
GH_1	3.5	62.2	15.0	3.6	1.8	3.3
DS_2	27.0	20.2	4.5	0.5	15.7	1.9

4.1.2 Asphalt Binders and Mastics

Two asphalt binders commonly used in the Midwest region of the United States are selected in this study: Flint Hills (FH) PG 64-22 and Valero (VAL) PG 64-16. The current design standard for asphalt mixtures provides guidance on selection of aggregates and asphalt binder and includes requirements for the amount of mineral filler to be added in the mix. The amount of filler that can be included is limited to a ratio in the range of 0.6–1.2 by mass of the filler to the binder (i.e., approximately 20-40% by volume of mastic). For this paper, the mastic preparation was selected by considering the maximum value of the dust to binder ratio allowed in the mix design. Therefore, 40% by volume of mineral filler was selected to produce the asphalt mastics. The asphalt mastic preparation consisted of mixing binder and fillers at 150°C for 30 minutes.

4.2 Methods

4.2.1 Gel Permeation Chromatography

Gel Permeation Chromatography (GPC) is used to measure the molecular size distribution of both binders and mastics. The detector used for the GPC analysis is the differential refractometer, which is a concentration sensitive detector that measures the difference in refractive index (dRI) between the eluent in the reference side, and the sample + eluent in the sample side. The GPC columns used are PolyPore 5 μm mixed, 300 x 7.5 mm. The flow rate was 1 mL/min and the testing temperature was 40°C. The tetrahydrofuran (THF) solvent was selected for the GPC analysis since previous researchers (Jennings et al., 1992; Kim et al., 2006; Shen et al. 2006) had successfully selected it for GPC applications in asphalt. According to Jennings et al. (1992), the asphalt binders' strongest associative bonds (polar and nonpolar) remain unaltered in THF solution.

The GPC samples are prepared in accordance to the following procedure:

1. Approximately 30 mg of sample is dissolved in 1 mL of THF;
2. The solution is then filtered through a 0.20 μm phobic PTFE syringe filter;
3. After the filtration, the solution is placed at ambient temperature for 8 hours, until the complete solvent evaporation;
4. The remaining asphalt film is weighted and THF solvent is added to achieve the concentration of 5 mg / 2 mL;
5. In the final step, each sample dissolved into THF is filtered through a 0.45 μm syringe filter prior to the injection.

Please note that this procedure is followed for preparation of both binder and mastic samples to assure that the asphalt film used to perform the GPC analysis would be equivalent. Figure 5 depicts the developed GPC testing procedure.

In the Gel Permeation Chromatography procedure the separation is based strictly on the size of the sample in solution, not chemistry (Yau and Malone, 1967). The mode of separation is not based on molecular weight, but on the size of the material being analyzed in solution. Once the sample has been suitably dissolved, it is introduced via an injection mechanism onto a set of columns which act as a molecular filtration system. The columns are packed with a gel which contains surface pores. These pores vary in size and act as molecular filters. Thus, the larger size molecules will not fit into the smaller pores. Conversely, the smaller molecules will fit into most of the pores and be retained longer. A typical GPC chromatogram is shown in Figure 6.

Figure 5. *GPC sample preparation: (a) solution of mastic sample and THF solvent; (b) solution filtered through a 0.20 µm phobic filter; (c) filtered solution left for evaporation of the solvent; and (d) remaining asphalt film (Moraes, 2014).*

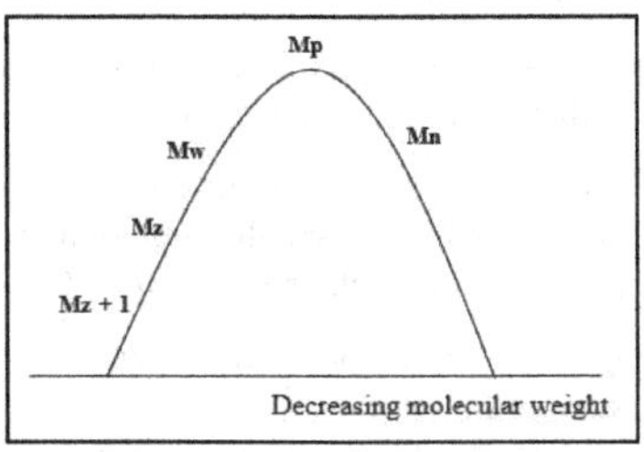

Figure 6. *Typical GPC chromatogram curve with M_w (weight-average molecular weight), M_n (number-average molecular weight), M_z (z-average molecular weight), M_{z+1} ((z+1)-average molecular weight), and M_p (peak molecular weight).*

Figure 6 shows that M_n value marks the value at which there are equal numbers of molecules on each side of the chromatogram, at higher and lower molecular weight. The value of M_n influences the thermodynamic properties of the molecule (Dawkins, 1979). M_w is defined as the value at which there are equal masses of molecules on each side, at higher and lower molecular weight. M_w is large-molecule sensitive and influences the bulk properties and toughness of the material. M_w affects many of the physical properties of material, and is the most often quoted molecular weight average (Dawkins, 1979). As well as M_n and M_w, there are other molecular weight averages that take increasing account of the higher molecular weight components of the sample, such as z-average molecular weight (M_z) and M_{z+1}. M_z is sensitive to even larger molecules and influences viscoelasticity and melt flow behavior. The ratio of M_w to M_n is used to calculate the polydispersity index (PDI) of a polymer, which provides an indication of the material's range of molecular mass. The broader the molecular weight distribution, the larger the PDI. The molecular weights are of the order $M_n < M_w < M_z < M_{z+1}$. The GPC molecular weight parameters used in this study are calculated in accordance with Equation 1 through Equation 4:

$$Mn = \frac{\Sigma N_i M_i}{\Sigma N_i} \quad [1]$$

$$Mw = \frac{\Sigma N_i M_i^2}{\Sigma N_i M_i} \quad [2]$$

$$Mz = \frac{\Sigma N_i M_i^3}{\Sigma N_i M_i^2} \quad [3]$$

$$M_{z+1} = \frac{\Sigma N_i M_i^4}{\Sigma N_i M_i^3} \quad [4]$$

Where: M_n = number-average molecular weight (daltons)

M_w = weight-average molecular weight (daltons)

M_z = z-average molecular weight (daltons)

M_{z+1} = (z+1)-average molecular weight (daltons)

M_p = peak molecular weight (daltons)

M_i = molecular weight (daltons)

N_i = number of molecules of molecular weight M_i

In this study, the GPC chromatogram was divided into 13 equal elution time areas (Jennings et al., 1980), as shown in Figure 7. These areas were then separated into three fractions: large molecular size (LMS, areas 1 to 5), medium molecular size (MMS, areas 6 to 9), and small molecular size (SMS, areas 10 to 13).

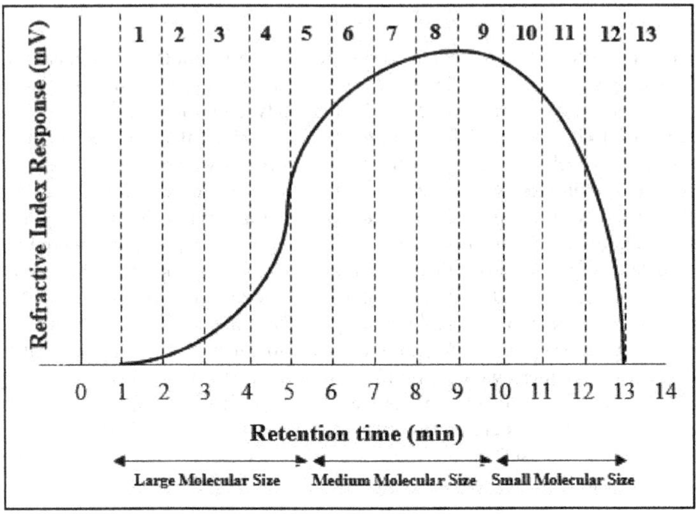

Figure 7. *GPC spectrum divided into 13 equal elution time areas (Moraes, 2014).*

4.2.2 Modified Pressure Aging Vessel Test Procedure

The Pressure Aging Vessel (PAV) aging procedure is used to simulate long-term field oxidative aging of asphalt binders in accordance to AASHTO R 28, "Accelerated Aging of Asphalt Binder Using a Pressurized Aging Vessel (PAV)" (2009). In this procedure, samples were aged in a vessel pressurized to 2.10 MPa (305 psi or 20.7 atmospheres). The materials in this study were heated to 100°C and aged for 24 and 48 h in PAV. Unaged control samples ("0 h aging") were also prepared for all mentioned materials.

4.2.3 Dynamic Shear Rheometer (DSR) Test Procedure

4.2.3.1 Frequency Sweep. In this study, a 1% strain amplitude is applied using a frequency sweep over the range of 0.1 to 30 Hz and temperatures of 10°C and 40°C. All tests are performed using 8 mm diameter parallel plate geometry with a fixed 2 mm sample height. The resulting complex shear modulus ($|G^*|$) data are used to calculate aging indexes. The complex shear modulus ($|G^*|$) can be considered the sample's total resistance to deformation when repeatedly sheared. It is expected that the effects of oxidative aging and mineral filler on binder mechanical properties will be observable using the complex shear modulus.

5.0 Analysis of GPC Testing Results

5.1 After 48 hours of Oxidative Aging

The oxidative aging in the Pressure Aging Vessel (PAV) involves a diffusion mechanism; therefore, film thickness is expected to be an important factor during aging in the PAV. In order to differentiate the diffusion effect from the mineral surface effect, different mastic thicknesses were selected and aged for 48 hours.

Figure 8 shows the GPC chromatograms of the neat binders and mastics after 48 hours in the PAV, where both base binders and mastics presented 3.15 mm (standard film thickness in a PAV pan) of film thickness during the aging process. In Figure 9, both base binders and mastics presented 1.00 mm of film thickness during the aging process in PAV. For each chromatogram, the area under the curve represents 100% of the binder molecules injected into the GPC system.

It can be seen that the GPC is a useful method for determining differences in molecular size distribution (MSD) of different asphalts and mastics concerning oxidative aging. It is visible that there are differences among the GPC curves, especially in identifying that the fillers are affecting the MSD portions of each binder differently.

In order to better investigate these molecular changes after oxidation, aging indexes were generated based on molecular size areas obtained by GPC (Figure 7). The aging indexes were calculated according to Equations 5 through 7.

$$\text{LMS Aging Index} = \frac{\text{Large Molecular Size Molecules after PAV}_{modified}}{\text{Large Molecular Size Molecules before PAV}_{modified}} \quad [5]$$

$$\text{MMS Aging Index} = \frac{\text{Medium Molecular Size Molecules after PAV}_{modified}}{\text{Medium Molecular Size Molecules before PAV}_{modified}} \quad [6]$$

$$\text{SMS Aging Index} = \frac{\text{Small Molecular Size Molecules after PAV}_{modified}}{\text{Small Molecular Size Molecules before PAV}_{modified}} \quad [7]$$

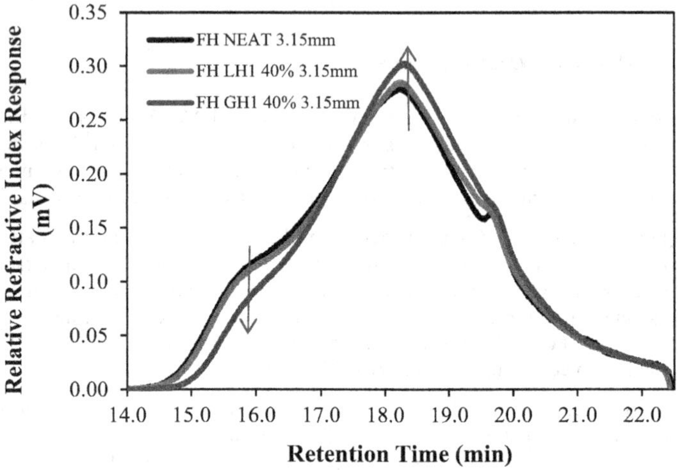

(a)

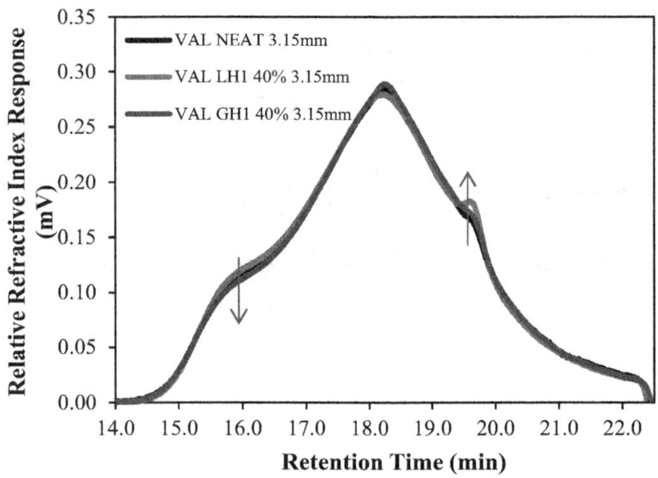

(b)

Figure 8. *GPC chromatograms after 48 hours of PAV aging for: (a) FH binder and 40% mastics, and (b) VAL binder and 40% mastics. Film thickness of 3.15 mm.*

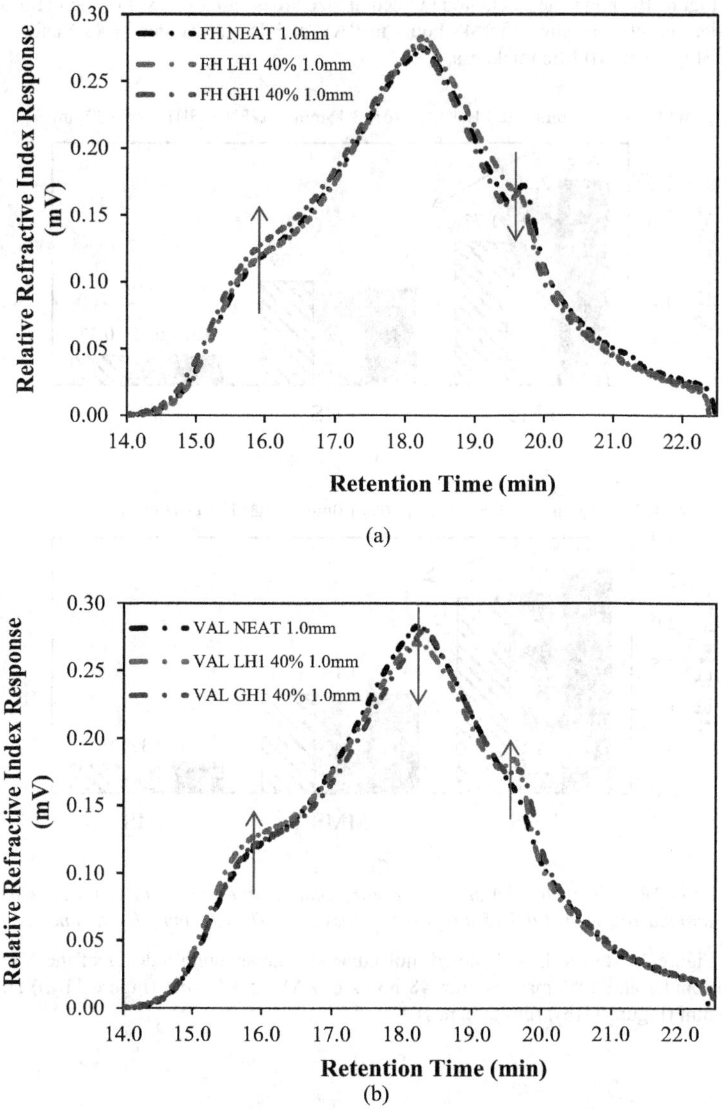

Figure 9. *GPC chromatograms after 48 hours of PAV aging for: (a) FH binder and 40% mastics, and (b) VAL binder and 40% mastics. Film thickness of 1.0 mm.*

Figure 10 shows the calculated molecular size areas aging indexes of the FH neat binder and 40% mastics after 48 hours in PAV at 3.15 mm (Figure 10(a)) and 1.0 mm (Figure 10(b)) film thickness.

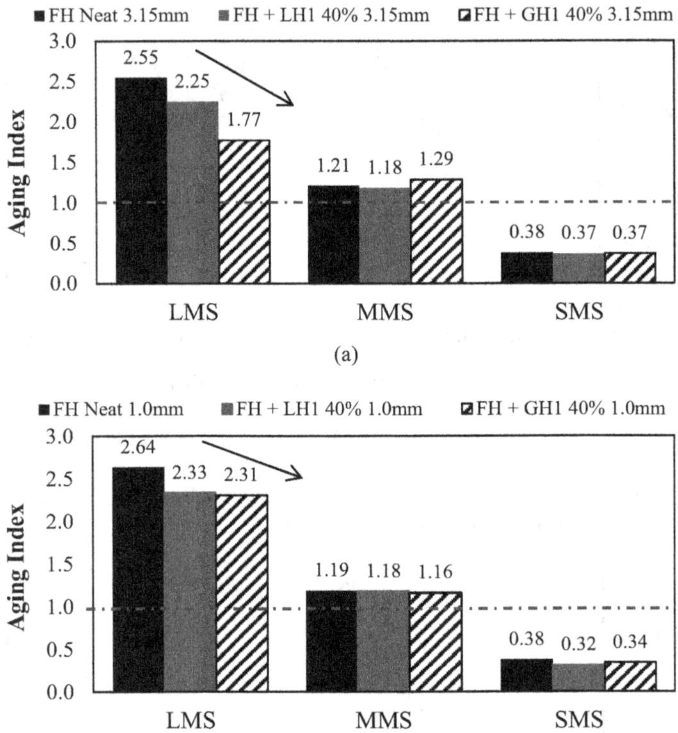

(b)

Figure 10. *GPC molecular size areas aging indexes after 48h in PAV of FH neat and mastics with: (a) 3.15 mm film thickness, and (b) 1.00 mm film thickness.*

Figure 11 shows the calculated molecular size areas aging indexes of the VAL neat binder and 40% mastics after 48 hours in PAV at 3.15 mm (Figure 11(a)) and 1.0 mm (Figure 11(b)) film thickness.

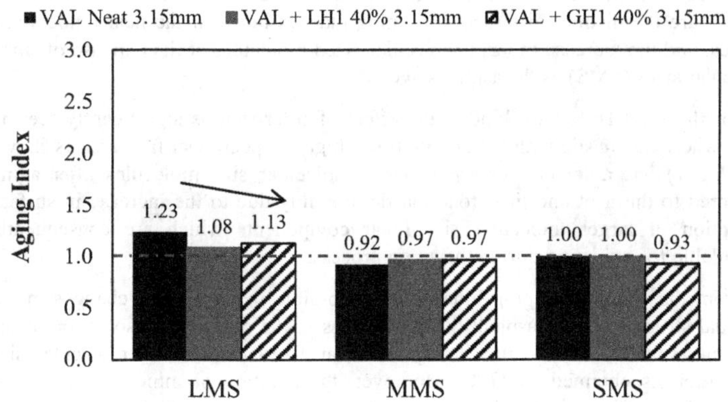

(a)

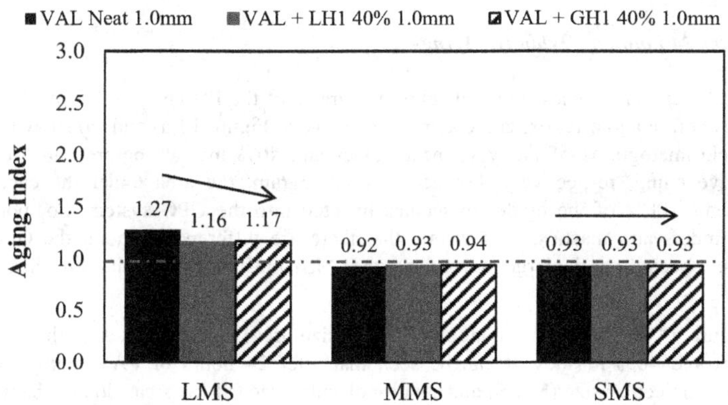

(b)

Figure 11. *GPC molecular size areas aging indexes after 48h in PAV of VAL neat and mastics with: (a) 3.15 mm film thickness, and (b) 1.00 mm film thickness.*

As can be seen in Figures 10 and 11, the addition of filler affects the LMS, MMS, and SMS molecular size areas of the base binders. For all materials (with and without filler), the oxidative aging increases LMS, decreases SMS, but does not change MMS significantly. Regardless of both binder type and film thickness during oxidative aging in PAV, the fillers are decreasing the production of the large molecular size molecules when compared to the base binders. Two noticeable features are the distinct differences in aging indexes between the neat binders and mastics, and the increase in large molecular sized molecules at the expense of small molecular sizes (SMS) as the asphalts age.

For the Flint Hills base binder, the effect of filler type is significantly seen in LMS, where the mastic with the granite filler (highest specific surface area as shown in Table 1) has a lower increase in large molecular size molecules after aging compared to the neat and limestone mastic, possibly due to the increase in surface adsorption of larger molecular size polar components which are consequently unavailable for oxidation during the aging process.

From the calculated aging indexes, it is possible to verify that changes in the colloidal structure of the asphalt due to changes in the degree of association of the different asphalt fractions after aging are clearly reflected in the molecular size distribution as obtained by GPC. Moreover, the addition of mineral filler to the asphalt phase affects the binder response to aging, and consequently its molecular distribution.

5.2 After 24 hours of Oxidative Aging

Figure 12(a) and (b) show the GPC chromatograms of the FH neat binder and 40% mastics before and after oxidative aging, respectively. Figure 13(a) and (b) show the GPC chromatograms of the VAL neat binder and 40% mastics before and after oxidative aging, respectively. For each chromatogram, the area under the curve represents 100% of the binder molecules injected into the GPC system. For both base binders and mastics, it is visible that there are differences among the GPC curves, especially in the large molecular size (LMS) and small molecular size (SMS) regions.

Figure 14 shows the calculated molecular size areas aging indexes of the neat binders and 40% mastics. It can be seen that after 24 hours of PAV aging, the medium molecular size (MMS) and small molecular size (SMS) areas do not change due to aging for any of the tested materials (with or without filler). For the Flint Hills base binder, a higher decrease in aging index of large molecular size (LMS) area is observed for the granite (GH_1) mastic. Regarding the Valero base binder, equal behavior is observed in the aging index of LMS for limestone and granite mastics. Also, the LMS aging index difference was less pronounced for VAL than that of the FH base binders and mastics.

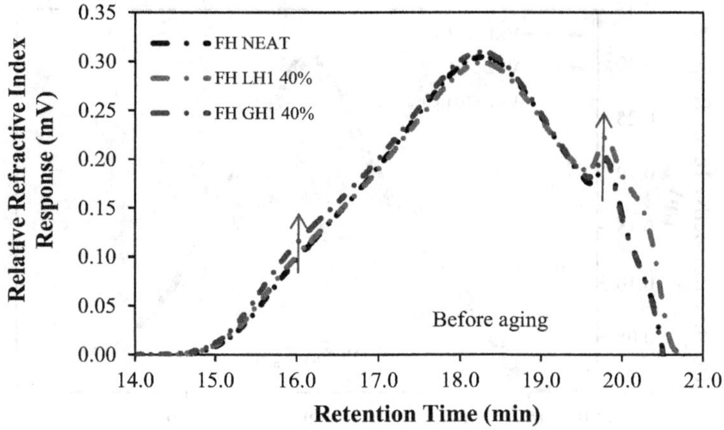

(a)

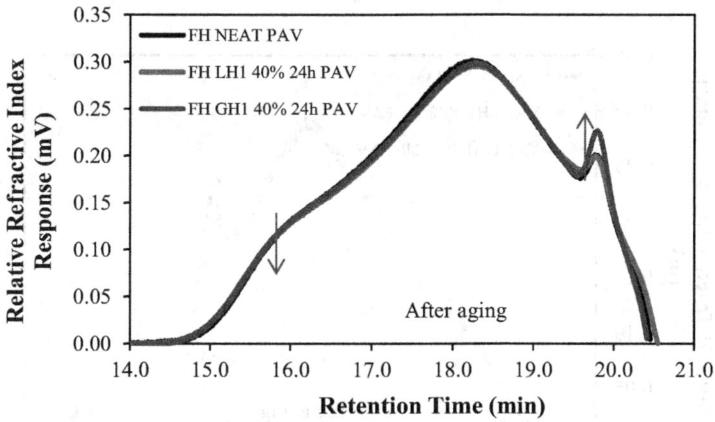

(b)

Figure 12. *GPC chromatograms of FH binder and mastics: (a) before, and (b) after 24 hours of PAV aging.*

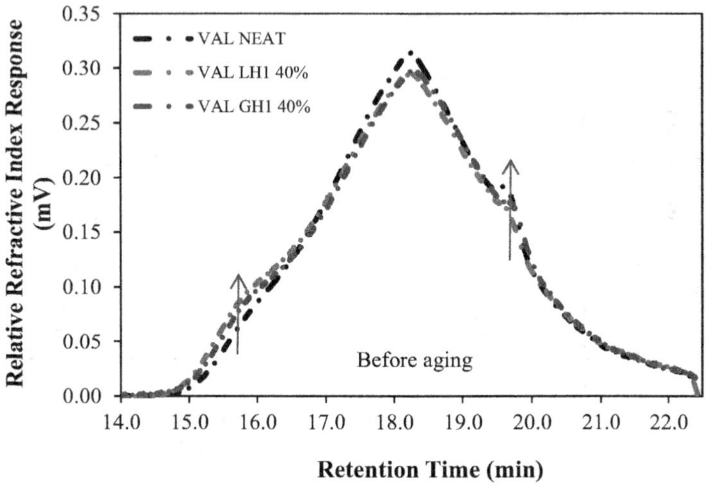

(a)

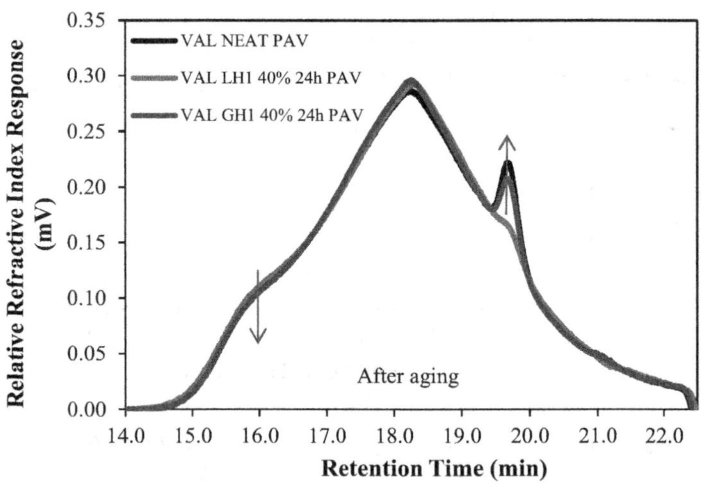

(b)

Figure 13. *GPC chromatograms of FH binder and mastics: (a) before, and (b) after 24 hours of PAV aging.*

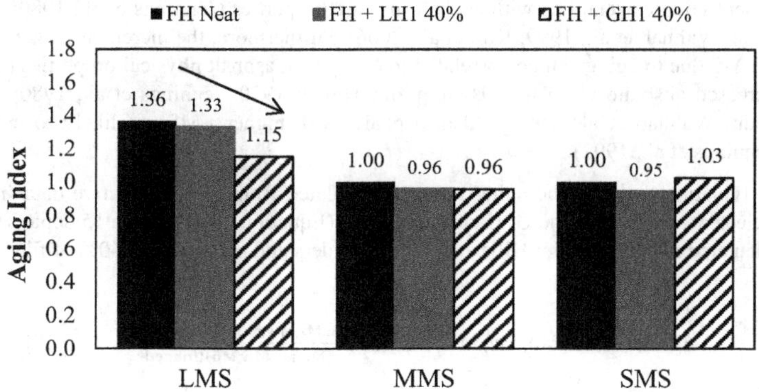

(a)

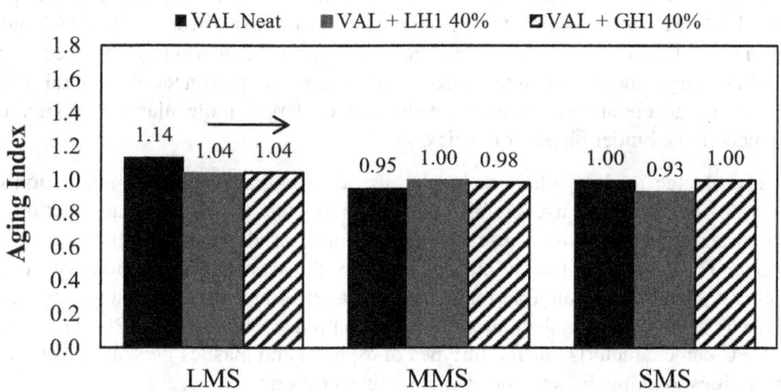

(b)

Figure 14. *GPC molecular size areas aging indexes of: (a) FH base binder and mastics, and (b) VAL base binder and mastics.*

The LMS is the percentage of high molecular weight molecules in a binder (Jennings et al., 1980). In asphalt binders, aging results in a shift towards the higher molecular weight molecules (Hagos, 2008), since oxidative aging causes formation of more polar molecules at the expense of the lower weight molecules. Many studies have indicated that the large molecular size (LMS) molecules of asphalt binder present good correlations with asphalt mixture properties (Jennings et al., 1980; Al-Abdul Wahhab et al., 1999; Kim et al., 2006). Furthermore, the incremental increase of LMS due to aging can be correlated to changes in asphalt physical properties (e.g. increased absolute viscosity) (Bynum and Traxler 1970; Jennings et al., 1980; Al-Abdul Wahhab et al., 1999). Also, asphalts with higher LMS are likely to crack (Jennings et al., 1992).

Equation 8 shows the aging indexes calculated from the quantitative data (i.e., molecular weights) of the GPC chromatogram (Equations 1-4). Figure 15 depicts the calculated results for Flint Hills and Valero binders and mastics with 40% of filler.

$$\text{Aging Index}\,(Mn, Mz, Mw, Mp) = \frac{(Mn,Mz,Mw,Mp)\text{After 24h PAV}}{(Mn,Mz,Mw,Mp)\text{Unaged}} \qquad [8]$$

It can be seen in Figure 15 that the molecular weight aging indexes of mastics were smaller than those observed for the aged neat binder. The most significant difference between mastics and base binder is observed for the M_z index (average molecular size, roughly corresponding to the LMS), where the aging index dropped from 1.54 (FH neat binder) to 1.13 (granite mastic), and from 1.39 (VAL neat binder) to 1.12 (granite mastic). These results support the previously observed trend in LMS (large molecular size molecules), where the presence of mineral filler appears to decelerate the rate of production of larger molecular size oxidation products in the binder phase of mastics.

The change in the asphalt molecular size distribution due to the interaction of mineral filler causing irreversible adsorption of the asphaltic polar fraction is hypothesized as the main cause of the softening of the mastics after the aging process. The changes in the colloidal structure of the asphalt binders and mastics due to changes in the asphalt fractions (i.e., asphaltenes and maltenes) upon aging is clearly reflected in the molecular weight distribution as obtained by GPC. Therefore, the GPC can characterize different types of asphalts and mastics presenting different sensitivities to aging, as well the changes caused by aging.

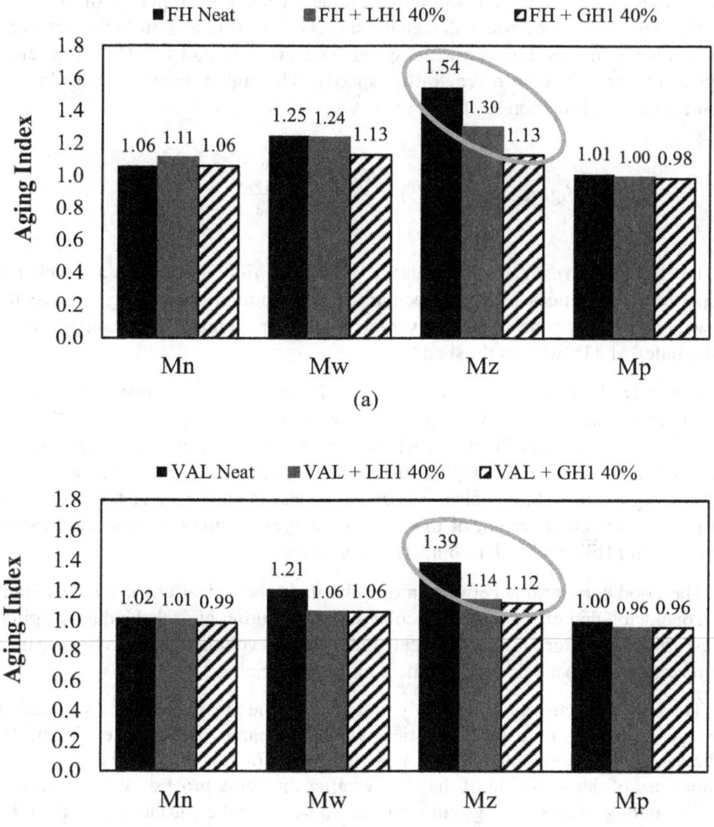

Figure 15. *Aging indexes of GPC molecular weight of: (a) FH base binder and mastics, and (b) VAL base binder and mastics.*

6.0 Relating Effect of Aging on GPC Results to Rheology

The changes in the colloidal structure of the asphalts due to oxidative aging, and the mineral fillers' effects on it, can be reflected on the rheological properties of the asphalts, where the transformation of the molecular size distribution seems to be the reason behind the changes in the mechanical properties of asphalts.

In order to show the effects of aging on the mechanical properties of binder and mastics, an aging index was calculated using |G*| at 10°C and 30 Hz (representing lower temperature or higher traffic speed) and at 40°C and 0.1 Hz (representing higher temperature or slower traffic speed). The aging index is calculated in accordance with Equation 9.

$$\text{Aging Index } (|G^*|) = \frac{(|G^*|)_{\text{After 24h PAV}}}{(|G^*|)_{\text{Unaged}}} \qquad [9]$$

To further investigate the relation between filler's effect on mechanical properties and changes in the molecular size distribution after aging, a correlation between the aging index of M_z (average molecular weight) (Equation 8) and the aging index of |G*| was established.

Figure 16 shows the correlations between M_z and |G*| aging indexes for the base binders and mastics. It can be seen that the M_z and |G*| aging indexes correlate with good R^2 values for both FH and VAL binders, especially at high temperatures. The observed trend shows that a higher increase in M_z leads to a higher stiffness at both low and high temperatures. Also, it is observed that the presence of fillers decreased the gain in molecular weight of mastics due to aging. Same behavior was observed for both Flint Hills and Valero binders and mastics.

The good correlations between molecular and mechanical characteristics lead to the conclusion that changes in molecular size distribution of asphalts during aging—a controlling factor in mechanical and thermo-volumetric behavior—can be engineered by proper selection of fillers in the mastic phase.

The results discussed in this study indicate that the presence of fillers decelerates the age hardening of asphalt mastics, when compared to the base binder. It is believed that the adsorption of the asphaltic polar fraction (i.e., asphaltenes) is the main cause of the softening of the mastics after the aging process. When comparing results among fillers, the granite has a higher specific surface area than both limestone and dolomite (as shown in Table 1), therefore, it is expected that the aging indexes for all granite mastics would have lower values than those of the limestone mastics, as is supported by the presented data.

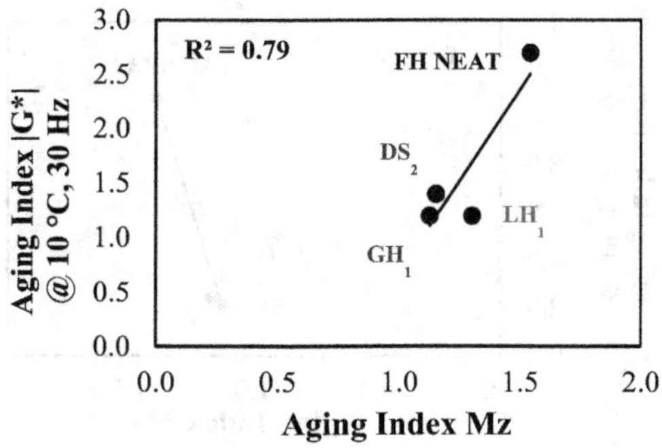
3.0
2.5
2.0
1.5
1.0
0.5
0.0
$R^2 = 0.79$
FH NEAT
DS 2
LH 1
GH 1
Aging Index |G*| @ 10 °C, 30 Hz
0.0
0.5
1.0
1.5
2.0
Aging Index Mz

(a)

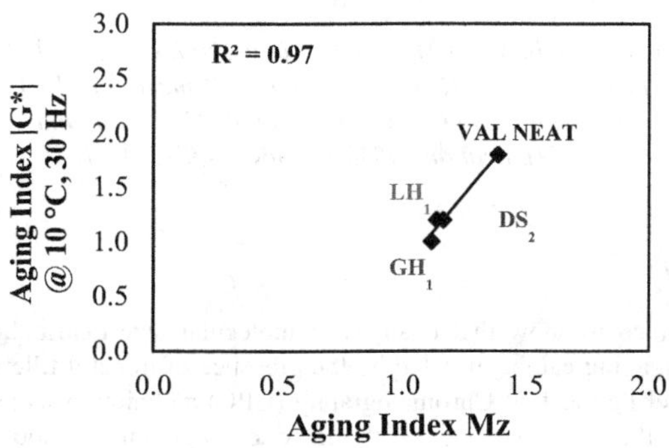
3.0
2.5
2.0
1.5
1.0
0.5
0.0
$R^2 = 0.97$
VAL NEAT
LH 1
DS 2
GH 1
Aging Index |G*| @ 10 °C, 30 Hz
0.0
0.5
1.0
1.5
2.0
Aging Index Mz

(b)

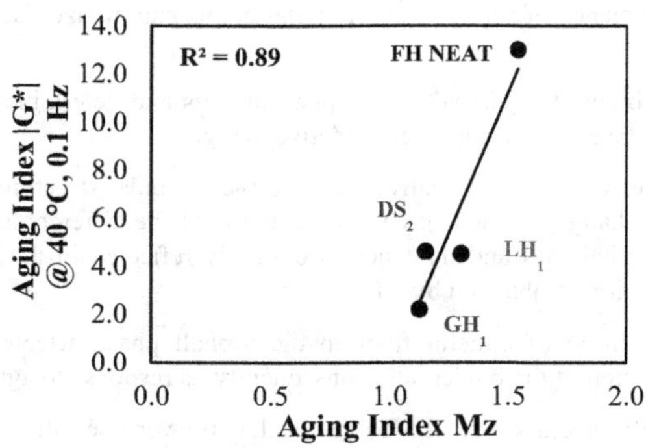
14.0
12.0
10.0
8.0
6.0
4.0
2.0
0.0
$R^2 = 0.89$
FH NEAT
DS 2
LH 1
GH 1
Aging Index |G*| @ 40 °C, 0.1 Hz
0.0
0.5
1.0
1.5
2.0
Aging Index Mz

(c)

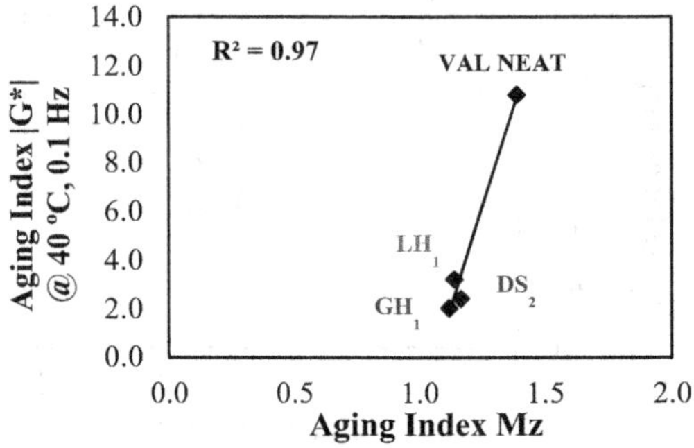

(d)

Figure 16. *Correlation between M_z and |G*| aging indexes of: (a)FH neat and 40% mastics at 10°C, 30 Hz, (b)VAL neat and 40% mastics at 10°C, 30 Hz, (c)FH neat and 40% mastics at 40°C, 0.1 Hz, and (d)VAL neat and 40% mastics at 40°C, 0.1 Hz.*

7.0 Conclusions

This study aimed to show that changes in molecular size distribution of asphalts binders during aging can be affected by the presence of mineral fillers in the mastic phase. The Gel Permeation Chromatography (GPC) technique was used to monitor the changes in the molecular size distribution due to aging in the laboratory using the PAV. Since these changes can significantly influence the asphalt binder consistency and consequently its mechanical properties, rheological measurements were made to confirm the influence of fillers. The following points summarize the major findings of this study:

- GPC can be used to characterize asphalt binders and determine changes at the molecular level before and after oxidative aging.
  - Changes caused by oxidative aging in the colloidal structure of the asphalt due to changes in the degree of association of the different asphalt fractions (i.e., asphaltenes and maltenes) are clearly reflected in the molecular size distribution as obtained by GPC.
  - The addition of mineral filler to the asphalt phase affects the molecular distribution of the binder, and consequently its response to aging.
- The large molecular size (LMS) molecules are very sensitive to aging, with asphalts showing an increase in their LMS contents upon aging. Therefore, the GPC technique can be used to monitor the aging process of asphalt binders with

regards to the production of high molecular weight molecules, and can clearly show the influence of fillers in mastics on the rate of aging of binders.

- Mineral filler appears to decelerate the rate of production of larger molecular size oxidation products in the binder phase of mastics, reducing the aging effects on mechanical properties.

- It is found that by selection of type and, consequently, the concentration of filler to be included in mastics used in asphalt pavements and roofing shingles, the impact of oxidative aging can be changed significantly. Reducing the impact of aging, enhances the service life of the asphaltic materials.

  - The simple procedures used in this study can be effectively used to determine how much each mineral filler can affect the aging rate of a specific asphalt binder. Thus, the durability of asphalt paving mixtures or roofing shingles can be engineered to provide better service life.

8.0 References

AASHTO R 28, "Accelerated Aging of Asphalt Binder Using a Pressurized Aging Vessel (PAV)," American Association of State Highway Transportation Officials, 2009.

Al-Abdul Wahhab, H. I., Asi, I. M., Ali, F. M., and AL-Dubabi, I. A., "Prediction of Asphalt Rheological Properties Using HP-GPC," *Journal of Materials in Civil Engineering,* Vol. 11, 1999, pp. 6–14.

Altgelt, K. H., "Gel Permeation Chromatography of Asphalts and Asphaltenes," Die Makromolekulare Chemie, Vol. 88, 1965, pp. 75–89.

Anderson, D. A., "Guidelines for Use of Dust in Hot-Mix Asphalt Concrete Mixtures," *Journal of the Association of Asphalt Paving Technologists*, Vol. 56, 1987, pp. 492–516.

Anderson, D. A., Bahia, H. U. and Dongre, R., "Rheological Properties of Mineral Filler-Asphalt Mastics and Their Relationship to Pavement Performance," *Effects of Aggregates and Mineral Fillers on Asphalt Mixture Performance*, ASTM STP 1147, Richard C. Meininger, editor, American Society for Testing Materials, Philadelphia, 1992.

Barbour, F. A., Barbour, R. V., and Petersen, J. C., "A Study of Asphalt-Aggregate Interactions Using Inverse Gas-Liquid Chromatography," *Journal of Applied Chemistry and Biotechnology*, Vol. 24, 1974, pp. 645–654.

Bowers, B. F., *Investigation of Asphalt Pavement Mixture Blending Utilizing Analytical Chemistry Techniques*, PhD Thesis, University of Tennessee-Knoxville, Tennessee, 2013.

Braunauer, S., Emmett, P. H., and Teller, E., "Adsorption of Gases in Multimolecular Layers," *Journal of the American Chemical Society*, Vol. 60, No. 2, 1938, pp. 309–319.

Bynum, D., and Traxler, R. N., "Gel Permeation Chromatography Data on Asphalt before and After Service in Pavement," *Association of Asphalt Paving Technologists Procedures*, Vol. 39, 1970, pp. 683–702.

Churchill, E. V., Amirkhanian, S. N., Burati Jr, J. L., "HP-GPC Characterization of Asphalt Aging and Selected Properties," *Journal of Materials in Civil Engineering*, Vol. 7, 1995, pp. 41–49.

Clopotel, C. S., *Filler Reinforcement Mechanisms in Asphalt Mastic*, PhD Thesis, University of Wisconsin-Madison, Madison, Wisconsin, 2012.

Corbett, L. W., "Composition of Asphalt Based on Generic Fractionation Using Solvent Deasphalteneing, Elution-Adsorption Chromatography and Densiometric Characterization," Analytical Chemistry, Vol. 41, 1969, pp. 576–579.

Curtis, C. W., Ensley, K., and Epps, J., *Fundamental Properties of Asphalt-Aggregate Interactions Including Adhesion and Absorption*, Strategic Highway Research Program: Report No. SHRP-A-341, National Research Council, Washington, DC, 1993.

Dawkins, J. V., "Theory of Gel Permeation Chromatography: Mechanism of Separation and the Influence of Polymer-Sorbent Interaction," *Pure and Applied Chemistry*, Vol. 51, 1979, pp. 1473–1481.

Dehouche, N., Kaci, M., and Mokhtar, K. A., "Influence of Thermo-Oxidative Aging on Chemical Composition and Physical Properties of Polymer Modified Bitumens," *Construction and Building Materials*, Vol. 26, 2012, pp. 350-356.

Delaporte, B., Di Benedetto, H., Chaverot, P. and Gauthier, G., "Linear Viscoelastic Properties of Bituminous Materials: from Binders to Mastics," *Journal of the Association of Asphalt Paving Technologists*, Vol. 76, 2007, pp. 455–494.

Delaporte, B., Di Benedetto, H., Chaverot, P. and Gauthier, G., "Effect of Ultrafine Particles on Linear Viscoelastic Properties of Mastics and Asphalt Concrete," *Transportation Research Record: Journal of the Transportation Research Board*, Vol. 2051, Transportation Research Board of the National Academies, Washington, D.C.2008, pp. 41–48.

Enustum, B. V., Kim, S. S., and Lee, D. Y., *Correlation of Locally-Based Performance of Asphalts with Their Physicochemical Parameters*, Iowa DOT Project HR-298, Engineering Research Institute Project 1942, 1990.

Faheem, A., Bahia, H. U., Hintz, C., Al-Qadi, I., Reinke, G., Dukatz, E., *Test Methods and Specification Criteria for Mineral Filler Used in HMA*, Final Report 9-45, National Cooperative Highway Research Program, Transportation Research Board, National Research Council, 2010.

Garrick, N. W., "Use of Gel Permeation Chromatography in Predicting Properties of Asphalt," *Journal of Materials in Civil Engineering*, Vol. 6, 1994, pp. 376–389.

Gehrke, C. W., and Ruyle, C. D., "Gas-Liquid Chromatographic Analysis of Nucleic Acid Components," *Journal of Chromatography*, Vol. 38, 1968, pp. 473–491.

Glover, C. J., Davison, R. R., Bullin, J. A., Button, J. W., and Donaldson, G. R., "Chemical Characterization of Asphalt Cement and Performance Related Properties," *Transportation Research Record: Journal of the Transportation Research Board*, No. 1171, Transportation Research Board of the National Academies, Washington, D.C., 1988, pp. 71–81.

Hagos, E. T., *The Effect of Aging on Binder Properties of Porous Asphalt Concrete*, PhD Thesis, Delft University of Technology, Delft, Netherlands, 2008.

Haley, G. A., "Molecular and Unit Sheet Weights of Asphalt Fractions Separated by Gel Permeation Chromatography," *Analytical Chemistry*, Vol. 43, 1971, pp. 371–375.

Jacobson, M., O'Brien, J. F., and Hedgcoth, C., "Determination of Nucleoside Composition of Ribonucleic Acid by Gas-Liquid Chromatography," *Analytical Biochemistry*, Vol. 25, 1968, pp. 363–369.

Jennings, P. W., Pribanic, P. W., Campbell, W., Dawson, K., Shane, S., and Taylor, R., *High Pressure Liquid Chromatography as a Method of Measuring Asphalt Composition*, FHWA-MT-79-30 Final Report, Bozeman, Montana, 1980.

Jennings, P. W., Pribanic, P. W., Smith, J. A., Mendes, T. M., "HP-GPC Analysis of Asphalt Fractions in the Study of Molecular Self-Assembly in Asphalt," Preprints of Papers-American Chemical Society, Division of Fuel Chemistry, Vol. 37, 1992, pp. 1312.

Kim, K. W., and Burati, J. L., "Use of GPC Chromatograms to Characterize Aged Asphalt Cements," *Journal of Materials in Civil Engineering*, Vol. 5, 1993, pp. 41–52.

Kim, K. W., Kim, K., Doh, Y. S., and Amirkhanian, S. N., "Estimation of RAP's Binder Viscosity Using GPC without Binder Recovery," *Journal of Materials in Civil Engineering*, Vol. 18, 2006, pp. 561–567.

Kriz, P., Stastna, J., Zanzotto, L., "Glass Transition and Phase Stability in Asphalt Binders," *Road Materials and Pavement Design*, Vol. 9/SI, 2008, pp. 37–65.

Lesueur, D., "The Colloidal Structure of Bitumen: Consequences on the Rheology and on the Mechanisms of Bitumen Modification," *Advances in Colloid and Interface Science*, 145(1-2), 2009, pp. 42–82.

Liu, M. M., Lin, M. S., Chaffin, J. M., Davison, R. R., and Glover, C. J., "Oxidation Kinetics of Asphalt Corbett Fractions and Compositional Dependence of

Asphalt Oxidation," *Petroleum Science and Technology*, Vol. 16, 1998, pp. 827–850.

Lu, X., and Isacsson, U., "Artificial Aging of Polymer Modified Bitumens," *Journal of Applied Polymer Science*, Vol. 76, 2000, pp. 1811–24.

Moraes, R., *Investigation of Mineral Filler Effects on the Aging Process of Asphalt Mastics*, PhD Thesis, University of Wisconsin-Madison, Madison, Wisconsin, 2014.

Noureldin, A. S., and Wood, L. E., "Variation in Molecular Size Distribution of Virgin and Recycled Asphalt Binders Associated with Aging," *Transportation Research Record: Journal of the Transportation Research Board,* No. 1228, Transportation Research Board of the National Academies, Washington, D.C., 1989, pp. 191–197.

O'Sullivan, K. A., *Rejuvenation of Reclaimed Asphalt Pavement (RAP) in Hot Mix Asphalt Recycling with High RAP Content*, MSc Thesis, Worcester Polytechnic Institute, Massachusetts, 2011.

Petersen, J. C, Barbour, F. A. and Dorrence, S. M., "Catalysis of Asphalt Oxidation by Mineral Aggregate Surfaces and Asphalt Components," *Association of Asphalt Paving Technology*, Vol. 43, 1974, pp. 162–177.

Petersen, J. C., "A Review of the Fundamentals of Asphalt Oxidation - Chemical, Physicochemical, Physical Property, and Durability Relationships," Transportation Research Circular E-C140, Transportation Research Board, Washington, DC, 2009.

Petersen, J. C., Harnsberger, P. M., and Robertson, R. E., "Factors Affecting the Kinetics and Mechanisms of Asphalt Oxidation and the Relative Effects of Oxidation Products on Age Hardening," American Chemical Society Division of Fuel Chemistry Preprints, Vol. 41, No. 4, 1996, pp. 1232–1244.

Plancher, H., and Petersen, J. C., "Reduction of Oxidative Hardening in Asphalts by Treatment with Hydrated Lime: A Mechanistic Study," *Association of Paving Technologists*, Vol. 45, 1976, pp. 1–24.

Plummer, M. A., and Zimmerman, C. C., "Asphalt Quality and Yield Predictions from Crude Oil Analyses," *Association of Asphalt Paving Technologists*, Vol. 53, 1984, pp. 138-159.

Read, J., and D. Whiteoak, *The Shell Bitumen Handbook*, Fifth Edition, 2003.

Robertson, R. E., *Chemical Properties of Asphalts and their Relationship to Pavement Performance,* Strategic Highway Research Program: Report No. SHRP-A/UWP-91-510, National Research Council, Washington, DC, 1991.

Rostler, F. S., and White, R. M., "Composition and Changes in Composition of Highway Asphalts, 85-100 Penetration Grade," *Association of Asphalt Paving Technologists*, Vol. 31, 1962, pp. 35–89.

Shen, J., Amirkhanian, S., and Xiao, F., "High-Pressure Gel Permeation Chromatography of Aging of Recycled Crumb Rubber-Modified Binders with Rejuvenating Agents," *Transportation Research Record: Journal of the Transportation Research Board*, No. 1962, Transportation Research Board of the National Academies, Washington, D.C., 2006, pp. 21–27.

Snyder, L. R., "Determination of Asphalt Molecular Weight Distributions by Gel Permeation Chromatography," *Analytical Chemistry*, Vol. 41, 1969, pp. 1223–1227.

Tarefder, R. A., and Arisa, I., "Molecular Dynamic Simulations for Determining Change in Thermodynamic Properties of Asphaltene and Resin because of Aging," *Energy & Fuels*, Vol. 25, 2011, pp. 2211–2222.

Xiao, F., Amirkhanian, S. N., and Shen, J., "Effects of Various Long-Term Aging Procedures on the Rheological Properties of Laboratory Prepared Rubberized Asphalt Binders," *ASTM Journal of Testing and Evaluation*, Vol. 37, 2009, pp. 129–138.

Yau, W. W., and Malone, C. P., "An Approach to Diffusion Theory of Gel Permeation Chromatographic Separation," *Journal of Polymer Science Part B: Polymer Letters*, Vol. 5, 1967, pp. 663–669.

Zenewitz, J. A., and Tran, K. T., "A Further Statistical Treatment of the Expanded Montana Asphalt Quality Study," *Public Roads*, Vol. 57, 1987, pp. 72–81.

9.0 Discussion

JOHN HARVEY: Very interesting and great presentation. A question: you were discussing the literature and the role of the absorptivity of the aggregate. But then when I was looking at the results for the three aggregates that you had, which were a limestone and a granite that had similar absorptivity and another limestone that had a different absorptivity, it seemed, at least if I remembered the numbers right, that the granite and the limestone with similar absorptivity separated quite a bit in your comparison of the aging with the fillers versus the aging with the neat binder. So my question is, if I'm correct in what I saw, can you give us any insight, if you have any, with regard to something other than absorptivity that might be explaining those results, such as the chemistry of the aggregate? That was a very long question, so I can repeat the gist of it if you would like.

RAQUEL MORAES: I hope I understood. The fillers selected for this study have similar specific area. Therefore, it was seeing that the two main characteristics of fillers that would play a role were the mineralogy and the surface area. If you see some of the data, the trend that was observed, for example, for the granite with one base binder is different than that observed when using another base binder. Thus, even though we can control some of the filler properties, when changing the base binder, it also changes how the filler will react with it.

JOHN HARVEY: So is the same thing true for the aggregate, that if had you similar surface area and if you change the chemistry of the aggregate, it's the same idea as changing the chemistry of the base binder?

RAQUEL MORAES: Yes, I believe so.

JOHN HARVEY: How important do you think changing the chemistry of the aggregate is?

RAQUEL MORAES: I think it's important because, as I said, it will change how it will interact with the base binder. So, according to the data collected, the mineralogy of the mineral filler played a major role in the asphalt-mineral filler interaction.

JOHN HARVEY: Thank you.

RAQUEL MORAES: Thank you, Dr. Harvey. I will definitely take this into consideration.

GEOFFREY ROWE: Thank you, a very nice presentation. I enjoy these studies that combine some chemistry and some physical measurements. I find it always very encouraging. I've got a couple of comments which I'd just like to make. The first thing, your last comment that you made was that the when considering the aging, just looking at the rheology of the binder alone is misleading. Let's just put that in context; it's the aging of the binder that we'd use for looking at binder purchase specification. So the aging of the binder alone is just the aging of the binder alone! As you go towards mixtures and fillers and other materials, we definitely need to understand the interactions with the aggregates to get a better picture of the aging and the mixture, but maybe this is misleading if it is just for the purchase specification. The other comment I'd like to make is that when you looked at your aging index, you showed a master curve on slide number 19, you showed a part of a master curve, and you took the slope from two parts of that master curve?

RAQUEL MORAES: Correct.

GEOFFREY ROWE: You're using that as your aging index. I would like to suggest that you fit that data to the Christensen Anderson or the CAM model and use the R value as the descriptor of the shape. What you'll notice is that if you track two properties, if you look at the crossover frequency or some value of where you have some characteristic of that. Crossover frequency is where phase angle equals 45 degrees or G' equals G". If you plot that value at a given reference temperature and then you also look at the evolution or the change in the R value as the material ages, you will find that would better capture what you're describing as a slope. Because what we're doing here on a slope, we're just taking an arbitrary interval, you know, so it's a bit of an arbitrary thing, it's not really a real characteristic, whereas the R value is something which I believe is a quite nice characteristic. It's related to the relaxation spectrum, it's related to cracking, and it really is a useful tool to understand how these materials are changing their properties as they age. So I would like to suggest that as a possible way forward!

RAQUEL MORAES: Thank you, Dr. Rowe. I will definitely take this into consideration.

DAVID ANDERSON: I emphasize what Geoff had to say. If I had gotten here first, I would have been able to say it, but Geoff took it out of my mouth. But, you know, you really can't say that the slope of the curve is temperature susceptibility. It's the phase angle, really. And I think you said that that was a measure of temperature susceptibility? It's not. That slope is the phase angle. So that's something you really need to correct in the final paper.

RAQUEL MORAES: Thank you.

DAVID ANDERSON: And it's very misleading to talk about soft and hard limestone. Those mineral fillers are composed of different granules of mineral, and it's what's in there that needs to be.... I've said this before. This same body of data has been reported other places, and I've made the same comment. You need to characterize the mineral filler according to its mineralogy, the different minerals. The granite is composed of many different things, and we don't really know what components in that granite are reacting with the binder. So it's too late to do this now, but for others contemplating this kind of work, you really need to characterize the mineralogy of the mineral filler. That was my comment, but otherwise it was fine. Thank you.

RAQUEL MORAES: Thank you, Professor.

ROBERT KLUTTZ: Very nice presentation and some really interesting work. I have a couple of questions and comments, though. First of all, when you're attributing adsorption of the asphaltenes as reducing oxidation of the asphaltenes, well, what you're seeing is an increase in the high molecular weight peak. And pretty much all of the asphaltene molecules are there already, so what you're seeing is oxidation not of the asphaltenes but of the other components of the bitumen. So how are you getting inhibition of oxidation if what you're adsorbing is the asphaltenes? And the second comment – it's been widely demonstrated that you see, for instance by SARA analysis, an increase of asphaltenes, while the saturates stay about the same, and loss of aromatics and resins. The other comment is, is there any way you could separate a reduced oxidation due to inhibited chemistry versus inhibited diffusion since you're adding a particulate to the asphalt? And then after you respond, I have a couple of suggestions on experiments.

RAQUEL MORAES: Thank you, Bob. First regarding the main hypothesis, not only the asphaltenes are being adsorbed but different polar asphalt fractions (i.e., asphaltenes and maltenes). Therefore, through adsorption the fillers decrease the production of large molecular size molecules in the asphalt. So, from that, the binder is less susceptible for oxidation, due to increase in surface adsorption of larger molecular size polar components by fillers, which are consequently unavailable for oxidation during aging process. Regarding the other question about diffusion?

ROBERT KLUTTZ: Diffusion.

RAQUEL MORAES: The diffusion. Can you give me some suggestions?

ROBERT KLUTTZ: Sorry?

RAQUEL MORAES: You told me you have some suggestions on how to measure that.

ROBERT KLUTTZ: Oh, yes. I have two. The first was, as far as a particulate, add glass microspheres, which are not absorptive at all. And then if you see the same sort of effect, you know it's due to diffusion. And the second would be to use something like a talc or a mica, something with a very high aspect ratio but perhaps similar chemistry, and there you're going to have a much greater influence of diffusion than on a more or less spherical particle. So that might help you distinguish between the two types of behavior.

RAQUEL MORAES: I have on my thesis one study that was performed by establishing a comparison between the aging index of mastic produced with glass beads and mastic produced with granite mineral filler. With this study it was possible to show that even though the filler volume and particle size were similar in both mastics, the effect of the filler on diffusion path length is assumed to be similar, thus the additional decrease in the aging index is believed to indicate the presence of a surface interaction through adsorption mechanisms. This adsorption results in changes in molecular size distributions. These results are believed to confirm the existence of an adsorption effect in addition to any effect on diffusion in the binder..

ROBERT KLUTTZ: Okay, so the glass beads did not show a reduction in oxidation?

RAQUEL MORAES: Yes.

ROBERT KLUTTZ: Oh, okay. Good.

RAQUEL MORAES: Yes.

ROBERT KLUTTZ: Okay, thank you.

RAQUEL MORAES: Thank you, Bob.

NATHAN MORIAN: Bob and Dave already covered most of my questions, so those have been answered very well. If you could go to the very last figure that you had relating your GPC results to the $|G^*|$ aging index. I was looking at the hard limestone LH results, because it had different surface characteristics, the lowest surface area I believe, and that tends to be the one with the greatest Aging Index, but I see not in every case. Just to follow up, my question was, this particular information showing both the Flint Hills and Valero binder with 40% filler, is this from the 3 mm aging film or the 1 mm?

RAQUEL MORAES: The 3 mm.

NATHAN MORIAN: These are the 3 mm. Okay, so you potentially have limiting conditions based upon diffusion like Bob was talking about, but there may also be a counteracting reduction due to that lower surface area. I mention the diffusion again because you see a larger difference in the GPC results with the 3 mm aging film compared to the 1 mm. Is that correct?

RAQUEL MORAES: Yes, but this is for 24 hours, the results that I showed before for 3.15 mm was 48 hours of PAV.

NATHAN MORIAN: Okay, thank you.

RAQUEL MORAES: Thank you very much.

Influence of Asphalt Binder Oxidative Aging on Critical Thermal Cracking Characteristics of Asphalt Mixtures

Mohammad Zia Alavi[a], Nathan E. Morian[b], Elie Y. Hajj[c], and Peter E. Sebaaly[c]

[a]*Pavement Research Center, Department of Civil and Environmental Engineering, University of California, Davis, California, 95616*

[b]*Nevada Department of Transportation, Carson City, NV, 89712*

[c]*Department of Civil and Environmental Engineering, University of Nevada, Reno, Nevada, 89557*

ABSTRACT. *The influence of oxidative aging of asphalt binders on the thermal cracking characteristics of asphalt mixtures was evaluated through analyses of thermo-volumetric (Coefficient of Thermal Contraction [CTC]), thermo-viscoelastic (stiffness-temperature relationship), crack initiation and fracture, and damage characterization parameters. These properties were evaluated utilizing laboratory-prepared specimens aged for multiple durations in laboratory ovens. The thermo-viscoelastic characterization was founded upon the linear viscoelastic solution of thermally stressed mixture characterization determined though the Uniaxial Thermal Stress and Strain Test (UTSST). A critical aspect of the UTSST measurements was determined to be the crack initiation stress and temperature which was additionally supported by the Viscoelastic Continuum Damage (VECD) methodology. Validation of the evolution of the thermal cracking characteristics of the mixtures with aging was obtained through corresponding measurements of the core specimens from the nearly 20 year in-service WesTrack field sections.*

KEYWORDS: *stiffness, crack initiation, fracture, coefficient of thermal contraction, continuum damage characteristics, oxidative aging.*

The oral presentation was made by Mohammad Zia Alavi.

1.0 Introduction

Thermal cracking has been associated with the volumetric change in the asphalt concrete layer as the pavement temperature drops. Due to the continuous nature of the asphalt concrete layer initially after construction, this contraction is restrained and thus causes tensile stresses to be induced within the asphalt concrete layer. The developed thermal stresses are generally greater at the surface where the pavement experiences the lowest overall temperature and/or the highest temperature fluctuations (Jung and Vinson, 1994). The magnitude of the induced tensile stress increases with decreasing temperature to a point where it reaches the cracking limit (i.e., tensile strength) of the asphalt mixture resulting in initiation of micro-cracks which quickly propagate through the depth of the asphalt concrete layer. For extremely cold temperatures or very fast cooling rates, thermal cracks may develop through the entire depth of the asphalt concrete layer after one or only a few cooling cycles; this is generally referred to as a single-event thermal cracking. For milder or slower cooling conditions, micro-cracks advance at a slower rate with the full depth cracking only occurring after several thermal cycles due to the accumulation of damage; this mechanism is usually referred to as thermal-fatigue cracking.

The resistance of asphalt pavements to thermal cracking is greatly influenced by the properties of the asphalt mixture. Particularly, the modulus (which is rate and temperature dependent), the thermal contraction coefficient, and the strength property of the mixture all have an impact on the resistance of an asphalt pavement to thermal cracking. These critical thermal cracking properties are highly dependent on the asphalt binder type and quantity, the aggregate source and gradation, and the asphalt mixture volumetrics (i.e., air void, film thickness, etc.). With time, the asphalt binder within the mixture becomes harder and more brittle as a result of formation of highly polar compounds such as carbonyl functional groups and other chemical species associated with the oxidation reaction. Oxidative aging can significantly influence the stiffness, contraction, and strength properties of the asphalt mixture such that thermal cracking becomes more prevalent with the age of pavement. The rate of asphalt binder aging in the mixture (e.g., increased carbonyl), has been observed to be dependent upon the asphalt binder composition and some significant mixture characteristics (Morian et al., 2013).

Precise evaluation of the thermal cracking resistance of an asphalt mixture could be achieved if the mixture critical characteristics were obtained by a suitable thermally induced loading condition. Historically, the thermal stress restrained specimen test (TSRST) was one of the few test methods which could fundamentally describe fracture behavior of an asphalt mixture subjected to a constant cooling rate. The effect of oxidative aging on increasing fracture temperature and decreasing fracture stress was observed by Jung and Vinson (1994). A following study by Kliewer et al. (1996) also observed the impact of aging temperature and duration on the fracture properties of asphalt mixtures. The changes in the low temperature fracture properties (i.e., stress and temperature) of asphalt mixtures were found to reach limiting values with increased aging time at 50°C. However, continuous changes in low temperature fracture properties of asphalt mixtures were observed with increased aging duration at 85°C. Moreover, it was found that no significant differences were observed in the fracture properties between mixtures aged with 4 and 8% air voids. The dependency of the temperature and duration on the effect of aging may be related to the concentration of polar components within the asphalt binder (Jung and Vinson, 1994; Kliewer et al., 1996).

Overall, fracture stress and temperature alone may not provide sufficient information for a comprehensive evaluation of an asphalt mixture. A more complete characterization of the mixture properties can be obtained with the measurements of both the thermal strain and thermal stress, and through the examination and interpretation of the relationship between these two measurements. For this purpose the Uniaxial Thermal Stress and Strain Test (UTSST) has recently been developed through enhancement of the traditional TSRST setup. The test allows reliable and repeatable measurements of thermal stress and strain from restrained and unrestrained asphalt mixture specimens, respectively. The UTSST specimens can be easily obtained from Superpave gyratory compacted specimens or field cores with a lift thickness of at least 65 mm. More detailed information regarding the test setup and sample fabrication can be found in the literature (Alavi et al., 2013a, Hajj et al., 2013).

The UTSST test has been used in several recent studies to investigate the thermo-viscoelastic (i.e., stiffness-temperature relationship) and fracture properties of asphalt mixtures including the mixtures with high recycled asphalt pavement (RAP) contents (Mensching et al., 2014) and select warm-mix asphalt mixtures (Hajj et al., 2013). Moreover, the evolution of the stiffness-temperature curve calculated using the UTSST results with oxidative aging of asphalt binder has been observed and described recently (Morian et al., 2014; Morian, 2014).

2.0 Research Objective

The main objective of this study was to quantify the influence of oxidative aging on the critical thermal cracking characteristics of asphalt mixtures as obtained from the UTSST measurements. These critical characteristics include the asphalt mixture thermo-volumetric (i.e., coefficient of thermal contraction [CTC]), crack initiation and fracture properties (i.e., respective temperature, stress, and energy values). The crack initiation behavior of the materials was supported by continuum damage characterization under thermal loading. The influence of oxidative aging was observed and quantified for the mixtures at various aging levels produced and aged in the laboratory. Subsequently, the findings were validated with field core samples from three representative WesTrack sections.

3.0 Materials

The materials used in this study were the original WesTrack materials that had been used during the two paving operations of the project; the first in 1995 and the second in 1997. Both aggregates and binder samples were obtained from the Federal Highway Administration (FHWA) Materials Reference Library (MRL), where they have been stored in a warehouse since construction. Both asphalt binders were originally supplied as PG 64-22 materials but were from different sources. The laboratory prepared mixture from the 1995 materials (WT95F22) contained aggregates from the Dayton Pit in a fine gradation with binder from West Coast Refinery. The laboratory-prepared mixture from the 1997 materials (WT97C22) contained aggregates from the Lockwood Pit in a coarse gradation with binder supplied from Idaho Asphalt. More specific information regarding the parent material properties and mixture characteristics can be found in literature (Epps et al., 2002; Morian, 2014).

4.0 Experimental Design

4.1 Aging Procedure

The aging process used in this study included both short- and long-term aging. Short-term aging was conducted on all mixtures in accordance with AASHTO R 30 (AASHTO, 2011) by placing the loose mixtures in a 135°C oven for four hours prior to compaction. The long-term aging was conducted on the asphalt mixture samples compacted to the desired air void level ±0.5% using the Superpave Gyratory Compactor. The long-term aging subjected the uncut, compacted specimens to forced-draft oven aging at 60°C for periods of 3, 6, and 9 months (28 days per month). The 60°C temperature was primarily selected to reduce the overall duration of the aging time but also to minimize the potential for unrealistic changes in the chemical properties of asphalt binders aged at higher temperatures, as noted in the literature (Petersen, 2009). After the designated aging, two cylindrical specimens were cored perpendicular to the compaction direction; with a diameter of 57 mm and height of approximately 130 mm as has been outlined elsewhere (Alavi et al., 2013a; Hajj et al., 2013).

4.2 Uniaxial Thermal Stress and Strain Test (UTSST)

Figure 1 presents the layout of the UTSST apparatus. The thermal stress and thermal strain were measured, respectively, from the restrained and unrestrained asphalt mixture specimens using the UTSST device. Briefly, the measurements were obtained while the restrained and unrestrained specimens were simultaneously subjected to a cooling rate of 10°C/hr starting from an initial temperature of 20°C. Detailed information regarding the effect of cooling rate on UTSST results can be found in the literature (Alavi and Hajj, 2014). A total of two replicates for the restrained specimen were tested for each evaluated mixture. In the case of thermal strain, the same unrestrained specimen was tested twice, once for each of the corresponding restrained specimen replicate tests.

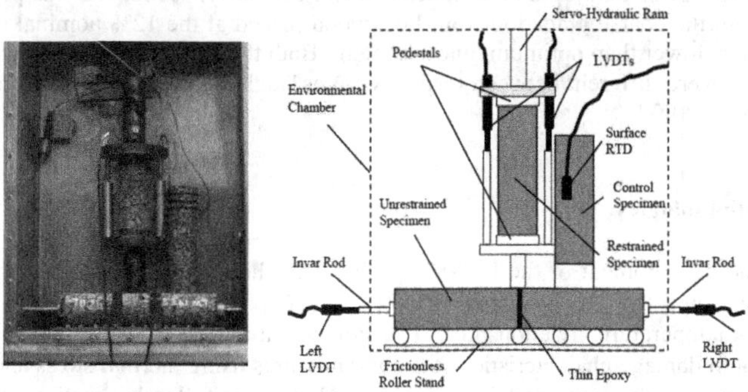

Figure 1. *Uniaxial thermal stress and strain test (UTSST) setup.*

4.3 Carbonyl Area (CA) Measurements

Following the testing, replicate mixture samples at the corresponding aging conditions were subjected to extraction and recovery to obtain representative asphalt binder samples that were tested by Fourier-Transform Infrared (FT-IR) Spectroscopy to determine an average level of oxidation represented by the carbonyl area (CA). A minimum of two replicate FT-IR spectra measurements were used to determine the CA, which is an indicator of oxygen content of the asphalt binder as an average of the entire extracted specimen. The CA is calculated as the area, in arbitrary units, between the IR absorption spectrum and the magnitude of the absorption at 1820 cm^{-1} used as a baseline. This area is integrated from 1650 to 1820 cm^{-1} wave numbers. Details regarding FT-IR testing procedure and measurements are available in the literature (Morian et al., 2013; Morian, 2014).

4.4 Field Core Samples

Field core samples were obtained in the spring of 2014 from three representative WesTrack sections. The specific sections of interest were those which roughly correspond to the mixtures evaluated in the laboratory aging studies. Specifically, the laboratory-prepared mixture designated as WT95F22 is analogous to the mixture used in section 15. Section 15 was constructed in 1995 as one of the fine gradation mixtures from Dayton placed at the 8% nominal air void level, near the optimum binder content.

The WT97C22 laboratory produced mixture with 7 and 11% air void represented the mixtures used in section 38 and section 56, respectively. Section 38 was placed in 1997 with the coarse gradation from Lockwood placed at the 8% nominal air void level, with lower than optimum binder content. Similarly, section 56 was placed in 1997 with the coarse gradation from Lockwood placed at the 12% nominal air void level, with lower than optimum binder content. Both the aggregate source and binder suppliers were different between respective WesTrack paving operations completed in 1995 and 1997.

5.0 Methodologies

With the development of the UTSST, different methodologies have been proposed to determine: (a) thermo-volumetric (i.e., CTC), (b) thermo-viscoelastic (i.e., stiffness-temperature relationship), (c) crack initiation, (d) fracture, and (e) continuum damage characteristics of asphalt mixtures using thermal stress and strain measurements. These methodologies are described in the following sections.

5.1 Determination of Thermo-Volumetric Properties

The thermal contraction strain of the unrestrained specimen caused by constant cooling rate, as shown in Figure 2, can be fitted with the proposed model presented in Equation 1 (Bahia, 1991) to obtain the thermo-volumetric property of the asphalt mixture.

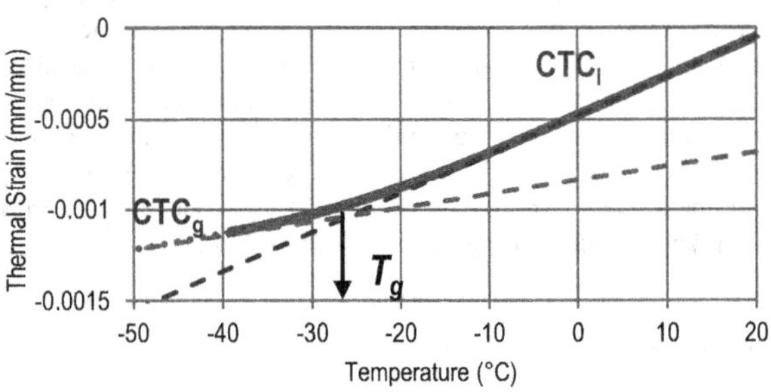

Figure 2. *Determination of thermo-volumetric properties from thermal strain measurements.*

$$\varepsilon_{th} = \frac{\Delta l}{l_0} = C + CTC_g(T - T_g) + ln\left\{\left[1 + e^{\frac{(T-T_g)}{R}}\right]^{R(CTC_l - CTC_g)}\right\} \quad [1]$$

Where $\Delta l/l_0$ is the relative change of length or thermal strain; C is the fitted intercept; CTC_l and CTC_g are the liquid and glassy coefficients of thermal contraction, respectively; T_g is glass transition temperature; and R is the parameter representing the curvature between the two linear asymptotes. T_g is determined by the intersection of the two linear portions of the curve with respect to the CTC_l and CTC_g, i.e., above and below T_g.

5.2 Determination of Thermo-Viscoelastic Properties

The measured thermal build-up stress can be related to the corresponding measured thermal strain using the uniaxial constitutive equation for linear viscoelastic materials, i.e., Boltzmann's equation (Christensen, 2003). These conditions are considered to be met for uniform and undamaged specimens.

By considering the synchronized thermal stress and thermal strain measurements, the UTSST modulus of an asphalt mixture at each temperature (corresponding to testing time t_n) can be calculated from the discrete form of the Boltzmann's equation as presented in Equation 2.

$$E(T(t_n)) = \frac{\left(\sigma_{Th}(t_{n+1}) - \sum_{i=2}^{n+1} E(t_{n+1} - t_i)(\varepsilon_{Th}(t_i) - \varepsilon_{Th}(t_{i-1}))\right)}{\varepsilon_{Th}(t_1)} \quad [2]$$

Where: E = asphalt mixture UTSST modulus
σ_{Th} = thermal stress
ε_{Th} = thermal strain
T = testing temperature
t_n =: testing time
n and I = time indices (*i* changes from 0 to *n+1*)

Five characteristic stages of material behavior were identified from the developed stiffness-temperature curve relationship and thermal build-up stress curve (Figures 3a and 3b). A brief description of these behaviors is listed below. In this study, the main evaluation was performed by examining the stresses and temperatures corresponding to fracture, crack initiation, and viscous softening stages. Further information regarding the thermo-viscoelastic properties and the stiffness-temperature relationship can be found in the literature (Alavi et al., 2013a; Hajj et al., 2013; Morian, et al., 2014; Alavi and Hajj, 2014; Mensching et al., 2014).

- Viscous softening: From this stage the relaxation modulus of the asphalt mixture increases rapidly with decreasing temperature.

- Viscous-glassy transition: At this stage the glassy properties of the material overcome the viscous properties.
- Glassy hardening: At this stage the behavior of the material is pure glassy.
- Crack initiation: In this stage micro-cracks occur in the specimen due to the induced thermal stresses when the material behavior is glassy.
- Fracture: At this stage the asphalt mixture specimen breaks due to the propagation of micro-cracks by the induced thermal stresses, i.e. macro failure.

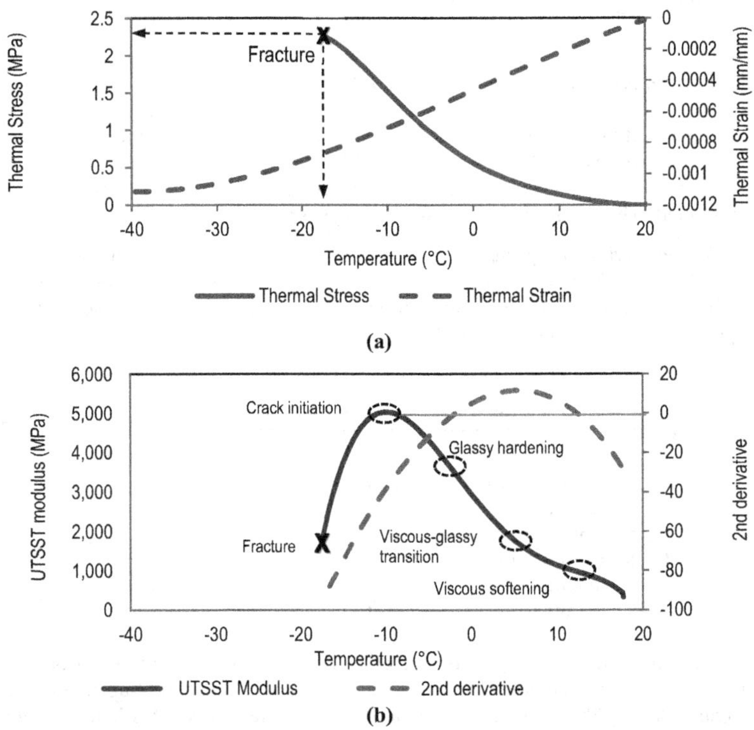

Figure 3. *(a) Measured thermal stress and strain; (b) Calculated UTSST modulus and associated characteristic stages.*

5.3 Determination of Failure, Crack Initiation, and Fracture Energies

In addition to the proposed characteristic terms for thermo-viscoelastic and fracture behaviors of asphalt mixtures, other critical properties of asphalt mixture under thermal loading can be determined from the area under the thermal stress (σ_{Th})-thermal strain (ε_{Th}) curve. The absolute value of the thermal strain is used in order to express both the thermal stress and strain as positive values. The area under the stress-strain curve is the definition of the energy per unit volume.

$$U = \frac{1}{V}\int F\, dL = \int_0^L \frac{F}{A_0}\frac{dL}{L_0} = \int_0^\varepsilon \sigma\, d\varepsilon \qquad [3]$$

Where U = energy per volume
F = thermal load
V = specimen volume
A_0, and L_0 = original cross section and length of the specimen, respectively.

From the area under the thermal stress-thermal strain curve, presented in Figure 4, three characteristic energy values are defined, which can be used to describe the toughness and brittleness of an asphalt mixture under thermal loading.

- Failure energy (U_f): this is defined as the total area under the thermal stress-thermal strain curve from the viscous softening stage to the fracture stage. Total failure energy (U) may indicate the amount of work or energy required by a mixture to fracture under thermal loading.
- Crack initiation energy ($U_{C.I.}$): this is defined as the area under the thermal stress-thermal strain curve from the viscous softening stage up to the crack initiation stage. This energy generally indicates the toughness of an asphalt mixture under thermal loading. In other words, crack initiation energy may be interpreted as the ability of the material to withstand a certain level of associated thermal stress and strain before the occurrence of micro cracks.
- Crack propagation energy ($U_{C.P.}$): this is determined as the difference between the failure energy (U_f) and the crack initiation energy ($U_{C.I.}$). This energy value is proposed as an indication of the mixture brittleness under thermal loading after the initiation of micro cracks.

The viscous softening stage was selected as the starting point in the calculation of the area under the thermal stress-thermal strain curve as opposed to the start temperature of the test in order to determine more fundamental energy values. Further, the viscous softening temperature is a characteristic temperature of asphalt materials as defined earlier. The crack initiation and propagation energy values can provide a better indication of the ductile or brittle behavior of a given asphalt mixture compared to merely crack initiation and fracture stress values.

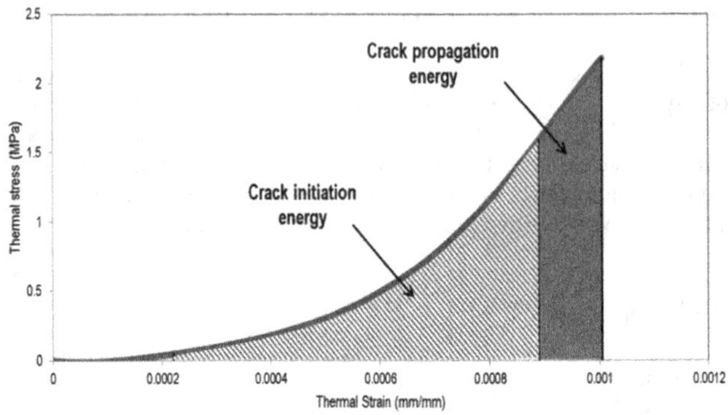

Figure 4. *Crack initiation and crack propagation energies determined by the area under the thermal stress-thermal strain curve.*

5.4 Determination of Continuum Damage Characteristics

Due to the heterogeneous nature of an asphalt mixture (i.e., a composite made with granular aggregate and asphalt binder), several defects do exist within this continuum mostly in the form of various sized air voids. Upon loading (either by traffic or temperature), damage may initiate and grow from these potential defect locations. Propagation or growth of such micro damage upon loading may not necessarily be physically visible, but on the macroscopic scale the influence can be detected with a continuous reduction in the stiffness. This complex behavior of asphalt mixtures upon traffic or thermally associated loading has been modeled through the proposed viscoelastic continuum damage (VECD) approach. In the VECD model, the viscoelastic portion of the model accounts for the time and temperature dependency of the materials while considering the history of the material upon loading through the time-temperature superposition principle. Additionally, the continuum damage portion of the model considers the reduction in the stiffness of the material due to the generation and growth of micro damage upon loading according to the work potential theory (Kim, 2009).

As a common practice, two types of testing are required to develop a one-dimensional (1-D) VECD model for an asphalt mixture. First, testing of the asphalt mixture in the undamaged linear viscoelastic range of behavior is needed to determine the time dependent stiffness (i.e., relaxation modulus) of the mixture. Second, a monotonic tensile test at low temperature using a single crosshead rate is required to determine the continuum damage characteristics of the material. The basis for the approach to determine the VECD model parameters can be found in the literature (Kim, 2009).

The application of VECD and more generally viscoelastoplastic continuum damage (VEPCD) models for predicting thermal build-up stress at various cooling rates in TSRST specimens were examined by Chehab and Kim (2005). It was determined that calculation of thermal stress without considering damage (i.e., using a linear viscoelastic (LVE approach) may result in noticeable error in the predictions. On the other hand, the accuracy of prediction was found to be improved considerably if either the VECD or VEPCD model were used. The application of VEPCD model might be limited since additional monotonic tests at high temperatures are needed for the determination of the VEPCD model parameters.

Based on the findings from the study (Chehab and Kim, 2005), it was proposed that damage may grow in the UTSST restrained specimen due to stress development upon cooling. Hence, the current effort was initiated to evaluate the potential for determining the continuum damage characteristics of asphalt mixtures using the UTSST thermal stress and strain measurements instead of the stress measurements from the mechanical monotonic testing at constant low temperatures. The fundamentals of the proposed methodology are presented in the following section.

5.4.1 Elastic-Viscoelastic Correspondence Principle

The thermal build-up stress in a UTSST restrained specimen neglecting damage can be calculated based on the elastic-viscoelastic correspondence principle as shown in Equation 4.

$$\sigma_{Th}(t) = E_R \times I \times \varepsilon^R_{Th}(t) \qquad [4]$$

$\varepsilon^R_{Th}(t)$ is the pseudo-thermal strain, which is calculated from the following convolution integral knowing the measured thermal strain from the unrestrained specimen ($\varepsilon_{Th}(t)$) and the undamaged relaxation modulus of asphalt mixture ($E_r(\xi)$). In this study, the undamaged relaxation modulus, represented in the form of a Prony series was determined from the complex modulus at the specific aging level based on the proposed approach by discretizing the continuum relaxation spectrum (Alavi et al., 2013b). Further, E_R is the reference modulus which assumed to be 1 and has the same unit as the undamaged relaxation modulus of the asphalt mixture.

$$\varepsilon^R_{Th}(t) = \frac{1}{E_R}\int_0^t - E_r(\xi(t) - \xi(t'))\frac{\partial \varepsilon_{Th}(t')}{\partial t'}dt' \qquad [5]$$

For the calculation of pseudo-thermal strain, the finite difference solution of the convolution integral was used (Equation 6), considering relaxation modulus in terms of a Prony series (Equation 7).

$$\varepsilon^R_{Th}(t) = \frac{-1}{E_R}\sum_{i=1}^{m}[e^{-\Delta\xi/\rho_i}\varepsilon^R_{Th}(t-\Delta t) + \Delta\varepsilon_{Th}E_i\frac{\rho_i}{\Delta\xi}(1 - e^{\frac{-\Delta\xi}{\rho_i}})] \qquad [6]$$

$$E_r = \sum_{i=1}^{m} E_i e^{-\xi/\rho_i} \quad [7]$$

Where: $\Delta\varepsilon_{Th}$ = change in thermal strain over time $t - \Delta t$ to t
$\Delta\xi$ = change of the reduced time over time $t - \Delta t$ to t
E_i and ρ_i = Prony coefficients of asphalt mixture relaxation modulus
m = number of Maxwell arms in the Prony series of relaxation modulus

Since the temperature changes continuously with time upon cooling during a UTSST test, the reduced time (ξ) is calculated from Equation 8. In this study, the WLF shift function (Equation 9) was employed for the calculation of reduced time at the reference temperature (T_r). The reference temperature was set to be 25°C in this study.

$$\xi(t) = \int_0^t \frac{dt'}{a_T(T(t'))} \quad [8]$$

$$a_T(T) = \frac{C_1 \times (T - T_r)}{C_2 + T - T_r} \quad [9]$$

Where: C_1 and C_2 = Coefficients of the WLF shift function equation
T_r = Reference temperature (°C)
T = testing temperature (°C); varies over time with the UTSST measures

The relationship between stress and pseudo-thermal strain follows the line of equality when no damage exists. Due to the fact that the complex modulus was determined on a separate set of representative specimens with different orientation relative to the axis of compaction, the obtained relaxation modulus determined from the complex modulus test may not be exactly the modulus of UTSST asphalt mixture specimens from which thermal stress and thermal strain were measured. Thus, the thermal stress and pseudo-thermal strain ratio may not be exactly one (i.e., follow the line of equality). In order to minimize the error associated with sample to sample variability, typically an adjustment factor (*I*) has been introduced to the equation. This factor was found to be 0.9 to 1.1 for samples with the same orientation (Kim, 2009). In this particular study the complex modulus specimen for measurement of the undamaged relaxation modulus has the same orientation as compaction, while the UTSST specimens were cored perpendicular to the axis of compaction (Alavi et al., 2013a). Hence, the *I* factor for these particular mixtures were found to be in range of 0.4 to 0.6 due to the aforementioned sample variability and anisotropic effect of the mixtures.

5.4.2 Implementing Continuum Damage Theory

As mentioned earlier, in reality due to the heterogeneous nature of asphalt mixtures, there are possible locations available in the asphalt mixture continuum where micro damage can grow due to thermal loading. Based upon the continuum damage theory

(CDT), the effect of damage growth can be qualified in the macroscopic view with a continuous decrease in the material stiffness. If damage exists, non-linearity will be observed in the curve of thermal stress–thermal pseudo-strain relationship as illustrated in Figure 5a for an example asphalt mixture. The temperature at which the non-linearity exists is defined as the damage initiation temperature (DIT) in this study (Figure 5b).

Subsequently, in the case of a viscoelastic material with growing continuum damage the thermal stress can be calculated from the equation expressed using the damage function (C). The damage function varies from a value of 1 for when there is no damage to 0 for a condition of complete damage in the continuum.

$$\sigma_{Th}(t) = C(S) \times E_R \times I \times \varepsilon^R_{Th}(t) \qquad [10]$$

According to the work potential theory, the damage function (C) is defined as a function of an internal state variable, i.e., the so-called damage parameter (S). The damage parameter is defined as any microstructural change that results in an observed stiffness reduction. Several mathematical functions were represented to calculate the damage parameter (S). In this study the damage parameter was determined numerically based on the equation proposed by Lee and Kim (1998). The coefficient "α" in this equation is the inverse of the maximum slope of the relaxation modulus versus reduced time curve in the log-log scale.

$$S(t + \Delta t) = S(t) + \left[-\frac{1}{2}(C(t + \Delta t) - C(t))(\varepsilon^R_{Th}(t))^2\right]^{\frac{\alpha}{1+\alpha}} \Delta t^{\frac{1}{1+\alpha}} \qquad [11]$$

5.4.3 Determination of C-S Relationship

The relationship between damage functions and damage parameter (S) can be defined through a decaying exponential function (Equation 12). The parameters a and b are determined by fitting the C-S exponential function to the measured values of C obtained from Equation 13 and the corresponding S values from the numerical form of Equation 11. The regression parameters a and b were estimated by minimizing the sum of the squared errors of the predicted values using the Solver feature in Microsoft Excel®. Figure 5c depicts the developed C-S relationship derived from the example asphalt mixture.

$$C(S) = e^{a \times S^b} \qquad [12]$$

$$C = \frac{\sigma_{Th}}{E_R \times I \times \varepsilon^R_{Th}} \qquad [13]$$

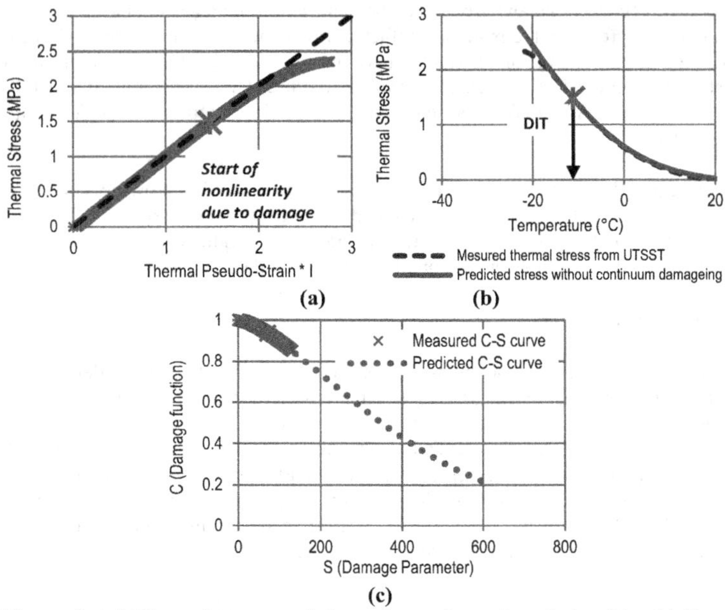

Figure 5. *(a) Thermal stress and thermal pseudo-strain relationship, (b) Damage initiation temperature determination, (c) Damage function (C-S relationship).*

6.0 Results and Discussion

The oxidative aging properties and evolution of critical thermal cracking characteristics of asphalt mixtures produced in the laboratory will be discussed in this section.

6.1 Carbonyl Growth

Figure 6 shows the measured CA values for the recovered asphalt binder from the respective aged mixtures as a function of aging time. As expected, the CA of the asphalt binder increased with aging time. Furthermore, the rate of increase in the CA values with aging time is relatively similar for the evaluated mixtures. However, the measured CA for the 11% air void mixture was noted to be larger than those of the 4 and 7% air void levels with the WT97C22 mixture. This is attributed to the fact that the asphalt binder in the higher air void mixture was more readily exposed to fresh air and thus higher concentrations of oxygen. The measured CA for the asphalt binder from the WT95F22 mix was noted to be even higher than those of the coarse mixtures, although it must be recognized that the asphalt binder sources differ between these mixtures. Thus, the overall oxidation magnitude should not be compared directly.

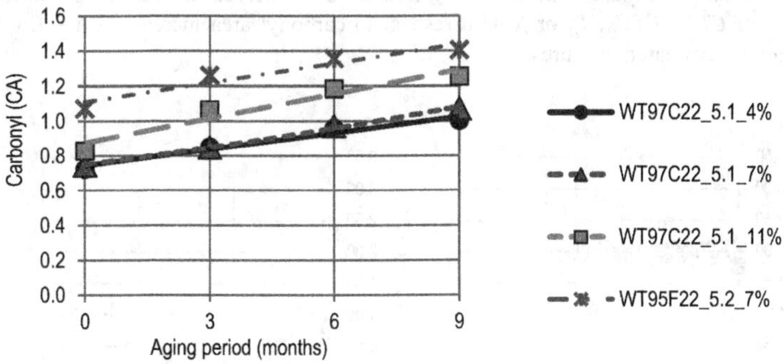

Figure 6. *Carbonyl area (CA) of extracted asphalt binder from the various mixtures aged at 60°C for 0, 3, 6, and 9 months.*

6.2 Influence of Oxidative Aging on Thermo-Volumetric Properties

The thermo-volumetric properties of the laboratory-produced mixtures made from WesTrack raw materials at various aging levels were calculated by fitting Equation 1 to the thermal strain versus temperature curve. Figure 7 shows the relationship between thermo-volumetric properties (i.e., CTC_l, CTC_g, T_g, and R) with the change in carbonyl area ($CA - CA_0$) measured in the asphalt binder for the laboratory-produced mixtures aged at 60°C. CA_0 is the initial carbonyl area at the age of 0 month for each mixture, i.e., after short-term aging only.

Overall, the CTC_l value was found to be more than twice the CTC_g value. The ratio between the CTC_l and CTC_g values was observed be as high as 7.5 for the case of the WT97C22_5.2-7% mixture subjected to 3 months of aging at 60°C. The glassy transition temperature (T_g) ranged between -20 and -26.5°C, corresponding to the WT97C22_5.2-11% mixture aged for 9 months and the WT97C22_5.2-11% with 0 month aging at 60°C, respectively. The values of R were observed to be within the range of 6 to 8. The differences between CTC_l and CTC_g values corresponding to the temperatures above and below T_g were determined to be substantial. This finding becomes critical given the fact that the value of CTC is commonly assumed to be a single and constant value regardless of the temperature, which can potentially lead to significant errors in the estimation of the thermal stresses in thermal cracking models and prediction analyses. Hence, the temperature dependency of CTC must be considered to obtain more reasonable computations of a given pavement response exposed to thermal loading regimes.

According to Figure 7, no consistent trends were observed for the changes in the values of CTC_l, CTC_g, T_g or R with respect to carbonyl area increase ($CA - CA_0$), among the evaluated mixtures.

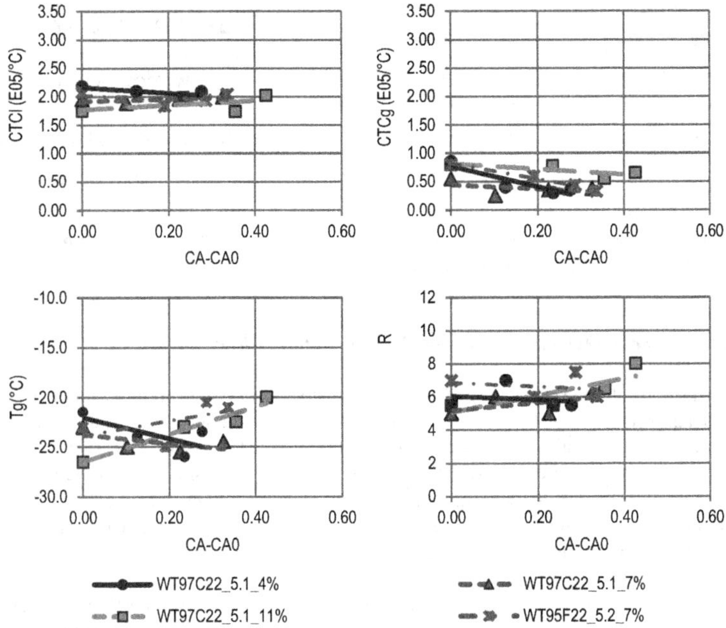

Figure 7. *Thermo-volumetric properties as a function of the increase in carbonyl ($CA - CA_0$).*

Statistical analyses were also performed to examine the sensitivity of thermo-volumetric properties to the aging (i.e., in terms of CA), the air void content (i.e., 4, 7, and 11%), and qualitative aggregate gradation (i.e., fine versus coarse). A mixture parameter was found to be significant if the P-value from t-statistics test was less than the significance level of 0.05. The analysis was completed using Minitab 16 statistical tool package. Accordingly, it was found that CTC_l for the evaluated mixtures was not sensitive to the measured levels of aging or the qualitative aggregate gradation. However, it was found to be sensitive to the air void level in the mixture (i.e., P-value = 0.01).

6.3 Influence of Oxidative Aging on Thermo-Viscoelastic and Fracture Properties.

According to the established methodology, the stiffness-temperature relationships of the asphalt mixtures were determined including the corresponding viscous softening, glassy transition, glassy hardening, crack initiation, and fracture stages. The temperature, stress, and modulus values corresponding to these stages of behavior are reported in Table 1. However, the main consideration to identify the influence of aging focused primarily on fracture, crack initiation, and viscous softening stresses and temperatures.

Table 1. *Critical cracking stress, temperature, and energy properties for WesTrack aggregate mixtures.*

ID	Air Void	Dur. (mo)	CA	Temperature (°C)			Stress (MPa)			Energy (Pa.mm/mm)			
				Fr.[a]	C.I.[b]	V.S.[c]	Fr.	C.I.	V.S.	F[d]	C.I.	C.P.[e]	C.P./F (%)
WT97C22_5.1_60°C	4%	0	0.723	-25.3	-24.6	11.6	4.0	3.9	0.0	696	662	34	5
		3	0.849	-19.7	-19.6	14.3	3.4	3.4	0.0	591	587	4	1
		6	0.959	-19.4	-17.2	13.7	3.4	2.9	0.0	567	511	56	10
		9	1	-20.6	-16.5	12.9	3.5	2.7	0.0	634	437	197	31
	7%	0	0.745	-26.2	-23.9	10	3.4	3.0	0.0	520	442	77	15
		3	0.846	-19.8	-19.7	14.1	3.0	3.0	0.0	436	430	6	1
		6	0.97	-22.7	-15.7	12.3	3.3	2.3	0.0	670	369	302	45
		9	1.071	-18.3	-14.1	14	2.8	2.1	0.0	522	358	164	31
	11%	0	0.829	-26.3	-26.2	12.2	2.3	2.3	0.0	369	366	3	1
		3	1.063	-21.4	-21.4	15	2.3	2.2	0.0	344	341	3	1
		6	1.184	-17.4	-16.1	14.7	2.1	2.0	0.0	330	294	35	11
		9	1.256	-15.9	-14	13.9	1.7	1.5	0.0	283	237	46	16
WT95F28 5.2_60°C	7%	0	1.069	-24.2	-23.2	11.1	3.0	2.7	0.0	505	462	43	9
		3	1.26	-19.7	-14.3	12.2	2.8	2.1	0.0	439	289	150	34
		6	1.356	-14.1	-10.9	14.9	2.5	2.0	0.1	351	294	57	16
		9	1.405	-12.8	-7.1	14.8	2.3	1.7	0.1	373	199	174	47

a- *Fracture*
b- *Crack initiation*
c- *Viscous softening*
d- *Failure*
e- *Crack propagation*

Figure 8 provides the relationship between the fracture, crack initiation, and viscous softening stresses and temperatures with $CA - CA_0$ for the laboratory-produced mixtures aged at 60°C and for their respective durations.

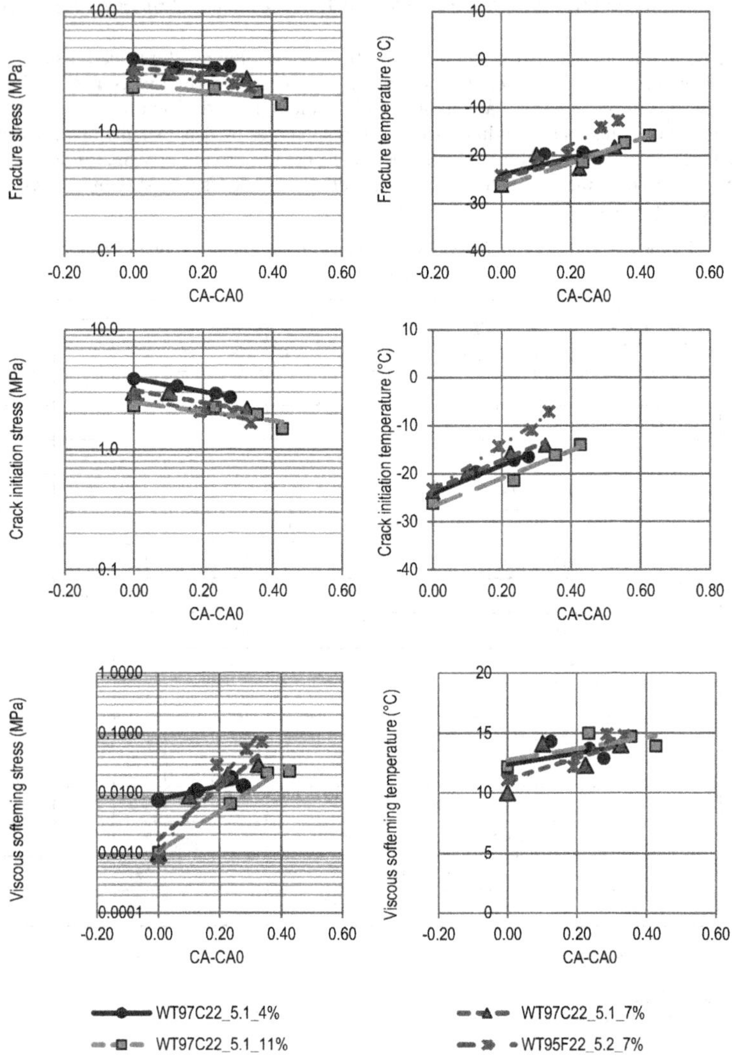

Figure 8. *Evolution of fracture, crack initiation and viscous softening stresses and temperatures with change in carbonyl (CA-CA_0).*

Overall, fracture and crack initiation stresses were observed to decrease with increased aging. The relationship between the viscous softening stresses with aging followed an opposite trend, i.e., increased with aging. The fracture, crack initiation, and viscous softening stages were exhibited at a warmer temperature with the increased aging level. These observations indicate a reduction in the ductility of the asphalt mixture with the increase in the asphalt binder oxidation. As the asphalt binder becomes more brittle with oxidative aging, the thermal stresses started to develop at warmer temperatures and with larger magnitudes (i.e., observed by warmer viscous softening temperature and greater viscous softening stress); due to a reduction in the relaxation potential of the asphalt mixture. Subsequently, the crack initiation and fracture limits occurred at relatively warmer temperatures. The crack initiation and the fracture stress limits of the material have also been reduced with the oxidative aging of the asphalt binder, due to a potential residual damage that may have been induced in the asphalt mixture causing a reduction in the tensile strength tolerance of the material.

The crack initiation and fracture stress and temperature values were also compared within the respective mixtures in order to detect the influence of air void and qualitative aggregate gradation. Comparisons of the mixtures at different air void levels revealed lower values of fracture and crack initiation stresses for those with higher air void contents. Similarly, the fracture and crack initiation points occurred at warmer temperatures for the mixtures with higher air void contents. The crack initiation and fracture stress for the fine gradation mixture was observed to be within the range of values for the coarse gradation mixture at the same air void level (i.e., 7%). The crack initiation and fracture temperatures for WesTrack fine mixture were observed to be higher (i.e., warmer) than other mixtures when compared at a higher level of increase in carbonyl ($CA - CA_0$), again acknowledging the differences in the binder and aggregate sources

6.4 Influence of Oxidative Aging on Failure, Crack Initiation, and Crack Propagation Energy Values

The mixtures were also evaluated in terms of the proposed energy values associated with the failure, crack initiation, and crack propagation stages as discussed previously. Figure 9 illustrates the evolution of the aforementioned energies with the increased oxidation level of the asphalt binder.

For all evaluated asphalt mixtures, the failure and crack initiation energies decreased with the increase in CA. The opposite trend was observed for the crack propagation energy. The influence of the asphalt binder oxidation on the properties of the mixtures can more fundamentally be described through the examination of the crack initiation and propagation energy values. The crack initiation energy particularly describes the toughness of the asphalt mixture, defined as its ability to withstand thermal loading (in a cooling event) before yielding to cracking. The higher crack initiation energy suggests an improved performance of the mixture under thermal loading. As the asphalt binder oxidation increased, the crack initiation energy decreased, indicating a decrease in the toughness of the asphalt mixture. After the crack occurrence, its propagation though the mixture can be observed by means of the crack propagation energy. For the evaluated mixtures, the crack propagation energy increased with increasing levels of oxidation. However, it must be noted that the interpretation of the crack propagation energy results for the overall performance of the mixture might not be a completely appropriate measure. Due to the loss of integrity in the specimen after the crack initiation, the crack propagation measures after this point are not representative of the global properties of asphalt mixture and are localized to the tested specimen. In other words, the specimen is no longer a continuous medium and thus the results should be viewed with caution.

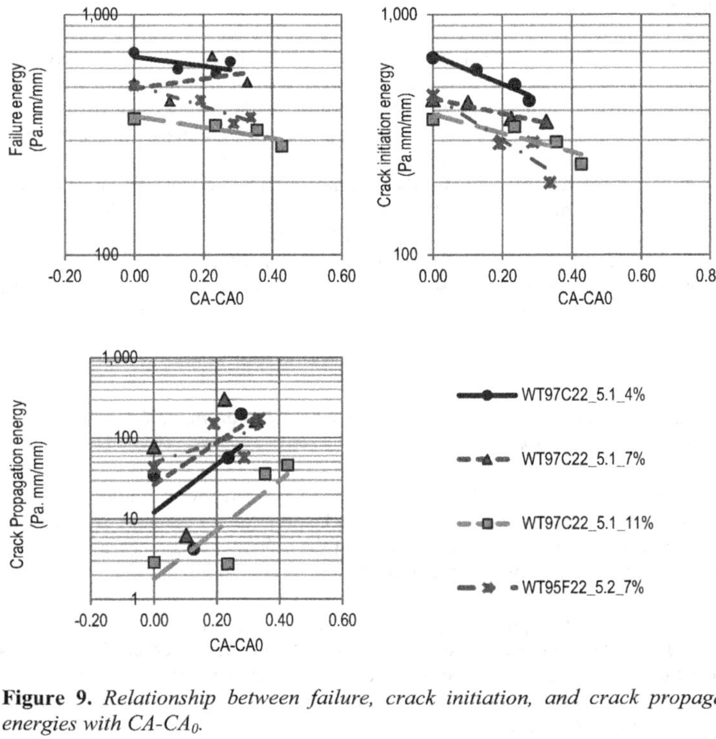

Figure 9. *Relationship between failure, crack initiation, and crack propagation energies with CA-CA*$_0$.

In general, the relationship between the crack initiation stress and energy with oxidation can be expressed with a decaying exponential function. Equation 14 expresses the generic form of the exponential function used to describe the evolution of the crack initiation stress and energy with oxidation. Similar decaying exponential function may be used to relate fracture stress and failure energy to the oxidation measures. An increasing exponential function could also be used to describe the increase in the viscous softening stress with oxidation. However, the main focus of this study was on the properties at crack initiation stage. Table 2 provides the values of the exponential function describing the evolution of the crack initiation stress and energy with increasing oxidation for the evaluated asphalt mixtures aged at 60°C for the respective durations.

$$Stress_{Crack\ ini.}\ or\ Energy_{crack\ ini.} = A \times e^{B(CA-CA_0)} \quad [14]$$

Table 2. *Coefficients of exponential regression equations relating crack initiation stress and energy to CA-CA₀.*

Mixture ID	Crack Init. Stress			Crack Init. Energy		
	A (MPa)	B	R^2	A (Pa.mm/mm)	B	R^2
WT19C22_5.1_4%_60°C	3.9	-1.260	0.99	678	-1.390	0.93
WT19C22_5.1_7%_60°C	3.1	-1.168	0.89	448	-0.722	0.93
WT19C22_5.1_11%_60°C	2.5	-0.886	0.67	385	-0.913	0.79
WT19F28_5.2_7%_60°C	2.7	-1.252	0.91	463	-2.174	0.88

6.5 Influence of Oxidative Aging on Continuum Damage Characteristics

In accordance with the methodology explained previously, the continuum damage characteristics of the respective mixtures were calculated using the thermal stress and strain measurements from the UTSST results. Figure 10 provides a comparison of the defined damage initiation temperature with the crack initiation temperature for evaluated mixtures.

A reasonable correlation was observed between the crack initiation temperatures from the stiffness-temperature relationship to the damage initiation temperatures obtained from the continuum damage analysis. This finding substantially indicates that the continuum damage analysis supports the findings observed by stiffness-temperature relationship with respect to the definition of the crack initiation stage. It has been reported that the continuum damage of the specimen initiates and grows mainly due to development of micro-cracks in the asphalt mixture continuum (Kim, 2009). It is worth mentioning that continuum damage analyses were also performed with other mixtures composed of dissimilar aggregate source, binder types, and mix designs with generally similar agreement being noted between the crack initiation and damage initiation temperatures (Alavi, 2014).

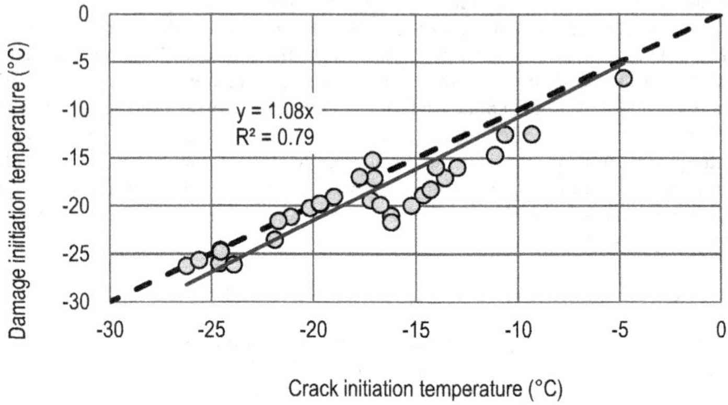

Figure 10. *Comparison of crack initiation and damage initiation temperatures.*

7.0 Evaluation of Critical Thermal Cracking Characteristics of WesTrack Field Cores

To evaluate the level of oxidation between the laboratory- and field-aged mixtures, the comparisons of the measured CA of both the laboratory-prepared mixtures and the field cores were conducted. However, due to the lack of established CA measures for the field sections (i.e., representative sample of the short-term aging condition), the CA measures comparing the field and lab measures represent the actual measures of the carbonyl area as opposed to the $(CA - CA_0)$ measures utilized previously. In this manner, the comparisons of the oxidation levels of the respective laboratory-aged and field-aged mixtures are considered in Figure 11.

Figure 11 presents the comparison of the CA measures between the field mixtures and the corresponding laboratory-prepared and -aged mixtures. Accordingly, the measured CA values for the field mixtures were generally observed to indicate a higher level of oxidation than those of the laboratory-prepared mixtures. The WT95F22 mixture obtained CA values that were quite similar between the measured core and the 9-months aging duration in the lab. However, the discrepancy in the measured CA values increased with the air void level for the WT97C22 mixtures. Thus, these findings generally support the conclusion that oxidative aging in the field may not only be material dependent, but also can be influenced by other factors leading to the noted discrepancy.

Additional considerations of the material properties comparing the field- and laboratory-aged mixtures were conducted by evaluating the crack initiation stress and crack initiation energy values as presented in Figure 12.

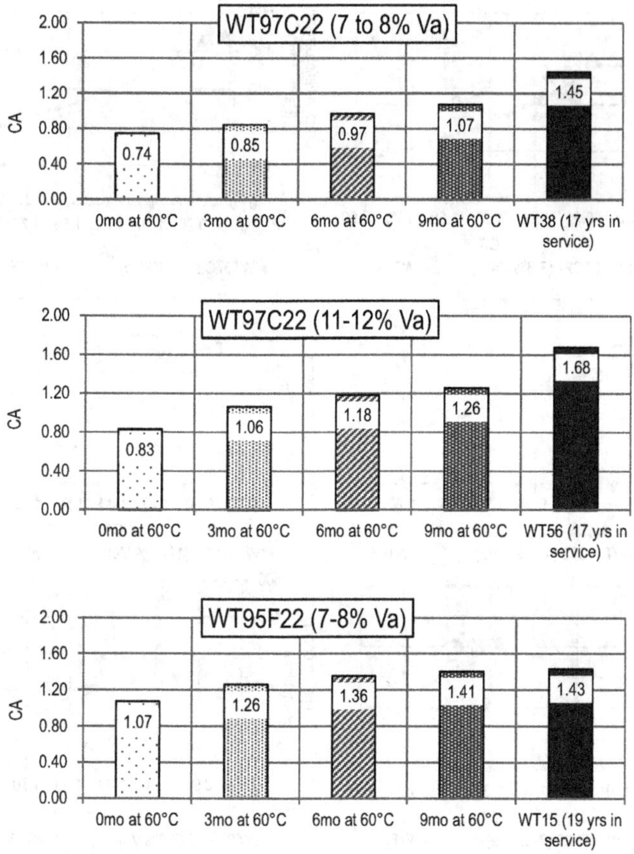

Figure 11. *Comparison of carbonyl area (CA) for the extracted binder from the field core mixture with the corresponding laboratory-produced samples.*

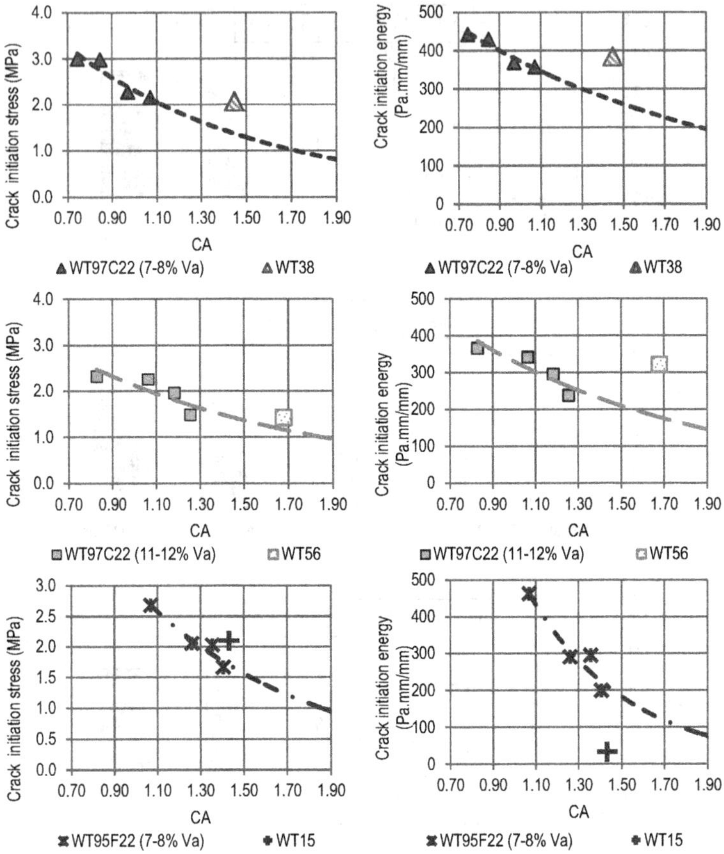

Figure 12. *Comparison of critical crack initiation stress and energy of the field and lab mixtures.*

Consideration of the crack initiation properties of the field core specimens relative to the laboratory-aged specimens with respect to the measured oxidation level indicates a moderate level of agreement with the crack initiation stress. A somewhat reduced level of agreement was observed with the crack initiation energy measures for the respective mixtures.

The general level of agreement was a beneficial finding, especially when viewed within the context of predictive modeling efforts of field performance based upon laboratory determined mixture properties. Given the fact that the relative indication of the crack initiation characteristics of the mixtures was reasonably quantified though the laboratory measures was a general indication that predicted field performance may be an achievable goal. Utilizing the appropriate mixture characterization methodologies coupled with appropriately simulated aging conditions provided a reasonable estimate of a given mixture after considerable in-service aging of the field mixture.

8.0 Summary and Recommendations

In summary, this paper provided a fundamental approach to characterize the thermo-volumetric and thermo-viscoelastic material properties, which were further supported by the continuum damage techniques to characterize the thermal cracking performance of asphalt mixtures. The thermo-volumetric (CTC) properties of the mixtures were not found to significantly change with increased levels of aging. However, they were found to be affected by the air void level of each respective mixture. The thermo-viscoelastic properties tended toward warmer temperatures, lower stress values, and generally lower fracture and crack initiation energies, but typically increased levels of crack propagation energies with increased levels of oxidation. These properties were again found to be dependent upon the respective mixture air void levels. The general agreement between the crack initiation temperatures provided by the UTSST measures and the point of damage identified through the VECD analyses, provides supplemental support of the appropriate material characterization of the micro-damage concept proposed in the UTSST methodology.

The increased reliability in the UTSST characterization should permit the expansion of the UTSST methodology into more in-depth evaluations of mixture characterization and performance prediction efforts. Having the validity of the material characterization supported should permit the additional comparison of laboratory- and field-prepared mixtures in an effort to further examine the complex changes in material properties as asphalt mixtures age in-service.

9.0 Acknowledgements

The content of this paper is part of the overall effort in the Asphalt Research Consortium (ARC). However, the contents of this report reflect the views of the authors and do not necessarily reflect the official views and policies of the FHWA. The authors gratefully acknowledge the FHWA support.

10. References

Alavi, M. Z., *Comprehensive Methodologies for Analysis of Thermal Cracking in Asphalt Concrete Pavements*, Ph.D. Dissertation, University of Nevada, Reno,

UMI Number: 3638109, ProQuest UMI Dissertation Publishing, Ann Arbor, MI, August 2014.

Alavi, M. Z., and Hajj, E. Y., "Effect of Cooling Rate on the Thermo-Volumetric, Thermo-Viscoelastic, and Fracture Properties of Asphalt Mixtures," *Asphalt Pavements*, Taylor and Francis Group, London, 2014, pp. 405–416.

Alavi, M. Z., E. Y. Hajj, N. E., Morian, and Sebaaly, P. E., "Low Temperature Characterization of Asphalt Mixtures by Measuring Visco-Elastic Properties under Thermal Loading," *ISCORD 2013: Proceeding of International Symposium on Cold Regions Development*, American Society of Civil Engineering, 2013a, pp. 404–415.

Alavi, N. E., Hajj, E.Y., and Morian, N. E., "Approach for Quantifying the Effect of Binder Oxidative Aging on the Viscoelastic Properties of Asphalt Mixtures," *Transportation Research Record*, No. 2373, Vol. 4, 2013b, pp. 109–120.

American Association of State Highway and Transportation Officials (AASHTO), *Standard Specifications for Transportation Materials and Methods of Sampling and Testing*, 31st Edition, Washington, D.C., 2011.

Bahia, H. U., *Low-Temperature Isothermal Physical Hardening of Asphalt Cements.*, Ph.D. thesis, Pennsylvania State University, University Park, Pa, 1991.

Chehab, G. R., and Kim, R. Y., "Viscoelastoplastic Continuum Damage Model Application to Thermal Cracking of Asphalt Concrete," *Journal of Materials in Civil Engineering*, ASCE, Vol. 17, No. 4, 2005, pp. 384–392.

Christensen, R., *Theory of Viscoelasticity*, Dover Publication Incorporation, New York, 2003.

Epps, J. A., Hand, A. J. T., Seeds, S., Schulz, T., Alavi, S., Ashmore, C., Monismith, C. L., Deacon, J. A., Harvey, J. T., and Leahy, R., *Recommended Performance-Related Specification for Hot-Mix Asphalt Construction: Results of the WesTrack Project*, NCHRP Report 455, Transportation Research Board, National Academies Press, Washington, D.C., 2002.

Hajj, E. Y., Alavi , M. Z., Morian, N. E., and Sebaaly, P. E., "Effect of Select Warm-Mix Additives on Thermo-Viscoelastic Properties of Asphalt Mixtures," *Road Materials and Pavement Design*, Vol. 14, Special Issue: EATA 2013, 2013, pp. 175–186.

Jung, D.H., and Vinson, T. S., *Low-Temperature Cracking: Test Selection*, Report No. SHRP A-400, Strategic Highway Research Program, Washington DC, 1994.

Kim, Y. R., *Modeling of Asphalt Concrete,* ASCE Press, American Society of Civil Engineers, 2009.

Kliewer, J. E., Zeng, H., and Vinson, T. S., "Aging and Low-Temperature Cracking of Asphalt Concrete Mixture," *Journal of Cold Regions Engineering*, Vol. 10, Issue 3, 1996, pp. 134–148.

Lee, H. J., and Kim, Y. R., "Viscoelastic Constitutive Model for Asphalt Concrete under Cyclic Loading," *Journal of Engineering Mechanics*, ASCE, Vol. 124, No. 1, 1998, pp. 32–40.

Morian, N. E., *Influence of Mixture Characteristics on the Oxidative Aging of Asphalt Binders*," Ph.D. Dissertation, University of Nevada, Reno, UMI Number: 3626103, ProQuest UMI Dissertation Publishing, Ann Arbor, MI, May 2014.

Morian, N. E., Hajj, E.Y., and Sebaaly, P.E., "Significance of Mixture Parameters on Binder Aging in HMA Mixtures," *Transportation Research Record*, No. 2370, Vol. 1, 2013, pp. 116–127.

Morian, N. E., Alavi, M. Z., Hajj, E. Y., and Sebaaly, P. E., "Evolution of Thermo-Viscoelastic Properties of Asphalt Mixtures with Oxidative Aging," *Transportation Research Record,* In press, Washington, D.C., 2014.

Mensching, D. J., Daniel, J. S., Bennert, T., Medeiros Jr., M. S., Elwardany M. D., Mogawer, W., Hajj, E. Y., and Alavi, M. Z., "Low Temperature Properties of Plant-Produced RAP Mixtures in the Northeast," *Road Materials and Pavement Design*, Vol. 15, Supplement 1, Special Issue: 89th Association of Asphalt Paving Technologists, 2014, pp. 1–27.

Petersen, J. C., *A Review of the Fundamentals of Asphalt Oxidation*, Transportation Research Circular E-C140, Transportation Research Board, Washington, D.C., 2009.

11.0 Discussion

<u>ESHAN DAVE</u>: Excellent presentation, Zia.

<u>MOHAMMAD ZIA ALAVI</u>: Thank you.

<u>ESHAN DAVE</u>: It's really good. Just a quick question on two things. First of all, you are calculating stresses and strain, so you are assuming a certain geometry and that's providing you with your curve that shows when you damage and then…

<u>MOHAMMAD ZIA ALAVI</u>: That's right.

<u>ESHAN DAVE</u>: Yeah. One of the early ones that you have. So the assumption here is that the linearity of your viscoelasticity still holds true after the damage?

<u>MOHAMMAD ZIA ALAVI</u>: After the damage?

ESHAN DAVE: Yeah, after your peak point.

MOHAMMAD ZIA ALAVI: That's a good question. We see the instantaneous drop in modulus curve after the peak point because the material is not anymore linear viscoelastic due to the initiation of a crack. In other words, we calculate the modulus based on the linear viscoelastic theory. Thus, when the crack initiates, this theory is no longer valid and the drop in modulus curve will be observed.

ESHAN DAVE: That's excellent. And I think that has a lot of potential, as you showed compared to, but my only worry is that other things that you are trying to extract out of it beyond that point I don't know if you'll be able to see a consistent trend...

MOHAMMAD ZIA ALAVI: That's right.

ESHAN DAVE: ...because, you know, you could have multiple types of failure. Just a caution there. The other question is have you looked at the effect of rate and has that shown any differences?

MOHAMMAD ZIA ALAVI: That's really a good question., Basically the main concern here was the measured properties before crack initiation point. Yes, you are right. We don't know what really happens after crack initiation since the material is no longer linear viscoelastic. You may even see some unusual trends in the behavior of material after initiation of crack. Therefore, only the calculated properties up to this point are valid and can be used for modeling.

ESHAN DAVE: Absolutely. But did you see any differences even up to that point if you changed rates?

MOHAMMAD ZIA ALAVI: In terms of cooling rates, we measured thermal strain and stress at three different cooling rates: 2.5, 10, and 17.5°C per hour. The thermal stress curves and calculated values of coefficient of thermal contraction were found to be different with regards to the cooling rate. Overall, based on the differences in thermal stress and thermal strain values, the shape of calculated modulus versus temperature was also found to change with the rate of cooling. Therefore, I guess the next step is to find a shift factor among various cooling rates.

ESHAN DAVE: Thanks.

MOHAMMAD ZIA ALAVI: You're welcome.

Quantification of Biasing Effects during Fatigue Tests on Asphalt Mixes: Non-Linearity, Self-Heating and Thixotropy

S. Mangiafico[a], C. Sauzéat[a], H. Di Benedetto[a], S. Pouget[b], F. Olard[b], and L. Planque[c]

[a]*Université de Lyon, École Nationale des TPE (LGCB), LTDS (CNRS UMR 5513) rue Maurice Audin, 69518 Vaulx-en-Velin, France*

[b]*EIFFAGE Travaux Publics - Direction R&D, 8 rue du Dauphiné BP 357, 69960 Corbas Cedex, France*

[c]*BP France - Services bitumes, Immeuble Le Cervier A, 12 av. des Béguines, 95866 Cergy Pontoise Cedex, France*

ABSTRACT. *Various phenomena other than fatigue (so called "biasing effects") occur during laboratory fatigue tests on asphalt mixes because of cyclic loading applications, thus altering experimental results and leading to misleading conclusions. The purpose of this study is to isolate and quantify biasing effects, therefore isolating real fatigue damage. In particular, non-linearity, self-heating and thixotropy (defined as a recoverable viscosity reduction after shear application) were evaluated.*

Six different mixes were produced using three distinct asphalt binders. Tests were performed in tension/compression mode on cylindrical samples. A particular test procedure was followed, consisting of two parts. In the first part, complex modulus measurements were performed at temperatures from 8°C to 14°C and strain amplitudes from 50µm/m to 110 µm/m, at 10 Hz. Regression equations were fitted in order to evaluate variations of norm of complex modulus and phase angle caused by temperature and strain level changes around common fatigue test conditions (10°C, 100 µm/m).

In the second part of the test, five partial fatigue tests (each one consisting of 100,000 cycles at a 100 µm/m strain amplitude) were performed at 10°C, 10 Hz. After each fatigue lag, a 24 hour rest period was imposed. During rest periods, short complex modulus measurements were performed (10°C, 10 Hz) in order to monitor recovery of mechanical properties.

Surface and internal temperatures of samples were constantly measured throughout the entire test in order to monitor self-heating due to repeated loading. A significant temperature increase was observed during each fatigue lag, while, during rest periods, temperature rapidly decreased to the initial value. Self-heating was observed to be correlated to viscoelastic energy dissipation.

The procedure used in this study allowed quantitatively estimating biasing effects. Therefore, unrecovered mechanical properties, due to damage accumulation, were obtained. 90% of

The oral presentation was made by Dr. Mangiafico.

This is a reproduction of an article whose final and definitive form has been published in *Road Materials and Pavement Design*© 2015 Taylor & Francis. The article is available online at: http://dx.doi.org/10.1080/14680629.2015.1077000

total complex modulus and phase angle variations observed during each fatigue lag were found to be completely reversible. Non-linearity and thixotropy appear to influence mechanical properties variations more importantly than self-heating.

KEYWORDS: asphalt mix, fatigue, biasing effects, non-linearity, heating, thixotropy.

1.0 Introduction

A widely accepted definition of fatigue is the accumulation of damage under the effect of repeated loading. Damage is intended as a deterioration of material properties. For this reason, common test procedures used to assess fatigue resistance of asphalt mixes require monitoring material properties, usually complex modulus, when applying repeated loading on samples.

Fatigue distress observed in real pavements is due to repeated traffic loading over time. Laboratory tests commonly performed need to simulate the phenomenon in a shorter time, therefore cyclic loading is applied at a fixed frequency and without rest periods.

One of major concerns about laboratory fatigue test procedures is the apparition of phenomena (so called "biasing effects") other than fatigue, triggering material property changes during tests, thus altering experimental results and leading to false conclusions (Di Benedetto et al., 2004). Most common biasing effects related to fatigue testing of viscoelastic material (such as asphalt mixes) are non-linearity, self-heating and thixotropy (Di Benedetto et al., 2011).

The objective of this study is to isolate and quantify biasing effects occurring during fatigue tests on asphalt mix samples because of cyclic loading application. For this purpose, a test protocol, called ALFABET (Advanced Laboratory Fatigue And Biasing Effects Test), has been developed at the LGCB laboratory of ENTPE. Moreover, calculating biasing effects allows estimating the effective variations of LVE (Linear ViscoElastic) properties due to damage accumulation.

This study is part of a wider research effort in the framework of a PhD thesis (Mangiafico, 2014) in collaboration between Ecole National des Travaux Publics de l'État (ENTPE), EIFFAGE Travaux Publics and BP. Materials for the study were provided and produced by EIFFAGE Travaux Publics and BP.

Fatigue testing of both asphalt binders and mixes is a widely common research topic (Boussad et al., 1996; Reese, 1997; Tayebali et al., 1994; Bahia et al., 1999, among others). A lot of effort is made, especially in the USA, to evaluate fatigue performances of bituminous materials using S-VECD (Simplified ViscoElastic Continuum Damage) models (Underwood et al., 2012). However, these models attempt to simulate fatigue test results without distinguishing between biasing effects and irreversible damage. Less research has been attempted about a more detailed quantification of real damage and biasing effects occurring during fatigue tests.

It is well known that bituminous materials subjected to repeated loading can exhibit a significant recovery of LVE properties if sufficiently long rest periods are maintained (Bonnaure et al., 1982; Deacon and Monismith, 1967; Raithby and Sterling, 1970; Hsu and Tseng, 1996; Smith and Hesp, 2000, among others). This phenomenon is usually referred to as "healing" (Bazin and Saunier, 1967; Pronk, 1997; Phillips, 1999, among others). However, a distinction should be made between recuperation of LVE properties after real damage occurrence (which is considered by the authors as the true "healing") and recovery of viscoelastic property transient variations due to other phenomena (Santagata et al., 2013, among others). The term "healing" cannot be properly used for materials that did not experience cracking. For this reason, the concept of biasing effects occurring during fatigue tests has been proposed (Di Benedetto et al., 2004; Di Benedetto et al., 1996), in order to differentiate real damage due to material deterioration and recoverable changes of mechanical properties. Since most of the research works characterize fatigue resistance of bituminous materials according to stiffness properties, some procedures have been proposed in order to correct experimental results of fatigue tests on samples subjected to cyclic loading (Di Benedetto et al., 1996; Tapsoba et al., 2012).

Biasing effects generally treated in literature are non-linearity, self-heating and thixotropy (although the latter is still a hypothesis).

It is widely accepted that asphalt binders exhibit a linear viscoelastic behavior (that is, strain amplitude is proportional to stress amplitude (Lakes, 2009), provided that, under strain-control mode, the applied strain level is not higher than a threshold value, called the LVE limit, characteristic of the material (Delgadillo et al., 2006, among others). It is complicated to determine a LVE limit for asphalt mixes. A generally accepted value is 100μm/m (Airey et al., 2003, among others). However, because of the internal structure, the binder located within inter-aggregate spaces can be subjected to strain levels significantly higher than the overall mix strain (Kose et al., 2000). Thus, mix behavior can show sensible non-linearity for relatively low strain levels applied on samples. Since this effect could be confused with damage accumulation, non-linearity should be taken into account when performing fatigue tests on asphalt mixes (Di Benedetto et al., 2011; Gauthier et al., 2010; Nguyen, 2011).

Viscoelastic and elasto-plastic materials show a temperature increase (usually referred to as "self-heating") when subjected to, respectively, repeated loading or a solicitation higher than the yielding limit of the material (Farren and Taylor, 1925; Oldyrev, 1967; Taylor and Quinney, 1934). Significant self-heating has been observed during fatigue tests performed on asphalt mixes (Pronk et al., 1996; Lundström et al., 2004, among others). Given the temperature dependency of mechanical properties of asphalt mixes, the self-heating effect cannot be disregarded in the context of fatigue testing. Nevertheless, different conclusions are reached about the influence of this phenomenon on fatigue test experimental results.

According to International Union of Pure and Applied Chemistry (IUPAC), "the application of a finite shear to a system after a long rest may result in a decrease of viscosity or the consistency. If the decrease persists when the shear is discontinued, this behavior is called work softening (or shear breakdown), whereas if the original viscosity or consistency is recovered this behavior is called thixotropy" (IUPAC). The accepted explanation of thixotropy consists of a slow build-up of a microstructure as a result of weak attractive forces between particles. Flow application causes the particle network to break (Mewis and Wagner, 2009). Several evidences of thixotropy occurrence have been found in bituminous materials (Shan et al., 2011; Mouillet et al., 2012).

Previous research was conducted on the quantification of biasing effects appearing during fatigue tests on asphalt mixes (Di Benedetto et al., 2011; Nguyen et al., 2012). The concept of these works has been used here, although the test procedure has been slightly modified. In particular, both fatigue solicitations and recovery times have been prolonged in order to increase damage occurrence and reduce the possibility of incomplete recovery of material properties.

2.0 Materials

Six different mixes were used for the study (Table 1). All mixes were produced using the same aggregate nature (limestone, from Haut-Lieu quarry, France) and the same grading curve (continuous 0/14, shown in Figure 1).

Table 1. *Mixtures tested for the study.*

mix	fresh binder	RAP content [%]	void content [%]	
35/50	35/50	0%	2.7%	
35/50 + 20% RAP	35/50	20%	4.0%	(sample A)
			5.3%	(sample B)
35/50 + 40% RAP	35/50	40%	3.4%	
35/50 + 2.5% SBS + 20% RAP	35/50 + 2.5% SBS	20%	5.4%	
35/50 B + 20% RAP	35/50 B	20%	3.7%	
Orthoprène® + 20% RAP	Orthoprène®	20%	4.0%	

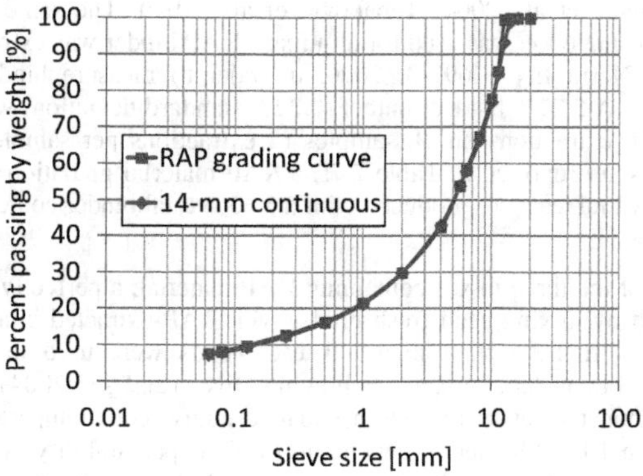

Figure 1. *Continuous 14-mm grading curve used for all mixes and RAP material grading curve.*

Four different binders were used (Table 1): a 35/50 pen grade straight run bitumen, the same binder modified with 2.5% SBS, a 35/50 B (so called "multigrade") pen grade bitumen issued from air rectification, and a soft paving grade bitumen modified with a high SBS content, cross-linked, called Orthoprène®. Results of classical semi-empirical tests (penetration, EN 1426:2007; Ring-Ball temperature, EN 1427:2007; Fraass, EN 12593:2007), SARA (Saturates, Aromates, Resins and Asphaltenes) fractioning and colloidal index I_C (calculated as the ratio of Saturates and Asphaltenes over Resins and Aromates) obtained for all binders are reported in Table 2. SARA fractioning was not performed on the modified binders.

Table 2. *Binder characteristics (BP laboratory, Gelsenkirchen, Germany).*

Bitumen		Penetration [dmm]	$T_{R\&B}$ [°C]	T_{Fraass} [°C]	SARA				I_C [-]
					Saturates [%]	Aromates [%]	Resins [%]	Asphaltenes [%]	
35/50		34	54.2	-7	5.2	59.4	24.8	10.7	0.19
35/50 + 2.5% SBS		30	58.4	-19	-	-	-	-	-
35/50 B		37	61.0	-17	9.6	54.1	18.6	17.7	0.38
Orthoprène®		56	94.0	-15	-	-	-	-	-
RAP binder	1st extraction	9	72.0	1	3.6	44.2	34.1	18.1	0.28
	2nd extraction	15	65.2	-1	3.5	46.6	31.7	18.2	0.28

Five mixes were produced with the addition of 20% RAP material by weight of the final mix without fresh binder. One mix was produced with the addition of 40% of the same RAP material. The influence of RAP on fatigue resistance of asphalt mixes has been investigated by several authors, although different conclusions have been drawn (Tapsoba et al., 2013; Mangiafico et al., 2015; Baaj et al., 2013; Hajj et al., 2009; Huang et al., 2004; Tabaković et al., 2010). The same RAP material, issued from a single lot, was used for all mixes. RAP binder was extracted according to European Norm EN 12697-3: 2006, in order to measure binder content and characteristics. A 5.22% binder content (0.22% standard deviation) was observed by performing 36 extractions on 18 samples (2 extractions per sample). RAP binder characteristics are reported in Table 2. The RAP material gradation curve is shown in Figure 1. All mixes were produced with a 5.35% total binder content by weight of the final mix.

For each of the three mixes containing RAP material, a perfectly blended binder was produced by mixing each fresh binder with RAP-extracted binder in the same proportions as in the asphalt mixes. These blends were used to fix production temperatures of asphalt mixes. Following EN 12697-35:2004+A1:2007, mix production temperatures were chosen equal to equiviscous temperatures of binder blends (with a 10°C tolerance), depending on their penetrability. Various research works have been carried out about actual degree of blending between binder in RAP material with fresh added binder in the final mix. A certain degree of blending has been observed for high RAP contents (Mangiafico et al., 2013a and 2013b; McDaniel et al., 2001; Stephens et al., 2001; Bonaquist, 2007; Eddhahak-Ouni et al., 2012; Al-Qadi, 2009). Preparation and mixing of aggregates, RAP material and fresh binder was performed following EN 12697-35:2004+A1:2007.

Compaction of materials was performed using an LCPC-type slab compactor (EN 12697-33:2003+A1:2007). Samples were cored from slabs and their void contents were checked before carrying out tests. Void content of samples is reported in Table 1. As explained in Section 3, because of the long duration of the test procedure used, for every material one sample was used, except for mix 35/50 + 20% RAP, for which two samples were tested.

3.0 Experimental Procedure

Tests on mixes were performed in tension/compression configuration on cylindrical samples (150 mm high with a 75 mm diameter). Cyclic sinusoidal loading was applied by using a hydraulic press in strain-control mode (Figure 2), using three extensometers placed at a 120° distance in the central part of the sample. Axial stress was measured with a load cell. In order to measure external surface temperature, a temperature probe was placed on the surface of the sample. Internal temperature was also measured. For this purpose, a 1 mm diameter thermocouple was also inserted inside the sample at a 30 mm depth, in a 1.7 mm radial hole drilled in the sample. After the insertion of the thermocouple, the hole was filled with bitumen in order to ensure material continuity.

Temperature control and sample conditioning was performed by means of a thermal chamber. Temperature probes used to measure test temperature (surface and internal) and to regulate the thermal chamber (ambient temperature) were independent, therefore a slight difference (approximately 0.5°C to 0.7°C) was observed between imposed temperature in the chamber and effective sample temperature. However, the effective measured temperature was always taken into account during analysis.

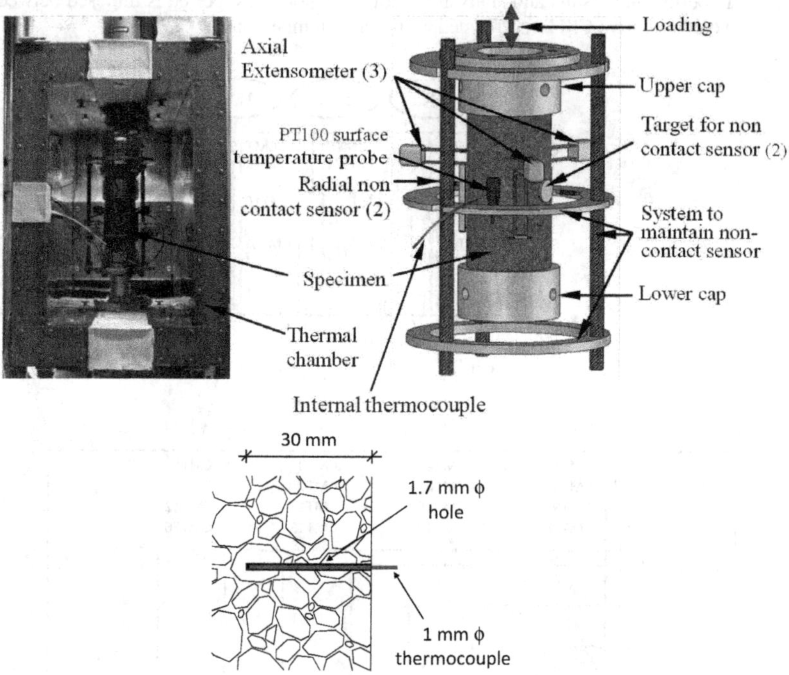

Figure 2. *(top, left) Tension-compression test apparatus; (top, right) Detailed schematic of measurement devices and sample [ENTPE laboratory, Vaulx-en-Velin, France]; (bottom) Schematic of internal thermocouple placement.*

The ALFABET test procedure is divided in two parts. In the first part, Complex Modulus Tests (CMT) are performed at four different temperatures and four different strain amplitudes (a schematic description of these tests is shown in Figure 3). The objective of this part of the procedure is to monitor LVE behavior of asphalt mixes in undamaged conditions at temperatures and strain levels close to the values used for fatigue tests (10°C, 100 µm/m). In particular, temperatures used are 8°C, 10°C, 12°C and 14°C, while the applied strain amplitudes are 50 µm/m, 75 µm/m, 100 µm/m and 110 µm/m. Frequency is always fixed at 10 Hz. For every temperature, a six-hour conditioning time is imposed, in order to be sure that the temperature is uniform in the whole sample. For every CMT, 200 loading cycles are applied. Because of hydraulic press characteristics, a certain number of cycles (approximately between 40 and 50 cycles) are necessary before the imposed strain level is reached. Measurements obtained during loading cycles performed at growing strain levels are used in the analysis (details are reported in the "Experimental results and analysis" section). A 900 s rest period is imposed between two consecutive CMTs performed at the same temperature.

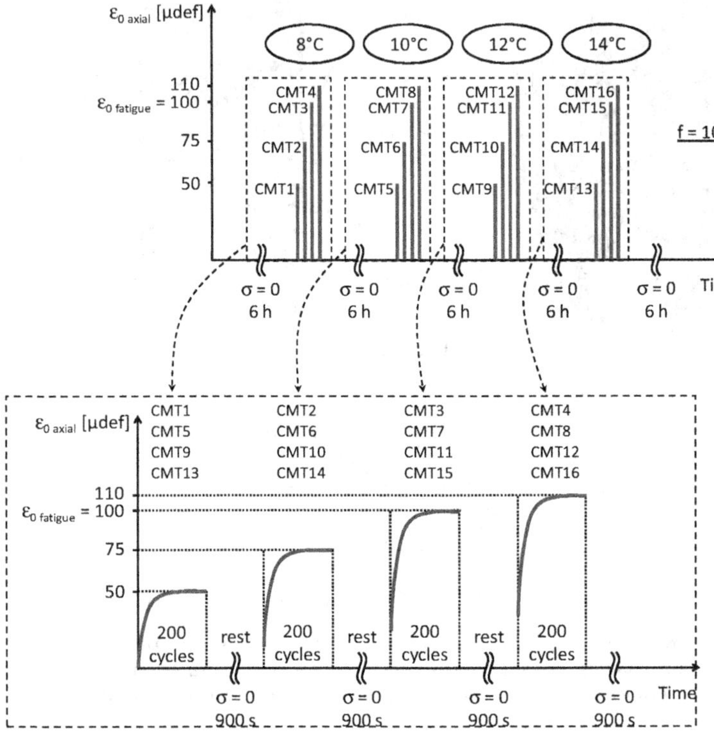

Figure 3. *Schematic procedure of preliminary Complex Modulus Tests (CMT) of ALFABET test (first part).*

The second part of the test, consisting of Partial Fatigue Tests (PFT), is performed immediately after all CMTs on the same sample. The objective is to quantitatively estimate different biasing effects and unrecovered LVE parameter variation during fatigue tests. A schematic description of PFTs is reported in Figure 4. Five fatigue lags are performed by applying a 100 μm/m sinusoidal tension/compression cyclic loading at 10 Hz. Imposed temperature within the thermal chamber is 10°C. For every fatigue lag, 100,000 cycles are applied. Also in this case, as for CMTs, between 40 and 50 cycles are necessary to reach the desired strain amplitude. Between two consecutive fatigue lags, a 24 hour rest period is imposed (stress within the sample is nil). In order to monitor the recovery of LVE properties, during every rest period, 16 short complex modulus measurements (100 cycles) are performed after 10, 20 and 30 minutes and 1, 2, 4, 6, 8, 10, 12, 14, 16, 18, 20, 22 and 24 hours. All complex modulus measurements are performed at 10°C, 10 Hz, by applying a 100 μm/m strain level.

It must be noted that the whole test procedure has a non negligible duration (approximately 7 days). For this reason, the experimental campaign was limited to six materials. For every material, one sample was used, except for mix 35/50 + 20% RAP, for which two samples were tested.

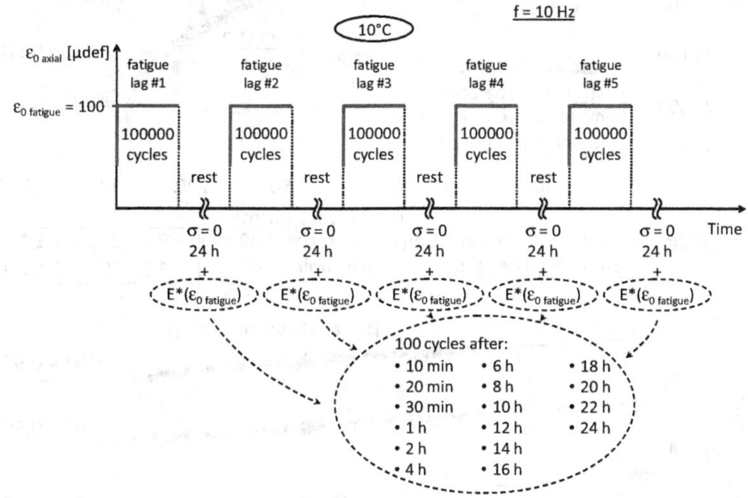

Figure 4. *Scheme of Partial Fatigue Tests (PFT) of ALFABET test (second part).*

4.0 Experimental Results and Analysis

4.1 Complex Modulus Tests (CMTs)

Complex modulus values measured during all CMTs were plotted against corresponding strain level. The first six cycles of every CMT were discarded. As an example, Figures 5 and 6 show CMT results regarding, respectively, norm of complex modulus and phase angle for mix 35/50 + 40% RAP.

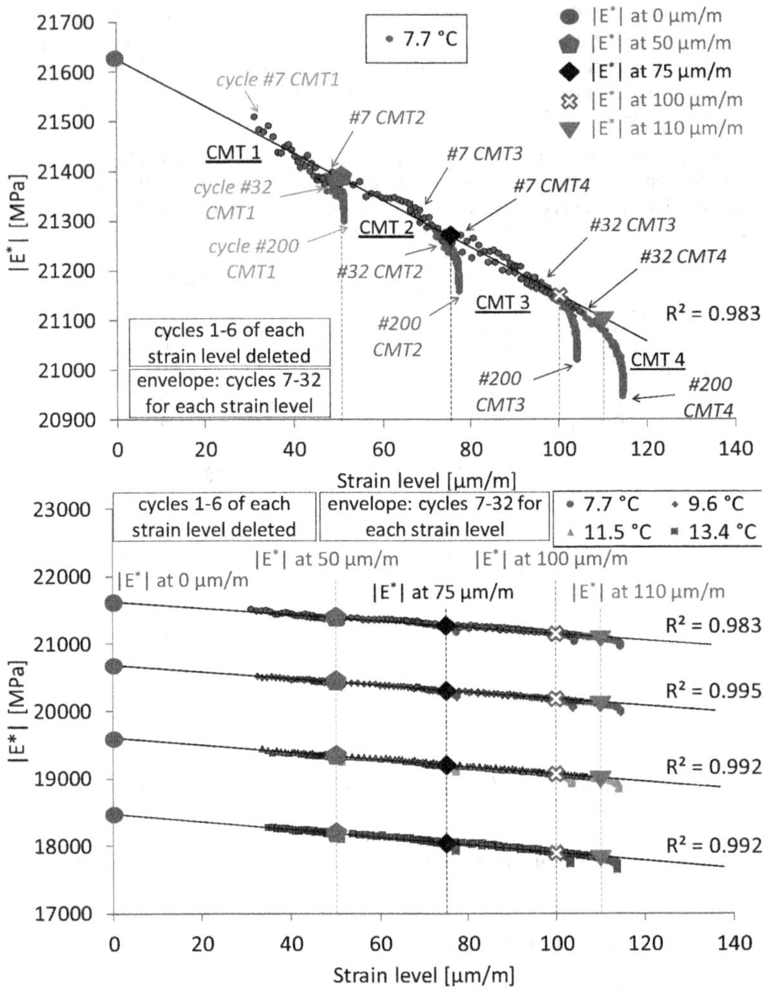

Figure 5. *CMT (Complex Modulus Test) results for mix 35/50 + 40% RAP: $|E^*|$ data and non-linearity envelope lines (cycles 7-32) at different temperatures (top, zoom on data obtained at 7.7°C; bottom, data obtained at all temperatures).*

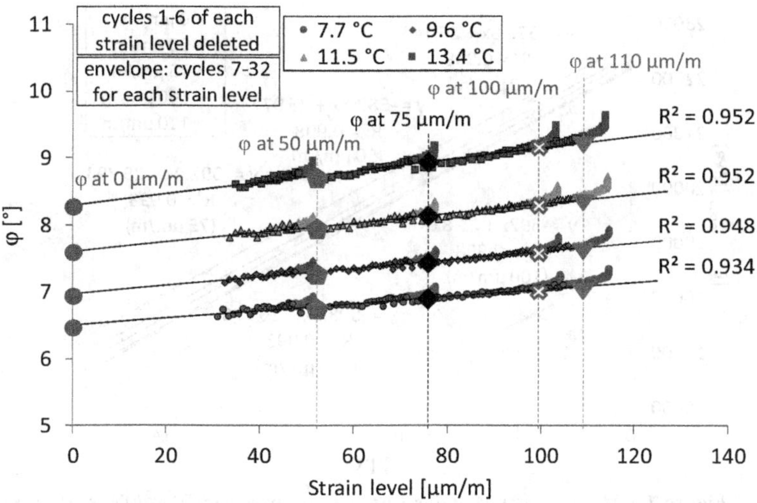

Figure 6. *CMT (Complex Modulus Test) results for mix 35/50 + 40% RAP: φ data and non-linearity envelope lines (cycles 7-32) at different temperatures.*

As described in Section 3, during the first part of all CMTs, strain level gradually increases to the imposed value. As can be noticed, the norm of complex modulus and phase angle follow a linear trend with strain level, depending on test temperature. Moreover, no linear viscoelastic limit can be observed for strain levels as low as 30 μm/m. This observation is in contrast with literature findings (Airey et al., 2003). For each μm/m of increase of applied strain level, the rate of variation of |E*| and φ is, respectively, between approximately -5 MPa and -6 MPa and between $5 \cdot 10^{-3\circ}$ and $9 \cdot 10^{-3\circ}$. After the imposed strain level is reached, prolonged cyclic solicitation appears to cause LVE parameters to slightly change. For this reason, only results obtained during cycles 7-32 of each CMT were fitted with linear regression equations (cycles 1-6 could be affected by transient effects) in order to obtain envelope lines of norm of complex modulus (Figure 5) and phase angle (Figure 6) as functions of strain level.

Values obtained using non-linearity envelope lines for all different temperatures at strain levels equal to 0 μm/m, 50 μm/m, 75 μm/m, 100 μm/m and 110 μm/m were plotted against corresponding temperatures. As an example, resulting plots for norm of complex modulus and phase angle of mix 35/50 + 20% RAP are shown, respectively, in Figures 7 and 8.

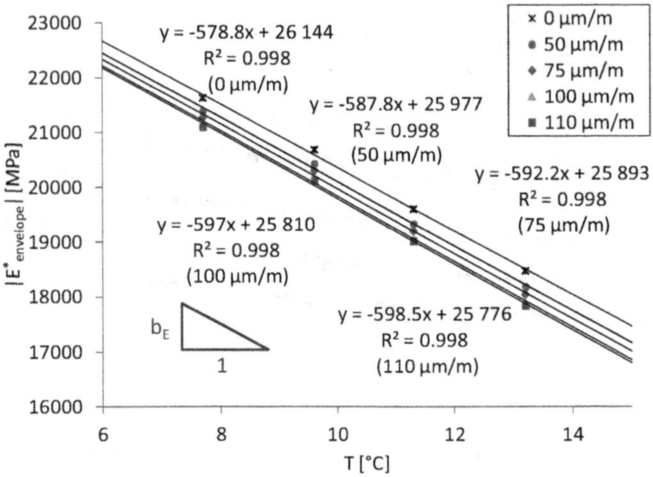

Figure 7. *CMT (Complex Modulus Test) results for mix 35/50 + 40% RAP: $|E^*|$ (obtained from non-linearity envelope lines) vs. temperature at different strain levels.*

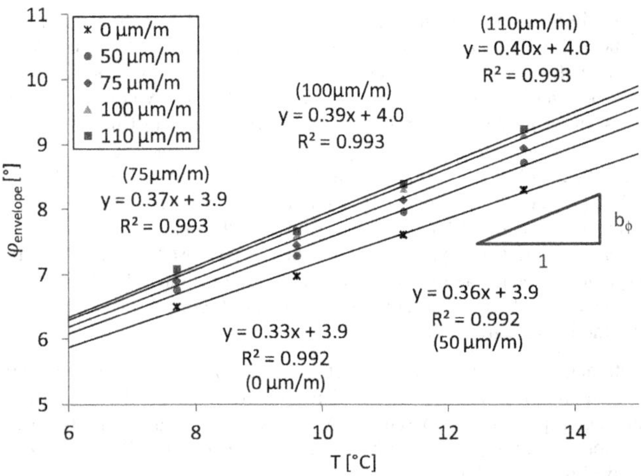

Figure 8. *CMT (Complex Modulus Test) results for mix 35/50 + 40% RAP: φ (obtained from non-linearity envelope lines) vs. temperature at different strain levels.*

Since all tests were performed at 10 Hz, it was impossible to apply the Time-Temperature Superposition Principle (Nguyen et al., 2009; Nguyen et al., 2013; Williams et al., 1955). Instead, it was observed that, within a limited temperature domain, both complex modulus and phase angle values follow linear trends with temperature. Regression Equations 1 and 2 were successfully fitted in order to estimate LVE parameters evolution with temperature T (in °C), depending on applied strain level:

$$|E^*|(\varepsilon, T) = b_E(\varepsilon)\,T + |E^*_{0°C}|(\varepsilon) \quad [1]$$

$$\varphi(\varepsilon, T) = b_\varphi(\varepsilon)\,T + \varphi_{0°C}(\varepsilon) \quad [2]$$

Where b_E and b_φ are the coefficients expressing the variation of, respectively, $|E^*|$ and φ with temperature and $|E^*_{0°C}|$ and $\varphi_{0°C}$ are, respectively, theoretical values of $|E^*|$ and φ at 0°C. Values of b_E and b_φ vary between, respectively, approximately -590 MPa/°C and 600 MPa/°C and 0.33°/°C and 0.39°/°C. Coefficients b_E and b_φ are used in the analysis of the PFTs to estimate the influence of non-linearity and self-heating phenomena during fatigue tests.

4.2 Partial Fatigue Tests (PFTs)

Figures 9 and 10 show, respectively, norm of complex modulus and phase angle evolution during all five PFT fatigue lags and recovery periods for mix 35/50 B + 20% RAP. The time axis is not to scale, in order to immediately visualize LVE variations due to both fatigue lags and rest periods. As already explained, applied strain level reaches the imposed value only after a certain number of cycles. In order to avoid undesired non-linearity effects, complex modulus and phase angle values considered during rest periods are equal to the average of values measured during the last five cycles (cycles 96-100). For the same reason, in Figures 9 and 10, the first 95 cycles of every fatigue lag are plotted in different size and color. Several observations can be made on $|E^*|$ and φ evolution:

- If the first 95 cycles of every fatigue lag are neglected, variations of $|E^*|$ and φ between cycles 96 and 100,000 are almost completely recovered during rest periods.
- $|E^*|$ and φ values measured during the 96th cycle of every fatigue lag are approximately the same as the ones measured at the end of the previous rest period.
- $|E^*|$ and φ values measured during the 96th cycle of consecutive fatigue lags appear to, respectively, slightly decrease and increase.

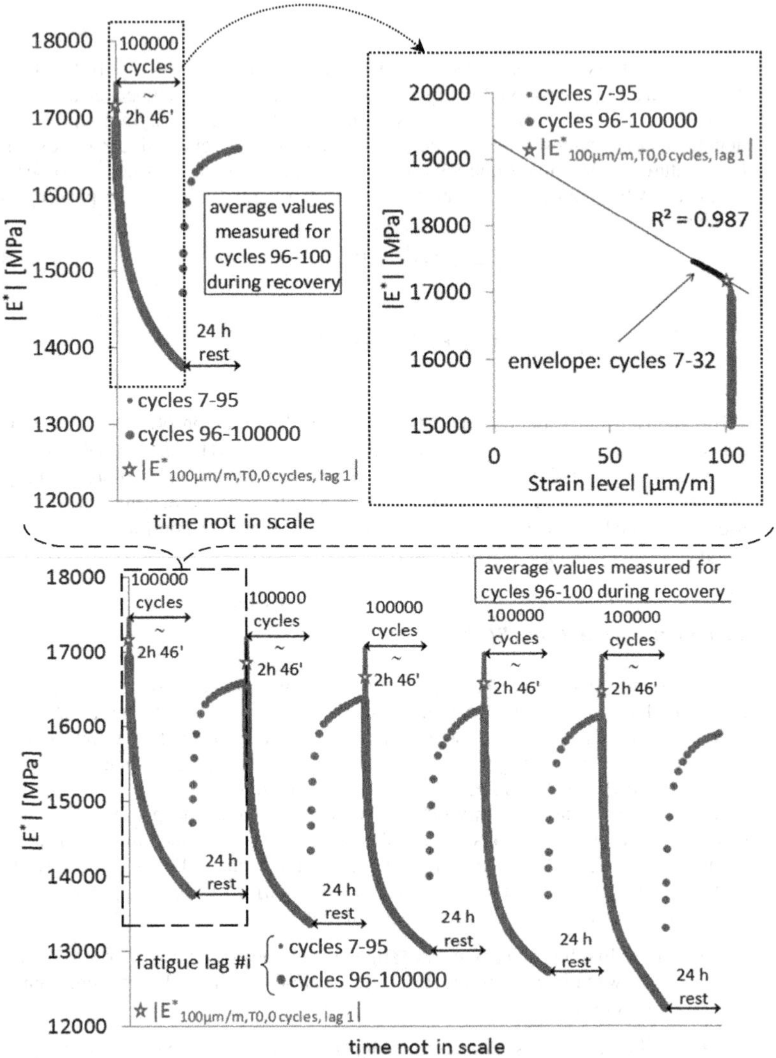

Figure 9. *PFT results for mix 35/50 B + 20% RAP: |E*| evolution during fatigue lags and recovery periods. Stars indicate values of* $|E^*_{100m/m,T0,0\ cycles,lag\ i}|$ *calculated for each ith fatigue lag with its own non-linearity envelope line (based on data obtained for cycles 7-32).*

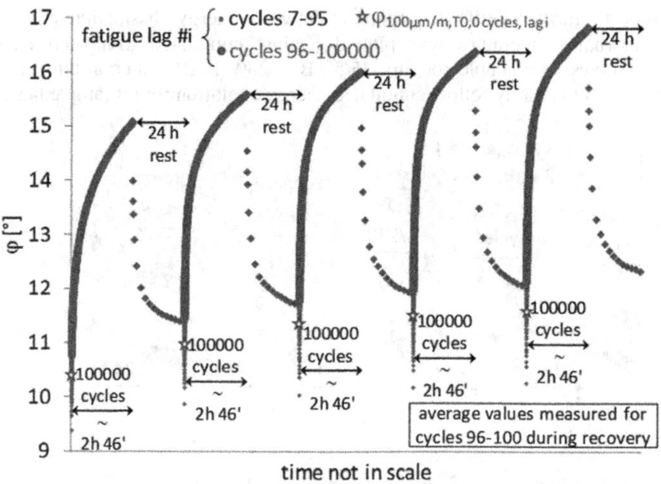

Figure 10. *PFT results for mix 35/50 B + 20% RAP: φ evolution during fatigue lags and recovery periods. Stars indicate values of $\varphi_{100m/m,T0,0\ cycles,lag\ i}$ calculated for each ith fatigue lag with its own non-linearity envelope line (based on data obtained for cycles 7-32).*

Surface and internal temperatures of samples were constantly measured in order to monitor the self-heating phenomenon. Figure 11 shows temperature measurements for mix 35/50 B + 20% RAP. No significant internal or surface temperature changes could be observed during the first 100 loading cycles. For this reason, average temperature values were considered for complex modulus tests during rest periods. During fatigue lags, both surface and internal temperatures of samples increase significantly. The increase of internal temperature is obviously more important than the surface temperature because of heat dispersion in the thermal chamber. It can be observed that, during fatigue lags, after a rapid temperature increase, a quasi-stationary condition is reached, due to the thermal equilibrium between self-heating occurring inside the sample and heat exchange on surface. During the 24 hour rest periods, temperature rapidly decreases to the initial value, equal to the thermal chamber temperature. Cooling of the sample is possible because no cyclic loading is applied, with the exception of the short complex modulus measurements. As self-heating is the consequence of viscous energy dissipation during cyclic loading, dissipated energy per cycle (W_N) and cumulated dissipated energy (ΣW) were monitored. Cumulated dissipated energy ΣW was calculated as follows:

$$\Sigma W = \sum_N W_N = \sum_N \pi\, \varepsilon_A\, \sigma_A \sin\varphi \qquad [3]$$

Where ε_A and σ_A are, respectively strain and stress amplitudes at cycle N.

In order to further verify the relation between energy dissipation and self-heating, internal temperature was plotted against cumulated dissipated energy (Figure 12 shows an example for mix 35/50 B + 20% RAP). Internal temperature increase appears to strictly follow cumulated energy evolution for all fatigue lags.

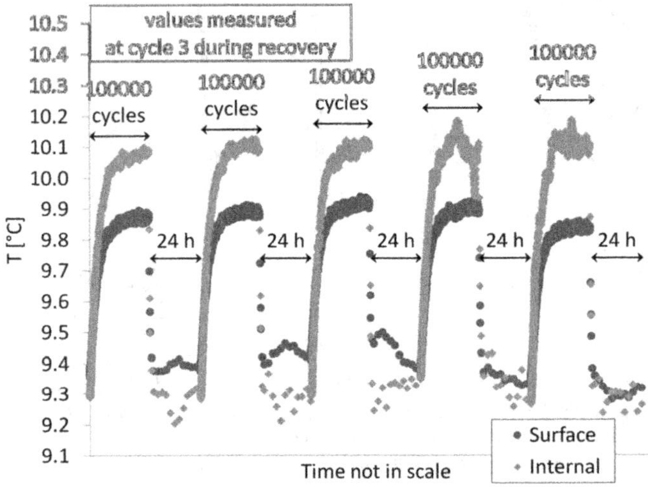

Figure 11. *PFT results for mix 35/50 B + 20% RAP: internal and surface temperature evolution during fatigue lags and recovery periods.*

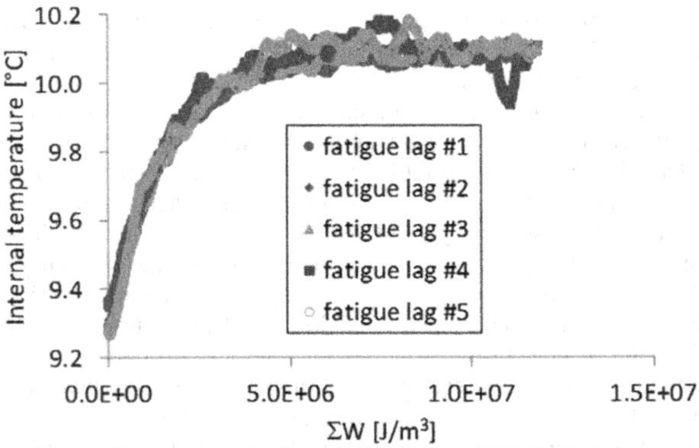

Figure 12. *PFT results for mix 35/50 B + 20% RAP: internal temperature vs. cumulated dissipated energy (ΣW) during fatigue lags.*

Based on all described observations, quantitative estimation of different biasing effects occurring during fatigue lags was performed. Each fatigue lag (and the following rest period) was considered separately. In order to do so, similarly to what was done for CMTs, a non-linearity envelope line was calculated for every fatigue lag during PFT (Figure 9 and Figure 10 show an example for mix 35/50 B + 20% RAP, for fatigue lag #1) . By using these envelope lines, for any ith fatigue lag, $|E^*|$ and φ values corresponding to 0 µm/m ($|E^*_{0\mu m/m,T0,0cycles,lag\ i}|$ and $\varphi_{0\mu m/m,T0,0cycles,lag\ i}$) and to 100 µm/m ($|E^*_{100\mu m/m,T0,0cycles,lag\ i}|$ and $\varphi_{100\mu m/m,T0,0cycles,lag\ i}$) could be calculated. T_0 is the internal temperature measured at the beginning of every fatigue lag (at temperature equilibrium after conditioning). This temperature does not show significant variations between all fatigue lags. Regarding the norm of complex modulus, the influence of non-linearity ($\Delta|E^*_{nonlinearity}|$) during any ith fatigue lag was calculated as the difference between values corresponding to 0 µm/m and 100 µm/m for each fatigue lag, as in Equation 4.

$$\Delta|E^*_{nonlinearity}|=|E^*_{0\mu m/m,T0,0cycles,lag\ i}|-|E^*_{100\mu m/m,T0,0cycles,lag\ i}| \quad [4]$$

A similar procedure was used to estimate the non-linearity effect for phase angle, $\Delta\varphi_{nonlinearity}$ during any ith cycle (Equation 5):

$$\Delta\varphi_{nonlinearity}=\varphi_{0\mu m/m,T0,0cycles,lag\ i}-\varphi_{100\mu m/m,T0,0cycles,lag\ i} \quad [5]$$

where $\varphi_{0\mu m/m,T0,0cycles,lag\ i}$ is the value of φ at 0 µm/m and $\varphi_{100\mu m/m,T0,0cycles,lag\ i}$ is the value of φ at 100 µm/m (both values are obtained from the non-linearity envelope line at any ith cycle).

For any ith fatigue lag, the unrecovered part of both $|E^*|$ and φ variations (respectively, $\Delta|E^*_{unrecovered\ 24h}|$ and $\Delta\varphi_{unrecovered\ 24h}$) after a 24 hour rest period is estimated. In particular, $\Delta|E^*_{unrecovered\ 24h}|$ and $\Delta\varphi_{unrecovered\ 24h}$ are estimated as in Equations 6 and 7:

$$\Delta|E^*_{unrecovered\ 24h}|=|E^*_{24h,lag\ i}|-|E^*_{100\mu m/m,T0,0cycles,lag\ i}| \quad [6]$$

$$\Delta\varphi_{unrecovered\ 24h}=\varphi_{24h,lag\ i}-\varphi_{100\mu m/m,T0,0cycles,lag\ i} \quad [7]$$

After taking into account the unrecovered part of $|E^*|$ and φ variations, recoverable biasing effects, i.e. self-heating and thixotropy, are estimated. The effect of self-heating on LVE parameter variations during fatigue lags is evaluated as a function of internal temperature decrease during rest periods, using coefficients b_E and b_φ evaluated with CMTs. For the norm of complex modulus and phase angle, respectively, Equations 8 and 9 were used:

$$\Delta|E^*_{heating}| = b_E\ \Delta T \quad [8]$$

$$\Delta\varphi_{heating} = b_\varphi\ \Delta T \quad [9]$$

Where $\Delta|E^*_{heating}|$ and $\Delta\varphi_{heating}$ are, respectively, norm of complex modulus and phase angle variations during any ith fatigue lag due to self-heating of the material, b_E and b_φ are, respectively, the coefficients determined during CMTs and ΔT is the difference of internal temperature between the value measured at the end (100,000th cycle) of the fatigue lag and the value measured during the last complex modulus test at the end of the rest period (after a 24 hours rest).

The effect of thixotropy on $|E^*|$ and φ variations (respectively $\Delta|E^*_{thixotropy}|$ and $\Delta\varphi_{thixotropy}$) during any ith fatigue lag was estimated equal to the portion of, respectively, $|E^*|$ and φ recovery not explained by self-heating. Therefore, $\Delta|E^*_{thixotropy}|$ and $\Delta\varphi_{thixotropy}$ are estimated using, respectively, Equations 10 and 11:

$$\Delta\left|E^*_{thixotropy}\right| = \left|E^*_{100000cycles,lag\,i}\right| - \left|E^*_{24h,lag\,i}\right| - \Delta\left|E^*_{heating}\right| \qquad [10]$$

$$\Delta\varphi_{thixotropy} = \varphi_{100000cycles,lag\,i} - \varphi_{24h,lag\,i} - \Delta\varphi_{heating} \qquad [11]$$

Where $|E^*_{100000cycles,lag\ i}|$ and $\varphi_{100000cycles,lag\ i}$ are, respectively, norm of complex modulus and phase angle measurements at the end (100,000th cycle) of the ith fatigue lag and $\Delta|E^*_{24h,lag\ i}|$ and $\Delta\varphi_{24h,lag\ i}$ are, respectively, norm of complex modulus and phase angle value measured during the last complex modulus test at the end of the rest period (after a 24 hours rest).

Finally, the five different fatigue lags were compared with each other. The following tendencies were observed regarding their envelope lines:

$$\left|E^*_{100\mu m/m,T0,0cycles,lag\,1}\right| > \left|E^*_{100\mu m/m,T0,0cycles,lag\,2}\right| >_{...} \left|E^*_{100\mu m/m,T0,0cycles,lag\,5}\right| \qquad [12]$$

$$\varphi_{100\mu m/m,T0,0cycles,lag\,1} < \varphi_{100\mu m/m,T0,0cycles,lag\,2} <_{...}< \varphi_{100\mu m/m,T0,0cycles,lag\,5} \qquad [13]$$

The opposite tendencies were found for extrapolated $|E^*|$ and φ values at 0 μm/m:

$$\left|E^*_{0\mu m/m,T0,0cycles,lag\,1}\right| < \left|E^*_{0\mu m/m,T0,0cycles,lag\,2}\right| <_{...}< \left|E^*_{0\mu m/m,T0,0cycles,lag\,5}\right| \qquad [14]$$

$$\varphi_{0\mu m/m,T0,0cycles,lag\,1} > \varphi_{0\mu m/m,T0,0cycles,lag\,2} >_{...}> \varphi_{0\mu m/m,T0,0cycles,lag\,5} \qquad [15]$$

It must be noted that, being obtained from regression equations, $|E^*|$ and φ values corresponding to a 100 μm/m strain level ($|E^*_{100\mu m/m,T0,0cycles,lag\ i}|$ and $\varphi_{100\mu m/m,T0,0cycles,lag\ i}$) are remarkably close to real experimental data, while values corresponding to 0 μm/m ($|E^*_{0\mu m/m,T0,0cycles,lag\ i}|$ and $\varphi_{0\mu m/m,T0,0cycles,lag\ i}$) are simply extrapolated. For this reason, envelope line differences between any ith fatigue lag and fatigue lag #1 are estimated by using $|E^*_{100\mu m/m,T0,0cycles,lag\ i}|$ and $\varphi_{100\mu m/m,T0,0cycles,lag\ i}$, as in Equations 16 and 17:

$$|E^*|envel.\,diff. = \left|E^*_{100\mu m/m,T0,0cycles,lag\,i}\right| - \left|E^*_{100\mu m/m,T0,0cycles,lag\,1}\right| \qquad [16]$$

$$\varphi envel.\,diff. = \varphi_{100\mu m/m,T0,0cycles,lag\,i} - \varphi_{100\mu m/m,T0,0cycles,lag\,1} \qquad [17]$$

As an example, Figure 13 shows calculations performed for $|E^*|$ of mix 35/50 B + 20% RAP, for fatigue lags #1 and #2.

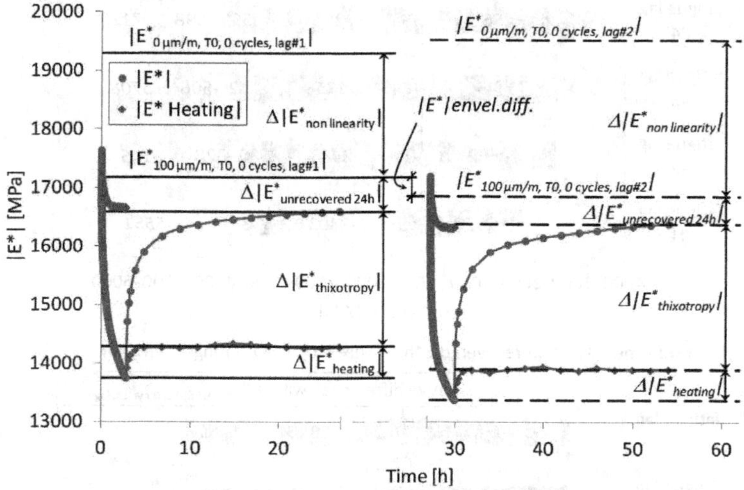

Figure 13. *Quantification of different contributions to $|E^*|$ evolution during fatigue lags #1 and #2 for mix 35/50 B + 20% RAP: a different envelope line is used for each fatigue lag.*

The relative importance of different biasing effects and unrecovered LVE parameter variations during fatigue lags with respect to total LVE parameter variations was evaluated. $|E^*|$ and φ total variations (respectively, $\Delta|E^*_{total}|$ and $\Delta\varphi_{total}$) are calculated as in, respectively, Equations 18 and 19:

$$\Delta|E^*_{total}| = -\Delta|E^*_{nonlinearity}| + \Delta|E^*_{unrecovered\ 24h}| + \Delta|E^*_{heating}| + \Delta|E^*_{thixotropy}| = \\ = |E^*_{100000cycles,lag\ i}| - |E^*_{0\mu m/m,T0,0cycles,lag\ i}| \quad [18]$$

$$\Delta\varphi_{total} = -\Delta\varphi_{nonlinearity} + \Delta\varphi_{unrecovered\ 24h} + \Delta\varphi_{heating} + \Delta\varphi_{thixotropy} = \\ = \varphi_{100000cycles,lag\ i} - \varphi_{0\mu m/m,T0,0cycles,lag\ i} \quad [19]$$

As an example, Figure 14 shows histograms of absolute values and relative importance of biasing effects and unrecovered $|E^*|$ variations during fatigue lags, for mix 35/50 B + 20% RAP.

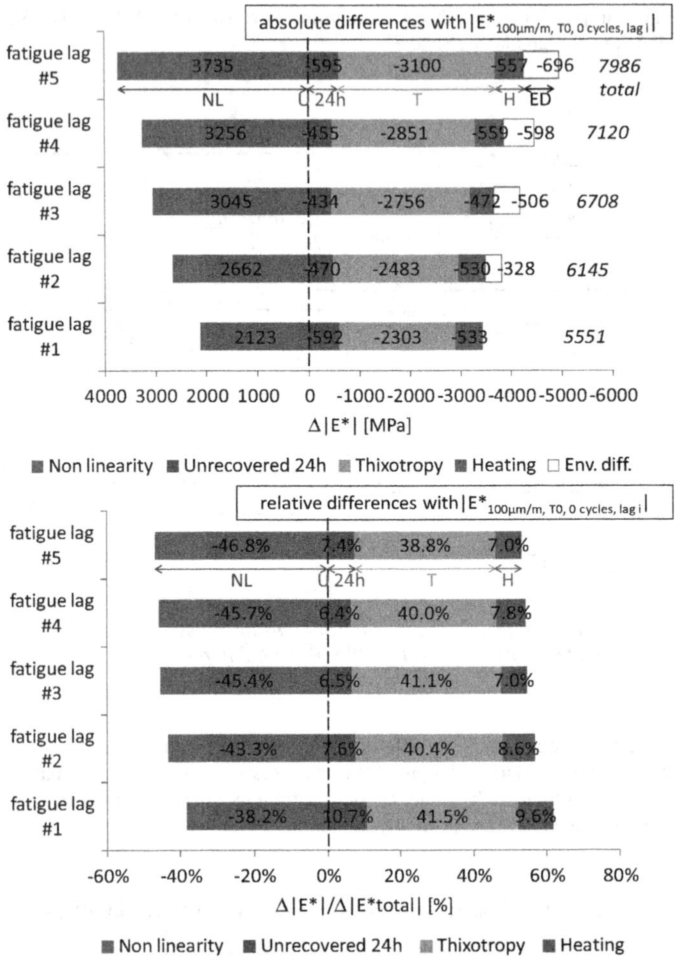

Figure 14. *Quantification of different contributions to* $|E^*|$ *evolution for mix 35/50 B + 20% RAP, calculated using a different envelope line for each fatigue lag: (top) absolute values; (bottom) relative values.*

A second approach was also used to quantify biasing effect and unrecovered $|E^*|$ and φ variations. It consists in neglecting non-linearity envelope lines of fatigue lags #2, #3, #4 and #5, and holding the non-linearity envelope line of fatigue lag #1 as a reference for the calculation of biasing effects occurring in all fatigue lags. Therefore, for any ith fatigue lag, the effects of non-linearity ($\Delta|E^*_{nonlinearity,\ ref:lag\ 1}|$ and $\Delta\phi_{non\ linearity,\ ref:lag\ 1}$) remain constant throughout the various fatigue lags and are calculated, respectively, as in Equations 20 and 21:

$$\Delta|E^*_{nonlinearity,ref:lag\ 1}|=|E^*_{0\mu m/m,T0,0cycles,lag\ 1}|-|E^*_{100\mu m/m,T0,0cycles,lag\ 1}| \quad [20]$$

$$\Delta\varphi_{nonlinearity,ref:lag\ 1}=\varphi_{0\mu m/m,T0,0cycles,lag\ 1}-\varphi_{100\mu m/m,T0,0cycles,lag\ 1} \quad [21]$$

Unrecovered variations ($\Delta|E^*_{unrecovered\ 24h,\ ref:lag\ 1}|$ and $\Delta\phi_{unrecovered\ 24h,ref:lag\ 1}$) are calculated, respectively, as in Equations 22 and 23:

$$\Delta|E^*_{unrecovered\ 24h,ref:lag\ 1}|=|E^*_{24h,lag\ i}|-|E^*_{100\mu m/m,T0,0cycles,lag\ 1}| \quad [22]$$

$$\Delta\varphi_{unrecovered\ 24h,ref:lag\ 1}=\varphi_{24h,lag\ i}-\varphi_{100\mu m/m,T0,0cycles,lag\ 1} \quad [23]$$

Estimations of self-heating and thixotropy effects are the same as for the first approach, since they do not depend on non-linearity envelope lines. By definition, $|E^*|$ and φ envelope line differences are not calculated. As an example, Figure 15 shows calculations performed with this approach for $|E^*|$ of mix 35/50 B + 20% RAP, for fatigue lags #1 and #2.

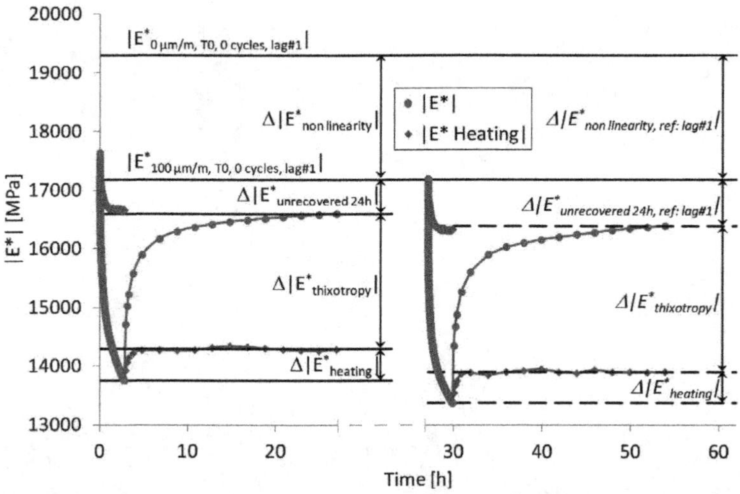

Figure 15. *Quantification of different contributions to $|E^*|$ evolution during fatigue lags #1 and #2 for mix 35/50 B + 20% RAP: the envelope line of fatigue lag #1 is used for all fatigue lags.*

$|E^*|$ and φ total variations were also calculated with this method. $|E^*|$ and φ total variations estimated using envelope line of fatigue lag #1 for all fatigue lags ($\Delta|E^*_{total,ref:lag\ 1}|$ and $\Delta\varphi_{total,ref:lag\ 1}$) are calculated as in Equations 24 and 25, respectively.

$$\Delta\left|E^*_{total,ref:lag\ 1}\right| = \left|E^*_{100000cycles,lag\ i}\right| - \left|E^*_{0\mu m/m,T0,0cycles,lag\ 1}\right| \quad [24]$$

$$\Delta\varphi_{total,ref:lag\ 1} = \varphi_{100000cycles,lag\ i} - \varphi_{0\mu m/m,T0,0cycles,lag\ 1} \quad [25]$$

According to this second approach, the relative importance of biasing effects and unrecovered $|E^*|$ and φ variations is evaluated with respect to $\Delta|E^*_{total,ref:lag\ 1}|$ and $\Delta\varphi_{total,ref:lag\ 1}$.

As an example, Figure 16 shows histograms of absolute values and relative importance of biasing effects and unrecovered $|E^*|$ variations during fatigue lags, estimated with respect to the non-linearity envelope line of lag #1, for mix 35/50 B + 20% RAP.

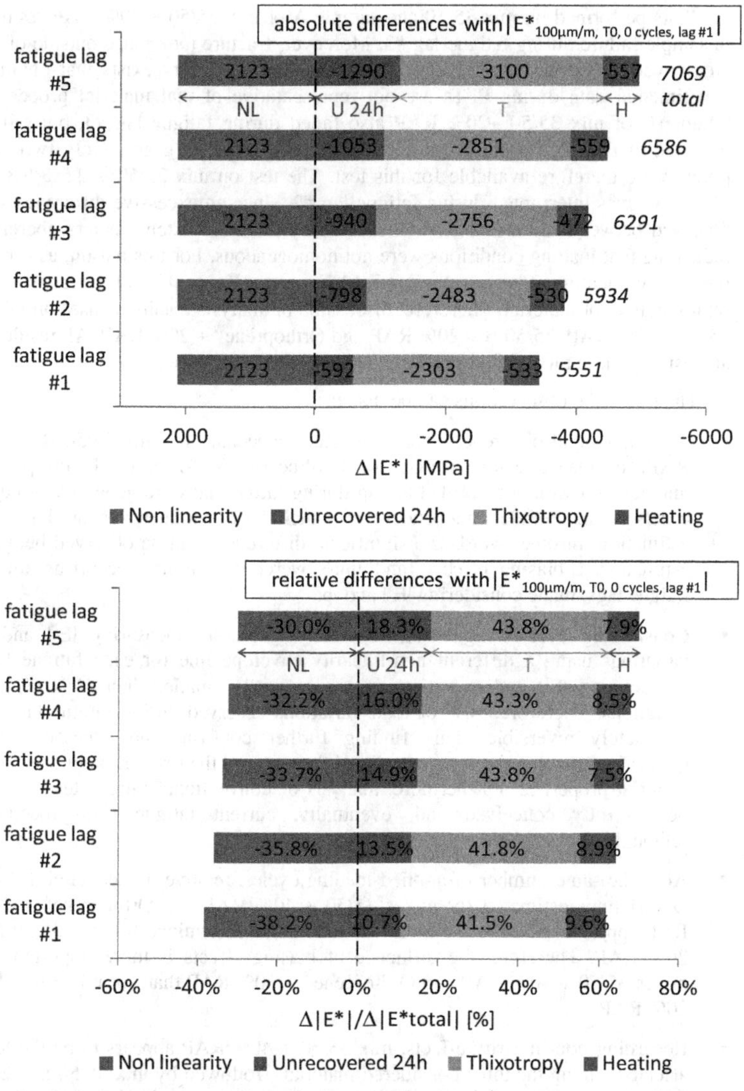

Figure 16. *Quantification of different contributions to* $|E^*|$ *evolution for mix 35/50 B + 20% RAP, calculated using the envelope line of fatigue lag #1: (top) absolute values; (down) relative values.*

Tests performed on mix 35/50 and sample A of mix 35/50 + 20% RAP resulted in sample failure during fatigue lag #3. Moreover, fracture took place outside of the area covered by extensometer gages, therefore the possibility exists that obtained experimental data during PFTs are not representative of real material properties. Sample B of mix 35/50 +20% RAP also failed during fatigue lag #3, but failure occurred at the edge of the area covered by extensometer gages. Only two data points were therefore available for this test. The test on mix 35/50 + 2.5% SBS + 20% RAP was interrupted during fatigue lag #2, since an excessive difference was observed between strain amplitude measured by the three extensometers, therefore indicating that loading conditions were not homogeneous. For this reason, as for the previously mentioned tests, measured LVE properties could not be considered representative of material. Therefore, discussion of analysis results focuses on mixes 35/50 + 40% RAP, 35/50 B + 20% RAP and Orthoprène® + 20% RAP. All results of analysis for these mixes are reported in Tables 3 to 10, in Annex.

The following observations can be made:

- Although small differences can be noticed (especially for mix 35/50 B + 20% RAP), estimated relative weights of non-linearity, self-heating, thixotropy and unrecovered variations of $|E^*|$ and φ during fatigue lags are generally of equal orders of magnitude, independent of which LVE parameter is used for the estimation. In other words, no significant differences can be observed between estimates of biasing effects (and unrecovered variations), occurring during fatigue lags, made considering $|E^*|$ and φ.
- Considering every fatigue lag separately (therefore measuring $|E^*|$ and φ variations using a different non-linearity envelope line for each fatigue lag), unrecovered $|E^*|$ and φ variations are generally smaller than 10% of total variations. Therefore, 90% of total variations observed during fatigue lags are completely reversible. This finding further confirms the necessity of a distinction between the recovery of real damage and the restoration of reversible material properties. Furthermore, analysis of conventional fatigue tests should be carefully conceived and, eventually, current fatigue test procedures rethought.
- After the same number of applied loading cycles, relative unrecovered $|E^*|$ and φ variations estimated for mixes 35/50 + 40% RAP and Orthoprène® + 20% RAP appear to be of less importance than those determined for mix 35/50 B + 20% RAP. Therefore, the influence of biasing effects is more important for mixes 35/50 + 40% RAP and Orthoprène® + 20% RAP than for mix 35/50 B + 20% RAP.
- Regarding non-linearity effects, mix 35/50 + 40% RAP appears to be the least affected among the three considered materials, followed by mix 35/50 B + 20% RAP then mix Orthoprène® + 20% RAP. In particular, more than 50% of total $|E^*|$ and φ variations determined for mix Orthoprène® + 20% RAP are explained by non-linearity effects.

- Thixotropy effects are most important for mix 35/50 + 40% RAP, while they have a smallest influence for mix Orthoprène® + 20% RAP.
- Thixotropy and non-linearity appear to contribute to $|E^*|$ and φ variations substantially more than self-heating. In particular, self-heating effects determined for mixes 35/50 B + 20% RAP and Orthoprène® + 20% RAP are approximately equivalent and explain less than 10% of the total variations of LVE properties. Mix 35/50 + 40% RAP appears to be more affected by variations of $|E^*|$ and φ due to self-heating.

From Equations 6, 16 and 22, the relationship reported in Equation 26 can be obtained:

$$\Delta\left|E^*_{unrecovered\ 24h,ref:lag\ 1}\right|=\Delta\left|E^*_{unrecovered\ 24h}\right|+\left|E^*\right|envel.diff. \quad [26]$$

showing that the unrecovered complex modulus loss occurring at any ith fatigue lag, estimated with respect to the non-linearity envelope line of fatigue lag #1 ($\Delta|E^*_{nonlinearity,\ ref:lag\ 1}|$), is equivalent to the sum of the loss estimated with respect to the envelope line of the ith cycle ($\Delta|E^*_{unrecovered\ 24h}|$) and the difference between considered envelope lines ($|E^*|$ envel. diff.).

Similarly, from Equations 7, 17 and 23, the analogous relationship for phase angle, reported in Equation 27, can be obtained:

$$\Delta\varphi_{unrecovered\ 24h,ref:lag\ 1}=\Delta\varphi_{unrecovered\ 24h}+\varphi envel.diff. \quad [27]$$

Equation 26 was divided by initial $|E^*_{100\mu m/m,T0,0cycles,lag\ 1}|$ value (Equation 28), in order to evaluate the unrecovered $|E^*|$ variation of different mixes with respect to their initial complex modulus (in undamaged condition).

$$\frac{\Delta\left|E^*_{unrecovered\ 24h,ref:lag\ 1}\right|}{\left|E^*_{100\mu m/m,T0,0cycles,lag\ 1}\right|}=\frac{\Delta\left|E^*_{unrecovered\ 24h}\right|}{\left|E^*_{100\mu m/m,T0,0cycles,lag\ 1}\right|}+\frac{\left|E^*\right|envel.diff.}{\left|E^*_{100\mu m/m,T0,0cycles,lag\ 1}\right|} \quad [28]$$

As an example, Figure 17 shows plots of unrecovered $|E^*|$ and φ variations according to relationships described, respectively, by Equations 28 and 26 for mixture 35/50 B + 20% RAP.

Considering unrecovered $|E^*|$ and φ variations occurring at every fatigue lag (estimated considering every fatigue lag separately, therefore measuring $|E^*|$ and φ variations using a different non-linearity envelope line for each fatigue lag), it can be noticed that $\Delta|E^*_{unrecovered\ 24h}|$ and $\Delta\varphi_{unrecovered\ 24h}$ values estimated for fatigue lag #1 are usually more important than values evaluated for all other fatigue lags. Also, $|E^*|$ and φ envelope line differences (calculated with respect to envelope line of fatigue lag #1) increase at a lower rate as fatigue lags come into succession. More tests and a wider experimental campaign are necessary to validate these results and to look for explanations about observed tendencies.

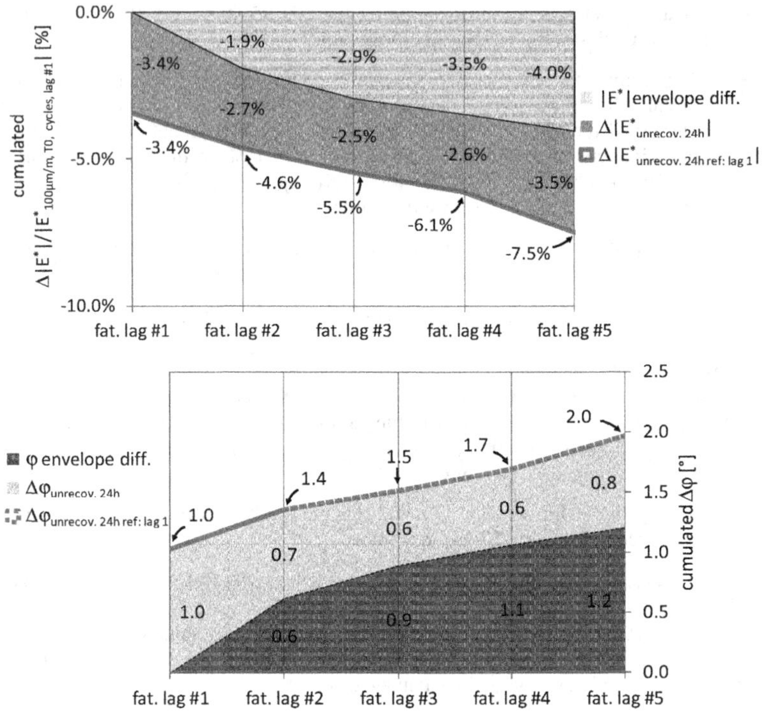

Figure 17. *Quantification of unrecovered* $|E^*|$ *(top) and* φ *(bottom) variations for mix 35/50 B + 20% RAP.*

5.0 Conclusions

Six different mixes were tested for the study, produced using four different binders.

1. Implemented test procedure, called ALFABET (Advanced Laboratory Fatigue and Biasing Effects Test), consists of two parts. The first part consists of complex modulus tests at temperatures and strain levels ranging, respectively, from 8°C to 14°C and from 50µm/m to 110 µm/m. No linear viscoelastic limit could be observed for mixes for applied strain amplitude values as low as 30 µm/m. Based on complex modulus results, regression equations were successfully fitted in order to estimate $|E^*|$ and φ evolution with temperature T and applied strain level ε_0. In the second part of the test procedure, five 100,000 cycle fatigue lags were carried out at 10°C, with a 100 µm/m applied

strain. After each fatigue lag, a 24 hour rest period was maintained. During rest periods, short complex modulus tests were performed at fixed intervals in order to monitor $|E^*|$ and φ recovery.

2. During fatigue lags, internal temperature of samples increased significantly. During the 24 hour rest periods, temperature rapidly decreased to the initial value, reaching thermal chamber temperature. Obtained results confirm that internal temperature increase is related to viscoelastic energy dissipation.

3. Based on all described observations, quantitative estimation of different biasing effects (non-linearity, self-heating and thixotropy) and unrecovered LVE properties observed during fatigue lags was performed. The following conclusions were drawn for mixes 35/50 + 40% RAP, 35/50 B + 20% RAP and Orthoprène® + 20% RAP:

   - Estimation of biasing effects does not show significant differences depending the LVE parameter used for the analysis ($|E^*|$ and φ).
   - 90% of total variations observed during each fatigue lag (considered separately) are completely reversible. For this reason, it is of primary interest to distinguish restoration from real damage (after crack occurrence) and recovery of reversible material properties when focusing on the so called "healing properties" of asphalt mixes.
   - Self-heating appears to influence $|E^*|$ and φ variations significantly less than thixotropy and non-linearity (less than 10%).

4. The procedure described in the study allows successfully isolating biasing effects from unrecovered LVE parameter variations occurring during fatigue tests. Moreover, quantitative estimation of different biasing effects is possible, and it should be considered when measuring fatigue performances by means of classical fatigue test laboratory procedures.

5. Results of ALFABET tests should be validated on a larger number of samples and mixtures. Comparisons with equivalent tests on corresponding bitumens and mastics are worth considering. Further development of the test procedure can also be considered, taking into account different rest period durations. Mechanisms of damage occurrence appear to be of primary importance to understand the fatigue phenomenon. Differences between self-healing of cracks and recovery of LVE properties should be studied.

6.0 Acknowledgements

The authors would like to express their sincere gratitude to S. Dupriet (EIFFAGE Travaux Publics) and R. Van Rooijen (BP) for their precious help preparing materials for the study, and to M.-D. Nguyen and J.-F. Roldan Agudelo for their useful contribution with the experimental campaign and data analysis.

7.0. References

Al-Qadi, I., Carpenter, S.H., Roberts, G.L., Ozer, H., Aurangzeb, Q., Elseifi, M.A. and Trepanier, J., *Determination of Usable Residual Asphalt Binder in RAP*, Research Report FHWA-ICT-09-031, Illinois Center for Transportation, 2009.

Airey, G.D., Rahimzadeh, B. and Collop, A.C., "Viscoelastic linearity limits for bituminous materials," *Materials and Structures*, Vol. 36, 2003, pp. 643–647, DOI: 10.1617/14081.

Baaj, H., Ech, M., Tapsoba, N., Sauzéat, C. and Di Benedetto, H., "Thermomechanical characterization of asphalt mixtures modified with high contents of Asphalt Shingle Modifier (ASM®) and Reclaimed Asphalt Pavement (RAP)," *Materials and Structures*, Vol. 46, 2013, pp. 1747-1763, DOI: 10.1617/s11527-013-0015-7.

Bahia, H.U., Zhai, H., Bonnetti, K. and Kose, S., "Non-Linear Viscoelastic and Fatigue Properties of Bitumen", *Journal of the Association of Asphalt Paving Technologists*, Vol. 68, 1999, pp. 1–34

Bazin, P. and Saunier, J., "Deformability, Fatigue and Healing Properties of Asphalt Mixes," *Proceedings of the 2nd International Conference on The Structural Design of Asphalt Pavements*, Ann Arbor, MI, August 1967, pp. 438–451.

Bonaquist, R., "Can I Run More RAP?," *HMAT: Hot Mix Asphalt Technology*, Vol. 12, No. 5, 2007, pp. 11–13.

Bonnaure, F.P., Huibers, A.H.J.J. and Boonders, A., "A Laboratory Investigation of the Influence of Rest Periods on the Fatigue Characteristics of Bituminous Mixes," *Journal of the Association of Asphalt Paving Technologists*, Vol. 51, 1982, pp.104–128.

Boussad, N., DesCroix, P. and Dony, A., "Prediction of Mix Modulus and Fatigue Law from Binder Rheology Properties," *Journal of the Association of Asphalt Paving Technologists*, Vol. 65, 1996, pp. 40–72.

Deacon, J.A. and Monismith, C.L., "Laboratory Flexural-Fatigue Testing of Asphalt Concrete with Emphasis on Compound Loading Tests," *Highway Research Record*, Vol. 158, Highway Research Board, 1967, pp. 1–31.

Delgadillo, R., Cho, D.W. and Bahia, H.U., "Nonlinearity of Repeated Creep and Recovery Binder Test and Relationship with Mixture Permanent Deformation," *Transportation Research Record: Journal of the Transportation Research Board*, No. 1962, 2006, pp. 3-11, DOI: 10.3141/1962-01.

Di Benedetto, H., Ashayer Soltani, A. and Chaverot, P., "Fatigue Damage for Bituminous Mixtures: A Pertinent Approach," *Journal of the Association of Asphalt Paving Technologists*, Vol. 65, 1996, pp. 142–158.

Di Benedetto, H., De la Roche, C., Baaj, H., Pronk, A.C., and Lundstrom, R., "Fatigue of bituminous mixtures," *Materials and Structures*, Vol. 37, 2004, pp. 202-216, DOI: 10.1007/BF02481620.

Di Benedetto, H., Nguyen, Q.T., and Sauzéat, C., "Nonlinearity, heating, fatigue and thixotropy during cycling loading of asphalt mixtures," *Road Materials and Pavement Design*, Vol. 12, 2011, pp. 129–158, DOI: 10.1080/14680629.2011.9690356.

Eddhahak-Ouni, A., Dony, A., Colin, J., Mendez, M., Navaro, J., Drouadaine, I. and Bruneau, D., "Experimental investigation of the homogeneity of the blended binder of a high rate recycled asphalt," *Road Materials and Pavement Design*, Vol. 13, No. 3, 2012, pp. 566-575, DOI: 10.1080/14680629.2012.700269.

Farren, W.S. and Taylor, G.I., "The Heat Developed during Plastic Extension of Metals," *Proceedings of the Royal Society of London. Series A, Containing Papers of a Mathematical and Physical Character*, Vol. 107, Issue 743, 1925, pp. 422–451, DOI: 10.1098/rspa.1925.0034.

Gauthier, G., Bodin, D., Chailleux, E. and Gallet, T., "Non linearity in bituminous materials during cyclic tests," *Road Materials and Pavement Design*, Vol. 11, Supplement 1: Special Issue on Asphalt Technology (EATA 2010), 2010, pp. 129–158, DOI: 10.1080/14680629.2010.9690339.

Hajj, E., Sebaaly, P.E. and Shrestha, R., "Laboratory evaluation of mixes containing Recycled Asphalt Pavement (RAP)," *Road Materials and Pavement Design*, Vol. 10, No. 3, 2009, pp. 495–517, DOI: 10.1080/14680629.2009.9690211.

Hsu, T-W and Tseng, K-H., "Effect of Rest Periods on Fatigue Response of Asphalt Concrete Mixtures," *Journal of Transportation Engineering*, Vol. 122, No. 4, 1996, pp. 316-322, DOI: 10.1061/(ASCE)0733-947X(1996)122:4(316).

Huang, B., Zhang, Z. and Kinger, W., "Fatigue crack characteristics of HMA mixtures containing RAP," *Proceedings of the 5th international RILEM conference on reflective cracking in pavements*, 2004, pp. 631–638.

IUPAC, *Compendium of Chemical Terminology—The Gold Book*, electronic version: http://goldbook.iupac.org/index.html.

Kose, S., Guler, M., Bahia, H.U. and Masad E., "Distribution of Strains within Binders in HMA: Applying Imaging and Finite-Element Techniques," *Transportation Research Record: Journal of the Transportation Research Board*, Issue 1728, 2000, pp. 21–27, DOI: 10.3141/1728-04.

Lakes, R.S., *Viscoelastic Materials*, Cambridge University Press, New York, 2009.

Lundström, R., Ekblad, J. and Isacsson, U., "Influence of Hysteretic Heating on Asphalt Fatigue Characterization," *Journal of Testing and Evaluation*, Vol. 32, Issue 6, 2004, pp. 484–493, DOI: 10.1520/JTE12284.

Mangiafico, S., *Linear Viscoelastic Properties and Fatigue of Bituminous Mixtures Produced with Reclaimed Asphalt Pavement and Corresponding Binder Blends*, (Unpublished doctoral dissertation), ENTPE, Lyon, 2014.

Mangiafico, S., Di Benedetto, H., Sauzéat, C., Olard, F., Pouget, S., Dupriet, S., Planque, L. and Van Rooijen, R., "Statistical analysis of the influence of RAP and mix composition on viscoelastic and fatigue properties of asphalt mixes", *Materials and Structures*, Vol. 48, Issue 4, 2015, pp. 1187–1205, DOI: 10.1617/s11527-013-0225-z

Mangiafico, S., Di Benedetto, H., Sauzéat, C., Olard, F., Pouget, S. and Planque, L., "Influence of RAP content on complex modulus of asphalt binder blends and corresponding mixes: experimental results and modeling", *Road Materials and Pavement Design*, Vol. 14, Supplement 1: EATA 2013, 2013a, pp. 132–148, DOI: 10.1080/14680629.2013.774751.

Mangiafico, S., Di Benedetto, H., Sauzéat, C., Olard, F., Pouget, S. and Planque, L., "New method to obtain viscoelastic properties of bitumen blends from pure and Reclaimed Asphalt Pavement binder constituents," *Road Materials and Pavement Design*, Vol. 15, 2013b, pp. 312–329, DOI: 10.1080/14680629.2013.870639.

McDaniel, R.S., Soleymani, H., Anderson, R.M., Turner, P. and Peterson, R., *Recommended Use of Reclaimed Asphalt Pavement in the Superpave Mixture Design Method*, NCHRP Web Document 30, Final Report on NCHRP Project 9–12, 2001.

Mewis, J. and Wagner, N.J., "Thixotropy," *Advances in Colloid and Interface Science*, Vol. 147-148, Special Issue, 2009, pp. 214–227, DOI: 10.1016/j.cis.2008.09.005.

Mouillet, V., De la Roche, C., Chailleux, E. and Coussot, P., "Thixotropic Behavior of Paving-Grade Bitumens under Dynamic Shear," *Journal of Materials in Civil Engineering*, Vol. 24, 2012, pp. 23–31, DOI: 10.1061/(ASCE)MT.1943-5533.0000354.

Nguyen, Q. T., *Comportement thermomécanique des enrobés bitumineux sous sollicitations cycliques dans les domaines linéaire et non-linéaire*, (Unpublished doctoral dissertation), ENTPE, Lyon, 2011 [in French].

Nguyen, H.M., Pouget, S., Di Benedetto, H. and Sauzeat, C., "Time-temperature superposition principle for bituminous mixtures," *European Journal of Environmental and Civil Engineering*, Vol. 13, 2009, pp. 1095–1107, DOI: 10.3166/EJECE.13.1095-1107.

Nguyen, Q. T., Di Benedetto, H., Sauzéat, C. and Tapsoba, N., "Time Temperature Superposition Principle Validation for Bituminous Mixes in the Linear and Nonlinear Domains," *Journal of Materials in Civil Engineering*, Vol. 15, 2013, pp. 1181–1188, DOI: 10.1061/(ASCE)MT.1943-5533.0000658.

Nguyen, Q.T., Di Benedetto, H. and Sauzéat, C., "Determination of thermal properties of asphalt mixtures as another output from cyclic tension-compression test," *Road Materials and Pavement Design*, Vol.13, 2012, p 85–103, DOI: 10.1080/14680629.2011.644082.

Oldyrev, P.P., "Self-heating and failure of plastics under cyclic loading," *Mekhanika Polimerov*, Vol. 3, No. 3, 1967, pp. 322–328, DOI: 10.1007/BF00858775.

Phillips, M.C., "Multi-step models for fatigue and healing, and binder properties involved in healing," *Proceedings of Eurobitume Workshop on Performance Related Properties for Bituminous Mixtures*, Luxembourg, May 1999, Paper No. 115.

Pronk, A.C., "Comparison of 2 and 4 Point Fatigue Tests and Healing in 4 Point Dynamic Bending Test Based on the Dissipated Energy Concept," *Proceedings of the 8th International Conference on Asphalt Pavements*, Seattle, WA, August 1997, pp. 987–994.

Pronk, A.C., Krans, R.L. and van Gogh, F., "Temperature increase in an asphalt beam during fatigue. Theory and practice," *Proceedings of Wegbouwkundige Werkdagen, CROW*, Ede, June 1996, ISBN 90-6628-221-5.

Raithby, K.D. and Sterling, A.B., "The Effect of Rest Periods on the Fatigue Performance of a Hot-Rolled Asphalt under Reversed Axial Loading," *Journal of the Association of Asphalt Paving Technologists*, Vol. 39, 1970, pp.134–152.

Reese, R., "Properties of Aged Asphalt Binder Related to Asphalt Concrete Fatigue Life," *Journal of the Association of Asphalt Paving Technologists*, Vol. 66, 1997, pp. 604–632.

Santagata, E., Baglieri, O., Tsantilis, L. and Dalmazzo, D., "Evaluation of self healing properties of bituminous binders taking into account steric hardening effects," *Construction and Building Materials*, Vol. 41, 2013, pp. 60-67, DOI: 10.1016/j.conbuildmat.2012.11.118.

Shan, L.Y., Tan, Y.Q., Underwood, B.S. and Kim, Y.R., "Separation of Thixotropy from Fatigue Process of Asphalt Binder," *Transportation Research Record: Journal of the Transportation Research Board*, Issue 2207, 2011, pp. 89–98, DOI: 10.3141/2207-12.

Smith, B.J. and Hesp, S.A.M., "Crack pinning in asphalt mastic and concrete: effect of rest periods and polymer modifiers on the fatigue life," *Proceedings of 2nd Eurasphalt and Eurobitume Congress*, Barcelona, September 2000, pp. 539–546.

Stephens, J.E., Mahoney, J. and Dippold, C., *Determination of the PG Binder Grade to Use in a RAP Mix*, Report No. JHR 00-278, Connecticut Department of Transportation, 2001.

Tabaković, A., Gibney, A., McNally, C. and Gilchrist, M.D., "Influence of Recycled Asphalt Pavement on Fatigue Performance of Asphalt Concrete Base Courses,"

Journal of Materials in Civil Engineering, Vol. 22, 2010, pp. 643–650, DOI: 10.1061/(ASCE)MT.1943-5533.0000093.

Tapsoba, N., Sauzéat, C. and Di Benedetto, H., "Analysis of Fatigue Test for Bituminous Mixtures," *Journal of Material in Civil Engineering*, Vol. 25, 2012, pp. 701–710, DOI: 10.1061/(ASCE)MT.1943-5533.0000636.

Tapsoba, N., Sauzéat, C., Di Benedetto, H., Baaj, H. and Ech, M., "Behavior of asphalt mixtures containing RAP and shingles," *Road Materials and Pavement Design*, Vol. 15, pp. 330–347, 2013, DOI: 10.1080/14680629.2013.871091.

Tayebali, A.A., Deacon, J.A., Coplantz, J.H., John, T. and Monismith, C.L., "Mixture and Mode-of-Loading Effects on Fatigue Response of Asphalt-Aggregate Mixtures," *Journal of the Association of Asphalt Paving Technologists*, Vol. 63, 1994, pp. 118–151.

Taylor, G.I. and Quinney, H., "The Latent Energy Remaining in a Metal after Cold Working," *Proceedings of the Royal Society of London. Series A, Containing Papers of a Mathematical and Physical Character*, Vol. 143, Issue 849, 1934, pp. 307–326, DOI: 10.1098/rspa.1934.0004.

Underwood, B.S., Baek, C., and Kim, Y.R., "Simplified Viscoelastic Continuum Damage Model as Platform for Asphalt Concrete Fatigue Analysis," *Transportation Research Record: Journal of the Transportation Research Board*, Issue 2296, 2012, pp. 36–45, DOI: 10.3141/2296-04.

Williams, M. L., Landel, L. F., and Ferry, J. D., "The Temperature Dependence of Relaxation Mechanisms in Amorphous Polymers and Other Glass-Forming Liquids," *Journal of the American Chemical Society*, Vol. 77, 1955, pp. 3701–3707, DOI: 10.1021/ja01619a008.

8.0 Discussion

PAVEL KRIZ: Thank you for your presentation. I have two questions. First, you're looking into nonlinearity and projected the lines to the zero strain. Would you expect linear viscoelastic range at very small strains where properties are independent of the strain? That's the first question. And the second question is related to the shear heating. Shear heating is also dependent on the gradient between the temperatures inside your sample and the environment. If you increase the strain, the heat flux will change as well, because there will be a larger difference between the temperatures. Would there be more variables impacting the process?

SALVATORE MANGIAFICO: Yes. Regarding the linear viscoelastic limit, I just plotted the results we obtained. I actually expected to see something, but as low as 30 microstrains, we could not see a linear viscoelastic limit. Maybe there is; I'm not saying there isn't. I just said that for sure it's lower than the amplitude that we worked at. I'm not saying that there is no linear viscoelastic limit; I just say that we could not observe it. Regarding heating, of course the flux will change. Still, if you change your strain amplitude, you will simply generate more viscous dissipation and therefore more heat. Sure, the flux of heat will change, the great end of temperature might change, but eventually you will have a stationary condition. You will have a stable flux from the inside of your sample to the thermal chamber.

PAVEL KRIZ: Well, thank you.

SALVATORE MANGIAFICO: No problem.

EMIN KUTAY: The maximum level of strain you ran the test at was 110 microstrain?

SALVATORE MANGIAFICO: Yes. Now we measure strain as a semi-amplitude, so if you take peak-to-peak it might be 210, 220, but still, yes.

EMIN KUTAY: So peak-to-peak 210.

SALVATORE MANGIAFICO: Yes.

EMIN KUTAY: So at that strain level, you have seen only about 10% is due to irreversible cracking?

SALVATORE MANGIAFICO: At that strain level, depending on the material, I could see that if you take the total reduction of complex modulus that you have during a fatigue period, which in our case was 100,000 cycles, at the end of the 24-hour rest period, you recovered 90% of the variation.

EMIN KUTAY: Okay. I guess my question is really if you increase the strain level to, let's say, 500-800 microstrain levels, do you have any comment on what would be the relative percentage of recovery?

SALVATORE MANGIAFICO: Well, if you increase the strain amplitude, of course you will see much more heating in proportion, then you might, I think, damage more the material, so maybe you will see a higher unrecovered variation. Still, I cannot project these results for different strain amplitudes. But, yes, it would be surely an interesting point to investigate.

JEAN-PASCAL PLANCHE: Great work, Salvo, and great presentation.

SALVATORE MANGIAFICO: Thanks.

JEAN-PASCAL PLANCHE: Just a question, if 90% of the effect that you are measuring is not fatigue, so there is 10% left, as you said. Is that enough to distinguish different asphalt mixtures' behavior, as if it was truly fatigue related?

SALVATORE MANGIAFICO: Well, I'm showing here the results obtained for three different mixtures produced with different binders. We see different trends. Let me show you the histogram for example, right. For example, complex modulus. We see different proportions between biasing effects and unrecovered variations. We do not explain the different proportions according to the binder. We just say that we observe different trends for different materials. So I think this would be another point to investigate the capability of the test to distinguish between materials. The purpose of the test at this point is just to highlight how much of the variation of complex modulus could be considered as fatigue. So maybe for different materials, probably we will see different things, yes, but I cannot guess what the outcome would be for the time being.

JEAN-PASCAL PLANCHE: Thank you.

9.0 Annex

Table 3. *Quantification of different contributions to* $|E^*|$ *evolution for mixes 35/50 + 40% RAP, 35/50 B + 20% RAP and Orthoprène®+ 20% RAP: values are calculated using a different envelope line for each fatigue lag.*

absolute differences with $\lvert E^*_{100\,\mu m/m,\,T0,\,0\,cycles,\,lag\,i}\rvert$		$\Delta\lvert E^*_{total}\rvert$ [MPa]	$\Delta\lvert E^*_{non\,linearity}\rvert$ [MPa]	$\Delta\lvert E^*_{heating}\rvert$ [MPa]	$\Delta\lvert E^*_{thixotropy}\rvert$ [MPa]	$\Delta\lvert E^*_{unrecov.\,24h}\rvert$ [MPa]	$\lvert E^*\rvert$ envel.diff. [MPa]
35/50 + 40% RAP	fat. lag #1	-2322	557	-309	-1320	-136	-
	fat. lag #2	-2423	533	-394	-1354	-143	-53
	fat. lag #3	-2591	600	-368	-1521	-102	-114
	fat. lag #4	-2705	610	-440	-1527	-127	-124
	fat. lag #5	-2846	654	-405	-1680	-107	-155
35/50 B + 20% RAP	fat. lag #1	-5551	2123	-533	-2303	-592	-
	fat. lag #2	-6145	2662	-530	-2483	-470	-328
	fat. lag #3	-6708	3045	-472	-2756	-434	-506
	fat. lag #4	-7120	3256	-559	-2851	-455	-598
	fat. lag #5	-7986	3735	-557	-3100	-595	-696
Orthoprène® + 20% RAP	fat. lag #1	-6543	3638	-541	-1941	-422	-
	fat. lag #2	-6345	3590	-527	-1889	-339	-216
	fat. lag #3	-6558	3843	-578	-1885	-252	-349
	fat. lag #4	-6496	3785	-547	-1927	-237	-381
	fat. lag #5	-6630	3925	-541	-1910	-254	-425

Table 4. *Quantification of different relative contributions to* $|E^*|$ *evolution for mixes 35/50 + 40% RAP, 35/50 B + 20% RAP and Orthoprène®+ 20% RAP: values are calculated using a different envelope line for each fatigue lag.*

relative differences with $\lvert E^*_{100\,\mu m/m,\,T0,\,0\,cycles,\,lag\,i}\rvert$		$\Delta\lvert E^*_{total}\rvert$ [%]	$\Delta\lvert E^*_{non\,linearity}\rvert$ [%]	$\Delta\lvert E^*_{heating}\rvert$ [%]	$\Delta\lvert E^*_{thixotropy}\rvert$ [%]	$\Delta\lvert E^*_{unrecov.\,24h}\rvert$ [%]
35/50 + 40% RAP	fat. lag #1	100.0%	-24.0%	13.3%	56.9%	5.8%
	fat. lag #2	100.0%	-22.0%	16.2%	55.9%	5.9%
	fat. lag #3	100.0%	-23.2%	14.2%	58.7%	4.0%
	fat. lag #4	100.0%	-22.6%	16.3%	56.5%	4.7%
	fat. lag #5	100.0%	-23.0%	14.2%	59.0%	3.8%
35/50 B + 20% RAP	fat. lag #1	100.0%	-38.2%	9.6%	41.5%	10.7%
	fat. lag #2	100.0%	-43.3%	8.6%	40.4%	7.6%
	fat. lag #3	100.0%	-45.4%	7.0%	41.1%	6.5%
	fat. lag #4	100.0%	-45.7%	7.8%	40.0%	6.4%
	fat. lag #5	100.0%	-46.8%	7.0%	38.8%	7.4%
Orthoprène® + 20% RAP	fat. lag #1	100.0%	-55.6%	8.3%	29.7%	6.4%
	fat. lag #2	100.0%	-56.6%	8.3%	29.8%	5.3%
	fat. lag #3	100.0%	-58.6%	8.8%	28.7%	3.8%
	fat. lag #4	100.0%	-58.3%	8.4%	29.7%	3.7%
	fat. lag #5	100.0%	-59.2%	8.2%	28.8%	3.8%

Table 5. *Quantification of different contributions to* $|E^*|$ *evolution for mixes 35/50 + 40% RAP, 35/50 B + 20% RAP and Orthoprène® + 20% RAP: values are calculated using the envelope line of fatigue lag #1.*

absolute differences with $|E^*_{100\ \mu m/m,\ T0,\ 0\ cycles,\ lag\ \#1}|$		$\Delta|E^*_{total,\ ref:\ lag\#1}|$ [MPa]	$\Delta|E^*_{non\ linearity,\ ref:\ lag\#1}|$ [MPa]	$\Delta|E^*_{heating}|$ [MPa]	$\Delta|E^*_{thixotropy}|$ [MPa]	$\Delta|E^*_{unrecov.\ 24h,\ ref:\ lag\#1}|$ [MPa]
35/50 + 40% RAP	fat. lag #1	-2322	557	-309	-1320	-136
	fat. lag #2	-2500	557	-394	-1354	-196
	fat. lag #3	-2662	557	-368	-1521	-216
	fat. lag #4	-2776	557	-440	-1527	-251
	fat. lag #5	-2904	557	-405	-1680	-262
35/50 B + 20% RAP	fat. lag #1	-5551	2123	-533	-2303	-592
	fat. lag #2	-5934	2123	-530	-2483	-798
	fat. lag #3	-6291	2123	-472	-2756	-940
	fat. lag #4	-6586	2123	-559	-2851	-1053
	fat. lag #5	-7069	2123	-557	-3100	-1290
Orthoprène® + 20% RAP	fat. lag #1	-6543	3638	-541	-1941	-422
	fat. lag #2	-6609	3638	-527	-1889	-555
	fat. lag #3	-6701	3638	-578	-1885	-601
	fat. lag #4	-6731	3638	-547	-1927	-619
	fat. lag #5	-6768	3638	-541	-1910	-679

Table 6. *Quantification of different relative contributions to* $|E^*|$ *evolution for mixes 35/50 + 40% RAP, 35/50 B + 20% RAP and Orthoprène® + 20% RAP: values are calculated using the envelope line of fatigue lag #1.*

relative differences with $|E^*_{100\ \mu m/m,\ T0,\ 0\ cycles,\ lag\ \#1}|$		$\Delta|E^*_{total,\ ref:\ lag\#1}|$ [%]	$\Delta|E^*_{non\ linearity,\ ref:\ lag\#1}|$ [%]	$\Delta|E^*_{heating}|$ [%]	$\Delta|E^*_{thixotropy}|$ [%]	$\Delta|E^*_{unrecov.\ 24h,\ ref:\ lag\#1}|$ [%]
35/50 + 40% RAP	fat. lag #1	100.0%	-24.0%	13.3%	56.9%	5.8%
	fat. lag #2	100.0%	-22.3%	15.7%	54.2%	7.8%
	fat. lag #3	100.0%	-20.9%	13.8%	57.1%	8.1%
	fat. lag #4	100.0%	-20.1%	15.9%	55.0%	9.0%
	fat. lag #5	100.0%	-19.2%	13.9%	57.9%	9.0%
35/50 B + 20% RAP	fat. lag #1	100.0%	-38.2%	9.6%	41.5%	10.7%
	fat. lag #2	100.0%	-35.8%	8.9%	41.8%	13.5%
	fat. lag #3	100.0%	-33.7%	7.5%	43.8%	14.9%
	fat. lag #4	100.0%	-32.2%	8.5%	43.3%	16.0%
	fat. lag #5	100.0%	-30.0%	7.9%	43.8%	18.3%
Orthoprène® + 20% RAP	fat. lag #1	100.0%	-55.6%	8.3%	29.7%	6.4%
	fat. lag #2	100.0%	-55.0%	8.0%	28.6%	8.4%
	fat. lag #3	100.0%	-54.3%	8.6%	28.1%	9.0%
	fat. lag #4	100.0%	-54.1%	8.1%	28.6%	9.2%
	fat. lag #5	100.0%	-53.8%	8.0%	28.2%	10.0%

Table 7. *Quantification of different contributions to φ evolution for mixes 35/50 + 40% RAP, 35/50 B + 20% RAP and Orthoprène®+ 20% RAP: values are calculated using a different envelope line for each fatigue lag.*

absolute differences with $\varphi_{100\,\mu m/m,\,T0,\,0\,cycles,\,lag\,i}$		$\Delta\varphi_{total}$ [°]	$\Delta\varphi_{non\,linearity}$ [°]	$\Delta\varphi_{heating}$ [°]	$\Delta\varphi_{thixotropy}$ [°]	$\Delta\varphi_{unrecov.\,24h}$ [°]	φ envel.diff. [°]
35/50 + 40% RAP	fat. lag #1	2.8	-1.0	0.2	1.4	0.2	0.0
	fat. lag #2	2.9	-1.0	0.3	1.5	0.2	0.1
	fat. lag #3	3.0	-1.0	0.2	1.6	0.2	0.2
	fat. lag #4	3.1	-1.0	0.3	1.6	0.2	0.2
	fat. lag #5	3.2	-1.0	0.3	1.7	0.2	0.2
35/50 B + 20% RAP	fat. lag #1	8.7	-4.0	0.4	3.2	1.0	0.0
	fat. lag #2	9.4	-4.8	0.4	3.5	0.7	0.6
	fat. lag #3	10.0	-5.3	0.4	3.7	0.6	0.9
	fat. lag #4	10.5	-5.6	0.5	3.9	0.6	1.1
	fat. lag #5	11.1	-5.8	0.5	4.0	0.8	1.2
Orthoprène® + 20% RAP	fat. lag #1	15.7	-9.4	0.8	4.6	0.9	-
	fat. lag #2	15.1	-9.1	0.7	4.6	0.7	0.5
	fat. lag #3	15.5	-9.6	0.8	4.6	0.6	0.7
	fat. lag #4	15.5	-9.6	0.8	4.6	0.5	0.8
	fat. lag #5	15.7	-9.8	0.8	4.6	0.5	0.9

Table 8. *Quantification of different relative contributions to φ evolution for mixes 35/50 + 40% RAP, 35/50 B + 20% RAP and Orthoprène®+ 20% RAP: values are calculated using a different envelope line for each fatigue lag.*

relative differences with $\varphi_{100\,\mu m/m,\,T0,\,0\,cycles,\,lag\,i}$		$\Delta\varphi_{total}$ [%]	$\Delta\varphi_{non\,linearity}$ [%]	$\Delta\varphi_{heating}$ [%]	$\Delta\varphi_{thixotropy}$ [%]	$\Delta\varphi_{unrecov.\,24h}$ [%]
35/50 + 40% RAP	fat. lag #1	100.0%	-34.5%	7.1%	50.8%	7.5%
	fat. lag #2	100.0%	-32.9%	8.9%	51.8%	6.5%
	fat. lag #3	100.0%	-33.3%	8.1%	53.4%	5.3%
	fat. lag #4	100.0%	-32.8%	9.3%	51.9%	6.0%
	fat. lag #5	100.0%	-32.4%	8.3%	53.8%	5.6%
35/50 B + 20% RAP	fat. lag #1	100.0%	-46.1%	4.9%	37.2%	11.8%
	fat. lag #2	100.0%	-51.1%	4.5%	36.6%	7.8%
	fat. lag #3	100.0%	-52.6%	3.8%	37.3%	6.2%
	fat. lag #4	100.0%	-52.8%	4.3%	36.9%	6.0%
	fat. lag #5	100.0%	-52.7%	4.1%	36.3%	6.9%
Orthoprène® + 20% RAP	fat. lag #1	100.0%	-59.7%	4.8%	29.5%	6.0%
	fat. lag #2	100.0%	-60.2%	4.9%	30.3%	4.6%
	fat. lag #3	100.0%	-61.8%	5.2%	29.4%	3.6%
	fat. lag #4	100.0%	-61.7%	4.9%	29.9%	3.5%
	fat. lag #5	100.0%	-62.3%	4.8%	29.6%	3.3%

Table 9. *Quantification of different contributions to φ evolution for mixes 35/50 + 40% RAP, 35/50 B + 20% RAP and Orthoprène®+ 20% RAP: values are calculated using the envelope line of fatigue lag #1.*

absolute differences with $\varphi_{100\ \mu m/m,\ T0,\ 0\ cycles,\ lag\ \#1}$		$\Delta\varphi_{total,\ ref:\ lag\#1}$ [°]	$\Delta\varphi_{non\ linearity,\ ref:\ lag\#1}$ [°]	$\Delta\varphi_{heating}$ [°]	$\Delta\varphi_{thixotropy}$ [°]	$\Delta\varphi_{unrecov.\ 24h,\ ref:\ lag\#1}$ [°]
35/50 + 40% RAP	fat. lag #1	2.8	-1.0	0.2	1.4	0.2
	fat. lag #2	3.0	-1.0	0.3	1.5	0.3
	fat. lag #3	3.1	-1.0	0.2	1.6	0.3
	fat. lag #4	3.2	-1.0	0.3	1.6	0.4
	fat. lag #5	3.4	-1.0	0.3	1.7	0.4
35/50 B + 20% RAP	fat. lag #1	8.7	-4.0	0.4	3.2	1.0
	fat. lag #2	9.3	-4.0	0.4	3.5	1.4
	fat. lag #3	9.6	-4.0	0.4	3.7	1.5
	fat. lag #4	10.0	-4.0	0.5	3.9	1.7
	fat. lag #5	10.5	-4.0	0.5	4.0	2.0
Orthoprène® + 20% RAP	fat. lag #1	15.7	-9.4	0.8	4.6	0.9
	fat. lag #2	15.8	-9.4	0.7	4.6	1.2
	fat. lag #3	16.0	-9.4	0.8	4.6	1.3
	fat. lag #4	16.1	-9.4	0.8	4.6	1.4
	fat. lag #5	16.2	-9.4	0.8	4.6	1.4

Table 10. *Quantification of different relative contributions to φ evolution for mixes 35/50 + 40% RAP, 35/50 B + 20% RAP and Orthoprène®+ 20% RAP: values are calculated using the envelope line of fatigue lag #1.*

relative differences with $\varphi_{100\ \mu m/m,\ T0,\ 0\ cycles,\ lag\ \#1}$		$\Delta\varphi_{total,\ ref:\ lag1}$ [%]	$\Delta\varphi_{non\ linearity,\ ref:\ lag\#1}$ [%]	$\Delta\varphi_{heating}$ [%]	$\Delta\varphi_{thixotropy}$ [%]	$\Delta\varphi_{unrecov.\ 24h,\ ref:\ lag\#1}$ [%]
35/50 + 40% RAP	fat. lag #1	100.0%	-34.5%	7.1%	50.8%	7.5%
	fat. lag #2	100.0%	-32.4%	8.5%	49.6%	9.4%
	fat. lag #3	100.0%	-31.2%	7.7%	50.8%	10.3%
	fat. lag #4	100.0%	-30.1%	8.9%	49.3%	11.7%
	fat. lag #5	100.0%	-29.1%	7.9%	51.0%	12.1%
35/50 B + 20% RAP	fat. lag #1	100.0%	-46.1%	4.9%	37.2%	11.8%
	fat. lag #2	100.0%	-43.4%	4.6%	37.4%	14.6%
	fat. lag #3	100.0%	-41.6%	4.0%	38.8%	15.7%
	fat. lag #4	100.0%	-40.0%	4.5%	38.7%	16.8%
	fat. lag #5	100.0%	-38.4%	4.3%	38.5%	18.8%
Orthoprène® + 20% RAP	fat. lag #1	100.0%	-59.7%	4.8%	29.5%	6.0%
	fat. lag #2	100.0%	-59.1%	4.6%	28.8%	7.5%
	fat. lag #3	100.0%	-58.4%	5.0%	28.5%	8.1%
	fat. lag #4	100.0%	-58.1%	4.7%	28.8%	8.4%
	fat. lag #5	100.0%	-57.9%	4.7%	28.7%	8.7%

Adaptation and Validation of Stochastic Limiting Strain Distribution and Fatigue Ratio Concepts for Perpetual Pavement Design

Mary M. Robbins[a], Nam H. Tran[a], David H. Timm[b], and J. Richard Willis[a]

[a]*National Center for Asphalt Technology at Auburn University, Auburn, AL 36830*

[b]*Department of Civil Engineering, Auburn University, Auburn, AL 36849*

ABSTRACT. *Traditional perpetual pavement thickness design is based, in part, on controlling strain levels at the bottom of the asphalt concrete layer below an endurance limit to prevent bottom-up fatigue cracking. A field-based limiting strain threshold was developed from cumulative distributions of field-measured tensile strains in the 2003 and 2006 research cycles at the National Center for Asphalt Technology Pavement Test Track to understand the limiting strain necessary to control fatigue cracking. Additionally, the fatigue ratio, the ratio of the nth percentile strain to the fatigue endurance limit, was developed. Both the tensile strain distributions and fatigue ratios showed a clear difference between sections that experienced bottom-up fatigue cracking and those that did not. However, it is necessary to adapt these thresholds to strains predicted by perpetual pavement design tools. PerRoad, a stochastic perpetual pavement design program, was used to predict strains for the same 2006 sections. Previously developed strain distributions and fatigue ratios were adjusted to reflect observed differences in predicted and measured strains. Cumulative distributions and fatigue ratios based on predicted strains for the 2009 research cycle validated the updated limiting strain distribution and maximum fatigue ratios for designing perpetual pavements to resist bottom-up fatigue cracking.*

KEYWORDS: *perpetual pavement design, limiting strain, asphalt, layered elastic*

1.0 Introduction

The Asphalt Pavement Alliance (APA) provided an early definition of perpetual pavements as "an asphalt pavement designed and built to last longer than 50 years without requiring major structural rehabilitation or reconstruction and needing only periodic surface renewal in response to distresses confined to the top of the

The oral presentation was made by Dr. Robbins.

This is a reproduction of an article whose final and definitive form has been published in *Road Materials and Pavement Design*© 2015 Taylor & Francis. The article is available online at: http://dx.doi.org/10.1080/14680629.2015.1077001

pavement" (APA, 2002). Therefore, perpetual pavements are designed to resist structural distresses, such as fatigue cracking that originates at the bottom of the pavement structure and rutting that occurs in the subgrade and base layers. Critical strains/stresses within the pavement structure are controlled to limit pavement distresses to the surface layers. Historically, a perpetual pavement is designed to have the tensile strain at the bottom of the asphalt concrete (AC) layer below its fatigue endurance limit. The fatigue endurance limit is defined as a strain level below which the pavement will have infinite fatigue life. In other words, if a pavement is designed such that the tensile strain levels at the bottom of the AC layer incurred due to loading are less than the fatigue endurance limit, fatigue damage originating at the bottom of the AC layer will not occur (Prowell et al., 2010). Thus, the pavement should perform for an unlimited number of years without experiencing structural failure (Prowell et al., 2010). Similarly, the pavement is designed to have the vertical strain at the top of the subgrade below a pre-determined limit to prevent structural rutting. The fatigue endurance limit used in perpetual pavement design is typically determined from laboratory fatigue testing and has ranged in magnitudes from early estimates of 70 microstrain (Monismith and McLean, 1972) to more recent estimates of up to 200 microstrain (Nishizawa et al., 1997; Prowell et al., 2010). A value of 200 microstrain has been proposed for the vertical strain limit (Monismith et al., 2004, Walubita and Scullion, 2010).

While there has been some debate about how high the fatigue endurance limit can be and still maintain a perpetual pavement, there has also been some concern raised as to whether one limiting strain value, under a fixed set of conditions, can control for fatigue. In the NCHRP Report 646 on validating the fatigue endurance limit, the authors conclude that "future efforts to incorporate the endurance limit into the Mechanistic-Empirical Pavement Design Guide (MEPDG) should consider a distribution of acceptable strains or endurance limits that vary as a function of temperature" (Prowell et al., 2010). Recent studies (Willis and Timm, 2009; Willis et al., 2011) at the National Center for Asphalt Technology (NCAT) Pavement Test Track used field-measured strain to determine the strain distribution over the service life of the pavement in a heavily loaded environment. Tensile strains measured at the bottom of the AC layer in 15 instrumented sections from the 2003 and 2006 NCAT Pavement Test Tracks were used in conjunction with known traffic and hourly average pavement temperatures to develop cumulative distributions of tensile strain. As a result, the cumulative strain distribution captures all strain levels measured under heavy loading for the temperatures experienced throughout the entire traffic period. Traffic periods for each test cycle are two years long; therefore, the developed cumulative strain distributions capture the strain due to a wide range of temperatures. In the cumulative strain distribution, the 100th percentile strain represents the maximum measured strain value and the minimum strain value is represented by the 0 percentile strain. Comparisons were drawn between the cumulative distribution of field-measured strain and the fatigue endurance limit as determined from laboratory testing of the base AC mixture. Researchers found that the fatigue endurance limit represented a wide range of percentiles on the cumulative distribution of strain for each section. Table 1 (Willis et al., 2011) shows the percentiles at which the fatigue endurance limits intersect the cumulative strain distributions from field-measured strain for five sections. It was found that these percentiles ranged greatly.

Table 1. *Comparison of cumulative strain distributions and fatigue endurance limits (after Willis et al., 2011).*

Section	**Percentile at Fatigue Endurance Limit**	**Performance**
N1 2006	10th	No Cracking
N2 2006	8th	No Cracking
N8 2006	50th	Cracked
N9 2006	88th	No Cracking
N10 2006	8th	Cracked
S11 2006	3rd	Cracked

For those sections that experienced fatigue cracking, the percentile ranged from the 3rd to the 50th, indicating that 3% to 50% of the strains measured in the field fell below the section's fatigue endurance limit. One would expect that these percentiles would be much greater for those sections that remained in good condition with no fatigue cracking; however, they too ranged greatly, such that the fatigue endurance limit was between the 8th and 88th percentile of field-measured strains. Based on the fatigue endurance limit concept, for sections that did not experience fatigue cracking the percentile at which the fatigue endurance limit intersects the cumulative distribution of strain should be relatively high, indicating that the majority of the strains fell below this limit and thus no damage was incurred. However, this was not the case, sections N1 and N2 did not experience fatigue cracking yet only 8-10% of the measured strains were less than the fatigue endurance limit. In this case, the concept of fatigue endurance limit would lead the designer to believe these sections were significantly under-designed. Whereas, section N8, a section that failed by fatigue, showed more promise of being a perpetual pavement with 50% of measured strains less than its fatigue endurance limit. As a result, the researchers were not able to develop correlations between fatigue endurance limits, pavement performance and field-measured strains at a single strain value, (Willis and Timm, 2009; Willis et al., 2011). Based on these results the application of a single fatigue endurance limit may not be an effective design criterion in perpetual pavement design.

Additionally, Willis and Timm (2009) found a noticeable difference between cumulative distributions of field-measured strains for sections that experienced bottom-up fatigue cracking and for those that did not. As a result, a limiting cumulative strain distribution, as shown in Table 2, was determined to identify the threshold for bottom-up fatigue cracking. The limiting cumulative distribution was based on field-measured strains for pavement sections in the 2003 and 2006 research cycles that did not experience fatigue cracking. It was recommended that the cumulative distributions of tensile strain be used as limiting criteria for fatigue cracking in the design of perpetual pavements either alone or in conjunction with the fatigue endurance limit, rather than using the fatigue endurance limit alone (Willis and Timm, 2009; Willis et al., 2011). Table 2 lists the limiting strain at each percentile for the field-based cumulative strain distribution.

Table 2. *Fatigue criteria based on field-measured strain (after Willis and Timm, 2009, and Willis et al., 2011).*

	Field-Based Cumulative Strain Distribution	
Percentile	**Limiting Strain (με)**	**Maximum Fatigue Ratio**
1%	49	
5%	72	
10%	90	
15%	101	
20%	112	
25%	122	
30%	132	
35%	143	
40%	155	
45%	168	
50%	181	1.27
55%	193	1.35
60%	205	1.44
65%	218	1.53
70%	232	1.63
75%	247	1.74
80%	263	1.85
85%	282	1.98
90%	310	2.18
95%	346	2.45
99%	394	2.83

From the developed cumulative distributions of strain for each section, the researchers developed another new concept, the fatigue ratio. The fatigue ratio is determined by dividing the cumulative strains at incremental percentiles by the fatigue endurance limit determined in the laboratory on the base layer asphalt mixture. The limiting fatigue ratios, also shown in Table 2, were a result of the distinct difference between sections that experienced fatigue cracking and those that did not. These studies showed that the fatigue ratios of the pavement sections that did not experience fatigue cracking fell below the limiting fatigue ratios shown in Table 2 (Willis and Timm, 2009; Willis et al., 2011).

These limiting strain criteria show promise for perpetual pavement design, however, the limiting cumulative strain distribution and maximum fatigue ratios were determined based on field-measured strains. Other studies at the NCAT Pavement Test Track have shown discrepancies between field-measured strains and predicted tensile strains at the bottom of the AC layer (Robbins and Timm, 2009; Timm et al., 2011). Differences between measured and predicted strains at the bottom of the AC layer have also been reported at the MnRoad facility (Nagarajan, 2012). Thus, the limiting cumulative strain distribution and maximum fatigue ratios may not be readily applicable to predicted strains that would result from perpetual pavement design. Therefore, there is a need to adjust the limiting strain distribution and fatigue ratios shown in Table 2 to reflect predicted strains determined in perpetual pavement design and to validate the adjusted criteria for future implementation.

The objectives of this study were (1) to determine the difference between the field-measured tensile strains and corresponding predicted tensile strains; and (2) to adjust the limiting strain distribution and maximum fatigue ratios (shown in Table 2) developed by Willis and Timm (2009) to reflect predicted strain values for future implementation in perpetual pavement design and if possible, (3) to validate the adjusted limiting strain distribution and maximum fatigue ratios.

2.0 Scope

Six sections of the 2006 NCAT Pavement Test Track were simulated in PerRoad version 3.5 to predict tensile strain values at the bottom of the AC layer. These sections were used to evaluate the application of the limiting strain distribution developed from field-measured strain. These sections were also used to develop a new limiting strain distribution based on predicted strain values for use in perpetual pavement design. Additionally, six sections from the 2009 NCAT Pavement Test Track were used to validate the newly developed limiting design distribution for predicted strain. Layer modulus values for each cross-section were determined through three-layer backcalculation using EVERCALC version 5.0. Average hourly pavement temperatures specific to each section were utilized to account for seasonal variation in the AC layer moduli. Bending beam fatigue testing (BBFT) was conducted to determine the fatigue endurance limit of the base AC mixtures for the 2009 sections.

3.0 Test Facility

The NCAT Pavement Test Track is a 1.7 mile Test Track located in Opelika, Alabama. It is a full-scale accelerated loading facility, with 46 200-foot test sections. Each research cycle operates on a three-year period, with two years designated for trafficking and one year split between construction and forensic evaluation at the conclusion of the traffic period. Approximately 10 million ESALs are applied over a two-year traffic period with a fleet of five triple trailer trucks operating at 45 mph for 16 hours a day, five days a week. The triple trailer trucks consist of a 12-kip steer axle, a 40-kip tandem axle and five trailing 20-kip single axles. Weekly performance evaluations that include crack mapping, rut-depth measurement and ride-quality measurements are augmented by frequent falling-weight deflectometer testing and strain-response measurements.

3.1 Test Sections

A total of twelve pavement sections from two research cycles at the NCAT Pavement Test Track were selected for this study and are shown in Figure 1. Six pavement sections, N3, N4, N8, N9, N10 and S11, were utilized from the 2006 NCAT Pavement Test Track research cycle. The remaining six sections were from the 2009 cycle and included sections N10, N11, S8, S9, S10 and S11. The 2006 sections were also used in the previous study in which the fatigue criteria shown in Table 2 were developed (Willis and Timm, 2009). Sections N3 and N4, originally constructed in 2003 and left in place in 2006 for additional trafficking, as well as section N9 from the 2006 research cycle did not exhibit bottom-up fatigue cracking. The cumulative field-measured strain distributions of Section N3 and N4 from both research cycles (2003 and 2006) were very similar to one another, and it was these four cumulative distributions that created the boundary between cracked and uncracked sections. As a result, the average of these four cumulative distributions was selected as the limiting field-measured strain distribution shown in Table 2 (Willis and Timm, 2009). Given the similarities in cumulative strain distributions of Sections N3 and N4 measured in the 2003 and 2006 research cycles, only the sections from the 2006 research cycle were used to evaluate the application of the limiting strain distribution on predicted strain values.

Sections N3 and N4 were placed as part of the 2003 NCAT Pavement Test Track and were left in place through the end of the 2009 research cycle. These sections were designed with 9 in. of AC over 6 in. of granite base material and the Test Track subgrade material, classified as an AASHTO A-4(0) soil (Timm, 2009). These two sections were designed to replicate each other with the exception of the binder type. Section N3 utilized a Performance Grade (PG) 67-22 unmodified binder throughout the AC layer, and section N4 used a PG 76-22 modified with styrene-butadiene-styrene (SBS) throughout. Section N8 was designed with 10 in. of AC including a stone-mastic asphalt (SMA) surface lift, conventional AC layers using PG 76-28 and PG 64-22 binders, and a rich-bottom layer designed at 2% air voids with PG 64-22 binder. Section N9 was the complement to N8, although it was designed to be perpetual at 14 in. of AC; N8 and N9 used the same mixtures throughout. Using a fatigue endurance limit of 100 $\mu\varepsilon$, section N9 was designed to maintain 90% of the predicted strains below this threshold (Timm et al., 2009). Both sections N8 and N9 utilized the compacted Test Track soil as a base material and a compacted soft subgrade material over the existing Test Track soil subgrade. Section N9 was left in place through the 2009 Test Track and as part of the still active 2012 Test Track. Section N10 was designed at 8 in. of AC consisting of mixes with PG 70-22 binder in the surface and intermediate layer and PG 64-22 binder in the base course mix.

Six in. of a Type-5 base material from Missouri was used as a granular base layer over the Test Track subgrade. Lastly, section S11 was designed at 7 in. of AC, featuring a PG 76-22 SBS-modified binder in the top two AC lifts and PG 64-22 in the bottom two lifts, on top of 6 in. of granite base material and the track soil subgrade. Detailed information about these sections can be found in a previous report (Timm, 2009).

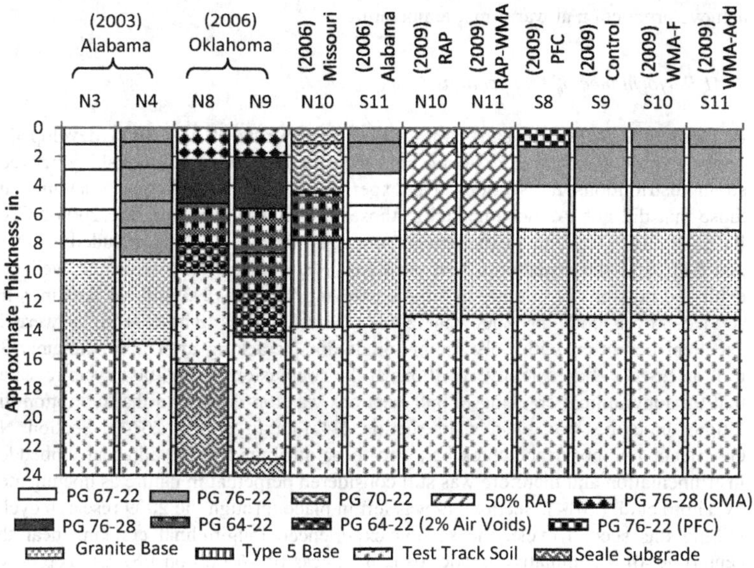

Figure 1. *Cross sections and materials of test sections used in this analysis.*

Test sections placed as part of the 2009 NCAT Pavement Test Track included in this study were six of the eight-section group experiment: N10, N11, S8, S9, S10, and S11. These sections were all designed at 7 in. of asphalt on top of 6 in. of granite base, placed on top of the Test Track soil subgrade material. These sections were selected due to the range in mix types used and although many were unconventional, they used commonly-available technology. Additionally, all of these sections experienced bottom-up fatigue cracking, and it was believed they would serve well for validation of the newly developed limiting design distribution for predicted strain. Section S9 served as the control section for the group experiment and featured conventional asphalt concrete mixtures including a surface course utilizing PG 76-22 SBS-modified binder, an intermediate layer also using a PG 76-22 SBS-modified binder and a base course using conventional PG 67-22 binder. Section S8 mirrored section S9 with the exception of the surface layer which replaced the conventional mix with a porous friction course (PFC) designed with 15% reclaimed asphalt pavement (RAP) at the same thickness of 1.25 in. Sections N10 and N11 shared the same mix designs which included 50% RAP in all three asphalt concrete layers, with the difference being that N11 utilized the foaming technique to produce it as a warm-mix asphalt (WMA) while the other was produced at typical production temperatures for hot-mix asphalt (HMA). Sections S10 and S11 shared mix designs with section S9 but were produced using warm-mix technologies, such that section S10 was produced with foaming technologies and S11 incorporated an additive to achieve production at warm-mix temperatures.

3.1.1 Performance of Test Sections

Previous studies (Willis and Timm, 2009; and Willis et al., 2011) developed a limiting strain distribution based on field-measured strain by distinguishing between strain distributions for sections that experienced bottom-up fatigue cracking and those that did not. As documented in those studies, sections from the 2006 NCAT Pavement Test Track that experienced bottom-up fatigue cracking included N8, N10 and S11; and the remaining 2006 sections, N3, N4 and N9 performed well and exhibited no signs of bottom-up fatigue cracking. Sections N3 and N4 remained in-service during both the 2006 and 2009 research cycles. These sections were in excellent condition with only minor longitudinal cracking after approximately 30 million ESALs. Forensic investigations at the conclusion of the 2009 cycle revealed that the longitudinal cracking in both sections was top-down and that no bottom-up fatigue cracking was evident in section N4 (Timm et al., 2014). Section N3 experienced a subsurface crack believed to be attributed to the adjacent embedded instrumentation and therefore was still considered perpetual in nature as documented by Timm et al. (2014). Section N9 was left in place through the 2009 research cycle. During this second cycle, the section experienced longitudinal cracking near the centerline after cumulative traffic loads in excess of 16 million ESALs. Top-down cracking at the Test Track has historically appeared as longitudinal cracking. Cores were cut at the location of the longitudinal crack from which it was confirmed that the cracking was top-down in nature. Section N9 was left in-place and continues to be trafficked as part of the 2012 research cycle, and after approximately 28 million ESALs, it only exhibits evidence of top-down cracking.

Six sections (N10, N11, and S8 through S11) placed as part of the 2009 NCAT Pavement Test Track performed well during the two-year trafficking period with little to no distresses evident and no fatigue cracking at the conclusion of the research cycle and application of 10 million ESALs. The six sections were left in-place for the 2012 research cycle during which time distresses became visible. Cracking was observed in all six sections.

Cores were extracted from four sections (S8 through S11). The cores confirmed that for sections S8 through S11 the transverse cracking was bottom-up and the longitudinal cracking was top-down. Extracted cores from areas of transverse cracking for these four sections are shown in Figure 2, where the cracking has been highlighted to make it more visible. Transverse cracking in S8 was first observed in the spring of 2013 after just over 12 million cumulative ESALs and later longitudinal cracking was evident as well. Figure 3 shows cracking in both the longitudinal and transverse directions (visible by the white paint marks on the surface) and the location of two cores that were extracted in December of 2013. These cores were taken at a transverse crack and longitudinal crack, from which it was confirmed that the longitudinal crack was top-down and the transverse crack was bottom-up cracking, as shown in Figure 2a. In a similar fashion, section S9 also exhibited longitudinal and transverse cracking during the 2012 research cycle. Cracking was first observed in March of 2013 in the longitudinal direction after approximately 12 million cumulative ESALs and transverse cracking was observed shortly thereafter. Figure 2b shows an extracted core from section S9 at the location of a transverse crack that was confirmed to be bottom-up fatigue cracking. Section S10 experienced both transverse and longitudinal cracking which was first observed in January of 2013. Figure 2c shows the extracted core from S10 for observed bottom-up fatigue cracking. Longitudinal cracking was first observed in Section S11 in December of 2012 after 10.7 million cumulative ESALs and transverse cracking was evident by the following spring (2013).

Section N10 was the last of the six sections continued from 2009 to crack. Transverse cracking was observed in section N10 in November of 2013 after a combined 15.5 million ESALs had been applied. However, this initial transverse cracking was in an area immediately adjacent to a patch that was placed to correct an area of severe but localized distress. Transverse cracking was observed in other areas in the section by February of 2014, as shown in the crack map from March 31, 2014, in Figure 4. The patched area is highlighted at the far right of the map, at approximately 170 feet and transverse cracking was observed near 55 feet and 120 feet in the outside wheelpath and near the edge of pavement, respectively. Cores were not extracted in this section; however, it is believed that the transverse cracking that has been observed originated at the bottom of the AC layer and propagated to the surface as has been the case in sections S8 through S11. Likewise, section N11 experienced transverse cracking and although cores were not extracted it is believed that the transverse cracking propagated from the bottom of the AC to the surface. Longitudinal cracking was first observed in N11 in February of 2013 and transverse cracking was evident shortly thereafter. As shown in Figure 5, a crack map from April 21, 2014, there is extensive cracking which has become interconnected in the outside wheel path (at the far right of the map). This area of interconnected cracking was the site of the initial observations of transverse cracking. Table 3 summarizes the performance of the 2006 and 2009 test sections included in this study.

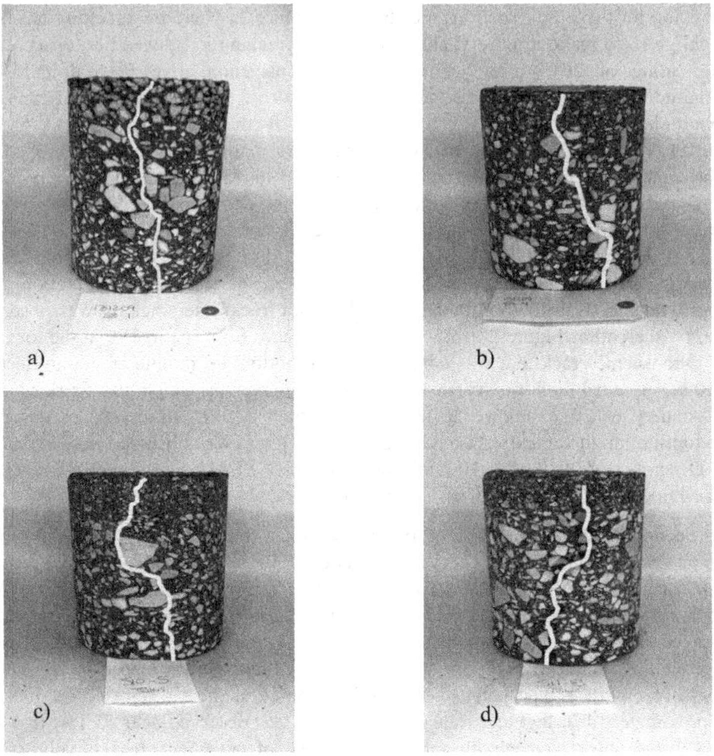

Figure 2. *Cores from transverse cracks in a) S8 b) S9 c) S10 and d) S11.*

Figure 3. *Evidence of transverse and longitudinal cracking in section S8.*

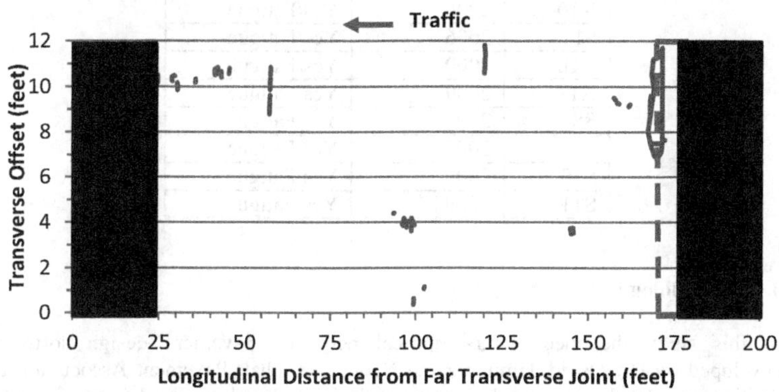

Figure 4. *N10 crack map, March 31, 2014.*

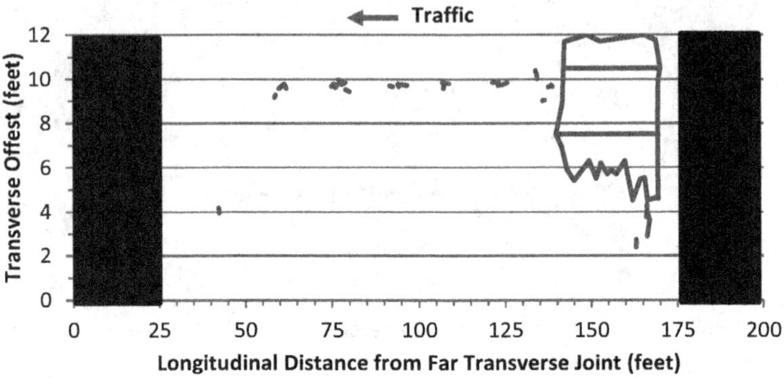

Figure 5. *N11 crack map, April 21, 2014.*

Table 3. *Summary of section performance.*

Section	Year Placed	Cracking/Type
N3	2003	No
N4	2003	No
N8	2006	Yes/Fatigue
N9	2006	Yes/Top-down
N10	2006	Yes/Fatigue
S11	2006	Yes/Fatigue
N10	2009	Yes/Fatigue
N11	2009	Yes/Fatigue
S8	2009	Yes/Fatigue
S9	2009	Yes/Fatigue
S10	2009	Yes/Fatigue
S11	2009	Yes/Fatigue

4.0 Methodology

In this study the mechanistic-empirical perpetual pavement design software developed by Dr. David Timm for the National Asphalt Pavement Association to facilitate perpetual pavement design and analysis, PerRoad version 3.5, was utilized. PerRoad was employed to predict horizontal tensile strain at the bottom of the AC layer for the six pavement sections from the 2006 Test Track and the six sections from the 2009 Test Track. The software utilizes Monte Carlo simulations to allow for the consideration of known variability associated with material properties and construction as well as seasonal variation effects on material moduli. PerRoad predicts stress and strain within the pavement structure through the use of WESLEA, a layered elastic analysis program developed by Don Alexander at the Waterways Experiment Station. Using the software, pavement responses at the bottom of the AC layer were determined from which strain distributions were computed and fatigue ratios determined. The strain distributions and fatigue ratios that resulted from the strain predictions were compared with the field-based limiting strain distribution and field-based maximum fatigue ratios shown in Table 2.

4.1 Predicted Strain Distributions

For each section, PerRoad was used to complete Monte Carlo simulations, generating 5,000 predictions of tensile strain at the bottom of the AC layer. Cumulative distributions were developed from the predicted strain for each section for comparing with the previously-developed cumulative distributions from field-measured strain. To predict tensile strain at the bottom of the AC layer using PerRoad, load spectra, pavement layer moduli and thickness as well as the associated coefficient of variation (COV) for each were necessary.

4.1.1 Load Spectra

Axle weights for each of the five triple trailer trucks used to apply traffic at the NCAT Test Track were determined previously (Taylor, 2008). Based on the total number of axles in the fleet, it was determined that single axles represent 71.42% of the total number of axles applied and the steer and tandem axles each accounted for 14.29% of applied axles. The axle weights for each axle type (steer, tandem and single) were entered into PerRoad to characterize the traffic loadings in the form of load spectra: 20% of the steer axles weighed 8-12 kips, with the remaining 80% falling into the 10-12 kip range; 80% of the tandem axles weighed between 38 and 40 kips, with the remaining percentage weighing between 40 and 42 kips; and 100% of the single axles weighed between 20 and 22 kips.

4.1.2 Pavement Cross-Section

A three-layer structure was selected for each section with layer one representing the asphalt concrete, and layers two and three representing the unbound granular base and subgrade, respectively. Four random locations (RL's) were identified at the start of each test cycle, and it was at these four locations in the outside, between, and inside wheelpaths that layer thickness was surveyed during construction. In each 200-ft test section, 25 ft on each end were identified as transition areas to allow for mixture and elevation changes. The middle 150 ft of each section was subdivided into three 50-ft subsections. Within each subsection, one location was randomly selected, corresponding to RL 1, 2, and 3 in Figure 6. RL 4 was centered in the instrumentation array. The schematic for the random locations is shown in Figure 6. From these measurements, the average layer thicknesses were determined for each section, as well as the COV of the layer thickness, calculated using Equation 1 and listed in Table 4. Sections N3 and N4 were originally constructed in 2003, and therefore did not have surveyed thicknesses of the unbound granular base layer. Therefore, the sections were assumed to have a thickness equivalent to their design layer thickness of 6 in. and a COV equivalent to the average COV for the unbound granular base layer in the 2009 sections. A normal distribution was assigned to the layer thickness variability.

$$COV = \sigma/\mu \quad [1]$$

Where: σ = standard deviation

μ = mean

Table 4. *Layer thickness and associated variability by section.*

Section	h_{AC} (in.)	COV_{AC} (%)	h_{GB} (in.)	COV_{GB} (%)
N3 (2006)	9.17	2.4	6	8.8
N4 (2006)	8.89	3.8	6	8.8
N8 (2006)	9.92	6.2	6.38	5.1
N9 (2006)	14.40	4.0	8.44	9.7
N10 (2006)	7.71	3.2	6.00	8.4
S11 (2006)	7.61	7.5	6.08	14.2
N10 (2009)	7.09	3.3	3.98	12.6
N11 (2009)	7.12	2.6	4.22	7.9
S8 (2009)	7.04	3.0	5.51	7.2
S9 (2009)	7.00	2.3	5.80	4.9
S10 (2009)	7.00	3.6	6.35	6.4
S11 (2009)	6.90	2.3	6.17	7.1

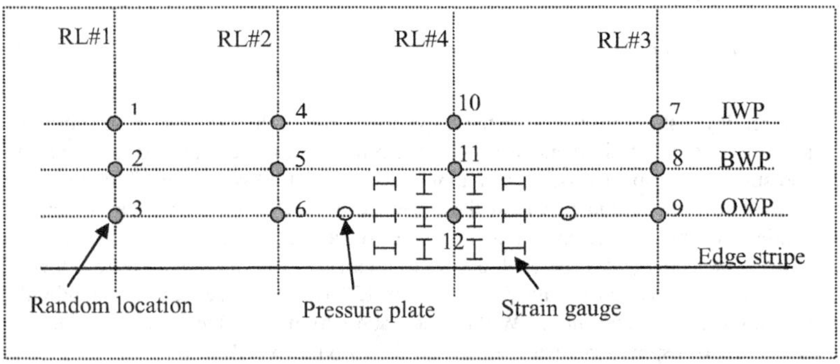

Figure 6. *Random locations schematic.*

Falling weight deflectometer (FWD) testing was also conducted at the four random locations in all three wheelpaths (Figure 6) throughout the duration of the test cycles. FWD testing included three replicates at four drop heights (load levels). Measured deflections were used to conduct a three-layer (AC, unbound base and subgrade layers) backcalculation in EVERCALC version 5.0. Section-specific unbound base and subgrade layer inputs were determined by calculating the average modulus for each layer over the entire two-year trafficking period. Average backcalculated layer moduli were selected for the 9-kip load level (load corrections were not applied) with root mean square error (RMSE) less than 3%. At the 9-kip load level, actual FWD loads ranged from 7.5 to 11 kips. The unbound materials used in the cross-sections of the 2006 and 2009 test sections have been explored through laboratory testing and extensive backcalculation in a previous study (Taylor, 2008). Based on the findings of the 2008 study characterizing these materials, the values shown in Table 5 are believed to be reasonable estimates of the material properties. The COV for the backcalculated base and subgrade moduli at the 9-kip load level was determined for each section following Equation 1 and are listed in Table 5 with the average backcalculated base and subgrade moduli. For the PerRoad simulations, a normal distribution was used in conjunction with the coefficients of variation listed in Table 5 for the backcalculated layer moduli of both the unbound base layer and subgrade material in each section.

Table 5. *Unbound layer moduli and associated variability.*

Section	Base (ksi)	Subgrade (ksi)	COV_{GB} (%)	COV_{SG} (%)
N3 (2006)	6.34	34.25	59.5	14.5
N4 (2006)	4.63	32.90	57.7	16.0
N8 (2006)	3.70	32.29	32.6	13.7
N9 (2006)	3.24	56.56	39.1	25.9
N10 (2006)	2.80	46.93	39.2	11.8
S11 (2006)	2.46	31.12	35.6	10.9
N10 (2009)	2.11	44.75	43.4	12.1
N11 (2009)	3.27	38.52	38.5	8.4
S8 (2009)	2.85	23.25	58.4	11.9
S9 (2009)	2.08	26.16	39.5	15.1
S10 (2009)	1.64	26.19	36.0	14.3
S11 (2009)	1.66	26.32	31.7	17.2

Backcalculation provides an estimate of the in situ AC modulus by considering one representative AC layer. Although these sections are comprised of multiple layers of AC, it is difficult to differentiate which deflections under FWD testing are associated to each AC layer. Furthermore, PerRoad, the layered elastic analysis software chosen for this study, is limited to only five pavement layers which would require the combination of AC layers for those test sections with more than three AC layers. Therefore, it was elected to use the AC modulus which results from a three-layer backcalculation as a modulus representative of the entire AC layer.

PerRoad allows for AC moduli to be entered for up to five seasons. To do so, first section-specific modulus-temperature relationships of the form listed in Equation 2 were developed with backcalculated AC moduli and the mid-depth pavement temperatures recorded throughout the two-year testing cycles. The backcalculated moduli selected for the modulus-temperature relationship were at the 9-kip load level and RMSE less than 3%. Hourly average temperatures were recorded for each section during the entire two-year trafficking period. The average hourly mid-depth pavement temperatures were then used in the modulus-temperature relationship in Equation 2 to develop a cumulative distribution of AC moduli experienced throughout the trafficking period. Once developed, the cumulative distribution of the AC moduli was used to select seasonal moduli in each section. The mid-point of each quintile was selected as the representative AC modulus for each of the five seasons: summer (10th percentile), spring 2 (30th percentile), spring (50th percentile), fall (70th percentile) and winter (90th percentile). These moduli and the corresponding mid-depth pavement temperatures are listed in Table 6 for each test section.

$$E_1 = \alpha_1(e^{\alpha_2 T}) \quad [2]$$

Where: E_1 = AC modulus (psi)

α_1, α_2 = regression coefficients

T = mid-depth pavement temperature (°F)

PerRoad requires the user to input the number of weeks in each season; therefore 10 weeks were assigned to spring, fall and winter, while summer and spring 2 were each assigned 11 weeks to be conservative. The COV for the AC modulus was calculated from temperature-corrected AC moduli. To do so, the backcalculated AC moduli, E_1, were corrected to 68°F, using Equation 3, and then the average and standard deviation of the corrected moduli were calculated to determine the COV. The COV for the corrected AC moduli are also listed in Table 6 for each test section. A lognormal distribution of AC moduli was used for each section.

$$E_{T_r} = E_1(e^{\alpha_2(T_r - T)}) \quad [3]$$

Where: E_1 = AC modulus at T (psi)

E_{Tr} = AC modulus at Tr (psi)

T = mid-depth pavement temperature (°F)

T_r = reference temperatures, 68°F

Table 6. *AC variability and AC moduli by season for PerRoad simulations.*

		Design Modulus, ksi				
Section	**COV (%)**	**Summer**	**Spring 2**	**Spring**	**Fall**	**Winter**
N3 (2006)	36.5	302.29	517.86	733.32	1022.28	1536.61
N4 (2006)	22.5	306.32	516.40	852.54	1309.14	1925.96
N8 (2006)	19.3	155.01	287.76	476.78	833.04	1333.62
N9 (2006)	22.4	240.45	384.75	654.68	977.31	1361.62
N10 (2006)	17.6	178.10	315.85	539.74	821.80	1197.28
S11 (2006)	17.2	139.64	249.47	451.50	760.88	1159.49
N10 (2009)	12.0	351.80	575.77	882.59	1432.38	2240.17
N11 (2009)	8.8	312.14	499.75	771.00	1288.27	2009.92
S8 (2009)	16.6	254.10	394.25	579.65	911.68	1386.25
S9 (2009)	12.5	249.20	397.99	632.82	1109.39	1794.56
S10 (2009)	11.5	232.54	390.86	597.95	996.86	1599.14
S11 (2009)	14.6	234.02	379.66	610.51	1018.44	1616.85
		Mid-Depth Pavement Temperature, °F				
N3 (2006)		87.30	68.56	56.43	44.85	30.65
N4 (2006)		96.20	80.70	65.82	53.09	41.64
N8 (2006)		102.0	87.20	75.12	61.77	50.51
N9 (2006)		99.30	86.30	71.60	60.52	51.35
N10 (2006)		98.40	84.50	71.50	61.30	52.17
S11 (2006)		101.8	88.40	74.7	62.65	52.92
N10 (2009)		106.3	90.50	76.80	61.27	46.90
N11 (2009)		106.6	91.50	77.59	61.12	46.85
S8 (2009)		106.5	91.00	77.4	61.42	46.60
S9 (2009)		106.5	92.60	78.83	62.16	47.87
S10 (2009)		108.7	91.90	78.15	61.61	46.32
S11 (2009)		105.7	91.20	76.99	61.63	47.78

4.2 Predicted Fatigue Ratio

To evaluate the field-based maximum fatigue ratios previously developed by Willis and Timm (2009), Equation 4 was applied to the predicted strain distributions and the fatigue endurance limits of each section. The fatigue endurance limits were determined by first conducting bending beam fatigue testing in accordance with AASHTO T 321. As documented by Willis and Timm (2009), the fatigue endurance limits for N3 and N4 were determined as part of the NCHRP 9-38 project (Prowell et al., 2010) for BBFT samples compacted to 7.0% target air voids, by applying a three-stage Weibull equation and were taken as the 95% one-sided lower prediction limit. The fatigue endurance limits for the sections placed in 2006 were also determined in the same manner using samples compacted to 5.5% target air voids at two strain levels, 400 and 800 microstrain (Willis et al., 2011). The fatigue endurance limits reported previously for the 2006 sections are listed in Table 7. Bending beam fatigue testing was also conducted for the base AC mixtures in the 2009 sections following AASHTO T 321 using specimens compacted to target air voids of 7.0% and tested at strain levels of 200, 400, and 800 microstrain. The fatigue endurance limits taken as the 95% one-sided lower prediction for the 2009 sections are also listed in Table 7.

$$R_n = \varepsilon_n / \varepsilon_f \qquad [4]$$

Where: R_n = fatigue ratio at the nth percentile

ε_n = microstrain at the nth percentile (developed using field-measured strain)

ε_f = fatigue endurance limit, microstrain

Table 7. *Fatigue endurance limit (after Willis and Timm, 2009; Willis et al., 2011).*

Section	Endurance Limit - 95th one-sided lower prediction limit, $\mu\varepsilon$	Target Air Voids of BBFT Specimens
N3 (2006)	151	7.0%
N4 (2006)	146	7.0%
N8 (2006)	203	5.5%
N9 (2006)	203	5.5%
N10 (2006)	130	5.5%
S11 (2006)	118	5.5%
N10 (2009)	100	7.0%
N11 (2009)	134	7.0%
S8 (2009)	92	7.0%
S9 (2009)	92	7.0%
S10 (2009)	99	7.0%
S11 (2009)	84	7.0%

5.0 Results and Discussion

To address the need to establish limiting strain criteria for improved perpetual pavement designs, previously developed field-based limiting strain distributions were evaluated using predicted tensile strain values using PerRoad, a stochastic design software.

5.1 Predicted Strain Cumulative Distributions

As a result of the PerRoad simulations, tensile strains were predicted at the bottom of the AC layer in each test section under the applied load spectra that accounted for seasonal variation in AC moduli and measured variability in layer moduli and thickness. Using the predicted tensile strains, cumulative distributions were calculated for each section. Figure 7 compares the predicted cumulative distributions from the 2006 Test Track with the limiting cumulative distribution (Table 2) developed based on field measurements from the previous study (shown in Figure 7 as "Field Meas Limit") (Willis and Timm, 2009).

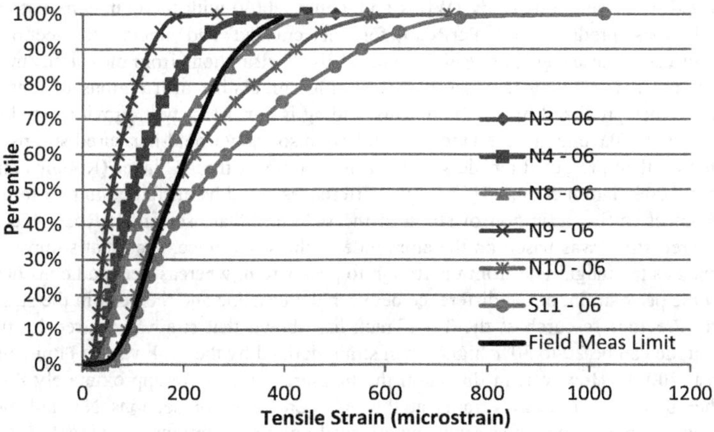

Figure 7. *Cumulative distributions for predicted tensile strain, 2006 sections.*

As previously mentioned, bottom-up fatigue cracking was observed in sections N8, N10 and S11 from the 2006 Test Track cycle, and the remaining 2006 sections (N3, N4, and N9) performed well with no fatigue cracking evident. In the previous study (Willis and Timm, 2009), the field-based limiting strain distribution was determined from the strain data measured in sections N3 and N4. This field-based limiting strain distribution (Table 2) can separate the uncracked sections from the cracked sections, such that the field-measured strain distributions of the uncracked sections fell to the left of the field-based limiting strain distribution.

If predicted strain matches field-measured strain, it would be expected that the field-based limiting strain distribution would be close to the predicted strain distributions of sections N3 and N4 in Figure 7. However, as shown in Figure 7, the field-based limiting strain distribution (shown as the "Field Meas Limit") does not fall near the predicted cumulative strain distributions of Sections N3 and N4, the same sections on which the field-based limiting strain distribution was based. Also, in Figure 7, the predicted strain distribution of section N8 that exhibited fatigue cracking falls to the left of the field-based limiting strain distribution along with the uncracked sections, N3, N4 and N9. Although the predicted cumulative distribution of section N8 does eventually cross the field-based limiting strain distribution, 80% of the predicted strain levels fall below the limiting distribution, and beyond this point there is little difference between the two distributions. This is evidence of large differences between predicted tensile strain and field-measured tensile strain at the bottom of the AC. These discrepancies in strain values will certainly influence design and reinforce the need to adapt this concept of strain distributions to predicted strains for use in perpetual pavement design.

Figures 8 and 9 compare the field-measured cumulative strain distributions reported in the previous study (Willis and Timm, 2009) with the cumulative strain distributions predicted by PerRoad for the cracked and uncracked sections, respectively. These figures reveal that cumulative distributions from measured strain values are much higher than the corresponding cumulative distributions of tensile strain values predicted by PerRoad. This finding is consistent with previous studies at the NCAT Pavement Test Track which have also shown field-measured strains to be higher than predicted tensile strains at the bottom of the AC layer (Robbins and Timm, 2009; Timm et al., 2011). These differences could be attributed in part to the differences in the definition of strain in the field and that used in PerRoad. Field-measured strain was based on the amplitude of the strain trace, such that strain was defined as the magnitude from the trough to peak strain, whereas PerRoad considers only the peak strain, or the difference between the baseline and the peak in the strain trace. Previous research at the Test Track has shown that strain measured by the amplitude can be 20 to 30% higher than strain defined by the peak value (Timm and Priest, 2008). However, in this case, the measured strains are approximately 80% higher than the predicted strains at the 50th percentile for sections N3 and N4. Although three of the six sections experienced bottom-up fatigue cracking and it is difficult to distinguish at what point fatigue cracking began as it takes time for the crack to propagate to the surface, measures were taken to minimize the use of measured strains that could be associated with fatigue damage, as documented by Willis and Timm (2009). The differences between measured and predicted strain could be due to materials and structural behaviors that are not simulated in the layered elastic analysis.

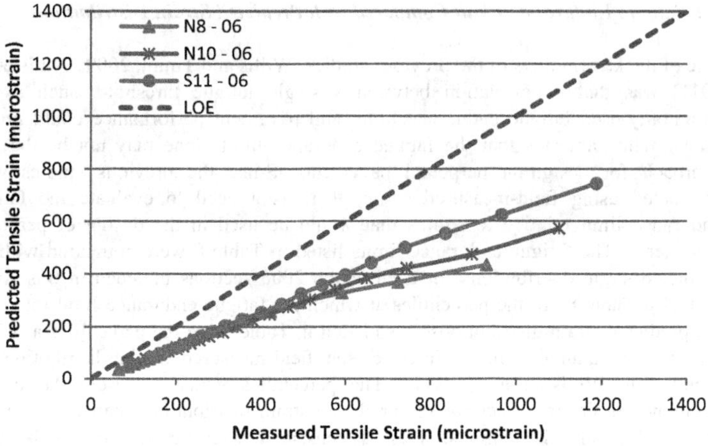

Figure 8. *Measured versus predicted cumulative strain distributions for cracked sections.*

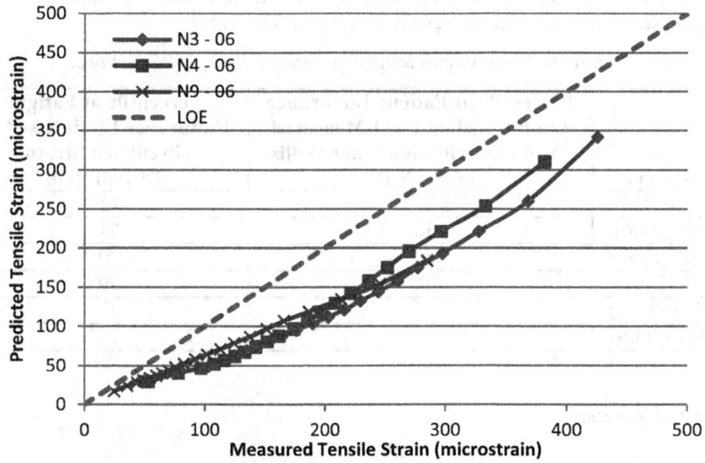

Figure 9. *Measured versus predicted cumulative strain distributions for uncracked sections.*

5.2 Fatigue Endurance Limit Compared with Predicted Strain Distributions

One of the key findings of the previous studies (Willis and Timm, 2009; Willis et al., 2011) was that a correlation between a single fatigue threshold, such as the commonly-used fatigue endurance limit, and pavement performance could not be drawn. This indicates that the fatigue endurance limit alone may not be the best approach for designing perpetual pavements. Since the previous studies were conducted using field-measured strain, there is a need to evaluate the fatigue endurance limit relative to strains that would be used in the design of perpetual pavements. The fatigue endurance limits listed in Table 8 were compared with the predicted strain distributions for each of the 2006 sections included in this study. Listed in Table 8 are the percentiles at which the fatigue endurance limit intersects the predicted strain distributions. Also listed in Table 8 are the percentiles at which the fatigue endurance limits intersect the field-measured strain distributions as reported in Willis et al. (2011). The percentiles which represent the fatigue endurance limit are higher for the predicted strain distributions than for the field-measured strain distributions. This is consistent with the differences between predicted and field-measured strain distributions discussed above. Section N8, which exhibited bottom-up fatigue cracking, ranked among sections N3 and N4 that received 30 million ESALs and performed well with no signs of fatigue cracking. This has obvious implications as based on the fatigue endurance alone one might expect section N8 to perform as well as sections N3 and N4.

Table 8. *Percentile at fatigue endurance limit, 2006 sections.*

Section	Percentile at Fatigue Endurance Limit based on Field-Measured Strain Distribution (*after* Willis et al., 2011)	Percentile at Fatigue Endurance Limit based on Predicted Strain Distribution
N3 (2006)	33rd	72nd
N4 (2006)	38th	71st
N8 (2006)	50th	70th
N9 (2006)	88th	99th
N10 (2006)	8th	33rd
S11 (2006)	3rd	19th

5.3 Fatigue Ratio

The fatigue ratios shown in Table 9 were calculated by dividing the predicted strain distributions by the corresponding fatigue endurance limits listed in Table 7. Table 9 also lists the maximum fatigue ratios determined from field-measured strains recommended by Willis and Timm (Willis and Timm, 2009). The ratios that do not meet the maximum fatigue ratios are shown in italics. Although cracked sections, N10 and S11, do not meet the field-based fatigue ratio criteria, N8 does for all percentile levels, despite exhibiting fatigue cracking during the 2006 Test Track cycle. This indicates that the field-based maximum fatigue ratios may not be a good indicator for failure when applied to predicted strain levels. As shown here, N8 passes the criteria for resistance to fatigue cracking, a critical aspect of perpetual pavement design; however, it experienced fatigue cracking after a year and a half of traffic (less than 10 million ESALs).

Table 9. *Fatigue ratio for 2006 sections compared to field-based maximum fatigue ratio.*

Percentile	Field-based Maximum Fatigue Ratio	N3	N4	N8	N9	N10[1]	S11[1]
50%	1.27	0.69	0.67	0.66	0.29	*1.37*	*1.93*
55%	1.35	0.74	0.74	0.74	0.32	*1.52*	*2.15*
60%	1.44	0.80	0.81	0.82	0.35	*1.70*	*2.41*
65%	1.53	0.88	0.89	0.90	0.39	*1.90*	*2.74*
70%	1.63	0.95	0.97	1.00	0.43	*2.10*	*3.02*
75%	1.74	1.04	1.08	1.11	0.48	*2.32*	*3.35*
80%	1.85	1.16	1.20	1.26	0.53	*2.61*	*3.75*
85%	1.98	1.28	1.34	1.43	0.59	*2.95*	*4.24*
90%	2.18	1.46	1.52	1.62	0.66	*3.25*	*4.69*
95%	2.45	1.72	1.74	1.82	0.76	*3.64*	*5.30*
99%	2.83	2.26	2.12	2.14	0.91	*4.40*	*6.31*

[1] *The ratios that do not meet the maximum fatigue ratios are shown in italics.*

As shown in Table 9, the predicted fatigue ratios for Sections N3 and N4 are much lower in magnitude than the field-based maximum fatigue ratio criteria that were developed based on the same sections. This is consistent with the differences illustrated in Figure 7, which shows the field-based limiting strain distribution is much higher than the predicted strain distributions for N3 and N4. This again indicates that there are differences in measured and predicted tensile strain levels.

5.4 Updating Limiting Strain Criteria for Use in Perpetual Pavement Design

Due to the differences between measured and predicted strains, the field-based limiting strain distribution and maximum fatigue ratios are not appropriate for design when predicted strain values are used. Therefore, the field-based limiting strain distribution and field-based maximum fatigue ratios need to be adjusted to implement this concept for the design of perpetual pavements.

As shown in Figure 7, there is an obvious separation at the predicted cumulative strain distributions of Sections N3 and N4, as all sections whose predicted strain distributions fall to the right of these curves exhibited bottom-up fatigue cracking and all sections whose predicted strain distributions fall to the left did not. This is consistent with the observations made from field-measured strains in the previous study (Willis and Timm, 2009). Since the cumulative distributions for predicted strains in Sections N3 and N4 lie nearly on top of one another and because these sections have the highest strain levels without experiencing fatigue cracking, they were selected for updating the limiting cumulative distribution of predicted tensile strain. To do so, consistent with the same methodology followed for developing the field–based limiting strain distribution, the average of the N3 and N4 predicted strain values at each percentile level were determined and plotted with the cumulative distributions of predicted strains of the other sections in Figure 10. For comparison, the field-based distribution is also shown in Figure 10 as a dashed line. Values for this new limiting distribution based on predicted strains are listed in Table 10.

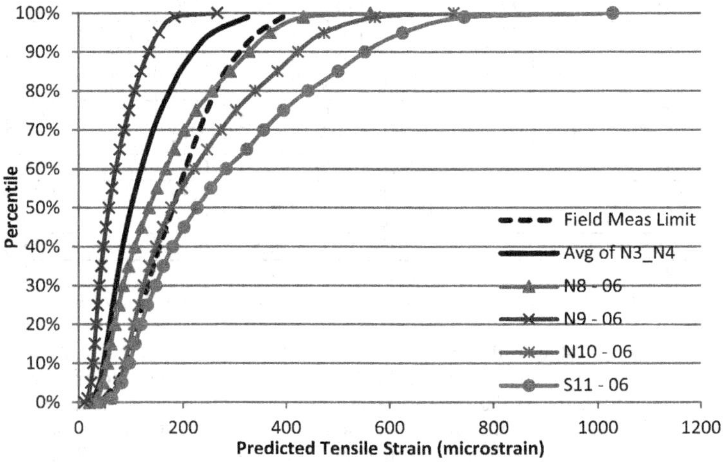

Figure 10. *Cumulative distributions for predicted tensile strain and new fatigue limit, 2006.*

In the development of the field-based limiting strain distribution, the variability of the strain measurements with the gauges used at the NCAT Test Track was reported as 30 microstrain (Willis and Timm, 2009). In that study, a tolerance of +/- 15 microstrain was applied to the limiting strain distribution to confirm that the distributions used to define the limiting strain distribution (sections N3 and N4) fell within the precision of the gauge. If this same variability were applied to the limiting design distribution here, it could also be used to determine if the average of N3 and N4 will suffice, or if the design distribution needs to be adjusted to be sure that the other distributions clearly fall outside of this band. A tolerance of 15 microstrain on either side of the limiting design distribution is plotted in Figure 11. The distributions for the remaining 2006 sections do not cross either the upper or lower bound of the limiting design distribution on the high end (above 50%) of the distribution, indicating that the limiting design distribution for predicted strains adequately defines the limiting strain criteria.

Table 10. *Updated limiting design strain distribution for predicted tensile strain.*

Percentile	Limiting Design Distribution for Predicted Strain (με)
1%	29
5%	41
10%	48
15%	54
20%	60
25%	66
30%	71
35%	78
40%	84
45%	91
50%	100
55%	110
60%	120
65%	131
70%	143
75%	158
80%	175
85%	194
90%	221
95%	257
99%	326

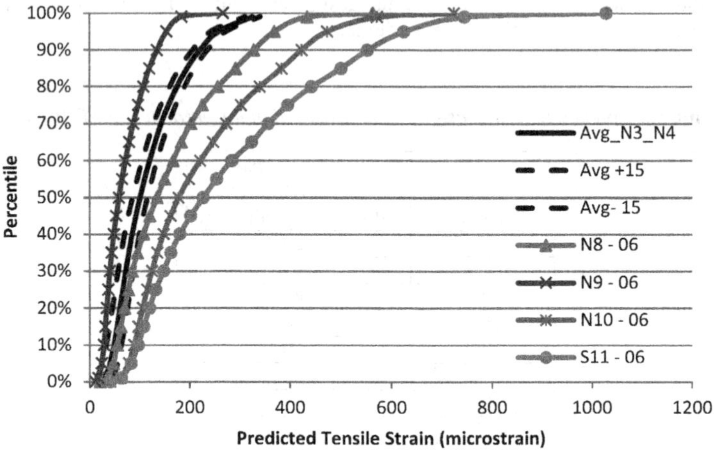

Figure 11. *Limiting design distribution for predicted strain with variability.*

To update the field-based maximum fatigue ratios to reflect tensile strain values predicted in PerRoad, the average ratio of N3 and N4 for each percentile was calculated and listed in Table 11. These new maximum fatigue ratios are compared in Table 11 to the ratios of the other 2006 sections investigated. Sections are denoted in Table 11 by the presence of fatigue cracking (FC) or no fatigue cracking (NC). Ratios that fail the criterion at that percentile level are shown in italics. As expected, sections that did not experience fatigue cracking passed these criteria. Sections N10 and S11 fail these criteria as expected given the fatigue cracking observed in these sections. However, section N8, a section with fatigue cracking, passes the criterion at the 50th, 55th and 99th percentile. Therefore, for these criteria to serve as in indicator of the ability of a perpetual pavement design to withstand bottom-up fatigue cracking, criterion at all percentiles from the 50th through the 99th percentile should be met.

Table 11. *Fatigue ratios from predicted tensile strain and updated fatigue ratio.*[1]

Percentile	Maximum Design Fatigue Ratio	N8 (FC)	N9 (NC)	N10 (FC)	S11 (FC)
50%	0.68	0.66	0.29	*1.37*	*1.93*
55%	0.74	0.74	0.32	*1.52*	*2.15*
60%	0.81	*0.82*	0.35	*1.70*	*2.41*
65%	0.88	*0.90*	0.39	*1.90*	*2.74*
70%	0.96	*1.00*	0.43	*2.10*	*3.02*
75%	1.06	*1.11*	0.48	*2.32*	*3.35*
80%	1.18	*1.26*	0.53	*2.61*	*3.75*
85%	1.31	*1.43*	0.59	*2.95*	*4.24*
90%	1.49	*1.62*	0.66	*3.25*	*4.69*
95%	1.73	*1.82*	0.76	*3.64*	*5.30*
99%	2.19	2.14	0.91	*4.40*	*6.31*

[1] *The ratios that do not meet the maximum fatigue ratios are shown in italics*

Despite a predicted strain distribution that is clearly far to the right of the design distribution, N8 had fatigue ratios that barely failed. This can be attributed to its very high fatigue endurance limit of 203 microstrain. In looking at the fatigue endurance limits listed in Table 7, sections N8 and N9, which shared the same base AC mixture and therefore had the same fatigue endurance limit, had the highest fatigue endurance limit of all of the sections in this study. However, N8 was 4 inches thinner and PerRoad simulations resulted in predicted strains significantly higher than its perpetual counterpart, N9. To account for scenarios such as this, reliability could be built into the fatigue ratios by using the variability associated with the strain gauge measurements (30 microstrain) as an initial assumption of strain variability. To do so, the lower bound (Avg – 15) shown in Figure 11 could be used. Fatigue ratios using the more conservative lower bound, or the average less 15 microstrain at each percentile, are shown in Table 12 with the other fatigue ratios for N8, N9, N10 and S11. By incorporating some reliability into the fatigue ratios those sections that experienced fatigue cracking in the field now clearly fail the fatigue ratio criteria. While the variability associated with strain gauge measurements was used to explore the effect of reliability on these ratios, more work should be done to determine the actual variability associated with predicted strain values to improve the lower bound fatigue ratios.

Table 12. *Lower bound fatigue ratio compared with ratios for each section.*[1]

Percentile	Lower Bound Design Fatigue Ratio	N8 (FC)	N9 (NC)	N10 (FC)	S11 (FC)
50%	0.57	*0.66*	0.29	*1.37*	*1.93*
55%	0.64	*0.74*	0.32	*1.52*	*2.15*
60%	0.70	*0.82*	0.35	*1.70*	*2.41*
65%	0.78	*0.90*	0.39	*1.90*	*2.74*
70%	0.86	*1.00*	0.43	*2.10*	*3.02*
75%	0.96	*1.11*	0.48	*2.32*	*3.35*
80%	1.08	*1.26*	0.53	*2.61*	*3.75*
85%	1.21	*1.43*	0.59	*2.95*	*4.24*
90%	1.39	*1.62*	0.66	*3.25*	*4.69*
95%	1.63	*1.82*	0.76	*3.64*	*5.30*
99%	2.09	*2.14*	0.91	*4.40*	*6.31*

[1] *The ratios that do not meet the maximum fatigue ratios are shown in italics*

5.5 Validation of Updated Limiting Strain Criteria

To determine if the updated limiting strain criteria (design distribution for predicted strains and maximum design fatigue ratios) were valid, six sections from the 2009 Test Track were simulated in PerRoad for comparison. The resulting predicted tensile strain values were used to develop cumulative strain distributions as was done with the 2006 test sections. Figure 12 shows the predicted strain distributions for each of the 2009 sections included in this study with the original field-based limiting strain distribution as well as the updated limiting design distribution. All six sections from 2009 are believed to have bottom-up fatigue cracking (four of the sections have been confirmed with field cores and the other two exhibit cracking consistent with bottom-up cracking historically observed at the Test Track). As was the case with predicted strain distributions for the 2006 sections, the original field-based limiting strain distribution did not effectively categorize sections prone to fatigue cracking based on predicted strain values. Predicted strain distributions for sections N10 and N11 both fall to the left of the field-based limiting strain distribution, which would indicate they passed this criteria. This is again an artifact of the underprediction of tensile strains. However, field-measured strains are not available during the design process; therefore the limiting strain distribution based on predicted strains from PerRoad is necessary to improve perpetual pavement design. The design limiting strain distribution, shown in Figure 12 as "Avg of N3_N4," accurately categorizes all six sections as prone to fatigue cracking. All six predicted strain distributions fall to the right of the design limiting distribution and all six sections experienced bottom-up fatigue cracking. These predicted strain distributions validate the updated design limiting strain distribution for predicted strain values for use in perpetual pavement design.

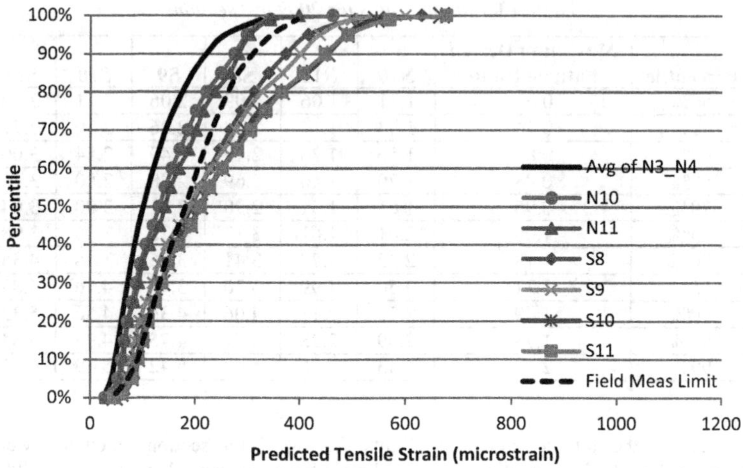

Figure 12. *Limiting design distribution compared with distributions for 2009 sections.*

Fatigue ratios using the predicted strain distributions shown in Figure 12 and associated fatigue endurance limits listed in Table 7 were determined for the 2009 test sections. The 2009 test sections serve as a validation dataset as they were not used to update the field based strain distribution and maximum fatigue ratio to incorporate predicted strain values. Table 13 lists the fatigue ratios for each of the 2009 test sections included in this study as well as the maximum design fatigue ratios. The maximum fatigue ratios were updated to reflect predicted strains for use in design. The maximum design fatigue ratios are meant to serve as criteria for perpetual pavement design such that ratios at all percentiles from 50 to 99 should be less than the maximum design fatigue ratios to achieve a section that will behave perpetually, as did sections N3, N4 and N9. It is expected that since all six sections from the 2009 research cycle experienced bottom-up fatigue cracking the fatigue ratios for all six sections should be greater than the maximum design fatigue ratios. That is the case as all sections clearly exceed the maximum design fatigue ratio at each percentile from 50 to 99, thus validating these updated fatigue ratios for perpetual pavement design.

Table 13. *Fatigue ratios for 2009 test sections.*

Percentile	Maximum Design Fatigue Ratio	N10	N11	S8	S9	S10	S11
50%	0.68	1.31	1.06	2.09	2.06	2.11	2.50
55%	0.74	1.43	1.14	2.28	2.26	2.30	2.71
60%	0.81	1.56	1.26	2.49	2.49	2.54	3.00
65%	0.88	1.70	1.40	2.69	2.78	2.80	3.34
70%	0.96	1.85	1.51	2.90	3.01	3.00	3.64
75%	1.06	2.01	1.61	3.13	3.25	3.26	3.98
80%	1.18	2.22	1.77	3.41	3.56	3.68	4.35
85%	1.31	2.50	1.99	3.70	3.99	4.16	4.83
90%	1.49	2.75	2.11	4.06	4.36	4.56	5.32
95%	1.73	2.99	2.28	4.51	4.75	4.93	5.84
99%	2.19	3.43	2.56	5.48	5.42	5.49	6.77

Lastly, the fatigue endurance limits for the 2009 sections included were compared with their predicted strain distributions, as was done with the 2006 sections. Table 14 lists the percentile on the predicted strain distribution that represents fatigue endurance limit for each section. Section N11 had the highest percentile, such that 45% of the strains predicted under Test Track traffic were less than or equal to the fatigue endurance limit of 134 microstrain. Section S11 was the lowest, with only 6% of the predicted strains falling at or below 100 microstrain. These percentiles rank among the percentiles reported in Table 7 for sections N10 and S11 from 2006. However, there is no obvious break point that relates fatigue endurance limit to pavement performance and predicted strain levels.

Table 14. *Percentile at fatigue endurance limit, 2009 test sections.*

Section	Percentile at Fatigue Endurance Limit based on Predicted Strain Distribution
N10 (2009)	37th
N11 (2009)	45th
S8 (2009)	9th
S9 (2009)	18th
S10 (2009)	16th
S11 (2009)	6th

6.0 Conclusions and Recommendations

Based on the findings of this study, the following conclusions and recommendations are made:

- A relationship between fatigue endurance limits, predicted strain levels and pavement performance was not clear. Therefore, fatigue endurance limits should not be the only consideration in perpetual pavement design. Cumulative distributions of predictive strain were found to be related to the presence of bottom-up fatigue cracking. Thus, it is suggested that cumulative distribution of the predicted tensile strain values be considered when designing perpetual pavements to resist bottom-up fatigue cracking.
- Significant differences exist between measured and predicted strain values, such that predicted strains were consistently lower than strains measured in the field. Although some of the differences observed are due to differences in definition of strain, there are inherent differences between predicted tensile strain and field-measured tensile strain. These differences must be accounted for during perpetual pavement design and when applying criteria developed from field measurements.
- The field-based limiting strain distribution and field-based maximum fatigue ratios were updated to reflect the difference between the field-measured and predicted tensile strain values and may serve as a threshold for designing perpetual pavements. Cumulative distributions of predicted tensile strain from the 2009 Test Track research cycle validated both the new design limiting strain distribution and maximum design fatigue ratios.
- The limiting cumulative strain distribution developed from field-measured tensile strain did not effectively categorize cross-sections when used in conjunction with predicted strain values. Therefore, the limiting design distribution that was adjusted to reflect predicted tensile strain values from PerRoad and validated with predicted tensile strains for sections that experienced fatigue cracking should be utilized for perpetual pavement design. It is recommended that in the perpetual pavement design, the designer develops a cumulative strain distribution from the predicted strains and compares that distribution with the limiting design distribution presented in Table 10. If the trial strain distribution does not meet these criteria, the pavement cross-section should be modified until the criteria are met.
- The maximum fatigue ratios determined based on field-measured strains does not adequately identify pavement cross-sections that can withstand fatigue cracking when applied to predicted tensile strain values. Therefore, the validated maximum fatigue ratios that have been adjusted for the difference in magnitude between measured and predicted strain levels should be used in design.
- The maximum design fatigue ratios serve as control points; therefore, for these fatigue ratios to be used for new designs, ratios should be less than those at the 50th through the 99th percentiles. It is recommended that additional work be done to incorporate reliability to guard against high fatigue endurance limits that could lead to erroneous acceptance of a design cross-section. Therefore, it is

recommend that the maximum design fatigue ratios only be used in conjunction with the limiting design distribution for predicted strains.

7.0 Acknowledgements

The authors thank the National Asphalt Pavement Association and the State Asphalt Pavement Associations for sponsoring this research.

8.0 References

Asphalt Pavement Alliance (APA), *Perpetual Pavements: A Synthesis*, APA 101, APA, Lanham, MD, 2002.

Monismith, C. and McLean, D., "Structural Design Considerations," *Proceedings of the Association of Asphalt Paving Technologists Technical Sessions*, Vol. 41, Cleveland, Ohio, 1972.

Monismith, C., Harvey, J., Bressette, T., Suszko, C. and St. Martin, J., "The I-710 Freeway Rehabilitation Project: Mix and Structural Section Design, Construction Considerations and Lessons Learned," *Proceedings of International Symposium on Design and Construction of Long Lasting Asphalt Pavements*, National Center for Asphalt Technology, Auburn, AL, 2004, pp. 217–262.

Nagarajan, V.V., *Comparison of Measured and Predicted Response of Six MnRoad Asphalt Test Sections*, Master's Thesis, University of Texas at Arlington, Arlington, TX, 2012. http://www.mrr.dot.state.mn.us/research/pdf/2012MRRDOC007.pdf.

Nishizawa, T., Shimeno, S., and Sekiguchi, M., "Fatigue Analysis of Asphalt Pavements with Thick Asphalt Mixture Layer," *Proceedings of the 8th International Conference on Asphalt Pavements*, Vol. 2, University of Washington, Seattle, WA, August 1997, pp. 969–976.

Prowell, B., Brown, E.R., Anderson, E.M., Daniel, J.S., Von Quintus, H., Shen, S., Carpenter, S., Bhattacharjee, S. and Maghsoodloo, S., *Validating the Fatigue Endurance Limit for Hot Mix Asphalt*, NCHRP Report 646, NCHRP, 2010.

Robbins, M.M. and Timm, D.H., "Effects of Strain Pulse Durations on Tensile Strain in a Perpetual Pavement," *Proceedings of the International Conference on Perpetual Pavement*, Columbus, Ohio, 2009.

Taylor, A. J., *Mechanistic Characterization of Resilient Moduli for Unbound Pavement Layer Materials*, M.S. Thesis, Auburn University, 2008.

Timm, D.H., *Design, Construction and Instrumentation of the 2006 Test Track Structural Study*, Report No. 09-01, National Center for Asphalt Technology, Auburn University, 2009.

Timm, D.H. and Priest, A.L., "Flexible Pavement Fatigue Cracking and Measured Strain Response at the NCAT Test Track," *Proceedings of the 87th Annual Meeting of the Transportation Research Board*, Transportation Research Board, National Research Council, Washington, D.C., 2008.

Timm, D.H., Gierhart, D., and Willis, J.R., "Strain Regimes Measured in Two Full Scale Perpetual Pavements," *Proceedings of the International Conference on Perpetual Pavement*, Columbus, Ohio, 2009.

Timm, D.H., Robbins, M.M., Willis, J.R., Taylor, A.J. and Tran, N.H., *Evaluation of Mixture Performance and Structural Capacity of Pavements Utilizing Shell Thiopave ®; Phase II: Construction, Laboratory Evaluation and Full-Scale Testing of Thiopave ® Test Sections – One Year Report*, Report No. 11-03, National Center for Asphalt Technology, Auburn University, 2011.

Timm, D.H., Robbins, M.M., and Willis, J.R., "Characteristics of Two Perpetual Pavements at the NCAT Test Track," *Proceedings of the 3rd International Conference on Transportation Infrastructure*, Pisa, Italy, 2014.

Walubita, L. and Scullion, T., *Texas Perpetual Pavements – New Design Guidelines*, Product 0-4822-P6, Texas Transportation Institute, College Station, TX, 2010.

Willis, J.R. and Timm, D.H., *Field-Based Strain Thresholds for Flexible Perpetual Pavement Designs*, Report No. 09-09, National Center for Asphalt Technology, Auburn University, 2009.

Willis, J.R., Timm, D.H., Taylor, A.J., Tran, N.H. and Kvasnak, A., "Correlating Laboratory Fatigue Endurance Limits to Field-Measured Strains," *Journal of the Association of Asphalt Paving Technologists*, Volume 80, 2011, pp. 135–160.

9.0 Discussion

ELIE HAJJ: Thank you for a good presentation. I have a question regarding two aspects, and I hope to get some insight from you. The first one is, typically, when we look at the endurance limits, you have that rest period. So in your case, there's a rest period between the axles and there's some kind of a different rest period between the trucks. How was that taken into consideration in the study? Because that would affect your endurance limit eventually. And the other one, the age of the material, as well, will affect your endurance limits. So you're looking at the data for two years. I understand it is still fresh compared to the life of the pavement, but still as the pavement ages, that endurance limit level is going to change. So you have two significant variables related to the test track. In your modeling, that was not taken into consideration. I'm not sure in the lab testing if it was. And as we're trying to relate these two together, how I am going to take that data and try to apply it in real-live simulations?

MARY ROBBINS: Thank you for the question. First I want to mention, we're not looking at fatigue endurance limit in the field, we're looking at a limiting cumulative strain distribution. So we're considering the entire distribution of strains that we would see under traffic for the entire two-year period. So we're looking at that full range of strain relative to temperature and to the traffic that we have at the test track. Now using PerRoad, we're not able to incorporate the axle spacings and thereby we're not able to incorporate rest periods, but we are looking at, again, this distribution of strain. So it's not necessarily looking at how it behaves when it rests, but we are looking at the different strain values that we have over the life of the pavement in the two-year period. Now I will mention in terms of your next question, which was aging, that the sections N3 and N4 were down. They were first placed in 2003, and they were taken up in 2012, so they were down for quite a long period in terms of the test track. They did experience 30 million ESALs. Although we're not able to account for all the aging that might occur in those sections, we were able to look at some of the aging. In fact, when I looked at the predicted strain values, we took the backcalculated values – moduli – from 2006, so this is already incorporating a couple years of aging at this point.

ELIE HAJJ: Right. But if I want to take that and apply in an actual traffic – if I have a specific material that was going to go in a given location, going all the way, let's say from Auburn to Reno, Nevada, now I have to incorporate – there a different climate, probably a different traffic, so for that specific location, I have to develop my limiting strain level.

MARY ROBBINS: Yeah. And simply taking our values and looking at the climate that you would have, so you can incorporate the seasonal variation. Again, it allows for up to five different seasons, and then looking at the traffic that you have, you can incorporate that load spectra and look at the cumulative distribution and compare it back to what we have. Now, I would say to calibrate it for different sections, we would probably want to again look at how is it performing. You need some sections to kind of indicate was it perpetual, what was not, in order to identify the criteria or that breakpoint so you can determine, "Well, I have the strain distribution but what does it mean in terms of performance?" So you would need to have a couple sections that did experience bottom-up fatigue cracking and a couple sections that behave perpetually.

ELIE HAJJ: All right. Thank you.

MARY ROBBINS: Thank you.

JOHN D'ANGELO: Mary, extremely difficult. We've talked about this many times, and I felt like I was trying to take a drink of water from Niagara Falls here.

MARY ROBBINS: Sorry. I tried.

JOHN D'ANGELO: Mary, very nice presentation. What I would like to ask you is about the issue with combining the two fatigue concepts presented. The first is fatigue endurance determined from a beam fatigue test, and fatigue ratio concept. Your presentation clearly showed we've got some issues with fatigue endurance limit and the fatigue ratio also seems to have limits. I really think that your approach using fatigue ratio based on strain distributions is the better way to analyze the pavement, but I think the issue Elie brought up is critical. How are you going to take into account aging? When we design perpetual pavement, we don't design it to last two years, three years, or four years. We would expect it to last 50 years. The distributions that you're analyzing at the track are all based on short time periods, with little aging. What we need to know what is going to happen 25 years down the road. It's going to be critical that you take that into account some aging concept to estimate how you expect those strains to change time, so that we can backcalculate critical levels. This will be needed so we can properly calculate critical strains to ensure truly long life pavements.

MARY ROBBINS: Certainly. Thank you for the comment, and I agree. We would hope to maybe look at how these pavements perform later in their life. But, you know, we have the great fortune of having the test track and the embedded strain gauges, but we do only operate on a two-year cycle, so.

GEOFFREY ROWE: Thanks, Mary. Nice presentation.

MARY ROBBINS: Thank you.

GEOFFREY ROWE: PerRoad uses a single value of modulus for the asphalt layer. You don't use any temperature depth distribution of moduli, do you?

MARY ROBBINS: Correct. We looked at the mid-depth pavement temperature, and that's how we determined the seasonal asphalt modulus based on the cumulative distribution of the temperatures that we see over the entire two-year period.

GEOFFREY ROWE: I would like to make a suggestion that you start to subdivide your layers. And the way I do this, and I found it to be quite useful, is I use a 20 mm top layer and then I double it and use a 40 mm next layer and an 80 mm and so. So I capture the temperature distribution with depth. I calculate the temperature depth profile and this can be done in several ways. For example, you can use the BELLS equations which developed as part of the LTPP program. My most recent way for considering this analysis is to develop master curves of material stiffness and then put them through aging models. It is then possible to develop from that a relationship between the FWD loading time and the stiffness of the material at different temperatures. I use a finite-element analysis and to compute the temperature depth profile in the pavement using the thermal properties and then factor into those layers the relative proportional differences, so you can back analyze with a temperature depth gradient, and you can also then do your forward analysis with temperature depth gradients. We did some work a number of years back looking at some pavements in China and trying to compare some of the analyses that were done using temperature depth gradients (I will include the figure in the discussion) and what we see is that the variation in strain that you calculate using layered elastic analysis isn't a singular point for a particular season. You almost get like a circular path depending on which hour of day you're on. As the top of the pavement warms and goes through its heating cycle and then it cools again, you get this sort of variation in strain that you're calculating at the bottom of the pavement. I would suspect that some of the effect that you see in your data with a downward shift there from where your calculated theoretical values might not be to what you're actually measuring but might be some of the variation due to temperature depth gradients. Layered elastic analysis is, of course, an old theory now and we can use viscoelastic analysis, but maybe there's more that we can do with the layered elastic analysis that you're running, and it might be quite interesting to look at that.

MARY ROBBINS: Yeah. Thank you for your comment, and, certainly, some of the differences that we have between predicted strain and measured strain are likely due to the fact that we're using layered elastic analysis. But I will mention that the intent with this project and this research was really to develop something that could be implemented right away in terms of perpetual pavement design. And so by calibrating to the field-measured strains and looking at that and using that as criteria and then backing up and using a tool that is very easy for pavement designers to use, uses all information that they can readily get. Unfortunately, they probably don't have temperature gradients to input, and so we wanted to work with the information that they would have ready to them, and so that is why we looked at PerRoad and we kept the layered elastic analysis approach and looked at just the mid-depth pavement temperature.

GEOFFREY ROWE: Yes. I'm just thinking there might be a few tricks you can do to better calibrate. I know WESLEA is restricted to the number of layers you can do, but in layered elastic analysis I can put 50 layers in if I need to, so there are ways of getting around that. And, you know, maybe it's worth doing sort of like a side study to see if you can better calibrate what you're seeing in some of those things.

MARY ROBBINS: And, certainly, I agree. I mean there's always room for improvement and we could look at the temperature gradient. We could also look at the rest periods, as John mentioned, but, again, we were really trying to create something that could be implemented for perpetual pavement design at an agency level that they could easily use for design and looking at tools that they would have readily available to them.

GEOFFREY ROWE: Thank you.

MARY ROBBINS: Thank you.

GEOFFREY ROWE: (Written Discussion). The figure I referenced in my discussion above which I indicated I would submit is shown below. The data represents four cases for which I have computed the strain at the underside of the pavement using layered elastic analysis (Rowe and Huber, Proceedings RMPD ICAM 2009). Two soil conditions are used and the S1 pavement is a weaker structure than S5. Each case has 5760 data points which represent a temperature depth profile for 20 years using 288 profiles for each year (12 months and 24 hours) using the average conditions. The data is clearly scattered and a degree of this is attributable to aging whereas the temperature depth profile accounts for the remainder.

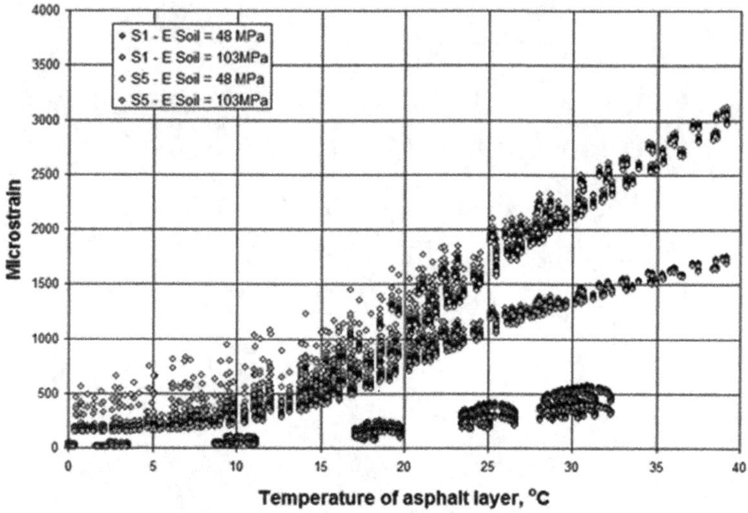

Figure 1A. Strain (micro) computed at the bottom of a pavement layer using layered elastic analysis for four pavement cases.

AUTHOR'S CLOSURE:

The authors approve the inclusion of the written addition and appreciate the comments provided.

Evaluation of Strain Relieving Interlayer to Retard Thermally-Induced Reflective Cracking

Hao Yin[a] and Charles Ishee[b]

[a]*Gemini Technologies, Egg Harbor Township, NJ, 08234*

[b]*Federal Aviation Administration, Atlantic City International Airport, NJ, 08405*

ABSTRACT. *To protect the structure and restore a smooth riding surface, many airports repair deteriorating concrete pavements with an overlay of hot mix asphalt (HMA). Reflective cracking is a serious concern associated with the use of thin overlays but is not addressed in the current Federal Aviation Administration (FAA) Advisory Circular for asphalt concrete (AC) overlaid rigid pavements. This paper presents a comprehensive study to examine the crack initiation and propagation of a test HMA overlay with and without a strain relieving interlayer, and to evaluate the interlayer performance to retard thermally-induced reflection cracks. To achieve these objectives, three-dimensional finite element analyses (FEA) were first conducted to assess key structural parameters controlling the tension stresses at the overlay bottom. Next, laboratory tests were performed to characterized fatigue, fracture, and viscoelastic performance of the interlayer mixture. Finally, a test pavement was built, instrumented, and tested at the FAA National Airport Pavement Test Facility (NAPTF). Full-scale test data suggested that the strain relieving interlayer considerably improved the reflective cracking resistance of the HMA overlay. Inclusion of a 1-in.-thick interlayer between existing concrete slabs and the overlay extended overlay service life up to 15%. The intact interlayer had prevented spalling and moisture infiltration at the joint and therefore prolonged the structural integrity of the pavement. In addition, mixed-mode fracture and channeling were observed in the crack propagation. This paper provides extensive information in hopes of increasing awareness of the FAA's contributions to the airport pavement research on reflective cracking.*

KEYWORDS: *Thermally-induced reflective cracking, full-scale test, instrumentation, finite element analysis.*

The oral presentation was made by Dr. Yin.

1.0 Introduction

At airfields, the potential for reflective cracking presents a major challenge for rigid pavement rehabilitation involving asphalt overlays. Reflection cracks most commonly result from temperature cycling-induced contraction (opening) and expansion (closing) in the concrete slab. Traffic loading exacerbates the thermally-induced reflection cracks, but the likelihood of spalling at the cracks and the potential for foreign object debris (FOD) to aircraft is a greater concern on airfields.

The LEDFAA overlay design model incorporated an estimate of reflection crack growth (1 in. per year), but this was believed unsatisfactory or illogical; and therefore was removed from FAARFIELD. The current version of FAA Advisory Circular does not explicitly consider this important distress mode (FAA, 2009). To incorporate HMA overlays as an alternative to increasing initial slab thickness, a reliable performance prediction model based on reflection crack growth is essential. The performance model is expected to relate the required thickness of HMA overlay to several input variables, including aircraft loads, climatic effects, and the condition of the existing pavement. One key element of developing such a sophisticated model is to acquire full-scale test data, particularly crack propagation rates, under controlled loading conditions to mimic the temperature cycles occurring in nature. Failure in this case would be the appearance of bottom-up cracks on the overlay surface that initiate from the underlying concrete joints.

An ongoing project sponsored by the FAA led to the development of the Temperature Effect Simulation System (TESS). The latest TESS upgrade was the installation of load cells that can measure actual forces being applied to a test HMA overlay. The primary objectives of the Phase I test were to examine the feasibility of substituting temperature load mechanically and understand the mechanism of thermally-induced reflective cracking. Through numerical simulations, Minhoto et al. (2005, 2008) demonstrated that climatic temperature variations in pavements lead to an increase of the reflective cracking phenomenon, due to the stress and strain states created by temperature, resulting in the premature distress of the asphalt overlay. It was hypothesized that thermally-induced reflective cracking would be critical to general aviation (GA) airports carrying small and light-weight aircraft that experienced significant temperature variations (Yin and Barbagallo, 2013). The Phase II test was designed to assess if bottom-up reflection cracks propagated roughly 1 in. per year (Yin and Barbagallo 2014). This paper describes the finite element analyses, laboratory characterization, and full-scale tests involved in Phase III.

2.0 Objective

The objective of this study was to evaluate the effectiveness of strain relieving HMA interlayers to retard thermally-induced reflective cracking under full-scale test conditions.

3.0 Reflective Cracking Mitigation

Existing reflective cracking mitigation techniques vary from simply increasing the asphalt concrete (AC) overlay thickness to a crack-arresting interlayer to a three-ply composite that is placed only over the joint/crack area. Although some of these techniques have been successful for mitigating reflection cracks in certain applications, many have performed poorly, particularly in colder climates (Lytton, 1989; Button and Lytton, 2007). Commonly used mitigation techniques include break/crack and seat, as well as rubblization, saw and seal, thick HMA overlay, mixture modification, reinforcement of HMA overlay, and interlayer system. Of all of these techniques, rubblization, when done correctly, is the only direct method for the complete elimination of reflective cracking potential. On the other hand, rubblization is also the method that most closely resembles reconstruction, since a relatively thick overlay is required to compensate for the reduction in structural support resulting from the elimination of concrete slabs.

One interesting observation from the Phase II test was that the crack length progressed aggressively once the crack reached the middle of the overlay. This observation implies that, given a specific pavement structure and materials, the critical zone to retard bottom-up cracks is the lower portion of the overlay. Logically, a layer of strain relieving HMA at the overlay bottom could be an effective control for thermally-induced reflective cracking.

4.0 Strain Relieving Interlayer

A strain relieving interlayer is a soft layer that is usually positioned at the bottom of an HMA overlay to absorb a large portion of the energy that would otherwise be part of the crack propagation process. The typical thickness of a strain relieving interlay in airfields is either less than 2 in. or greater than 3 in. (Von Quintus et al., 2009). The thin layers dissipate only the horizontal movements, while the thicker layers are hypothesized to dissipate both horizontal and differential vertical movements at joints and cracks. Two problems associated with thicker interlayers include 1) the total overlay thickness is generally much greater than some of the other mitigation strategies employ, and 2) if an open-graded HMA mixture is used, the interlayer can be a potential water conduit or reservoir between the overlay and existing pavement.

4.1 Parametric Analysis Using Finite Element Analysis

The Phase I test (Yin and Barbagallo, 2013) revealed that the driving mode for bottom-up reflection cracks was fracture Mode I; therefore, the focus of finite element analysis (FEA) was placed on the horizontal tensile stresses at the overlay bottom. Yin (2013a) demonstrated a three-dimensional finite element (FE)-based pavement simulation model to analyze thermally-induced reflective cracking. A haversine function describing the relationship between the loading time and the joint opening was used to approximate the temperature variations. In this study, the same full-scale FE model was used to assess two key structural parameters of the interlayer: thickness and stiffness.

According to Von Quintus et al. (2009), 4-in. dense-graded mixtures with adequate stability should be sufficient on GA facilities, while 6 in. should be sufficient for heavier aircraft at larger commercial aviation airports. For full-scale test purposes, the constructability and construction quality are critical. In addition, the overlay should not survive to a number of repetitions beyond the project time frame. Figure 1a shows three overlay thickness combinations proposed for the interlayer section. The control section contained 5-in. thick standard FAA P-401 material. The main purpose of a strain relieving interlayer is to isolate relative displacements between the overlay and existing pavement, such that the longitudinal horizontal tensile stresses in the overlay caused by thermal cycling can be significantly reduced. To accomplish these characteristics at low temperatures (i.e., 32°F), the interlayer material must possess excellent strain relieving and have the ability to relax stresses rapidly (Dave et al., 2007). Two reduction levels were selected to represent the interlayer's softness, 75% and 50% of the overlay's relaxation modulus, *E(t)*. The relaxation modulus was obtained from the creep compliance test on standard FAA P-401 material (Chiawat and Buttlar, 2012). The relaxation modulus master curves at the reference temperature of 32°F are provided in Figure 1b.

Figure 1c shows the tensile stress distributions in the longitudinal direction. The tensile stress sharply declined from its peak value atop the joint toward the middle of the slab. For the control section, full (100%) HMA relaxation modulus resulted in the maximum tensile stress of 355 psi. With the inclusion of an interlayer, the stress magnitude was considerably reduced. At 75% of the HMA relaxation modulus, the maximum tensile stresses at the interlayer bottom ranged from 277 to 300 psi. When the relaxation modulus was further reduced to 50%, the maximum tensile stress was only 40% of that in the control section, 150 psi. It is clear that, for the interlayer section, the interlayer stiffness had a much more pronounced influence on the tensile stresses compared to the thickness. The tensile stress distributions in the transverse direction are important for instrumentation (Yin, 2013b). As shown in Figure 1d, the tensile stresses reached peak values at 6 in. from the outer and inner edges for both the north and south strips. However, once the *E(t)* dropped to 50%, the tensile stresses at the interlayer bottom were nearly uniform. Like the longitudinal direction, the contribution to relieving tensile stress was predominantly from the reduction of HMA stiffness.

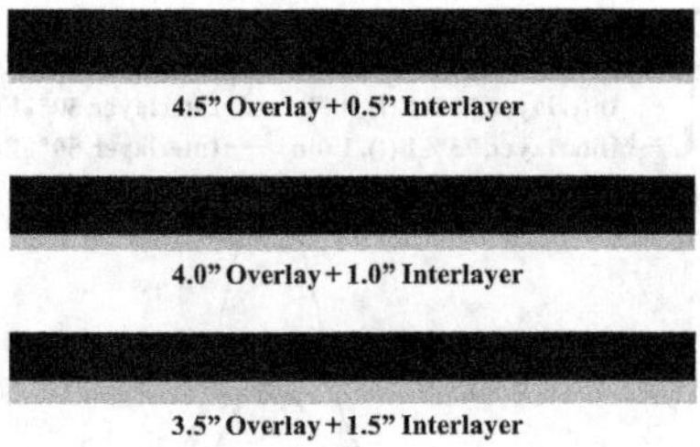

(a) Thickness combination.

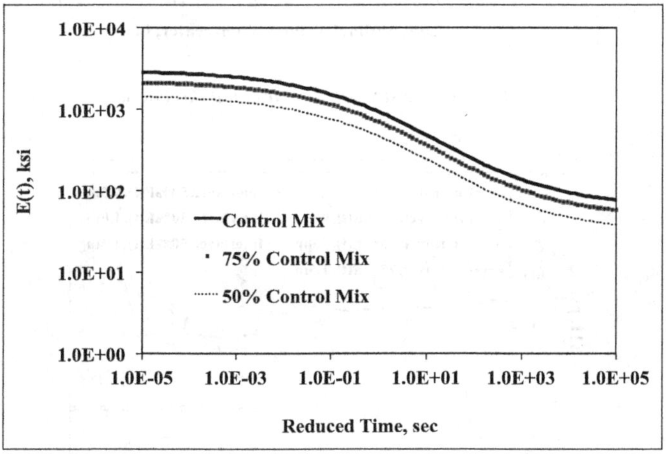

(b) Relaxation modulus master curves at a reference temperature of 32°F.

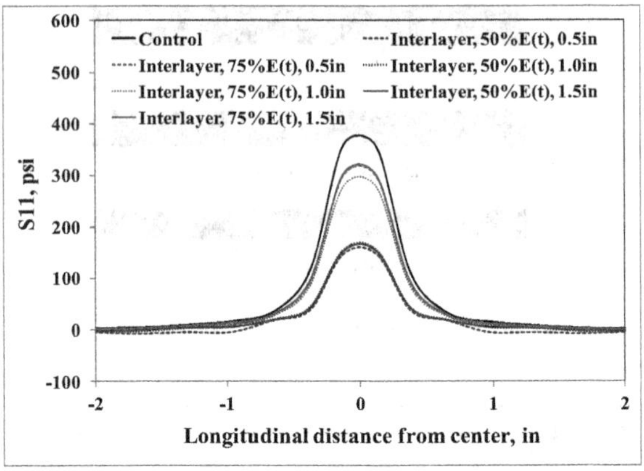

(c) Tensile stress in longitudinal direction.

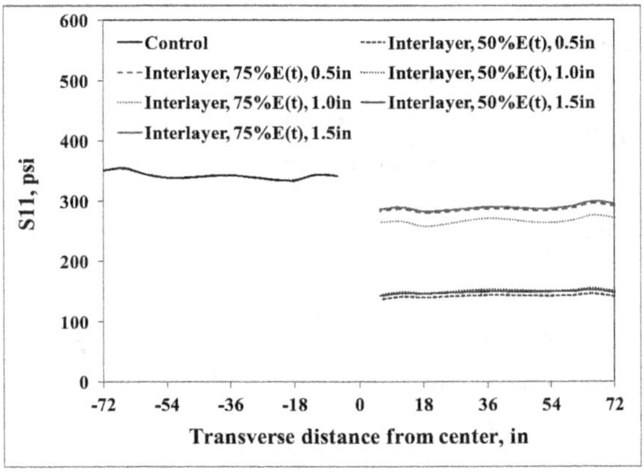

(d) Tensile stress in transverse direction.

Figure 1. *Summary of FE simulations.*

4.2 HMA Mix Design

The strain relieving interlayer is a fine graded, polymer modified asphalt (PMA), asphalt-rich mixture. The PMA is a cross-linked, elastomeric styrene-butadiene block copolymer system that is less susceptible to temperature and has a higher viscosity at ambient temperature compared to unmodified or neat asphalt mixtures. In addition, the PMA provides the elasticity to withstand and partially absorb the tension, shear, and bending exerted on the pavement. As FEA simulations demonstrated, a 1-in.-thick soft interlayer seemed promising to reduce the maximum tensile stress directly above the joint. Dave et al. (2010) reported that a thin strain tolerant interlayer mix can be incorporated in an HMA overlay to delay reflection cracks. For the Phase III test, two PMA-based interlayer mix designs were developed meeting Superpave PG 76-22 requirements. The aggregate source, gradation, and key volumetrics of each mix are summarized in Table 1.

Table 1. *Aggregate source, gradation, and volumetric of interlayer HMA mixes.*

Sieve Size	S-1	S-2	S-3	Cumulative Percent	
	Dolomite	Dolomite	Argillite	Blend A	Blend B
	#9	#10	#10	(40% S-1 and 60% S-2)	(35% S-1, 20% S-2, and 45% S-
⅜ in. (9.5 mm)	100.0	100.0	100.0		
No. 4 (4.75 mm)	85.5	99.0	99.6	92.0	92.3
No. 8 (2.36 mm)	18.0	60.5	83.3	46.8	55.9
No. 16 (1.18 mm)	2.8	30.1	54.1	25.3	34.7
No. 30 (0.600 mm)	0.3	16.5	34.7	16.7	23.7
No. 50 (0.300 mm)	0.0	10.1	23.2	12.4	16.8
No. 100 (0.150 mm)	0.0	4.5	16.6	8.3	11.8
No. 200 (0.075 mm)	0.0	0.0	12.1	4.6	8.2
Pan	0.0	0.0	0.0	0.0	0.0
Volumetrics				Mix A	Mix B
Optimum Asphalt Content (%)				8.2	7.9
Maximum Specific Gravity				2.532	2.450
Bulk Specific Gravity				2.455	2.326
Air Voids (%)				3.08	5.05
VMA (%)				16.36	20.48

4.3 Evaluation of Interlayer Mix

Laboratory tests were conducted to evaluate fatigue, fracture, and viscoelastic performance of the two interlayer mixes. These included Texas Overlay Tester, disk-shaped compact tension, and complex modulus tests. All test samples were 6-in. in diameter and laboratory-prepared with target air voids of 7%. An ideal strain relieving HMA should exhibit higher fracture energy and stiffness at low temperatures without compromising fatigue performance.

4.3.1 Texas Overlay Tester

Following the protocol established for the Texas Department of Transportation (2009), samples were cycled at 0.1 Hz in displacement control from 0 to 0.025 in. at 77°F. The crack growth, as observed on the sides or top of the sample, was manually recorded. Five samples were tested for each mix design. The peak load ranged from 608 to 736 lb and 349 to 473 lb for Mix A and Mix B samples, respectively. The test was terminated when either a 93% load reduction was reached or 1200 cycles. All samples reached 1200 cycles without failure.

4.3.2 Disk-Shaped Compact Tension

The disk-shaped compact tension tests were performed at 10°F, in accordance with ASTM D7313-13 (2013). A constant crack mouth opening displacement (CMOD) rate of 0.039 in./min was used. The specimens' fracture energies were determined by calculating the normalized area under the Load-CMOD curve. As shown in Figure 2a, the fracture energy of Mix B on average was 20% higher than Mix A. The higher fracture energy might have resulted from the tighter gradation of Mix B.

4.3.3 Complex Modulus

In addition to fatigue and fracture performance, it is also important to measure the bulk material viscoelastic properties, because the overall material response (stresses and strains) are governed by the bulk material characteristics. Three samples were tested for each mix design. Four temperatures (10°, 40°, 70°, and 100°F) and three loading frequencies (10, 1, and 0.1 Hz) were considered. Figure 2b shows the dynamic modulus ($|E^*|$) master curves at a reference temperature of 70°F for all test samples. It appeared that the difference between mixes varied, but Mix B was slightly stiffer than Mix A at higher frequencies and much softer at lower frequencies. AC mixtures that exhibited lower stiffness properties at high frequencies (low temperatures) tended to be more crack resistant.

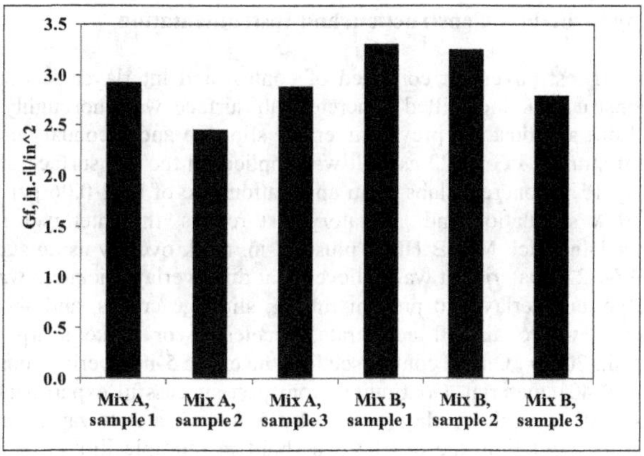

(a) Fracture energy.

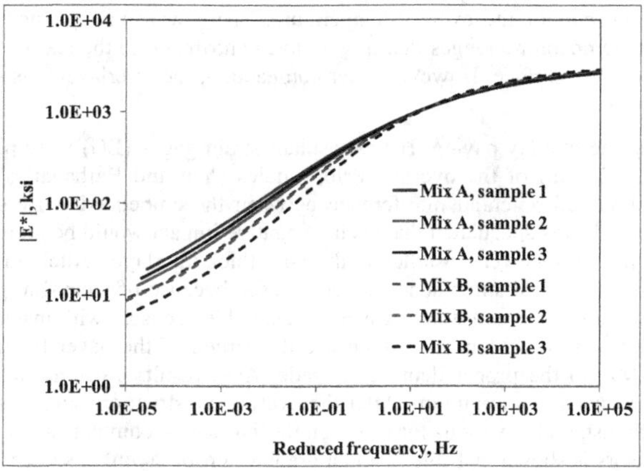

(b) Dynamic modulus master curves at a reference temperature of 77°F.

Figure 2. *Laboratory test results.*

5.0 Overlay Structure, Construction, and Instrumentation

The Phase III test pavement consisted of control and interlayer sections. Prior to overlay construction, the milled concrete slab surface was thoroughly washed to remove all dirt and dust. To prevent interface slippage and secondary cracks, a thin tack coat of straight PG 64-22 asphalt was applied on the dry surface of two 12-in. thick, 15 by 15 ft concrete slabs at an application rate of 0.04-0.06 gal/yd^2. On the basis of FEA simulation and laboratory test results, the interlayer section was designed as 1-in. thick Mix B HMA plus a 4-in. thick overlay using standard FAA P-401 (PG 64-22) material. It was believed that this overlay thickness was sufficient above the softer interlayer to prevent rutting, slippage cracks, and shoving in the overlay areas where aircraft accelerate, decelerate, or make sharp turns (Von Quintus et al., 2009) . On the control section, the entire 5-in. overlay consisted of the same FAA P-401 material. Continuing from past successful experiments (Yin and Barbagallo, 2014), each overlay was 30 by 5 ft with a 2 ft gap in between. To facilitate instrumentation, the overlay was built in multiple lifts (1.0, 1.5, and 2.5 in.). Between lifts, time was cautiously balanced to allow for the application of a tack coat, placement of instrumentation sensors, and an adequate mix temperature to achieve the desired density. Thermocouples were embedded at various depths and locations to monitor the overlay temperature. There were unexplained spikes and dips in the temperature ranges that might have contributed to the tearing appearance in the interlayer surface. However, after compaction, the interlayer was smooth and stable.

During the overlay paving, H-type asphalt strain gages (EG) were placed at the mid-depth (2.5-in.) of the overlay. Prior studies (Yin and Barbagallo, 2014; Yin, 2013b) suggested potential interference between the embedded sensors and crack propagation. It was speculated that such a negative impact would be more prominent if the H-type sensors were installed in the 1-in.-thick interlayer. Alternatively, Fiber-Bragg Grating optical strain sensors were chosen because of their slim profile. One installation lesson learned was to cover the embedded sensors with material close to the lift thickness (i.e., 1-in.) and then use the screed of the paver to strike off the excess HMA to the proper depth and grade. As a result, instrumentation damage could be reduced to a minimum. After the overlay construction, surface strain gages (SG) were installed at various locations on the surface. A complete array of installed strain sensors is shown in Table 2, and a graphic sensor layout is shown in Figure 3. Note that all sensors were installed on the "best guess" crack path, which was directly above and perpendicular to the concrete joint.

6.0 Full-scale Test

Full-scale tests began on June 3, 2014. The Phase III test employed the same test protocol as the Phase II test. The overlay bottom temperature was targeted at 32°F. The temperature variations were approximated by a haversine load waveform describing the relationship between the joint opening (0.012 in.) and cycle time (150 sec). At the end of each loading cycle, a rest (nonload) period of 600 sec was included to allow the overlay to relax. After 6350 cycles, the test concluded on July 30, 2014.

Table 2. *Summary of sensor location and cycles to failure.*

Overlay Section	Location in Overlay	Sensor ID	Cycles to Failure
Control	Bottom	FG6	2490
Control	Bottom	FG7	2026
Control	Bottom	SG14	4108
Control	Bottom	SG23	892
Control	Mid-depth	EG10	3004
Control	Mid-depth	EG12	2741
Control	Mid-depth	FG10	2477
Control	Mid-depth	EG14	2063
Control	Mid-depth	EG16	1967
Control	Mid-depth	SG15	4219
Control	Mid-depth	SG22	2955
Control	Surface	SG17	5173
Control	Surface	SG18	2815
Control	Surface	SG19	2254
Control	Surface	SG21	3196
Interlayer	Bottom	FG2	*
Interlayer	Bottom	FG3	*
Interlayer	Bottom	SG2	3257
Interlayer	Bottom	SG11	365
Interlayer	Mid-depth	EG2	2953
Interlayer	Mid-depth	EG4	2702
Interlayer	Mid-depth	FG9	2385
Interlayer	Mid-depth	EG6	1747
Interlayer	Mid-depth	EG8	1634
Interlayer	Mid-depth	SG3	3611
Interlayer	Mid-depth	SG10	2088
Interlayer	Surface	SG4	5047
Interlayer	Surface	SG6	2878
Interlayer	Surface	SG7	1882
Interlayer	Surface	SG9	3117**

** Malfunctioned after 1350 cycles*
*** From visual examination*

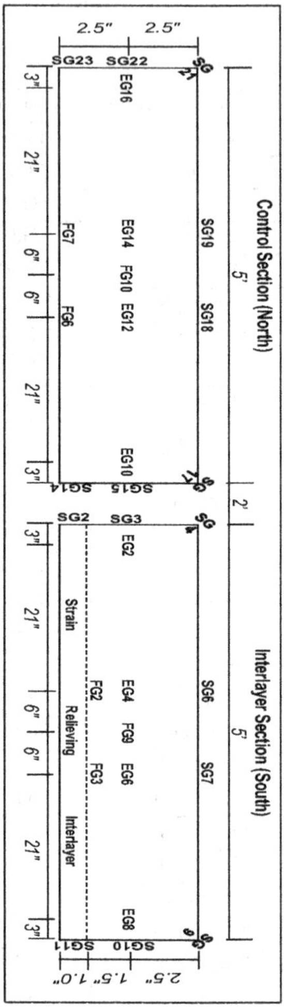

Figure 3. *Strain sensor layout.*

6.1 Performance of Fiber Optic Strain Sensor

As mentioned before, fiber optic strain sensor (FG) was introduced in the Phase III test. The FG provides active temperature compensation that is critical to temperature-dependent materials, such as AC. As illustrated in Figure 4a, the most attractive feature of FG is its thin profile. In addition, a stainless steel carrier was used to secure and protect the fiber. Figure 4b shows the strain responses from both EGs and the FG during one loading cycle. In a side-by-side comparison with H-type EGs, the FG responses were equally sensitive and reasonable. However, the FG's slimness may result in unexpected failure. As shown in Table 2, FGs 2 and 3 malfunctioned after only 1350 loading cycles.

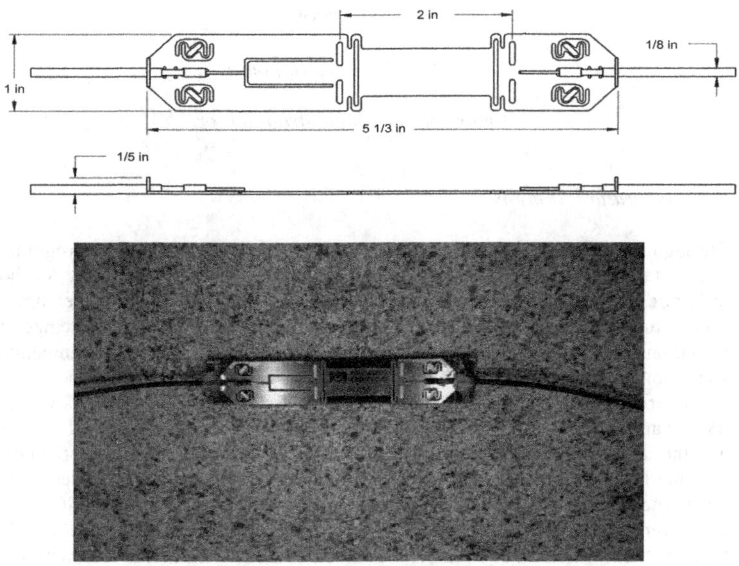

(a) Diagram.

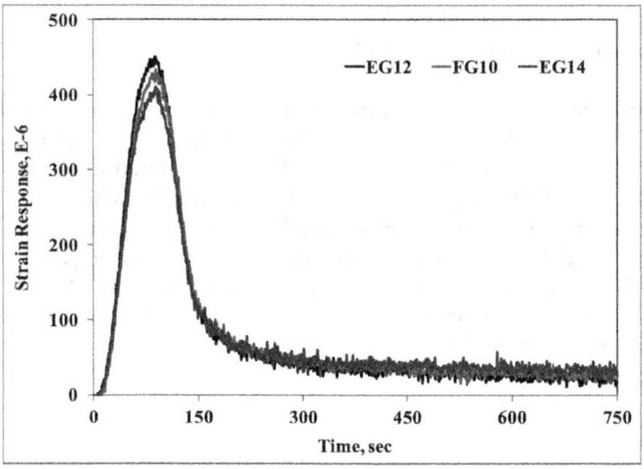

(b) FG vs. EG responses.

Figure 4. *Fiber optic strain sensor.*

6.2 Temperature Analysis

Through the course of full-scale testing, the TESS temperature unit was programmed to maintain the AC-PCC interface (overlay bottom) temperature at 32°F. As shown in Figure 5, except for the beginning of the Phase III test, the target temperature was successfully achieved through the course of testing. There were some temperature deviations in the upper portion of the overlay while the overlay bottom temperatures were almost identical (solid and dash lines lay on top of each other) for both control and interlayer sections. This dissimilarity in the first 2000 loading cycles was most likely caused by the different thermal conductivity of the standard P-401 PG 64-22 and the interlayer PG 76-22 mixes. Afterwards the divergence between the control and interlayer temperatures became more noticeable. This change coincided with the occurrence of the first through bottom-up crack which was captured by SG 7 on the interlayer section after 1882 loading cycles. Therefore, it is reasonable to believe that the heat transfer in a fractured structure had a significant impact on the pavement temperature. Statistical analysis (t-Test) was performed to determine if the control and interlayer temperatures had equal population means. Both populations were assumed to follow a normal distribution. Temperature records collected from the control and interlayer sections at the same depth and same time were treated as a paired sample and the variances of the two samples was assumed to be equal. A complete summary of t-Test results is provided in Table 3. The t value and critical t value suggested that the difference of bottom temperatures between the control and interlayer sections was not statistically significant. But for the mid-depth and surface temperatures, the difference between the control and interlayer sections was highly statistically significant (extremely small two-tailed P-value).

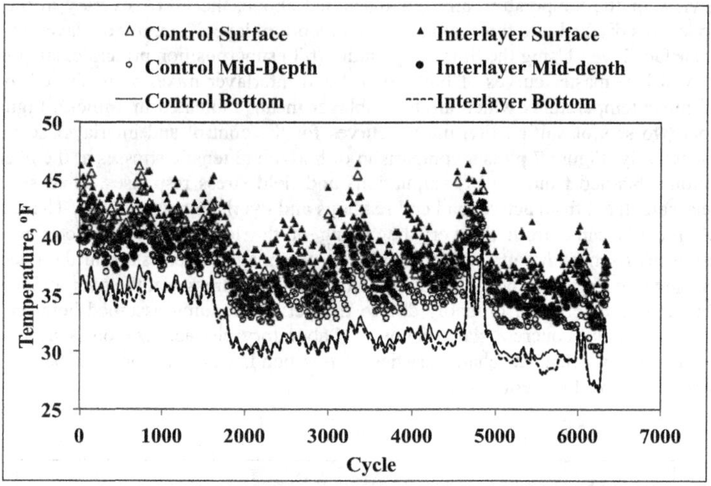

Figure 5. *Overlay temperature records.*

Table 3. *Summary of t-test results.*

Overlay Temperature	Surface		Mid-depth		Bottom	
Statistics	Control	Interlayer	Control	Interlayer	Control	Interlayer
Mean	38	40	35	37	32	32
Variance	7.7	7.4	6.9	6.1	6.4	6.4
Observations	6350					
Pooled Variance	7.5		6.5		6.4	
Hypothesized Mean Difference	0					
Df	12698					
t Stat	-8.29		-8.46		2.02	
P(T<=t) one-tail	0.00		0.00		0.02	
t Critical one-tail	1.65		1.65		1.65	
P(T<=t) two-tail	0.00		0.00		0.04	
t Critical two-tail	1.96		1.96		1.96	

6.3 Comparison of FE and Field Responses

In view of the temperature analysis discussed above, the HMA overlay in the FE model was divided into three sublayers: 1-in. bottom layer, 2-in. middle layer, and 2-in. surface layer. Using the time temperature (t-T) superposition principle, laboratory derived E(t) master curves of both control and interlayer mixes were shifted to the reference temperatures which are the sublayer means tabulated in Table 3. Figure 6 plots two sets of shifted E(t) master curves for the control and interlayer sections, respectively. Figure 7 plots a comparison of horizontal tensile stresses at the overlay bottom obtained from the FE simulations and field stress responses. Field stresses were calculated from active load cell readings and overlay cross sections. These load cell records came from the very beginning of the loading cycles; therefore, the responses from both sources represented an undamaged stress state in the overlay. As shown in Figure 7, the FE model predicted smaller stresses at the overlay bottom. This underprediction is probably due to the perfect bonding assumed between the overlay and the concrete slabs below it. Laboratory characterization is needed to evaluate the mechanical characteristics of this bond, which in turn can be used to further improve FE predictions.

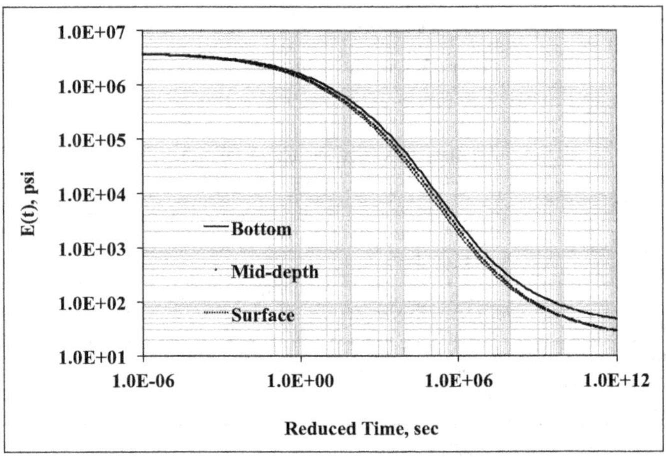

(a) Control section

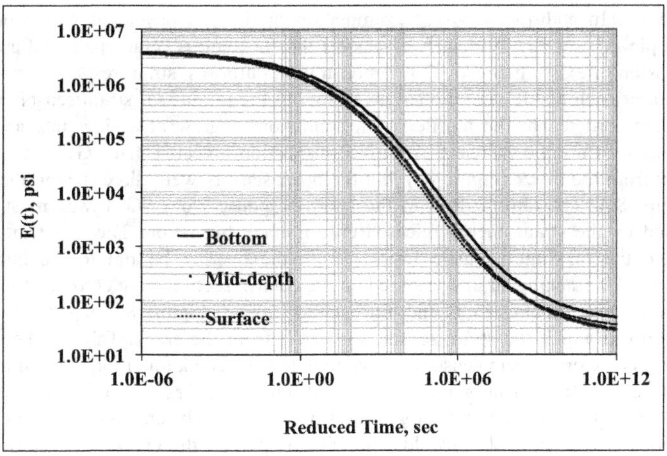

(b) Interlayer section

Figure 6. *E(t) master curves shifted to field temperatures.*

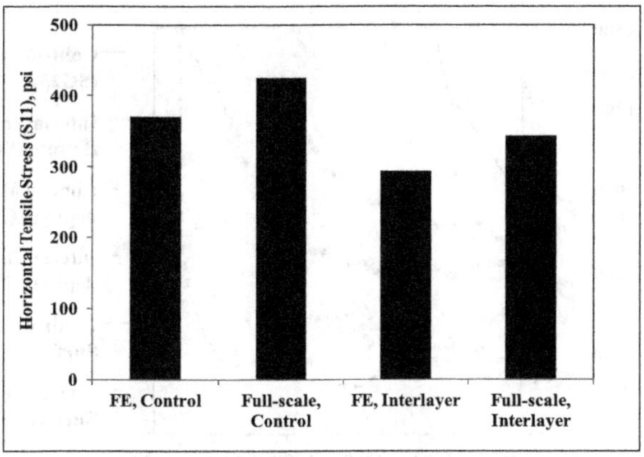

Figure 7. *Comparison of FE and field responses.*

6.4 Crack Initiation

In view of previous successful experiments (Yin and Barbagallo, 2013; Yin and Barbagallo, 2014), strain responses were used as the primary tool to determine crack initiation. In addition, a visual examination of the test pavement was conducted multiple times a day to identify cracks off the "best guess" crack path and trace the extension of existing cracks. For demonstration purposes, strain responses from one sensor at each depth were plotted, as shown in Figure 8a. All strain sensors except SG11 recorded that the tensile strain continuously grew at a slow rate and then experienced a rapid increase. A surface inspection revealed that SG11 was 0.5 in. away from the crack (Figure 8b), but the other sensors were directly atop the crack (Figure 8c). The abrupt drop in the strain response from SG11 was most likely caused by the strain energy released during crack formation. The intact interlayer shifted the maximum tensile stresses from the overlay bottom to the interlayer bottom. All cracking strains recorded on the control section were consistent with the previous experiments (Yin and Barbagallo, 2014). Figure 8d depicts the third scenario in which the strain sensor was offset from the crack. Table 2 shows that SG9 was the only strain censor not able to detect the crack initiation. Monismith and Coetzee (1980) and Button and Lytton (1987) reported that, due to its low stiffness, the interlayer will exhibit large deformations, which are accompanied by a dissipation of energy that would otherwise be part of the crack propagation. This concept is demonstrated in Figure 8a, as the strain sensor at the overlay bottom registered much higher tensile strains on the interlayer section. When the bottom-up cracks propagated into the upper portion of the overlay, the soft interlayer effect on the strain responses became less noticeable.

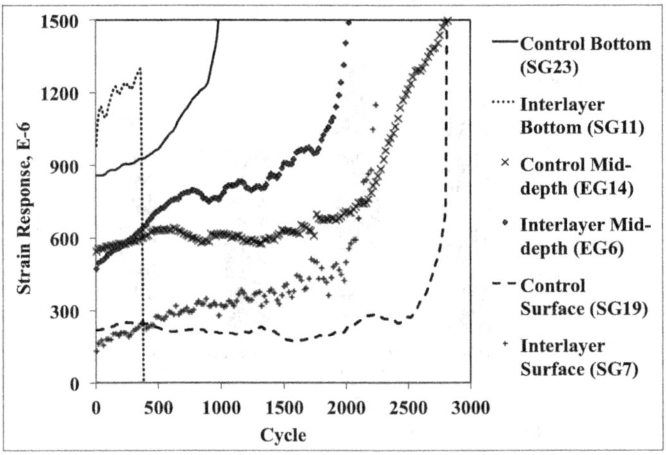

(a)

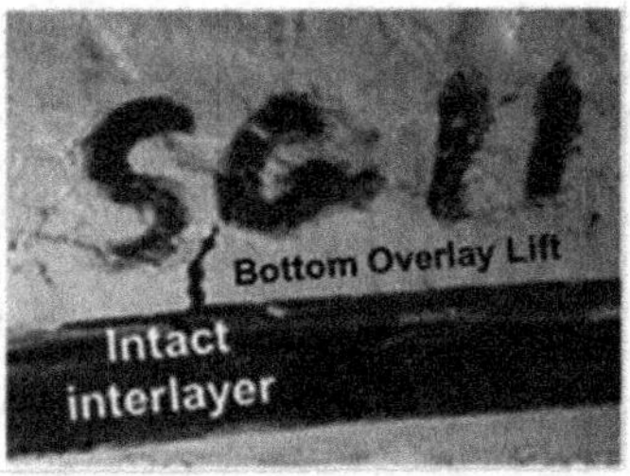

(b)

(c)

(d)

Figure 8. *Crack initiation.*

6.5 Crack Propagation

The first through crack on the control section was captured by SG23, EG16, and SG19. On the interlayer section, strain responses from SG11, EG8, and SG7 indicated the earliest bottom-up crack arriving at the surface. Because the control and interlayer sections had different overlay thicknesses (5-in. vs. 4-in.), the relationship between normalized crack length and number of cycles is shown in Figure 9a. For both control and interlayer sections, the crack length developed gradually at the beginning, and its propagation rate became higher and higher with the crack growth. At the middle of the overlay, the crack propagation stage underwent a transition from quiescent to aggressive. It is also evident in Figure 9a that the interlayer section required constantly more load repetitions to penetrate the crack through the overlay. To penetrate reflection cracks through the full-depth of overlay, additional 155 loading cycles were required for the interlayer section. This extension was roughly 15% of the overlay service life. Note that this increased life still needs to be justified by the cost increase of using an interlayer instead of standard P-401 materials. The effectiveness of the strain relieving interlayer was more pronounced at an early stage of crack propagation and slowly diminished as the crack length increased.

Figures 9b and 9c show the overlay failure for the interlayer and control sections, respectively. On the interlayer section, the crack started above the intact interlayer. At first, the crack propagated upward, indicating fracture Mode I dominance. Once this vertical crack reached the 0.25-in. benchmark, it progressed in the horizontal direction, departing from the concrete joint until it reached the 1.0-in. benchmark. In the final stage, the vertical crack progressed gradually on its initial path, and the horizontal crack started to deviate at an angle toward the surface. After 3117 cycles, the diagonal crack made an appearance on the overlay surface, 3.5 in. offset from the joint. At the same time, the vertical crack barely reached the 0.5-in. benchmark above the intact interlayer. It can be concluded that the fracture on the interlayer section was a mixed mode. The interlayer did not completely fail after 3232 cycles; therefore the intact interlayer had prevented spalling and moisture infiltration at the joint and therefore prolonged the structural integrity of the pavement. On the control section, the reflection crack initiated from the overlay bottom and progressed on its upright track. The horizontal tensile stresses seemed to be the principal driving force for the overlay fracture in this case.

The above paragraph describes the bottom-up crack propagation in a vertical direction. It is worth noting that the reflection cracks had a tendency to develop across the overlay (along the joint) as well. When a Mode I crack formed and developed to a critical length, the energy release rate decreased. This decreasing driving force was due to the presence of compressive stresses in the inner portion of the overlay. As a result, in addition to the advance in the through-thickness direction, a bottom-up crack can advance in other orientations. This channeling phenomenon is reflected by the cycles to failure listed in Table 2. Garzon et al. (2010) demonstrated this channeling in an AC overlay structure with FE simulations. Regardless of the depth, the crack occurrence was constantly observed earlier in the outer portion of the overlay than the inner portion.

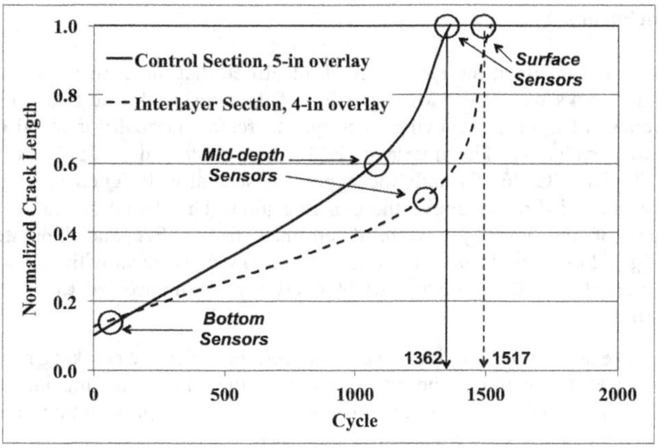

(a) Crack evolution.

(b) Failure edge of interlayer section.

(c) Failure edge of control section.

Figure 9. *Crack propagation.*

7.0 Conclusions

Previous experiments at the FAA NAPTF identified that the critical zone to control bottom-up cracks was at the lower portion of the overlay. In the Phase III test the effectiveness of a strain relieving interlayer to retard thermally-induced reflection cracks was evaluated. The interlayer HMA mix contained a highly polymerized asphalt binder (PG 76-22) and fine aggregate specially designed to absorb local strains generated directly above the concrete joint. The thickness and viscoelastic properties of the interlayer were determined from FEA and laboratory tests accordingly. Full-scale tests were conducted on HMA overlays with and without an interlayer. Side-by-side comparisons of overlay performance led to the following conclusions:

1. The strain relieving interlayer enhanced the reflective cracking resistance of an HMA overlay. The effectiveness of the strain relieving interlayer was more pronounced at an early stage of crack propagation and slowly diminished as the crack length increased.
2. Inclusion of a 1-in.-thick interlayer between existing concrete slabs and the overlay extended overlay service life up to 15%. The intact interlayer had prevented spalling and moisture infiltration at the joint and therefore prolonged the structural integrity of the pavement.
3. To realistically characterize the development of bottom-up reflection cracks, both mixed-mode fracture and crack channeling should be considered.

8.0 Acknowledgement

This work was supported by the FAA Airport Technology Research and Development Branch.

9.0 References

ASTM D7313-13, "Standard Test Method for Determining Fracture Energy of Asphalt-Aggregate Mixtures Using the Disk-Shaped Compact Tension Geometry," 2013.

Button, J.W., and Lytton, R.L., "Evaluation of Fabrics, Fibers and Grids in Overlays," *6th International Conference on Structural Design of Asphalt Pavements*, Vol. 1, 1987, pp. 925-934.

Button, J.W., and Lytton, R.L., "Guidelines for Using Geosynthetics with Hot-Mix Asphalt Overlays to Reduce Reflective Cracking," *Transportation Research Record: Journal of the Transportation Research Board*, No. 2004, 2007, pp. 111-119.

Chiawat, N., and Buttlar, W., *Interim Progress Report: Laboratory Evaluation of FAA Reflective Cracking Rig Materials*, University of Illinois at Urbana-Champaign, 2012.

Dave, E.V., Song, S.H., Buttlar, W.G., and Paulino, G.H., "Reflective and thermal cracking modeling of asphalt concrete overlays," *Proc., Advance Characterization of Pavement and Soil Engineering Materials,* 2007.

Dave, E.V., Ahmed, S., Buttlar, W.G., Bausano, J.P., and Lynn, T., "Investigation of Strain Tolerant Mixture Reflective Crack Relief Systems: an Integrated Approach," *Journal of the Association of Asphalt Paving Technologists, Vol. 79,* 2010, pp. 119-156.

Federal Aviation Administration, *Airport Pavement Design and Evaluation*, AC 150/5320-6E, 2009.

Garzon, J., Duarte, C.A., and Buttlar, W., "Analysis of Reflective Cracks in Airfield Pavements using a 3-D Generalized Finite Element Method," *Road Materials and Pavement Design,* Vol 11, Issue 2, 2010.

Lytton, R.L., "Use of Geotextiles for Reinforcement and Strain Relief in Asphalt Concrete," *Geotextiles and Geomembranes*, Vol. 8, 1989, pp. 217-237.

Minhoto, M.J.C., Pais, J.C., Pereira, P.A.A., Picado-Santos, L.G., "The Influence of Temperature Variation in the Prediction of the Pavement Overlay Life," *Road Materials and Pavement Design*, Vol 6, Issue 3, 2005, pp. 365-384.

Minhoto, M.J.C., Pais, J.C., and Pereira, P.A.A., The Temperature Effect on the Reflective Cracking of Asphalt Overlays, *International Journal of Road Materials and Pavement Design*, Vol 9, Issue 4. 2008, pp. 615-632..

Monismith, C.L., and Coetzee, N.F., "Reflection Cracking: Analysis, Laboratory Studies and Design Considerations," *Journal of the Association of Asphalt Paving Technologists*, Vol. 49, 1980.

Texas DOT (2009) *Tex-248-F: Test Procedure for Overlay Test.*

Von Quintus, H.L., Mallela, J., Weiss, W., and Shen, S., *Techniques for Mitigation of Reflective Cracks*, Technical Guide AAPTP 05-04, 2009.

Yin, H., and Barbagallo, D., "Development of Full-Scale Reflective Cracking Test at FAA NAPTF," *Transportation Research Record: Journal of the Transportation Research Board,* No.2368, 2013, pp. 81-91.

Yin, H., and Barbagallo, D., "Full-Scale Test of Thermally-Induced Reflective Cracking," *TRB 93rd Annual Meeting Compendium of Papers*, 2014.

Yin, H. "Finite Element Modeling and Analysis of Full-Scale Reflective Cracking Test," *International Journal of Roads and Airports,* 3(1), 2013a, pp. 1-15.

Yin, H., "The Impact of Strain Gage Instrumentation on Localized Strain Responses in Asphalt Concrete Pavements," *International Journal of Pavement Research and Technology*, Vol. 6, No. 3, 2013b, pp. 225-234.

Comparison of Laboratory Cracking Test Results and Field Performance

Wangyu Ma[a], Nam H. Tran[a], Adam J. Taylor[a], J. Richard Willis[a] and Mary M. Robbins[a]

[a]*National Center for Asphalt Technology at Auburn University, Auburn, AL 36830*

ABSTRACT. *Several tests have been developed over the years to evaluate the fatigue cracking resistance of asphalt mixtures. Two of these tests are the four-point bending beam fatigue test (BBFT) and overlay test (OT). The BBFT has been used to determine the fatigue cracking resistance of asphalt mixtures. The OT was originally developed to evaluate the resistance of asphalt overlays to reflective cracking but was further refined and evaluated for determining the fatigue cracking resistance of asphalt mixtures. The objective of this study was to evaluate the two laboratory tests based on the field performance of the base-layer mixtures used in five structural sections as part of the fourth (2009) research cycle of the NCAT Pavement Test Track. The BBFT and OT results were first used to rank the five asphalt mixtures for comparison. Then, for each test, a transfer function was fitted to the test results to determine the number of cycles to failure at the temperature-corrected strain measured in the field. Finally, the number of cycles to failure determined at the temperature-corrected field strain was used to re-rank the five mixtures for comparison with the field performance at the NCAT Pavement Test Track. The results for this study showed that the rankings of the five mixtures based on the OT results at each maximum opening displacement (0.381, 0.318, or 0.254 mm) were similar to those by the BBFT results at 800 and 400 microstrain but different from those at 200 microstrain. The rankings of the five mixtures based on the number of ESALs applied until the first crack was observed on the surface was the same as the rankings based on the OT results but different from those based on the BBFT results, both of which were determined at the temperature-corrected strain measured in the field. These findings indicate the potential use of the OT to characterize the relative fatigue cracking resistance of asphalt mixtures. Further evaluation using different pavement structures and mixtures is necessary to validate these findings.*

KEYWORDS: *fatigue cracking, field performance, cracking test, transfer function, ranking*

The oral presentation was made by Wangyu Ma.

1.0 Introduction

Several tests have been developed over the years to evaluate the fatigue cracking resistance of asphalt mixtures. Two of these tests are the four-point bending beam fatigue test (BBFT) and overlay test (OT). The BBFT has been used to determine the fatigue cracking resistance of asphalt mixtures. The OT was originally developed by Germann and Lytton (1979) to evaluate the resistance of asphalt overlays to reflective cracking. It was further refined by the researchers at the Texas Transportation Institute (Zhou and Scullion, 2003, Zhou and Scullion, 2005) and evaluated for determining the fatigue cracking resistance of asphalt mixtures Zhou et al., 2007b).

In a bending beam fatigue test, a beam specimen (380 ± 6 mm long by 63 ± 2 mm wide by 50 ± 2 mm thick) can be trimmed from a laboratory- or field-compacted beam specimen. The loading device is kept in an environmental chamber set at 20 ± 0.5°C. The loading frequency ranges from 5 to 10 Hz. Two waveforms can be applied: sinusoidal in accordance with AASHTO T 321 and haversine in compliance with ASTM D7460. The maximum deflection at the center of the specimen is maintained by a closed-loop control system, resulting in the maximum tensile strain typically ranging from 200 to 800 microstrain. The result of this test is the number of cycles to failure, which can be multiplied by a shift factor to determine the fatigue life of the asphalt pavement under repeated traffic loads (Prowell, 2010). Several shift factors have been suggested by researchers (Harvey et al., 1997; Pierce and Mahoney, 1996) depending on the pavement condition.

The OT has been used to evaluate the cracking resistance of asphalt mixtures in the mix design process and in the proposed asphalt overlay thickness design and analysis tool in Texas (Hu et al., 2011; Hu et al., 2010; Walubita et al., 2012; Hu et al., 2008). As specified in the Texas Department of Transportation (TxDOT) procedure (Tex-248-F-09), a test specimen (150 ± 2 mm long by 76 ± 0.5 mm wide by 38 ± 0.5 mm thick) is subjected to repeated tensile loads until failure. The test is conducted at 25°C and 0.1 Hz using a maximum opening displacement (MOD) of 0.635 mm. The number of cycles at which the applied peak load reduces by 93% is defined as the number of cycles to failure.

In a previous study Zhou et al., 2007b), the rankings of six asphalt mixtures based on their performance in the Federal Highway Administration's Accelerated Loading Facility (FHWA-ALF) were similar to the those determined by the OT results. The authors also compared the test results of the BBFT (under 374 uε) and the OT for three asphalt mixtures and found that the rankings of the three mixtures were the same for the two tests. They suggested that the OT could be an alternative test for determining the fatigue cracking resistance of asphalt mixtures and that further evaluation was needed.

In 2009, the National Center for Asphalt Technology (NCAT) Pavement Test Track located in Opelika, Alabama was reconstructed for the fourth research cycle. In this research cycle, five 200-foot sections were fully instrumented for a structural study. Instrumentation in each section of the test track included an array of strain gauges, temperature probes, and pressure plates. The strain gauges installed at the bottom of asphalt layer measure the longitudinal and transverse tensile strain under traffic loading. The temperature probe measures the mid-depth temperature. The pressure plates measure the vertical pressure on the base and subgrade (Timm et al., 2011). These sections were built on the two tangents of the 1.7-mile closed-loop accelerated pavement research facility and subjected to approximately 10 million ESALs in the two-year trafficking period. The five base-layer mixes placed in these sections were good candidates for evaluating the BBFT and OT.

2.0 Objectives and Scope

The objective of this study was to evaluate the two laboratory tests based on the field performance of the base-layer mixtures used in the five structural sections as part of the fourth (2009) research cycle of the NCAT Pavement Test Track.

In this study, the BBFT and OT results were first used to rank the five asphalt mixtures for comparison. Then, for each test, a transfer function was fitted to the test results to determine the number of cycles to failure at the temperature-corrected strain measured in the field. Finally, the number of cycles to failure determined at the temperature-corrected field strain was used to re-rank the five mixtures for comparison with the field performance at the NCAT Pavement Test Track.

3.0 Experimental Plan

3.1 Material Properties

Five mixtures used in the bottom asphalt layers of the five test sections (S9-3, S10-3, S11-3, N10-3, and N11-3) of the Group Experiment in the 2009 research cycle of the NCAT Test Track were selected for this study. These sections were all designed with 7 in. of asphalt layers on top of 6 in. of granite base atop of the test track soil subgrade. Table 1 shows the as-built properties of all the mixtures used for preparing overlay test and bending beam fatigue test specimens in this study. The mixtures were plant-produced and sampled during the construction of these test sections. The S9-3 mixture served as the control mixture in this study and was produced hot without reclaimed asphalt pavement (RAP). The S10-3 and S11-3 mixtures, which used the same mix design without RAP as the S9-3 mixture, were produced warm using foaming and additive WMA, respectively. The N10-3 and N11-3 mixtures were produced based on a mix design with 50% RAP. The N10-3 mixture was produced hot, and the N11-3 mixture was produced warm using foaming WMA. The 50% RAP in the mixtures included 20% fine fraction and 30% coarse fraction material. More information about the mixtures used in this study is available elsewhere (Powell, 2013).

Table 1. *As-built mixture properties.*

Properties	S9-3	S10-3	S11-3	N10-3	N11-3
NMAS (mm)	19	19	19	19	19
Virgin Asphalt PG Grade	67-22	67-22	67-22	67-22	67-22
% RAP	0	0	0	50	50
WMA	No	Foam	Additive	No	Foam
Air Voids, %	3.6	4.1	3.0	4.2	4.1
Total Combined Binder (P_b), % wt	4.9	4.7	5.0	4.7	4.6
Effective Binder (P_{be}), % wt	4.4	4.2	4.5	4.1	4.0
Dust Proportion (DP)	1.2	1.2	1.2	1.4	1.3
Maximum Specific Gravity (G_{mm})	2.532	2.533	2.522	2.537	2.544
VMA, %	14.0	14.0	13.7	13.8	13.7
VFA, %	75	71	78	70	70

3.2 Laboratory Testing Plan

In this study, the overlay test was conducted in the AMPT using an overlay test kit. A test specimen can be cut from a gyratory compacted specimen or a field core, epoxied to a set of two steel plates, and loaded vertically in the testing chamber. During the test, the top plate remained fixed while a cyclic triangle-wave load was applied to the bottom plate to achieve the target maximum opening displacement (MOD) between the two base plates at the peak of the waveform. The recommended MOD (0.635 mm) in the TxDOT procedure was originally selected for evaluating reflective crack resistance of asphalt mixtures used in overlays on top of old concrete pavements in Texas. However, the testing in this study was used to evaluate the fatigue cracking resistance of base-layer asphalt mixtures. Therefore, the maximum opening displacement needed to be lower so the number of cycles to failure for each mixture would not be too small to make effective comparisons. As shown in Table 2, three MOD levels (i.e., 0.381, 0.318, and 0.254 mm) were selected to evaluate the relationships between MOD and the number of cycles to failure. At lower MOD levels, the overlay test takes considerably longer to perform (especially at the MOD of 0.254 mm). Therefore, the overlay test was conducted at a higher frequency (i.e., 1 Hz) to reduce the testing time (Ma et al., 2014). Bending beam fatigue testing was conducted following the AASHTO T 321-07 standard (except haversine loading was utilized) at three maximum tensile strains (i.e., 800, 400, and 200 uε). Each mixture was tested using three replicates (or tests) for both the beam fatigue and the overlay test. All specimens (both beam and overlay) were produced to a target air void content of 7.0 ± 1.0% after the cut. The beam fatigue test was performed at 20°C (AASHTO T 321) and the overlay test was performed at 25°C (Tex-248-F). It should be noted that the BBFT results were completed and reported in another project (Timm, et al., 2011). The test results based on the testing plan are presented later in this paper.

Table 2. *Summary of laboratory testing plan.*

Section	Mixture	OT (at 1 Hz, 25 °C)		BBFT (at 10 Hz, 20 °C)	
		Max. Opening Displacement (mm)	**No. of Tests**	**Max. Tensile Strain (uε)**	**No. of Tests**
S9-3	Control (HMA)	0.381/0.318/0.254	3/3/3	800/400/200	3/3/3
S10-3	WMA Foam	0.381/0.318/0.254	3/3/3	800/400/200	3/3/3
S11-3	WMA Additive	0.381/0.318/0.254	3/3/3	800/400/200	3/3/3
N10-3	50% RAP	0.381/0.318/0.254	3/3/3	800/400/200	3/3/3
N11-3	50% RAP+Foam	0.381/0.318/0.254	3/3/3	800/400/200	3/3/3
Total Number of Tests		90			

4.0 Laboratory Testing and Field Evaluation Results

4.1 Overlay Test Results

The TxDOT procedure defines the test result (i.e., the number of cycles to failure) as the number of cycles at which the applied peak load is reduced by 93% compared to the applied peak load measured at the beginning of the test. However, cracks were often seen propagating through the entire thickness of the specimen long before the applied load was reduced by 93% (Ma et al., 2014). Therefore, an alternative method, referred to as the "normalized load × cycle" or "NLC", was used to determine the number of cycles to failure in this study. This method was developed based on the "normalized stiffness × cycle" method described in the ASTM D7460 standard for beam fatigue testing. The number of cycles to failure determined from this method was found right before the moment when crack just propagates through the thickness of specimen. Development and analysis of this method were documented elsewhere (Ma, 2014). Typically, this method determines the number of cycles to failure in the following three steps:

1. The "normalized load × cycle" is calculated using Equation 1 for each loading cycle;
2. The NLC is plotted against the number of cycles;
3. The number of cycles to failure is determined corresponding to the peak of the NLC curve as shown in Figure 1.

$$NLC = \frac{P_i \times N_i}{P_1 \times N_1} \qquad [1]$$

Where: NLC = normalized load × cycles, kN/kN
P_i = peak load at cycle number i, kN
P_1 = peak load at 1st cycle, kN
N_i = cycle number i
N_1 = 1(first cycle)

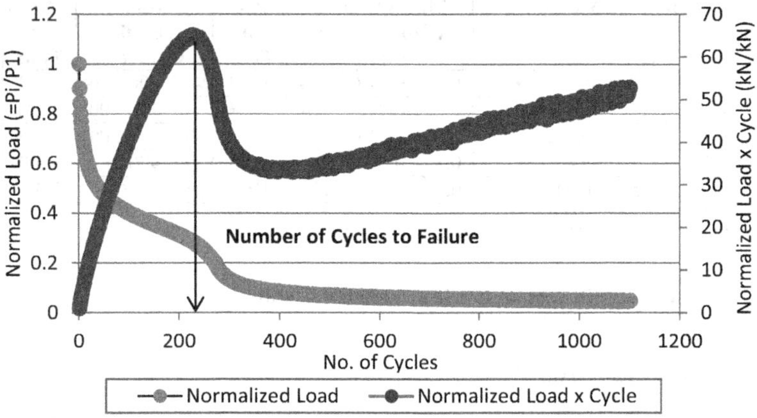

Figure 1. *Determination of the number of cycles to failure based on NLC method.*

By using the NLC method, the results of the individual overlay tests for the five mixtures were determined and summarized in Table 3. A statistical summary of the results is provided in Table 4. All the tests were conducted at 1-Hz loading frequency and the 25°C testing temperature. As shown, the number of cycles to failure increases when MOD decreases from 0.381 mm to 0.254 mm.

Table 3. *Overlay test results.*

MOD (mm)	Replicate ID	No. of Cycles to Failure				
		S11-3	S10-3	S9-3	N11-3	N10-3
0.381	1	1,372	1,110	895	512	96
	2	807	761	669	548	71
	3	931	784	722	499	253
0.318	1	2,028	1,876	1,672	530	276
	2	2,680	2,406	2,108	686	301
	3	3,672	1,804	1,372	806	467
0.254	1	4,044	5,520	5,756	4,948	936
	2	5,132	3,428	7,204	4,988	1,824
	3	5,132	4,300	9,076	2,848	2,072

Table 4. *Statistical summary of overlay test results.*

Mixture	MOD (mm)	Number of Cycles to Failure		
		[1]Average	[1]Std. Dev.	COV
S11-3	0.381	1,037	297	29%
	0.318	2,793	828	30%
	0.254	4,769	628	13%
S10-3	0.381	885	195	22%
	0.318	2,029	329	16%
	0.254	4,416	1,051	24%
S9-3	0.381	762	118	16%
	0.318	1,717	370	22%
	0.254	7,345	1,665	23%
N11-3	0.381	520	25	5%
	0.318	674	138	21%
	0.254	4,261	1,224	29%
N10-3	0.381	140	99	70%
	0.318	348	104	30%
	0.254	1,611	597	37%

[1]*Average and standard deviation with the unit of cycle were rounded up to the nearest integer.*

4.2 Bending Beam Fatigue Test Results

Bending beam fatigue tests for these five mixtures were completed in accordance with AASHTO T 321-07 standard. Beam specimens were tested at three strain levels (i.e., 800, 400, and 200 uε) until the stiffness of each specimen reached 50% of the initial value (i.e., stiffness at 50th loading cycle). The individual specimen results are given in Table 5 and a statistical summary of the results is provided in Table 6. As shown in Table 5, the number of cycles to failure determined from BBFT is much larger than those from OT listed in Table 3. This was expected because the applied maximum tensile strain on the beam specimen is much smaller than that on overlay test specimen. The relationship between MOD and applied strain in overlay test is discussed later in this paper.

Table 5. *Bending beam fatigue test results.*

Strain (uε)	Rep. ID	Number of Cycles to Failure				
		S11-3	S10-3	S9-3	N11-3	N10-3
800	1	9,090	3,610	7,890	2,560	3,320
	2	14,670	6,980	4,260	3,100	2,950
	3	7,720	16,850	17,510	2,100	680
400	1	114,370	157,270	201,060	101,150	13,480
	2	185,490	261,010	216,270	146,210	31,460
	3	299,680	135,930	141,250	124,920	112,630
200	1	1,746,710	4,233,170	6,953,800	53,100,000	15,200,000
	2	4,789,300	7,244,350	5,994,840	54,500,000	9,797,400
	3	4,594,330	365,870	2,165,480	4,501,250	3,328,290

Table 6. *Summary of BBFT results at three strain levels for the five mixtures.*

Mixture	Strain (με)	Number of Cycles to Failure		
		[1]Average	[1]Std. Dev.	COV
N11-3	800	2,587	501	19%
	400	124,094	22,542	18%
	200	37,367,084	28,471,254	76%
S9-3	800	9,887	6,874	70%
	400	186,194	39,659	21%
	200	5,038,040	2,533,496	50%
N10-3	800	2,317	1,430	62%
	400	52,524	52,825	101%
	200	9,441,897	5,943,834	63%
S10-3	800	9,147	6,881	75%
	400	184,737	66,911	36%
	200	3,947,797	3,448,108	87%
S11-3	800	10,494	3,682	35%
	400	199,847	93,486	47%
	200	3,710,114	1,703,150	46%

[1]*Average and standard deviation with the unit of cycle were rounded up to the nearest integer.*

4.3 Cracking Performance on the Test Track

Pavement performance on the test track has been routinely examined and recorded, providing great benefit for performance-related investigations. The five sections focused on in this study were placed in the 2009 research cycle and performed well during the two-year trafficking period. No fatigue cracking was evident when the research cycle was completed. Therefore, these sections were left in place for the following 2012 research cycle, during which fatigue cracking became visible at the pavement surface. In this study, two factors were considered to represent the cracking performance on the test track. The first factor was the number of equivalent single-axle loads (ESALs) applied when crack appeared on the pavement surface (1% crack area of the section was assumed as the starting point). The sections exhibiting cracking earlier are more susceptible to fatigue cracking and tend to have mixtures with less cracking resistance in the bottom asphalt layer. Table 7 lists the exact date and the number of ESALs applied when 1% crack area was exhibited on each of the five sections. Section S11 started to exhibit cracking at the beginning of 2012 research cycle when an additional 584,991 ESALs was applied after 2009 research cycle was completed. Then, cracks started to appear in other sections (with 1% crack area) following the order of S10, N11, S9, and finally N10 on November 18, 2013.

Table 7. *Date and ESALs when 1% crack area was exhibited on the sections.*

Section	Date (mm/dd/yyyy)	No. ESALs Applied[1]
S11	12/03/2012	10,727,131
S10	01/14/2013	11,178,452
S9	03/11/2013	11,952,663
N11	02/20/2013	11,736,631
N10	11/18/2013	15,547,240

[1] *Cumulative ESALs since August 27, 2009*

The second factor used to represent the cracking performance was the cracking area measured in the summer 2013. All the measurements were conducted on the same day (August 5, 2013). Crack maps of the five test track sections are shown in Figures 2 through 6 (Powell, 2013). Each map indicates the distribution and the amount of cracking within a 200-foot section. The dashed line shows the traffic direction and the red lines represent the cracking. A rectangle plotted at the far right of the map in Figure 6 indicates a patch on the pavement. Also, an observation can be made that the transverse fatigue cracking dominates within each cracked section. Since this study was focused on transverse fatigue cracking, these crack maps were appropriate to use for comparison with fatigue test results in the laboratory. Later measurements in the spring of 2014 showed extensive cracks on these sections except for N11. Longitudinal cracks (recently found to be top-down cracking) became interconnected with the transverse fatigue crack by that time. Therefore, crack maps recorded in the spring 2014 were not used for the comparisons in this study. As shown in Figures 2 through 6, sections S11 and S10 showed the heaviest cracking; sections S9 and N11 exhibited intermediate cracking; and section N10 showed no cracking. Table 8 summarizes the crack area of the five sections showed in the crack maps (Figures 2 through 6).

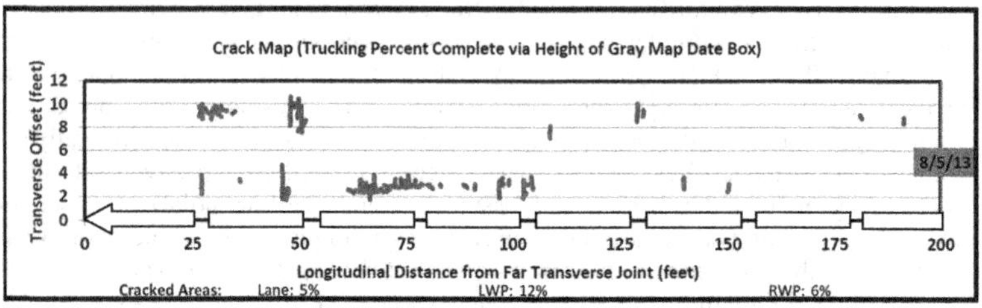

Figure 2. *S11 section crack map - 8/5/13.*

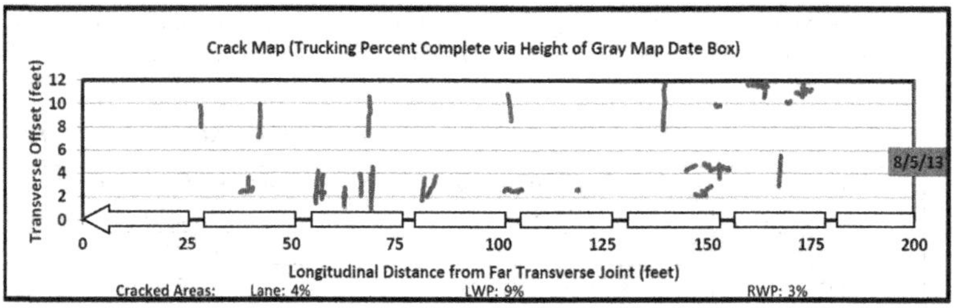

Figure 3. *S10 section crack map - 8/5/13.*

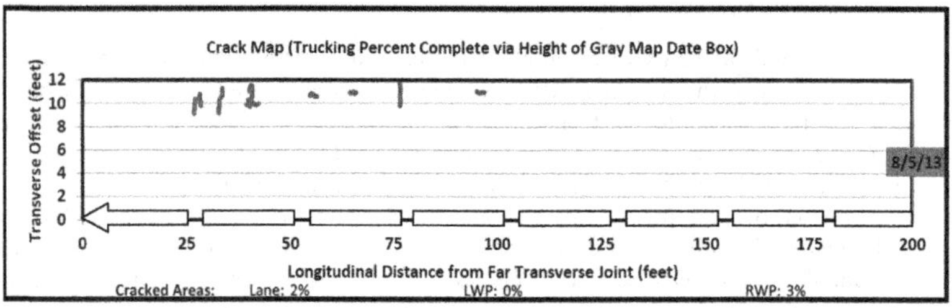

Figure 4. *S9 section crack map - 8/5/13.*

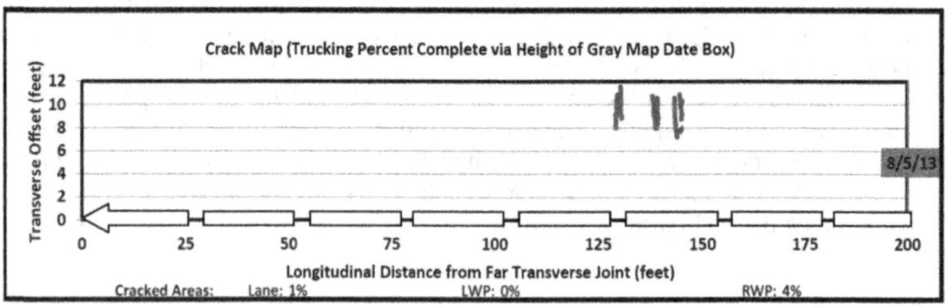

Figure 5. *N11 section crack map - 8/5/13.*

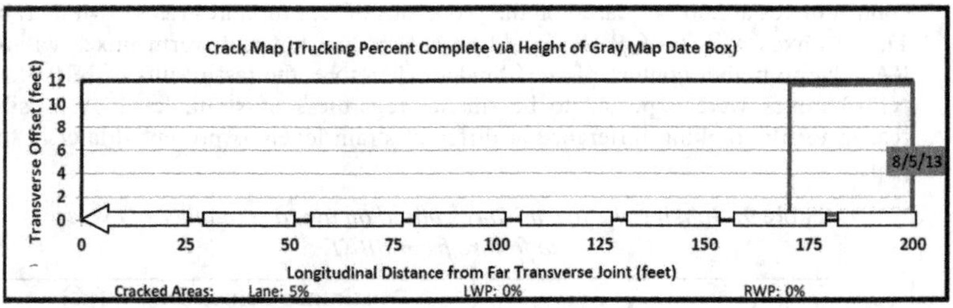

Figure 6. *N10 section crack map - 8/5/13.*

Table 8. *Approximate crack area in five test sections – 8/5/13.*

Section	Lane (%)[1]	LWP (%)[2]	RWP (%)[3]
S11	5	12	6
S10	4	9	3
S9	2	0	3
N11	1	0	4
N10	0	0	0

[1]*crack area percentage of the entire lane (or section)*
[2]*crack area percentage of the left wheel path (or inner wheel path)*
[3]*crack area percentage of the right wheel path (or outer wheel path)*

5.0 Ranking Mixtures Based on Overlay Test and Bending Beam Fatigue Test Results

Both the overlay test and bending beam fatigue test have been studied to determine the fatigue cracking resistance of asphalt mixture (Zhou et al., 2007a). The results of the two tests were used to rank the crack resistance of the five mixes. The ranking by bending beam fatigue test results is discussed first, followed by the ranking by overlay test and their comparisons.

5.1 Ranking Based on Bending Beam Fatigue Test Results

The five mixtures were ranked based on the average results of three replicates tested at each applied maximum strain. Then the correlation analysis of rankings was performed to determine the level of similarity among these rankings.

Table 9 ranks the mixtures based on the average number of cycles to failure from the largest to the smallest (i.e., 1st indicates the largest and 5th indicates the smallest) at each strain level. The ranking at 800 and 400 με are the same. However, the ranking at 200 με is much different from those of the other two strain levels because the mixtures with 50% RAP (N11-3 and N10-3) rank higher than the WMA and control mixes at 200 με. Based on the properties of five mixtures (as shown in Table 1), the mixes with 50% RAP should be stiffer than hot and warm mixes without RAP due to higher content of aged binder. Therefore, the test results of N10-3 and N11-3 mixes were expected to be smaller regardless of strain level. A possible reason for the ranking difference at different strain levels is presented later in this paper.

Table 9. *Ranking the five mixtures based on average number of cycles to failure from BBFT.*

Mixture	Processing Information	Maximum Tensile Strain (με)		
		800	**400**	**200**
N10-3	50%RAP+HMA	5th	5th	2nd
N11-3	50%RAP+WMA Foam	4th	4th	1st
S9-3	0%RAP+HMA (Control)	2nd	2nd	3rd
S10-3	WMA Foam	3rd	3rd	4th
S11-3	WMA Additive	1st	1st	5th

5.2 Ranking Based on Overlay Test Results

The overlay test has been suggested as an alternative to determine the fatigue cracking resistance of asphalt mixture (Zhou et al., 2007a). Table 10 shows the ranking of these mixtures based on the average test result in descending order (i.e., 1st indicates the largest and 5th indicates the smallest) at each MOD. The rankings at 0.381 and 0.318-mm MOD are the same, but the rankings at 0.254-mm MOD for the S9-3, S10-3, and S11-3 mixtures are different. Also, the mixtures with 50% RAP (N11-3 and N10-3 mixes) are ranked lower than the two warm mixes and HMA without RAP at all MOD levels. Unlike with the beam fatigue testing, the lower strain levels in the overlay test did not benefit the 50% RAP and 50% RAP+WMA mixture in terms of cycles to failure. This ranking seems reasonable since the aged binder in the RAP could make the N11-3 and N10-3 mixes much stiffer.

Table 10. *Ranking the five mixtures based on average number of cycles to failure from OT.*

Mixture	Processing Information	Maximum Opening Displacement (mm)		
		0.381	**0.318**	**0.254**
N10-3	50%RAP+HMA	5th	5th	5th
N11-3	50%RAP+WMA Foam	4th	4th	4th
S9-3	0%RAP+HMA (Control)	3rd	3rd	1st
S10-3	WMA Foam	2nd	2nd	3rd
S11-3	WMA Additive	1st	1st	2nd

5.3 Rank-Order Correlation

Spearman's Rank-Order Correlation Coefficient (r_s) was calculated to determine the strength and direction of correlation (or similarity) among six rankings from Table 9 and Table 10. The closer r_s is to zero, the weaker the correlation. The positive or negative sign of r_s value shows whether or not the two rankings are in the same direction (Royal Geographical Society). Table 11 shows the correlation coefficient matrix among these six rankings. This matrix compares the ranking at each overlay test MOD to the ranking at each BBFT maximum tensile strain. Also, it gives the correlations among three rankings within each test.

Table 11. *Spearman's rank-order correlation coefficients.*

	OT at 0.381 mm	**OT at 0.318 mm**	**OT at 0.254 mm**	**BBFT at 800 με**	**BBFT at 400 με**	**BBFT at 200 με**
OT at 0.381 mm	1	1	0.7	0.9	0.9	-0.9
OT at 0.318 mm		1	0.7	0.9	0.9	-0.9
OT at 0.254 mm			1	0.9	0.9	-0.6
BBFT at 800με				1	1	-0.8
BBFT at 400με					1	-0.8
BBFT at 200με						1

A hypothesis test was performed to determine if the ranking correlation truly existed (in this case, the average number of cycles to failure of five mixes was ranked). For a significance level of 0.05, the critical value for r_s is 0.9. That is, the correlation would occur at 95% chance if the coefficient is equal or above 0.9 (Gautheir, 2001). As shown in Table 11, the rankings of the mixes tested at three overlay test MOD levels correlate well with the rankings of the mixes tested at 800 με and 400 με in BBFT ($r_s = 0.9$). However, the rankings of the mixes based on the OT results were almost opposite compared to those based on the BBFT results at 200 με ($r_s = -0.9$ or -0.6). The rankings of the mixes at the three overlay test MOD levels are very similar ($r_s \geq 0.7$). Even though 0.7 is smaller than the critical value of 0.9, it still indicates a positive correlation. However, the BBFT rankings at 200 με are in a reversed order compared to those at the other strain levels. Overall, the OT rankings at each MOD are similar to the BBFT rankings at 800 με and 400 με but different from those at 200 με. The good correlations between the two tests indicate the potential of using the overlay test to characterize fatigue cracking resistance of asphalt mixtures.

6.0 Transfer Function for Predicting the Number of Cycles to Failure

6.1 Transfer Function for BBFT

Figure 7 illustrates the relationship between maximum tensile strain and number of cycles to failure from BBFT for the five asphalt mixtures. Trendlines were fitted to the testing data for each mixture. In the log-log scale, each trendline becomes linear. The slopes of trendlines for S9-3, S10-3, and S11-3 mixtures are different from those of N11-3 and N10-3 mixtures. The number of cycles to failure of high RAP mixes (i.e., N11-3 and N10-3) is higher at the smallest strain level (i.e., 200 με) but lower at the largest strain level (i.e., 800 με) compared to other three mixes. Therefore, Table 9 shows the different rankings of the mixes at 200 με and at 800 με in BBFT. High RAP mixes tend to last longer at lower strain levels but fail quicker than the virgin mixes at higher strain levels. A possible explanation of this difference is that the fatigue resistance of the high RAP mixes is more sensitive to the change in the strain level used in the test.

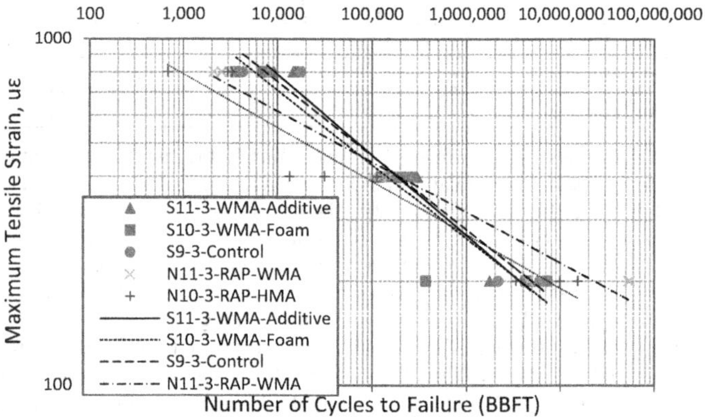

Figure 7. *Maximum tensile strain versus the number of cycles to failure from BBFT for the five mixtures.*

The number of cycles to failure at a given maximum tensile strain in the BBFT can be determined by a fatigue transfer function shown in Equation 2 (Huang, 1993). Table 12 includes the power model (hereafter referred to as transfer function) coefficients determined by fitting Equation 2 to the BBFT testing data for the five mixtures.

$$N_f(BBFT) = \alpha_1 \left(\frac{1}{\varepsilon}\right)^{\alpha_2} \qquad [2]$$

Where: $N_f(BBFT)$ = number of cycles to failure from BBFT

ε = maximum tensile strain in BBFT, uε
α_1, α_2 = transfer function coefficients

Table 12. *Fitted transfer function coefficients from BBFT testing data.*

Mixture	Processing Information	Transfer Function Coefficients		
		α_1	α_2	R^2
S11-3	WMA Additive	1×10^{16}	4.1923	0.9743
S10-3	WMA Foam	7×10^{15}	4.1092	0.8792
S9-3	0%RAP+HMA (Control)	1×10^{17}	4.5321	0.9686
N11-3	50%RAP+WMA Foam	3×10^{22}	6.5846	0.9600
N10-3	50%RAP+HMA	4×10^{20}	6.0192	0.9288

6.2 Transfer Function for OT

Similarly, Figure 8 shows the maximum opening displacement (MOD) versus the number of cycles to failure in OT for the five mixtures. The transfer function in Equation 3 was used to fit the testing data. As shown in Figure 8, the slopes of S10-3 and S11-3 trendlines are different from the slopes of the S9-3, N10-3, and N11-3 mixes. As MOD level changes, the rate of change of number of cycles to failure is higher for the high RAP mixes and the virgin control mix than for the virgin warm mixes. The data suggest that the high RAP mixes and virgin control mix are more sensitive than the virgin warm mixes to the change of MOD level.

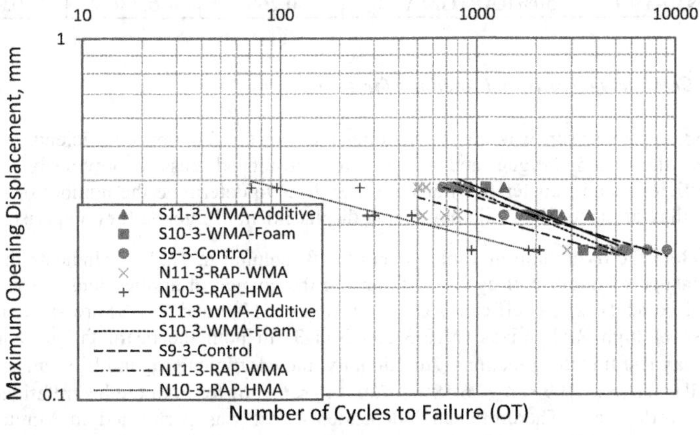

Figure 8. *Maximum opening displacements versus number of cycles to failure from OT for the five mixtures.*

Equation 3 shows the fitted transfer function between number of cycles to failure and the reciprocal of MOD (i.e., 1/MOD). The fitted function can be used to predict the number of cycles to failure at the desired maximum opening displacement. Table 13 lists the transfer function coefficients for the five mixtures. All the R^2 values are larger than 0.80, indicating a good model fit.

$$N_f(OT) = \beta_1 \left(\frac{1}{MOD}\right)^{\beta_2} \qquad [3]$$

Where: $N_f(OT)$ = number of cycles to failure from OT
MOD = maximum opening displacement in OT, mm
β_1, β_2 = transfer function coefficients

Coefficient β_1 is a scaling factor, moving the trendline up or down (e.g., larger β_1 moves the trendline up). Coefficient β_2 governs the rate of growth or decline of the trendline (e.g., smaller absolute value of β_2 corresponds to a lower rate of change). The mixture with smaller absolute value of β_2 in the fitted transfer function is less sensitive to MOD change.

Table 13. *Fitted transfer function coefficients from OT testing data.*

Mixture	Processing Information	Transfer Function Coefficients		
		β_1	β_2	R^2
S11-3	WMA Additive	29.956	3.760	0.87
S10-3	WMA Foam	20.478	3.934	0.93
S9-3	0%RAP+HMA (Control)	3.158	5.601	0.96
N11-3	50%RAP+WMA Foam	2.595	5.233	0.85
N10-3	50%RAP+HMA	0.269	6.286	0.87

6.3 Comparing Transfer Functions for Two Tests

Next, a comparison was made regarding the transfer function coefficients derived from the beam fatigue and overlay test results discussed previously. These coefficients and transfer functions are used later to determine the number of cycles to failure at the tensile strain or opening displacement of interest for comparison.

Both the α_2 column in Table 12 and the β_2 column in Table 13 indicate the rate of change of number of cycles to failure to the change of applied tensile strain (or displacement). The coefficients of warm mixes (S10-3 and S11-3) are smaller than those of high RAP mixes (N10-3 and N11-3) in both the beam fatigue and the overlay test transfer functions. Additionally, the coefficients α_2 and β_2 seem to have similar range ($4.109 < \alpha_2 < 6.585$, $3.760 < \beta_2 < 6.286$) for both the beam fatigue and the overlay test. Therefore, an independent-t test was performed in Minitab 15 software to test the equality of the means of two coefficients for the five mixes. The Minitab output is documented elsewhere (Ma, 2014). The analysis result showed that the mean of α_2 and the mean of β_2 are statistically no different at significance level of 0.05 (p-value = 0.724). Even though the assumption of equal variances was satisfied, the power of t-test was weakened due to the small sample size (n=5)

(Montgomery, 2009). To obtain a more reliable conclusion for the comparison of the slope coefficients from the beam fatigue and the overlay tester, more types of mixtures need to be tested in the future.

7.0 Comparing Ranking Orders Based on Lab Results and Field Performance

For each unique transfer function, the number of cycles to failure at the temperature-corrected field strain was determined. The predictions were compared to the cracking performance on the test track, which were represented by the two selected factors: the number of ESALs applied until 1% crack area appears in the section and the crack area percentage of the section measured on August 5, 2013.

7.1 Determining the Strain Input for the Transfer Function

The transfer function for each test can be used to predict the number of cycles to failure at the strain level of interest. To compare with the test track performance, the measured strain at the bottom of the asphalt layer corrected for the testing temperature (i.e., 20°C for BBFT or 25°C for OT) was applied. Previous studies have showed the transverse strains were only about 66% of longitudinal strains (Priest and Timm, 2006; Timm and Priest, 2008). Also, a recently conducted investigation by coring on these sections indicated the dominant form of transverse cracking in these sections was bottom-up in nature. Therefore, the longitudinal strains which induce transverse cracking were the focus of this study. The measured longitudinal strains for the five sections are documented elsewhere (Timm et al., 2011). The dependence of strain on temperature determined by Equation 4 was applied in this study to determine the strain at the corresponding testing temperature (Timm et al., 2011). The temperature-corrected strains measured at the bottom of the asphalt layer for each section at the Test Track are provided in Table 14.

$$\varepsilon = k_1 e^{k_2 T_m} \qquad [4]$$

Where: ε = longitudinal strain, uε
T_m = measured mid-depth temperature, °F
k_1, k_2 = section-specific regression constants

Table 14. *Temperature-corrected strains measured at the bottom of asphalt layer (after Timm et al., 2011).*

Section	k_1	k_2	Temperature-Corrected Tensile Strain (uε)	
			at 20 °C	at 25 °C
S11	62.7602	0.0236	312	386
S10	68.9107	0.0226	320	393
S9	67.5669	0.0238	341	422
N11	73.6127	0.0182	254	299
N10	77.3254	0.0174	252	295

7.2 Prediction of N_f from Overlay Test Transfer Function

The transfer function for overlay test is the relationship between the number of cycles to failure (N_f) and MOD instead of maximum tensile strain. Therefore, the field strain listed in the last column of Table 14 needs to be converted to the applied MOD on the bottom of specimen. Figure 9a shows the steel plates prior to gluing where a 6.09-mm wide tape is covered on the gap between two plates. Figure 9b shows the specimen being glued and cured on the plates. During the test, the bottom of the specimen above the tape is not glued to the plates, and is therefore free to be extended as the plates move apart. Based on the definition of engineering strain, change of length to the original length, the original length for OT becomes the width of the tape and the change of length becomes the MOD. In this study the method to determine MOD is based on Equation 5, assuming the strain is constant during the test.

$$\text{MOD} = \varepsilon \times w_{\text{tape}} \qquad [5]$$

Where: MOD = maximum opening displacement, mm
ε = temperature-corrected field strain at 25 °C
w_{tape} = tape width without glue, 6.09 mm

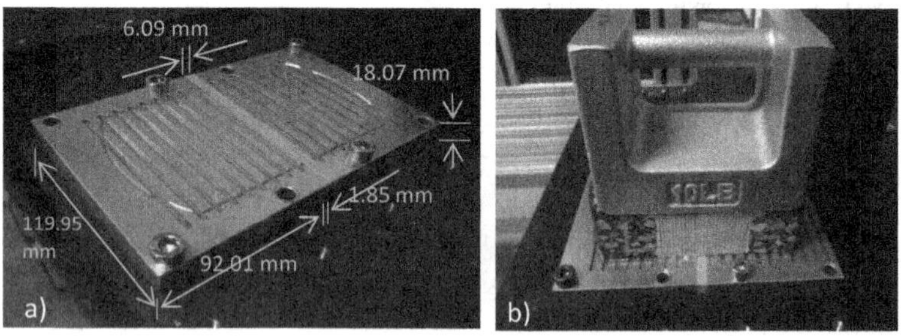

Figure 9. *a) Prepared base plates; and b) glued overlay test specimen.*

The MOD determined using Equation 5 was applied to predict the number of cycles to failure from the overlay test. The results provided in Table 15 show that the MOD is too small to be applied in practice. The determined cycles to failure from the fitted transfer function become very large. It should be noted that the strain on the specimen is difficult to measure and the value may also be changing as the crack grows. Therefore, the MOD is only an approximation. The relationship between MOD and strain in overlay test needs further investigation in the future.

Table 15. *Overlay test MOD and N_f at temperature-corrected field strains.*

Mixture	MOD at 25°C (mm)	N_f (OT) at 25°C
S11-3	0.00235	2.29×10^{11}
S10-3	0.00239	4.19×10^{11}
S9-3	0.00257	1.01×10^{15}
N11-3	0.00182	5.64×10^{14}
N10-3	0.00180	4.88×10^{16}

7.3 Prediction of N_f from Bending Beam Fatigue Test Transfer Function

For the bending beam fatigue test, the strain can be directly used in the transfer function in Equation 2 without conversion. Table 16 lists the determined number of cycles to failure from BBFT under the strain. The prediction is much smaller compared to the prediction from overlay test, and the magnitude of the cycles to failure is in line with typical beam fatigue test results. Also, the determined result can be multiplied with a certain factor (within the range between 6 and 51) for the five different sections to be equal to the number of ESALs applied when 1% crack area occurs on the sections (as shown in Table 7).

Table 16. *Temperature-corrected field strain and corresponding BBFT N_f.*

Mixture	Strain at 20°C (uε)	N_f (BBFT) 20°C
S11-3	312	349,748
S10-3	320	355,579
S9-3	341	332,127
N11-3	254	4,387,864
N10-3	252	1,404,591

7.4 Comparing the Ranking from Laboratory Test Results to the Ranking of Test Track Performance

Fatigue cracking often initiates and propagates from the bottom of the asphalt layer. In the group experiment of the Test Track in 2009 research cycle, the bottom asphalt layers of the five sections (i.e., S11, S10, S9, N11, and N10) were designed to resist cracking under increased tensile strains. Therefore, fatigue cracking resistance of the mixtures in the bottom asphalt layer determined from the laboratory tests (i.e., bending beam fatigue test and overlay test) should govern the fatigue cracking performance of the entire section. Ranking of these five bottom layer mixtures based on the determined number of cycles to failure were expected to be similar to the ranking of cracking performance in these sections. The rankings based on the predictions from the two tests and cracking performance on the test track are summarized in Table 17.

Table 17. *Comparison of the rankings from laboratory test results to the ranking of test track pavement performance.*

Section[3]	**N_f (OT) at 25°C**[1]	**N_f (BBFT) at 20°C**[1]	**ESALs to Exhibit 1% Crack**[1]	**Crack Area**[2]
S11	5th	4th	5th	5th
S10	4th	3rd	4th	4th
S9	2nd	5th	2nd	3rd
N11	3rd	1st	3rd	2nd
N10	1st	2nd	1st	1st

[1]*For N_f or the number of ESALs to exhibit 1% crack, 1st indicates the longest and 5th indicates the shortest*

[2]*For crack area, ranking was based on percentage of crack on the section measured on the same day in 2013 summer, 1st indicates the smallest and 5th indicates the largest*

[3]*Results used for ranking are listed in Table 7, 8, 15, and 16.*

As indicated in Table 17, the two laboratory tests provide different rankings for the bottom asphalt layer mixtures in the five sections. However, the ranking of test track performance based on the number of ESALs until 1% cracking is exactly the same as the overlay test rankings and different from the bending beam fatigue test rankings. The ranking based on cracking area of total lane is also similar to the ranking by the determined overlay test results. Only section S9 and section N11 were ranked differently. The same ranking obtained by the determined overlay test cycles to failure and the measured number of ESALs to exhibit crack is reasonable because as long as the crack propagates through the bottom asphalt layer, it would be expected to appear on the pavement surface. Overall, the rankings based on cracking performance on the test track were different from the rankings determined by beam fatigue results but correlated well with the rankings the determined by overlay test results.

8.0 Conclusions and Recommendations

Based on the results of this study, the following conclusions and recommendations can be offered:

- The correlation analysis between two laboratory tests indicated that the mixture rankings by the AMPT-based overlay test results (i.e., the number of cycles to failure determined by the NLC method) at each maximum opening displacement (MOD) was similar to the bending beam fatigue test result rankings at 800 and 400 microstrain but different from that at 200 microstrain.
- The power model used in the transfer function for the bending beam fatigue test (BBFT) can also be used to fit the transfer function for overlay test (OT) to characterize the relationship between the number of cycles to failure and MOD. In both tests, the fatigue cracking resistance (indicated by the number of cycles to failure) of warm mixes was less sensitive to the change of strain or MOD level than the high RAP mixes (i.e., mixes containing 50% RAP). The number of cycles to failure from both BBFT and OT at a temperature-corrected field strain level was determined using fitted transfer functions.
- Cracking performance on the test track used for the lab-to-field comparisons in this study was represented by two factors: (1) the number of ESALs applied until 1% crack area appeared on the section; and (2) the total cracking area percentage measured in the summer of 2013. Comparisons indicated that the ranking based on the number of ESALs until 1% cracking was exactly the same as the ranking by the strain-corrected OT results, but different from that by the strain-corrected BBFT results. The ranking based on the measured cracking area percentage was also found similar to the ranking by the strain-corrected OT results.
- The same (or similar) rankings between the strain-corrected OT results and cracking performance described above were reasonable because as long as the crack propagated through the bottom asphalt layer, it would be expected to appear on the pavement surface soon. The test track sections with more crack-resistant mixtures in the bottom layer were also expected to exhibit smaller crack areas after the same trafficking period. These findings indicate the potential of the overlay test to determine the relative cracking resistance of asphalt mixtures. However, these findings need to be further validated in other pavement sections with different structures and mixtures.
- The determined OT results were much larger than the number of ESALs applied until 1% crack area appears on the section. This was because the small MOD, converted from the temperature-corrected field strain, was used for the prediction. The equation used for converting the field strain to the overlay test MOD assumed that the strain is constant during the test. Therefore, the MOD was only an approximation. The relationship between MOD and strain in the overlay test needs further investigation.

9.0 Acknowledgements

The authors thank Federal Highway Administration for supporting this research.

10.0 References

Gautheir, T.D., "Detecting Trends using Spearman's Rank Correlation Coefficient," *Environmental Forensics, 2*(4), 2001, pp. 359–362.

Germann, F.P., and Lytton, R.L., *Methodology for Predicting the Reflection Cracking Life of Asphalt Concrete Overlay*, Interim Report, Sep. 1974-Mar. 1979, Texas A&M University, College Station, 1979.

Harvey, J.T., Deacon, J.A., Taybali, A.A., Leahy, R.B., and Monismith, C.L., "A Reliability-Based Mix Design and Analysis System for Mitigating Fatigue Distress," *Eighth International Conference on Asphalt Pavements (Vol. I)*, 1997, pp. 301–323.

Hu, S., Zhou, F., and Scullion, T., "Factors that Affect Cracking Performance in Hot-Mix Asphalt Mix Design," *Transportation Research Record: Journal of the Transportation Research Board, 2210*(1), 2011, pp. 37–46.

Hu, S., Zhou, F., and Scullion, T., "Reflection Cracking-Based Asphalt Overlay Thickness Design and Analysis Tool," *Transportation Research Record: Journal of the Transportation Research Board, 2155*(1), 2010, pp. 12–23.

Hu, X., Zhou, F., and Scullion T., *Pilot Implementation of the Overlay Tester and Double-Blade Saw*," Report No. FHWA/TX-08/5-4467-01-1, Texas Transportation Institute, Texas A&M University System, 2008.

Huang, Y.H., *Pavement Analysis and Design*, Prentice Hall, Inc., New Jersey, 1993.

Ma, W., Tran, N.H., Taylor, A.J., and Li, X., "Proposed Improvements to Overlay Test for Determining Cracking Resistance of Asphalt Mixtures," Paper No. 14-3183, *Presented at 2014 Annual Meeting of the Transportation Board*, Washington, D.C., 2014.

Ma, W., *Proposed Improvements to Overlay Test for Determining Cracking Resistance of Asphalt Mixtures*, M.S. Thesis, Auburn University, 2014.

Montgomery, D.C., *Design and Analysis of Experiments*, 7th Edition, John Wiley & Sons, Inc., New York, 2009.

Pierce, L.M., and Mahoney, J.P., "Asphalt Concrete Overlay Design Case Studies," *Transportation Research Record: Journal of the Transportation Research Board, 1543*(1), 1996, pp. 3-9.

Powell, R.B., "NCAT Pavement Test Track: Construction," retrieved from http://www.pavetrack.com/construction.htm.

Priest, A.L., and Timm, D.H., *Methodology and Calibration of Fatigue Transfer Functions for Mechanistic-Empirical Flexible Pavement Design*" Report 06-03, National Center for Asphalt Technology, Auburn University, Alabama, 2006.

Prowell, B.D., "Estimate of Fatigue Shift Factors between Laboratory Tests and Field Performance," *Transportation Research Record: Journal of the Transportation Research Board, 2181*(1), 2010, pp. 117–124.

Royal Geographical Society, "Spearman's Rank Correlation Coefficient - Excel Guide," retrieved from http://www.rgs.org/NR/rdonlyres/4844E3AB-B36D-4B14-8A20-3A3C28FAC087/0/OASpearmansRankExcelGuidePDF.pdf.

Timm, D.H. and Priest, A.L., "Flexible Pavement Fatigue Cracking and Measured Strain Response at the NCAT Test Track," *Presented at 87th Annual Meeting of the Transportation Board*, Washington, D.C., 2008.

Timm, D.H., Willis, J.R., and Kvasnak, A., "Full-Scale Structural Evaluation of Fatigue Characteristics in High Reclaimed Asphalt Pavement and Warm-Mix Asphalt," T*ransportation Research Record: Journal of the Transportation Research Board, 2208*(1), 2011, pp. 56–63.

Walubita, L.F., Faruk, A.N., Das, G., Tanvir, H.A., Zhang, J., and Scullion, T., *The Overlay Tester: A Sensitivity Study to Improve Repeatability and Minimize Variability in the Test Results*," Report No. FHWA/TX-12/0-6607-1, Texas Transportation Institute, Texas A&M University System, 2012.

Zhou, F., and Scullion, T., *Upgraded Overlay Tester and Its Application to Characterize Reflection Cracking Resistance of Asphalt Mixtures*, Report No. FHWA/TX-04/0-4467-1, Texas Transportation Institute, Texas A&M University System, 2003.

Zhou, F., and Scullion, T., *Overlay Tester: A Rapid Performance Related Crack Resistance Test*, Report No. FHWA/TX-05/0-4467-2, Texas Transportation Institute, Texas A&M University System, 2005.

Zhou, F., Hu, S., and Scullion, T., *Development and Verification of the Overlay Tester Based Fatigue Cracking Prediction Approach*, Report No. FHWA/TX-07/9-1502-01-8, Texas Transportation Institute, Texas A&M University System, 2007a.

Zhou, F., Hu, S., Chen, D.H., and Scullion, T., "Overlay Tester: Simple Performance Test for Fatigue Cracking," *Transportation Research Record: Journal of the Transportation Research Board, 2001*(1), 2007b, pp. 1–8.

11.0 Discussion

JIM SCHEROCMAN: I've followed the overlay tester for many, many years. I have some concerns about it.

WANGYU MA: Okay.

JIM SCHEROCMAN: My first concern is related to the accuracy of the gap and the accuracy of the glue application. Do those two items affect any of the test results that you've gotten?

WANGYU MA: Thank you for the question. The glue amount is pretty consistent. I applied about 10 grams uniformly on the bottom of each specimen. Then, a tape was used to cover the gap of the two base-plates so there was no glue underneath the tape of fixed width. This procedure controls the original length when calculating tensile strain. Other procedures in the lab were also consistent.

JIM SCHEROCMAN: Second, when you compare your test results to actual field performance, since the overlay tester is horizontal displacement only and pavements also deflect vertically under load, are you really measuring the total performance of the roadway pavement when you're only measuring one of the two directions that the pavement moves?

WANGYU MA: For the field performance, we only used the longitudinal strain in order to characterize the transverse cracking, which we found is the major issue.

JIM SCHEROCMAN: But don't pavements undergo vertical deflection also?

WANGYU MA: Yes.

JIM SCHEROCMAN: So what's the correlation between the overlay tester test results and the vertical movement in terms of the long-term performance of a roadway?

WANGYU MA: The overlay test is basically to characterize the Type 1 cracking, which is mainly due to horizontal movement. You are right, the road also has vertical deflection. Probably it's better to consider that in lab testing as well, but we need to know which type of strain is more critical to the distress so we can use the lab testing method to simulate that. We did not characterize the vertical movement in the scope of this study.

JIM SCHEROCMAN: But that other method, the two things are happening at the same time so you have to coordinate the two and somehow come up with a combined answer, correct?

WANGYU MA: Right.

JIM SCHEROCMAN: Good luck.

WANGYU MA: Thank you.

JOHN D'ANGELO: Very interesting work. You've put a lot of effort into this. I'm just wondering, though, when you talked about relating the strains in the mixture in your test to actual field conditions, which field conditions are you talking about? When we start talking about the real world, not even the test track but the real world, things become extremely more complicated. In the real world today, we're dealing mostly with overlays, not full-depth new construction, and so how do we tie some of these strains with strains developed in the overlay tester? If I'm talking about a surface mix, I have one type of strain, say surface tension at the edge of a tire, which is completely different if I have debonded layers. So there's a lot of different things that have to be addressed. I'm just wondering if you have started to look at how you are going to approach some of these different types of real world mechanisms that we're dealing with one test.

WANGYU MA: Okay, thank you for the comment. I think your question is to ask me if I can use this test method to simulate the field, which has many different issues in terms of conditions and strain levels. But here in this study, we have to control the parameters in the lab, so based on our finding at this step, the only thing we can do is to use the longitudinal strain and mid-depth temperature, as the simulation reference. It would be better if we can get more measurements from the field, so we can use these measurements and upgrade the overlay test to get better performance result.

GEOFFREY ROWE: Nice presentation. I was very interested to see that you had applied that method to the overlay cracking, looking at the cycles x N_f. The nice thing about that technique and the reason I adopted it is you get a nice peak that very clearly defines a point of something changing in a test. I also have used that technique with repeated load deformation tests where I consider the compliance in that test, and I use the same technique to pick out sort of like an inflection failure in a test like that. In the work that we recently did – and I can see Phil standing up here. On bending beam fatigue, Phil was part of that work – we used that to try and reduce the coefficient of variation in that test result, so we felt using that technique helped us with that. And one of the criticisms against the overlay test has been the coefficient of variation or the variability of the test, and I'll be particularly interested to hear your comments about how much you think using a technique such as that to define the failure point will maybe affect or change or reduce the coefficient of variation that you have in that test device.

WANGYU MA: Thank you very much. The overlay test and most fatigue tests have high variability, and we want to control it as low as possible. Based on our findings, we found the coefficient of variation is typically below 30% when using AMPT to perform overlay test. If we want to reduce that variability and improve that consistency, we need to make the material much more consistent. Also, some studies in Texas DOT or TTI already investigated how to reduce the variability. They have many recommendations for the glue amount, tape width, and how to better determine the fatigue life. In the future, I think it's better to reduce the displacement levels, so at higher frequency, you won't have higher variability.

GEOFFREY ROWE: Did you compare the variability that you had when you had the 93% reduction, such as the original Texas method to your modified method?

WANGYU MA: Yes.

GEOFFREY ROWE: What was that reduction in variability?

WANGYU MA: There is a reduction in variability. We can see the 93% load reduction criterion has higher variability.

GEOFFREY ROWE: Okay. Thank you. I mean it would be nice just to quantify what that reduction is for the folks.

TOM HARMAN: You're going to have an opportunity to edit the comments. One of the things that Geoff's asking you to do is when you edit them, put the actual data in there. That would be excellent.

WANGYU MA: Thanks. Our test results shows the average coefficient of variation using normalized load × cycle method is 28.1%, slightly less than 28.4% by Texas original method.

PHILLIP BLANKENSHIP: The variability that you had on your fatigue test seems very high.

WANGYU MA: Right.

PHILLIP BLANKENSHIP: I am used to seeing fatigue test variability in the 30 to 50% range. Did you post analyze the fatigue data or use the result that is provided by the manufacture's software which assumes 50% stiffness failure?

WANGYU MA: Thank you for the question. We used the AASHTO Standard, which is 50% stiffness reduction, as the number of cycle to failure in the bending beam fatigue test. We determined the result right after test and didn't perform any other procedure to trim the data.

PHILLIP BLANKENSHIP: That's the classic in analysis.

WANGYU MA: Yes, original data.

PHILLIP BLANKENSHIP: That explains a lot. Thank you so much.

Unified Failure Criterion for Asphalt Binder under Cyclic Fatigue Loading

Chao Wang[a,b], Cassie Castorena[b], Jinxi Zhang[a], Y. Richard Kim[b]

[a]*Department of Road and Railway Engineering, Beijing University of Technology, Beijing, P. R. China, CN 100124*

[b]*Department of Civil, Construction, and Environmental Engineering, North Carolina State University, NC 27695-7908*

ABSTRACT. Defining failure and developing a unified failure criterion for the fatigue testing of asphalt materials remain a challenge. This study seeks to develop a failure criterion for the fatigue testing of asphalt binders under cyclic loading in the dynamic shear rheometer. Newly developed pseudo strain energy (PSE) - based failure analysis is introduced for both the time sweep fatigue test and the accelerated linear amplitude sweep (LAS) test (AASHTO TP 101). The presented methodology builds upon recent advances in the simplified viscoelastic continuum damage (S-VECD) modeling of asphalt mixtures. Trends in stored PSE have been proven to be effective in defining failure for the LAS tests of asphalt binders. This new proposed failure definition is material-dependent and, thus, is effective in capturing the benefits of asphalt modification for binder fatigue resistance. In addition, it is found that a unique relationship that is independent of loading history exists between the PSE release rate and fatigue life. The fatigue life predictions using this relationship and the S-VECD model are in reasonable agreement with the laboratory-measured fatigue life data and also generally relate well with the field fatigue performance measured in the FHWA-ALF pooled fund study.

KEYWORDS: failure criterion, asphalt fatigue, pseudo strain energy, damage, S-VECD, fatigue life

1.0 Introduction

Fatigue cracking caused by repeated traffic loading is a critical distress in asphalt pavements. The accurate evaluation and prediction of the fatigue performance of asphalt materials is required for material selection, asphalt mixture design, and pavement structural design. Fatigue cracking is a complex problem, affected by both material characteristics and structural design. It is well known that the thickness of the layers comprising a pavement greatly impacts fatigue resistance. However, it is also recognized that the asphalt mixture volumetric properties play a significant role in determining performance. In addition, asphalt binder properties can significantly

The oral presentation was made by Dr. Castorena.

This is a reproduction of an article whose final and definitive form has been published in *Road Materials and Pavement Design*© 2015 Taylor & Francis. The article is available online at: http://dx.doi.org/10.1080/14680629.2015.1077010

impact fatigue resistance of pavements as asphalt binder comprises the weakest component of asphalt concrete and thus, fatigue cracks generally initiate and propagate within the binder phase of asphalt concrete. Therefore, if asphalt binder fatigue resistance can be quantified, material selection and consequently pavement performance can be improved.

Several different fatigue performance tests have been proposed for evaluating both asphalt binder and mixture fatigue resistance. Data interpretation, modeling, and consequently performance predictions can be accomplished using phenomenological observations and/or mechanical modeling. The definition of fatigue failure during testing is a critical component of the fatigue characterization of asphalt materials. A proper fatigue failure criterion must be applicable to a wide range of loading conditions so that it can be characterized from a set of material tests and then can be applied to the prediction of pavement performance under complex loading and environmental conditions.

A significant amount of research has been conducted to develop fatigue failure definitions and criteria for asphalt materials. The traditional and simplest definition of failure used in fatigue analysis of asphalt materials is the 50% loss in stiffness/pseudo stiffness criterion (Hicks et al., 1993; Kim et al., 1997). However, this approach has been challenged because it is based on an arbitrary definition, without theoretical or phenomenological justification. Dissipated energy approaches offer an alternative for defining failure and have been widely used for asphalt material fatigue analysis. Different forms of dissipated energy criteria, including the dissipated energy ratio (DER), ratio of dissipated energy change (RDEC), and dissipated pseudo strain energy (DPSE), have been proposed (Van Dijk, 1975; Van Dijk and Visser, 1977; Rowe, 1993; Ghuzlan and Carpenter, 2000; Bonnetti et al., 2002; Shen and Carpenter, 2005; Carpenter and Shen, 2006; Masad et al., 2008). Dissipated energy approaches have an advantage over the arbitrary 50% loss in initial stiffness criterion because they are based on mechanistic principles and are material-dependent. Fracture mechanics-based indices also have been employed for failure analysis of asphalt materials (Branco et al., 2008; Hintz et al., 2011). In addition, phenomenological parameters related to marked changes in the progression of damage growth have been used. These parameters include the peak in phase angle (Reese, 1997) and peak in $S \times N$ where S is the stiffness and N is the number of loading cycles (Rowe and Bouldin, 2000). These parameters have been shown to correlate with the point of damage localization in asphalt mixtures and, hence, the onset of macro-cracking.

Recently, a pseudo strain energy (PSE) - based failure criterion, G^R (i.e., pseudo-strain energy (PSE) release rate) has been developed and incorporated into the simplified viscoelastic continuum damage (S-VECD) model for asphalt mixtures (Zhang, 2012; Zhang et al., 2013; Sabouri and Kim, 2014; Norouzi et al., 2014). This new failure criterion is independent of mode of loading and is applicable for asphalt mixture to a range of temperatures where viscoplastic strains are negligible. Therefore, this PSE-based criterion allows the prediction of fatigue failure at different temperatures, loading modes, and stress/strain amplitudes using limited test results.

The linear amplitude sweep (LAS) test (AASHTO TP 101) has been proposed for fatigue specifications of asphalt binders. The LAS test consists of repeated cyclic loading in the dynamic shear rheometer (DSR) and utilizes systematically increasing strain amplitudes to accelerate damage growth. The test procedure takes approximately thirty minutes, including time for thermal equilibration. The S-VECD model has been shown to be applicable for the analysis of the LAS test results, thereby allowing the prediction of damage evolution under variable loading conditions (Hintz et al., 2011; Safaei et al., 2014). However, accurately defining the point of failure in the LAS test and developing a fatigue failure criterion are still needed for accurate fatigue performance predictions. This paper attempts to apply the fatigue modeling and failure concepts developed from mixture characterization to binder fatigue tests.

2.0 Objectives

The objectives of this study are to:

- Develop a means to detect failure from asphalt binder fatigue test results.
- Develop a material-dependent failure criterion that can be incorporated into the S-VECD model for performance predictions.
- Evaluate the predictive capabilities of the developed approach by comparing the predicted results to field performance data.

3.0 Materials and Test Methods

3.1 Materials

This study was conducted in two phases. The first phase included testing a single neat PG 64-22 binder from North Carolina (hereinafter termed NC binder) to investigate the applicability of a unified failure definition and criterion. For the NC binder, both short-term aging (STA) and long-term aging (LTA) conditions were evaluated. STA and LTA samples were produced using the universal simple aging test (USAT) developed at the Western Research Institute (WRI) (Farrar et al., 2012). The USAT utilizes the aging of thin films of asphalt binder in an oven and offers an efficient alternative to conventional rolling thin-film oven (RTFO) and pressurized aging vessel (PAV) methods that generate only limited amounts of aged material. Details of the USAT procedure and its development can be found elsewhere (Farrar et al., 2012; WRI, 2013). The STA samples were produced using the recommended USAT procedure to produce samples that were aged to the standard RTFO condition. The LTA samples were produced using the USAT procedure to produce samples aged to the standard RTFO+PAV condition.

The second phase included verification and assessment of performance prediction capabilities using four different binders from the Federal Highway Administration - Accelerated Loading Facility (FHWA-ALF) pooled fund study TPF-5(019). These binders are referred to as Control (PG 70-22), CR-TB (PG 76-28), Terpolymer (PG 70-28), and SBS-LG (PG 70-28). The ALF binders were subjected to standard RTFO short-term aging prior to testing. Both the mixture performance test results obtained from cyclic direct tension tests and field performance data were available for the ALF mixtures. Thus, the binder performance prediction results could be compared to the corresponding mixture and field performance data to judge the predictive capabilities of the proposed approach.

3.2 Test Methods

All tests were conducted using a TA Instruments AR-G2 DSR. The newly developed 4-mm parallel plate geometry (Sui et al., 2010; Sui et al., 2011) was employed in this study for both the linear viscoelastic frequency sweep tests and fatigue tests. Linear viscoelastic characterization was required to obtain a measure of the undamaged material response so that the damage and viscoelastic effects could be separated when interpreting the fatigue test results.

3.2.1 Linear Viscoelastic Characterization

Frequency sweep tests were conducted to obtain the linear viscoelastic properties of the binders at the intermediate temperature range used for fatigue testing. Dynamic shear modulus ($|G^*|$) mastercurves were generated to provide a fingerprint of the undamaged material responses. In this study, the machine compliance of the TA AR-G2 rheometer was measured using a strain sweep from 1×10^{-5} rad to 1×10^{-4} rad using a thin film (50 micro meters) of epoxy to weld the 4-mm parallel plates together to allow for an accurate measurement of the asphalt binder properties. By plotting the displacement versus measured oscillatory torque, the slope of the best-fit line was determined as the machine compliance. The final machine compliance was determined to be 1.5×10^{-2} rad/N· m. This machine compliance correction method follows the procedure recommended by the WRI (WRI, 2012).

3.2.2 Fatigue Tests

Two types of cyclic fatigue tests, the LAS test and the time sweep (TS) test, were conducted in this study to evaluate the damage resistance of asphalt binder. NC binder STA and LTA samples were tested at the intermediate temperature of 18°C. This temperature was selected based on the recommendation by Sabouri et al. (2014) to avoid the effects of viscoplasticity. The ALF RTFO binders were tested at 19°C, consistent with the temperature used in the ALF project.

3.2.2.1 Linear Amplitude Sweep (LAS) Test. The standardized LAS procedure (AASHTO TP 101) consists of two steps. First, a frequency sweep is conducted as a fingerprint test to determine the undamaged material response of the sample being tested. The frequency sweep is nondestructive and is followed by a linear oscillatory strain sweep with strain amplitudes ranging from 0.1% to 30% to measure the asphalt binder damage tolerance (Hintz et al. 2011). In this study, in addition to the specified LAS test, further amplitude sweep procedures were tried to evaluate the effects of different rates of increase in strain amplitude. The generated constant strain-amplitude rate (CSR) that resulted from running the standard LAS strain sweep for 300 seconds was approximately 30%/300 s, which equals 0.001 s^{-1}. Increasing or decreasing the designed sweep time results in different CSRs given the same strain amplitude range is utilized. The different CSRs used in this study are shown in Figure 1, and the employed CSR levels for each tested binder are detailed in Table 1.

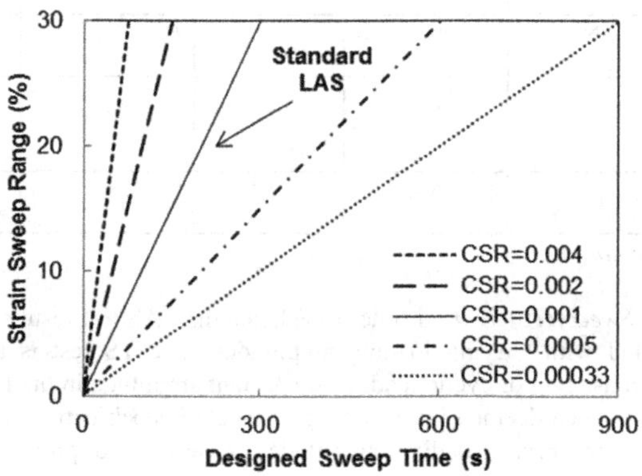

Figure 1. *Multiple loading sequences used in LAS tests.*

For the NC binders, two replicates were conducted for all LAS tests under different CSRs to verify the observed binder failure behavior are repeatable, whereas for the FHWA-ALF binders only one replicate was used to verify the developed fatigue failure definition and criterion. Agreement between data was assessed through overlap of S-VECD damage characteristic curves obtained from tests with varying CSRs. Damage characteristic curves, as will be discussed, are assumed to be loading history independent and thus, deviance between curves could be used to identify poor results.

Table 1. *Summary of LAS test procedure.*

Binder	LAS - Sweep Time (minutes) and Corresponding CSR (s^{-1})				
	LAS-1.25	LAS-2.5	LAS-5 (Standard)	LAS -10	LAS -15
	CSR=0.004	CSR=0.002	CSR=0.001	CSR=0.0005	CSR=0.00033
NC STA	×	×	×	×	NA*
NC LTA	×	×	×	×	NA
Control	NA	NA	×	×	×
CR-TB	NA	NA	×	×	×
Terpoly-mer	NA	NA	×	×	×
SBS-LG	NA	NA	×	×	×

*Note: *Not available.*

3.2.2.2 Time Sweep (TS) Test. In order to validate the LAS test results, TS tests were also conducted with varying loading amplitudes. The TS test is a conventional fatigue test, consisting of cyclic loading at constant amplitude in the DSR. However, the TS test has been deemed impractical for standard specifications due to the need to select an appropriate loading amplitude for testing to produce failure in a reasonable amount of time, which requires knowledge of the material's damage resistance a priori. In this study, the NC binder STA samples and ALF Control binder were tested using both constant displacement amplitude in the TS test (CD-TS) and constant stress amplitude in the TS test (CS-TS). The NC binder LTA samples and the rest of the ALF binders were tested using only the CD-TS test. Details regarding the TS testing for each material tested in this study are included in Table 2. One replicate was conducted for each loading amplitude. Agreement between results was assessed through agreement of S-VECD damage characteristic curves. If the damage characteristics curves did not align, a replicate test was conducted.

3.3 S-VECD Analysis

Both LAS and TS test data were analyzed using the S-VECD model. Details regarding the S-VECD model and its recent application for asphalt binder fatigue analysis are provided elsewhere (Underwood et al., 2010; Underwood, 2011; Safaei et al., 2014). Key aspects of this modeling approach are summarized as follows.

Table 2. *Summary of time sweep procedures.*

Binder	CD-TS	CS-TS
	Strain Amplitude	Stress Amplitude
NC STA	2%, 3% and 4%	500 kPa
NC LTA	2%, 2.5% and 3%	NA*
Control	3%, 5% and 7%	650 kPa
CR-TB	5% and 7%	NA
Terpolymer	5% and 7%	NA
SBS-LG	5% and 7%	NA

*Note: *Not available*

Damage evolution is based on Schapery's work potential theory (Schapery, 1984). In work potential theory, the rate-dependent damage evolution rate is expressed as:

$$\frac{dS}{dt} = \left(-\frac{\partial W^R}{\partial S}\right)^{\alpha} \qquad [1]$$

Where S is the internal state variable representing damage; W^R is the work performed; α is the undamaged material-dependent constant; and t is time. In this study, α=1/m, where m is the fitting slope parameter of the linear viscoelastic dynamic shear modulus ($|G^*|_{LVE}$) mastercurve. W^R is quantified using PSE density:

$$W^R = \frac{1}{2}C(S)\left(\gamma^R\right)^2 \qquad [2]$$

Where C(S) is pseudo stiffness and can be determined as:

$$C(S) = \frac{\tau_p}{\gamma_p{}^R \times DMR} \qquad [3]$$

Where τ_p is the effective (measured) peak shear stress in a given cycle. $\gamma_p{}^R$ is the peak pseudo strain for that given cycle, defined as:

$$\gamma_p{}^R{}_i = \frac{1}{G_R}\gamma_{pi} \times |G^*|_{LVE} \qquad [4]$$

Where $|G^*|_{LVE}$ is the linear viscoelastic dynamic shear modulus at a given temperature and loading frequency, and γ_{pi} is the peak shear strain in the given cycle. G_R is the arbitrary reference modulus, selected to be 1, which implies that the pseudo strain is equivalent to the linear viscoelastic stress response to a given loading input. Hence, for the case of cyclic loading in the DSR with zero mean displacement, Equation 4 can be reduced to the following for cycle i:

$$\gamma_{p\ i}^{R} = \gamma_{pi} \times |G^*|_{LVE} \qquad [5]$$

Because pseudo strain is equivalent to the linear viscoelastic stress response for a given loading history, it can be observed that pseudo stiffness, defined in Equation 3, is equal to 1 when the material is behaving linear viscoelastically and decreases as the sample undergoes damage.

In Equation 3, DMR = dynamic modulus ratio $=|G^*|_{fingerprint} / |G^*|_{LVE}$, which is introduced to eliminate sample-to-sample variability. For LAS test analysis, $|G^*|_{fingerprint}$ is determined using the frequency sweep step in the LAS test, whereas for TS test analysis, it is obtained by subjecting the specimen to several cycles of loading within the linear viscoelastic range prior to TS testing.

Equations 1 to 5 are combined, and Equation 1 is numerically integrated to solve for damage, S, as a function of time:

$$S = \sum_{j=1}^{N} \left[\frac{DMR}{2} \left(\gamma^{R}\right)^{2} \left(C_{j-1} - C_{j}\right) \right]^{\frac{\alpha}{1+\alpha}} \left[\left(t_{j} - t_{j-1}\right) \right]^{\frac{1}{1+\alpha}} \qquad [6]$$

Where j represents the time steps.

The power of the S-VECD model is that the relationship between C(S) and S is loading history independent and can be used to back calculate fatigue performance under any loading condition of interest. To allow for performance predictions, the data from the fatigue test are used to calculate C(S) and S, and the model is then fitted as follows:

$$C = 1 - C_1 (S)^{C_2} \qquad [7]$$

Where C_1 and C_2 are the model parameters that best fit the data.

Combining Equations 1, 2, 5, and 7 allows for the derivation of a model between loading cycles and strain amplitude, γ_p.

$$N = \frac{f \cdot 2^{\alpha} \cdot S^{1-\alpha C_2 + \alpha}}{(1 - \alpha C_2 + \alpha)(C_1 C_2)^{\alpha} (|G^*|_{LVE} \cdot \gamma_p)^{2\alpha}} \qquad [8]$$

Where f is the loading frequency used in fatigue testing and t is the loading time. If S is set to S at failure, Equation 8 can be used to predict the fatigue life in terms of the numbers of cycle to failure (N_f) at any strain amplitude.

4.0 Failure Definition Used in LAS and TS Tests

The goal in establishing a failure criterion is to capture the transition from stable damage growth, where the material can still be considered a continuum, to unstable damage growth, where there is macro-crack propagation, typically inferred from a significant increase in the rate of damage growth. Thus, the establishment of a failure criterion is a critical component of the S-VECD modeling framework.

4.1 LAS Test Failure Characterization

4.1.1 Material Yield and Failure Definition

Representative LAS test results of NC binder STA samples at multiple CSRs are shown in Figure 2(a). Figure 2(a) displays trends in phase angle and peak stress with increasing peak strain (and, hence, number of loading cycles). It can be observed that both the measured shear stress and phase angles exhibited peaks during the test. Typical replicate results in Figure 2(b) showed these are repeatable material behavior with the mean absolute error (MAE) less than 10% for both shear stress and phase angle responses. The peak stress can be easily understood as the yield threshold of the material under increasing loading, whereas the phase angle drop, which occurred in the tests after the material yielded, can reasonably be defined as the ultimate failure or damage tolerance. The peak in phase angle corresponds to the point at which the material could not resist additional loading, as indicated by the onset of a rapid decrease in the stress response that coincides with the phase angle peak. Thus, it is felt that the phase angle peak is a more logical failure definition than the yield point. This interpretation is consistent with that of Kim et al. (2003) who demonstrated that yield proceeded ultimate failure in strain sweep testing of asphalt-sand mixtures. In addition, the use of the peak in phase angle to define failure has been utilized extensively in defining asphalt mixture fatigue failure (Reese, 1997; Zhang, 2012; Zhang et al., 2013; Sabouri et al., 2014; Norouzi et al., 2014).

Another phenomenological failure definition that has been applied to the LAS test is the maximum of C×N (Safaei et al., 2014). It was found in this study that C×N always reached its peak value simultaneously with the time that the material yielded during the LAS test, as typically illustrated in Figure 2(c). This coincidence can be explained by Equation 9, which demonstrates that C×N is mathematically dependent only on the shear peak stress (τ_p) in the S-VECD analysis of the LAS test results, because the $|G^*|_{LVE}$, CSR, DMR, and loading frequency (*f*) are all constants and predefined for a given test. Thus, the C×N approach in the LAS test will always give the point of yield as the failure, which is interpreted here as proceeding failure. It should be noted that the C×N derivation in Equation 9 is specific to the LAS test and C×N remains a valid parameter to assess failure in TS tests.

$$C \times N = \frac{\tau_p}{\gamma^R{}_p \cdot DMR} \times t \cdot f = \frac{\tau_p}{\left|G^*\right|_{LVE} \cdot \gamma_p \cdot DMR} \times t \cdot f$$
$$= \frac{\tau_p}{\left|G^*\right|_{LVE} \cdot (CSR \cdot t) \cdot DMR} \times t \cdot f = \frac{\tau_p}{\left|G^*\right|_{LVE} \cdot CSR \cdot DMR} \times f \qquad [9]$$

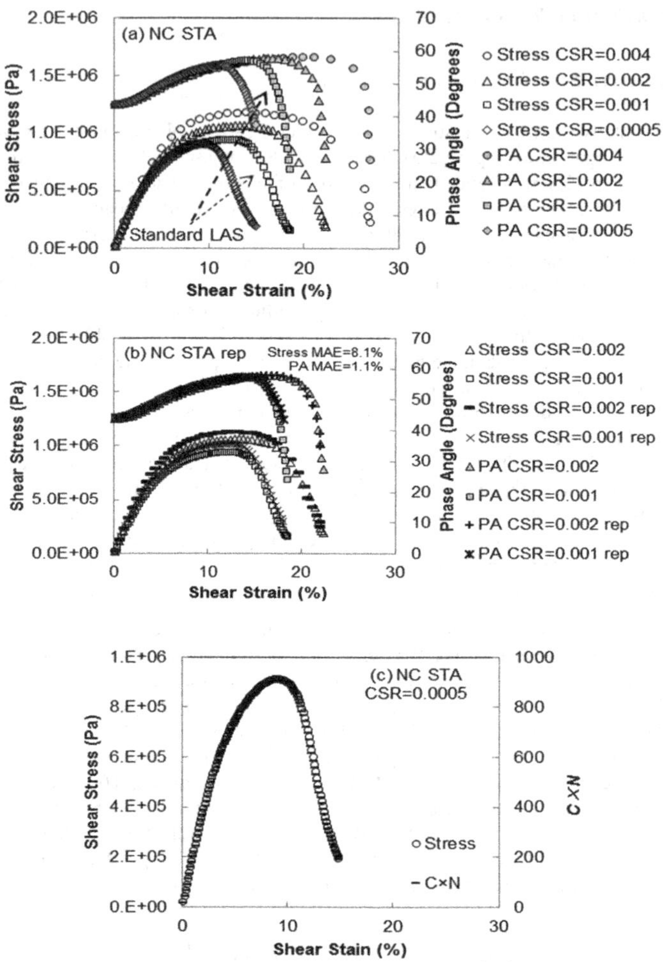

Figure 2. *LAS test results under multiple CSRs: (a) phase angle evolution (b) typical replicate results and (c) typical C×N evolution.*

4.1.2 Pseudo Strain Energy (PSE) Variables-Based Analysis

The relative change in phase angle during fatigue testing is sometimes small and gradual, especially for modified systems, which makes identification of fatigue failure based on phase angle trends difficult in some cases, particularly for modified binders. The reason behind phase angle increases in fatigue testing is also poorly understood and thought to be largely related to loading waveform distortion as a result of damage rather than a true change in material phase angle (Reese, 1997). In addition, the phase angle is not included in the S-VECD model framework, and therefore, the phase angle-based failure definition could be used for determining the failure cycle from experimental data but it cannot be used for predicting fatigue life. Thus, a mechanistic-based parameter to use in place of the phase angle to define failure is desirable. In order to develop a mechanistic approach to define failure, the LAS test results were analyzed again in the pseudo strain domain where the time-dependent behavior of the asphalt material could be eliminated, thereby allowing for the true assessment of damage effects.

Figure 3(a) depicts the initial damage evolution in the LAS test. The 'undamaged' line is included as a reference to show the response if the material were behaving linear viscoelastically. The measured material response deviates from the undamaged line as loading progresses, indicating that damage has occurred. For a particular data point, the stored and released PSE can be calculated. Stored PSE, or W^R_s, represents the PSE that is stored with each loading cycle applied, as calculated using Equation 10 and can be represented as the triangular area with dashed lines in Figure 3(a). The total PSE input is the total amount of PSE that goes into the material; it can be approximately defined as Equation 11. Hence, the cumulative released PSE, W^R_r, can be derived approximately by subtracting the stored PSE, W^R_s, from the total PSE, as shown in Equation 12. The cumulative released PSE is depicted in Figure 3(a) by the triangular area with solid vertical lines. Thus, using Equations 10 through 12, stored and cumulative released PSE can be calculated for each data point collected during LAS testing.

$$W^R_s = \frac{1}{2}\cdot\tau_p\cdot\gamma^R_p / DMR = \frac{1}{2}\cdot C\cdot(\gamma^R_p)^2 \qquad [10]$$

$$W^R_{total} = \frac{1}{2}\cdot\tau_{undamaged}\cdot\gamma^R_p = \frac{1}{2}\cdot(\gamma^R_p)^2 \qquad [11]$$

$$W^R_r = W^R_{total} - W^R_s = \frac{1}{2}\cdot(1-C)\cdot(\gamma^R_p)^2 \qquad [12]$$

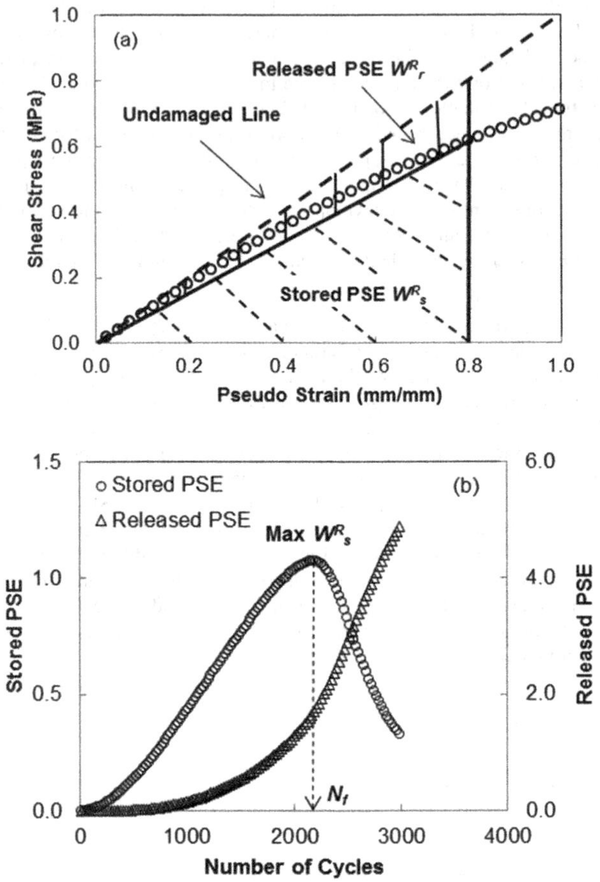

Figure 3. *LAS-based PSE definition and evolution: (a) schematic representation of PSE distribution and (b) evolution of stored and cumulative released PSE.*

The calculated W^R_s and W^R_r data for the entire LAS testing for the NC STA binder under CSR 0.0005 s^{-1} are shown in Figure 3(b). The results demonstrate that there is a maximum W^R_s during the test, whereas W^R_r continually increases. An increasing W^R_s indicates that the material retains the ability to store additional energy as the loading amplitude (and hence energy input) increases in the LAS test. However, a decrease in W^R_s indicates that the material is losing the ability to store PSE as the loading input increases, indicating that failure has occurred. Thus, the maximum of W^R_s, i.e., Max W^R_s, is a logical energy-based failure definition. Normally, a general failure definition is specified in terms of some energy characteristics that are easily found from the stress–strain curves (Christensen, 2013). Additionally, Figure 4 shows the correlations between fatigue life data (N_f), which have been determined based on the phase angle drop, and maximum W^R_s for all the binders tested in this study. The fatigue life data determined from the two approaches are in good agreement, demonstrating that the peak in stored PSE is a reliable measure for defining fatigue failure.

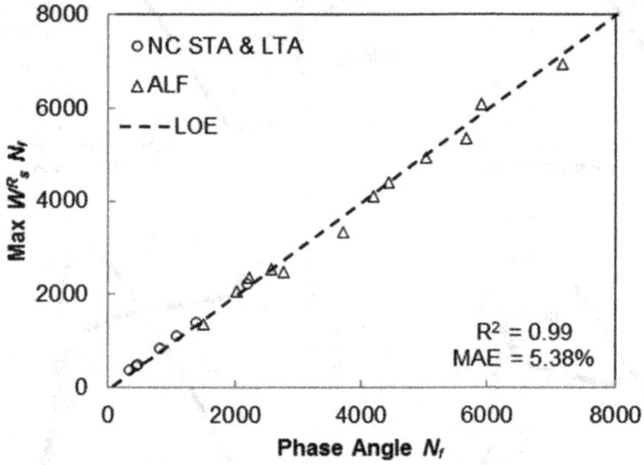

Figure 4. *Failure definition comparison between phase angle drop and Max W^R_s.*

A further derivation of Max W^R_s expressed in Equation 13 shows that this parameter in LAS testing is mathematically dependent not only on the shear stress, τ_p, but also on time, t (and hence, the number of loading cycles). Similar to the previously presented C×N analysis, this derivation is specific to the LAS test loading scheme.

$$W^R{}_s = \frac{1}{2}\cdot\tau_p\cdot\gamma^R{}_p = \frac{1}{2}\cdot\tau_p\cdot|G^*|_{LVE}\cdot\gamma_p = \frac{1}{2}\cdot\tau_p\cdot|G^*|_{LVE}\cdot(\text{CSR}\cdot\text{t}) \qquad [13]$$

4.2 TS Test Failure Definition

4.2.1 Phase Angle and C×N Evolution in TS Test

Figure 5 presents typical C×N and phase angle trends with respect to the number of loading cycles in the TS tests. For the neat binders (NC binders and ALF-Control), the results presented in Figures 5(a) and (c) show that the C×N and phase angle curves both reached peaks during the test and that the peaks occurred at roughly the same time for the controlled displacement tests (CD-TS). For the stress-controlled tests (CS-TS), the phase angle dramatically increased rather than decreased upon failure. However, the maximum C×N remains a logical failure definition.

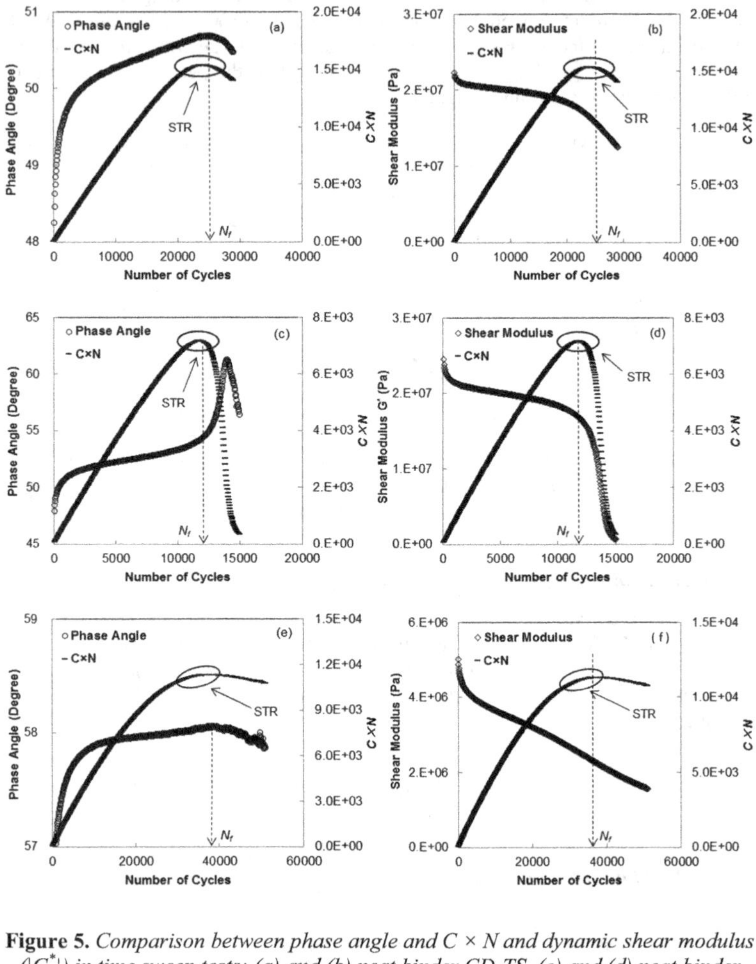

Figure 5. *Comparison between phase angle and C × N and dynamic shear modulus (|G*|) in time sweep tests: (a) and (b) neat binder CD-TS, (c) and (d) neat binder CS-TS, and (e) and (f) PMB CD-TS.*

In the verification study using the ALF binders, it was found that for some polymer-modified binders (PMBs), C×N in the CD-TS did not display an obvious peak but only exhibited a slope transition range (STR), as shown in Figure 5(e). This behavior can be explained by the different stiffness/pseudo stiffness evolutions of the neat binders and PMBs after failure, which are thought to coincide with the onset of macro-crack propagation. For the neat binders, the material integrity always decreased rapidly after the phase angle peak, indicating rapid crack propagation. However, for the PMBs, the stiffness did not experience as rapid a transition due to the toughening of the binder by the polymer. These phenomena are illustrated in Figure 5(b), (d) and (f) for the neat binders and PMBs. Thus, based on the limited test results of the five binders used in this study, the phase angle drop remains a logical failure definition for the displacement-controlled tests of all the binders and, for the controlled-stress tests, the peak in C×N is logical. However, it should be noted that the relative change in phase angle can be small in the TS tests, requiring close examination of the data.

4.2.2 PSE-Based Analysis

This research attempted to develop a PSE-based approach for defining CD-TS test failure. However, the PSE input history in the TS tests differed from that in the LAS tests because the loading amplitude was constant in the TS tests. Hence, the PSE input increased with every loading cycle in the LAS test whereas for the CD-TS it decreased. Figure 6 shows that, using the same definition of W^R_s and W^R_r in Equations 10 to 12, the evolutions differed significantly between the LAS and TS tests. The stored PSE decreased throughout the CD-TS testing because the loading input was constant and the material integrity decreased. Thus, there was no obvious means to define failure based on W^R_s for the TS test.

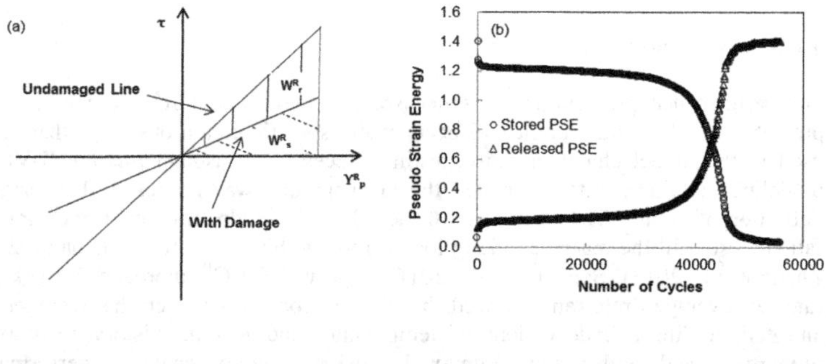

Figure 6. *CD-TS-based PSE definition and evolution: (a) schematic representation of PSE distribution and (b) evolution of stored and cumulative released PSE.*

It is worthwhile exploring the W^R_s and C×N definitions for the displacement-controlled TS tests, which are similar to the LAS tests with the exception of increasing strain amplitudes in the LAS test. Equation 14 shows the calculation of

W^R_s for the CD-TS tests based on the pseudo strain definition given in Equation 5. It is interesting to observe that W^R_s in the CD-TS tests is not dependent on $\tau_p \times t$ but is determined only from τ, because the loading input is constant and explains the observed trends in Figure 6(b) as the stress response continually decreases in the CD-TS tests. Further derivation of C×N in Equation 15 shows that the C×N in the CD-TS tests can be reduced to being dependent on $\tau_p \times t$, which is the same parameter on which W^R_s was found to be dependent in the LAS tests. For the CS-TS case, Equation 16 illustrates that C×N is determined from $(1/\gamma_p) \times t$, which is similar to the CD-TS case, because, in the stress-controlled test, strain is the dependent measurement and the stress level is fixed. Thus, the maximum C×N remains a reasonable means to define failure for the TS tests and can be used in place of a maximum phase angle in the case of binders that exhibit a clear peak. Because the TS test is not proposed for specifications, it is not anticipated to be problematic that a unified failure definition based on PSE could not be derived.

$$W^R{}_s = \frac{1}{2} \cdot \tau_p \cdot \gamma^R{}_p = \frac{1}{2} \cdot \tau_p \cdot \left|G^*\right|_{LVE} \cdot \gamma_p \quad \text{(for CD-TS)} \qquad [14]$$

$$C \times N = \frac{\tau_p}{\gamma^R{}_p \cdot DMR} \times t \cdot f = \frac{\tau_p}{\left|G^*\right|_{LVE} \cdot \gamma_p \cdot DMR} \times t \cdot f \quad \text{(for CD-TS)} \qquad [15]$$

$$C \times N = \frac{\tau_p}{\gamma^R{}_p \cdot DMR} \times t \cdot f = \frac{\tau_p}{\left|G^*\right|_{LVE} \cdot \gamma_p \cdot DMR} \times t \cdot f \quad \text{(for CS-TS)} \qquad [16]$$

5.0 Unified Failure Criterion: G^R Approach

5.1 G^R Background

For performance predictions, it is necessary to have a failure criterion that can predict when the failure point will occur under specific conditions other than those used in the model characterization testing. Recently, in asphalt mixture S-VECD modeling, a failure criterion called the G^R approach was proposed. The original definition of G^R is: "The stable rate of the released pseudo strain energy during the fatigue test until the macro-cracking localization within the material" (Zhang, 2012; Zhang et al., 2013). Sabouri et al. (2014) improved the G^R approach from 'stable rate' to 'averaged rate' and showed that the relationship between the averaged G^R and fatigue life is independent of temperature and loading history for asphalt mixtures tested under cyclic uniaxial tension, thereby enabling performance predictions with only limited testing. The G^R concept is similar to the ratio in dissipated energy change (RDEC) criterion employed by other researchers (Shen and Carpenter, 2005; Carpenter and Shen, 2006) but is consistent with the S-VECD modeling approach and easily implementable within the current analysis framework for the LAS tests.

5.2 Binder G^R Verification

The averaged G^R can be calculated easily from binder fatigue test results once the failure point is defined, using the following two steps:

Step 1: Calculate the average total released PSE ($\overline{W^R_r}$) during the test until failure.

Step 2: Calculate the G^R using Equation 17, where A is the area under the W^R_r curve until failure, as shown in Figure 3(b) and Figure 6(b).

$$G^R = \frac{\overline{W^R_r}}{N_f} = \frac{A / N_f}{N_f} = \frac{A}{(N_f)^2} \quad [17]$$

Figure 7 demonstrates the relationship between G^R and fatigue life (N_f) for NC binder STA and LTA samples subjected to both TS tests at different loading levels and LAS tests at different CSRs. It can be observed that a unique relationship exists between G^R and N_f that is independent of loading history. These results suggest that G^R versus N_f is a fundamental relationship for a given material, which is consistent with the asphalt mixture findings. This strong link between the LAS and TS tests is promising evidence that the accelerated LAS test can be used for performance predictions. The relationship between G^R and N_f can be derived by running several LAS-type tests using different CSRs, which is much more efficient than TS tests in which the testing time is highly variable.

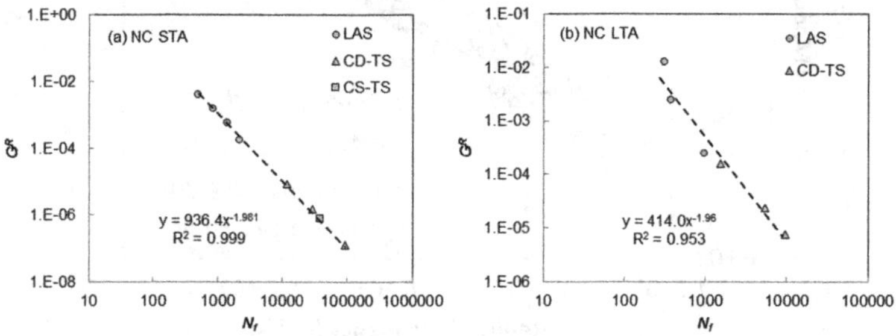

Figure 7. *NC binder* G^R *versus* N_f*: (a) STA and (b) LTA.*

6.0 G^R Verification of FHWA-ALF Binders

The four ALF RTFO-aged binders were tested to verify the G^R based failure criterion using additional materials, including polymer-modified asphalts. Each binder was subjected to LAS testing at three CSRs. For all the LAS tests, the strain amplitudes ranged from 0.1% to 30% strain, and the CSRs were varied. The CSRs were chosen so that the tests could be completed in 5 minutes (standard LAS test), 10 minutes, and 15 minutes. For each binder, TS tests also were conducted, as detailed in Table 2. The results were analyzed to assess the predictive capabilities of the S-VECD model combined with the G^R failure criterion.

6.1 Linear Viscoelastic Dynamic Modulus Mastercurves

The dynamic shear modulus mastercurves at the reference temperature of 19°C for the ALF binders are shown in Figure 8. It can be seen that the Control binder is the stiffest of the four materials tested. The three modified binders show similar stiffness at high loading frequency but distinguished modulus within low loading frequency.

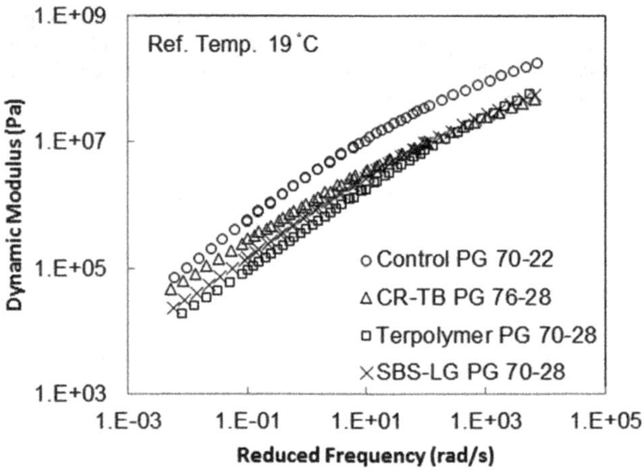

Figure 8. *Dynamic modulus mastercurves of ALF binders.*

6.2 Damage Characteristic Curves (DCCs)

Figure 9(a) shows the C versus S damage characteristic curves (DCCs) for the four asphalt binders tested using the standard LAS test with five minutes of loading. Previous DCCs of the FHWA-ALF asphalt mixtures obtained from cyclic direct tension tests are shown in Figure 9(b) (Kim and Underwood, 2010). The results demonstrate similar trends in the DCCs obtained from the binders and the mixtures, suggesting that the binder plays a critical role in determining the damage resistance of the mixtures.

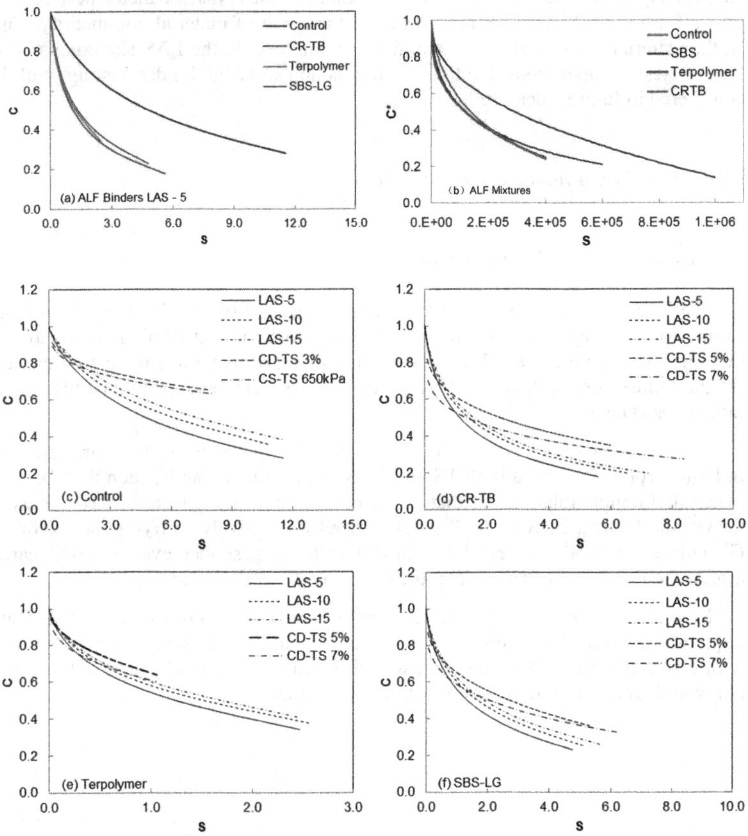

Figure 9. *Damage characteristic curves (DCCs) of ALF binders: (a) comparison of standard LAS-5 DCCs for all binders, (b) DCCs of the ALF asphalt mixtures, and (c) - (f) DCCs under different loading histories for each binder.*

Figure 9(c) to (f) present the DCCs for each binder derived from the LAS and TS tests. In Figure 9(c), the DCCs from the controlled-displacement test and the controlled-stress test collapse well, indicating the mode-of-loading independency of the DCC. It can be observed that the DCCs from the LAS tests under three CSRs collapse relatively well for all the binders. In addition, the DCCs from the TS tests with different strain amplitudes collapse fairly well. However, the discrepancies between the DCCs from the TS tests and the LAS tests are greater than those among the different loading conditions within the same test (i.e., LAS or TS test). This discrepancy is the greatest for the Control binder. These discrepancies between the DCCs from the LAS and TS tests could be the result of material nonlinearity. That is, the material is strained to a much higher strain level in the LAS test compared to the TS test. Proper consideration of the nonlinearity in binder testing will be considered in future research.

6.3 Fatigue Performance Characterization

6.3.1 Failure Criteria Determination

The fatigue life data for all ALF binders under different loading modes are summarized in Figure 10(a) (obtained directly from the test data). It is found that multiple loading modes exhibit a consistent ranking of materials, indicating that the fatigue failure defined here for the LAS and TS tests can be used to quantify binder fatigue resistance.

The characteristic G^R versus N_f relationships for the ALF materials are provided in Figure 10(b), including both TS and LAS test results. It can be seen that for each material, a unique relationship that is independent of loading history exists between the G^R and N_f (i.e., LAS versus TS tests), thereby effectively verifying the use of the G^R failure criterion. Interestingly, these results suggest that even if nonlinearity affects the DCCs, it does not affect the G^R versus N_f relationship.

Note that in order to derive the relationship, several data points are needed. Thus, implementation of the failure criterion would require conducting several LAS tests with differing CSRs. This approach would increase the total characterization time, but would greatly improve the predictive capabilities.

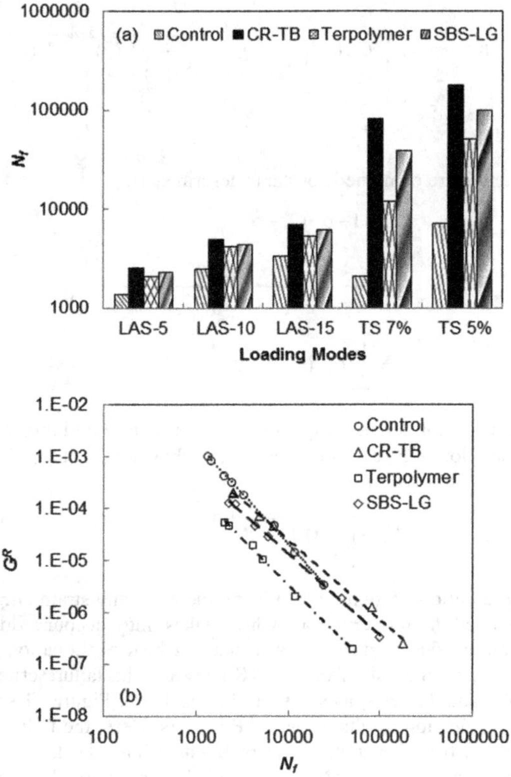

Figure 10. *ALF binder fatigue characterization: (a) N_f ranking based on different loading modes and (b)* G^R *versus* N_f *characteristic relationships.*

6.3.2 Fatigue Prediction Assessment

If a model is fit to the relationship between G^R and N_f, this failure criterion can be incorporated into the S-VECD model, thereby enabling performance predictions. The power law model is used in this study, as shown in Equation 18.

$$G^R = a \cdot (N_f)^b \qquad [18]$$

Based on Equations 1, 2 and 5, A in Equation 17 can be expressed as a function of γ and N_f:

$$A=\int_0^{N_f} W^R{}_r dN=\left[\frac{1}{2}\cdot C_1\cdot(|G^*|_{LVE})^2\cdot q^{(-\frac{C_2}{p})}\cdot\frac{1}{\frac{C_2}{p}+1}\right]\cdot(\gamma)^{2+2\alpha(\frac{C_2}{p})}(N_f)^{\frac{C_2}{p}+1}$$
$$=k\cdot(\gamma)^{2+2\alpha(\frac{C_2}{p})}(N_f)^{\frac{C_2}{p}+1} \quad [19]$$

Where p, q and k are combined constants described as:

$$p=1-\alpha\cdot C_2+\alpha \quad [20]$$

$$q=\frac{f\cdot 2^{\alpha}}{(1-\alpha\cdot C_2+\alpha)(C_1\cdot C_2)^{\alpha}(|G^*|_{LVE})^{2\alpha}} \quad [21]$$

$$k=\frac{1}{2}\cdot C_1\cdot(|G^*|_{LVE})^2\cdot q^{(-C_2/p)}\cdot\frac{1}{(C_2/p)+1} \quad [22]$$

Substituting Equation 19 into Equation 17, and then combining with Equation 18 derives the solution of N_f only in terms of γ, as shown in Equation 23.

$$N_f=\left[\frac{k}{a}\cdot(\gamma)^{2+2\alpha(\frac{C_2}{p})}\right]^{\frac{1}{b+1-\frac{C_2}{p}}} \quad [23]$$

Equation 23 allows fatigue life to be predicted at any strain amplitude using the newly developed failure criterion, which takes into account differences in the damage at failure due to differences in loading history. Equation 23 was verified using the C versus S results from the TS tests and the failure criterion determined from the LAS and TS tests, as shown in Figure 10(b). Figure 11 shows the TS test fatigue life verification results of the ALF binders. Here, the failure criterion model constants, a and b in Equation 18, were fitted to both the LAS and TS test data, whereas the damage model constants, C_1 and C_2 in Equation 7, were fitted to the TS test data only. The R^2 value of 0.95 and mean absolute error (MAE) of 22% demonstrate good agreement between the measured and predicted fatigue life data, indicating that the S-VECD model combined with the G^R failure criterion can be effectively applied for fatigue performance predictions.

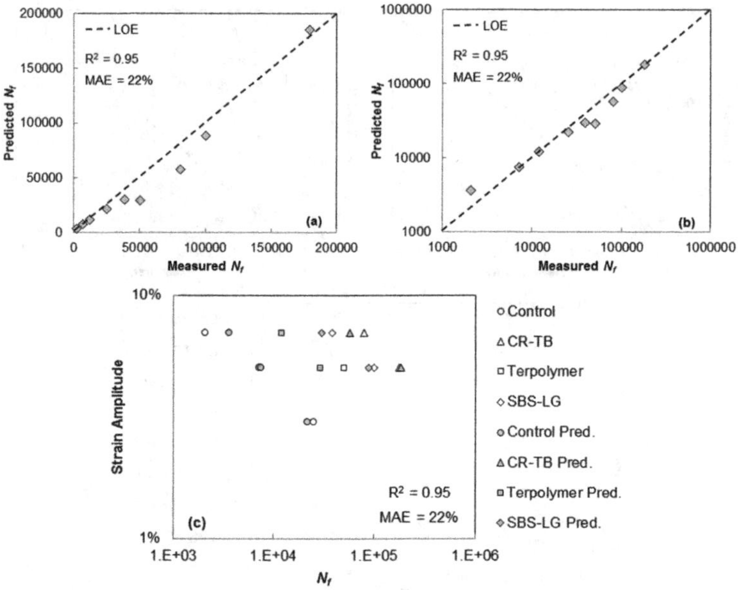

Figure 11. *Verification results for TS fatigue life: (a) arithmetic scale, (b) logarithmic scale, and (c) simulation for constant-displacement TS tests.*

The performance prediction ability of the combined S-VECD model and failure criterion only derived from LAS testing is of great interest as it could be practically implemented into standard specifications. In order to assess the performance prediction capability of the LAS tests, the C versus S curves derived from the LAS tests conducted at differing CSRs were fitted with one curve. Also, only LAS test results were used to develop the G^R failure criterion. The DCC parameters obtained (C_1 and C_2) were incorporated into Equation 23 combined with the G^R failure criterion relationship and used to predict fatigue life measured from the TS tests. The corresponding prediction results are shown in Figure 12, which indicates that the predictions are of only slightly lower accuracy than the verification results presented in Figure 11. This outcome is due to the differences in the DCCs of the TS and LAS test results and also due to the differences between the failure criterion obtained from the LAS and TS tests together and the failure criterion obtained from the LAS test alone. However, the predicted and measured fatigue life data are in reasonable agreement, with an R^2 value of 0.89 and MAE of approximately 30%. In addition, the predictions maintain a consistent ranking of the materials with the measured values, suggesting that the failure criterion combined with the S-VECD model applied to the LAS testing can be used to distinguish material fatigue performance.

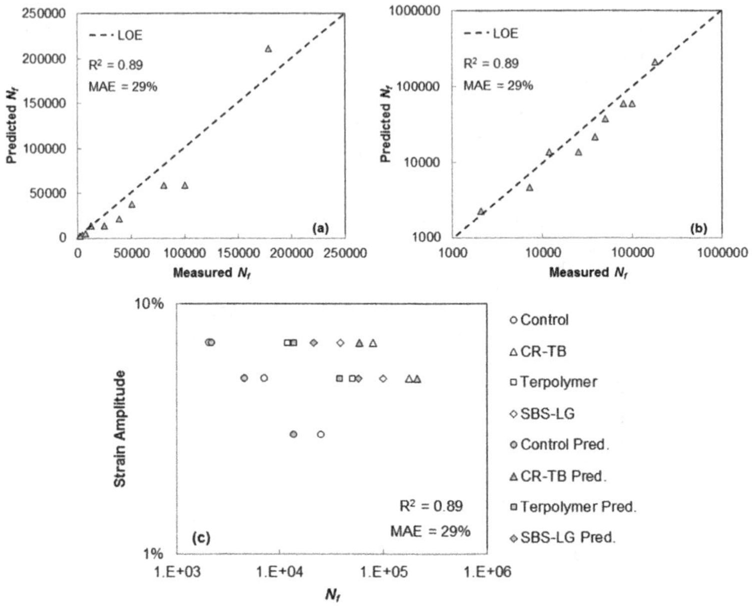

Figure 12. *LAS-based ALF binder fatigue prediction results: (a) arithmetic scale, (b) logarithmic scale, and (c) simulation for constant-displacement TS Tests.*

In order to implement combined S-VECD and G^R failure criterion approach, at least two additional LAS tests are required at varying CSRs in addition to the standard LAS test (5 minutes). Additional LAS test procedures that are designed to cover the same strain amplitude range (0.1% to 30%) but last 10 minutes and 15 minutes are recommended. These additional LAS tests could be used in place of replicate tests in order to improve efficiency, because any discrepancy in the results could be detected by observing C versus S trends, which should agree if the results are consistent. The aforementioned approach is practically implementable for fatigue specification of binders as it requires the use of standard DSR equipment and short test durations. It is envisioned that the specification could be implemented within the AASHTO MP 19 framework, which includes specification limits for the Multiple Stress Creep and Recovery (MSCR) test that are traffic dependent through having higher fatigue life requirements at certain strain amplitudes which increase with increasing traffic designation.

6.3.3 Field Validation

The FHWA-ALF study focused on evaluating modified asphalt pavements (Gibson, 2012). The study included lanes that were constructed with identical structures but varying asphalt concrete mixtures. The different mixtures were prepared using the same aggregate source and gradation, same air void content, and same asphalt content, but with different binders, including the ones tested in this study. The facility was kept at a constant temperature of 19°C during the fatigue evaluation. Loading consisted of 71kN wheel loads applied using tires inflated to 827kPa. Wheel wander was incorporated into the loading. The field cracking performance was closely monitored at the test site, and the fatigue life was quantified based on the number of load cycles (wheel passes) until a crack length reached 25m (Gibson, 2010).

The contribution of the asphalt binder to the fatigue performance of pavements is a function of the asphalt's ability to resist both damage and strain within the asphalt. Thus, it is important to consider both the loading that a given material will undergo in a given pavement structure as well as the material's damage resistance. Past research using the ALF mixtures that contained the binders that are under consideration in this study modeled the undamaged mixture response to a single moving wheel load that was equivalent to the conditions at the ALF with respect to pressure, speed, and temperature using layered viscoelastic pavement structural analysis (Kim et al., 2010). Based on this analysis, the tensile strain kernel at the bottom of the asphalt concrete layer of each ALF section under consideration in this study was determined. In order to use this information for binder fatigue life predictions, the peak tensile strain obtained from the strain kernel analysis of the asphalt mixtures was multiplied by 80, because it is well known that the strain experienced within the asphalt binder phase of asphalt mixtures is much greater than that experienced in bulk mixtures. The strain ratio of 80 was selected based on previous findings (Safaei et al., 2014) that suggested a binder-to-mix strain ratio of 80 works well for conducting identical fatigue life predictions between asphalt binders and mixtures. This information was input into Equation 23 for performance prediction using the combined S-VECD model and failure criterion.

Table 3. *Summary of the field performance validation.*

Materials	Kernel Strain	Strain Ratio	Binder Strain	LAS Test-based Binder N_f	Field N_f at 25-*m* Cracking
Control	3.60E-04	80	2.88E-02	14,987	45,000
CR-TB	5.15E-04	80	4.12E-02	441,755	100,000
Terpolymer	5.99E-04	80	4.79E-02	42,789	140,000
SBS-LG	5.50E-04	80	4.40E-02	85,649	207,000

The validation results in this study are displayed in Table 3, which shows that the rankings of materials based on binder and field observations are consistent, with the exception of the CR-TB material for which the asphalt binder results indicate a fatigue life that is much longer than that of the other binders, but the field performance indicates that the Terpolymer and SBS-LG materials performed better. Considering only the binders, it is not surprising that the CR-TB material performed the best because it contains both terminal blend ground tire rubber and styrene-butadiene-styrene, thereby including more modification than the other binders under consideration (Kluttz et al., 2014). It is speculated that this high level of modification may have resulted in the difference in the compaction properties of the corresponding mixture and, hence, difference in the aggregate structure formation, which cannot be captured in a binder test alone. Future research is needed to investigate the reason for the outlier behavior of the CR-TB material in the binder tests. Figure 13 shows the correlations between the measured field fatigue life data and the predicted binder fatigue life data for all the binders, with the exclusion of the CR-TB material. The results demonstrate a high correlation between the field performance and the predictions based on the binder test results. However, this information was derived from a very limited data set, and future research is needed to confirm these trends.

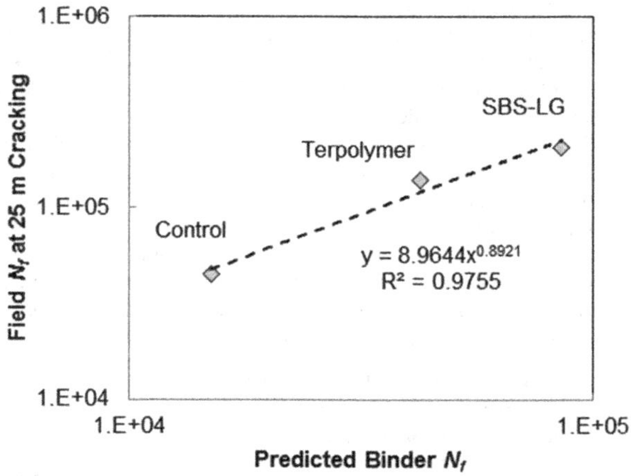

Figure 13. *Comparison between predicted binder* N_f *and field load passes using 25 m cracking formulation.*

7.0 Conclusions

This study proposes a PSE-based method for defining failure using LAS tests. In addition, a new fatigue failure criterion is proposed that can be incorporated into the S-VECD modeling framework for fatigue performance predictions of asphalt binders. The specific findings of this study are:

- The maximum stored PSE (Max W^R_s) is proposed for defining failure in the LAS test. This definition is based on logical trends in energy evolution and is consistent with phase angle trends that have been used extensively in defining asphalt mixture fatigue failure. For a "purchase specification" for asphalt binder, it is only needed to run the LAS test under one CSR, then identify the failure occurrence using the proposed Max W^R_s definition, which aims to quickly rank different binder materials.
- A unified failure criterion based on PSE release rates (G^R) can be applied to both LAS and TS tests to provide a unique relationship with the number of cycles to failure (N_f) that is independent of loading mode/history, which aims to establish a full fatigue characterization for one asphalt binder material. Deriving this fundamental relationship requires testing under several conditions. It is proposed that several LAS-type tests should be conducted with differing rates of increase in strain amplitude in order to derive the failure criterion efficiently.
- In this study, the fatigue life predictions using the newly developed failure criterion combined with the S-VECD model were in reasonable agreement with the laboratory-measured values and also generally related well with the field fatigue performance.

8.0 Acknowledgements

The authors would like to gratefully acknowledge the sponsorship from China Scholarship Council. The supports from National Natural Science Foundation of China under Grant No. 51478021 and Beijing Natural Science Foundation under Grant No. KZ201510005002 are also acknowledged.

9.0 References

Bonnetti, K. S., Nam, K. and Bahia H. U., "Measuring and Defining Fatigue Behavior of Asphalt Binders," *Transportation Research Record: Journal of the Transportation Research Board*, No. 1810, 2002, pp. 33–43.

Branco, V. T. F. C., Masad, E. A., Bhasin, A. and Little, D. N., "Fatigue Analysis of Asphalt Mixtures Independent of Mode of Loading," *Transportation Research*

Record: Journal of the Transportation Research Board, No. 2057, 2008, pp. 149–156.

Carpenter, S. H. and Shen, S., "A Dissipated Energy Approach to Study HMA Healing in Fatigue," *Transportation Research Record: Journal of the Transportation Research Board*, No. 1970, 2006, pp. 178–185.

Christensen, R. M., "Defining Yield Stress and Failure Stress (Strength)," Chapter 9 in *Theory of Materials Failure*, Oxford University Press, 2013, pp. 1–14.

Farrar, M. J., Grimes, R. W., Sui, C., Planche, J-P., Huang, S-C., Turner, T. F. and Glaser, R., "Thin Film Oxidative Aging and Low Temperature Performance Grading Using Small Plate Dynamic Shear Rheometer: An Alternative to Standard RTFO, PAV, and BBR," *Presented at the Fifth Europe Bitumen Congress, Istanbul, Turkey, 2012.*

Ghuzlan, K. A. and Carpenter, S. H., "Energy-derived, Damage-based Failure Criterion for Fatigue Testing," *Transportation Research Record: Journal of the Transportation Research Board*, No. 1723, 2000, pp. 131–141.

Gibson, N., "*Summary* of Binder Fatigue Findings from FHWA-ALF," *Presented at the Binder Expert Task Group Meeting, Irvine, California, 2010.*

Gibson, N., "Performance Testing for Superpave and Structural Validation," *Report No. FHWA-HRT-11-045*, Federal Highway Administration, 2012.

Hicks, R. G., Finn, F. N., Monismith, C. L. and Leahy, R. B., "Validation of SHRP Binder Specification through Mix Testing," *Journal of Association of Asphalt Paving Technologists*, Vol. 62, 1993, pp. 565–614.

Hintz, C., Velasquez, R. A., Johnson, C. M. and Bahia, H. U., "Modification and Validation of the Linear Amplitude Sweep Test for Binder Fatigue Specification," *Transportation Research Record: Journal of the Transportation Research Board*, No. 2207, 2011, pp. 99–106.

Kim, Y. R., Lee, H. J. and Little, D. N., "Fatigue Characterization of Asphalt Concrete using Viscoelasticity and Continuum Damage Theory," *Journal of Association Asphalt Paving Technologists*, Vol. 66, 1997, pp. 520–569.

Kim, Y., Little, D. N. and Lytton, R. L.., "Fatigue and Healing Characterization of Asphalt Mixtures," *Journal of Materials in Civil Engineering*, Vol. 15, No. 1, 2003, pp. 75–83.

Kim, Y. R. and Underwood, B. S., "S-VECD Fatigue Test Protocol and Analysis Software," *Presented at the Asphalt Mixture and Construction Expert Task Group Meeting, Irvine, California*, 2010.

Kluttz, B., Castorena, C., Puchalski, S. and Andriescu, A., "Recent Developments in Asphalt Binder Linear Amplitude Sweep (LAS) Test," *Presented at the 51st Peterson Asphalt Research Conference, Laramie, Wyoming, 2014.*

Masad, E. A., Branco, V. T. F. C., Little, D. N. and Lytton, R. L., "A Unified Method for Analysis of Controlled-Strain and Controlled-Stress Fatigue Testing," *International Journal of Pavement Engineering*, Vol. 9, No. 4, 2008, pp. 233–246.

Norouzi, A., Sabouri, M. and Kim, Y. R., "Evaluation of the Fatigue Performance of High RAP Asphalt Mixtures," *Presented at 12th International Society of Asphalt Pavement (ISAP) Conference, June 1-5, 2014, Raleigh, North Carolina, U.S.A.* In press.

Reese, R., "Properties of Aged Asphalt Binder Related to Asphalt Concrete Fatigue Life," *Journal of Association Asphalt Paving Technologists*, Vol. 66, 1997, pp. 604–632.

Rowe, G. M., "Performance of the Asphalt Mixtures in the Trapezoidal Fatigue Test," *Journal of Association Asphalt Paving Technologists*, Vol. 62, 1993, pp. 344–384.

Rowe, G. M. and Bouldin, M. G., "Improved Techniques to Evaluate the Fatigue Resistance of Asphalt Mixtures," *Proceeding of 2nd Euroasphalt & Eurobitumen Congress, Barcelona, Spain*, 2000, pp. 754–763.

Sabouri, M. and Kim, Y. .R., "Development of a Failure Criterion for Asphalt Mixtures under Different Modes of Fatigue Loading," *Transportation Research Record: Journal of the Transportation Research Board No. 2447*, 2014, pp. 117–125.

Safaei, F., Lee, J., Nascimento, L. A. H., Hintz, C. and Kim, Y. R., "Implications of Warm-Mix Asphalt on Long Term Oxidative Aging and Fatigue Performance of Asphalt Binders and Mixtures," *Journal of the Association of Asphalt Paving Technologists*, 2014, pp. 143–170.

Schapery, R. A., "Correspondence Principles and a Generalized J-Integral for Large Deformation and Fracture Analysis of Viscoelastic Media," *International Journal of Fracture*, Vol. 25, No. 3, 1984, pp. 195–223.

Shen, S. and Carpenter, S. H., "Application of Dissipated Energy Concept in Fatigue Endurance Limit Testing," *Transportation Research Record: Journal of the Transportation Research Board*, No. 1929, 2005, pp. 165–173.

Sui, C., Farrar, M. J., Harnsberger, P. M., Tuminello, W. H. and Turner, T. F., "New Low-Temperature Performance-Grading Method: Using 4-mm Parallel Plates on a Dynamic Shear Rheometer," *Transportation Research Record: Journal of the Transportation Research Board*, No. 2207, 2011, pp. 43–48.

Sui, C., Farrar, M. J., Tuminello, W. H. and Turner, T. F., "New Technique for Measuring Low-Temperature Properties of Asphalt Binders with Small Amounts of Material," *Transportation Research Record: Journal of the Transportation Research Board*, No. 2179, 2010, pp. 23–28.

Underwood, B. S., "Multiscale Constitutive Modeling of Asphalt Concrete," *Ph.D. Dissertation*, North Carolina State University, Raleigh, N.C., 2011.

Underwood, B. S., Kim, Y. R. and Guddati, M. N., "Improved Calculation Method of Damage Parameter in Viscoelastic Continuum Damage Model," *International Journal of Pavement Engineering*, Vol. 11, No. 6, 2010, pp. 459–476.

Van Dijk, W., "Practical Fatigue Characterization of Bituminous Mixes," *Journal of Association Asphalt Paving Technologists*, Vol. 44, 1975, pp. 38–74.

Van Dijk, W. and Visser, W., "The Energy Approach to Fatigue for Pavement Design," *Journal of Association Asphalt Paving Technologists*, Vol. 46, 1977, pp. 1–40.

Western Research Institute, "Standard Method of Test for Determining the Low Temperature Rheological Properties of Asphalt Binder Using a Dynamic Shear Rheometer (DSR)," *Draft AASHTO Standard Method,* 2012.

Western Research Institute, "Standard Method of Test for the Universal Simple Aging Test (USAT): Simulating Asphalt Binder Short and Long Term Hot and Warm Mix Oxidative Aging in the Laboratory," *Draft AASHTO Standard Method*, 2013.

Zhang, J., "Development of Failure Criteria for Asphalt Concrete Mixtures under Fatigue Loading," *Master Thesis*, North Carolina State University, Raleigh, NC., 2012.

Zhang, J., Sabouri, M., Guddati, M. N. and Kim, Y. R., "Development of a Failure Criterion for Asphalt Mixtures under Fatigue Loading," *Journal of the Association of Asphalt Paving Technologists*, Vol. 82, 2013, pp. 1–22.

10.0 Discussions

DAVID ANDERSON: Did you make any accommodation for, or maybe there was no difference, but the difference in temperature between the field loading and the laboratory loading? Are they are the same temperature?

CASSIE CASTORENA: We did this at 19°C, which was the temperature at which the....

DAVID ANDERSON: So we don't have a temperature effect between the field and the lab then?

CASSIE CASTORENA: No.

DAVID ANDERSON: That's something that most people don't seem to grasp. We go into the laboratory and do beam fatigue tests at 19° and then try to correlate it with something in the field where the temperatures run anything from minus whatever to plus whatever, and any correlation there is strictly fortuitous not rational. So I was curious about that. And then we also have the rate of loading effect, as well. And I wonder if your loading rates correlate to what we have in the field? The ALF is a pretty low loading rate. Are we going to have the same fatigue damage in the field if we are on an Interstate highway or if we're running at relatively low speeds that the ALF has?

CASSIE CASTORENA: That's something we're currently looking into. In this study, we just use the strain amplitude that we scaled from the mix, but we have the strain kernel from the pavement structural analysis, so we could input the actual loading form into the S-VECD model and can model the fatigue response in place of sinusoidal loading used herein. We are looking into that as it would account for the rate effects. Thank you.

AUTHOR'S CLOSURE

CHAO WANG: The fatigue failure analysis of asphalt binder in this paper was established based on the LAS test results under a single temperature regarding the possibility of arriving at a more effective binder fatigue specification. Nonlinearity, temperature/loading rate and aging effects as well as the healing potential should be further addressed and characterized for a comprehensive understanding of binder contribution to pavement fatigue resistance.

Modifying Laboratory Mixture Design to Improve Field Compaction

Ali Hekmatfar[a], Ayesha Shah[b], Gerald Huber[c], Rebecca McDaniel[b], John E. Haddock[a]

[a]*Purdue University, Lyles School of Civil Engineering, 550 Stadium Mall Drive, West Lafayette, IN*

[b]*North Central Superpave Center, Purdue University, Lyles School of Civil Engineering, West Lafayette, IN*

[c]*Heritage Research Group, 7901 West Morris Street, Indianapolis, IN*

ABSTRACT. Field data collected in Indiana suggests that if asphalt mixtures were designed to be more compactable in the field, they could be compacted to a field density equivalent to the laboratory mixture design density, potentially increasing pavement durability. The objective of this research was to modify the mixture design in order to increase in-place mixture density without sacrificing permanent deformation characteristics. Three 100-gyration mixtures were used, each meeting applicable specifications and designed according to AASHTO M 323. For each mixture, additional designs were completed using 30, 50, and 70 gyrations. Optimum binder content for these mixtures was chosen at 5% air voids, rather than 4%, and the effective binder content was held constant. Results indicate that the mixture produced using 30, 50, and 70 gyrations had permanent deformation characteristics equal to or better than the original 100-gyration mixtures. A trial field project confirmed that mixtures designed at 5% air voids could be compacted to this level, showing that the approach is feasible from a construction viewpoint.

KEYWORDS: hot-mix asphalt; permanent deformation; durability; mixture design

1.0 Introduction

Asphalt pavements in Indiana currently have nominal service lives of 15-20 years. Generally, these pavements reach the end of their service lives because of durability issues associated with asphalt binder aging. As the asphalt binder ages, stiffness and embrittlement increase. In terms of surface condition, aging is manifested as cracking that is usually non-load associated, but may also include load associated cracking. Asphalt binder aging is predominantly related to oxidation, which is

The oral presentation was made by Professor Haddock.

This is a reproduction of an article whose final and definitive form has been published in *Road Materials and Pavement Design*© 2015 Taylor & Francis. The article is available online at: http://dx.doi.org/10.1080/14680629.2015.1077003

controlled by access of air into the asphalt pavement. Reducing the air permeability of asphalt mixtures decreases the rate of asphalt binder aging, and thus allows for a longer pavement life.

The current Indiana Department of Transportation (INDOT) Superpave asphalt mixture design method targets a 4% air void (V_a) content in the laboratory compacted (design and quality assurance) specimens. On the road, the density acceptance criterion for INDOT quality assurance mixtures is a percent-within-limits specification with a lower threshold limit of 91% of maximum theoretical specific gravity (G_{mm}). The average density needed to comply with this specification is approximately 93% (7% V_a). The result is that when the in-place density specification is met, 10% of the pavement area may have a density of less than 91%, meaning a V_a content of more than 9%. Test data has shown that when in-place densities exceed 8%, air permeability begins to increase dramatically (Cooley et al., 2002).

Increasing average in-place asphalt mixture densities to 95% (V_a of 5%) would significantly decrease asphalt binder aging. Conservatively, the pavement life increase would be two to three years, representing a 12 to 20% increase in pavement life. The key to achieving this in-place density is to modify the relationship between laboratory and field compaction. Therefore, the objective of the research summarized here is to modify asphalt mixture laboratory design compaction as it relates to field compaction in order to increase in-place asphalt pavement durability without sacrificing permanent deformation characteristics of the mixtures.

2.0 Background

Since its adoption in the 1940s, the Marshall mixture design method has been widely used throughout world. In the 1960s and 70s, the Laboratoire Central des Ponts et Chaussées (LCPC) developed a mixture design method still used in France. Today, most agencies in the United States make use of the Superpave mixture design method to design asphalt mixtures. This mixture design method contains similarities to both the Marshall and LCPC methods.

2.1 Marshall Mixture Design Method

The Marshall mixture design philosophy involves selecting an aggregate gradation and a compaction level. Aggregate is mixed with varying percentages of asphalt binder and then compacted using a drop hammer. The air voids in the compacted samples are then determined and compared to the specification values. Normally, 4% air voids is desired with a voids in the mineral aggregate (VMA) requirement that is based on the nominal maximum size of the aggregate blend. If the specified air voids cannot be achieved by merely varying the asphalt content, a new aggregate gradation, or even the use of new aggregate materials, must be examined. A design is selected when the gradation is within a specified range and the volumetrics meet specified criteria (Asphalt Institute, 1997).

The philosophy of Marshall design compaction is that the density of specimens in the mold should match the ultimate density that will be obtained after the mixture has been in service for some period of time. Typically air voids in the Marshall mold were designed to be from 3 to 5% (McFadden and Ricketts, 1948; Griffith, 1949). The mixture was to be compacted to 8 to 10% air voids on the road. After being trafficked—some believe for two to four years, others believe at the end of in-service life (15 years)—the density would increase, leaving only 3 to 5% air voids in the pavement.

Early developmental research for the method sought to correlate the density of field test sections to the number of blows of a Marshall hammer. As a result, the design compaction level is related to the expected amount of traffic. Three levels of compaction were recommended, 35, 50 and 75 blows on each side of a specimen to be used for light, medium and heavy traffic, respectively (McFadden et al., 1948).

2.2 LCPC Mixture Design Method

In the 1960s and 70s, LCPC developed a new method of mixture design. The design method is based on the principle that an asphalt mixture should be designed and, during construction, compacted to its ultimate density. Therefore, no post construction densification is anticipated. The density at the end of the mixture's in-service life is the same as the density immediately after construction.

Developmental research for the LCPC method sought to establish a design compactive effort to match the compaction that occurs during construction. LCPC defined a standard rolling train as 16 passes of a 3 tonne (6,600 lbs) pneumatic-tired roller and having a tire pressure of 600 kPa (87 psi).

A full-scale laboratory compaction bench was built and a range of different mixtures was compacted. The compaction bench was 2.1 m (7 feet) wide and 4.2 m (14 feet) long with one pneumatic tire. It was discovered that, for a specific mixture, the density achieved under a standard rolling train was related to the lift thickness. When lift thickness was too thin, density was lower (Moutier, 1977).

As a result, a standard lift thickness was established based on maximum aggregate size. The required lift thickness is five times the maximum aggregate size. This is equivalent to six times the nominal maximum aggregate size as defined by the Superpave mixture design method. So for a 0/14 mixture, the maximum aggregate size is 14 mm (0.55 in.) and the lift thickness is 80 mm (3 in.), about six times the nominal maximum aggregate size of 12.5 mm (0.5 in.). The method results in the use of thick lifts in France. In the United States lifts tend to be thinner; about four times the nominal maximum aggregate size.

For laboratory compaction, LCPC chose a gyratory compactor. They had prior experience with the Marshall hammer and with static compaction. (The Duriez method of mixture design, used before the LCPC method, used static compaction.) In 1959, a French delegation visited Texas and witnessed the Texas gyratory compactor. Studies were done, and by 1968 a gyratory compaction protocol was developed in France.

The Texas compactor uses an angle of nearly six degrees and applies gyrations at the rate of one per second in bursts of three gyrations. After each set of gyrations the specimen is checked for resistance to further compaction. If the specimen is not yet resistant enough, an additional set of gyrations is applied. Compaction is complete when the specimen is sufficiently resistant to compaction (Ortolani and Sandberg, 1952).

LCPC designed and constructed a gyratory compactor that applied gyrations continuously instead of in bursts. Studies were done to investigate the effect of gyratory angle and vertical pressure for different types of mixtures. Studies were also done to select specimen size. As a result of these studies LCPC standardized the gyratory angle at one degree with a vertical pressure of 600 kPa (87 psi). LCPC also monitored the rate of increase in density and discovered a nearly linear relationship between density and the log of the number of gyrations (Moutier, 1974).

After correlating compaction curves in the gyratory to densification curves from the laboratory compaction bench, LCPC set the design number of gyrations to equal the lift thickness in millimeters. Therefore, a 14-mm maximum size mixture (12.5-mm nominal maximum size) would be constructed 80 mm (3 in.) thick (five times maximum size) and the design compaction would be 80 gyrations.

In the LCPC method, the design asphalt binder content is fixed for each mixture type; there are adjustment factors for asphalt absorption, aggregate specific gravity, and gradation or surface area. Since the effective asphalt binder volume is fixed, the design method becomes one of selecting an aggregate structure to provide the correct range of air voids at the design compaction. The range of allowable air voids is 4 to 8%. Most designers target 5% air voids. In the field, the required density is 95% of maximum theoretical gravity. LCPC has documented that there is little or no increase in density under traffic during the pavement in-service life.

2.3 Superpave Mixture Design Method

The Superpave mixture design method and procedures were developed as part of the Strategic Highway Research Program (SHRP) to be a comprehensive system for the design and modeling of asphalt materials. Asphalt binder testing was implemented to relate the performance of the binder to the climate and traffic level. Aggregate quality specifications were established in an effort to improve the performance of the resulting mixtures. The gyratory compactor was developed as a laboratory tool that more closely simulates field compaction of asphalt mixtures.

During SHRP, developmental studies focused on relating gyratory compactive effort to the density of pavements at the end of their in-service lives. The underlying principle in SHRP was carried over from the Marshall method. The design air voids were to be selected at a low level (air voids were 4% instead of the range of 3 to 5%) and construction air voids were expected to be 8%. The N-design experiment during SHRP determined the number of gyrations required to match in-place density for pavements that were at least 12 years old and had been subjected to various levels of traffic as measured by Equivalent Single Axle Loads (ESAL) (Huber et al., 1994). The decision to fix design air voids at 4%, instead of using the range of 3 to 5% came about because of an issue dealing with VMA.

In the late 1980s the Asphalt Institute was evaluating issues dealing with VMA. The reason for VMA, since its inclusion in Marshall design in 1962, was to ensure that a mixture had a minimum volume of effective asphalt binder and a minimum volume of air voids. Minimum VMA was fixed according to the aggregate nominal maximum aggregate size and was constant regardless of the design air voids. As a result, the minimum effective volume of asphalt binder could vary. For example, a 12.5-mm nominal maximum size mixture required 14% VMA. The volume of effective asphalt binder could range from 11 to 9% depending on whether the design air voids were chosen at 3 or 5%. As a result, in 1989 the Asphalt Institute changed the design criteria in MS-2, the *Manual of Mix Design Methods*, from a range of 3 to 5% to a single value of 4%.

During SHRP, consideration was given to discontinuing the use of VMA and instead specifying a minimum effective asphalt volume, for example, specifying a minimum effective asphalt volume of 10% for a 12.5-mm nominal maximum mixture. Then, the air void content could be specified as a range of 3 to 5%. Instead, the decision was made to keep VMA as a design criterion. To prevent the problem of effective asphalt volume varying depending on design air voids, the decision was made to adopt the Asphalt Institute recommendation of a constant 4% design air void criteria.

3.0 Study Methodology

INDOT uses the standard Superpave method to design asphalt mixtures. Optimum binder content for a mixture is chosen at 4% air voids. The compaction level depends on the amount of traffic the asphalt mixture is expected to encounter over its service life. Table 1 shows the asphalt pavement traffic categories specified by INDOT along with the number of Superpave Gyratory Compactor (SGC) gyrations (N_{design}) required for compacting laboratory specimens during the mixture design process. The category reflects the anticipated traffic on the roadway where the asphalt mixture will be used over a 20-year design life.

Table 1. *Asphalt pavement categories.*

Category	Equivalent Single Axle Load	N_{design}
1	< 300,000	50
2	300,000 to < 3,000,000	75
3	3,000,000 to < 10,000,000	100
4	10,000,000 to <30,000,000	100
5	≥ 30,000,000	125

In order to complete the objective of the study, the following tasks were undertaken:

- Appropriate asphalt mixtures were chosen for testing. This involved considering various combinations of factors such as pavement categories, aggregate types, mixture types (gradations), and asphalt binder types.
- For each of the asphalt mixture designs chosen for inclusion in the research, a standard asphalt mixture design was completed. This standard mixture design is referred to as the "original" mixture design.
- For each original mixture design, three new mixture designs were completed using lower N_{design} levels. Optimum binder contents for these mixtures were selected to achieve 5% air voids with the same effective binder volume as their respective original designs. The effective binder content was kept the same to ensure good durability of the mixtures. These are referred to as "re-designed" mixtures.
- Dynamic modulus and flow number test specimens were prepared and tested. Specimens for the original mixtures were compacted to 7% air voids because that is the anticipated air void content to which these mixtures would be compacted at the time of construction. The re-designed mixture test specimens were compacted to 5% air voids because this corresponds to their anticipated field density at the time of construction. It is known that the air void level affects asphalt mixture stiffness and flow, but the point is to explore how mixtures designed at these different laboratory compaction levels will perform when compacted to the anticipated field densities.
- The data was analyzed to determine how the dynamic moduli and flow number values for the re-designed mixtures compare to those of the standard mixtures.

In order to keep the experiment to a manageable level, three asphalt mixtures were chosen from the many possibilities; these are shown in Table 2. The three mixtures represent two traffic pavement categories (3 and 4, as shown in Table 1), both requiring an N_{design} of 100 gyrations. The mixtures are of two sizes (9.5 and 19.0 mm) and incorporate three coarse aggregate types (limestone, dolomite, and air cooled blast furnace (ACBF) slag).

Table 2. *Experimental design matrix.*

<table>
<tr><td colspan="5" rowspan="2"></td><td colspan="2">Mixture Type</td></tr>
<tr><td>9.5-mm</td><td>19.0-mm</td></tr>
<tr><td rowspan="6">Pavement Category</td><td rowspan="3">3</td><td rowspan="6">ESAL</td><td rowspan="3">3,000,000 to <10,000,000</td><td rowspan="6">Number of Gyrations</td><td>30</td><td>X</td><td></td></tr>
<tr><td>50</td><td>X</td><td></td></tr>
<tr><td>70</td><td>X</td><td></td></tr>
<tr><td rowspan="3">4</td><td rowspan="3">10,000,000 to <30,000,000</td><td>30</td><td>X</td><td>X</td></tr>
<tr><td>50</td><td>X</td><td>X</td></tr>
<tr><td>70</td><td>X</td><td>X</td></tr>
</table>

3.1 Materials

The gradations of the aggregates used in the study are shown in Table 3 along with their bulk specific gravities (G_{sb}). Dolomite is defined as carbonate rock containing at least 10.3% elemental magnesium. A PG 64-22 asphalt binder was used for all mixtures. No recycled materials were included in the designs.

Table 3. *Aggregate gradations (percent passing) and specific gravities.*

Sieve Size, mm	Limestone				Blast Furnace Slag	Dolomite			Natural Sand	Bag House Dust
	#8	#11	#12	#24 Sand	#11	#11	#12	#24 Sand		
25.0	100.0									
19.0	90.3									
12.5	49.7	100.0	100.0		100.0	100.0	100.0			
9.5	25.8	83.3	99.6	100.0	84.4	89.1	99.9	100.0	100.0	
6.3	13.0	44.2	63.8	99.6	41.5	43.3	79.9	99.7	98.6	
4.75	6.8	25.2	46.4	99.4	21.7	22.1	70.7	99.6	97.9	
2.36	2.6	4.7	11.1	90.6	6.0	6.8	20.8	94.9	90.8	
1.18	2.0	2.2	4.1	56.0	4.4	2.9	6.3	63.4	75.3	
0.600	1.7	1.9	2.5	33.4	3.7	1.8	2.6	36.7	57.3	100.0
0.300	1.4	1.7	2.1	16.1	3.2	1.4	1.6	20.1	23.4	99.4
0.150	1.2	1.5	1.9	7.8	2.6	1.2	1.2	8.9	2.4	97.8
0.075	1.0	1.4	1.7	5.0	2.0	0.9	1.0	4.4	1.1	94.6
G_{sb}	2.664	2.664	2.655	2.671	2.442	2.681	2.688	2.748	2.612	2.700

3.2 Mixture Designs

As mentioned previously, the three original mixture designs were used as the starting points for this experiment. Each mixture was re-designed in order to reach 5% air voids, instead of 4%, at lower gyration levels with the same effective asphalt binder content. Of course as is often the case with asphalt mixture designs, these air void and effective binder content values were nominal. The same aggregates were used in different proportions to achieve the designs. The Bailey method (Vavrik et al., 2002) was used to suggest how to change the proportions of the various aggregate components to make the mixtures more compactable. Table 4 shows the aggregate gradations of the original and re-designed mixtures. The optimum binder contents and volumetric properties of the mixtures are shown in Table 5. N100, N70, N50, and N30 refer respectively to 100, 70, 50, and 30, the number of SGC gyrations used to compact the mixture specimens during performance of the mixture design. In each case, the N100 mixtures are the original mixtures, while N70, N50, and N30 are re-designed mixtures.

The Category 4, 19.0-mm asphalt mixtures consisted of limestone coarse aggregate, limestone sand, and baghouse fines. Several trial gradations were tested before the final gradations for each of the mixtures (Figure 1) were selected.

The Category 3, 9.5-mm mixtures were produced from dolomite coarse aggregate, dolomitic and natural sands, and baghouse fines. The gradations of the four mixtures are shown in Figure 2.

Finally, the Category 4, 9.5-mm mixtures were made with dolomite and slag coarse aggregates, dolomite and natural sand, and baghouse fines. The gradations of these mixtures are shown in Figure 3. Only three mixture designs were completed for this mixture; the original (N_{design} of 100) and re-designed mixtures compacted with N_{design} values of 30 and 50. Mixture design data from both the Category 4, 19.0-mm and Category 3, 9.5-mm mixtures indicated little difference between the N100 and N70 mixtures when the optimum binder content was chosen at 4 and 5% air voids respectively. Lowering the number of design gyrations by 30 effectively raised the mixture air voids by one percent. Thus the gradation was either not changed or changed only slightly from the original; therefore the 70-gyration mix was eliminated.

Table 4. *Aggregate gradation of the mixtures and specific gravity.*

Mixture	Category 4, 19.0-mm				Category 3, 9.5-mm				Category 4, 9.5-mm		
Sieve Size, mm	N100	N70	N50	N30	N100	N70	N50	N30	N100	N50	N30
25.0	100.0	100.0	100.0	100.0							
19.0	97.4	97.4	96.1	95.3							
12.5	86.4	86.4	79.9	75.5	100.0	100.0	100.0	100.0	100.0	100.0	100.0
9.5	77.0	77.4	68.4	62.1	97.4	97.3	95.6	95.4	95.3	94.7	94.1
6.3	59.6	60.7	55.1	51.8	75.8	76.0	71.6	72.6	72.8	72.2	71.4
4.75	51.2	52.7	48.6	46.8	65.4	65.7	60.0	61.6	61.9	61.3	60.5
2.36	36.5	38.2	37.8	39.3	33.0	34.7	38.1	43.7	34.0	38.2	42.4
1.18	22.8	23.9	23.9	25.5	19.1	20.4	24.6	29.5	20.4	24.9	29.2
0.600	14.6	15.3	15.3	16.6	12.4	12.9	16.1	19.4	12.8	16.4	19.7
0.300	8.5	8.8	8.8	9.7	7.8	7.7	9.5	11.3	7.6	9.9	11.5
0.150	5.6	5.7	5.6	6.4	4.9	4.4	5.3	6.0	4.2	5.7	6.1
0.075	4.4	4.5	4.4	5.1	4.0	3.4	4.0	4.5	3.0	4.3	4.6
G_{sb}	2.665	2.665	2.650	2.651	2.692	2.692	2.692	2.694	2.631	2.630	2.626

Table 5. *Mixture design volumetrics.*

Category 4, 19.0-mm Mixture						
N_{design}	Avg. V_a, %	Avg. G_{sb}	Avg. VMA, %	Avg. VFA, %	P_{be}, %	Avg. Dust Ratio
100	4.0	2.665	13.6	70.6	4.1	1.1
70	4.9	2.665	14.5	66.3	4.1	1.1
50	4.9	2.650	14.4	66.0	4.1	1.1
30	4.9	2.651	14.9	67.2	4.3	1.2
Category 3, 9.5-mm Mixture						
N_{design}	Avg V_a, %	Avg. G_{sb}	Avg. VMA, %	Avg. VFA, %	P_{be}, %	Avg. Dust Ratio
100	4.1	2.692	15.0	72.9	4.6	0.9
70	5.1	2.692	16.0	67.9	4.6	0.7
50	4.9	2.692	15.8	68.9	4.6	0.9
30	5.3	2.694	16.3	67.6	4.7	0.9
Category 4, 9.5-mm Mixture						
N_{design}	Avg. V_a, %	Avg. G_{sb}	Avg. VMA, %	Avg. VFA, %	P_{be}, %	Avg. Dust Ratio
100	3.8	2.631	15.0	74.9	4.8	0.6
50	4.9	2.630	16.4	70.0	5.0	0.9
30	5.0	2.626	16.4	69.6	5.0	0.9

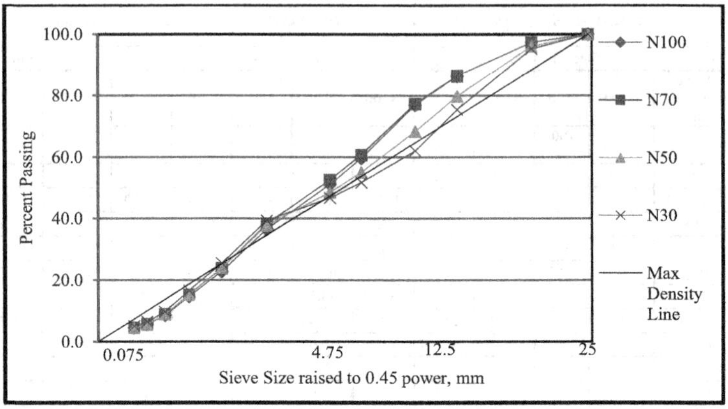

Figure 1. *Category 4, 19.0-mm mixture gradations.*

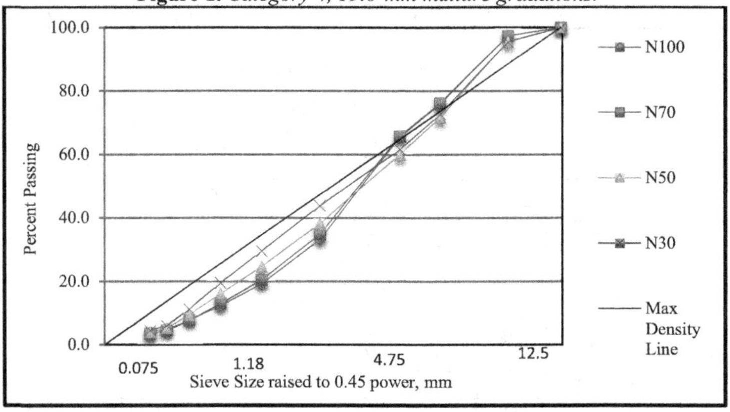

Figure 2. *Category 3, 9.5-mm mixture gradations.*

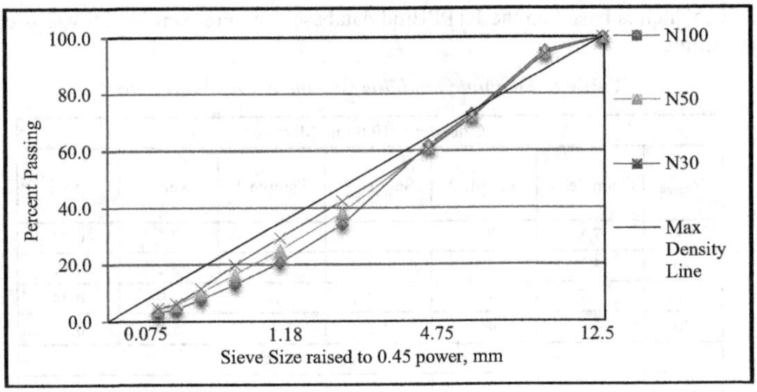

Figure 3. *Category 4, 9.5-mm mixture gradations.*

3.3 Test Methods

Once the mixture designs were complete, a set of four test specimens was produced, according to AASHTO PP 60, "Preparation of Cylindrical Performance Test Specimens Using the Superpave Gyratory Compactor (SGC)," to a specific air void content corresponding to the expected field compaction level. Samples of the original mixture (N100) were compacted to 7% air voids (after coring and sawing) to simulate field compaction at the start of pavement life. Test specimens for the re-designed mixtures were compacted, cored and drilled to produce 5% air voids, to simulate field compaction at the start of pavement life. Once the specimens were fabricated and air void content checked, the dynamic modulus and flow number tests were performed. Table 6 shows the air void contents for the specimens prior to testing.

The mixtures were characterized using the dynamic modulus test (AASHTO TP 79-10) and the flow number test (AASHTO TP 79-10). Both tests were completed in the Asphalt Mixture Performance Tester (AMPT). The dynamic modulus test determines the |E*| value, commonly referred to as the mixture stiffness. The flow number test denotes the onset of tertiary flow in an HMA mixture and is believed to be a reliable indicator of in-service rutting.

The dynamic modulus test was performed and master curves developed according to AASHTO PP 62-10 using the second order polynomial shift factor equation. The test was performed at six frequencies (25, 10, 5, 1, 0.5, and 0.1 Hz) and four temperatures (4, 21, 37, and 50°C). The reference temperature was 30°C.

The flow number test was unconfined, with a deviator stress (σ_d) equal to 600 kPa (87 psi) and contact stress of 30 kPa (4.35 psi) (5% σ_d) at a temperature of

50.5°C, which is based on the LTPPBind database. The Franken Model was used for curve fitting.

Table 6. *Dynamic modulus specimens' air void data.*

Category 4, 19.0-mm Mixture						
N_{design}	Sample 1	Sample 2	Sample 3	Sample 4	Average	SD
100	7.1	7.5	7.1	-[1]	7.2	0.23
70	5.5	5.3	4.7	-[1]	5.2	0.42
50	4.7	4.9	5.2	4.7	4.9	0.24
30	4.6	4.7	5.2	5.4	5.0	0.39
Category 3, 9.5-mm Mixture						
N_{design}	Sample 1	Sample 2	Sample 3	Sample 4	Average	SD
100	7.7	7.2	6.5	6.7	7.0	0.54
70	4.8	4.7	5.0	5.4	5.0	0.31
50	4.7	4.6	4.8	4.9	4.8	0.13
30	5.1	4.9	5.1	4.9	5.0	0.12
Category 4, 9.5-mm Mixture						
N_{design}	Sample 1	Sample 2	Sample 3	Sample 4	Average	SD
100	6.9	6.9	7.1	7.0	7.0	0.10
50	4.8	5.1	4.8	4.8	4.9	0.15
30	5.1	5.1	4.9	4.9	5.0	0.12

[1]—No data collected due to insufficient material

4.0 Results and Analysis

4.1 Category 4, 19.0-mm Mixture

The master curves for all four mixtures (gyration levels) are shown in Figure 4. For better clarity at high frequencies, the semi-log plot of this data is shown in Figure 5. The results indicate that the N70 mixture is stiffest of the four mixtures while the N100 mixture has the lowest stiffness. The stiffness of the N30 and N50 mixtures were approximately the same and were higher than that of the N100 mixture. When observing these results it should be remembered that two groups of mixtures are presented. The N100 mixture was designed for 4% air voids in the laboratory and the test specimens were produced at 7% air voids (original mixture). The second group of mixtures (N70, N50, N30) were designed for 5% air voids and test specimens produced at 5% air voids.

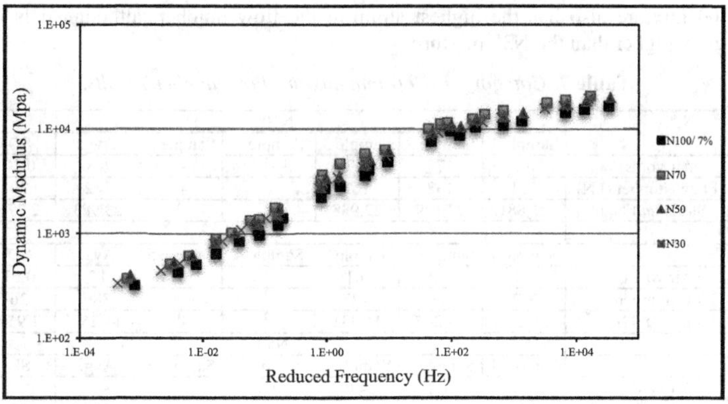

Figure 4. *Category 4, 19.0-mm master curves.*

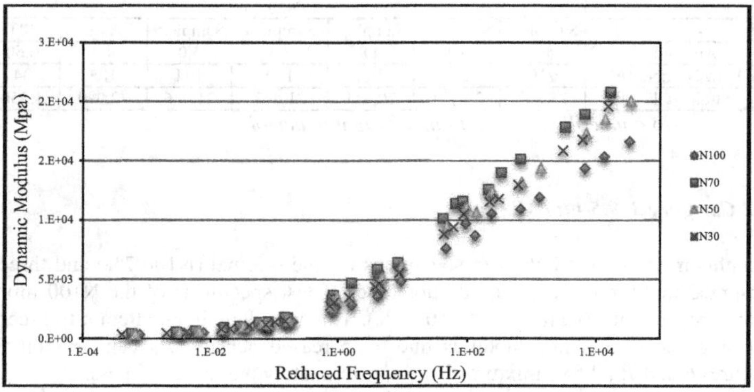

Figure 5. *Category 4, 19.0-mm master curves, semi-log plot.*

After the dynamic modulus test the same specimens were used for the flow number test because of insufficient material to compact additional specimens. Because all of the specimens were treated identically and the dynamic modulus test did not change the air void content of the specimens significantly, it is reasonable to make a relative comparison between two different designs, although it is not an absolute conclusion. Table 7 shows the flow number results for Category 4, 19.0-mm designs. All three of the re-designed mixtures have flow numbers higher than the original mixture. This suggests that these mixtures, when compacted to 5% air voids in the field, should have better rutting resistance than the original, conventionally designed mixture, when compacted in the field to 7% air voids. The

N100 mixture also has the highest strain at the flow number, although it is only slightly higher than the N30 mixture.

Table 7. *Category 4, 19.0-mm mixture flow number results.*

	N100						
	Sample 1	Sample 2	Sample 3	Sample 4	Sample 5	Avg.	SD
Air voids, %	6.7	7	6.7	-[1]	-[1]	6.8	0.17
Flow Number (FN)	191	132	164	-[1]	-[1]	162	30
Strain @ FN, με	21,881	27,129	22,938	-[1]	-[1]	23,983	2,776
	N70						
	Sample 1	Sample 2	Sample 3	Sample 4	Sample 5	Avg.	SD
Air voids, %	5.1	4.9	4.1	-[1]	-[1]	4.7	0.53
Flow number (FN)	371	360	427	-[1]	-[1]	386	36
Strain @ FN, με	20,230	18,245	16,331	-[1]	-[1]	18,269	1,950
	N50						
	Sample 1	Sample 2	Sample 3	Sample 4	Sample 5	Avg.	SD
Air voids, %	4.3	4.2	4.0	4.0	-[1]	4.1	0.15
Flow Number (FN)	220	337	424	410	-[1]	348	93
Strain @ FN, με	20,179	20,722	18,973	19,654	-[1]	19,882	747
	N30						
	Sample 1	Sample 2	Sample 3	Sample 4	Sample 5	Avg.	SD
Air voids, %	4.0	4.1	4.8	5.3	5.0	4.6	0.60
Flow Number (FN)	210	221	162	140	194	185	34
Strain @ FN, με	23,862	23,171	20,811	21,542	21,066	22,090	1,350

[1]*No data collected due to insufficient material.*

4.2 Category 3, 9.5-mm Mixture

For this mixture, in addition to specimens for the original (N100/7%) and three re-designed mixture designs, an additional set of test specimens of the N100 mixture were prepared at 5% air voids (N100/5%). This was done in an attempt to ascertain the increase in dynamic modulus due to increased density. As can be seen from Figures 6 and 7, all five mixtures have approximately the same stiffness.

Table 8 shows the flow number results for the Category 3, 9.5-mm designs. Again the original mixture has the lowest average flow number value, 91, while the three re-designed mixtures have flow numbers that average approximately 162. Again, the results indicate that the re-designed mixtures should have better rutting resistance when placed in service at 5% air voids than the original mixture, when placed at 7% air voids.

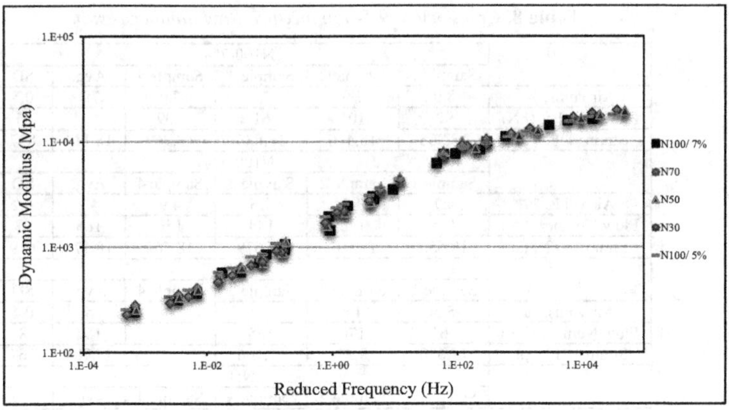

Figure 6. *Category 3, 9.5-mm master curves.*

Interestingly, the original design shows the highest average flow number (166) when the test specimens were compacted to 5% air voids (N100/5%). This is an increase of 75 over the N100 specimens compacted to 7% air voids (N100/7%), and is similar to the values for the three re-designed mixtures.

The N50 mixture has the highest strain at flow number, the N70 mixture the lowest. The N30 mixture has the second highest strain at flow number followed by the both N100 mixtures.

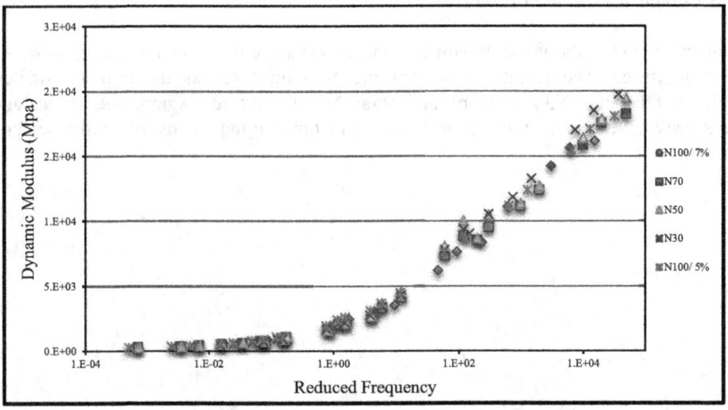

Figure 7. *Category 3, 9.5-mm master curves, semi-log plot.*

Table 8. *Category 3, 9.5-mm mixture flow number results.*

	N100-7%					
	Sample 1	Sample 2	Sample 3	Sample 4	Avg.	SD
Air voids, %	8.0	7.0	7.3	7.0	7.3	0.5
Flow Number (FN)	92	101	61	109	91	21
Strain @ FN, με	18,366	18,436	16,886	18,769	18,114	837
	N100-5%					
	Sample 1	Sample 2	Sample 3	Sample 4	Avg.	SD
Air voids, %	4.7	4.5	4.6	4.9	4.7	0.2
Flow Number (FN)	195	172	144	153	166	23
Strain @ FN, με	18,653	17,601	17,679	18,764	18,174	619
	N70					
	Sample 1	Sample 2	Sample 3	Sample 4	Avg.	SD
Air voids, %	5.5	4.8	4.8	-[1]	5	0.4
Flow Number (FN)	161	176	165	-[1]	167	8
Strain @ FN, με	17,650	18,296	17,166	-[1]	17,704	567
	N50					
	Sample 1	Sample 2	Sample 3	Sample 4	Avg.	SD
Air voids, %	4.5	4.3	4.6	-[1]	4.5	0.15
Flow Number (FN)	183	156	151	-[1]	163	17
Strain @ FN, με	18,008	18,861	24,030	-[1]	20,300	3258
	N30					
	Sample 1	Sample 2	Sample 3	Sample 4	Avg.	SD
Air voids, %	5.3	5	4.9	-[1]	5.1	0.21
Flow Number (FN)	141	162	164	-[1]	156	13
Strain @ FN, με	19,000	19,419	19,193	-[1]	19,204	210

[1]*No data collected due to insufficient material.*

4.3 Category 4, 9.5-mm Mixture

Figures 8 and 9 show the dynamic modulus master curves for the Category 4, 9.5-mm mixtures. The results show that the N50 mixture has the highest stiffness, followed by the N30 mixture and then N100 mixture. Again, the re-designed mixtures have equal or better stiffness than the original, conventionally designed mixture.

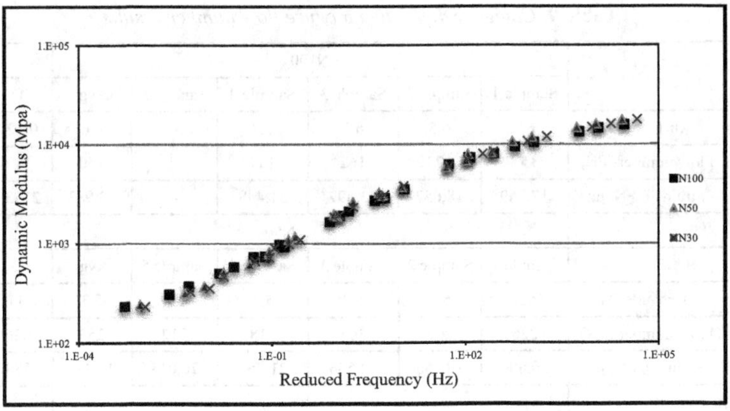

Figure 8. *Category 4, 9.5-mm master curves.*

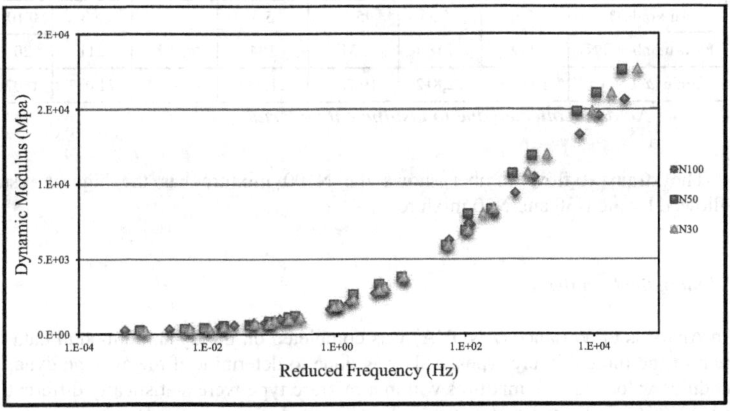

Figure 9. *Category 4, 9.5-mm master curves, semi-log plot.*

Table 9 contains the flow number results for the mixtures. The N30 mixture has the highest flow number, followed by the N70 mixture and the N100 mixture. This suggests that both the N30 and N50 mixtures could have better rutting performance in the field, when compacted to 5% air voids, than the N100 mixture compacted to 7% air voids.

Table 9. *Category 4, 9.5-mm mixture flow number results.*

	N100						
	Sample 1	Sample 2	Sample 3	Sample 4	Sample 5	Avg.	SD
Air voids, %	6.5	6.5	6.7	6.7	-[1]	6.6	0.12
Flow number (FN)	157	173	162	147	-[1]	160	11
Strain @ FN, με	17,269	18,657	23,497	21,448	-[1]	23,983	2793
	N50						
	Sample 1	Sample 2	Sample 3	Sample 4	Sample 5	Avg.	SD
Air voids, %	5.2	5.5	5.3	5.4	5.4	5.3	0.11
Flow number (FN)	226	207	303	318	212	253	53
Strain @ FN, με	20,600	19,756	22,556	21,751	20,013	20,935	1188
	N30						
	Sample 1	Sample 2	Sample 3	Sample 4	Sample 5	Avg.	SD
Air voids, %	5.3	5.3	5.1	5.3	-[1]	5.3	0.10
Flow number (FN)	209	218	231	184	-[1]	211	20
Strain @ FN, με	19,645	22,812	19,702	21,973	-[1]	21,033	1607

[1]*No data collected due to insufficient material.*

The strain at flow number shows the N100 mixture has the highest result, followed by the N30 and N50 mixtures.

4.4 Statistical Analysis

An Analysis of Variance (ANOVA) was completed on the dynamic moduli data for each of the three mixture types. This was done to determine if the average dynamic modulus values for the mixtures within a mixture type were statistically different. If the ANOVA indicated a statistically significant difference, the Bonferroni method was used to determine which mixtures were significantly different.

4.4.1 Category 4, 19.0-mm Mixture

The dynamic modulus data for the Category 4, 19.0-mm mixture is shown in Table 10. At 25 Hz and 6°C, the analysis leads to the conclusion there are no significant differences between the N30, N50, N70, and N100. For all other combinations of frequencies and temperatures, the ANOVA indicates there are significant differences between at least two of the groupings.

Table 10. *Category 4, 19.0-mm dynamic modulus data summary.*

@ 25 Hz	N30		N50		N70		N100	
	Avg. |E*|, MPa	CV, %	Avg. |E*|, MPa	CV, %	Avg. |E*|, MPa	CV, %	Avg. |E*|, MPa	CV, %
6°C	19,554	15.2	20,041	6.4	20,145	9.6	16,166	9
22°C	11,276	3.7	12,465	7.4	12,132	9.3	10,088	11
37°C	4,157	6.4	4,552	9.4	5,299	15	3,561	7.7
50°C	1,431	7.9	1,743	6.3	1,792	13.1	1,366	0.6
@ 10 Hz	N30		N50		N70		N100	
	Avg. |E*|, MPa	CV, %	Avg. |E*|, MPa	CV, %	Avg. |E*|, MPa	CV, %	Avg. |E*|, MPa	CV, %
6°C	16,770	1.8	18,477	5.2	18,277	7.2	14,898	5.2
22°C	10,328	3.3	11,097	4	11,219	5.4	9,207	11.3
37°C	3,236	7.6	3,547	6.4	4,251	10.7	2,841	6.7
50°C	1,089	7.9	1,299	9.1	1,341	10.8	1,052	2.1

Table 11 shows the Bonferroni groupings for the dynamic modulus data. As indicated earlier, this data shows that in all cases the re-designed mixtures have equal or higher dynamic moduli than the original, conventionally designed mixture. Again, for this to hold true, the re-designed mixtures would need to be put into service at 5% air voids, while the original would be at the typical 7% air voids.

4.4.2 Category 3, 9.5-mm Mixture

For the Category 3, 9.5-mm mixture data shown in Table 12, the ANOVA indicates that at a temperature of 37°C and frequencies of 25 Hz and 10 Hz, the average dynamic modulus values of all the designs are statistically equivalent. All other combinations do have at least one significantly different mixture. The Bonferroni groupings for these are shown in Table 13. For this data it is interesting to note that when the N100 mixture was tested at 5% air voids, it showed a higher average dynamic modulus value than when tested at 7% air voids. This seems to confirm that increasing the in-place density of asphalt mixtures increases the mixture stiffness.

Table 11. *Category 4, 19.0-mm dynamic modulus Bonferroni groups.*

Test Temperature, °C	Test Frequency, Hz	
	25	10
6	No differences	(N50=N70=N30) (N30=N100)
22	(N50=N70=N30) (N70=N30=N100)	(N70=N50=N30) (N30=N100)
37	(N70=N50=N30) (N50=N30=N100)	(N70=N50=N30) (N50=N30=N100)
50	(N70=N50) (N50=N30) (N30=N100)	(N70=N50=N30) (N50=N30=N100)

Table 12. *Category 3, 9.5-mm dynamic modulus data summary.*

@ 25 Hz	N30		N50		N70		N100	
	Avg. |E*|, MPa	CV, %	Avg. |E*|, MPa	CV, %	Avg. |E*|, MPa	CV, %	Avg. |E*|, MPa	CV, %
4°C	19715	4.4	19413	4	18226	8.2	16144	13.1
21°C	10529	6.3	10480	3.7	9504	8.8	8351	10.7
37°C	2756	8.1	2678	13.4	2529	11.2	3046	20.2
50°C	919	7.5	1114	5.3	1081	8.1	944	6.2
@ 10 Hz	N30		N50		N70		N100	
	Avg. |E*|, MPa	CV, %	Avg. |E*|, MPa	CV, %	Avg. |E*|, MPa	CV, %	Avg. |E*|, MPa	CV, %
4°C	18492	6.1	17843	3.3	17543	6.7	15639	6.4
21°C	9331	7.1	10042	5.6	8864	9.2	7628	11.5
37°C	2191	10.7	2136	15.4	2092	14	2461	20.4
50°C	671	7.9	811	4.4	789	7	707	6.1

Table 13. *Category 3, 9.5-mm dynamic modulus Bonferroni groups.*

Test Temperature, °C	Test Frequency, Hz	
	25	10
4	(N70=N50=N100/7%) (N30=N50=N100/5%)	(N70=N100/7%) (N70=N30=N50=N100/5%)
21	(N70=N100/7%) (N70=N30=N50=N100/5%)	(N70=N100/7%) (N70=N30=N50=N100/5%)
37	No differences	No differences
50	(N70=N100/7%) (N70=N50) (N30) (N100/5%)	(N70=N100/7%) (N70=N50) (N30) (N100/5%)

4.4.3 Category 4, 9.5-mm Mixture

The Category 4, 9.5-mm dynamic modulus data is shown in Table 14. ANOVA indicates that at 21°C and 25 Hz there is no statistically significant difference between the N100, N50, and N30 mixtures. All other mixture combinations have some significant differences as shown in the Bonferroni groupings of Table 15. As before, the results statistically indicate that the re-designed mixtures have average dynamic moduli values as high as or higher than the original mixture. This suggests their in-service performance should be as good as, if not better than, the original mixture. The reader is again cautioned to keep in mind the differences in the air voids between the re-designed and original mixtures.

Table 14. *Category 4, 9.5-mm dynamic modulus data summary.*

@ 25 Hz	N100		N50		N30	
	Avg. |E*|, MPa	CV, %	Avg. |E*|, MPa	CV, %	Avg. |E*|, MPa	CV, %
4°C	15,632	1.9	17,610	0.9	17,682	2.8
21°C	8,025	3.7	8,308	2.5	8,392	3.7
37°C	2,721	4.4	3,297	2.8	2,970	4.8
50°C	971	11.7	1171	4.1	1,093	8.8
@ 10 Hz	N100		N50		N30	
	Avg. |E*|, MPa	CV, %	Avg. |E*|, MPa	CV, %	Avg. |E*|, MPa	CV, %
4°C	14,591	2.3	16,055	1.4	16,107	3.0
21°C	7,364	1.1	6,897	2.6	6,957	4.0
37°C	2,105	6.8	2,632	2.7	2,334	7.4
50°C	731	12.1	829	3.1	910	8.9

Table 15. *Category 4, 9.5-mm dynamic modulus Bonferroni groups.*

Test Temperature, °C	Test Frequency, Hz	
	25	10
4	(N30=N50) (N100)	(N30=N100) (N50)
21	No differences	(N30=N100) (N50)
37	(N50=N30) (N100)	(N50) (N30=N100)
50	(N50) (N30=N100)	(N30=N50) (N100)

4.4 Field Trial

Based on the positive results of the laboratory study it was decided that a field trial should be completed to see if the laboratory findings were applicable to the field. As such, a search was made for asphalt paving contracts that would be suitable for such a field trial. Several locations were identified and then eliminated for various reasons. Eventually, INDOT settled on the SR-13 project near Ft. Wayne, Indiana. The project called for the surface to be milled and an overlay placed. The new asphalt overlay mixture was designed as a Category 4, 9.5-mm surface. This means the design was done using 100 gyrations of the SGC and optimum asphalt binder content chosen at 4% air voids. It was initially decided by the study advisory committee (SAC), in consultation with the research team, that the re-designed mixture for use on the project should be designed using 30 gyrations of the SGC and selecting the optimum asphalt binder content at 5% air voids. Later, the decision was made to change the 30-gyration design to a 50-gyration design instead.

Both the original and re-designed mixtures consisted of steel slag and limestone coarse aggregates, limestone and natural sands, recycled asphalt shingles (RAS), and a PG 70-22 asphalt binder. The aggregate gradations of both mixtures are given in Table 16 and shown graphically in Figure 10. Relatively small changes in the aggregate proportions were needed to allow the re-designed mixture to achieve the design air voids at reduced gyrations.

Table 16. *Aggregate gradation of the field mixtures and specific gravity.*

Sieve Size, mm	Original (N100)	Re-designed (N50)
	Percent Passing	
12.5	100.0	100.0
9.5	92.0	89.6
6.3	71.8	70.4
4.75	62.0	61.1
2.36	46.4	40.1
1.18	32.0	27.7
0.6	22.7	19.5
0.3	13.4	11.0
0.15	7.5	6.0
0.075	4.8	4.4
G_{sb}	2.953	2.972

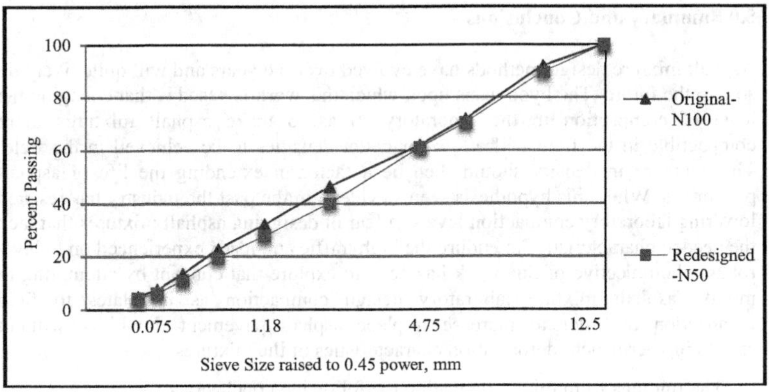

Figure 10. *Field trial mixture gradations.*

A total of nine sublots of the original mixture and three sublots of the re-designed mixture were placed. For a surface mixture, INDOT specifications define a sublot as 544 tonnes (600 tons) of mixture. As a standard quality assurance measure, INDOT extracted two cores from each sublot to establish the in-place mixture densities. The data is summarized in Table 17. The average in-place density of the original mixture was 91.8% for the re-designed mixture the average was 94.7%, close to the goal of 95%. The contractor reported that the re-designed mixture field density was achieved without making any changes in the rollers or rolling patterns. The work to determine the mechanical properties of both the original and re-designed mixtures is currently ongoing.

Table 17. *Field core densities.*

Mixture	Sublot Number	Density, Percent of G_{mm}		
		Core 1	Core 2	Average
Original (N100)	1	92.3	89.7	91.0
	2	90.3	94.6	92.5
	3	92.6	92.7	92.6
	4	92.4	93.9	93.1
	5	90.5	90.0	90.3
	6	90.2	90.0	90.1
	7	92.4	91.4	91.9
	8	92.6	92.4	92.5
	9	92.3	92.4	92.3
Re-designed (N50)	1	92.3	94.5	93.4
	2	93.6	94.7	94.1
	3	96.3	96.7	96.5

5.0 Summary and Conclusions

Asphalt mixture design methods have evolved over the years and will quite likely do so into the future. The hypothesis upon which this work is based is that changing the level of compaction in the laboratory so as to make asphalt mixtures more compactible in the field will allow for better densities to be achieved in the field. This increase in density should then be a factor in extending the life of asphalt pavements. While this hypothesis seems evident, in the past the industry has resisted lowering laboratory compaction levels in fear of designing asphalt mixtures that lack the needed characteristics to endure the high traffic and loads experienced on today's roads. The objective of this work has been to explore that concept by attempting to modify asphalt mixture laboratory design compaction as it relates to field compaction in order to increase in-place asphalt pavement durability without sacrificing permanent deformation characteristics of the mixtures.

The findings from this study lead to the following conclusions:

- With changes in gradation, mixtures can be designed at 5% air voids in the laboratory while keeping effective binder content constant.
- Mixtures designed at 5% air voids and compacted to the anticipated field air void content of 5% can have |E*| and flow number values as high as or higher than those designed at 4% and tested at anticipated field density corresponding to 7% air voids.
- Asphalt mixtures designed in the laboratory at 5% air voids can be compacted to 5% air voids in the field. The field test indicates this can be done without additional compaction effort.
- Results of the study suggest that if asphalt mixtures were designed at 5% air voids and placed in service at 5% air voids, they could outperform mixtures designed and placed in a more conventional manner.

6.0 Recommendations

- Fatigue and low-temperature testing, as well as moisture susceptibility testing, should be completed for the standard and re-designed mixtures used in the study to quantify the enhanced durability of the re-designed mixtures.
- Additional laboratory work should be done so as to include the effects of more traffic levels, mixtures containing RAP, RAS, or both, and additional binder grades and aggregate types.
- The performance of the trial project placed as part of this study should be monitored for performance.
- Additional trial projects should be placed and evaluated.

7.0 Acknowledgments

The authors wish to acknowledge the support of Indiana Department of Transportation and the Joint Transportation Research Program at Purdue University. Also, the authors acknowledge the valuable assistance of Mr. William Pine, who assisted the project by estimating trial aggregate blends for the asphalt mixtures. Without his help this project would have been much more difficult.

The contents of this report reflect the views of authors who are responsible for the facts and accuracy of the data presented herein. The contents do not necessarily reflect the official views or policies of the Indiana Department of Transportation or the Federal Highway Administration at the time of publication.

8.0 References

Asphalt Institute, *Mix Design Methods for Asphalt Concrete and Other Hot Mix Types*, Manual Series No. 2, 6th ed., Lexington, KY, 1997.

Cooley, L.A., Prowell, B.D., and Brown, E.R., "Issues Pertaining to the Permeability Characteristics of Coarse-Graded Superpave Mixes," *Journal of the Association of Asphalt Paving Technologists*, Vol. 71, 2002, pp. 1–29.

Griffith, J. M., "Evaluation of the Method of Asphalt Pavement Design Developed by The Corps of Engineers," *Proceedings of the Technical Sessions of the Association of Asphalt Paving Technologists*, Vol. 18, 1949, pp. 221–260.

Huber, G.A., Blankenship, P.B. and Mahboub, K.C., "Rational Method for Laboratory Compaction of Hot-Mix Asphalt," *Transportation Research Record*, No. 1454, 1994, pp. 144–153.

McFadden, G. and Ricketts, W.C., "Design and Field Control of Asphalt Paving Mixtures for Military Installations," *Proceedings of the Technical Sessions of the Association of Asphalt Paving Technologists*, 1948, pp. 93–113.

Moutier, F., "Le Banc De Compactage," *Bulletin Liaison Laboratoire Ponts et Chaussées*, Vol. 88, 1977, pp. 132-136.

Moutier, F., "La Presse à Cisaillement Giratorie: Modèle de Série," *Bulletin Liaison Laboratoire Ponts et Chaussées*, Vol. 74, 1974, pp. 137–148.

Ortolani, L. and Sandberg, H.A., "The Gyratory-Shear Method of Molding Asphaltic Concrete Test Specimens: Its Development and Correlation with Field Compaction Methods. A Texas Highway Department Standard Procedure," *Proceedings of the Technical Sessions of the Association of Asphalt Paving Technologists*, 1952, pp. 280–297.

Vavrik, W.R., Huber, G.A., Pine, W.J., Carpenter, S.H., and Bailey, R. (2002) "Bailey Method for Gradation Selection in Hot-Mix Asphalt Mixture Design," Circular E-C044, *Transportation Research Board of the National Academics*, 2002.

9.0 Discussion

MICHAEL DUNNING: On your N50 mix, did you increase the oil content?

JOHN HADDOCK: The effective binder contents are essentially the same for each combination of mixture type and size; it did not change. Now, there may be some differences in the total binder contents because of the gradations and the absorptivity of the aggregate, but the effective binder contents are the same.

MICHAEL DUNNING: Very good, thank you. By the way, great presentation. It was very good for us here at the county. Did you perform rut tests?

JOHN HADDOCK: We did not do any laboratory wheel tracking tests for these mixtures. We were doing flow number and dynamic modulus testing to indicate whether or not the mixtures might be susceptible to rutting problems. We do have some cores from a recently completed field project that will be tested in the Hamburg machine. In fact, those tests may have already started, correct, Ali? Yes, that testing has begun, but we do not yet have the data.

MICHAEL DUNNING: Very good. We don't have the sophistication of a gyratory compactor at the county level. Marshall is our design method. We have minimum 93% on our specifications, and we want to move it up like you've done. I'm glad to see you doing that. We have to adjust our Marshall mix design from a 4% air void to a 3.5 because we know our gradations right now are quite good because the aggregates are close together, effectively reducing further densification, therefore we feel the air voids will contain more binder which increases the density. In order to prevent a mistake of too high of a binder content, we do a break point on our rolling pattern. So I'm glad to see this. It gives me some encouragement on which way to go to help the process here. Thank you.

JOHN HADDOCK: All right. Thank you.

DAVE ANDERSON: Nice interesting study. I assume that what you're trying to do is to get a more durable pavement by virtue of minimizing the amount of oxidation you have in the binder. Is that correct?

JOHN HADDOCK: Yes, the idea is to lower the permeability in the pavement so that the…

DAVE ANDERSON: Okay, you just used the word that you didn't use in your presentation very much, and that's permeability. I would maintain that air voids are a secondary variable here, that what's really important is the permeability of the pavement, not the air voids, and if those air voids are disconnected, the oxygen isn't going to get down to the binder. So I really think that you should be emphasizing permeability. And I'm talking about air permeability, not in a turbulent manner but in the manner in which the air really gets to the pavement. That's a laminar flow type of thing. I think we've not measured permeability correctly and we don't understand it. I don't think you can address this in a laboratory because the mix design that we use in the laboratory doesn't simulate what we do in the field. We don't mix the same in the laboratory, we don't pack the same in the laboratory, the particles aren't coated in the same way. We have an artifact in the laboratory. We really need to study this with mixtures that are put down in the field, and it's the connected air voids that are important. And that is going to really give us a handle on the durability of the pavement. This is fine, but there's a long way to go to real life.

JOHN HADDOCK: You are absolutely right, permeability is important, and it's really not just the air void level, it's whether or not the air voids are interconnected, because if they are not, then you don't have such the…

DAVE ANDERSON: I would say it's more than important. It is the main, main variable, actually. It's not just important.

JOHN HADDOCK: We currently have some additional work underway in which we are trying to identify from cores if the air voids in the cores are interconnected. In the lab we are trying to do that just to see if there is maybe a way that we can do exactly what you say. Is there some kind of a test that we can run in the field that would really, really tell us what we have in the field from a permeability standpoint.

DAVE ANDERSON: There are techniques for measuring the air permeability, the permeability of air through the pavement, and those should be employed, and I suggest you incorporate those into your experiment.

JOHN HADDOCK: Thank you. You are absolutely right. Thank you.

LEE GALLIVAN: Consultant. That would be the first time I've said that.

JOHN HADDOCK: So it's nice to see you, Lee. Lee and I only see one another if we are not in Indiana.

LEE GALLIVAN: And we only work and live just 60 miles apart. Two questions. One is thank you very much for the good presentation. I think we're going to see more and more about that one. And you talked about some future work with some other mixes and you're going to go into RAS and RAP and all the other good stuff and other binders. And that's great. Have you thought about transportation of this process into another region of the US to give it another try?

JOHN HADDOCK: Well, I guess we could say we had thought about it, but at least in the position that I am, I don't have the authority to do that. But it's certainly something that could be tried, and I think there may be some other states out there that might be interested. We would certainly be willing to offer them any data or any help that we could, if they wanted to try it.

LEE GALLIVAN: In my travels, I see a lot of states that are already down in that 8% air voids area and they're struggling, and there's quite a few in the 6% air voids that you can start off with. So there's probably some states that should be interested in improving their densities.

JOHN HADDOCK: And we would be happy to work with anyone who was interested.

LEE GALLIVAN: The other question was, the other question was... I can't remember.

JOHN HADDOCK: It was probably an easy one, though. Right, Lee?

LEE GALLIVAN: It was an easy one. I just want to say thank you very much.

JOHN HADDOCK: Thank you. Brian?

BRIAN POWELL: Good job, John, as always. I always enjoy your presentations. For the author types, you know, look at John's slides. My old eyes, I could actually read them. That's something that a lot of us really appreciate. So think about that next year.

JOHN HADDOCK: Well, I too am getting older.

BRIAN POWELL: Yup. My first question for you, John, is in your conclusions you said that you could design mixes at 5% voids without lowering the effective binder content. In production or in the specification, would that not only be true if you also increased the minimal VMA 1% corresponding to that?

JOHN HADDOCK: Yes, you are correct. If the air voids go up a percent, the VMA goes up a percent when the effective binder content is held constant, right?

BRIAN POWELL: Right.

JOHN HADDOCK: So it does. If this method were adopted, it would be necessary to raise the minimum VMA specification by 1%.

BRIAN POWELL: So for a 9.5 you'd go from 15 to 16?

JOHN HADDOCK: Exactly.

BRIAN POWELL: And then you also mentioned as you decrease in design gyrations that roughly in 25-30 gyration increments that you saw a 1% increase in voids, which realistically was a 1% increase in VMA, was it not?

JOHN HADDOCK: Right.

BRIAN POWELL: It wasn't the voids that just went up by themselves; VMA increased.

JOHN HADDOCK: Right.

BRIAN POWELL: And that really corresponds to work that was done by Mike Anderson at the Asphalt Institute and Shane Buchanan at NCAT that showed for similar changes in gyrations, 25-30 gyrations, you got a 1% shift in VMA. So looking at that relationship, if you were to take your N30 gyrations and add 25-30 to it, you're at 55-60, so you could get kind of the same result by keeping the current VMA minimum requirements, say 15 for a 9.5-mm mix, and designing it at either 55 or 60 if you were trying to get N30 or 75 or 80 if you were trying to get N50 at 5%. So same VMA 4% voids versus 25-30 gyrations higher or the values you used 5% air voids a percent higher VMA.

JOHN HADDOCK: Right. Yes.

BRIAN POWELL: They work out to be the same thing, is that not correct?

JOHN HADDOCK: Yes, they probably do. And I think in Indiana, the DOT really hasn't settled on it, but I think we'll probably end up somewhere around N50. At least, that's kind of what we feel as a research team. When we did the first field project we were discussing using N30, which the DOT hesitated to use. They were not too sure about using N30 on a state highway project.

BRIAN POWELL: But for other people, then, using 4% air voids and current VMA, that would be about 75-80 gyrations for heavy traffic.

JOHN HADDOCK: Yes.

BRIAN POWELL: So I think your work at N70 really goes back then and confirms work that Ray Bonaquist and Don Christensen did when they looked at the volumetric properties for Superpave that did show really the value of better field compaction and how that could help stiffness and rutting.

JOHN HADDOCK: Right.

BRIAN POWELL: Thanks, John.

JOHN HADDOCK: Thanks, Brian. Erv?

ERV DUKATZ: Hi, John. Excellent presentation. Very thought provoking. One point of clarification on your slide. When you said increase in density, I'm assuming that's percent max Rice density?

JOHN HADDOCK: Yes, yes. I'm sorry if I did not mention that. It is the percent of maximum theoretical.

ERV DUKATZ: And then another question which is a follow-up to what Brian was just asking. Are you proposing that the mix designs be done at the original gyration level and then checked by backing off the level of gyrations, or are you suggesting in the future that you design at the lower gyration level and set the air void target at 5%?

JOHN HADDOCK: I get that question a lot because to this point what we have done is start with an N100 design and then change the gradation and lower the gyration level. What we would propose, for example if 50 gyrations was adopted as the design number of gyrations, is to begin from scratch and do the design at N50.

ERV DUKATZ: Very interesting. I think the other thing that is shown in your slides is that as you kept your asphalt content the same and increased your VMA, the gradation curves moved much closer to the maximum density line to achieve your higher VMA and higher air void values. These are changes which would be considered counter to normal aggregate gradation adjustments, where the gradation curve is moved further from the 0.45 power max density line to increase volumetrics.

JOHN HADDOCK: Well, that is what Bill does with the Bailey method.

ERV DUKATZ: My thought is adding more sand to the mix, but…

JOHN HADDOCK: It is interesting that you ask that question because Brian asked me yesterday, "Have you looked at the gradations that you are using now and compared them to gradations used in Indiana before Superpave?" We have not done so, but as I look at them, I think that they are probably more similar to what we used to do, prior to Superpave, under the old Marshall method with the exception that we have taken natural sand out. There is no sand hump in them. So you make a good point and I am sure Ali will be looking at the old gradations to make the comparisons.

ERV DUKATZ: Thank you.

HONGBIN XIE: Really interesting idea to make the transition load, make mix designs we've always done. I mean typically it's done at 4% and then we just change the compaction level, but that you kind of changed angle, changed lab design air voids, and then achieve a better compaction in the field. One clarification before I ask the question. Is the material for category 4, for example, like a 19 mm are those with the same exact material? I mean coarse and fine aggregate with four different compaction effort?

JOHN HADDOCK: For the three mixtures that we used, the two Category 4 mixtures would have the same aggregates because those are the higher traffic mixtures; they must have dolomite and blast furnace slag in them by specification. The Category 3 mixture did not contain blast furnace slag, rather it had limestone in it, so there are some slight differences, but two of the mixtures did have the same aggregates in them.

HONGBIN XIE: Okay. Because if you're comparing the compaction effort in the lab when you're doing the design, if the rock source is the same, what are you really changing? It's really just the composition of material, right? So lab design is how you put a mix together, aggregates and asphalt, combine them together to make the mix which has better workability. Just by looking at your gradation chart with different compaction effort, it seems like there's a really minimal difference the 100 and the 70 gyrations.

JOHN HADDOCK: Right. There is almost no difference between the gradations for the N100 and N70 mixtures.

HONGBIN XIE: Right. And you also mentioned you kept the AC content the same, so if the gradation is the same and the AC content is the same and the materials are the same, then really there's no real difference in terms of the material you're producing, right?

JOHN HADDOCK: Right.

HONGBIN XIE: So really what I want to say is since between 100 to 70 gyrations, there's minimal difference in the actual material composition, so if you're achieving better compaction at the end, that's kind of interesting to me, so I'm not really sure why. Do you see that? That's the first question, really. And the second one is that I do see you have a bigger difference in terms of gradation between the lower compaction effort to, you know, the base design. You have to go much coarser since you keep the same compaction effort. That gets back to Brian's question, you're going to have a higher VMA, but you didn't do that. So you kept the same AC content but you really changed the structure of the gradation.

JOHN HADDOCK: First of all, I am not sure about the difference in the compactability of the N100 and the N70 mixtures. The effective binder contents are the same, the gradations are approximately the same, but we did not put any N70 mixtures down in the field so we do not know from a field standpoint if they would be easier or more difficult to compact. We compacted them to 5% in the lab when we made specimens in the lab, but we did not put those down in the field. So I am not sure what their compactability would be.

HONGBIN XIE: Okay, I missed that point that you didn't compare the N100 and N70 in the field. I was expecting the field performance could be similar, really, because no matter how you design it, if you design it with the same gradation, same AC content, same rock, same binder, field performance will be the same....

JOHN HADDOCK: Yes. From a field standpoint, it stands to reason that if you have an N100 and an N70, they have pretty much the same gradation, they have the same binder content, the same materials, you would expect that N70 mixture to act just like the N100 mixture in the field. We could probably get down to maybe 7 or 8% air voids in the field. We do not know because we did not do that. We only used N50 mixtures in the field and so we only have that to refer to. And in the case of the N50 mixture, it does have a different gradation even though the effective binder contents are pretty much constant across the mixtures. However, we were able, or rather the contractor was able to compact those in the field to 5% air voids, so they do appear to be more compactable. I think the big question that we all have is, are they so compactable that the trucks will continue to compact them down to 2 or 3% air voids? We don't think so. We have not seen that on State Road 13. We have not seen any problems with that, and feel like at least from the results that we have gotten in lab testing the field mixtures, we do not expect that to happen. But until we see what happens in the field, I don't think any conclusions can be drawn, because what happens on the road is most important. Many different things can be done in the lab, but unless it works in the field, it does not really matter.

HONGBIN XIE: Yeah, okay. So that's why I feel there's really little difference between the 70 and the 100, the compaction effort really resulted in a different gradation. So that's really what actually matters. But that's a really good angle to think your design approach....

JOHN HADDOCK: Right. It is the gradations that are changing in these mixtures and that is what allows them to be compacted to lower air voids in the field.

HONGBIN XIE: Okay, thank you.

JOHN HADDOCK: You're welcome.

ROSS BENTSEN: Wonderful presentation, John. I certainly applaud the practicality of the research and being able to immediately move from the lab to the field. That's something that the industry certainly needs more of. You've done a good job of planting all your questions this morning because I've only got one more that hasn't been answered. You talked about the headaches in the field with the production of the new mix. Was that design related? It sounded to me is was more related of typical problems of receiving materials that aren't the same.

JOHN HADDOCK: The problems that we had in the field on State Road 13 were specifically related to the slag. It was extremely variable and it caused problems for the contractor. There were times when I wondered if they were upset with our trial. I tried to avoid the project site.

ROSS BENTSEN: Make Bill take it on.

JOHN HADDOCK: I sent Ali to the project site.

ROSS BENTSEN: That's Bill's job. Thank you.

Short-Term Aging of Asphalt Mixtures

Fan Yin[a], Amy Epps Martin[a], Edith Arambula[b], and David E. Newcomb[b]

[a]*Zachry Department of Civil Engineering, Texas A&M University, College Station, Texas 77843-3136*

[b]*Texas A&M Transportation Institute, College Station, Texas 77843-3136*

ABSTRACT: In the last two decades, changes in asphalt mixture components, production parameters, and plant design have occurred and raised the question of the validity of the current mix design procedures in adequately assessing the volumetric needs of asphalt mixtures and the physical characteristics required to meet performance expectations. A study of short-term aging of asphalt mixtures was performed considering the impacts of various asphalt mixture components and production parameters, including binder source, aggregate absorption, warm mix asphalt (WMA) technology, inclusion of recycled materials, plant type, and production temperature.

In this study, the laboratory short-term oven aging (STOA) protocols of two hours at 135°C for hot mix asphalt (HMA) and two hours at 116°C for WMA were used to fabricate laboratory mixed and laboratory compacted (LMLC) specimens for volumetric analysis and performance evaluation by Resilient Modulus (M_R) test, Dynamic Modulus (E) test, and Hamburg Wheel Tracking Test (HWTT). The simulation of asphalt aging and absorption during plant production and construction by the laboratory STOA protocols was evaluated by comparing mixture volumetrics and laboratory test results with those for plant mixed and plant compacted (PMPC) specimens and cores at construction. In addition, the laboratory test results were used to identify those mixture components and production parameters with significant effects on the performance of short-term aged asphalt mixtures.*

Correlations for LMLC specimens and PMPC specimens and cores at construction in terms of mixture volumetrics and laboratory test results indicated that the laboratory STOA protocols of two hours at 135°C for HMA and two hours at 116°C for WMA were able to simulate the asphalt aging and absorption that occurred during plant production and construction. According to the laboratory test results, among those factors investigated in this study; binder source, aggregate absorption, WMA technology, and inclusion of recycled materials had significant effects on the performance of short-term aged asphalt mixtures. However, no significant effects from mixture production temperature and plant type were observed.

KEYWORDS: short-term aging, asphalt absorption, mixture components, production parameters.

The oral presentation was made by Mr. Yin.

1.0 Introduction

Asphalt mixtures may be produced in either batch or drum mix plants and then compacted at temperatures ranging from 104°C to 163°C (Kuennen, 2004; Newcomb, 2005). The goal of asphalt mixture production is to ensure complete drying of the aggregate, proper coating and bonding of the aggregate with the binder, and adequate mixture workability for handling and compaction. These processes are important to the mixture's durability and resistance to permanent deformation, moisture susceptibility, and cracking. Advances in asphalt technology, including the use of polymer modified binders and use of more angular aggregate resulted in increased mixing and compaction temperatures up to a limit of approximately 177°C where polymer breakdown in the binder can occur. Conversely, the use of warm mix asphalt (WMA) technology led to reduced production and paving temperatures without sacrificing the quality of the final product. The result is a wider range of available production temperatures for asphalt mixtures.

Traditionally, asphalt mixtures have been designed on the basis of volumetric parameters (Asphalt Institute, 1984; 1995). Laboratory mixing and compaction temperatures were dependent upon the stiffness or viscosity of the asphalt binder. This system was refined with time, first for the Marshall and Hveem procedures, and then for Superpave. In the last two decades, changes have occurred in asphalt mixture components which are beneficial, but pose challenges in terms of how mixtures are designed and evaluated. Increased use of polymer modifiers, incorporation of recycled asphalt materials including reclaimed asphalt pavement (RAP) and reclaimed asphalt shingles (RAS), and the advent of WMA are departures from the norm under which the current volumetric mix design was developed.

Compounding this complexity is the evolution in asphalt plant design. The 1970s saw a rapid and persistent increase in the number of continuous plants, as drum mix plants (DMPs) replaced batch mix plants (BMPs). In a BMP, the aggregates are dried prior to being loaded into hot bins in a batching tower. Gates on the hot bins are opened, allowing for the proper proportion of aggregates to be weighed in a bin prior to dropping into the pugmill, where the asphalt binder is introduced and mixed with the aggregates. A DMP differs from a BMP in that cold aggregates are fed onto a weigh belt in the proper proportions prior to entering the elevated end of the drum for drying. The aggregates are dried as they tumble through the drum toward the lower end, where they are mixed with the asphalt binder prior to exiting to a slat conveyor for loading into a silo.

As discussed, the changes in asphalt mixture components, production parameters, and plant design have raised the question of the accuracy of the current mix design procedures in assessing the volumetric needs of asphalt mixtures and the physical characteristics required to meet performance expectations. Therefore, there is a need to address many of these issues, considering the impact of binder source, aggregate absorption, WMA technology, inclusion of recycled materials, plant type, and production temperature on the volumetric and performance characteristics of asphalt mixtures during production and construction.

The objectives of this study are to: 1) validate a laboratory short-term oven aging (STOA) protocol for asphalt loose mix prior to compaction to simulate the asphalt aging and absorption of asphalt mixtures produced in a plant and then loaded into a truck for transport, and 2) identify mixture components and production parameters with significant effects on the performance of short-term aged asphalt mixtures. Nine field projects were utilized including a wide spectrum of asphalt mixture components and production parameters. Plant mixed and plant compacted (PMPC) specimens (i.e., specimens compacted at the plant after stabilizing the mix to the compaction temperature) and cores at construction were acquired for each mixture from each field project, in conjunction with raw materials used for laboratory mixed and laboratory compacted (LMLC) specimen fabrication. According to previous research, the laboratory STOA protocols of two hours at 135°C for hot mix asphalt (HMA) and 116°C for WMA were used in this study for conditioning loose mix prior to compaction (Yin et al., 2013). Laboratory tests including the Resilient Modulus (M_R) test, the Dynamic Modulus (E*) test, and the Hamburg Wheel Tracking Test (HWTT) were utilized, and test results were used to validate the selected laboratory STOA protocols in simulating short-term aging and asphalt absorption during plant production and construction, and to identify those mixture components and production parameters with significant effects on the performance of short-term aged asphalt mixtures.

This paper first provides a brief review of written documentation of previous research on short-term aging of asphalt mixtures in the field and laboratory. Next, the experimental design is described, followed by test results and data analysis. Finally, conclusions and recommendations for future research are proposed.

2.0 Background

The aging of asphalt binders in mixtures has long been a concern to those in the pavement field. The standard practice for laboratory mix design of asphalt mixtures is to simulate the asphalt aging and absorption that occurs during production and construction by conditioning the loose mix prior to compaction for a specific amount of time at a specific temperature. For HMA, the recommended loose mix aging procedure per AASHTO R 30 for preparing specimens for volumetric mix design is two hours at the compaction temperature (T_c) and four hours at 135°C for preparing performance testing specimens. The implementation of WMA raised the question of the impact of lower plant production temperatures on the aging characteristics and absorption of asphalt by aggregates in the WMA mixtures and how to adequately simulate the differences in the laboratory.

A number of studies have been performed focusing on short-term aging characteristics of asphalt mixtures and simulation of plant aging by laboratory STOA protocols; a brief summary is presented in Table 1. In general, these studies have concluded that most short-term aging of asphalt mixtures occurs during production through placement and compaction and does not stop at plant load-out, and that binder source and type, including the presence of polymers, WMA technology, and plant type have substantial effects on short-term aging characteristics of asphalt mixtures. Additionally, a variety of laboratory STOA protocols have been evaluated to simulate short-term aging and asphalt absorption during plant production and construction when fabricating LMLC specimens for volumetric analysis and performance testing. A general trend has been observed that an increase in laboratory STOA temperature and/or time leads to more aging and asphalt absorption in asphalt mixtures. However, the validity of the standard short-term aging procedure per AASHTO R 30 in simulating aging during plant production and construction has been questioned because it produces a more significant level of binder/mixture aging. Also, a comprehensive study to establish a standard laboratory STOA protocol that encompasses the effects of binder source, aggregate absorption, WMA technology, inclusion of recycled materials, plant type, and production temperature has not been undertaken.

A previous study focusing on evaluation of laboratory conditioning protocols for WMA provided preliminary results toward understanding short-term aging through the plant and its simulation by laboratory STOA protocols (Yin et al., 2013). Different conditioning protocols were selected based on available literature for fabricating HMA and WMA LMLC specimens from two field projects, and these specimens were tested to determine the effect of each STOA protocol on mixture stiffness. Cores at construction and PMPC specimens were also incorporated in the study to represent asphalt mixtures experiencing short-term aging and asphalt absorption during production through compaction. Table 2 summarizes the M_R stiffness results for LMLC specimens with different STOA protocols compared through statistical analysis of Tukey's Honest Significant Differences (HSD) test at a 95% confidence level. In Table 2, solid shading indicates statistically higher M_R stiffness values for LMLC specimens as compared to either cores at construction or PMPC specimens, no shading/hatching indicates statistically equivalent performance for these specimen comparisons, and vertical hatching indicates that LMLC specimens exhibited statistically lower M_R stiffness values for the same comparisons. In summary, M_R stiffness results showed that mixture stiffness increased with higher temperature and longer time in the STOA protocol. Among the five selected STOA protocols for producing LMLC specimens, two hours at 135°C and two hours at T_c were more representative in terms of stiffness of HMA and WMA mixtures with plant aging, respectively. Considering the difficulty in accurately defining T_c in the field and the common range of T_c for WMA, two hours at 116°C instead of two hours at T_c was recommended as the standard laboratory conditioning protocol for WMA LMLC specimens. For HMA LMLC specimens, two hours at 135°C was recommended prior to compaction.

Therefore, the STOA protocols of two hours at 135°C for HMA and two hours at 116°C for WMA were selected in this study for simulating asphalt aging and absorption during plant production and construction, with a further validation by asphalt mixtures with a wider range of asphalt mixture components and production parameters.

Table 1. *Previous research on short-term aging of asphalt mixtures.*

Reference	Laboratory STOA Protocol	Major Findings
Heithaus and Johnson, 1958 Traxler, 1958 Chipperfield and Welch, 1967 Aschenbrener and Far, 1994	N/A	Most aging during production and construction through compaction
Traxler, 1961 Chipperfield and Welch, 1967 Lund and Wilson, 1984 & 1986 Topal and Sengoz, 2008 Zhao et al., 2009 Morian et al., 2011 Rashwan and Williams, 2011 Mogawer et al., 2012 Daniel et al., 2014	N/A	• Factors with significant effect: binder source, binder type, aggregate absorption, WMA technology, recycled material, plant type, production temperature, and silo storage time • Factors with no effect: aggregate gradation
Aschenbrener and Far, 1994	2-4h @ T_c 1-3h @ T_c	• Asphalt absorption and HWTT results comparable to cores at construction • Recommend 2h @ T_c
Estakhri et al., 2010	2h @ 104°C (WMA) 2 and 4h @ 135°C (WMA) 2 and 4h @ 121°C (HMA)	• Recommend 4h @ 135°C for WMA
Bonaquist, 2011	2h @ T_c 4h @ T_c	• Asphalt absorption and IDT strength comparable to cores at construction • Recommend 2h @ T_c for WMA and suggested additional longer aging period for evaluating rutting and moisture susceptibility
Clements et al., 2012	½, 2, 4, and 8h @ 135°C (HMA) ½, 2, 4, and 8h @ 114°C (WMA)	Recommend the same STOA protocol for WMA and HMA
Sharp and Malone, 2013	1h @150°C (WMA)	Recommend 1h @ 150°C for WMA
Yin et al., 2013	2h @135°C (HMA) 2h @116°C (WMA)	• M_R stiffness comparable to PMPC specimens and cores at construction • Recommend 2h @ 135°C for HMA and 2h @ 116°C for WMA

Table 2. *Summary trends of laboratory STOA protocols (Yin et al., 2013).*

STOA Protocol	2h@Tc		4h@Tc		2+16+2h@Tc		2h@275°F		4h@275°F	
Asphalt Mixture	C*	P*	C*	P*	C*	P*	C*	P*	C*	P*
Iowa HMA			■	■	■	■	■		■	
Iowa WMA I									■	
Iowa WMA II			■		■		■		■	
Texas HMA					N/A					
Texas WMA I			■	■			■	■	■	■
Texas WMA II	■	■	■	■			■	■	■	■

* *C = cores at construction; P = PMPC specimens*

3.0 Experimental Design

This section first provides an overview of the experimental design, including selection of field projects and materials, specimen fabrication procedures, and laboratory tests. Figure 1 presents the research methodology used in this study. First, the laboratory test results of LMLC specimens fabricated using the selected STOA protocols were compared against those of corresponding PMPC specimens and cores at construction. A second set of comparisons was also performed to evaluate the effect of each factor (asphalt mixture components or production parameters) on short-term aging of asphalt mixtures for each type of early-life specimens, i.e., LMLC specimens, PMPC specimens, and cores at construction.

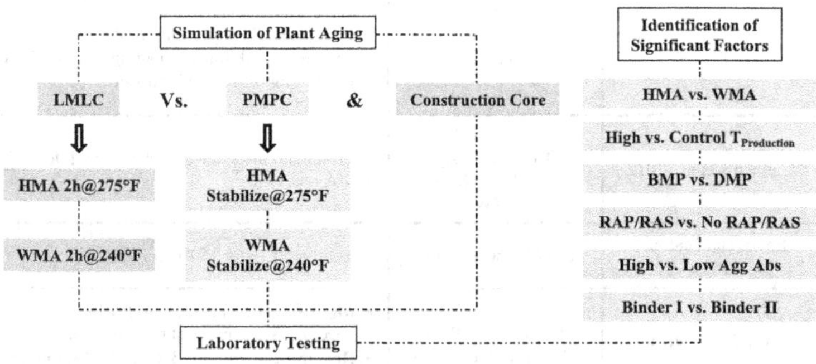

Figure 1. *Research methodology.*

3.1 Field Projects

Materials used in this study are from nine field projects located in the states of Texas, New Mexico, Connecticut, Wyoming, South Dakota, Iowa, Indiana, and Florida. Table 3 provides a summary of these field projects. The following factors were taken into consideration in the selection of field projects to include a wide spectrum of asphalt mixture components and production parameters: binder source, aggregate absorption, WMA technology, inclusion of recycled materials, plant type, and production temperature. During construction of these field projects, PMPC specimens were fabricated on-site in conjunction with acquisition of raw materials and cores at construction.

Table 3. *Summary of field projects used in this study.*

Location & Date	Asphalt	Aggregate & Additive	Mixture & $T_{Production}$	Factor(s)
Texas I FM 973 01/2012	PG 70-22 PG 64-22	Limestone	325°F HMA 325°F HMA+15%RAP/3%RAS 275°F Foaming 275°F Evotherm 270°F Evotherm+15%RAP/3%RAS	HMA vs. WMA RAP/RAS vs. No RAP/RAS
New Mexico IH 25 10/2012	PG 76-28 PG 64-28	Siliceous Gravel 1% Versabind	345°F HMA 315°F HMA+35%RAP 285°F Foaming+35%RAP 275°F Evotherm+35%RAP	HMA vs. WMA RAP/RAS vs. No RAP/RAS
Connecticut IH 84 08/2012	PG 76-22	Basalt	322°F HMA+20%RAP 312°F Foaming+20%RAP	HMA vs. WMA
Wyoming SR 196 08/2012	PG 64-28	Limestone 1% Lime	315°F HMA 275 & 295°F Foaming 255 & 275°F Evotherm	HMA vs. WMA High vs. Control $T_{Production}$
South Dakota SH 262 10/2012	PG 58-34	Quartize 1% Lime	310°F HMA+20%RAP 275°F Foaming+20%RAP 270°F Evotherm+20%RAP 280°F Advera+20%RAP	HMA vs. WMA
Iowa Fairgrounds 06/2013	PG 58-28	Limestone (0.9% AC) Limestone (3.2% AC) Field Sand	295 & 325°F Low Abs HMA+20%RAP 295 & 310°F High Abs HMA+20%RAP 265 & 295°F Low Abs Foaming+20%RAP 260 & 290°F High Abs Foaming+20%RAP	HMA vs. WMA High vs. Control $T_{Production}$ High vs. Low Aggregate Absorption
Indiana Indianapolis 08/2013	PG 64-22	Limestone	305°F HMA+25%RAP (BMP) 300°F HMA+25%RAP (DMP) 273°F Advera+25%RAP (BMP) 271°F Foaming+25%RAP (DMP)	HMA vs. WMA BMP vs. DMP
Florida Jacksonville 08/2013	PG 58-28	Granite (0.6% AC) Limestone (3.7% AC) 0.5% Anti-Stripping	306°F Low Abs HMA+25%RAP 308°F High Abs HMA+25%RAP 272°F Low Abs Foaming+25%RAP 267°F High Abs Foaming+25%RAP	HMA vs. WMA High vs. Low Aggregate Absorption
Texas II Odessa 04/2014	PG 64-22	Limestone	HMA (BMP & Binder A) HMA (DMP & Binder A) HMA (BMP & Binder V) HMA (DMP & Binder V)	BMP vs. DMP Binder I vs. Binder II

3.2 Specimen Fabrication and Short-Term Aging Protocols

To fabricate LMLC specimens, aggregates and binders were heated to the specific plant production temperature ($T_{production}$) independently and then mixed with a portable mixer. Afterwards, loose mix was short-term aged in the oven following the selected laboratory STOA protocols of two hours at 135°C for HMA and two hours at 116°C for WMA prior to compaction in the Superpave Gyratory Compactor

(SGC). Trial specimens were fabricated to assure specimens were obtained with air voids (AV) of 7±0.5%.

Cores were obtained for all field projects soon after construction, and the AV of the cores is summarized in Table 4. To fabricate PMPC specimens, plant produced loose mix was taken from the trucks before leaving the plant or at the paving site, if necessary, transported to an on-site laboratory and maintained in an oven for one to two hours at the field compaction temperature prior to compaction to stabilize the temperature of the mix. Trial specimens were fabricated to assure PMPC specimens achieved a target AV of 7±0.5%. In this study, PMPC specimens and cores at construction were included as representative of plant produced asphalt mixtures experiencing asphalt aging and absorption during production and construction.

Table 4. *Summary of air voids for cores at construction.*

Field Project	AV Range
Texas I	6.5% - 7.7%
New Mexico	3.6% – 8.3%
Connecticut	4.3% - 6.3%
Wyoming	5.1% - 9.3%
South Dakota	6.9% - 10.0%
Iowa	5.8% – 12.0%
Indiana	6.3% - 11.2%
Florida	5.2% - 9.4%
Texas II	4.7% - 10.4%

3.3 Laboratory Tests

Based on previous experience, M_R, E*, and HWTT tests were used in this study to compare the performance of asphalt mixtures fabricated following the selected laboratory STOA protocols versus the corresponding PMPC specimens and cores at construction and also to investigate the effects of mixture components and production parameters on the performance of short-term aged asphalt mixtures.

3.3.1 Resilient Modulus (M_R) Test

The M_R test was conducted through repetitive applications of compressive loads in a haversine waveform along a vertical diametral plane of cylindrical asphalt concrete specimens. The resulting horizontal deformations of the specimen were measured by two linear variable differential transducers (LVDT) aligned along the horizontal diametral plane. M_R stiffness at 25°C was measured in accordance with the current ASTM D7369 with a modification consisting of replacing on-specimen LVDTs with those aligned along the horizontal diametral plane (i.e., gauge length as a fraction of diameter of the specimen = 1.00). The calculation of M_R was based on vertical load, horizontal deformation, and the asphalt mixture's Poisson's ratio at the test temperature, as shown in Equation 1. The M_R test equipment used to perform the measurements is presented in Figure 2.

$$M_R = \frac{p(\upsilon+0.2732)}{t\Delta} \quad [1]$$

Where, M_R, resilient modulus of asphalt mixture;
p, vertical load;
υ, Poisson's ratio;
t, specimen thickness; and
Δ, horizontal deformation measured by LVDTs.

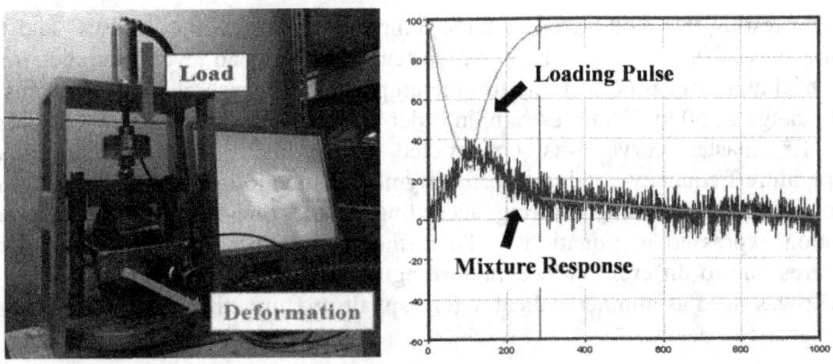

(*a*) (*b*)

Figure 2. *Resilient modulus (M_R) test equipment; (a) specimen setup in loading frame, (b) data acquisition system.*

3.3.2 Dynamic Modulus (E) Test*

The E* test was conducted under unconfined conditions using the Asphalt Mixture Performance Tester (AMPT) shown in Figure 3, following the test procedure specified in AASHTO TP 79-13. SGC compacted specimens were compacted to a height of 170mm, and then cored and trimmed to obtain test specimens with a diameter of 100mm and a height of 150mm.

Figure 3. *Asphalt Mixture Performance Tester (AMPT) for dynamic modulus testing.*

The testing was conducted at three temperatures of 4, 20, and 40°C and three frequencies of 0.1, 1, and 10 Hz for each temperature. Load levels were determined by a trial and error process to assure the amplitude of measured vertical strains was in the range of 50 to 75 microstrain, in order to prevent damage to the test specimen. The E* master curve was constructed by fitting the E* values at each temperature/frequency condition to the sigmoidal function described in Equation 2, followed by horizontally shifting according to the time-temperature shift factor function expressed in Equation 3. To further discriminate E* stiffness of asphalt mixtures due to different binder/mixture aging levels, the E* stiffness at 20°C and 10 Hz was used as another indicator for asphalt mixture stiffness in addition to the E* master curve.

$$log|E^*| = a + \frac{b}{1+\frac{1}{e^{d+g*log(f_R)}}} \quad [2]$$

Where, E^*, dynamic modulus of asphalt mixture;
f_R, reduced frequency; and
a, b, c, and g, fitting coefficients of the sigmoidal function.

$$loga_T = \alpha_1 T^2 + \alpha_2 T + \alpha_3 \quad [3]$$

Where, a_T, time-temperature shift factor; and
α_1, α_2, and α_3, fitting coefficients of time-temperature shift factor function.

3.3.3 Hamburg Wheel Tracking Test (HWTT)

The HWTT (AASHTO T 324) is a laboratory test commonly used for evaluating rutting resistance and moisture susceptibility of asphalt mixtures. The test consisted of submerging specimens in warm water at 50°C and subjecting them to 52 passes of a loaded steel wheel per minute. Each of two replicate specimens was loaded for a maximum of 20,000 load cycles or until the center of the specimen deformed by 12.5mm per Texas Department of Transportation specification Tex-242-F. The

HWTT equipment used to perform the measurements is shown in Figure 4. Typical results of the rut depth versus number of load cycles consist of three phases: post compaction, creep, and stripping (Solaimanian et al., 2003). The traditional test parameter for evaluating rutting performance of asphalt mixtures is rut depth at a specific number of load cycles. In this study, rut depth at 5,000 load cycles was used as a mixture rutting resistance parameter in the HWTT.

Figure 4. *Hamburg Wheel Tracking Test (HWTT) equipment.*

In addition to the traditional rutting analysis in the HWTT, a novel method developed by Yin et al. (2014) was also used in this study to discriminate asphalt mixtures with different rutting resistance, and the viscoplastic strain at the stripping number ($\Delta\varepsilon_{SN}^{vp}$) was employed as the alternative HWTT rutting resistance parameter. The determination of $\Delta\varepsilon_{SN}^{vp}$ is illustrated in Figure 5 and details of the analysis method can be found elsewhere (Yin et al., 2014). As compared to the traditional rutting parameter of rut depth at a specific number of load cycles, the determination of $\Delta\varepsilon_{SN}^{vp}$ isolated the viscoplastic strain during the creep phase and did not include contributions due to different specimen AV or stripping. Therefore, the novel parameter was expected to be able to better characterize mixture rutting resistance in HWTT. Specifically, asphalt mixtures with higher $\Delta\varepsilon_{SN}^{vp}$ values were expected to be more susceptible to rutting than those with lower $\Delta\varepsilon_{SN}^{vp}$ values.

According to the laboratory experience with the novel analysis method, early stripping had been frequently observed for short-term aged asphalt mixtures using softer asphalt binders and/or recycled materials, with LC_{SN} observed at less than 3,000 load cycles. These mixtures had a limited duration of the creep phase before stripping occurred, and as a consequence, the determination of the viscoplastic strain was unfeasible. Therefore in this study, the evaluation of rutting resistance of asphalt mixture by $\Delta\varepsilon_{SN}^{vp}$ value was only performed for asphalt mixtures having LC_{SN} greater than 3,000 load cycles.

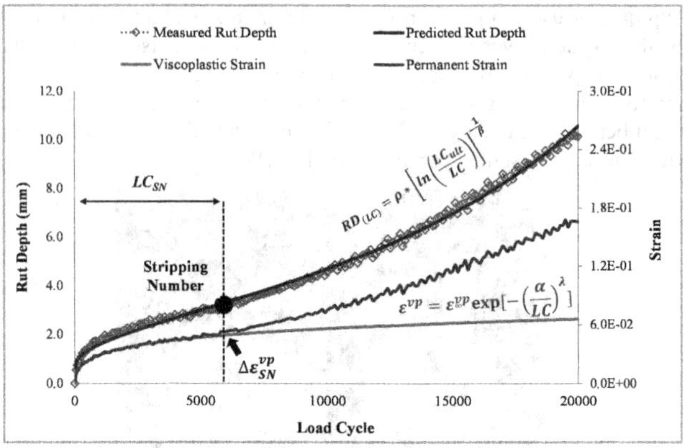

Figure 5. *Alternative rutting resistance parameter of* $\Delta\varepsilon_{SN}^{vp}$ *in the Hamburg Wheel Tracking Test (HWTT).*

4.0 Test Results and Data Analysis

This section provides the mixture test results for LMLC specimens fabricated following the selected laboratory STOA protocols of two hours at 135°C for HMA and two hours at 116°C for WMA, PMPC specimens, and cores at construction. Mixture volumetrics, M_R stiffness, E* stiffness, and HWTT rutting resistance results were analyzed toward simulating asphalt aging and absorption during plant production and construction by the selected laboratory STOA protocols, and identifying mixture components and production parameters with significant effects on the performance of short-term aged asphalt mixtures.

4.1 Mixture Volumetrics

Table 5 presents the comparison of volumetrics of LMLC specimens fabricated following the selected laboratory STOA protocols and PMPC specimens in terms of theoretical maximum specific gravity (G_{mm}), percentage of absorbed asphalt (P_{ba}), percentage of effective asphalt (P_{be}), and effective binder film thickness (FT_{be}). The volumetrics were calculated using the mix design aggregate gradation and asphalt content per Saskatchewan Highways and Transportation method STP 204-19.

Table 5. *Mixture volumetrics for LMLC specimens and PMPC specimens.*

Field Project	Mixture Type	PMPC				LMLC			
		G_{mm}	P_{ba}	P_{be}	FT_{be}	G_{mm}	P_{ba}	P_{be}	FT_{be}
Texas I	HMA	2.420	0.53	4.70	9.09	2.397	0.10	5.11	9.88
	Evotherm	2.408	0.30	4.91	9.50	2.399	0.13	5.07	9.81
	Foaming	2.400	0.15	5.06	9.77	2.407	0.28	4.93	9.53
	HMA+RAP/RAS	2.410	0.83	4.42	7.89	2.418	0.98	4.27	7.64
	Evotherm+RAP/RAS	2.420	1.02	4.24	7.58	2.417	0.96	4.29	7.67
New Mexico	HMA	2.342	0.41	5.01	10.21	2.329	0.16	5.25	10.70
	HMA+RAP	2.340	0.66	4.78	9.66	2.339	0.64	4.79	9.70
	Evotherm+RAP	2.343	0.72	4.72	9.55	2.333	0.52	4.91	9.93
	Foaming+RAP	2.335	0.56	4.87	9.85	2.349	0.84	4.61	9.32
Connecticut	HMA+RAP	2.676	1.26	3.71	8.55	2.652	0.90	4.04	9.33
	Foaming+RAP	2.675	1.24	3.72	8.59	2.658	0.99	3.96	9.14
Wyoming	HMA	2.470	0.76	4.28	8.81	2.491	1.13	3.93	8.09
	Evotherm High T	2.479	0.92	4.13	8.50	2.494	1.18	3.88	7.98
	Evotherm Ctrl T	2.487	1.06	3.99	8.22	2.501	1.30	3.76	7.75
	Foaming High T	2.485	1.03	4.03	8.29	2.497	1.24	3.83	7.88
	Foaming Ctrl T	2.470	0.76	4.28	8.81	2.505	1.37	3.70	7.61
South Dakota	HMA+RAP	2.441	0.58	4.75	7.28	2.441	0.58	4.75	7.28
	Evotherm+RAP	2.440	0.56	4.77	7.31	2.440	0.56	4.77	7.31
	Foaming+RAP	2.428	0.35	4.97	7.62	2.440	0.56	4.77	7.31
	Advera+RAP	2.432	0.42	4.90	7.51	2.432	0.42	4.90	7.51
Iowa	High Abs HMA+RAP High T	2.425	2.35	4.82	10.18	2.373	1.35	5.75	12.14
	High Abs HMA+RAP Ctrl T	2.439	2.61	4.57	9.67	2.373	1.35	5.75	12.14
	High Abs Foaming+RAP High T	2.435	2.54	4.64	9.81	2.365	1.19	5.89	12.45
	High Abs Foaming+RAP Ctrl T	2.437	2.57	4.61	9.74	2.373	1.35	5.75	12.14
	Low Abs HMA+RAP High T	2.481	0.61	4.42	9.28	2.482	0.63	4.40	9.25
	Low Abs HMA+RAP Ctrl T	2.476	0.52	4.50	9.46	2.479	0.58	4.45	9.35
	Low Abs Foaming+RAP High T	2.477	0.54	4.49	9.42	2.488	0.73	4.30	9.04
	Low Abs Foaming+RAP Ctrl T	2.474	0.49	4.54	9.53	2.489	0.75	4.29	9.00
Indiana	HMA+RAP BMP	2.451	1.31	4.77	8.12	2.458	1.32	4.65	7.92
	HMA+RAP DMP	2.446	1.48	5.00	8.55	2.443	1.43	5.05	8.64
	Advera+RAP BMP	2.448	1.29	4.84	8.24	2.456	1.43	4.71	8.02
	Foaming+RAP DMP	2.455	1.43	4.73	8.06	2.440	1.16	4.98	8.49
Florida	High Abs HMA+RAP	2.350	2.03	4.66	6.93	2.341	1.86	4.83	7.17
	High Abs Foaming+RAP	2.363	2.18	4.37	6.48	2.365	2.22	4.33	6.43
	Low Abs HMA+RAP	2.537	0.79	3.74	5.61	2.540	0.84	3.70	5.54
	Low Abs Foaming+RAP	2.548	1.09	3.64	5.46	2.540	0.96	3.76	5.65
Texas II	HMA BMP Alon	2.402	1.39	5.06	7.97	2.393	1.11	5.11	8.10
	HMA DMP Alon	2.415	1.34	4.65	7.28				
	HMA BMP Valero	2.395	1.26	5.19	8.17	2.392	1.16	5.21	8.20
	HMA DMP Valero	2.411	1.26	4.71	7.38				

Figure 6 and Figure 7 present the correlations for LMLC specimens versus PMPC specimens in terms of G_{mm} and P_{ba} values. As illustrated in Figure 6, most of the data points fell on the line of equality indicating equivalent G_{mm} values were achieved by PMPC specimens and LMLC specimens. A reasonable correlation in terms of P_{ba} values was also observed in Figure 7 when comparing the two specimen types regardless of a set of outliers being observed for Iowa mixtures that corresponded to the high absorptive aggregates. The P_{ba} values were mixed when ranking LMLC and PMPC specimens for all mixtures except those from the Connecticut and Wyoming field projects. Higher P_{ba} values for PMPC versus LMLC specimens were shown for the Connecticut mixtures, while the opposite trend was shown for the Wyoming mixtures. Based on mixture volumetrics summarized in Table 5, Figure 6, and Figure 7, equivalent mixture volumetrics were observed for PMPC specimens and LMLC specimens for a wide range of asphalt mixtures. Therefore, the selected laboratory STOA protocols of two hours at 135°C for HMA and 116°C for WMA were considered suitable to simulate the asphalt absorption during plant production and construction.

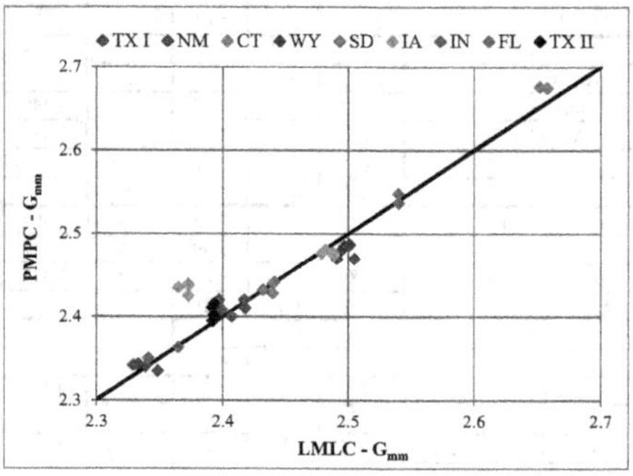

Figure 6. *G_{mm} correlation for LMLC specimens versus PMPC specimens.*

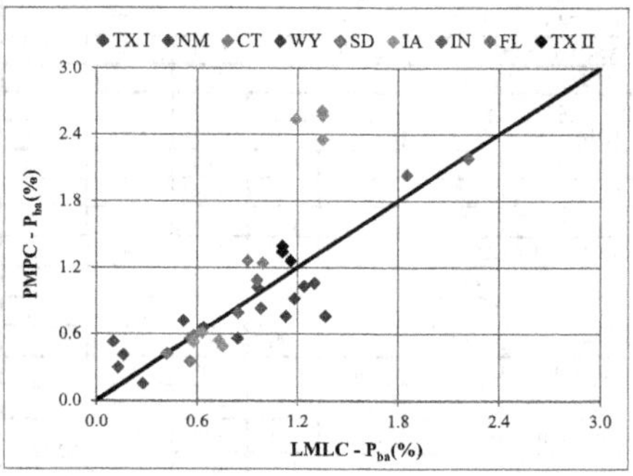

Figure 7. *P_{ba}(%) correlation for LMLC specimens versus PMPC specimens.*

4.2 Simulation of Plant Aging

As mentioned previously, the laboratory STOA protocols of two hours at 135°C for HMA and two hours at 116°C for WMA were used in this study to simulate asphalt aging during plant production and construction. In order to explore the correlation in binder/mixture aging induced by the selected laboratory STOA protocols versus that occurred during plant production, M_R stiffness, E* stiffness, and HWTT rutting resistance parameters for LMLC specimens from nine field projects were plotted against the corresponding results obtained for PMPC specimens and cores at construction. A detailed discussion of the results is presented in the following sections.

4.2.1 M_R Test Results

Figure 8 and Figure 9 present the M_R stiffness correlation of LMLC specimens versus PMPC specimens and cores at construction, respectively. As illustrated, most of the data points fell on the line of equality, which indicated that M_R stiffness for LMLC specimens with the selected laboratory STOA protocols of two hours at 135°C for HMA and two hours at 116°C for WMA closely mimicked the M_R stiffness of the PMPC specimens and cores at construction. In addition, there was a remarkable similarity in the pattern of M_R stiffness results for PMPC specimens and cores at construction indicating an insignificant difference between these two types of specimens.

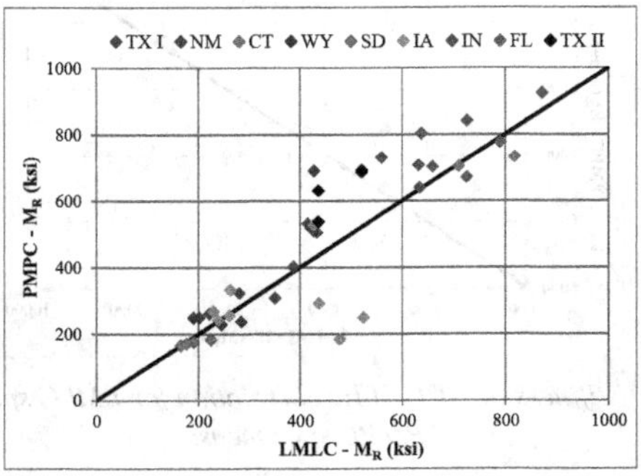

Figure 8. *M_R stiffness correlation for LMLC specimens versus PMPC specimens.*

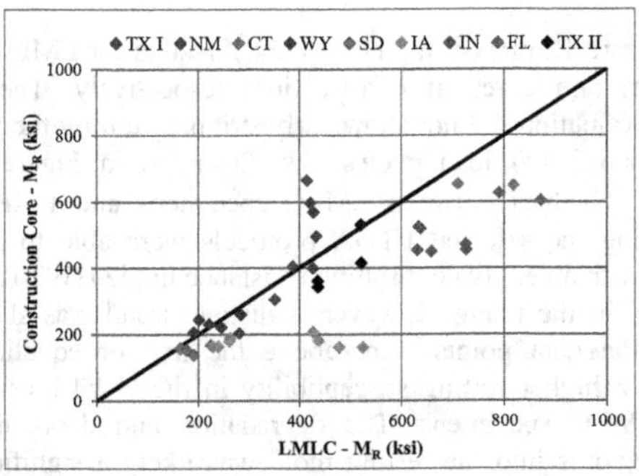

Figure 9. *M_R stiffness correlation for LMLC specimens versus cores at construction.*

4.2.2 E Test Results*

Figure 10 presents the correlation of E* stiffness results at 20°C and 10Hz for LMLC specimens versus PMPC specimens of asphalt mixtures from the Connecticut, Indiana, and Texas II field projects. Consistent with results shown in Figure 8, a good correlation in E* stiffness was observed for LMLC specimens versus PMPC specimens in Figure 10. Therefore, the laboratory STOA protocols of two hours at 135°C for HMA and two hours at 116°C for WMA were able to produce laboratory asphalt mixtures with an equivalent E* stiffness as compared to plant produced asphalt mixtures. The outlier shown in Figure 10 was the BMP PMPC specimen of HMA from the Indiana field project, which showed a significantly lower E* stiffness as compared to its corresponding LMLC counterpart.

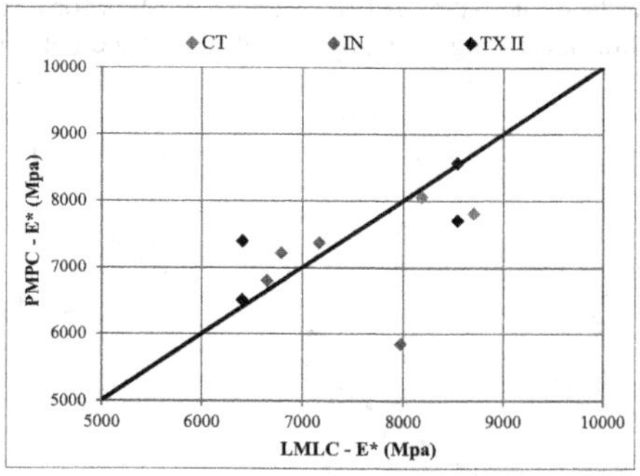

Figure 10. *E* stiffness at 20°C/10Hz correlation for LMLC specimens versus PMPC specimens.*

4.2.3 HWTT Results

Figure 11 and Figure 12 present the HWTT $\Delta\varepsilon_{SN}^{vp}$ results for LMLC specimens versus PMPC specimens and cores at construction, respectively. The asphalt mixtures included in this evaluation did not show early stripping during the tests and had LC_{SN} values greater than 3,000 load cycles. As illustrated in Figure 11, a reasonable correlation in $\Delta\varepsilon_{SN}^{vp}$ values between LMLC specimens and PMPC specimens was obtained, indicating the selected LTOA protocols were able to produce laboratory asphalt mixtures with an equivalent rutting resistance in the HWTT as compared to the mixture produced in the plant. However, a distinct trend was shown in Figure 12, where most of the data points were above the line of equality. Thus, cores at construction had a higher rutting susceptibility in the HWTT as compared to their corresponding LMLC specimens. The degradation and debonding of the plaster needed to fit the cores into the testing mold was likely a significant contributor to higher rut depths for cores at constructions and a consequent poor correlation with the LMLC results. Therefore, there is a need for appropriate modifications to the HWTT procedure for testing field cores in the future.

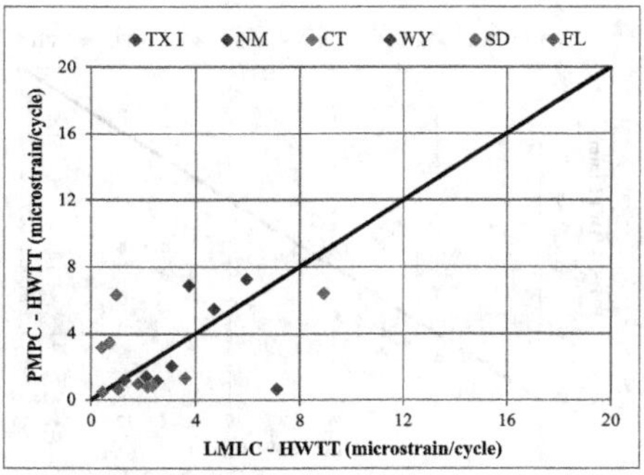

Figure 11. *HWTT $\Delta\varepsilon_{SN}^{vp}$ correlation for LMLC specimens versus PMPC specimens.*

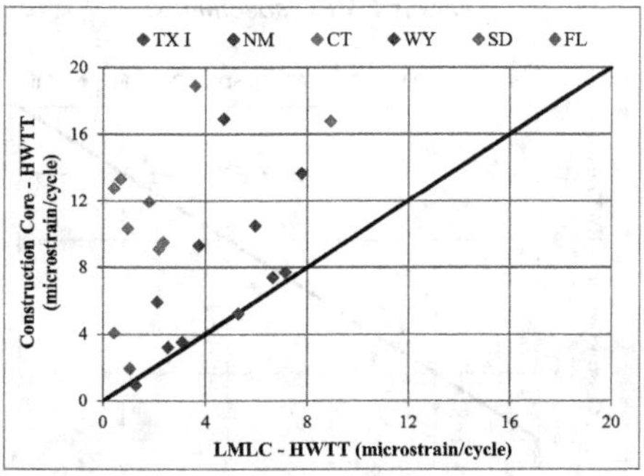

Figure 12. *HWTT $\Delta\varepsilon_{SN}^{vp}$ correlation for LMLC specimens versus cores at construction.*

In addition to the HWTT $\Delta\varepsilon_{SN}^{vp}$ results, the traditional rutting resistance parameter of rut depth at 5,000 load cycles was also used to evaluate the simulation of plant aging by the selected laboratory STOA protocols. The results for LMLC specimens versus PMPC specimens and cores at construction are presented in Figure 13 and Figure 14, respectively. Although a substantial variability in the rut depth measurements was exhibited in Figure 13, there was a reasonable correlation in terms of rutting resistance between LMLC specimens and their corresponding PMPC specimens. Similar to the test results shown in Figure 12, a higher rutting susceptibility as indicated by higher rut depths at 5,000 load cycles for cores at construction versus their corresponding LMLC specimens was also observed in Figure 14. Again, HWTT results for cores at construction were possibly biased due to the disintegration and debonding of the plaster needed to properly fit the cores into the molds during the test.

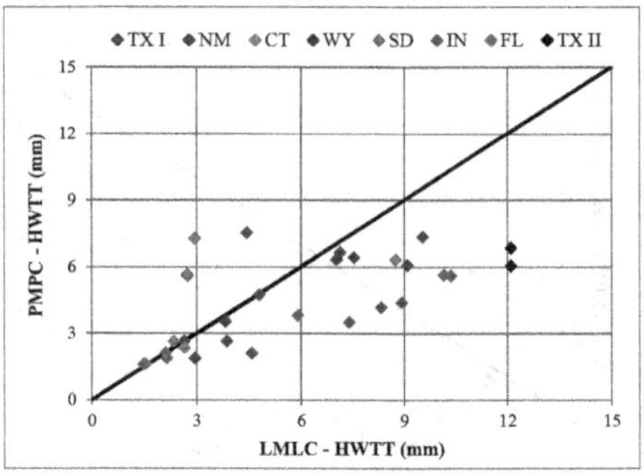

Figure 13. *HWTT rut depth at 5,000 load cycles correlation for LMLC specimens versus PMPC specimens.*

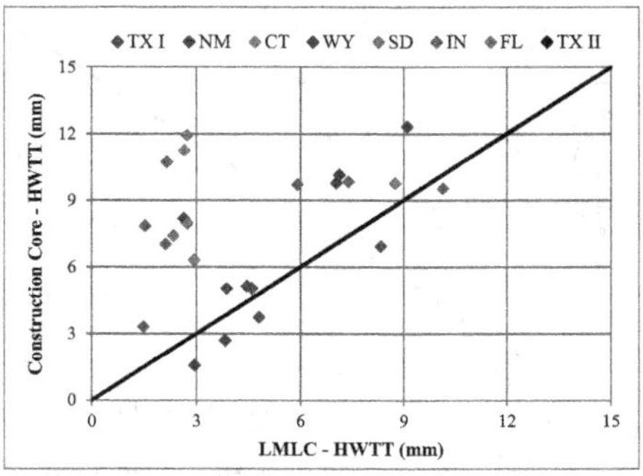

Figure 14. *HWTT rut depth at 5,000 load cycles correlation for LMLC specimens versus cores at construction.*

4.2.4 Summary

According to the M_R and E* test results, good correlations in mixture stiffness between LMLC specimens with the selected laboratory STOA protocols and PMPC specimens and cores at construction were obtained for a wide range of asphalt mixtures from nine field projects. In addition, an approximately equivalent rutting resistance was also observed for LMLC specimens and PMPC specimens in terms of HWTT $\Delta\varepsilon^{vp}_{LC_{SN}}$ values and rut depths at 5,000 load cycles. A higher rutting susceptibility in the HWTT was shown for cores at construction than the corresponding LMLC specimens, which was possibly caused by the need to plaster the cores to fit the height of the HWTT molds. Thus, the simulation of binder/mixture aging during plant production and construction by the laboratory STOA protocols of two hours at 135°C for HMA and 116°C for WMA was verified in this study for a wider range of asphalt mixtures.

4.3 Identification of Factors Affecting the Performance of Short-Term Aged Asphalt Mixtures

This section presents the laboratory test results for identifying mixture components and production parameters with significant effects on the performance of short-term aged asphalt mixtures. These factors include binder source, aggregate absorption, WMA technology, inclusion of recycled materials, plant type, and production temperature. Detailed discussions for each factor are presented in the following sections.

4.3.1 WMA Technology (HMA vs. WMA)

The M_R and HWTT results for LMLC specimens, PMPC specimens, and cores at construction from eight field projects are shown in Figures 15 to 17, with M_R stiffness and HWTT rutting resistance parameters of $\Delta\varepsilon_{SN}^{vp}$ values and rut depth at 5,000 load cycles for HMA mixtures plotted against those of corresponding WMA mixtures. The x-axis coordinate represents HMA test results, and the y-axis coordinate represents the corresponding WMA test results. The black solid line is the line of equality, and the red dashed line is the fitted trendline for M_R stiffness or rutting resistance parameters in the HWTT.

The M_R stiffness comparison for HMA versus WMA shown in Figure 15 illustrated that most of the data points were below the line of equality, indicating a higher M_R stiffness for HMA as compared WMA. According to the fitted trendline, the average ratio of WMA M_R stiffness over HMA M_R stiffness is approximately 0.85. Thus, the inclusion of WMA technology in the asphalt mixture is likely to produce asphalt mixtures with lower mixture stiffness.

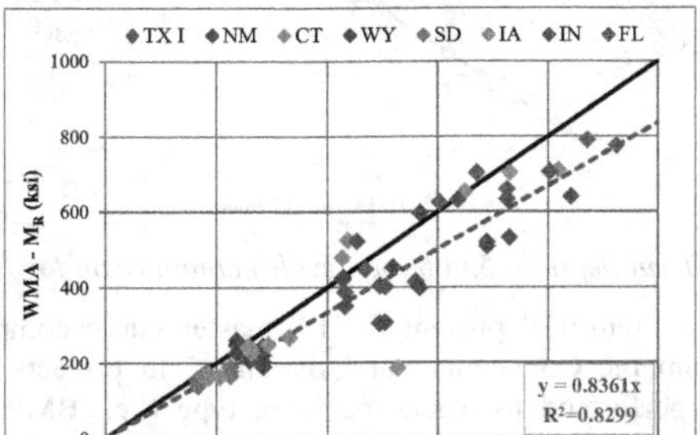

Figure 15. *M_R stiffness comparison for HMA versus WMA.*

Figure 16 presents the HWTT $\Delta\varepsilon_{SN}^{vp}$ comparison for HMA versus WMA. As illustrated, most of the data points were above the line of equality, indicating a better rutting resistance in the HWTT for HMA than WMA. Figure 17 presents the HWTT traditional rutting resistance parameter of rut depth at 5,000 load cycles. Similar to

Figure 16, most of the data points in Figure 17 aligned above the line of equality. According to the slope for the fitted trendline, the rut depth after the first 5,000 load cycles in the HWTT for WMA mixtures was approximately 1.26 times higher than that for HMA. Thus, the inclusion of WMA technology is likely to produce asphalt mixtures with a higher rutting susceptibility.

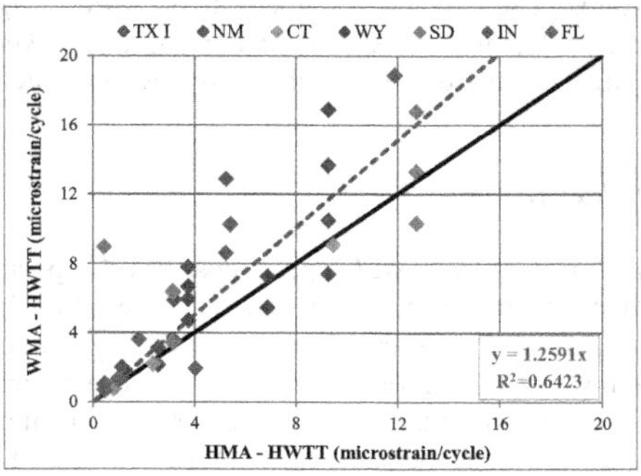

Figure 16. *HWTT* $\Delta\varepsilon_{SN}^{vp}$ *comparison for HMA versus WMA.*

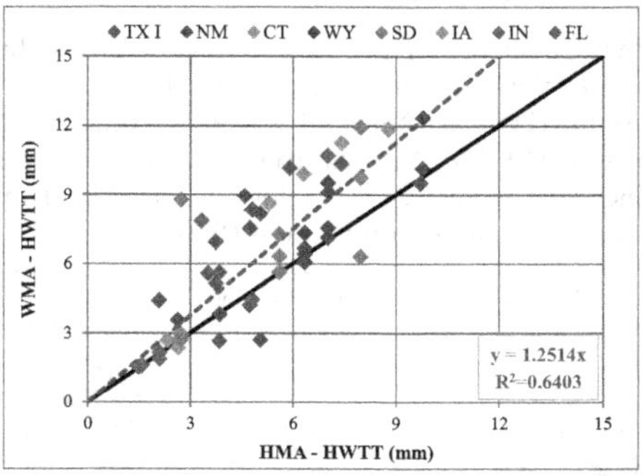

Figure 17. *HWTT rut depth at 5,000 load cycles comparison for HMA versus WMA.*

Figure 18 and Figure 19 present the E* master curve comparisons for HMA versus WMA from the Connecticut and Indiana field projects, respectively. The comparison was performed for each specimen type (i.e., BMP PMPC specimen, DMP PMPC specimen, and LMLC specimen). For all comparisons in terms of E* stiffness for HMA versus WMA (except for BMP PMPC specimens from the Indiana field project), the E* master curves for HMA mixtures were above or overlapping with those for WMA counterpart mixtures indicating higher or equivalent E* stiffness values over a wide range of testing temperatures and frequencies. For the exceptional case, slightly lower E* stiffness values were observed for HMA mixtures as compared to their WMA counterparts.

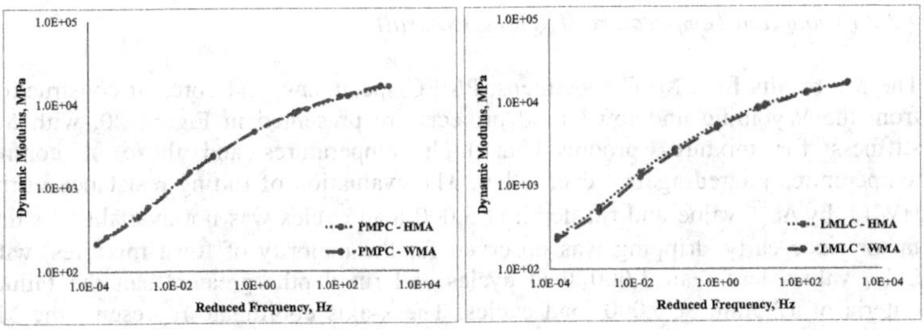

(*a*) (*b*)

Figure 18. *E* master curve comparison for HMA versus WMA for the Connecticut field project; (a) PMPC specimens, (b) LMLC specimens.*

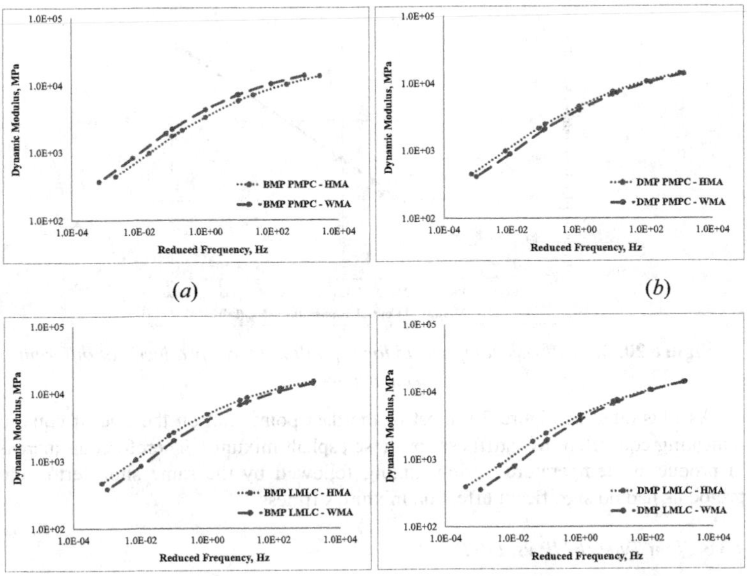

(*c*) (*d*)

Figure 19. *E* master curve comparison for HMA versus WMA for the Indiana field project; (a) BMP PMPC specimens, (b) DMP PMPC specimens, (c) BMP LMLC specimens, (d) DMP LMLC specimens.*

4.3.2 Production Temperature (High vs. Control)

The M_R results for LMLC specimens, PMPC specimens, and cores at construction from the Wyoming and Iowa field projects are presented in Figure 20, with M_R stiffness for mixtures produced at high temperatures and those at control temperatures plotted against each other. The evaluation of rutting resistance in the HWTT by $\Delta\varepsilon_{SN}^{vp}$ value and rut depth at 5,000 load cycles was not available for this factor since early stripping was observed for the majority of Iowa mixtures, with LC_{SN} values less than 3,000 load cycles and rut depths greater than the failure criteria of 12.5mm at 5,000 load cycles. The x-axis coordinate represents the M_R stiffness for mixtures produced at control temperatures, and the y-axis coordinate represents M_R stiffness for mixtures produced at high temperatures. The black solid line is the line of equality, and the red dashed line is the fitted trendline for M_R stiffness.

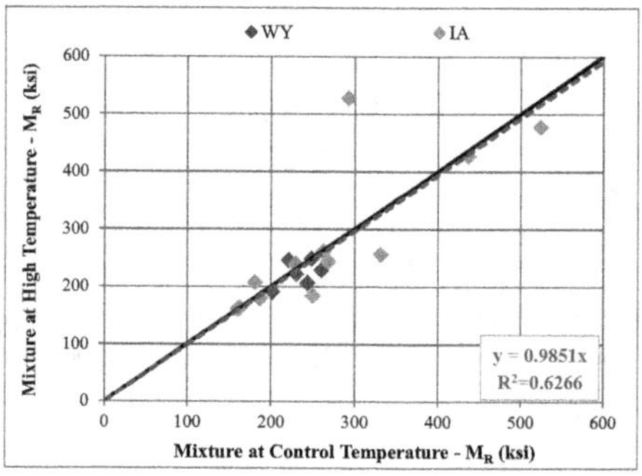

Figure 20. *M_R stiffness comparison for asphalt mixtures produced at different temperatures.*

As illustrated in Figure 20, most of the data points fell on the line of equality, indicating equivalent M_R stiffness for those asphalt mixtures. Therefore, an increase in production temperature during mixing followed by the same short-term aging protocols had no significant effect on mixture stiffness.

4.3.3 Plant Type (BMP vs. DMP)

The M_R and HWTT test results for LMLC specimens, PMPC specimens, and cores at construction from the Indiana and Texas II field projects are presented in Figure 21 and Figure 22, with M_R stiffness and the HWTT rutting resistance parameter of rut depth at 5,000 load cycles for BMP produced mixtures and DMP produced mixtures plotted against each other. The evaluation of rutting resistance in the HWTT by $\Delta\varepsilon_{SN}^{vp}$ value was not available for this factor since early stripping was observed for the majority of Indiana and Texas II asphalt mixtures. The x-axis coordinate represents the test results for BMP produced mixtures, and the y-axis coordinate represents corresponding test results for DMP produced mixtures. The black solid line is the line of equality, and the red dashed line is the fitted trendline for M_R stiffness or rut depth measurements in the HWTT.

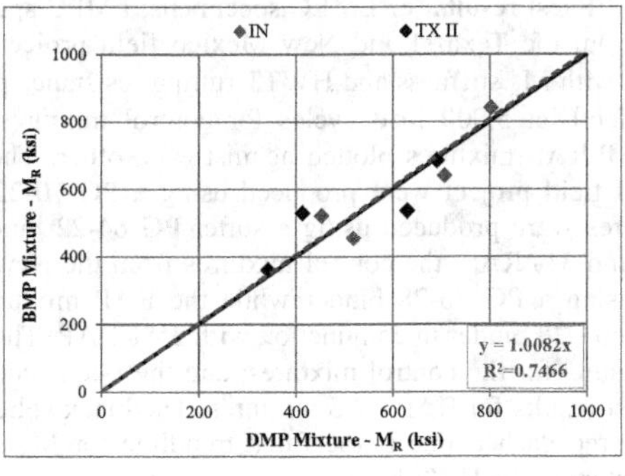

Figure 21. *M_R stiffness comparison for asphalt mixtures produced at BMP versus DMP.*

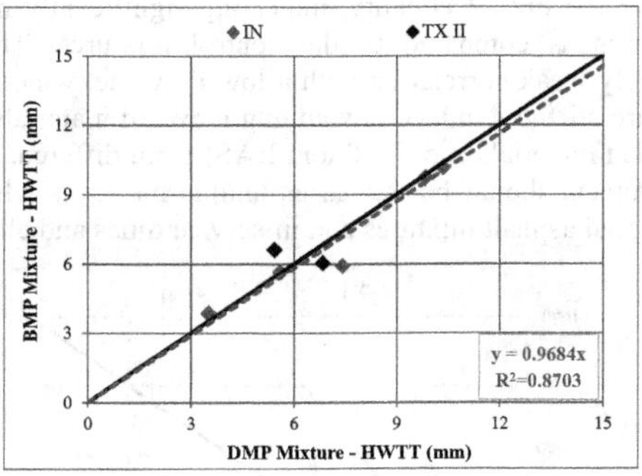

Figure 22. *HWTT rut depth at 5,000 load cycles comparison for asphalt mixtures produced at BMP versus DMP.*

The M_R results in Figure 21 illustrated that most of the data points were on the line of equality, indicating equivalent mixtures stiffness was achieved by asphalt mixtures produced in a BMP and the corresponding mixtures produced in a DMP. Figure 22 presents the traditional HWTT rutting resistance parameter of rut depth at 5,000 load cycles for the comparison of BMP produced mixtures versus DMP produced mixtures. Similar to Figure 21, most of the data points fell on the line of equality. Therefore, equivalent mixture stiffness and rutting resistance was observed for asphalt mixtures produced in a BMP and a DMP.

4.3.4 Inclusion of Recycled Material (RAP/RAS vs. No RAP/RAS)

The M_R and HWTT test results for LMLC specimens, PMPC specimens, and cores at construction from the Texas I and New Mexico field projects are presented in Figures 23 to 25, with M_R stiffness and HWTT rutting resistance parameters of $\Delta\varepsilon_{SN}^{vp}$ values and rut depth at 5,000 load cycles for control mixtures without recycled mixtures and RAP/RAS mixtures plotted against each other. The control mixtures from the Texas I field project were produced using a PG 70-22 binder while the RAP/RAS mixtures were produced using a softer PG 64-22 binder in conjunction with 15% RAP and 3% RAS; the control mixtures from the New Mexico mixtures were produced using a PG 76-28 binder while the RAP mixtures were produced using a softer PG 64-28 binder in conjunction with 35% RAP. The x-axis coordinate represents test results for the control mixtures, and the y-axis coordinate represents corresponding test results for RAP/RAS mixtures. The black solid line is the line of equality, and the red dashed line is the fitted trendline for M_R stiffness or rutting resistance parameters in the HWTT.

The M_R stiffness results shown in Figure 23 illustrated that most of the data points were above the line of equality, indicating significantly higher stiffness for RAP/RAS mixtures as compared to the control mixtures. The fitted trendline presents a relatively weak correlation with a low R^2 value, which indicated that the increase in mixture stiffness induced by adding recycled materials was inconsistent. Therefore, recycled materials (i.e. RAP and RAS) from different sources utilized in different field projects should be treated as unique materials whose properties are related to the original asphalt mixtures and in-service times and climatic conditions.

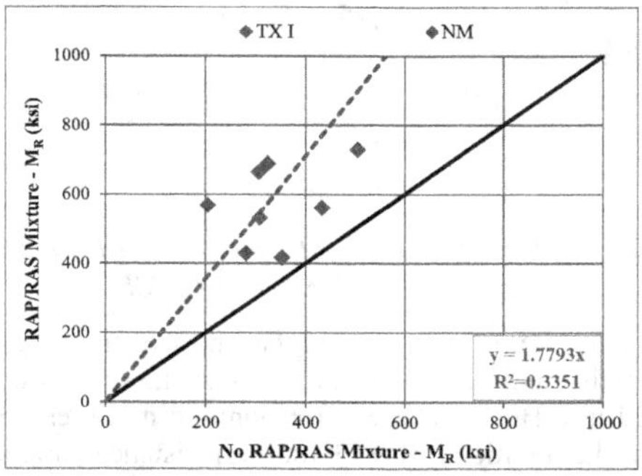

Figure 23. *M_R stiffness comparison for asphalt mixtures with and without RAP and RAS.*

Figure 24 presents the HWTT $\Delta\varepsilon_{SN}^{vp}$ comparison for RAP/RAS mixtures versus control mixtures. As illustrated, most points were above the line of equality, indicating decreased rutting resistance in the HWTT for RAP/RAS mixtures as compared to the control counterpart mixtures. According to the fitted trendline, the average ratio of $\Delta\varepsilon_{SN}^{vp}$ value for RAP/RAS mixtures over $\Delta\varepsilon_{SN}^{vp}$ value for the control mixtures is approximately 1.44. Figure 25 presents the traditional HWTT rutting resistance parameter of rut depth at 5,000 load cycles. No consistent trend in the comparison of RAP/RAS mixtures versus control mixtures was observed for this parameter, as indicated by a negative R^2 value for the fitted trendline.

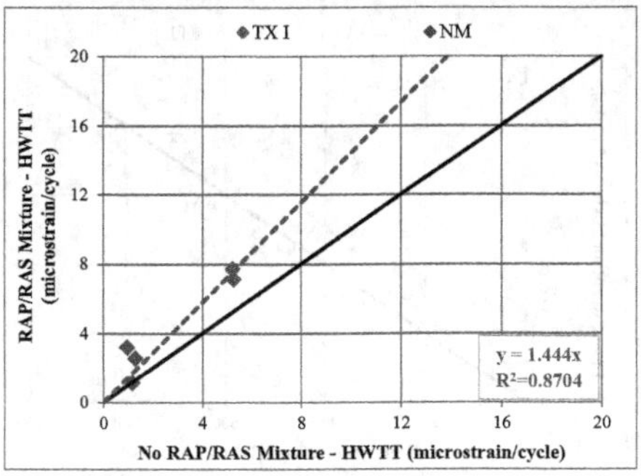

Figure 24. *HWTT* $\Delta\varepsilon_{SN}^{vp}$ *comparison for asphalt mixtures with and without RAP and RAS.*

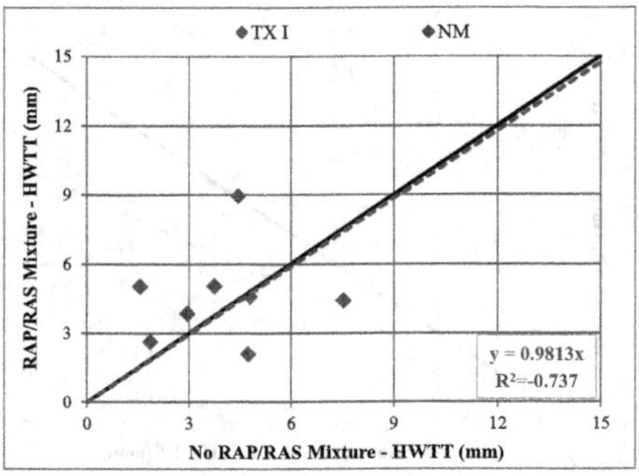

Figure 25. *HWTT rut depth at 5,000 load cycles comparison for asphalt mixtures with and without RAP and RAS.*

4.3.5 Aggregate Absorption (High vs. Low Absorptive Aggregate)

The M_R and HWTT test results for LMLC specimens, PMPC specimens, and cores at construction from the Iowa and Florida field projects are presented in Figure 26 and Figure 27, with M_R stiffness and the HWTT rutting resistance parameter of rut depth at 5,000 load cycles for asphalt mixtures using high absorptive aggregates versus low absorptive aggregates plotted against each other. The evaluation of rutting resistance in the HWTT by $\Delta\varepsilon_{SN}^{vp}$ value was not available for this factor since early stripping was observed for the majority of Iowa and Florida asphalt mixtures, with LC_{SN} values less than 3,000 load cycles. The x-axis coordinate represents test results for mixtures using high absorptive aggregates, and the y-axis coordinate represents corresponding test results for mixtures using low absorptive aggregates. The black solid line is the line of equality, and the red dashed line is the fitted trendline for M_R stiffness or rut depth measurements in the HWTT.

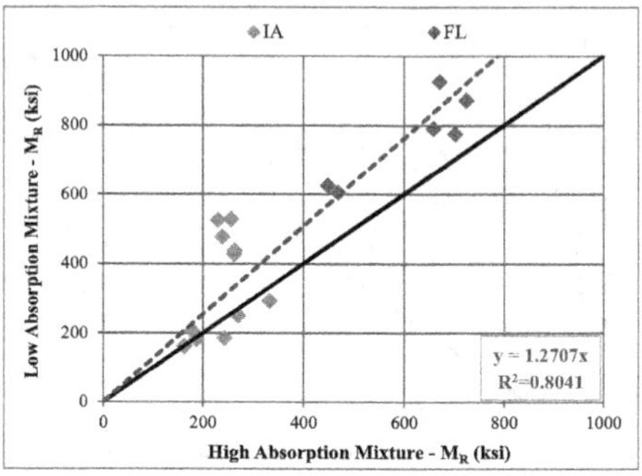

Figure 26. *M_R stiffness comparison for asphalt mixtures using high absorptive versus low absorptive aggregates.*

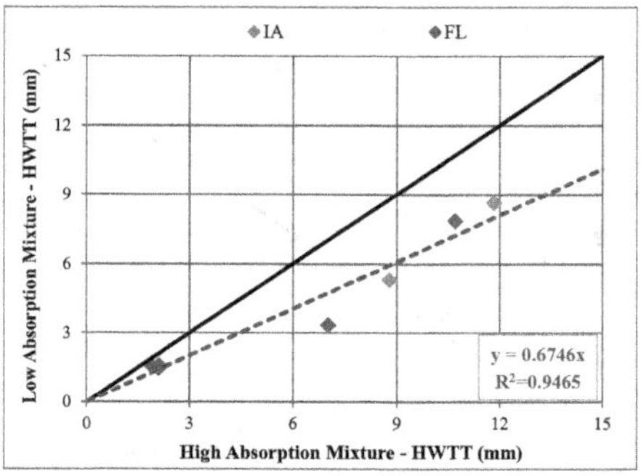

Figure 27. *HWTT rut depth at 5,000 load cycles comparison for asphalt mixtures using high absorptive versus low absorptive aggregates.*

The M_R stiffness comparison for mixtures using high versus low absorptive aggregates shown in Figure 26 illustrated that most of the data points were above the line of equality, indicating a higher M_R stiffness for mixtures using low absorptive aggregates as compared to the counterpart mixtures using high absorptive aggregates. Figure 27 presents the HWTT results in terms of rut depth at 5,000 load cycles. Similar to Figure 26, most of the data points in Figure 27 were below the line of equality, indicating better rutting resistance in the HWTT for mixtures using low versus high absorptive aggregates. The better mixture performance in terms of stiffness and rutting resistance observed for mixtures using low absorptive aggregates as compared to those using high absorptive aggregates may be attributed to thinner effective binder film thicknesses in these mixtures as indicated by lower P_{be} and FT_{be} values in Table 5.

4.3.6 Binder Source (Binder I vs. Binder II)

The M_R results for LMLC specimens, PMPC specimens, and cores at construction from the Texas II field project are presented in Figure 28, with M_R stiffness for mixtures using binder A plotted against those of corresponding mixtures using binder V. The evaluation of rutting resistance in the HWTT by $\Delta\varepsilon_{SN}^{vp}$ value and rut depth at 5,000 load cycles was not available for this factor since early stripping was observed for all Texas II mixtures with LC_{SN} values less than 3,000 load cycles and rut depths greater than the failure criteria of 12.5mm at 5,000 load cycles. The x-axis coordinate represents M_R stiffness for mixtures using binder A, and the y-axis coordinate represents M_R stiffness for mixtures using binder V. The black solid line is the line of equality, and the red dashed line is the fitted trendline for M_R stiffness.

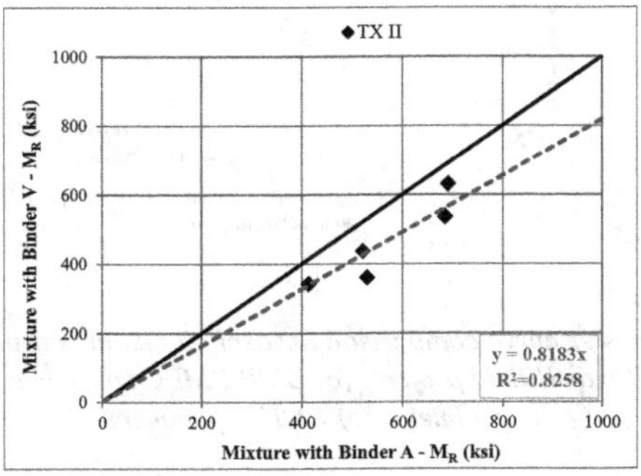

Figure 28. *M_R stiffness comparison for asphalt mixtures using different binder sources.*

As shown in Figure 28 for the M_R stiffness comparison for asphalt mixtures using two different binder sources, most of the data points were below the line of equality, indicating a higher M_R stiffness for asphalt mixtures using binder A than the counterpart mixtures using binder V.

Figure 29 presents the E* master curve comparisons for PMPC specimens and LMLC specimens of asphalt mixtures with binder A versus binder V. The comparison was performed for each specimen type (i.e., BMP PMPC specimen, DMP PMPC specimen, and LMLC specimen). As illustrated, the E* master curves for mixtures with binder A were consistently above the curves obtained for mixtures with binder V. Therefore, binder source exhibited a significant effect on the performance of short-term aged asphalt mixtures, and consequently, asphalt mixtures using the same PG graded binders from different sources would exhibit substantially different mixture performance in terms of stiffness and rutting resistance.

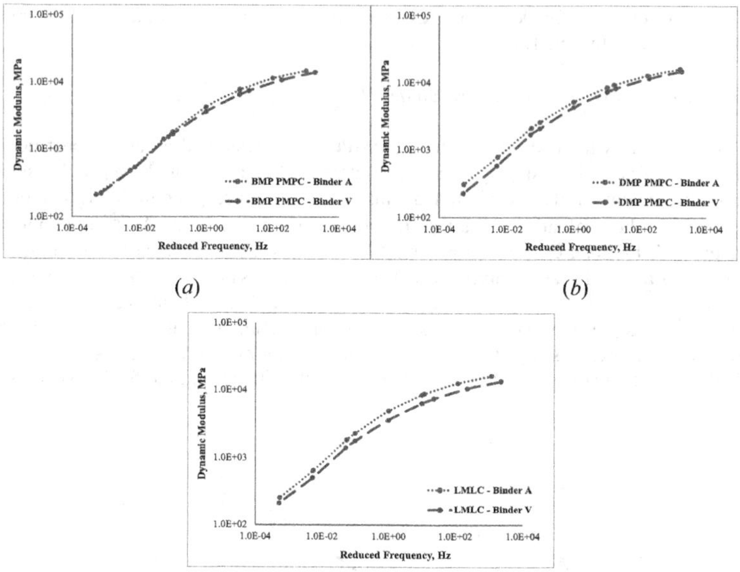

(*a*) (*b*)

(*c*)

Figure 29. *E* master curve comparison for asphalt mixtures using different binder sources for the Texas II field project; (a) BMP PMPC specimens, (b) DMP PMPC specimens, (c) LMLC specimens.*

4.3.7 Summary

In this section, the effects of various mixture components and production parameters including WMA technology, production temperature, plant type, inclusion of recycled materials, aggregate absorption, and binder source, on the performance of short-term aged asphalt mixtures were evaluated. The correlations in terms of M_R stiffness and HWTT rutting resistance parameters were performed for each factor, and the results are summarized in Table 6. Those mixture components and production parameters with significant effects were identified based on the magnitude of the slope for the fitted trendlines being greater than 1.05 or smaller than 0.95 (i.e., 5% off from the line of equality).

According to Table 6, binder source, aggregate absorption, WMA technology, and inclusion of recycled materials had significant effects on stiffness and rutting resistance of short-term aged asphalt mixtures. However, no significant effect from plant type and production temperature was observed.

Table 6. *Summary of the effects of mixture components and production parameters on the performance of short-term aged asphalt mixtures.*

Factor	M_R Stiffness		HWTT $\Delta\varepsilon_{SN}^{vp}$		HWTT Rut Depth	
	Slope Magnitude	**Significant Effect?**	**Slope Magnitude**	**Significant Effect?**	**Slope Magnitude**	**Significant Effect?**
WMA Technology	0.8361	Yes	1.2591	Yes	1.2514	Yes
Production Temperature	0.9851	No	N/A			
Plant Type	1.0082	No	N/A		0.9684	No
Inclusion of Recycled Materials	1.7793	Yes	1.444	Yes	0.9813	No
Aggregate Absorption	1.2707	Yes	N/A		0.6746	Yes
Binder Source	0.8183	Yes	N/A			

5.0 Conclusions and Recommendations

Asphalt mixtures have been traditionally designed on the basis of volumetric parameters in which the optimum asphalt content for a given aggregate gradation was dependent upon the compaction effort, air voids (AV) content, voids in mineral aggregate, and asphalt absorption into aggregate voids. These volumetric procedures worked well for agencies and contractors in an era when the components for asphalt mixtures were relatively constant. However, in the last two decades, changes in asphalt mixture components, production parameters, and plant design have occurred and, consequently, raised the question of the validity of the current mix design procedures in adequately assessing the volumetric needs of asphalt mixtures and the physical characteristics required to meet performance expectations. Therefore, there is a need for the current mix design procedure to consider the impact of these recent changes including binder source, aggregate absorption, WMA technology, inclusion of recycled materials, plant type, and production temperature on the volumetric and performance characteristics of asphalt mixtures during production and construction.

In this study, the laboratory short-term oven aging (STOA) protocols of two hours at 135°C for hot mix asphalt (HMA) and two hours at 116°C for warm mix asphalt (WMA) were used to fabricate laboratory mixed and laboratory compacted (LMLC) specimens for volumetric analysis and performance testing. The simulation of asphalt aging and absorption during plant production and construction by the selected laboratory STOA protocols was evaluated by comparing laboratory test results with those for plant mixed and plant compacted (PMPC) specimens and cores at construction. Additionally, the laboratory test results were used toward identifying mixture components and production parameters with significant effects on the performance of short-term aged asphalt mixtures. The following conclusions are made based on this study:

- The correlation between laboratory produced mixes and plant produced mixes in terms of theoretical maximum specific gravity (G_{mm}) and percentage of absorbed asphalt (P_{ba}) values indicated that the laboratory STOA protocols of two hours at 135°C for HMA and two hours at 116°C for WMA for fabricating LMLC specimens were able to simulate the asphalt absorption during production in the plant.

- The correlations between LMLC specimens and PMPC specimens and cores at construction in terms of Resilient Modulus (M_R) stiffness and Dynamic Modulus (E*) stiffness for a wide range of asphalt mixtures indicated that the laboratory STOA protocols for LMLC specimens were able to simulate the short-term aging during plant production and construction. The correlation between LMLC specimens and PMPC specimens in terms of Hamburg Wheel Tracking Test (HWTT) rutting resistance parameters of viscoplastic strain increment at the stripping number ($\Delta\varepsilon_{SN}^{vp}$) and rut depth at 5,000 load cycles also provided evidence that the laboratory STOA protocols produced representative specimens for performance testing. HWTT results for cores at construction did not correlate with those for LMLC specimens, possibly due to testing difficulties of thin lifts and the associated use of plaster to fit cores into the HWTT molds.

- Laboratory test results indicated a significant effect on the performance of short-term aged asphalt mixtures from WMA technology. Lower mixture stiffness and decreased rutting resistance were observed for WMA mixtures as compared to HMA mixtures, possibly due to the reduced production temperature.

- Laboratory test results on the effect of recycled materials indicated a significant increase in mixture stiffness even with the use of a softer asphalt binder, although the effect was inconsistent due to the variability of recycled materials in terms of original asphalt mixtures and in-service times and climatic conditions.

- A significant effect on the performance of short-term aged asphalt mixtures was observed for aggregate absorption. Asphalt mixtures using low absorptive aggregates exhibited higher stiffness and better rutting resistance than the counterpart mixtures using high absorptive aggregates, which was attributed to the thinner effective film thicknesses within the mixture.

- Binder source had a significant effect on the performance of short-term aged asphalt mixtures. Different mixture performance in terms of stiffness and rutting resistance should be expected from asphalt mixtures using the same PG graded binders from different sources.

- Laboratory test results on the effect of production temperature and plant type indicated no significant change in mixture stiffness and rutting resistance.

The following recommendations for future research can be made based on this study:

- Mixture stiffness and rutting resistance were utilized in this study to discriminate asphalt mixtures with different short-term aging characteristics. Additional mixture properties need to be considered for further validation.
- Early stripping in the HWTT was observed for a substantial portion of short-term aged asphalt mixtures in this study. As a consequence, the rut depth measurements for these mixtures were possibly biased from the stripping of binder from aggregates. Therefore, future research into a more accurate rutting evaluation for softer asphalt mixtures by performing the HWTT test at a lower temperature is necessary. Additionally, there is a need for appropriate modifications to the HWTT procedure for testing field cores.

6.0 Acknowledgements

The authors would like to acknowledge Dr. Jon Epps and Mr. Arif Chowdhury from the Texas A&M Transportation Institute, Dr. Nam Tran from the National Center for Asphalt Technology, and Mr. Chuck Paugh from the Federal Highway Administration for their contributions to this study. The assistance of Rick Canatella, Stephen Johnson, Daniel Furdock, Guillermo Gomez Salas, Adam Joel Taylor, Grant Julian, and Eyoab Zegeye in laboratory-related activities is greatly appreciated. This study was part of the National Cooperative Highway Research Program (NCHRP) Project 09-52, "Short-Term Laboratory Conditioning of Asphalt Mixtures."

7.0 References

Aschenbrener, T., and N. Far, *Short-Term Aging of Hot Mix Asphalt*, CDOT Public Report No. CDOT-DTD-R-94-11, 1994.

Asphalt Institute, *Mix Design Methods for Asphalt Concrete and Other Hot Mix Types*, MS-2, The Asphalt Institute, Lexington, KY, 1984.

Asphalt Institute, *Superpave Level I Mix Design*, SP-2, The Asphalt Institute, Lexington, KY, 1995.

Bonaquist, R., *Mix Design Practices for Warm Mix Asphalt*, NCHRP Report 691, National Cooperative Highway Research Program, Washington, DC, 2011.

Chipperfield, E.H. and T.R. Welch, "Studies on the Relationships between the Properties of Road Bitumens and Their Service Performance," *Proceedings of the Association of Asphalt Paving Technologists Annual Meeting*, Vol. 36, 1967, pp. 421–488.

Clements, T.M., P.B. Blankenship, and K. C. Mahboub, "The Effect of Loose Mix Aging on the Performance of Warm Asphalts," *Journal of the Association of Asphalt Paving Technologists*, Vol. 80, 2012, pp. 541–567.

Daniel, J.S., T. Bennert, Y. R, Kim, and W. Mogawer, *TPF 5(230): Evaluation of Plant Produced RAP Mixtures in the Northeast*, Pooled Fund Phase I Interim Report, Federal Highway Administration, 2014.

Estakhri, C., J. Button, and A. E. Alvarez, *Field and Laboratory Investigation of Warm Mix Asphalt in Texas*, Report No. FHWA/TX-10/0-5597-2, Texas Transportation Institute, College Station, TX, 2010.

Heithaus, J.J. and R.W. Johnson, "A Microviscometer Study of Road Asphalt Hardening in the Field and Laboratory," *Proceedings of the Association of Asphalt Paving Technologists Annual Meeting*, Vol. 27, 1958, pp. 17–34.

Kuennen, T., "Warm Mixes are a Hot Topic," *Better Roads*, James Informational Media, Inc., Des Plaines, Illinois, June 2004.

Lund, J.W. and J.E. Wilson, "Evaluation of Asphalt Aging in Hot Mix Plants," *Proceedings of the Association of Asphalt Paving Technologists Annual Meeting*, Vol. 53, 1984, pp. 1–18.

Lund, J.W. and J.E. Wilson, "Field Validation of Asphalt Aging in Hot Mix Plants," *Proceedings of the Association of Asphalt Paving Technologists Annual Meeting*, Vol. 55, 1986, pp. 92–119.

Mogawer, W., T. Bennert, J. S. Daniel, R. Bonaquist, A. Austerman, and A. Booshehrian, "Performance Characteristics of Plant Produced High RAP Mixtures," *Journal of the Association of Asphalt Paving Technologists*, Vol. 80, 2012, pp. 403–439.

Morian, N., E.Y. Hajj, C.J. Glover, and P. Sebaaly, "Oxidative Aging of Asphalt Binders in Hot-Mix Asphalt Mixtures," *Transportation Research Record 2207,* Transportation Research Board, Washington, DC, 2011, pp. 107–116.

Newcomb, D., "Warm Mix: The Wave of the Future?" *Hot Mix Asphalt Technology*, National Asphalt Pavement Association, Lanham, Maryland, July/August 2005, pp. 33-36.

Rashwan, M.H., and R.C. Williams, "An Evaluation of WMA Additives and RAP on Performance Properties of Asphalt Mixtures." *91st Annual Meeting of the Transportation Research Board*, Washington, D.C., 2011.

Sharp, K., and S. Malone, *Evaluation Protocol of Warm Mix Asphalt*, Austroads Publication No. AP-T231-13, 2013.

Solaimanian, M., H. John, T. Maghsoud, and T. Vivek, "Test Methods to Predict Moisture Sensitivity of Hot-Mix Asphalt Pavements," *Moisture Sensitivity of Asphalt Pavements A National Seminar*, San Diego, CA, February 4–6, 2003.

Topal, A. and B. Sengoz, "Effect of SBS Polymer Modified Bitumen on the Ageing Properties of Asphalt," 4th *Eurasphalt and Eurobitume Congress,* 2008.

Traxler, R. N., "Relation Between Asphalt Composition and Hardening by Volatization and Oxidation," *Proceedings of the Association of Asphalt Paving Technologists Annual Meeting*, Vol. 27, 1958 pp. 359–377.

Tseng, T.-H., and R.L. Lytton, "Prediction of Permanent Deformation in Flexible Pavement Materials" *Implication of Aggregates in the Design, Construction, and Performance of Flexible Pavements*, ASTM, West Conshohocken, PA, 1989, pp.154–172.

Yin, F., E. Arambula, R. Lytton, A. Epps Martin, and L. Garcia Cucalon, "A Novel Method to Evaluate Moisture Susceptibility and Rutting Resistance of Asphalt Mixtures using the Hamburg Wheel Tracking Test," Transportation Research Board, Washington, D.C, 2014, forthcoming.

Yin, F., L. Garcia Cucalon, A. Epps Martin, E. Arambula, and E. S. Park, "Laboratory Conditioning Protocols for Warm-Mix Asphalt." *Journal of the Association of Asphalt Paving Technologists*, Vol. 82, 2013, pp. 177–211.

Zhao, D., M. Lei, and Z. Yao, "Evaluation of Polymer-Modified Hot-Mix Asphalt: Laboratory Characterization." *Journal of Materials in Civil Engineering* Vol. 21, No. 4, 2009, pp. 163–170.

8.0 Discussion

EMIN KUTAY: From the theory of viscoelasticity, you can calculate the resilient modulus from the dynamic modulus master curve. I was wondering if there was a specific reason why you ran both of the tests?

FAN YIN: Okay. The resilient modulus test used in this study is a nondestructive test, and if we look at the specimen thickness, we can use the same specimen for the Hamburg testing. So before we run the Hamburg test on those specimens, we do the M_R first to determine mixture stiffness. And also the E* testing is very popular. Most of the pavement design programs need these numbers so that's why we included these two mixture stiffness testing in the study.

EMIN KUTAY: Did you run E* ON all of the mixes?

FAN YIN: No, E* is only run for selected mixtures, not for all the mixtures.

MICHAEL DUNNING: We use the APA machine, not the Hamburg. We've noticed that between the laboratory and the field there is about a 1 to 2 mm difference bias, towards more rut in the field. We also found, of course, no-brainer here, and the lower the compaction, the higher the rut tests because, of course, you have more movement from the low compaction. We also found there is a lower rut with higher compaction with low binder content, meaning; the binder was not in the specification range. The reason why this happens is because in the field, they could over-compact the mix to the point of rock to rock contact so their rut test is very good. For those reasons you have to be careful with that comparison.

FAN YIN: Okay.

MICHAEL DUNNING: On top of that, of course, Plaster of Paris is a key factor in your mold. If you don't see movement, it's probably okay. What it could be in field is the segregation of the aggregate in the field. You remove a core and then you've have some aggregate gradation changes which sometimes won't correlate very well with your rut in your laboratory.

FAN YIN: Yeah, I totally agree with you. The confinement in the specimen when we run the test has a significant effect on the result. And also you mentioned APA test. I assume the test was run at a dry condition?

MICHAEL DUNNING: Yes.

FAN YIN: Okay. I want to make a point that in Texas, we use hot water at 50° C for the Hamburg test.

MICHAEL DUNNING: Ok.

FAN YIN: This has an effect on the plaster.

MICHAEL DUNNING: Oh, you're right. That would. Good point.

FAN YIN: Sometime we observe the debonding issue for the cores.

MICHAEL DUNNING: Good point. By the way, very good presentation here for us, as well, in Clark County. One other observation here. I was a hot plant manager for a while. With very high absorptive aggregates and your comment about them adding more oil possibly causing a problem with the correlation between bulk and effective asphalt is very true. I know for us, we had to add the extra oil because we knew it would absorb and therefore by the time it got to the jobsite it would be okay. So in the mix itself, in the mix design, we'd soak it in the oven for a couple hours, to absorb the asphalt and get the proper oil content. But sometimes if you don't do that step, it will give the appearance of a high effective asphalt content, which it is, of course, because not enough absorption had occurred..

FAN YIN: Thank you. Thanks for the information.

LOUAY MOHAMMAD: Nice presentation. We're doing similar work on NCHRP Project 9-48. One of the things that caught my eye is the high air void on the lab made mix versus field cores. Did you make an attempt to fabricate laboratory samples to have similar air void levels as the field cores (known air voids). This way you can eliminate the different air voids variability in the comparison of production vs. construction.

FAN YIN: I agree with the direction. One thing we realized in our study is even if we look at just one field site, we take the cores after construction. If we have six or eight specimens, they have different air voids and the variability can be higher so…

LOUAY MOHAMMAD: Yes, but you could cluster them based on common air void levels for a given test. For example, you have four cores for LWT test. Then fabricate LWT lab samples to the cores' air voids. You want to minimize variability due to effects other than that you have control on.

FAN YIN: Yeah, that's right. I agree.

LOUAY MOHAMMAD: So that's what I would do. On the LWT analysis of the test results, you've done a lot of extensive work in terms of coming up with a model. Now to start with a model, you have to have a response curve from LWT test (rut depth vs number of passes). To have a curve, you've got to have rut depth measurement data along the sample. So what rut depth measurement data did you use from an LWT test in order to come up with that curve? For example, certain LWT equipment measures 11 rut depth points along the sample. My question is which one of those points did you use to come up with these elaborate expressions?

FAN YIN: I assume you're talking about the curve of air voids versus the stiffness, right?

LOUAY MOHAMMAD: No, I'm looking at the curve of rut depth versus number of passes.

FAN YIN: Okay, let me pull it up. This one?

LOUAY MOHAMMAD: Yes, that curve, right.

FAN YIN: Okay.

LOUAY MOHAMMAD: See those curves that come from some rut depth measurement data points.

FAN YIN: Yeah, that's right.

LOUAY MOHAMMAD: What data points did you use?

FAN YIN: The red dot here. I think it's really hard to see.

LOUAY MOHAMMAD: No. I am not talking about the red dots. Just the response curve from LWT test itself. Before you do anything, you need a curve (rut depth vs. number of passes). Therefore, you need data from your machine to construct the response curve, right?

FAN YIN: Yup.

LOUAY MOHAMMAD: So the question is which one of those data from your machine that you have used in the plotting of the response curve?

FAN YIN: The rut depth values.

LOUAY MOHAMMAD: Did you use all 11 points? Did you use the middle? Which one did you use?

FAN YIN: We used the middle point in this study.

LOUAY MOHAMMAD: Middle?

FAN YIN: Yeah, we used the middle LVDT reading.

LOUAY MOHAMMAD: Okay. Well, I caution from the use of the middle points only as this may not be the maximum rut depth values.

FAN YIN: Yeah.

LOUAY MOHAMMAD: The middle point may not be the maximum rut depth as we have observed from an ongoing NCHRP study.

FAN YIN: Okay.

LOUAY MOHAMMAD: It is important to calibrate the LWT tester in order to ascertain the maximum rut depth location. Therefore, one has to be careful of how you are going to analyze LWT test data because any subsequent analysis will depend on the data that are collected and how the LWT tester was calibrated.

FAN YIN: Yeah, that's right.

LOUAY MOHAMMAD: So I encourage you to look at the data prior to getting into this elaborate analysis.

FAN YIN: Thanks for the comments. I do remember there's a TRB paper presented by Scott Schramm from Iowa DOT talking about for the Hamburg, we get 11 data points for rut depth and which of the data points should be used for the data analysis. I don't have the numbers on top of my head, but if I remember correctly, he was recommending using different LVDT readings for mixtures with different performance. But I need to double check those values, but thanks for the comments. Dave, do you have some other comments to make?

DAVE NEWCOMB: I would like to go back to Louay's comments on the air voids. The reason that we did not adjust the lab air voids to match the field air voids was we were looking at lab mix, lab compacted as being representative of mix design process, and so from that standpoint for the typical protocol you target 7% air voids. So that was the reason. You're absolutely right about the effect, though, and it's clearly demonstrated here. It was just a matter of how we were approaching the comparison between the lab mix, lab compacted and the field.

FAN YIN: Thanks, Dave.

A Simplified Performance-Based Specification for Asphalt Pavements

Minkyum Kim[a], Louay N. Mohammad[b], Harshavardhan Challa[c], and Mostafa A. Elseifi[c]

[a]*Louisiana Transportation Research Center*

[b]*Louisiana Transportation Research Center and Dept. of Civil and Environmental Engineering, Louisiana State University*

[c]*Civil and Environmental Engineering, Louisiana State University*

ABSTRACT: *In order to guarantee quality construction of asphalt pavements, an effective construction specification is essential. Such a specification should clearly identify the quality goals to ensure that the as-built pavement meets the as-designed criteria. Louisiana's current construction specification for asphalt pavements adopts a Quality Control (QC) and Quality Assurance (QA) procedure, which describes the required quality goals in terms of the volumetric and physical properties of asphalt mixtures and roadways. However, there is no fundamental correlation to ensure that these volumetric properties are sufficient to provide satisfactory long-term performance of the asphalt pavements. Therefore, developing performance-based specifications (PBS), which rely on the fundamental mechanical asphalt mixture properties as performance predictors, is needed to complement current QC/QA specifications. The objective of this paper is to introduce a simplified performance-oriented specification approach developed for new and rehabilitated asphalt pavements in Louisiana. A total of nine field projects across Louisiana were selected, of which six were existing projects that have been in service for three to eight years, and three were new projects. Hamburg-type loaded wheel tracking (LWT) device and semi-circular bending (SCB) test were conducted to measure proposed performance indicators of asphalt pavements from the field core samples for rutting and cracking performances. In addition, indirect tensile dynamic modulus (IDT |E*|) test was conducted for viscoelastic characterization of asphalt mixtures, which were used in the performance predictions of AASHTOWare Pavement ME-Design. The laboratory-measured rutting and cracking performance indicators were then compared with the field distress data obtained from the Louisiana Pavement Management System (LA PMS) for the selected projects and compared with the laboratory test results. The LWT measured rut depths of 10 mm and 6 mm were established as the tentative target quality limits for Level 1 and Level 2 (High Traffic) asphalt pavements, respectively. Minimum SCB* J_c *values of 0.5 and 0.6* kJ/m^2 *were selected as the tentative performance criteria to avoid cracking related problems in Level 1 and Level 2 asphalt pavements, respectively.*

KEYWORDS: *performance specifications, loaded wheel tracking (LWT), rutting, semi-circular bending (SCB), cracking*

The oral presentation was made by Professor Mohammad.

This is a reproduction of an article whose final and definitive form has been published in *Road Materials and Pavement Design*© 2015 Taylor & Francis. The article is available online at: http://dx.doi.org/10.1080/14680629.2015.1077005

1.0 Introduction

In order to guarantee quality construction of asphalt pavements, an effective construction specification is essential. Such a specification should clearly identify the quality goals to ensure that the as-built pavement meets the as-designed criteria. Louisiana's current construction specifications for asphalt pavements adopt a Quality Control (QC) and Quality Assurance (QA) procedure, which describes the required quality goals in terms of the roadway density and surface roughness profile of asphalt pavements (State of Louisiana, 2006). Since its introduction in 1976, Louisiana's QC/QA specification has served well in numerous highway construction projects. In recent years, however, asphalt pavements built with acceptable levels of quality according to the current specifications have started to experience premature failures more frequently than before. Primary causes of the frequent premature failures of asphalt pavements can be attributed to the increased traffic volume on highways (Epps et al., 2002). Adaptation of unconventional asphalt paving technologies, such as the use of high percentages of recycled materials, polymer modified asphalt cements, and warm-mix asphalt technologies, can also make it challenging to adequately guarantee the long-term performance of pavements with conventional volumetric quality goals. Thus, enhancing the reliability of current asphalt pavement specifications is necessary.

Performance-oriented specifications, which adopt direct performance characteristics of pavements as the quality goals of the construction, such as the performance related specifications (PRS) and performance-based specifications (PBS), are gaining national popularity to overcome disadvantages of conventional QC/QA specifications (AASHTO, 2003). In order to implement these advanced specification approaches for local highway construction, however, clear understanding of the features and validation of the key performance specification criteria are needed.

2.0 Objective and Scope

The primary objective of this study was to introduce a simplified performance-oriented specification approach developed for new and rehabilitated asphalt pavements in Louisiana. To achieve this objective, a total of nine field projects across Louisiana were selected, of which six were existing projects that have been in service for three to eight years, and three were new projects completed during the 2013-2014 construction season. Existing projects provided field performance history and in-place asphalt mixture samples for proposed laboratory performance tests, which were used to establish simplified performance criteria of roadways. Level 1 (low traffic volume) and Level 2 (high traffic volume) mixtures were considered separately. New projects provided a verification set for the proposed criteria. A suite of laboratory tests was conducted to measure proposed performance indicators of asphalt pavements for field core samples based on the Hamburg-type loaded wheel tracking (LWT) device for rutting performance and the semi-circular bending (SCB) test for cracking performance. In addition, the indirect tensile dynamic modulus

(IDT |E*|) test was conducted for viscoelastic characterization of asphalt mixtures, which were used in the performance predictions in AASHTOWare Pavement ME-Design.

3.0 Literature Review

3.1 Asphalt Pavement Specifications

Different types of construction specifications have been used by state highway agencies (SHAs) in the US in order to guarantee quality construction of transportation infrastructures. These specifications range from method specifications to performance-based specifications. Table 1 summarizes the definitions, features, and advantages/disadvantages of various specifications throughout the history of pavement constructions (AASHTO, 2003). To address drawbacks of the earlier types of specifications (e.g., method specifications or end-result specifications), SHAs started to adopt QA specifications in the late 1960s. According to a survey performed by Ksaibati and Butts in 2003 (Ksaibati and Butts, 2003), a majority of the SHAs adopted some variation of QA specifications for their asphalt pavement constructions. Figure 1 shows their 2003 survey results. Of 45 SHAs that responded to the survey, 40 of them responded that they were using QA specifications. Only five states indicated that they did not have QA specifications for asphalt pavements.

In recent years, SHAs have been increasingly concerned about the performance of asphalt pavements and have put their efforts on implementing PRS and/or PBS. Clearly, a common benefit of implementing either of the two lately evolved types of specifications (PRS and PBS) is that they are both "performance-oriented" for the acceptance of the products. However, distinctions among different levels of performance-oriented specifications need to be understood as discussed in NCHRP Synthesis 212 (Chamberlin, 1995). Table 2 summarizes those distinctions in terms of the acceptance quality characteristics (AQCs), use of a mechanical property prediction model, use of a performance prediction model, and finally, level of knowledge on the performance of the final product.

Table 1. *Types of pavement construction specifications (AASHTO, 2003).*

Specification Type	Definition	Advantage/Disadvantage
Method Specifications	Require the contractor to produce and place a product using specified materials in definite proportions and specific types of equipment and methods under the direction of the Agency	• Leaving no opportunity to the contractor to implement innovative or economical procedures and equipment. • Difficult to determine whether to correct the process, shut down production, or reject/remove material • Increased chances of disputes and confrontations
End-Result Specification (ERS)	Require the Contractor to take the entire responsibility for producing and placing a product	• QC responsibility is often not clearly defined. • Difficult for timely identification and correction of work • Acceptance target values are often subjective
Quality Assurance (QA) Specification	Require contractor Quality Control (QC) and Agency Acceptance (QA) activities throughout production and placement of a product	• Clearly specified sampling schedule for both QC and QA • Easier QC activity decision based on real-time monitoring of production • Require multiple measurements within an entire lot for reliability • Acceptance is based on the percentage of material within specified limits. • Provide rational mechanism to determine incentive or reduced pay
Performance-Related Specifications (PRS)	Describe quantified Quality Characteristics that are related to product performance	• Acceptance based on key quality characteristics that correlate with fundamental engineering properties that predict performance • Models that predict the fundamental properties of materials using the key quality characteristics and the performance of pavements using the predicted fundamental properties are required. • Pay adjustments can be based on the expected life of the pavement.
Performance-Based Specifications (PBS)	Describe the desired levels of fundamental engineering properties that are predictors of performance and appear in primary prediction relationships	• Acceptance based on fundamental engineering properties that predict performance • Performance prediction model that uses fundamental engineering property is required. • Pay adjustments can be based on the expected life of the product.

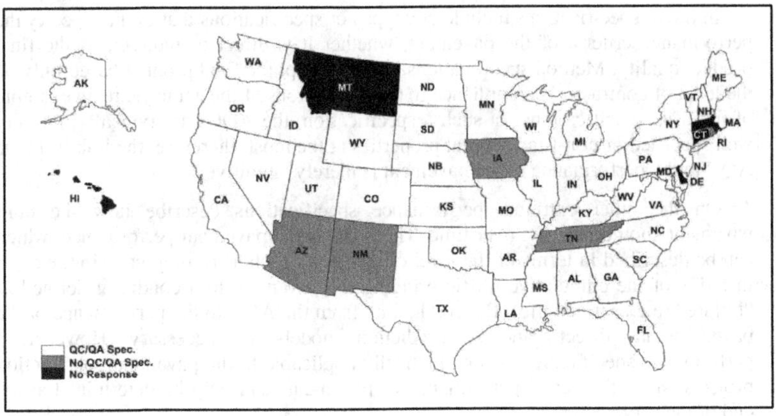

Figure 1. *Use of asphalt pavement specifications (Ksaibati and Butts, 2003).*

Table 2. *Level of knowledge on the performance in various specifications.*

Key Features	Intuitive Specifications	PRS	PBS	Performance Spec.
Acceptance Quality Characteristic (AQC)	Non-fundamental or None	Volumetric Properties	Mechanical Properties	Performance
Mechanical Property Prediction Model	No	Yes	No	No
Performance Prediction Model	No	Yes	Yes	No
Knowledge on Performance	Intuitive	Empirical	Mechanistic-Empirical	Direct

AQCs are the measurements of the product to be collected during the construction and the basis of QC and QA decisions. Mechanical or volumetric properties of asphalt concrete such as the modulus, strength, asphalt content, and density are examples of the AQCs. A mechanical property prediction model is used to estimate a mechanical property of asphalt mixtures (e.g., dynamic modulus) using other mechanical or non-mechanical properties. Witczak's predictive equation (Witczak et al., 2002) is an example of such a model, which predicts the dynamic modulus of asphalt mixtures given volumetric properties of the mixture composition. A performance prediction model is used to forecast the long-term performance of pavements based upon the mechanical behavior of the pavement structures under specific environmental and loading conditions. The mechanistic-empirical pavement design guide (NCHRP, 2004), a commercial version of which is available as 'Pavement ME,' is considered as the most widely accepted performance prediction models in recent years (McCarthy et al., 2014).

Intuitive specifications include any types of specifications that do not specify the performance criteria of the pavement, whether it is direct or indirect, as the final product quality. Method specifications, which accept the final product based only on the level of contractor's compliance to the instructions of the agency, are an example of intuitive specifications. In such a specification, the AQCs are typically not very well specified or non-fundamental properties at the most; therefore, the link from the AQC to the performance of the pavement is merely "intuitive."

On the other extreme, performance specifications describe how the built pavement should perform over time. The AQC is the pavement performance, which can be described in terms of the level of distresses such as rutting, cracking, etc., or in terms of the cumulative traffic bringing the pavement to a condition defined as "failure" (e.g., service life). Hence, the link from the AQC to the performance of the pavement is "direct" and no prediction models are necessary. However, a performance specification is not practically applicable to the pavement construction projects, since the actual performance of pavements can only be determined at the end of the service life.

PRSs describe the desired levels of AQC that are some volumetric properties such as air voids, asphalt cement content, voids in the mineral aggregate, etc. PRS needs two prediction models including a material model for the AQC to mechanical property relationship and a subsequent performance model for the mechanical property to pavement performance relationship. Witczak's Predictive Equation and the Pavement-ME are examples of the material model and the performance model, respectively. Thus, the linkage from the AQC to the performance is mostly empirical.

PBSs describe the desired levels of fundamental mechanical properties of materials as the AQC. Mechanical properties such as the dynamic modulus or creep compliance are the key material input variables for the performance prediction models (e.g., the Pavement-ME). The linkage from the AQC to the performance is mechanistic-empirical, since the predicted performance is based on the measured fundamental mechanical properties through empirical transfer functions between computed stress-strain and in-field performance of pavement structures. Compared to the PRS, a PBS has an advantage on the AQCs that are fundamental mechanical properties of materials, and thus, eliminates the need for a material property prediction model. Yet, unlike the performance specification, a standard PBS procedure still requires a performance prediction model to estimate the long-term pavement performance. The prediction usually requires an extensive amount of information on traffic, climate of the project area, structural pavement design, and material properties of sub-layers in addition to the asphalt surface course.

An alternative approach can be attempted to address the limitations of both PBS and performance specification. Table 3 compares the alternative approach (simplified PBS) to the standard PBS. In the simplified PBS, the AQCs can be selected as semi-direct performance indicators of the pavement to eliminate the need for a performance prediction model. As long as the reliability of the selected performance indicators is proven by comparison analyses between field and laboratory performances, the simplified PBS can be a practical tool to ensure the long-term performance of constructed asphalt pavements.

Table 3. *Alternative (simplified) PBS concept.*

Types of Specifications	Standard PBS	Simplified PBS	Performance Spec.
Acceptance Quality Characteristic (AQC)	Mechanical Properties	Performance Indicator	Performance
Mechanical Property Prediction Model	No	No	No
Performance Prediction Model	Yes	No	No
Link to Performance	Mechanistic-Empirical	Semi-direct	Direct

3.2 Pavement Performance Indicators

Many laboratory test methods have been proposed to measure the pavement performance indicators in simple and practical manners. Rutting at high service temperatures and cracking at intermediate service temperatures are the two major distresses for which test methods have been developed.

3.2.1 Rutting Performance Tests at High Temperature

At high in-service temperatures, asphalt pavements need to resist permanent deformation (or rutting) due to traffic loading. Rutting in asphalt pavements can occur as a combination of progressive densification and shear deformation. Rutting is defined as a mix stability-related problem and the stability is considered as the main property for asphalt mixture design to minimize rutting. Hubbard-Field, Marshall, and Hveem stability test methods are examples of the stability-related mixture design (Zhou et al., 2008).

More recently, NCHRP 9-19 (Witczak et al., 2002) compared various laboratory test methods and their response parameters as the simple performance tests (SPTs) for evaluating asphalt mixtures' resistance to rutting. Reliability of discrimination, repeatability, complexity of procedure, equipment cost, testing time, and so forth were considered to select the best candidate test methods and response parameters for evaluating the rut resisting performance of various asphalt mixtures. The analysis results led to a conclusion that the rut factor ($|E^*|/sin\delta$) measured at high temperatures (e.g., 38 to 54°C), flow time (F_t), and the flow number (F_n) from triaxial tests are the three best laboratory test methods and parameters for rutting resistance evaluation.

Other researchers have used the Hamburg type LWT device to evaluate the combined rutting and moisture damage potential (susceptibility) of asphalt mixtures (Izzo and Tahmorressi, 1999; Williams and Prowell, 1999; Cooley et al., 2000; Brown et al., 2001; Yildirim and Stokoe, 2005; Mohammad et al., 2006). According to an extensive literature review performed by Cooley et al. (2000), the correlations between the LWT measured rut depth (RD) and the actual field rutting performance are reasonably good when the loading and environmental conditions are considered, and thus they concluded that the test methods have a potential to be used as a 'pass/fail' type of criterion.

Some of the laboratory rutting performance indicators ($|E^*|/\sin\delta$, F_t, F_n, and RD) were recently evaluated by the Louisiana Transportation Research Center (LTRC), LTRC Project No. 04-6B (Mohammad et al., 2014), for the applicability to local asphalt mixtures. It was commonly observed with each of the four test parameters that the average values of the parameters calculated for different traffic level mixtures (i.e., level 1, level 2, and level 3 mixture groups with level 1 being the lowest volume highway mixture and level 3 being the highest volume highway mixture as specified in the 2000 edition of LA Specification) were able to discriminate the different mixture groups. The observation means that any of the four test parameters can be used as the asphalt mixtures' rutting performance indicators. However, their relationships to the actual pavement rutting performance have not been thoroughly investigated.

As a part of the literature review of this study, an extensive survey of the asphalt pavement specifications across the US was conducted in 2013. The latest version of construction specifications of all 50 Department of Transportations (DOTs) were collected and reviewed to identify the quality requirements for asphalt pavements. Figure 2 summarizes the survey results with the types of quality characteristics adopted in the construction specifications of the state DOTs. A majority of states specify mixture volumetric properties, roadway density, and roughness profile as the quality characteristics in their latest version of specifications. A number of states (11 out of 50), in addition to the volumetric properties, have included tensile strength ratio (TSR) of moisture conditioned specimens to unconditioned specimen as performance criteria to minimize the moisture damage potential. Among them, Nevada is the only state to further specify the minimum tensile strength of unconditioned specimens as the mixture quality requirements. Four states (Arkansas, Ohio, Texas, and Utah) have implemented the Hamburg type wheel tracking tests as the rutting performance criteria. Specific criteria or ranges of volumetric quality requirements vary from state to state as the historical performance of local pavements varies and as the specific performance goals of one state differ from those of another. Similarly, the wheel tracking test requirements need to be determined on the basis of historical rutting performance data of individual states.

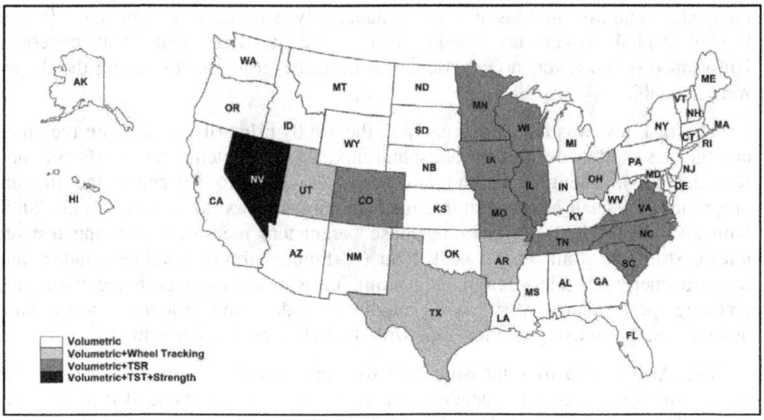

Figure 2. *Types of quality characteristics used by state DOTs.*

3.2.2 Cracking Resistance Tests at Intermediate Temperature

Intermediate temperature cracking is a major distress type in asphalt pavements and are caused by damage induced from repeated loadings. Formation of cracks on the pavement surface, coupled with moisture infiltration through the crack, generally expedites deterioration of the pavement structures and leads to eventual loss of serviceability. Many laboratory test methods, such as the beam fatigue, Texas overlay tester (OT), indirect tensile (IDT) strength, Asphalt Mixture Performance Tester (AMPT), and SCB tests, have been used to characterize the resistance of asphalt mixture to cracking.

One of the most widely accepted laboratory test methods to evaluate cracking resistance of asphalt concrete at intermediate temperatures is the repeated loading beam fatigue test (Tayebali et al., 1996; Carpenter et al., 2002; Tsai et al., 2004). Load repetitions to failure or a constant rate of dissipated strain energy per loading cycle is the typical response parameter of the test. Despite its popularity as a fundamental research tool among researchers, the test method has not been widely adopted by practitioners for routine use due to relatively large beam specimen size and longer testing time to complete (Chiangmai, 2010).

The Texas Overlay Test (OT) is one of the recently developed test methods for characterizing fatigue cracking resistance of asphalt mixtures (Zhou et al., 2007). Fatigue lives of asphalt pavements are estimated from fracture property parameters in Paris' Law, A and n obtained from the OT test, along with other material property inputs such as the dynamic modulus and Poisson's ratio of asphalt mixtures. Reliability of the test method was validated through comparisons with five FHWA-ALF fatigue test lanes and showed good agreements between the predicted and measured fatigue lives. While the test method looked promising in the literature, many steps and assumptions used in the data analysis process to estimate the fatigue life of asphalt pavements would need to be justified with local materials. Unfortunately, however, no experience or historical test database using the device were available.

The IDT test was developed early in the 1990s primarily to measure the creep compliance and tensile strength of asphalt mixtures at low temperatures (Buttlar and Roque, 1994). The test method has been also used to determine the fracture properties of asphalt mixtures at intermediate temperatures (Mohammad et al., 2008; Kim and Wen, 2002). Primary response parameters measured from the test are tensile strength, strain at the peak load (or failure strain), toughness index, and fracture energy. While strength and strain measures do not correlate with field cracking performance very well, toughness index and fracture energy have demonstrated promising relationships with the field cracking performance.

The AMPT primarily measures the dynamic modulus of asphalt mixtures at broad ranges of loading frequencies and temperatures as recommended by NCHRP 9-19 (Witczak et al., 2002). NCHRP 9-19 specifically recommended the $|E^*|$ measured at low temperature ranges (e.g., 4 to 15°C) for fatigue cracking prediction. Compatibility with the MEPDG fatigue cracking prediction model and consistency in test method for both rutting and cracking were suggested as the benefits of the AMPT measured $|E^*|$.

The SCB test method was originally used in rock mechanics and was adapted for asphalt mixture fracture property characterization about a decade ago (Mull et al., 2002). Fracture energy and critical *J*-integral (or critical strain energy release rate, J_c) are the two main response parameters of the test. Since its introduction, the SCB test method has been investigated by many researchers and was found to be a simple and reliable test method for estimating the fracture resistance of asphalt mixtures (Molenaar et al., 2002; Mohammad et al., 2004; Mull et al., 2006; Birgisson et al., 2006; Shu et al., 2008; and Kim et al., 2012). Some advantages of the SCB test method acknowledged by the researchers include the flexibility of specimen fabrication, quick testing time, and simplicity of the test procedure. Given these advantages, the SCB test was selected as the laboratory cracking procedure in this study.

3.3 Summary of the Literature Review

Key findings of the literature review are summarized as follows:

- The concept of performance-oriented specifications is promising, since it takes a more direct performance measure as the quality goal of pavement construction. However, applications of the concept to actual projects require use of complicated prediction models for material properties and pavement performance.

- A simplified approach to the standard performance-related or performance-based specification procedure can be attempted to improve the reliability of current QC/QA specifications, while avoiding the use of complicated prediction models.
- The LWT test can be a practical tool for evaluating rutting performance of asphalt mixtures and pavements. Some state highway agencies have already started to adopt the wheel tracking test in their latest specifications.
- The SCB test can measure the cracking performance indicator of asphalt mixtures. Key advantages of this test method include the simple specimen geometry, ease of specimen fabrication, quick testing time, and simple test procedure.

4.0 Methodology

4.1 Field Projects and Materials

Figure 3 shows the approximate locations of selected field projects across the state. A total of nine field projects were selected through consultation with Louisiana Department of Transportation and Development (LADOTD) construction and research personnel. All relevant design and construction records were collected including project design proposals and mixture design job-mix formula (JMF). Table 4 lists detailed information of the field projects and asphalt mixtures included in this study. Of the nine, six were existing projects constructed three to eight years before sample coring was conducted during the study. Two different traffic levels are included in the existing projects. I10 Egan (I10EG), I10 Vinton (I10VT), and LA964 Baker (964BK) consist of three pavement sections with five asphalt mixtures designed for Level 2 traffic volume in accordance with the 2006 LA Standard Specifications. US171 Shreveport (171SP), LA3121 Spearsville (3121SV), and LA116 Pineville (116PV) consist of seven pavement sections with eight asphalt mixtures designed for Level 1 traffic volume. These existing field projects provided historic performance data of various pavement sections along with the cored asphalt samples (or plant-produced and field-compacted, PF, samples), which were brought to the LTRC laboratory to measure the proposed performance indicators such as the LWT RD, SCB J_c, and dynamic modulus ($|E^*|$).

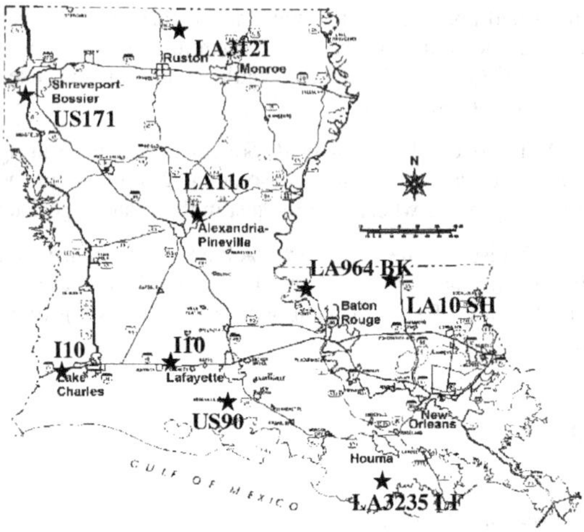

Figure 3. *Approximate locations of selected field projects.*

In addition to the six existing projects, three new field projects, i.e., LA10 St. Helena (10SH), LA3235 Lafourche (3235LF), and US90 Iberia (90IB), which consist of three pavement sections with six asphalt mixtures, were selected to provide construction data and asphalt mixtures necessary for the verification of the proposed PBS approach. These asphalt mixtures were designed for Level 1 traffic volume. Asphalt binders used to produce the mixtures ranged from PG70-22m to PG82-22rm where the letter 'm' stands for 'polymer modified' and 'rm' stands for 'crumb rubber modified' asphalt binders, respectively. Three nominal maximum aggregate sizes (NMAS), 12.5, 19.0, and 25.0 mm, were included. Varying percentages of reclaimed asphalt pavement (RAP) from 0 to 30 % were used in the mixtures. All mixtures were Superpave designed except L964W and LA964B, which were designed as Marshall mixtures. It is worth noting that a number of pavement sections included two types of asphalt mixtures while other sections had only one type of asphalt mixture; that is, three existing pavement sections (116-2, I10EG, and 964BK) consist of wearing course (WC) and binder course (BC) layers, while seven pavement sections included in 3121SV, 171SP, and I10VT projects only contain a WC layer. All three new pavement sections consist of both WC and BC.

Table 4. *Field projects and mixtures.*

	Traffic Level	Project Name (ID)	Built in	Mix Type	Section Code	Mix Layer	Mix ID	Binder Grade	NMAS (mm)	RAP (%)	Mix Design
Existing Projects	2	I10 Egan (I10EG)	2004	HMA	I10EG	WC	I10EW	PG76-22m	12.5	0	SP[1]
						BC	I10EB		25.0		
		I10 Vinton (I10VT)	2003	HMA	I10VT	WC	I10VW	PG76-22m	12.5	0	SP SMA[2]
		LA964 Baker (964BK)	2005	HMA	964BK	WC	L964W	PG76-22m	19.0	15	MS[3]
						BC	L964B		25.0	19	
	1	US171 Shreveport (171SP)	2010	HMA	171-1	WC	171H1	PG70-22m	12.5	15	SP
				WMA	171-2		171W1			15	
					171-3		171W2			30	
		LA3121 Spearsville (3121SV)	2009	HMA	3121-1	WC	3121H1	PG70-22m	12.5	14	SP
				WMA	3121-2		3121W1			14	
					3121-3		3121W2			29	
		LA116 Pineville (116PV)	2010	WMA	116-2	WC	116W1	PG70-22m	12.5	14	SP
						BC	116W2		19.0	19	
New Projects	1	LA10 St. Helena (10SH)	2013	WMA	10SH	WC	10W	PG82-22rm	12.5	15	SP
						BC	10B	PG70-22m	19.0	15	
		LA3235 Lafourche (3235LF)	2013	HMA	3235LF	WC	3235W	PG70-22m	12.5	19	SP
						BC	3235B		19.0	24	
		US90 Iberia (90IB)	2014	HMA	90IB	WC	90W	PG70-22m	12.5	15	SP
						BC	90B		19.0	19	

[1] *Superpave Mixture Design*
[2] *Superpave-SMA Design*
[3] *Marshall Mixture Design*

4.2 Laboratory and Field Experiments

Cored field samples were fabricated in the laboratory for the proposed laboratory performance tests. Standard test protocols followed are summarized in Table 5.

Table 5. *Laboratory performance test parameters and protocols.*

Test Method	Response Parameters	Test Temperature	Test Protocol
LWT	Rut Depth (mm)	50°C	AASHTO T 324
SCB	J_c (kJ/m^2)	25°C	Mohammad et al. (2004)
IDT |E*|	Dynamic Modulus	-10 to 30°C	Kim et al. (2002)

4.2.1 Loaded Wheel Tracking (LWT) Test

The LWT device, manufactured by PMW, Inc., of Salina, Kansas, was used to measure the rutting resistance of asphalt mixtures in accordance with AASHTO T 324. The test is performed by rolling a 703 N (158 lb.) steel wheel across the surface of two conjoined asphalt core samples (approximately 280 mm in total length), which are submerged in 50°C water, for 20,000 passes at approximately 56 passes per minute loading rate. The rutting accumulated after 20,000 loading passes is a measure of the asphalt mixtures' rut resistance after a considerably long-term in service. It was compared in this study to actual field rutting performance data of asphalt pavement sections. Another LWT test parameter used in the comparison analyses is the post consolidation. The post consolidation is the rut depth measured at 1000 passes, which primarily shows the rutting due to densification in the asphalt mixtures.

4.2.2 Semi-Circular Bending (SCB) Test

The SCB test characterizes the fracture resistance of asphalt mixtures at an intermediate temperature range (25°C) in terms of the critical strain energy release rate or the critical value of J-integral (J_c). The J_c is a function of the rate of strain energy change per notch depths (*dU/da*), as shown in Equation 1, and represents the consumed strain energy while a unit area of fractured surface is formed in a mixture. Therefore, the higher J_c values indicate that a material is tougher to resist cracking and crack propagation. Typically, SCB specimens are conditioned for five days at 85°C prior to testing to simulate the long-term aging effect in field, so that the measured J_c values represent the fracture resistance of asphalt mixtures after a considerable amount of time in service (Mohammad et al., 2004). To determine the J_c, semi-circular specimens with three notch depths (25.4, 31.8, and 38.0 mm) are tested under a constant rate of vertical ram displacement of 0.5 mm/min in a three-point bending configuration until the fracture takes place. The load and deformation are continuously recorded and the strain energy to failure (U) is calculated, which is used to determine the J_c using the Equation 1:

$$J_C = -\left(\frac{1}{b}\right)\frac{dU}{da} \qquad [1]$$

Where: b = sample thickness;
a = notch depth; and
U = strain energy to failure as the area under the stress-strain curve.

4.2.3 Indirect Tension Dynamic Modulus (IDT |E|) Test*

The indirect tension dynamic modulus test was conducted for viscoelastic characterizations of field-cored asphalt concrete samples following the test protocol described by Kim and Wen (2002). The dynamic modulus (|E*|) is determined as the absolute value of a ratio between the peak-to-peak amplitude of dynamic stress (σ_0) and that of the recoverable strain (ε_0) measured from steady state sinusoidal stress and strain waveforms, respectively:

$$|E^*| = \frac{\sigma_0}{\varepsilon_0} \quad [2]$$

Tests were performed at three temperatures (-10, 10, and 30°C) and six frequencies (10, 5, 1, 0.5, 0.1, and 0.01) for characterizing dynamic modulus spectrum of asphalt samples at broad ranges of temperatures and loading frequencies. A broad range of measured dynamic modulus values is required for the Level 1 material inputs in the AASHTOWare Pavement ME Design software for pavement response analysis and performance prediction. In fact, the standard format of dynamic modulus input in the software is a complete array of five temperatures by six frequencies ([5×6]) as specified in AASHTO T 342. Thus, the smaller set of dynamic modulus data (i.e., [3×6] array) measured under the indirect tension mode was expanded to the complete [5×6] array of |E*|. The expansion was performed using the extrapolation technique described in Kim et al. (2015).

4.2.4 Field Distress Survey: Louisiana PMS (LA PMS) Database

LADOTD surveys its highway network using the automatic road analyzer (ARAN) every two years and records distress data in the Louisiana PMS database (Khattak et al., 2008). The ARAN uses transverse LASER profiler mounted at the back of the survey van, which makes 1280 measurements across a lane width (perpendicular to the driving direction) to compute the average rut depth of a location. The transverse rutting profile is continuously measured along the driving direction of a pavement section, and then the average rutting in inches for a 0.16 km (0.1 mile) long unit segment of the pavement section is calculated.

For a crack performance survey, on the other hand, a digital pavement imaging system mounted at the back of the survey van records the planar view of pavement surface for every 6.4 m (21.1 ft) long pavement area while driving at highway speeds. The continuous aerial images of the pavement surface are then processed to detect types and severity of cracks in accordance with "Louisiana Cracking and Patching Protocol for Asphalt Surface Pavements." The LA PMS records alligator, longitudinal, transverse, block, and random cracking in low, medium, and high severity levels. Among them, alligator and random cracking are the two most commonly observed cracking types in the LA PMS database. Random cracking is the sum of longitudinal and transverse cracks. The extent of the cracks is reported every 0.16 km (0.1 mile) unit segment. Alligator cracks are calculated as the square footage per 0.16 km (ft^2/0.1mile) segment, while longitudinal, transverse, and random cracking are counted as the linear feet per 0.16 km (ft/0.1 mile) segment.

4.2.5 Distress Index System

The Louisiana PMS converts the distress survey data into distress indices scaled from 0 to 100, with 100 being the perfect condition. Table 6 shows the conversion between the rutting measured in inches and the corresponding index values. The rutting index stays at 100 until the measured rutting reaches 0.3175 cm, then decreases linearly as the measured rutting increases. Hence, the exact index value corresponding to a certain rutting should be calculated by the linear interpolation.

Cracking indices are determined by a 'deduct point' system as shown in Tables 7 and 8 for alligator and random cracking, respectively. These deduct points are functions of types, extent, and severity levels of cracking. As discussed earlier, the alligator and random cracking are the two most commonly observed cracking types in Louisiana asphalt pavements. Therefore, further investigation of the field cracking performance was conducted using these two types of cracking.

Table 6. *LA DOTD rutting index.*

No.	**Rut (cm)**	**Rutting Index**
1	0.000	100
2	0.3175	100
3	0.635	90
4	1.27	70
5	1.905	50
6	2.54	30
7	3.175	10
8	3.4925	0

Table 7. *Deduct points for alligator cracking index (ACI).*

	Extent (ft^2)					
Severity	0-51	51-701	701-1301	1301-2401	2401-3168	3168-9999.9
Low	0	1-16	16-21	21-25	25-28	28
Medium	0	1-21	21-29	29-36	36-49	49
High	0	1-29	29-43	43-50	50-61	61

Table 8. *Deduct points for random cracking index (RCI).*

	Extent (ft)					
Severity	0-31	31-301	301-1601	1601-5001	5001-6001	6001-9999.9
Low	0	1-3	3-16	16-18	18-20	20
Medium	0	1-16	16-21	21-30	30	30
High	0	1-26	26-28	28-42	42-48	48

The alligator cracking index (ACI) and random cracking index (RCI) used to make comparisons with laboratory measured SCB J_c values are calculated using Equations 3 and 4, respectively:

$$ACI = 100 - [(AD)_L + (AD)_M + (AD)_H] \quad [3]$$

$$RCI = 100 - [(RD)_L + (RD)_M + (RD)_H] \quad [4]$$

Where, ACI and RCI: Alligator Cracking and Random Cracking Indices;
$(AD)_L$ and $(RD)_L$= Alligator and Random deducts for low severity level;
$(AD)_M$ and $(RD)_M$= Alligator and Random deducts for medium severity;
$(AD)_H$ and $(RD)_H$= Alligator and Random deducts for high severity level.

A certain type of maintenance or rehabilitation action is triggered when the distress indices exceed the set limit for that action. For example, a 'thin overlay on interstate' is required if the rutting index of the pavement is less than 80 (corresponding to 9.5 mm). Likewise, a rutting index of 65 (corresponding to 14.3 mm) is the trigger for a 'thin overlay on arterial.' For the 'medium overlay' on interstate and arterial, RCI 90 and 80 are the respective triggers. For the 'structural overlay' on interstate and arterial, ACI 65 and 60 are the respective triggers.

5.0 Discussion of Results

5.1 Laboratory Performance Test Results

Tables 9 and 10 summarize the LWT and SCB test results, respectively. In this study, it was assumed that the asphalt layers are completely bonded and resist the deformation and fracture together as a whole layer, but with different contribution rates from the individual layers (e.g., different LWT and SCB test results of different asphalt mixtures). Thus, the average of wearing and binder course mixture test results (if present) should be representative of the entire asphalt pavement section's rutting or cracking resistance. Based on this assumption, laboratory test results were averaged for the wearing and binder course mixtures of I10EG, 964BK, 116-2, 10SH, 3235LF, and 90IB to represent the overall performance of these pavement sections. Moreover, these section averages of laboratory test results enabled direct section-by-section comparisons with field distress performance obtained from the PMS.

Table 9. *LWT test results.*

	Traffic Level	Layer Average					Section Average			
		Mix ID	Thickness[1] (mm)	DEN[2] (%)	RD[3] (mm)	PC[4] (mm)	Section ID	DEN (%)	RD (mm)	PC (mm)
Existing Projects	2	I10EW	50.8	95.3	2.6	1.8	I10EG	95.3	2.5	1.5
		I10EB	189.0	95.2	2.4	1.1				
		I10VW	50.8	95.4	2.7	1.3	I10VT	95.4	2.7	1.3
		L964W	40.0	93.8	4.1	2.0	964BK	94.1	4.1	2.0
		L964B	110.0	94.4	4.1	2.0				
	1	171H1	50.8	95.6	6.7	2.1	171-1	95.6	6.7	2.1
		171W1	50.8	96.4	4.3	2.0	171-2	96.4	4.3	2.0
		171W2	50.8	94.9	4.8	2.7	171-3	94.9	4.8	2.7
		3121H1	50.8	94.8	3.1	1.7	3121-1	94.8	3.1	1.7
		3121W1	50.8	95.3	4.5	1.4	3121-2	95.3	4.5	1.4
		3121W2	50.8	95.0	4.8	1.9	3121-3	95.0	4.8	1.9
		116W1	38.1	93.0	3.2	1.4	116-2	94.3	2.5	1.1
		116W2	50.8	95.5	1.7	0.7				
New Projects	1	10W	38.1	95.1	3.7	1.5	10SH	95.2	2.7	1.3
		10B	100.8	95.3	1.7	1.0				
		3235W	38.1	95.2	4.6	1.2	3235LF	94.1	3.8	1.3
		3235B	50.8	93.0	3.0	1.4				
		90W	38.1	92.1	9.5	2.5	90IB	93.6	6.8	2.2
		90B	50.8	95.0	4.0	1.8				

[1] Layer Thickness, [2]Density of roadway cores as percentage to the maximum density; [3] LWT measured rut depth in mm; and [4] LWT measured post consolidation in mm

Table 10. *SCB test results.*

	Traffic Level	Layer Average			Section Average		
		Mix ID	DEN (%)	J_c (kJ/m^2)	Section ID	DEN (%)	J_c (kJ/m^2)
Existing Projects	2	I10EW	94.3	0.39	I10EG	94.9	0.64
		I10EB	95.5	0.88			
		I10VW	95.6	0.93	I10VT	95.6	0.93
		L964W	94.3	0.30	964BK	93.9	0.37
		L964B	93.5	0.43			
	1	171H1	95.7	0.52	171-1	95.7	0.52
		171W1	96.2	0.73	171-2	96.2	0.73
		171W2	94.9	0.60	171-3	94.9	0.60
		3121H1	95.2	0.66	3121-1	95.2	0.66
		3121W1	94.8	0.91	3121-2	94.8	0.91
		3121W2	96.1	0.60	3121-3	96.1	0.60
		116W1	93.6	0.75	116-2	93.9	0.66
		116W2	94.2	0.57			
New Projects	1	10W	94.7	0.52	10SH	95.2	0.57
		10B	95.7	0.61			
		3235W	96.2	0.61	3235LF	95.7	0.69
		3235B	95.1	0.77			
		90W	93.6	0.28	90IB	94.5	0.30
		90B	95.3	0.31			

Good rutting performance of existing pavement sections were predicted by the LWT RD, ranging from 2.5 to 6.7 mm with an average of 4.0 mm, which are lower than the thin overlay trigger values of 9.5 and 14.3 mm for interstate and arterial roads, respectively. I10EG, I10VT, and 116-2 showed the lowest RD values of 2.5, 2.7, and 2.5 mm, respectively, which imply that these pavement sections will develop low rutting problems. 10SH showed the lowest RD among the three new sections. In-place densities of all pavement sections were well above the minimum requirement of 92%, which would indicate that there were no abnormalities in construction of these pavement sections. A reasonable amount of post consolidation around 2.0 mm was measured for all existing pavement sections. Similar density, rut depth, and post consolidation results were obtained from the three new pavement sections on average. Separate layer results, however, show that the 90W mixture may experience an excessive rutting problem with 9.5 mm of RD. The SCB J_c values of existing pavement sections ranged from 0.37 to 0.93 kJ/m^2. The 964BK section showed the lowest value of 0.37 kJ/m^2, while all other sections showed J_c values greater than or equal to 0.52 kJ/m^2. Among the three new pavement sections, 90IB section showed 0.30 kJ/m^2 of J_c value, while the other two sections showed reasonably higher values of J_c. These observations would imply that the two pavement sections (i.e., 964BK and 90IB) will likely experience cracking problems more than any other pavement sections investigated in this study.

Figure 4 presents the dynamic modulus master curves of all 19 individual asphalt mixtures included in this study. Master curves were constructed at the reference temperature of 10°C. A rule of thumb expectation from the master curve is that a stiffer asphalt mixture at the low reduced frequency range (approximately from 10^{-5} Hz to 10^{-3} Hz) would result in low rutting, while a softer asphalt mixture at the middle reduced frequency range (approximately from 1 Hz to 100 Hz) would result in low cracking. Among the existing pavement mixtures, L964W, L964B, 116W1, and 116W2 looked stiffer than the rest at the low reduced frequency range, while 171H1, 171W2, and I10EW looked softer than the rest at the middle reduced frequency range. Thus, 964BK and 116-2 existing pavement sections are expected to have minimal rutting, and 171-1, 171-3, and I10EG existing pavement sections are expected to have low cracking problems. Similarly, 10SH section is expected to develop the lowest rutting, and 90IB section is expected to experience the least cracking problems among the three new projects. Dynamic modulus data were also used as material input in the AASHTOWare Pavement ME Design for the field rutting performance predictions of the pavement sections.

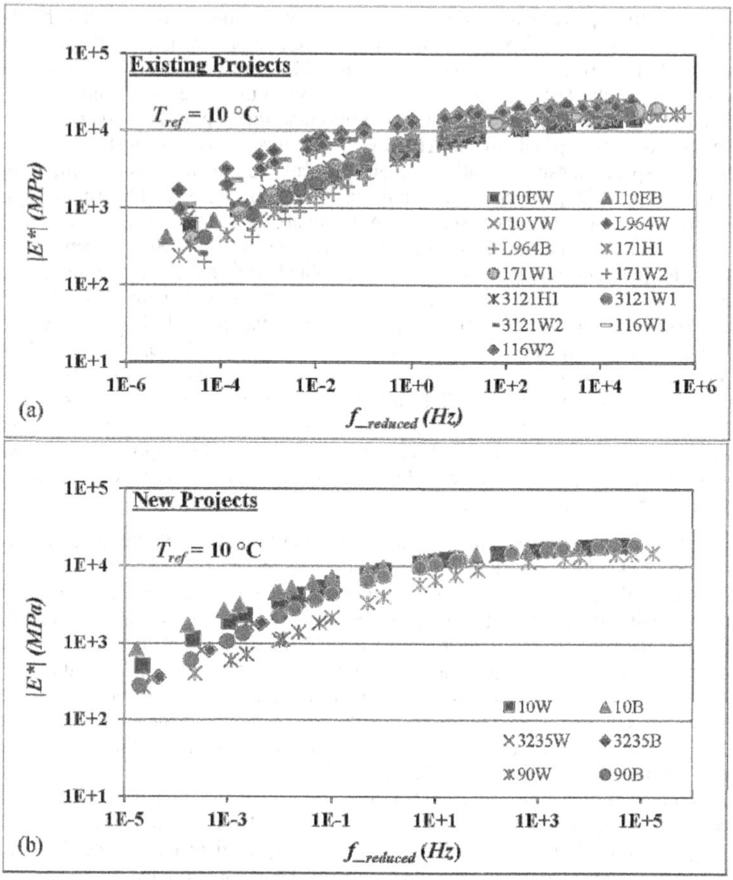

Figure 4. *Dynamic modulus master curves: (a) existing projects and (b) new projects.*

5.2 Field Performance

5.2.1 PMS Rutting Performance Record

Table 11 summarizes the field rutting performance data recorded in the Louisiana PMS. Specific locations of the pavement sections where the rutting performance data were collected are shown with the control section location (CSL). Older pavement sections (e.g., I10EG, I10VT, and L964BK) have longer records of rutting performance starting from 2007 for I10EG and I10VT and 2009 for L964BK. On the other hand, seven younger pavement sections have shorter performance records starting from the 2011 survey cycle.

Table 11. *Field rutting performance record.*

Pavement Sections	CSL[1] (km)	Age (month)	Rutting (mm), years measured				Terminal Rutting (mm)
			2007	2009	2011	2013	
I10EG	11.20 – 23.84	101	2.2	2.7	2.8	3.1	4.0
I10VT	3.52 – 16.10	115	2.3	3.0	2.8	3.6	3.7
L964BK	0.64 – 8.16	93	-	2.8	3.4	3.6	4.3
3121-1	0.00 – 3.52	39	-	-	2.2	2.1	3.9
3121-2	3.84 – 6.88	39	-	-	2.1	2.0	2.5
3121-3	7.04 – 8.00	39	-	-	2.2	2.1	2.7
116-2	7.52 – 10.40	33	-	-	2.2	1.5	3.1
171-1	4.48 – 5.92	36	-	-	2.7	5.6	7.1
171-2	6.72 – 10.24	36	-	-	2.6	4.6	5.2
171-3	1.28 – 3.20	36	-	-	2.6	5.9	8.1

[1] *control section location data used in PMS to identify specific locations of projects*
- *Not available*

In general, gradual increases of field rutting over time can be observed except for a few irregularities such as the reversed trends observed in 3121-2, 3121-3, and 116-2. Nonetheless, given the very small magnitude of rutting in these pavement sections (e.g., only about 2 mm), the reversal in trends (e.g., 0.7 mm drop from 2.2 mm in 2011 to 1.5 mm in 2013 for 116-2) may only indicate that the rutting in these pavements over the years has been very minimal. In fact, good rutting performance of the 10 existing pavement sections was consistently observed through the 2013 survey record. The highest rutting value of 5.9 mm was observed in 171-3 after 36 months in service, which is much lower than the thin overlay trigger value of 14.3 mm for arterial roads. Overall average rutting of all ten sections was only 3.4 mm. For better unbiased comparisons, however, these rutting data for different pavement sections at different ages needs to be projected to the terminal service life as shown in the last column of Table 11.

Projections of the 'terminal rutting' were made by the rutting prediction model of Pavement ME Design software as shown in Equation 5 through curve fittings to the rutting history data.

$$\frac{\varepsilon_p}{\varepsilon_r} = k_z\, \beta_{r1}\, 10^{k_1}\, T^{k_2\beta_{r2}}\, N^{k_3\beta_{r3}} \qquad [5]$$

Where, $\varepsilon_p/\varepsilon_r$ = ratio of accumulated permanent strain to elastic strain at N^{th} loading
N = number of traffic loadings
T = pavement temperature, °F
k_z = function of asphalt layer thickness for confining pressure adjustment
k_1, k_2, and k_3 = nationally calibrated model coefficients
β_{r1}, β_{r2}, and β_{r3} = local calibration constants

A trial-and-error based least square curve fitting approach was employed for the prediction of terminal field rutting using the Pavement ME software. First, the local calibration constants, namely, β_{r1}, β_{r2}, and β_{r3}, were initialized to a set of recommended values reported in the LRTC Project 04-6B (Mohammad et al., 2014). An initial simulation was run for a specific pavement structure with known climate, traffic, and asphalt mixture dynamic modulus data. Then, a sum of squared error (SSE) between the predicted and observed rut depths was computed. The calibration constants were adjusted to reduce the SSE for the second simulation. The process was iterated until the SSE was minimized. The final rutting prediction at this stage was regarded as the projected 'terminal rutting' of the pavement. Figure 5 shows an example plot of the Pavement ME-Design based curve fitting to the rutting history data for the projection. It should be also noted that the field rutting from the PMS database is the rutting measured at the surface of pavements, which may include rutting from non-bituminous layers. For fair comparisons of rutting performance, only the asphalt layer portion of the Pavement ME rutting prediction was further compared with the LWT measured rut depth of asphalt mixtures.

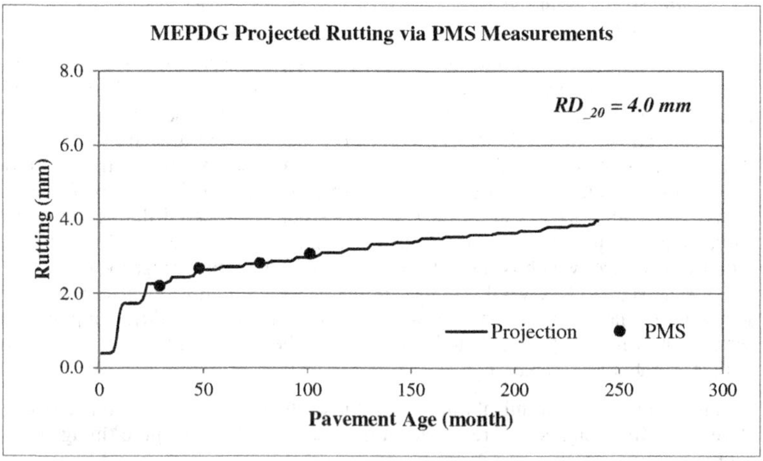

Figure 5. *Example rutting projection using Pavement ME Design.*

5.2.2 PMS Cracking Performance Record

Table 12 summarizes the field cracking performance records of the ten existing pavement sections. For a concise presentation of the data, three severity levels (e.g., low, medium, and high) of alligator, longitudinal, transverse, and random cracks are summed together without weights. Very little alligator cracking was recorded for six pavement sections (i.e., I10EG, I10VT, 116-2, 171-1, 171-2, and 171-3) and relatively small amounts of alligator cracking were observed in three 3121SV sections. On the other hand, significantly higher amounts of alligator cracks were observed in the 964BK section. Similarly, the 964BK section showed the highest amount of random cracks, while the rest of pavement sections showed some moderate amount of random cracks with reasonable growth patterns over time.

The significance of these crack counts can be more clearly demonstrated when the crack counts are converted into the index as previously discussed. Table 13 presents the RCI values converted from the random crack counts at the present time. It should be noted that the alligator cracking index (ACI) values are not presented nor further analyzed hereafter because most of the alligator crack counts remained at zero or very small resulting in the ACI values near 100 except the 964BK section. At present time, no pavement sections have crossed the medium overlay triggers at 90 and 80 for interstate and arterial roads, with 964BK being the only pavement section approaching the trigger. It is interesting to note that the 964BK section has been constantly rated as having a high chance of cracking problems by both laboratory and field performance indicators. The RCI values of the existing pavement sections were further compared with the SCB J_c values of corresponding asphalt mixtures to identify the potential relationship between the field and laboratory.

Table 12. *Field cracking performance record.*

Pavement Sections	Age (Month)	Alligator (ft^2/0.1 mile)	Longitudinal (ft/0.1 mile)	Transverse (ft/0.1 mile)	Random (ft/0.1 mile)
I10EG	29	0	2	0	2
	48	0	4	3	7
	77	0	4	2	7
	101	0	12	4	17
I10VT	47	0	3	1	4
	63	2	1	3	4
	89	1	7	4	12
	115	5	121	16	137
L964BK	42	0	1	3	3
	70	255	58	170	228
	93	674	81	441	522
3121-1	15	7	0	0	1
	39	11	3	2	5
3121-2	15	48	10	1	11
	39	131	34	25	59
3121-3	15	96	27	1	28
	39	122	59	7	66
116-2	9	5	2	4	6
	33	9	13	67	79
171-1	9	0	9	3	12
	36	0	262	3	265
171-2	9	0	10	3	13
	36	0	92	32	124
171-3	9	0	21	0	21
	36	0	212	0	212

Table 13. *Random cracking index (RCI).*

Pavement Sections	Random Cracking (ft/0.1mile)				RCI (0-100)
	Low	Medium	High	Total	
I10EG	7	7	0	13	100
I10VT	29	107	1	137	96
964BK	165	356	1	522	82
3121-1	3	2	0	5	100
3121-2	25	34	1	59	100
3121-3	15	46	5	66	99
116-2	52	27	0	79	99
171-1	24	242	0	265	89
171-2	27	97	0	124	96
171-3	1	211	0	212	90

* crack count numbers are rounded to nearest ones place

5.3 Field vs. Laboratory Performance Comparisons

Figure 6 presents quantitative comparisons between the field and laboratory measured performance indicators. Figure 6(a) shows a scatter plot of the projected field rutting versus LWT RD and Figure 6(b) shows a scatter plot of the field RCI versus SCB J_c. The coefficient of determination (R^2) value of the linear trend line for the field rutting vs. LWT RD relationship was 0.30. The low R^2 value means that the "one-to-one" relationship between the field and lab measured rutting performance indicators is very weak. Such a weak correlation can be expected, since field rutting is not only a function of the asphalt mixture's rutting resistance, but also, a function of many other factors such as the asphalt layer thickness, asphalt mixture density, asphalt content, etc. On the other hand, the R^2 value of the power law trend line for the RCI vs. SCB J_c was considerably higher (0.73) than the field rutting vs. LWT RD. It can be noted that the power law trend line crosses the RCI value of 80, the medium overlay trigger for arterial roads, when the J_c value is around 0.37. Based upon the strong correlation between the RCI and SCB J_c, it can be expected that the asphalt mixtures with SCB J_c values of 0.37 or higher may not experience severe cracking problems to trigger a rehabilitation action. However, it is noted that the field pavement performance is not only a function of a single performance indicator such as the LWT RD or SCB J_c, but also a function of many other potentially influencing factors.

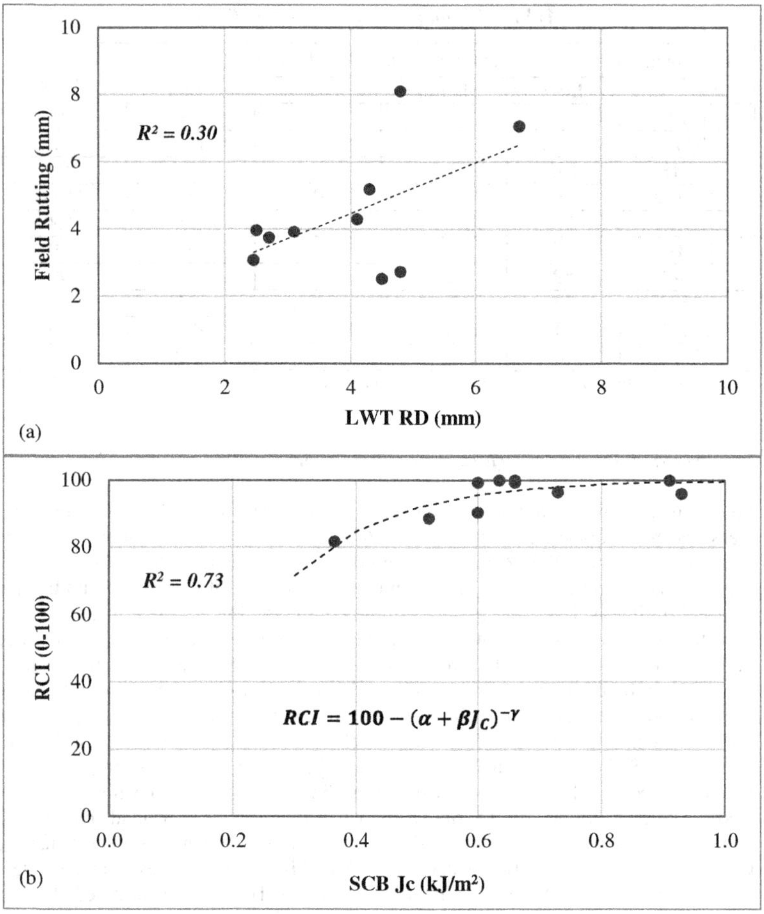

Figure 6. *Field versus lab performance indicators: (a) rutting vs. LWT RD and (b) RCI vs. SCB* J_c.

To further investigate the correlation between the field performance and some of these potentially influencing factors, a statistical analysis was conducted to build a linear regression model that considers these additional factors. These factors are presented in Tables 14 and 15 for field rutting and cracking, respectively. A stepwise regression method was employed to reduce a large set of independent variables to a smaller set through either a 'backward elimination' or a 'forward addition' algorithm (Weisberg, 1985) using a commercial statistical analysis software package. In backward elimination, the initial model includes all independent variables in the form of Equation 6:

$$Y = \beta_0 + \beta_1 X_1 + \beta_2 X_2 + \beta_3 X_3 + \cdots + \beta_n X_n + \varepsilon \qquad [6]$$

Where, Y = the dependent variable
β's = model coefficients
X's = the independent variables
ε = regression error

Table 14. *Potential factors affecting field rutting performance.*

Section ID	Density (%)	LWT RD (mm)	PC (mm)	AC (%)	VMA (%)	Thickness (mm)
I10-12	95.3	2.5	1.5	4.5	13.7	240
I10-3	95.4	2.7	1.3	6.0	16.6	226
964-12	94.1	4.1	2.0	4.2	13.3	150
3121-1	94.8	3.1	1.7	4.8	15.0	50.4
3121-2	95.3	4.5	1.4	4.8	15.0	50.4
3121-3	95.0	4.8	1.9	4.8	15.0	50.4
116-34	94.3	2.5	1.1	4.4	13.5	88.5
171-1	95.6	6.7	2.1	5.0	14.5	50.4
171-2	96.4	4.3	2.0	5.0	14.5	50.4
171-3	94.9	4.8	2.7	5.4	14.0	50.4

Table 15. *Potential factors affecting random cracking index.*

Section ID	AC (%)	Unit Peak Load (kN/mm)	Coarse Aggregate Slope (nCA)	Film Thickness (micron)	ESAL (million)	SCB J_c (kJ/m^2)
I10EG	4.5	0.032	0.41	6.5	48.0	0.64
I10VT	6.0	0.028	0.56	7.0	48.0	0.93
964 BK	4.2	0.019	0.38	8.3	1.4	0.37
3121-1	4.8	0.018	0.29	8.2	0.1	0.66
3121-2	4.8	0.016	0.29	6.0	0.1	0.91
3121-3	4.8	0.000	0.29	8.2	0.1	0.60
116-34	4.4	0.022	0.28	6.1	0.6	0.66
171-1	5.0	0.014	0.32	8.7	1.3	0.52
171-2	5.0	0.016	0.29	7.6	1.3	0.73
171-3	5.4	0.011	0.30	9.3	1.3	0.60

Goodness of fit was assessed through the Mallow's Cp. Mallow's Cp is a measure of the fit of a regression model based on the ordinary least squares, which is typically applied for model selection from multiple independent variables. The backward elimination algorithm takes out one independent variable at a time and computes the updated Mallow's Cp due to the removal of the variable and compares the updated statistics to that of the immediate previous step. If the Cp decreases, then the elimination process continues until no further improvement is observed. The resulting final model will include fewer independent variables than the initial model and the variables selected in the final model are supposed to be the factors affecting the field distresses. On the other hand, the forward addition algorithm starts from a constant function model with no variables (y = constant) and adds one independent variable in every iteration until the best combination of independent variables is chosen by the minimum Cp. The final models chosen by either of the two processes can be used to predict the field distresses. The final linear models selected for the field rutting from the factors presented in Table 14 and for the RCI from the factors in Table 15 are shown in Equations 7 and 8:

$$R_{field} = 13.02 + 4.27(AC) + 0.57(RD_{LWT}) - 2.19(VMA) \quad [7]$$

$$RCI = 100.4 - 5.8(AC) + 34.9(SCB\ Jc) \quad [8]$$

Where, R_{field} is the predicted 20-year field rutting, *RCI* is Random Cracking Index, *AC* is asphalt content, RD_{LWT} is the LWT RD, *VMA* is voids in the mineral aggregates, and *SCB* J_c is SCB measured J_c.

Figure 7 shows comparisons between the regression models predicted performance parameters, i.e., rutting by Equation 7 and RCI by Equation 8, and the projected field performance parameters. As shown in the figure, improved predictions of field performance indicators were achieved through the use of the linear models, which included more than one independent variable. The computed field rut depth using the linear regression model built with the three independent variables agreed with the field rutting performance very well, showing a good R^2 value of 0.87. Likewise, the computed RCI values using the two-variable linear regression model agreed reasonably well with the field RCI, showing a decent R^2 value of 0.60. It is worth noting that the R^2 value of 0.60 in Figure 7 (b) is lower than that shown in Figure 6(b). The reduction is due to the difference in trend line functions, i.e., power model versus linear model, but not due to the addition of AC variable in Equation 8. The strong relationships between field and laboratory performance indicators observed in Figures 7(a) and 7(b) may be translated into an expectation that the laboratory performance indicators such as the LWT RD and SCB J_c, in addition to some other relevant quality characteristics of asphalt mixtures and pavement construction, can be added into the current specification as supplemental requirements to better guarantee the ultimate performance of asphalt pavements.

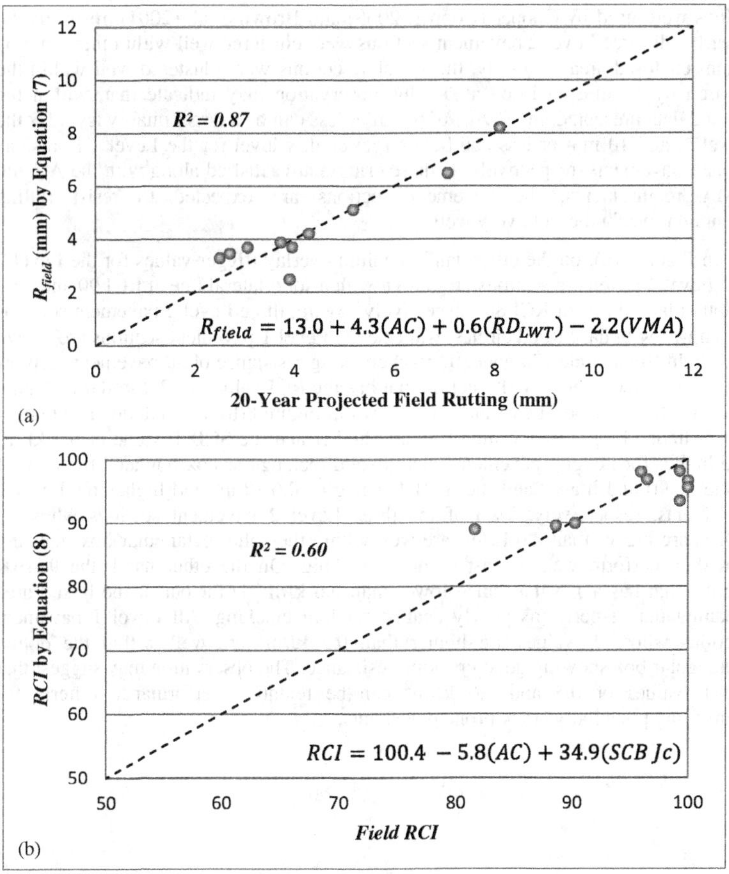

Figure 7. *Regression model vs. projected field performance: (a) rutting and (b) RCI.*

Figure 8 recaptures the relationship between the projected field performance indicators and the laboratory measured performance indicators previously presented in Figure 6. Although the one-to-one correlations between the field and laboratory performance indicators were not good enough, a further qualitative examination on the clustered data points showed some useful implications of their relationship. In Figure 8(a), data points of Level 2 mixtures are represented as dark filled circles, while the Level 1 mixtures are represented with light filled circles. A rectangular box with dark solid line is drawn in the plot to show an enclosed area under 6 mm of rutting on both axes, while another rectangular box with light dotted line encloses another larger area under 10 mm of rutting on both axes. These limits of rutting values were used by Cooper (Cooper, 2009) and Brown et al. (2001), respectively. Clearly, all three Level 2 pavement sections were clustered well within the 6 mm by 6 mm enclosed area; similarly, the Level 1 sections were clustered well within the 10 mm by 10 mm enclosed area. This observation may indicate that, within the limited data presented, the RD_{LWT} of 6 mm or less can be a target quality level for the Level 2, and 10 mm or less can be a target quality level for the Level 1 Louisiana asphalt pavements, respectively. If these criteria are satisfied along with the AC and VMA requirements, the pavement sections are expected to resist rutting accumulations in the field very well.

In Figure 8(b), on the other hand, medium overlay trigger values for the Level 2 and Level 1 pavement sections are shown with a solid lateral line at RCI 90 and with a dotted lateral line at RCI 80, respectively. Again, three Level 2 pavement sections are shown with dark filled circles, while seven Level 1 pavement sections are shown with light filled circles. In general, good cracking resistance of all pavement sections is evident, except the 964BK section that has the RCI value of 82. Similar to Figure 8(a), two boxed zones are created and shown in Figure 8(b), i.e., a dotted rectangular box bounded by the RCI value of 80 and higher and the SCB J_c value of 0.5 kJ/m^2 and higher for Level 1 pavements; and a solid rectangular box bounded by the RCI value of 90 and higher and the SCB J_c value of 0.6 kJ/m^2 and higher for Level 2 pavements, respectively. Two of the three Level 2 pavement sections whose J_c values are higher than 0.6 kJ/m^2 are well within the solid rectangular box meaning that they perform well against random cracking. On the other hand, the 964BK section that has a J_c value much lower than 0.6 kJ/m^2 is far out of the boxed area meaning that it performs poorly against random cracking. All Level 1 pavement sections whose J_c values are higher than 0.5 kJ/m^2 are well within the dotted rectangular box showing good cracking resistance. The observation may suggest that the J_c values of 0.5 and 0.6 kJ/m^2 can be tentative performance criteria for identifying potentially crack prone pavements.

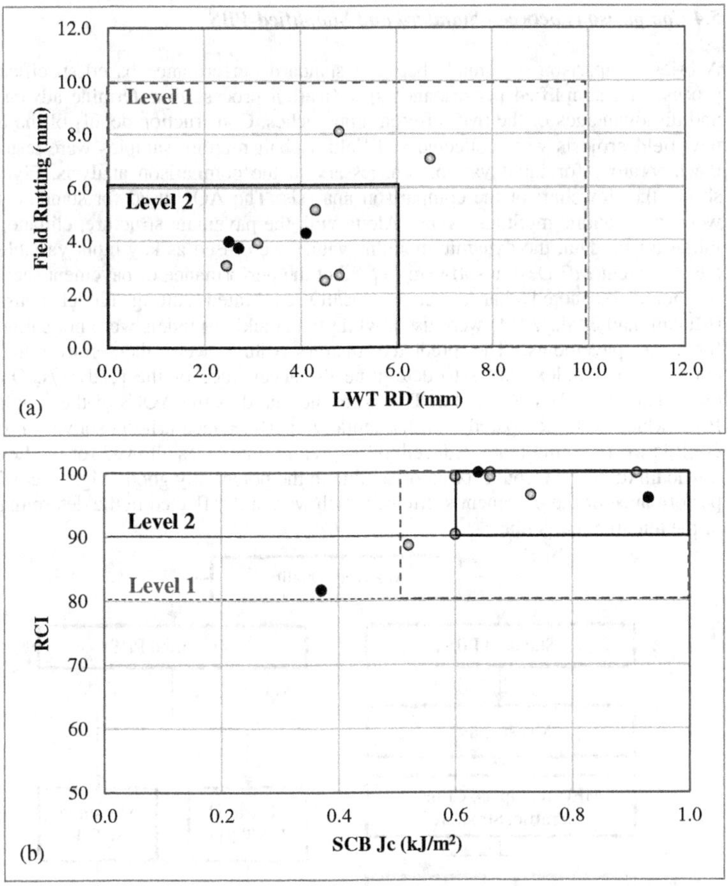

Figure 8. *Tentative guidelines of laboratory performance indicators: (a) rutting and (b) random cracking.*

5.4 Comparisons between Standard and Simplified-PBS

A brief comparison was made between standard performance based specification process and simplified performance specification process to determine advantages and disadvantages of the two different approaches. Construction details of the three new field projects were collected and field asphalt mixture samples were tested in the laboratory for input variables necessary in the comparison analysis. Figure 9 shows the flowchart of the comparison analysis. The AQC used for standard PBS were the dynamic modulus values. Along with the pavement structure, climate, and initial traffic data, the dynamic modulus values were used as key input variables in the Pavement ME Design software to predict the performance of pavement sections. It should be noted that a set of locally calibrated rutting model constants (Mohammad et al., 2014) were used, while the cracking models were not calibrated for local pavements. The predicted distress values were then converted into corresponding index values to determine the acceptance of the roadways. On the other hand, the LWT RD and SCB J_c were measured as the AQCs of the simplified PBS, which were compared with the tentative performance criteria shown in Table 16. It is worth noting that a reduced SCB J_c requirement was allowed for the Level 1 traffic mixtures to account for an offset due to the potentially good alligator cracking performance of the pavement sections, which were not reflected in the determination of the tentative guideline.

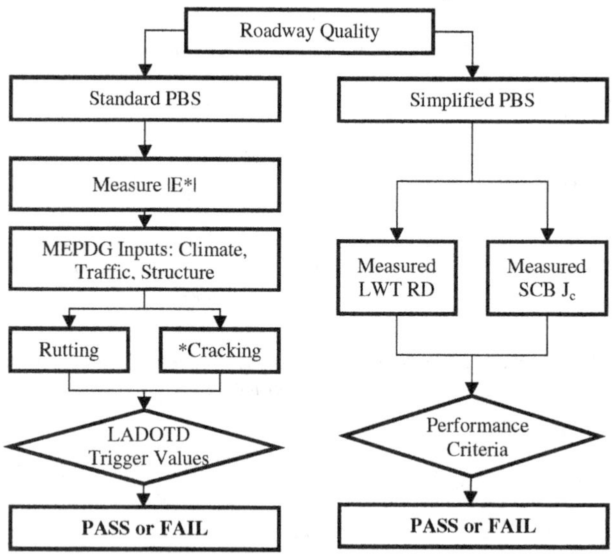

*MEPDG Cracking Models are not calibrated locally.

Figure 9. *Standard PBS vs. simplified PBS.*

Table 16. *Proposed specification limits for laboratory performance tests.*

Performance Based Tests	Level 1 Traffic	Level 2 Traffic
LWT RD @50°C, mm	10.0	6.0
SCB J_c @25°C, kJ/m^2	0.5	0.6

Table 17 presents a summary of the comparison results. The thin overlay trigger for the Level 1 mixtures is a rutting index (RI) of 65 or less. The medium overlay trigger for the Level 1 mixtures is RCI 80 or less. Meanwhile, the simplified performance criteria for rutting and cracking are 10 mm and 0.5 kJ/m^2, respectively. All three pavement sections passed the rutting performance criteria for both standard and simplified specifications. Moreover, the orders of performance measures by the two approaches match exactly. Such good agreement between the standard and simplified specification procedures in acceptance would support the applicability of the simplified procedure in lieu of a far more complicated and time-consuming approach. For the cracking performance acceptance, on the other hand, three pavement sections passed the standard specification performance criterion for the random cracking, while 90IB section failed the simplified cracking performance criterion of 0.5 kJ/m^2. In fact, it should be reminded that the un-calibrated cracking models of Pavement ME-Design were used in the performance predictions of these new pavement sections. Thus, the RCI values shown in Table 17 are not capable of discriminating different materials' performance, while the SCB J_c looks to suggest, with relatively reasonable reliability, that the 90IB section will experience a cracking related problem in the future.

Table 17. *Summary of standard vs. simplified specification comparison.*

Pavement Section	Rutting		Random Crack	
	Standard (RI>65)	Simplified (LWT RD>10 mm)	Standard (RCI>80)	Simplified (SCB J_c > 0.5 kJ/m^2)
10SH	96	2.7	100	0.57
3235LF	92	3.8	82	0.69
90IB	89	6.8	100	0.30

6.0 Conclusions

The primary objective of this research was to develop a simplified performance-oriented specification for new and rehabilitated asphalt pavements in Louisiana. Understanding the performance oriented specifications and establishing simplified performance guidelines for roadways were the two specific objectives of this study. A total of nine field projects across Louisiana were selected, of which six were existing projects that have been in service for three to eight years, and three were new projects. Hamburg type LWT device and SCB test were conducted to measure proposed performance indicators of asphalt pavements from the field core samples for rutting and cracking performance indicators, respectively. In addition, an IDT |E*| test was conducted for full viscoelastic characterization of asphalt mixtures, which was used in the performance predictions of AASHTOWare Pavement ME Design. Findings of the research are summarized as follows:

- The concept of performance-oriented specifications is promising, since it takes more direct performance measure as the quality goal of the pavement construction. However, applications of the concept to actual projects require use of complicated prediction models for material properties and pavement performance.
- A simplified approach to the standard performance-related or performance-based specification procedure can be attempted to improve the reliability of current QC/QA specifications, while avoiding the use of complicated prediction models.
- The loaded wheel tracking (LWT) device can be a practical tool for evaluating rutting performance of asphalt mixtures and pavements. The LWT measured rut depths of 6 mm or less and 10 mm or less can be the tentative target quality limits for the Level 2 and Level 1 Louisiana asphalt pavements, respectively.
- Semi-circular bending (SCB) device can measure the cracking performance indicator of asphalt mixtures and pavements. The minimum SCB Jc values of 0.6 and 0.5 kJ/m2 can be tentative criteria to avoid crack related problems in Level 2 and Level 1 asphalt pavements, respectively.

According to findings of this research, it can be concluded that a simplified performance-based specification (i.e., Simplified PBS) procedure, which includes LWT and SCB tests can be added to the current QC/QA specification to better guarantee the long term performance of Louisiana asphalt pavements. A continued research effort to collect further field and laboratory performance data is underway to validate the findings of the study discussed in this paper.

7.0 Acknowledgements

This work was supported by the Louisiana Transportation Research Center in cooperation with the Louisiana Department of Transportation and Development. The contribution of staffs in the asphalt laboratory and the Engineering Material Characterization and Research Facility to this project is acknowledged.

8.0 References

AASHTO, *Major Types of Transportation Construction Specifications, A Guideline to Understanding Their Evolution and Application*, a report of the AASHTO Highway Subcommittee on Construction, August 2003.

Birgisson, B., Montepara, A., Napier, J., Romeo, E., Roncella, R., and Tebaldi, G., "Micromechanical Analyses for Measurement and Prediction of Hot-Mix

Asphalt Fracture Energy," *Transportation Research Record: Journal of the Transportation Research Board, No. 1970*, 2006, pp. 186–195.

Brown, E. R., Kandhal, P. S., and Zhang, J., *Performance Testing for Hot Mix Asphalt*, NCAT Report 01-05A. NCAT, 2001.

Buttlar, W. G. and Roque, R., "Development and Evaluation of the Strategic Highway Research Program Measurement and Analysis System for Indirect Tensile Testing at Low Temperatures," *Transportation Research Record, No. 1454*, 1994, pp. 163–171.

Carpenter, S. H., Ghuzlan, K. A., and Shen, S., "Fatigue Endurance Limit for Highway and Airport Pavements," *Transportation Research Record: Journal of the Transportation Research Board, No. 1832*, 2002, pp. 131–138.

Chamberlin, W.P., *Performance-Related Specification for Highway construction and Rehabilitation,* NCHRP Synthesis 212, Transportation Research Board, Washington, DC, 1995.

Chiangmai, C. N., *Fatigue-Fracture Relation on Asphalt Concrete Mixtures,* Thesis, University of Illinois at Urbana-Champaign, Urbana, IL, 2010.

Cooley Jr., L. A., Kandhal, P. S., Buchanan, M. S., Fee, F., and Epps, A., *Loaded Wheel Testers in the United States: State of the Practice,* Transportation Research E-Circular. No. E-C016. Transportation Research Board, 2000.

Cooper, S. B., *Evaluation of HMA Mixtures Containing Sasobit®*, LTRC Report 06-1TA. Louisiana Transportation Research Center, 2009.

Epps, J. A., Hand, A., Seeds, S., Schulz, T., Alavi, S., Ashmore, C., Monismith, C. L., Deacon, J. A., Harvey, J. T., and Leahy, R., *Recommended Performance-Related Specification for Hot-Mix Asphalt Construction: Results of the WesTrack Project*," NCHRP Report 455, Transportation Research Board, Washington, D.C., 2002.

Izzo, R. P. and Tahmoressi, M., "Use of the Hamburg Wheel-Tracking Device for Evaluating Moisture Susceptibility of Hot-Mix Asphalt," *Transportation Research Record: Journal of the Transportation Research Board, No. 1681*, Transportation Research Board, Washington, D.C., 1999, pp. 76-85.

Khattak, M. J., Baladi, G. Y., Zhang, Z., and Ismail, S., "Review of Louisiana's Pavement Management System," *Transportation Research Record: Journal of the Transportation Research Board, No. 2084*, Transportation Research Board, Washington, D.C., 2008, pp. 18–27.

Kim, M., Mohammad, L., and Elseifi, M., "Characterization of Fracture Properties of Asphalt Mixtures As Measured by Semicircular Bend Test and Indirect Tension Test," *Transportation Research Record: Journal of the Transportation Research Board, No. 2296*, Transportation Research Board, Washington, D.C., 2012, pp. 115–124.

Kim, M., Mohammad, L., and Elseifi, M., "Effects of Various Extrapolation Techniques for Abbreviated Dynamic Modulus Test Data on the MEPDG Rutting Predictions," *Journal of Marine Science and Technology*, v.23 (3), 2015.

Kim, Y. R. and Wen, H., "Fracture Energy from Indirect Tension Testing," *Journal of the Association of Asphalt Paving Technologists,* Vol. 71, 2002, pp. 779-793.

Ksaibati, K. and Butts, N., *Evaluating the Impact of QC/QA Programs on Asphalt Mixture Variability*, MPC Report No. 03-146, Mountain-Plains Consortium, Fargo, ND, June 2003.

McCarthy, L. M., Guercio, M. C. Bennert, T. DeJarnette, V., "Comparing Flexible Pavement Performance Using Emerging Analysis Tools," *Journal of Transportation Engineering*, Vol.140 (5), 2014.

Mohammad, L. N., Wu, Z., and Aglan, M., "Characterization of Fracture and Fatigue Resistance on Recycled Polymer-Modified Asphalt Pavements," *5th RILEM Conference on Pavement Cracking,* France, 2004.

Mohammad, L. N., Wu, Z., Obulareddy, S., Cooper, S., and Abadie, C., "Permanent Deformation Analysis of Hot-Mix Asphalt Mixtures with Simple Performance Tests and 2002 Mechanistic-Empirical Pavement Design Software," *Transportation Research Record: Journal of the Transportation Research Board, No. 1970*, Transportation Research Board, Washington, D.C., 2006, pp. 133–142.

Mohammad, L. N., Kabir, M. D., and Saadeh, S., "Evaluation of Fracture Properties of Hot Mix Asphalt," *The 6th RILEM International Conference on Cracking in Pavements*, Chicago, IL, 2008, pp. 427–436.

Mohammad, L.N., Kim, M., Raghavendra, A., and Obulareddy, S., *Characterization of Louisiana Asphalt Mixtures Using Simple Performance Tests and MEPDG*, LTRC Report No. FHWA/LA 11/ 499, Louisiana Transportation Research Center, Baton Rouge, Louisiana, 2014.

Molenaar, A.A.A., Scarpas, A., Liu, X., and Erkens, S.M.J.G., "Semi-Circular Bending Test; Simple but Useful?," *Journal of the Association of Asphalt Paving Technologists,* Vol. 71, 2002, pp. 794–815.

Mull, M. A., Stuart, K., and Yehia, A., "Fracture Resistance Characterization of Chemically Modified Crumb Rubber Asphalt Pavement," *Journal of Materials Science,* Vol. 37, 2002, pp. 537–566.

Mull, A. M., Othman, A., and Mohammad, L. "Fatigue Crack Growth Analysis of Hot-Mix Asphalt Employing Semicircular Notched Bend Specimen" TRB 85th Annual Meeting Compendium of Papers CD-ROM, 2006.

NCHRP 1-37A, *Guide for Mechanistic-Empirical Design of New and Rehabilitated Pavement Structures*, National Cooperative Highway Research Program, Transportation Research Board, Washington, D.C., March 2004.

State of Louisiana Department of Transportation and Development, *Louisiana Standard Specifications for Roads and Bridges*, Section 502, 2006.

Shu, X., Huang, B., and Vukosavljevic, D., "Laboratory Evaluation of Fatigue Characteristics of Recycled Asphalt Mixture," *Construction and Building Materials*, Vol. 22, 2008, pp. 1323–1330.

Tayebali, A. A., Deacon, J. A., and Monismith, C. L., "Development and Evaluation of Dynamic Flexural Beam Fatigue Test System," *Transportation Research Record,* No. 1545, Transportation Research Board, Washington, D.C., 1996, pp. 89–97.

Tsai, B-W, Harvey, J. T., and Monismith, C. L., "Two-Stage Weibull Approach for Asphalt Concrete Fatigue Performance Prediction," *Journal of the Association of Asphalt Paving Technologists,* Vol. 73, 2004, pp. 623-655.

Weisberg, S., *Applied Linear Regression*, John Wiley & Son, 2nd. New York, NY, 1985.

Williams, R. C. and B. D. Prowell, "Comparison of Laboratory Wheel-Tracking Test Results with WesTrack Performance," *Transportation Research Record: Journal of the Transportation Research Board, No. 1681*, Transportation Research Board, Washington, D.C., 1999, pp. 121–128.

Witczak, M. W., Kaloush, K., Pellinen, T., El-Basyouny, M., and Von Quintus, H., *Simple Performance Test for Superpave Mix Design*, NCHRP Report 465, Transportation Research Board, Washington, D.C., 2002.

Yildirim, Y. and Stokoe, K. H., *Analysis of Hamburg Wheel Tracking Device Results in Relation to Field Performance*, CTR Technical Report 0-4185-5. Center for Transportation Research at the University of Texas at Austin, 2005.

Zhou, F., Hu, S., Scullion, T., Qi, X., Chen, D., Claros, G., "Development and Verification of Overlay Tester Based Fatigue Cracking Prediction Approach," Journal of Association of Asphalt Paving Technologists, 2007.

Zhou, F., Fernando, E., Scullion, T., *A Review of Performance Models and Test Procedures with Recommendations for Use in the TEXAS M-E Design Program*, Report FHWA/TX-08/0-5798-1, September 2008.

9.0 Discussion

MARTIN VAN DE VEN: Delft University. The cracking index is interesting to me. As we heard earlier, you have cracks that start from the bottom and cracks that start at the surface, and my first thinking was you do that SCB test on the surfacing mixtures, but I saw now that you do it on all mixes. For base mixtures we normally perform repeated load tests like the four point bending fatigue test. I assume the SCB test is a monotonic displacement controlled failure test. So you do not separate the surface cracking from the cracking coming from below I see.

LOUAY MOHAMMAD: Yes, the SCB test, that I described, is a monotonic test. It measures the resistance of asphalt mixtures to fracture. Currently, the recommended criterion of critical strain energy release rate, Jc for crack resistance mixtures is based on the traffic volume level. This criterion, as described, was developed based on measurement of field performance.

MARTIN VAN DE VEN: In Europe, we are using this test only for the surface layer to just rank mixtures with regards to cracks coming from the surface. Do we really have with this test a design method to have a handle on bottom up fatigue, because for the fatigue design based on bottom up, we perform stress/strain calculations, to relate them to the fatigue behavior of the mixture. We are talking here millions of load repetitions at a relatively low stress level

LOUAY MOHAMMAD: Our attempt is to incorporate a performance test as a part of asphalt mixture design process to minimize intermediate temperature cracking. .

MARTIN VAN DE VEN: Okay.

RAY BONAQUIST: I've got a question concerning how this would be used. I'm assuming that a contractor or producer is required to meet all binder selection requirements and normal specification, all volumetric requirements, and then these two tests get added on at the end.

LOUAY MOHAMMAD: Correct.

RAY BONAQUIST: And I'm assuming that if I don't pass the cracking test, I've got to do something to change the mixture.

LOUAY MOHAMMAD: Correct, such as redesign.

RAY BONAQUIST: In any of the work that you have done, do you have information or have you developed relationships between the things that I as a producer can change and its effect, let's say, on the SCB intermediate cracking temperature parameter?

LOUAY MOHAMMAD: We have been performing this SCB test for quite some time. As a result, a large database has been compiled for many mixtures' Jc with different asphalt binder type, mixture design parameters (such as aggregate types, VMA, VFA, film thickness, dust to asphalt ratio, etc). A preliminary statistical model has been developed to predict J_c values for a given asphalt mixture design. This model can be used as guidance for designers to predict J_c for a give mixture design. This approach of J_c prediction is similar to prediction models for dynamic modulus (Hirsch model or the Witczak model). In addition, we are are organizing a SCB test training workshop on April 16 for our contractors, design labs, and DOT to familiarize them with sample preparation, equipment setup, conduct of the test, and data analysis..

RAY BONAQUIST: Yes. Well, I think to continue to develop those relationships and improve them. Those are going to be important things for the people who ultimately have to design the mixes, so that they know what to change in case they don't pass one of these tests.

LOUAY MOHAMMAD: Yes. We will broadcast those models in the near future.

RAY BONAQUIST: Thanks.

IMAD AL-QADI: Excellent presentation, Louay.

LOUAY MOHAMMAD: Thank you.

IMAD AL-QADI: I will follow up on the previous question. But first, I have a comment. I did notice in the cracking relationship between field performance and laboratory results that the regression was pulled by just one point/number, so if you take that number out, I think it will be completely different relationship. The reason I'm bringing this up is because I think this test is mainly a quality control or a quality assurance test for material as it's produced. When you take cores from the field, you need to correct for the time it has been in the field. So, the values from the field cores are different than mix specimens you prepared. That's the first comment. The second is to follow up on the previous question. This SCB test is very simple, but there are many parameters that you can get from the test itself that will allow you to predict how ductile the material is for lab mixes and field cores. I expect developing a database to predict field performance from lab specimens.

LOUAY MOHAMMAD: Thank you for the comment. The relationship presented was based on field cores. However, this relationship was developed with mixture design samples in our database described earlier. I do expect that a factor has to be established and used by the designers. This is similar to what has been performed by NCHRP Project 9-48, "Field versus Laboratory Volumetrics and Mechanical Properties." Your comments are well taken and it's in the process. In terms of various other parameters, as described in the paper, a stepwise regression process was performed to correlate field performance to various parameters measured from the SCB test. However, a simple, practical, and fundamental value of J_c was ultimately recommended

IMAD AL-QADI: In the State of Illinois we're planning to use a suite of tests too in a similar way. This includes the wheel test, and the SCB is being considered. We are developing an index for the contractors to use—to answer Ray's question—a mix may include as much as possible of RAP or RAS or other materials as far as a criterion is met. A correction factor must be applied for field cores.

LOUAY MOHAMMAD: Yes. Similarly, in Louisiana two tests are recommended: LWT for high temperature and this SCB test for intermediate temperature (no index) to evaluate the designed mixture. These two tests will provide a tool for our DOT to evaluate mixtures with recycled materials such as RAP and/or RAS or any other materials as well....

ELIE HAJJ: Thank you, Louay, for the good presentation. My question is regarding the conditioning and laboratory aging of the asphalt mixtures. How are you planning on incorporating the aging effect in the implementation phase? Will the developed criteria be applicable for both HMA and WMA? Will there be any adjustment to the conditioning of the mixtures or another set of criteria will be developed for WMA mixtures? Can you please further elaborate on this matter?

LOUAY MOHAMMAD: The criteria in Louisiana for WMA is permissive, basically. Whatever applies to HMA applies to WMA except as long as they are in the different production/compaction temperatures schemes. Currently, mixtures are long-term aged and tested.

ELIE HAJJ: That is five days at 85°C?

LOUAY MOHAMMAD: Yes. Five days, 85°C, correct. However, we are working on reducing the aging time period using mixture aging prior to compaction.

ELIE HAJJ: Thank you.

LOUAY MOHAMMAD: Sure.

FRANK FEE: I applaud your work, obviously.

LOUAY MOHAMMAD: Thank you.

FRANK FEE: Louay, have you thought through how you're going to move this into reality and give it to the contractors to work with, where are you going to get the testing infrastructure? I assume right now you're doing all this, right?

LOUAY MOHAMMAD: Chris Abadie if he is around may assist with the reply to this question.

FRANK FEE: But you understand what I'm saying?

LOUAY MOHAMMAD: The research was completed and recommendations were made to be incorporated into Louisiana's' specifications. Further in-house research was completed to evaluate the recommendation made. The in-house research conducted by Samuel Cooper III confirmed, among other things, the feasibility of using a Marshall type of load frame to perform the SCB test. As mentioned earlier, we are conducting an SCB test training workshop on April 16 for our contractors, design labs, and DOT to familiarize them with sample preparation, equipment setup, conduct of the test, and data analysis. Several of our contractors are equipped to perform this test. I also do expect that private laboratories may assist in the conduct of the test.

FRANK FEE: Now you're going to do this across the board for all your mixes now, is that the idea?

LOUAY MOHAMMAD: Yes, all the mixes as presented in Table 502-7 of the Louisiana's specifications.

FRANK FEE: Okay. All right.

LOUAY MOHAMMAD: This is the table.

FRANK FEE: Got it. Okay. That's important.

LOUAY MOHAMMAD: Yes.

FRANK FEE: You've got to have the incentive to get the infrastructure. Thank you.

RAY BONAQUIST: I guess I heard two things, one from you, Louay, and one from Imad, and I'm going to have to ask this question and maybe Imad's going to have to get up and explain what's going on in Illinois. But as I understand it, in Louisiana, the idea is to do current design as you do it, all the criteria have to be met for binder selection, maximum RAP and RAS content, all volumetric properties, and you add two tests on. And so I would really see this as a tightening of a design specification. What I thought I heard Imad say was put what you want in there, meet the criteria, and the mix is okay. These are two really big different philosophies.

LOUAY MOHAMMAD: Yes. That is correct for Louisiana. Further, if mixtures containing higher percentages of RAP and/or RAS can be evaluated using the two recommended tests. Currently, the use of RAS is not allowed in the specifications.

RAY BONAQUIST: Yeah. And so, Imad, if you would kind of explain what the plan is because I thought I heard a couple of different philosophies here.

FRANK FEE: One thing I'm saying—I think we're saying here—is you're going to have to give some forgiveness to the volumetrics side. For example, if you have an issue where you want to get some improved cracking, we could probably just simply add some binder, but you'll have to adjust the voids. And that should be a permissible way to do this. You're going to have to give some forgiveness of the volumetrics.

LOUAY MOHAMMAD: Thank you for the comment. It is noted.

IMAD AL-QADI: Ray, I think I should explain myself a little better. When it comes to the volumetrics, this is something specified by the state. The way that I tried to explain is that whenever we're trying to add RAP or RAS, we need a test that will be able to distinguish. So if I run a test and I look at the fracture energy at low temperature, you definitely cannot see that because if I have a lot of RAP there, that's going to be increasing the peak, and you may end up with the same area underneath both materials with and without RAP and still you may end up having cracks. So what we are trying to do is to come up with an index, which is called the flexibility index. This flexibility index is based on two things. One is the fracture energy itself but also on the ductility of the material. As the material now is propagating the crack itself, it depends now how, I guess, quickly that crack is going to propagate. If you have a brittle material, it's going to propagate very, very quickly, which means that your material is not good. So we were able to capture that very accurately. And whenever you have a brittle material and you cannot distinguish between that in the fracture energy we're able to do that with the flexibility index. We create three categories for that: One, this material is really poor. One is the material is something you need to go ahead and work on it, do something, change in the design; and the other one is a material that is super. Now this is based on things that we're doing in the lab. When you are taking cores immediately, you should be able to use the same index. If these cores are in the field for a long time, this material is aged, this material it's definitely going to need the shift factor. And that shift factor should depend on the number of years this material is in the field. So we feel very comfortable with this approach right now, and we'll be glad to share with you the modification that we did with the SCB test, as well as the criteria that we have established. So the only thing, I guess, is we're trying to avoid the discussion about the percentage of RAP and RAS that you're putting. All we're interested in—is this material going to perform or not? And we were able to see a material that it has zero recycled material but still not performing and it can crack very quickly, and we all know that. I hoped that addresses it.

LOUAY MOHAMMAD: The use of fracture energy by itself is misleading as compared to critical strain energy. For example, if one is to compare a load of 1000 lbs to 6000 lbs; or is 10 psi to 50 psi.

IMAD AL-QADI: Let me add this. I think what we are looking at is the fracture energy and the tail of the curve, because that's where you learn most of the information from.

ERV DUKATZ: This is what I love about AAPT, the exchange of ideas. And to quote a very famous alumni and paraphrasing him somewhat, we at Mathy Construction Company have studied this issue. We hope to have a paper to present in front of this august body next year. To address the point that Imad has made about similar areas under the load-displacement curve, is that we've looked at changes to the SCB procedure to take that into account. We have similar results that show that just looking at the area underneath the curve, you can get similar results for very different performing mixes. However, there are ways of looking at that data which we are now exploring in a ruggedness study for a proposed ASTM specification on SCB testing at intermediate temperatures. The additional analysis procedure will be added in an appendix. The procedure looks at the rate of loading to calculate the stiffness at the peak load and divides the area under the loading curve into elastic and compliant areas to differentiate between differences in mixes. I think that's very important.

My other comment is on mix specifications. As previously mentioned numerous times today, the state sets the requirements for the mix. So if Indiana changes the specification to require 5% air voids, the contractors will figure out how to make mixes with 5% air voids. If the highway agency wants 3%, as Michigan specifies for some mixes, the contracting industry will figure out how to make mixes with 3% air voids. So, I think what we need to continue as we have been doing in this group, is have the discussions about the ins and outs of the test procedures. Then look at what actually happens in the road. This field data will show that the small details we have been discussing are important and cannot be ignored.

And the final thing I want to say is that I compliment Louay for studying aged mixes in Louisiana. We need a way of studying the aged materials in laboratory. Because when you put RAP and RAS into mixes the components will blend together over time. However, as Dr. Anderson likes to say, "(we) ignore the rate of diffusion between the aged materials and new asphalt materials." And, I have to compliment Dr. McDaniel on her work on "Black-Rocks". Our research has shown when we put recycled materials into mixes, they are black rock at that moment when they first come out of the asphalt plant. However, when analyzed over time, a week later, a year later, and so forth, they are no longer black rocks. They're an intregal part of the asphalt mix with age dependent properties.

LOUAY MOHAMMAD: Thank you for the comment.

ERV DUKATZ: There was a question in there someplace, but I forgot.

LOUAY MOHAMMAD: Yes. We selected a test that measures fundamental properties, simple, and can be implemented in contractor laboratories. The SCB test is based in fracture mechanics principles. A good text book that I recommend is titled *Fracture Mechanics Fundamentals and Applications* by T. L. Anderson. We need to stay with these concepts and violations to assumptions of theories. That's where I want to leave it at.

Effect of Mixing Sequence on the Workability and Performance of Asphalt Mixtures

Ebrahim Hesami[a], Bjorn Birgisson[b], Niki Kringos[a]

[a]*KTH Royal Institute of Technology, Stockholm, 10049, Sweden*

[b]*Aston University, Birmingham, B47ET, UK*

ABSTRACT. *In Sweden, during recent years, a new type of mixing protocol has been applied in which the order of mixing is changed from the conventional method. Improved workability and diminished mixing and compaction energy needs have been important drivers for this. Considering that it is the mastic phase which is modified by changing the mixing order, it provides an interesting case study for explaining the mechanisms of workability in connection with the mastic phase. To do so, an analytical viscosity framework was combined with a mixture morphology framework to scale up to the mixing level and tribology principles to explain the interaction between the mastic and the aggregates. From the mastic viscosity protocol it was found that the mixing order significantly affects the resulting mastic viscosity. To analyze the effect of this on the workability and resulting mixture performance, X-Ray Computed Tomography was used to analyze mixtures produced by the two different mixing sequences. Mechanical testing was utilized to determine the long-term mechanical performance. In this part of the study, mastic viscosity as a function of particle concentration and distribution was directly coupled to improved mixture workability and enhanced long-term performance.*

KEYWORDS: *asphalt mixtures, mixing method, sequential mixing, asphalt mastics, workability, mastic viscosity.*

1.0 Introduction

Mixing aggregates with bitumen is an important task in the production of asphalt mixtures. The quality of mixing can affect several characteristics of asphalt mixtures, such as film coating thickness, bitumen-aggregate adhesion and its homogeneity, which have significant impact on the performance of asphalt pavements. So far, the aggregate and bitumen characteristics, mix design and their effect on the performance of the asphalt pavements have been the subject of much research on asphalt mixtures (Kandhal and Parker, 1998; Petersen, 1984; Taylor and

The oral presentation was made by Professor Birgisson.

This is a reproduction of an article whose final and definitive form has been published in *Road Materials and Pavement Design*© 2015 Taylor & Francis. The article is available online at: http://dx.doi.org/10.1080/14680629.2015.1077008

Khosla, 1983; Brown and Cooley, 1999; Goodrich, 1988). The effect of mixing sequence, however, has been given considerably less attention in the literature.

Most research available for predicting or calculating the performance of asphalt pavements assumes asphalt mixtures are homogeneous materials. Even when assuming a homogeneous behavior, a similar and predictable behavior of the mixture throughout the pavement is crucial for its long-term performance. The only phase that guarantees the homogeneity of the material is the mixing process. Though homogeneity of asphalt mixtures can be disrupted after the mixing stage, utilizing suitable facilities for mixing, designing the material for a suitable workability level and adequate mixing time can help to have a more homogeneous mixture with all aggregates coated appropriately.

Conventionally, most asphalt mixtures are produced by first mixing all the mineral aggregates together (the 'dry mixing'), followed by a spraying of the bitumen on this aggregate mixture (the 'wet mixing') followed by further mixing, as described, for example, in Roberts et al. (2009). Though the risk for segregation and heterogeneity of the mixture are still noted as issues (Brown et al., 1989), this type of mixing is still most commonly used.

In Sweden, for the last few years, a new type of mixing protocol (Jacobsson, 2009), has been applied, in which the order of mixing is changed from this conventional method as follows:

Step 1: The load carrying aggregates (here named "primary structure") are mixed with the bitumen.

Step 2: The filler particles are added to the mixture. Filler in this case consists of the mineral aggregates with size 75μm or smaller.

Step 3: The fine aggregates, which are bigger than filler and smaller than the primary structure (here named "secondary structure"), are mixed in with the entire mixture.

Since 1998, about 1 million tons of asphalt mixtures have been produced with this method, mostly for use as a wearing layer on highways and secondary roads. However, the mixture design and mixing protocols have essentially consisted of an empirical "mixture recipe" approach, without a broader theoretical basis. Improved workability and decreased manufacturing temperature of asphalt mixture by 30°C have been important drivers for this (Viman et al., 2008; Viman, 2011). According to a field study which was performed during a three-year period after construction, the mixes produced sequentially showed good performance in terms of noise, friction, texture, wear resistance and rutting compared with conventionally mixed asphalt mixtures (Viman et al., 2008; Viman, 2011).

This paper presents a new framework that explains: (i) the effect of the mixing sequence on the workability of asphalt mixtures, focusing on the mastic phase, and (ii) the linkage between the mixing procedure and resulting mixture mechanical performance. The new framework is subsequently used to explain the special case associated with the sequential method. In particular, the effects of mix sequence on workability and asphalt mixture performance are investigated by comparing asphalt mixtures produced through sequential and conventional mixing methods.

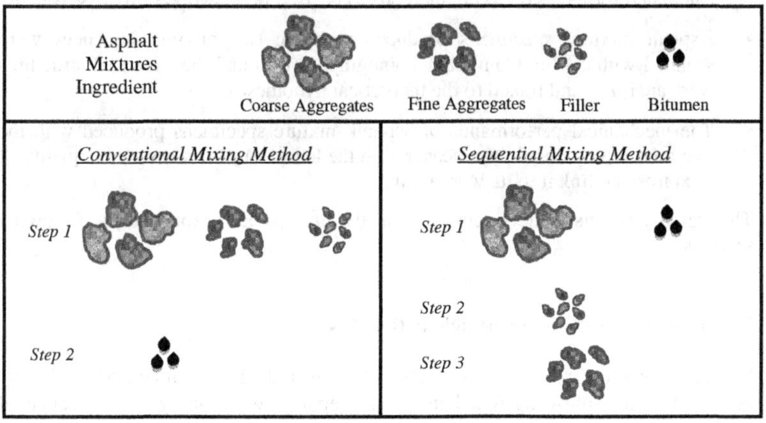

Figure 1. *Schematic of conventional vs. sequential mixing methods.*

2.0 Research Aim and Methodology

Though the application of the sequential method in Sweden has resulted in satisfactory results in terms of reduced energy needs and sufficient mechanical performance, it does pose a number of challenges. One is moving towards performance-based specifications using this method since it is based on an empirical recipe procedure; there is currently no clear analytically-based linkage between mixture design and performance. Secondly, in the case of specialized applications, there is currently very little guidance towards the needed changes in the mix design or production. Finally, as a community we are moving towards removing black boxes as much as possible as it will enhance control on long-term performance and reduced risk.

For this reason, the research described in this paper is focusing on linking the effect of mixing sequence on the workability and mechanical performance of asphalt mixtures. For this the following steps were taken:

- Hypotheses were formed to theoretically explain the reduction in mixing temperature, the enhanced coating thickness and the reduction of agglomeration when changing the mixing sequence.
- Rheology measurements were performed on asphalt mastics, which were produced with the sequential and normal mixing methods to assess the energy needs upon mixing.

- The binder coating thickness around the load carrying aggregate structure and the theoretical friction coefficient between coated particles were theoretically calculated for the two mixing methods to be used in the workability assessment.
- Asphalt mixture specimens produced with the two mixing sequences were scanned with X-Ray Computed Tomography (CT) and their internal structures were analyzed and linked to the theoretical hypotheses.
- The mechanical performance of asphalt mixture specimens produced with the two mixing sequences was measured in the laboratory to assess as the quality of mixtures and link it to its workability.

The details, results and conclusions of these steps are explained in following sections.

3.0 Discussion of the Sequential Mix Benefits

From previous research (Jacobsson, 2009; Viman et al., 2008; Viman, 2011; Hesami et al., 2013) the following four benefits are reported when applying the sequential method:

1. Lower manufacturing temperature.
2. More homogeneous mixture.
3. Thicker coating around the coarser particles.
4. Environmental and economical benefits.

The first benefit, the mixing temperature reduction is not only a benefit but also part of the instructions of the sequential mixing. Since the first step of mixing involves a relatively high amount of bitumen against the surface area of the coarse gradation component, there is a risk of the bitumen draining down. To avoid this, the bitumen viscosity should be higher than the viscosity of the bitumen in the conventional mix. Reducing the mixing temperature thus avoids this problem since it increases the viscosity of the bitumen.

Though the total aggregate surface area in this method is the same as in the conventional method, introducing the final fractions at a later stage in the process means that less adhesive forces have to be overcome during the mixing. This also explains why a lowering of the manufacturing temperature is possible.

Due to the high surface area of the filler particles, the filler grains have more interaction with each other and thus have a high risk of forming clusters (Hesami et al., 2013). These clusters can actually be quite hard to break, once formed, through the mixing process since they will still be relatively small compared to the mechanical mix arms. Additionally, the liquid bridges that bind the particles together, Figure 2, arrest part of the bitumen (Pietsch, 2008). This portion of the binder is therefore not available as free binder. This is frequently seen in the normal mixing process through the "dullness" of the loose mixture. In the sequential mixing method, filler is added gradually on the surface of the bitumen that is already partially adhered to the primary structure. As such, the agglomeration problem is significantly reduced in this case. This is also observed through the "shininess" of this type of mixture.

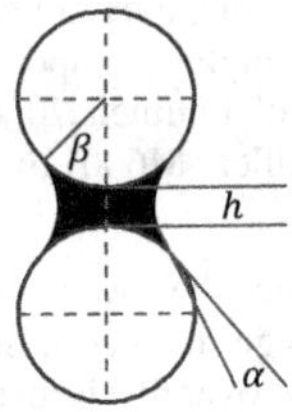

Figure 2. *Liquid bridge Between two monosized spheres.*

In conventional mixing procedures, all the aggregates are initially mixed dry. Considering that all asphalt mixtures either have a continuous or a gap graded aggregate mixture, there will always be aggregates of different sizes. This means that in the case of dry mixing of the aggregates, there is a tendency of the smaller particles to move to the top of the mixture. This segregation can thus result in an inhomogeneous mixture upon introduction of the binder. It would take a significant mixing effort to reduce this segregation at this stage and "over-mixing" could become a high risk. In the case of the sequential mixing method, the larger aggregate sizes are first coated with bitumen. Then, upon introduction of the filler there is no chance of segregation since the filler will immediately stick to the binder. This concept is also referred to as cohesive mixing (Figure 3) (Rhodes, 2008). This thus results in a more homogeneous mixture with less mixing effort.

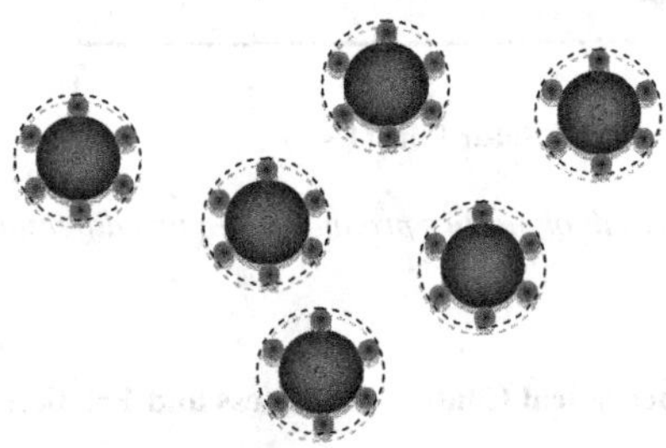

Figure 3. *Small particles on carrier particles.*

To assess the above described processes of agglomeration and its viscosity behavior, viscosity measurements with the shear rheometer were made in which the mixing order of particles and the bitumen was varied.

In the first case (mix I), filler grains were gradually dispersed on top of the bitumen while bitumen was agitated. In the second case (mix II) all mineral aggregates were first placed in the cup and then bitumen was mixed in.

The mixing time for both methods was kept the same (15 min) and done at 100°C. The preparation and test methods follow the protocol for measuring the viscosity of asphalt mastics at higher temperature developed by Hesami et al. (2014). For both samples 100g of bitumen 70/100 were mixed with 150g of fine aggregate (0.5-1 mm) and 40% filler (M600) by volume. In Figure 4 the results are shown.

Results show that the mix II has the lower viscosity, which is expected based on the above discussion since the agglomerated minerals formed larger particles in the bitumen matrix thus reducing the overall viscosity (Hesami et al., 2012; Hesami et al., 2014).

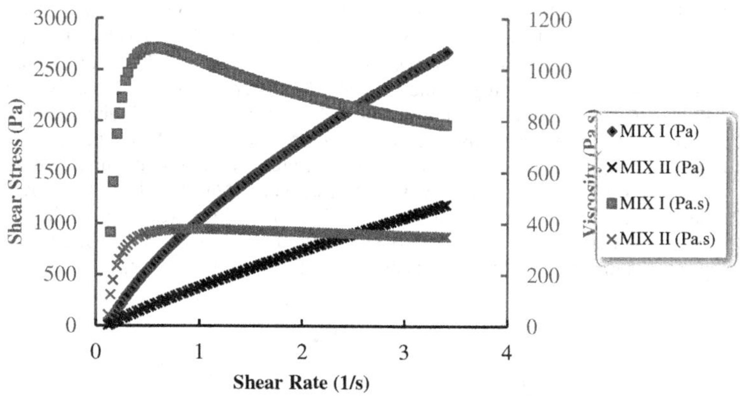

Figure 4. *Viscosity result of mastics produced with two different methods at 100°C.*

4.0 Workability: Theoretical Coating Thickness and Friction Coefficients

4.1 Quantification of the Effects of Mixture Morphology

To demonstrate the effect of the mixing sequence on the workability of asphalt mixtures, in the following, an example mixing recipe (Swedish ABS 11 (Andersson, 2011) was used. In this mixture, the bitumen content is 6.3% and Table 1 shows the aggregate gradation.

Table 1. *Aggregate gradation for calculating friction coefficient.*

Sieve size (mm)	16	11.2	8	5.6	4	2	1	0.5	0.063
Passing (%)	100	95	54	34	28	22	19	15	9

To calculate the theoretical workability of the two mixing sequences, the mixture morphology concepts developed by Das et al. (2014) were employed. In the mixture morphology framework (Das et al., 2014), a central assumption is that the portion of aggregate which has a significant role in transferring the load through the asphalt mixture is referred to as the primary structure and aggregates smaller than primary structure, but bigger than filler, are considered to form the mixture's "secondary structure." In the sequential mixing method it is important to determine the boundary limit between primary and secondary structures. To do so, the weighted average grain size ($D_{w,avg}$) of two contiguous sieve sizes can be calculated by utilizing Equation 1 (Yideti et al., 2013):

$$D_{w,avg} = D_1 D_2 \sqrt[3]{\frac{\varphi_1 + \varphi_2}{\varphi_1 D_2^3 + \varphi_2 D_1^2}} \qquad [1]$$

Where D_1 and D_2 are the sieve sizes and φ_1 and φ_2 are the concentration of each sizes. Two contiguous sieve sizes can be in the primary structure range only if the Equation 2 is fulfilled. The analysis starts from the biggest sieve size in the gradation and goes down until sieve size of 2.00 and 2.36 mm of European and American sieve size standard, respectively (Yideti et al., 2013).

$$\frac{1.1 \times D_1 D_2}{\sqrt[3]{D_2^3 + 2.36 \times D_1^3}} \leq d_{w,avg} \leq \frac{1.1 \times D_1 D_2}{\sqrt[3]{2.36 \times D_2^3 + D_1^3}} \qquad [2]$$

In this Equation $d_{w,avg}$ is the weighted average void diameter which is equal to $0.732 D_{w,avg}$.

The coating thickness of the primary structure at the production phase of asphalt mixture does not only show the quality of asphalt mixture but also gives a rough estimation of distance between aggregates. Once the primary and secondary structure's range are recognized, the asphalt film coating thickness around the primary structure can be calculated by using the morphology framework concept (Cooke et al., 1999; Das et al., 2014; Bernardita et al., 2013). In this, the film growth imposed by contact between particles based on the aggregate packing arrangement is taken into account (Cooke et al., 1999) and it is assumed that the aggregates in loose mixture are arranged in the simple cubic packing system (Figure 5). To calculate the film thickness in this method, the volume of a unit cell and its porosity must be calculated. A unit cell is a six-sided box which is formed by joining the center of eight spheres. For the simple cubic arrangement, the volume of the unit cell can be calculated by the following equation.

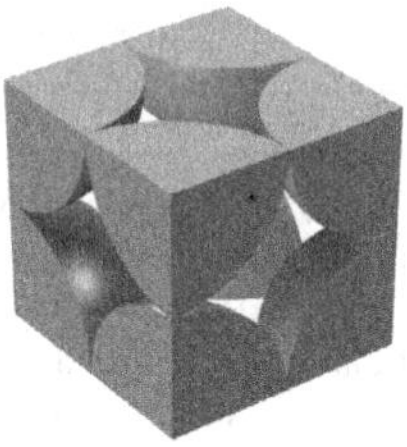

Figure 5. *Simple cubic packing system.*

$$V_{uc} = \alpha_m d_p^3 \qquad [3]$$

Where d_p is the diameter of a particle and α_m is a packing factor, which for the simple cubic is equal to 1. Porosity of the unit cell simply can be calculated by subtracting the total volume of the unit cell by the volume of the aggregates and bitumen within the unit cell. In the simple cubic unit cell there is only one spherical aggregate in the unit cell, so by having the average weighted aggregate size for each part of aggregate gradation, the volume of solid in the unit cell can be calculated. Volume of bitumen also can be determined by using the bitumen content in the mix, specific gravity of aggregate and specific gravity of bitumen.

The relationship between the porosity and coating thickness index ($\frac{2L_t}{d_p}$) can be determined from Equation 4, which is proposed by Taylor et al. (1990).

$$n = 1 - \frac{\pi}{\alpha_m}\left[\frac{2-m}{12}\left(\frac{2L_t}{d_p}\right)^3 + \frac{4-m}{8}\left(\frac{2L_t}{d_p}\right)^2 + \frac{1}{2}\left(\frac{2L_t}{d_p}\right) + \frac{1}{6}\right] \qquad [4]$$

Where n is the porosity, m is number of contact points and L_t is a film thickness. Figure 6 shows the modified Taylor relationship for different packing systems.

For quantifying the film thickness of the primary structure coating and the frictional coefficients will be calculated. Hence, as a first step calculations of the size of the course aggregate fractions are made, based on the aggregate gradation shown in Table 2 (shown later). Combining this with the bitumen percentages, and assuming a simplified packing arrangement, the theoretical film coating thickness of the primary structure can be determined. From this, the aggregate contact points and mixture porosity are determined. These then allow for the determination of the theoretical workability.

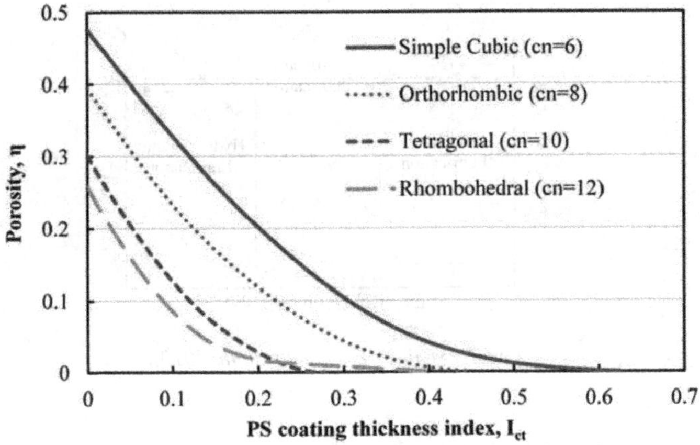

Figure 6. *The relationship of porosity and film coating thickness (Das et al., 2014).*

4.2 Quantification of the Workability of Sequential and Conventional Mixes

According to the developed viscosity framework (Hesami et al., 2012), the relative viscosity of a suspension, mastic or mixture can be calculated based on particle concentration. The relative viscosity is defined as the viscosity of the suspension divided by the viscosity of the binder. The relative viscosities of suspensions are divided into two regimes: hydrodynamic and frictional. If the particles are far from each other, the suspension behaves in a hydrodynamic regime otherwise it behaves in a frictional regime.

As the asphalt mixture in all mixing stages has a high aggregate concentration, asphalt mixtures in the loose stage and during mixing are dominated by frictional behavior. Hence, the most resistance to shear in loose asphalt mixtures comes from the frictional forces between aggregates. However, aggregates during mixing are not dry and a thin layer of the bitumen exists between aggregates. This layer of bitumen has the lubrication effect and causes a lower friction coefficient.

The Stribeck curve, Figure 7 (Stachowiak and Batchelor, 2013), is a tribology theory that explains the changes in friction coefficient due to change in properties of the film lubricant. In this context, the Sommerfeld number is a function of lubricant viscosity. Equation 5 describes the Sommerfeld number as a function of velocity of movement of two surface and load which is applied on the sliding surface.

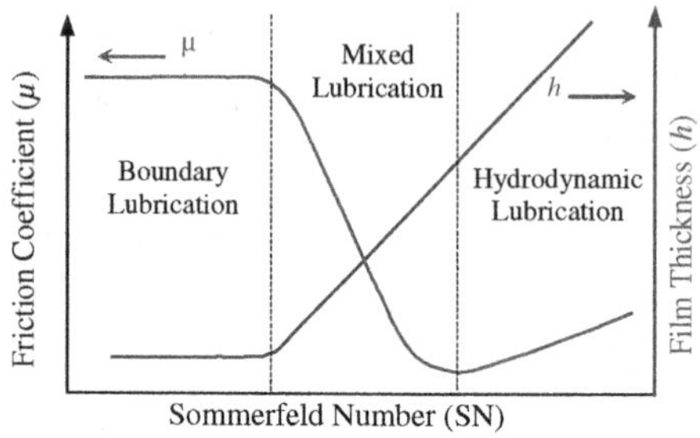

Figure 7. *Stribeck curve (Bhushan, 2010).*

$$SN = \frac{Velocity \times Viscosity}{Load} \qquad [5]$$

So, changing any of these variables changes the frictional coefficient, which may lead to changes in the lubrication regime. According to the Stribeck curve, three different lubrication regimes are distinguishable; boundary, mixed and hydrodynamic lubrication. The distance between two surfaces is increased from very narrow thickness in the boundary lubrication to more than 25μm in hydrodynamic regime (Bhushan, 2010).

Considering that during mixing a relatively high velocity can be expected for two sliding aggregates in the vicinity of each other, as well as a low load magnitude, the hydrodynamic regime is the dominant lubrication regime during the mixing of asphalt mixtures. Therefore, at a same job mix formula and mixing conditions, for a given velocity and load, the higher bitumen viscosity will give the higher frictional force.

Frictional forces of two dry aggregates in contact that have relative lateral motion can be calculated with the Coulomb Equation 6:

$$F = \mu \times N \qquad [6]$$

Where F is the frictional force, μ is the frictional coefficient and N is the vertical load on the contact point. For two bitumen coated aggregates, the relative lateral motion follows the viscosity law. With existence of a layer of Newtonian fluid between two surfaces, the resistance to the relative lateral motion is calculated by the following Equation:

$$\tau = \eta\,\dot{\gamma}$$

$$F = \frac{\eta . V . A}{h} \qquad [7]$$

Where τ is the shear stress, η is the viscosity of the liquid, $\dot{\gamma}$ is a shear rate, F is the lateral force, V is the relative velocity, A is the contact area and h is the distance between the surfaces of aggregates. Substituting Equation 6 into Equation 7 will give the friction coefficient as Equation 8.

$$\mu = \frac{\eta . V . A}{h . N} \quad [8]$$

This Equation can be used for relatively comparing the change of friction coefficient during mixing for both the conventional and the sequential mixing methods.

In the conventional mixing method, when the bitumen is added to all the mineral aggregates, it is assumed that filler, due to its high surface area, adsorbs the bitumen first and makes mastic. In this stage, the viscosity of mastic is a function of filler concentration and viscosity of the bitumen. Then the secondary structure (SS), which has a higher surface area than the primary structure, will mix with the mastic phase. According to Roscoe (1952), a suspension with smaller particles plays a media role for bigger particles. This means the viscosity of the mixture of secondary structure and mastic is a function of the viscosity of mastic and concentration of the secondary structure. The same applies to the primary structure, so its viscosity is a function of the combined viscosity of the secondary structure and mastic and concentration of the primary structure.

In the sequential mix, bitumen first coats the primary structure, so the mixture viscosity is a function of viscosity of bitumen and concentration of the primary structure. After adding filler gradually on top of the bitumen coating layer, the mastic forms. In this stage, the viscosity of the mixture is a function of mastic viscosity and the primary structure concentration. And finally, by adding secondary structure, the overall viscosity of the mixture is a function of the combined viscosity of the secondary structure and mastic and primary structure concentration. The above is summarized in Table 2.

Table 2. *Summary of viscosity as a function of particle concentration at different stages of mixing.*

Conventional Mix	Sequential Mix
1. Bitumen + Filler $\eta_{Mastic} = f(\phi_{Filler}, \eta_{binder})$	1. Bitumen + Primary Structure $\eta_{PS} = f(\phi_{PS}, \eta_{binder})$
2. Mastic + Secondary Structure $\eta_{SS} = f(\phi_{SS}, \eta_{mastic})$	2. PS Mix + Filler $\eta_{PS+F} = f(\phi_{PS}, \eta_{mastic})$
3. Mortar + primary Structure $\eta_{PS} = f(\phi_{PS}, \eta_{SS})$	3. PS/F Mix + Secondary Structure $\eta_{PS} = f(\phi_{PS}, \eta_{SS})$

The boundary limit between primary and secondary structures of the example aggregate gradation was calculated as 2 mm. By utilizing the Taylor relationship and the other calculated volumetric parameters, the film coating thickness can be calculated, Table 3.

Table 3. *Film coating thickness for sequential and conventional mixing methods.*

Sequential	Conventional
Stage 1 *PS coating thickness (Bitumen)* $T_{PS1} = 326\ \mu m$	Stage 1 *Filler particles in bitumen (Hydrodynamic Regime)*
Stage 2 *PS coating thickness (mastic)* $T_{PS2} = 718\ \mu m$	Stage 2 *SS coating thickness (mastic)* $T_{SS2} = 84\ \mu m$
Stage 3 *PS coating thickness (mastic + SS)* $T_{PS3} = 1044\ \mu m$	Stage 3 *PS coating thickness (Mastic + SS)* $T_{PS3} = 1044\ \mu m$

Combining the film coating thickness results and friction coefficient equation (Equation 8), shows the effect of mixing sequence on the friction coefficient. Table 4 shows friction coefficients at different stages for both methods.

Table 4. *Friction coefficients at different stages.*

Sequential	Conventional
Stage 1 $\mu_1 = \frac{\eta_b . A_{cPS}}{t.(652\mu m)}$	Stage 1 *Filler particles are not in contact and the resistance to flow comes from the viscosity of mastic.**
Stage 2 $\mu_2 = \frac{\eta_{mastic} . A_{cPS}}{t.(1436\mu m)}$	Stage 2 $\mu_2 = \frac{\eta_{mastic} . A_{cSS}}{t.(168\mu m)}$
Stage 3 $\mu_3 = \frac{\eta_{mastic+SS} . A_{cPS}}{t.(2088\mu m)}$	Stage 3 $\mu_3 = \frac{\eta_{mastic+SS} . A_{cPS}}{t.(2088\mu m)}$

**Viscosity of mastic (η_{mastic}) can be determined in hydrodynamic regime utilizing viscosity framework (Hesami et al., 2012).*

Assuming the same mixing velocity and contact area for both mixing methods, the equations in Table 4 illustrate that the frictional coefficient during the mixing in the sequential method is lower than that for conventional mixing. This implies that the mixture in the sequential mixing scheme is more workable during mixing compared to the conventional. However, frictional coefficients at the end of the mixing process are the same for both mixtures. In practice, however, due to lower levels of mixture workability in the conventional method, it will be difficult to achieve the ideal mixture without a higher mixing effort.

5.0 Effect of Mixing Sequence on the Internal Structure of Asphalt Mixtures

Changing the mixing method but keeping the recipe and materials constant may vary the internal structure of asphalt pavement. Changes in the internal structure of asphalt pavement can lead to changes in behavior of asphalt pavement. So, knowledge of the internal structure of the asphalt mixtures produced by sequential and conventional mixing methods could help explain the connection between mixing method on performance. Therefore, in this study three-dimensional X-ray digital images were analyzed using an image processing software (Avizo).

To have real field conditions, specimens were cored from a highway in Sweden, E18, between Slyte and Örebro, where there are two sections that were paved with asphalt mixtures produced by the sequential and the conventional method. Two asphalt specimens, which were produced with sequential and conventional mixing methods, were used for the analyses. Both specimens are Swedish ABS 11 mixtures in which the biggest aggregate size is 11 mm and 6.3% of weight of samples is bitumen 70/100 (Andersson, 2011). Figure 8 shows the aggregate gradation of the specimen.

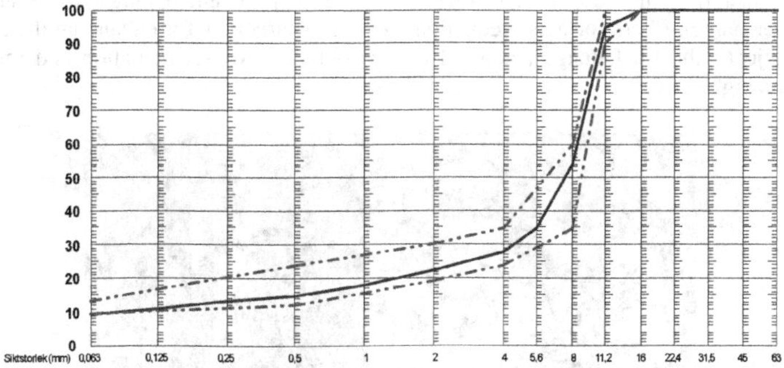

Figure 8. *Aggregate gradation of asphalt samples used for X-ray computed tomography.*

The KTH X5000 Computed Tomography X-ray scanner is used to capture digital images of asphalt mixture specimens for visualization. The specimens were scanned at energy intensity of 225kV without beam filtration. The scanning resolution is 1949 x 1799 with a slice thickness of 59 μm and a total of 400 slices. Figure 9 shows raw pictures of asphalt samples before image processing.

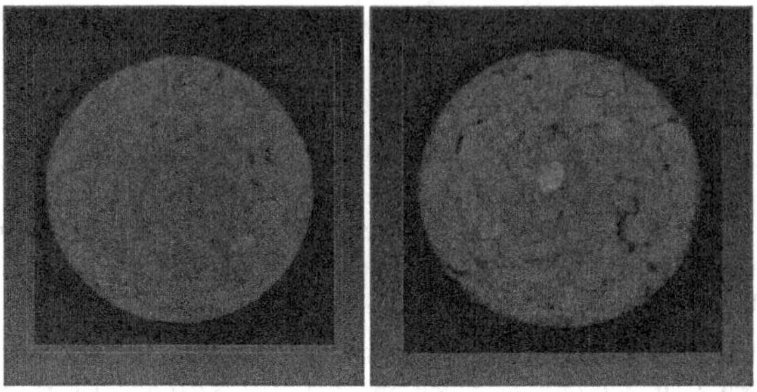

Figure 9. *Example of X-ray image before image processing.*

One of the artifacts of X-ray scanning is beam hardening, which refers to the loss of resolution in the center of the scanned images. In this study, the beam hardening is corrected by utilizing background and flat field correction feature in Avizo Fire, Figure 10. After applying some other image processing options, such as equalization for transferring the grayscale histogram from I shape (limited range) to O shape (wider range), filters for noise reduction and edge sharpening for enhancing the edge of objects, the final image (Figure 10(c)) is obtained for segmentation and image analyzing.

Figure 10. *a) Beam hardening, b) flat field image and c) corrected image.*

Figure 11(a) shows the segmented scanned image of the mixture with its three phases: aggregates, mastic and air voids. From these, it is possible to compute the volumetric properties and calculate the air void area, mean distance and contact area between adjacent aggregates with depth for both asphalt samples.

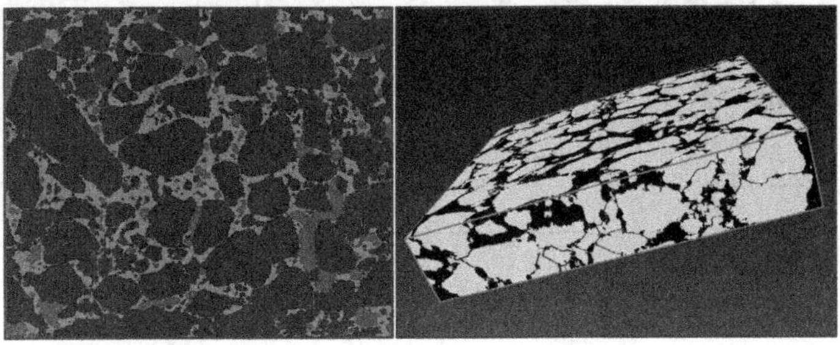

Figure 11. *a) Mixture segmentation and b) 3D volume rendered.*

Figure 12 shows the 2D area of air voids per slice along the depth. From the Figure it can be seen that the sequentially mixed asphalt mixture has an even air void distribution along the depth, unlike the conventionally mixed asphalt mixture. This indicates better compactability of the sequential asphalt mixture compared to the conventional one.

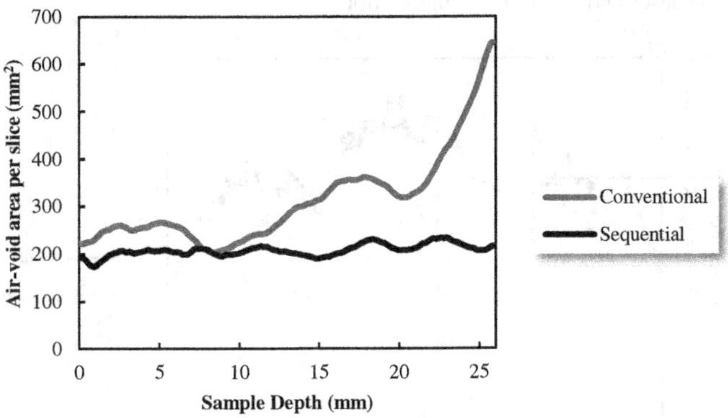

Figure 12. *Density distribution of air voids with depth.*

Correspondingly, the mean distance between adjacent aggregates shows the same trend (Figure 13) as the sequentially mixed asphalt shows a lower and more even distribution of mean distance after compaction along the depth of sample.

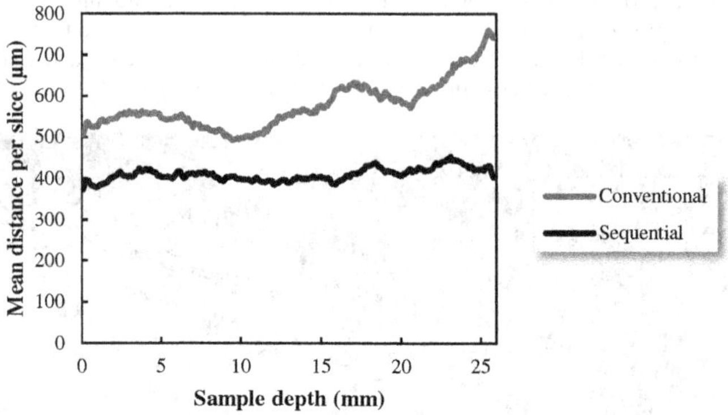

Figure 13. *Mean distance distribution with depth.*

Figure 14 gives the computed contact area of adjacent aggregates and shows that the sequential mix shows a slightly higher contact area in each slide. Lower and more even distribution of mean distance combined with higher contact area shows that the sequential mix has more homogeneity and better internal structure for transferring loads. To further characterize this, mechanical testing was performed, which is described in the following section.

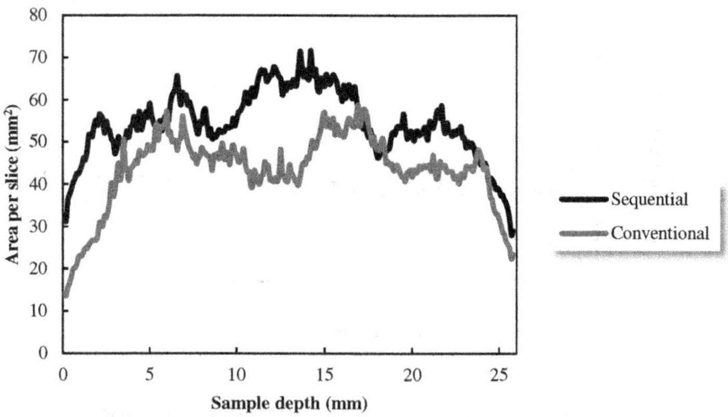

Figure 14. *Aggregate contact area with depth.*

6.0 Effect of Mixing Sequence on Performance

Six samples with ABT-16 (i.e. biggest aggregate size was 16 mm) (Andersson, 2011) and bitumen 70/100 mm/mm were produced with the sequential and conventional mixing methods. Figure 15 shows the aggregate gradation and Table 5 gives the other specifications of the mixtures.

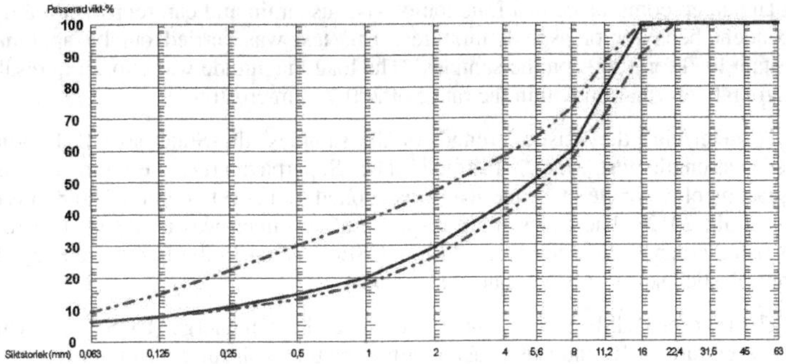

Figure 15. *Aggregate gradation.*

Table 5. *Mixture specifications.*

Mix Type	Aggregate Density (g/cm³)	Flakiness index (%)	Micro-Deval (%) (10-14mm)	Los Angeles (%) (10-14mm)	Bitumen density (kg/m³)	Bitumen Content (%)
Sequential	2.66	15	10	25	1020	6
Conventional	2.66	15	10	25	1020	6.3

Mix Type	Mixture Bulk Density (kg/m³)	Mixture Maximum Density (kg/m³)	Void content (%)
Sequential	2358	2430	3
Conventional	2340	2415	3.1

To have consistency and more realistic conditions for both types of mixing, the loose asphalt mixtures were taken from the Skanska Rockneby asphalt plant, which produces asphalt with both sequential and conventional mixing methods. The loose samples were warmed up in the laboratory and compacted with gyratory compactor (Servopac) in a 10 cm mold. The compaction force was 400 kN and the loading angle was 1.25°. These samples were trimmed according to the procedure developed by Roque et al. (1992) for Superpave Indirect Tension Testing (IDT). All IDT tests were performed at 10°C.

The Superpave IDT is utilized to determine resilient modulus (Mr), creep compliance, tensile strength (St), fracture energy limit (FE), dissipated creep strain energy limit (DCSE) and energy ratio (ER).

Resilient modulus is the ratio of applied stress to the recoverable strain under repeated load. The test was performed by applying a haversine load, each loading cycle contains 0.1s loading and 0.9s rest period. This loading represents the speed of 55 mph. In order to ensure the validity of linear viscoelasticity, the load magnitude was adjusted to keep horizontal deformation within the range of 200-300 microstrain (Das et al., 2012).

The creep compliance is a function of viscous strain and can represent the time-dependent behavior of asphalt mixtures. This test was carried out by applying a constant load for 1000s on the samples. The load magnitude was chosen to result in horizontal deformation within the range of 200-750 microstrain.

To determine the tensile strength of the samples, the Superpave IDT indirect tensile strength test was performed. The Superpave IDT strength test is a displacement controlled test and load was applied in a constant rate of 50.8 mm/min (Das et al., 2012). The stress-strain response of specimen was used to calculate the dissipated creep strain energy ($DCSE_f$). $DCSE_f$ is equal to the fracture energy (FE) minus elastic energy (EE) (Roque et al., 1992).

The ratio of $DCSE_f$ to the minimum dissipated creep energy ($DCSE_{min}$) is called the energy ratio (ER) and is a measure of the fracture resistance of mixtures:

$$ER = \frac{DCSE_f}{DCSE_{min}} \qquad [9]$$

The material parameters obtained from Superpave IDT for both sequential and conventional mix methods are summarized in Table 6. From the results one may observe that producing asphalt mixture with the sequential method resulted in stiffer mechanical behavior. Resilient modulus represents the material's capability to recover after releasing loading, so generally higher resilient modulus shows higher elastic recovery. The resilient modulus (Mr) of the sequential mix is 14% higher than the conventional mix. Furthermore, the creep compliance of the sequential mix at 1000s was found to be reduced by 19%, which shows the better performance of sequential mix in term of resistance to permanent deformation. Figure 16 compares the creep compliance of both mixtures.

Table 6. *Material parameters from Superpave IDT.*

Type of Mixture	Mr (GPa)	D0 (1/GPa)	D1 (1/GPa)	m-value	Creep Rate @ 1000s
Sequential	13.26	0.075	0.1828	0.710	0.018
Conventional	11.59	0.086	0.3203	0.659	0.020

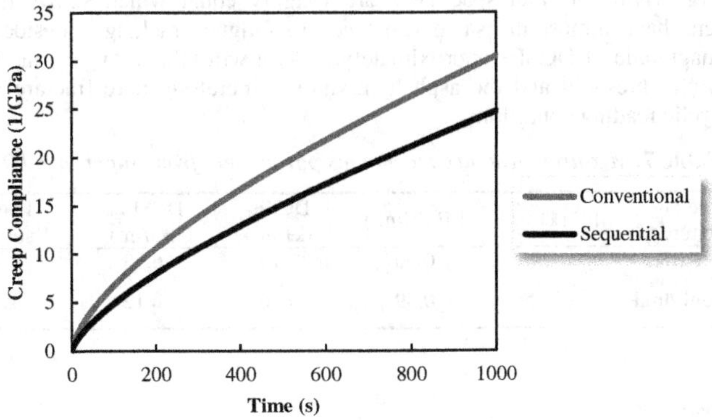

Figure 16. *Creep compliance at 10°C.*

As shown in Figure 17 the tensile strength of the sequential mix is significantly higher than for conventional mix, about 45%. This can be a result of a better distribution of the aggregates and bitumen in the asphalt mixture produced by the sequential method, providing thus a suitable structure for transferring loads.

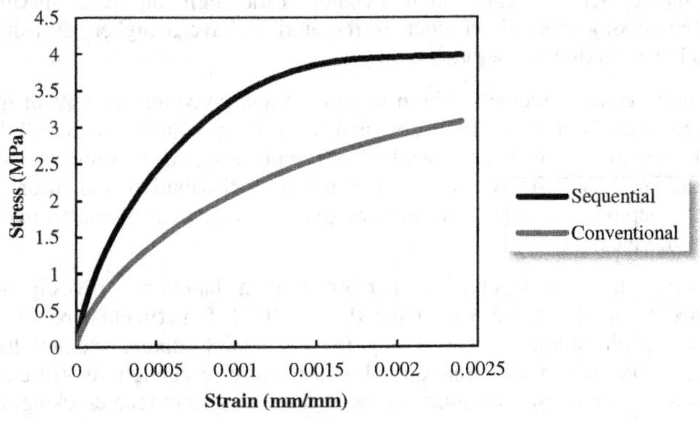

Figure 17. *Tensile strength at 10°C.*

The summary of the results from the asphalt mixture fracture mechanics analysis is presented in Table 7. Comparing the fracture energy of these two mixtures shows 22% lower FE for the sequential mix. This indicates that the sequential mixture seems more susceptible to brittle cracking such as thermal induced cracking or fracture due to a single passing of a very heavy vehicle (Butt et al., 2010). However, the energy ratios of both specimens are roughly equal which shows that both specimens have almost the same resistance to fatigue cracking. Considering the lower magnitude of $DCSE_f$, approximately 25%, it will take a longer time to reach the fracture threshold and the asphalt mixture is therefore more fracture resistant under cyclic loading conditions.

Table 7. *Mixtures' fracture mechanics parameters from Superpave IDT.*

Type of Mixture	FE (kJ/m^3)	EE (kJ/m^3)	$DCSE_f$ (kJ/m^3)	$DCSE_{min}$ (kJ/m^3)	Energy Ratio (ER)
Sequential	7.79	0.60	7.19	6.53	1.10
Conventional	10.05	0.49	9.56	8.12	1.18

7.0 Conclusion

In this paper, the possible reasons for differences in the rheological properties of sequential and conventional mixing methods were discussed. In particular, the viscosity framework (Hesami et al., 2012), morphology framework (Das et al., 2014; Bernardita et al., 2013) and concepts from tribology were used to evaluate the effect of mixing sequence on the workability of asphalt mixture during mixing. The results presented showed that asphalt mixtures produced with sequential mixing method are more workable with less segregation. Furthermore, due to lower specific surface area of the coarser aggregate which is coated at the beginning of the mixing in the sequential mixing method, bitumen is required to have a higher viscosity which leads to lower mixing temperatures.

Despite the same recipe for both sequential and conventional asphalt mixtures, the image analysis of 3D pictures obtained from X-ray scanner confirmed that their internal structures are not somewhat different. Asphalt mixture produced by sequential mix method showed an even air void distribution and mean distance along the depth of sample. This implies that the sequential asphalt mixture may compact more easily.

To evaluate their mechanical performance, a laboratory investigation was performed by applying the Superpave IDT at 10°C. Experimental results showed that the asphalt mixture produced by the sequential mixing method has more resistance to permanent deformation, while it is more susceptible to brittle cracking. Both asphalt samples show almost the same resistance to fatigue cracking, however the sequential mix is slightly more resistant to fracture under cyclic loading conditions.

For the continuation of this work, it is recommended to repeat some of the performed testing for multiple samples. From this, more generic conclusions can be drawn between the dominant parameters for mix design using different mixing procedures and resulting workability, compactability and long-term performance.

8.0 References

Andersson, P., "TRVKB 10 Bituminous stock," *Swedish Transport Administration*, TRV 2011:082, 2011, ISBN number: 978-91-7467-151-3.

Bernardita, L., Jelagin, D. and Birgisson, B., "Gradation-Based Framework for Asphalt Mixture," *Materials and Structures*, Vol 46(8), 2013, pp. 1401–1414.

Bhushan, B., *Modern Tribology Handbook*, Two Volume Set, CRC Press, 2010, ISBN 0-8493-8403-6.

Brown, E. R., Collins, R. and Brownfield, J. R., "Investigation of Segregation of Asphalt Mixtures in the State of Georgia," *Transportation Research Record No. 1217*, 1989, pp. 1–8, ISBN: 0309048117.

Brown, E. R. and Cooley, L. A., *Designing Stone Matrix Asphalt Mixtures for Rut-Resistant Pavements*, NCHRP Report No. 425, Transportation Research Board, 1999.

Butt, A. A., Jelagin, D., Tasdemir, Y. and Birgisson, B., "The Effect of Wax Modification on the Performance of Mastic Asphalt," *International Journal of Pavement Research and Technology,* Vol 3(2), 2010, pp. 86–95.

Cooke, A. J. and Rowe, R. K., "Extension of Porosity and Surface Area Models for Uniform Porous Media," *Journal of Environmental Engineering*, Vol 125(2), 1999, pp. 126–136.

Das, P. K., Birgisson, B., Jelagin, D. and Kringos, N., "Investigation of the Asphalt Mixture Morphology Influence on its Ageing Susceptibility," *Materials and Structures*, 2015, Vol. 48(4), pp. 987-1000, DOI: 10.1617/s11527-013-0209-z.

Das, P. K., Tasdemir, Y. and Birgisson, B., "Low Temperature Cracking Performance of WMA with the Use of the Superpave Indirect Tensile Test," *Construction and Building Materials*, Vol 30, 2012, pp. 643–649.

Goodrich, J. L., "Asphalt and Polymer Modified Asphalt Properties Related to the Performance of Asphalt Concrete Mixes (with discussion)," *Association of Asphalt Paving Technologists*, Vol 57, 1988, pp. 116–175.

Hesami, E., Bidewell, N., Birgisson, B. and Kringos, N., "Evaluation of Environmental Susceptibility of Bituminous Mastic Viscosity as a Function of Mineral and Biomass Fillers", *Transportation Research Record: Journal of the Transportation Research Board,* Vol 2371, 2013, pp. 23–31.

Hesami, E., Birgisson, B. and Kringos, N., "A New Protocol for Measuring Bituminous Mastic Viscosity as a Function of its Filler Concentration," *Road Materials and Pavement Design*, Vol 15(2), 2014, pp. 420–433, DOI:10.1080/14680629.2013.876926.

Hesami, E., Jelagin, D., Kringos, N. and Birgisson, B., "An Empirical Framework for Determining Asphalt Mastic Viscosity as a Function of Mineral Filler

Concentration," *Journal of Construction and Building Materials,* Vol 35, 2012, pp. 23–29.

Jacobsson, T., *VVKB Flödesblandad asfalt av typ KGO-III,* Swedish Transport Administration, 2009, ISBN number: 1401–9612.

Kandhal, P. S. and Parker, F., *Aggregate Tests Related to Asphalt Concrete Performance in Pavements*, NCHRP Report No. 405, Transportation Research Board, 1998.

Petersen, J. C., "Chemical Composition of Asphalt as Related to Asphalt Durability: State of the Art", *Transportation Research Record,* Transportation Research Board, Vol 999, 1984, pp. 13–30.

Pietsch, W., *Agglomeration Processes*, Wiley-VCH, 2008, SN: 9783527620678.

Rhodes, M., *Introduction to Particle Technology*, Wiley, 2008, SN: 978-0-470-01427-1.

Roberts, F. L., Kandhal, P. S., Brown, E. R., Lee, D. Y. and Kennedy, T. W., *Hot Mix Asphalt Materials, Mixture Design and Construction*, NAPA Education Foundation, 2009.

Roque, R. and Buttlar, W. G., "The Development of a Measurement and Analysis System to Accurately Determine Asphalt Concrete Properties Using the Indirect Tensile Mode (with discussion)," *Journal of the Association of Asphalt Paving Technologists*, Vol 61, 1992, pp. 304–332.

Roscoe, R., "The Viscosity of Suspensions of Rigid Spheres," *British Journal of Applied Physics*, Vol 3, 1952, pp. 267–269.

Stachowiak, G. and Batchelor, A. W., *Engineering Tribology*, Butterworth-Heinemann, 2013, ISBN 978-0-12-397047-3.

Taylor, M. A. and Khosla, N. P., "Stripping of Asphalt Pavements: State of the Art (discussion, closure)," *Transportation Research Record,* Vol 911, Transportation Research Board, 1983, pp. 150–158.

Taylor, S. W., Milly, P. C. D. and Jaffe, P., R., "Biofilm Growth and the Related Changes in the Physical Properties of a Porous Medium, 2, Permeability," *Water Resources Research*, Vol 26(9), 1990, pp. 2161–2169.

Viman, L., *Uppföljning av vägar utförda med KGO-III-metoden*, The Swedish National Road and Transport Research Institute (VTI), 2011.

Viman, L., Wendel, M. and Said, S., "Long-Term Performance of Flow Mixing Technique," *Proceedings of the 4th EurAsphalt and EuroBitume Congress,* 2008.

Yideti, T. F., Birgisson, B., Jelagin, D. and Guarin, A., "Packing Theory-Based Framework to Evaluate Permanent Deformation of Unbound Granular Materials," *International Journal of Pavement Engineering*, Vol 14, 2013, pp. 309–320.

9.0 Discussion

JAN VOSKUILEN: Recently we carried out a similar research work. We didn't carry out the CT scans, but we carried out indirect tensile strength tests. We were curious if we could improve the mechanical properties with another sequence of the mixing. The conclusion of our research work was that we couldn't improve the mechanical properties, they were the same, but from our gyratory compaction research work, we learned another sequence could improve the workability and compactability. So this is similar with your first conclusion, so it's good to hear that you have the same conclusions about it.

BJORN BIRGISSON: Thank you very much, Jan. That's very nice to hear, actually, that you've been looking at the same thing and actually found similar things to what we found. Thank you for that.

ANDREW BRAHAM: Bjorn, thank you for the presentation. I really like how you're thinking outside the box on one of the steps of specimen preparation that we often take for granted, the mixing procedure. I think that's fascinating. Specifically, I had a question about the Stribeck curve. You said that there are three regions within the Stribeck curve—the boundary lubrication, the mix lubrication, and the hydrodynamic lubrication—and those regions were defined by the SN, which is a function of velocity, viscosity, and load.

BJORN BIRGISSON: Velocity squared divided by load.

ANDREW BRAHAM: Okay. And you mentioned during mixing you saw an increase in viscosity with a decrease of load, and based on that you determined that you were in the hydrodynamic region?

BJORN BIRGISSON: No, actually not, no. There is a separate study we did on the mastics and the fines, which is written up in two separate papers where we consistently found that for the mastics and the fines we were actually in that region very consistently. So, sorry.

ANDREW BRAHAM: And that's fine. I browsed through the paper while you were talking so I probably didn't get all of it, but are those two references inside your paper?

BJORN BIRGISSON: Yes, they certainly are.

ANDREW BRAHAM: Okay, great.

BJORN BIRGISSON: Thank you very much.

ERV DUKATZ: Bjorn, very excellent presentation. And I have a question. Are you suggesting we have a new method of mix design or is this a call out to the equipment manufacturers that we need to have new advanced batch plants to replace all of our drum plants?

BJORN BIRGISSON: Very, very good question. You could do both. Because the morphology framework actually indicates that one can actually design mixtures quite nicely based on that concept. And actually we are looking at putting together the Bailey Method with the morphology framework, actually with Gerry Huber and Bill Pine and their team, to see how that might actually work. So certainly as a mix, advance mix guidelines, I would say. The method is quite simple. This is actually on a spreadsheet readily programmable, very easily used, so no black boxes in this. Related to the plant, the only issue with the plants is that one has to introduce a metal sheet essentially to rain the fine particles into the mix. It's quite straightforward, a very minor adjustment. And given, since this was actually developed in Sweden, where batch plants are the vast majority of the plants, we've not looked at what it would take for other types of production systems.

ERV DUKATZ: Thank you. I think it's an excellent idea for mix design, but at a 500-ton-an-hour drum plant, I was having a hard time seeing how you would sequentially mix the aggregate up. Thank you.

BJORN BIRGISSON: I agree with you. But I think it is probably like everything else, we could probably come up with a way to do that, but right now I have a hard time seeing it as well.

GEOFFREY ROWE: Very nice presentation, very interesting. I particularly like you're looking at the shape of the aggregate and how you related that to the porosity. I think that's some very nice work. And I think what you're doing builds so much upon what we've seen before where we've looked at various sort of empirical aggregate factors or empirical aggregate shapes and have shown that you get different permeabilities. One of my favorite papers back in the annals of AAPT is the Goode and Lufsey paper of 1965 where they looked at the air permeability with different surface areas of aggregates and they showed the effect. And I'm certain had they had the tools and applied what you're doing now, they'd be able to explain their results very neatly with your single curve there. So I just compliment you on that. I think this is a nice piece of work, and I think it goes to addressing some of the comments that Dave Anderson made early on on the importance of understanding permeability in the asphalt mixes, so this is a very nice piece of work, and I congratulate you on it.

BJORN BIRGISSON: Thank you very much, Geoff. I really appreciate that. I have to say that the old volumes of AAPT proceedings have been a real inspiration to me throughout my career. I would encourage this organization not to drop them, okay? We actually started this study, if I may reflect a little bit, trying to measure or establish the viscosity of the mix. So we did some measurements on that. In fact, we had a big rotational viscometer that we put together. It is doable. For unmodified mixtures, we get something that looks very much like a Bingham fluid, Bingham behavior. But I decided that would probably be too complicated, actually, to go down that road, so we backed off, focused on the mastics and the fillers, developed a method for measuring those, which is in separate papers, which gets back to my comments on the Stribeck curve and where we're at on the flow regime and used the morphology framework. Because the more I see of that, the more comfortable I am with that whole approach. It seems to work rather well.

GERALD REINKE: Did you do any moisture sensitivity comparisons between the conventional mix and the sequential mix?

BJORN BIRGISSON: No, that was not really the point. All these mixtures, you know, you've got to understand this is Sweden, so there are liquid antistripping additives in all these. These are plant mixtures, real mixtures that were produced for the road that we studied and they all have antistripping agents in there. But what we're seeing in the field, based on the history in Sweden, is at least as good a stripping performance with these sequential mixtures as the conventional, if not better. And there are reports on that, actually, unfortunately in Swedish, written by VTI, that's the Transport Research Institute in Linköping in Sweden. They've done those. But that's a very, very good point.

GERALD REINKE: Thank you.

BJORN BIRGISSON: Thank you.

LOUAY MOHAMMAD: Very nice presentation. I would like to hear your comment. You showed in one of the slides that you improve air void distribution in a specimen, yet there were no differences in mechanical properties. However, as documented in the literature, one would think that air void distribution is a partial contributor to some of those variations.

BJORN BIRGISSON: I had the same thoughts, Louay. Good question. You have to remember that both mixtures in the end were compacted to 3 and 3.1%. They are similar, you know, in the bulk. So I don't know that one can start measuring differences in performance and relate them to air voids with that. But the other issue that I should point out, the sequential mixture had 0.3% lower bitumen content as compared to the conventional one. And that may also have led to why we didn't see any difference in performance. Now with the analysis framework and understanding we have, one can actually start playing with this and look at mixtures from many different angles and look at the sequencing. It's kind of amazing we haven't done this before. Actually, thinking about it, this is darn near the way we put together cakes, very much in the food industry. They play with this all the time, so it would seem that this would be a logical step for us.

MARTIN VAN DE VEN: Bjorn, did you also look at the different types of mixtures with regards to the mixing order? One of them is stone skeleton mixes, and I have a feeling that the mixtures you're now talking about are more or less what we call sand skeleton mixture (or continuously graded mixtures). And definitely in sand skeleton mixtures, it's not only the larger aggregate, but also the sand part is really taken care of the load going through that mixture. So for a stone skeleton type of mixture, I think this approach is very interesting, specifically porous asphalt mixtures, because there you would like to have good resistance against raveling at the surface. But wouldn't you, in those cases, like to have the sand/filler also early mixed with the bitumen, not only those large stones but also the sand? So I'm a little bit surprised that this would be better than the mix sequence that we use at the moment in asphalt plants. From a production point of view, do you have an idea how much extra time it would take to do the new sequence? Normally, it's less than one minute. How much time would this new sequence approach take in the asphalt plant?

BJORN BIRGISSON: A couple of questions there. One is what does this mean for production? A plant that's set up and does this routinely, it means no change in production times. Plants that are starting, we found this can mean up to 30% slower production. So, yes, there can be that effect, but once they are humming, there is virtually no difference. Now, related to the types of structures, this is a little different question again because I think you're getting into the morphology issue.

MARTIN VAN DE VEN: Absolutely.

BJORN BIRGISSON: So, yes. With the morphology framework, we have applied it to all kinds of mixtures, and there is no problem with that. We can very nicely look at SMAs and open-graded mixtures with it. In that case, you have one or two stone sizes that dominate, but it doesn't become less important because then the importance of that supporting structure, the secondary structure, is not lessened. And actually understanding what needs to be done to that to get stability is something we can now start looking at, I think, much more systematically than we've done before, Martin.

MARTIN VAN DE VEN: I was referring to the stone skeleton mixtures, I think this will be a very interesting approach because then you're assured immediately of a thick film around the skeleton system.

BJORN BIRGISSON: Okay, you're talking mixing. Yes, absolutely, yeah. And, as you know, for certain types of stone skeleton mixes, one actually needs to introduce fibers and other things. This is capable of taking that into account because technically we could actually characterize the mastic, you know, the filler and the mastic viscosity up front and actually design the stability criterion based on that.

Development of an Image-Based Multi-Scale Finite Element Approach to Predict Mechanical Response of Asphalt Mixtures

Amir Arshadi[a] and Hussain Bahia[a]

[a]Department of Civil and Environmental Engineering, University of Wisconsin-Madison, Madison, WI 53706, USA

ABSTRACT. Image-based simulation of complex materials is a very important tool for understanding their mechanical behavior and an effective tool for successful design of composite materials. Asphalt concrete, as one of these multi-phase complex materials, is a composite of asphalt binder, air voids, and mineral aggregate particles. Simulation of asphalt concrete with numerical methods is not a new topic but is faced with many challenges. In addition to requiring tremendous computational cost, it is not clear yet how to effectively model the aggregate-to-aggregate contact behavior during loading and deformation. In this paper an image-based, multi-scale modeling is introduced to reduce the computational cost significantly by reducing the amount of elements in the numerical model. In this approach the "up-scaling" and homogenization of each scale to the next is critically designed to improve accuracy. In addition to this multi-scale efficiency, this study introduces an approach for consideration of particle contacts at each of the scales in which mineral particles exist. The FE analysis is performed in the study at four scales, asphalt binder, mastic, mortar, and asphalt mixture scale.

The well-known finite element (FE) software ABAQUS is used to conduct FE simulations at all scales. The inputs are based on the experimentally derived measurements for the binder viscoelastic properties from which the binder constitutive model is implemented into the software via the user material subroutine (UMAT). For the scales of mastic and mortar, the artificially 2-dimensional (2D) images of mastic and mortar scales were generated and used to characterize the properties of those scales. Finally, the 2D scanned images of asphalt mixtures after elimination of fine aggregate particles is used to study the asphalt mixture behavior under uniaxial creep and recovery loading. Comparison between experimental results and the results from the model shows that the model developed in this study is capable of predicting the effect of asphalt binder properties and aggregate micro-structure on mechanical behavior of asphalt concrete under repeated creep loading.

KEYWORDS: multi-scale modeling, finite element, homogenization, particle-to-particle contact.

This paper is for publication only.

This is a reproduction of an article whose final and definitive form has been published in *Road Materials and Pavement Design*© 2015 Taylor & Francis. The article is available online at: http://dx.doi.org/10.1080/14680629.2015.1077007

1.0 Introduction

Visualization techniques have helped many industries better analyze and understand components they produce or utilize, and have facilitated significant advancements in many industries. In the field of asphalt paving there have been numerous attempts to utilize this technique to better understand asphalt concrete (AC) behavior, especially for modeling material mechanical viscoelastic response and damage behavior. Two-dimensional (2D) scanning, and to some extent three-dimensional (3D) X-ray tomography, have been utilized in asphalt pavement research, but challenges regarding accuracy and precision of resulting visualization, and the cost issues with regards to 3D methodologies such as x-ray tomography persist. This paper is an attempt to enhance the accuracy of image analysis and visualization of AC components and a step toward further developing 2D visualization as an effective characterization and research tool.

It is well known that asphalt concrete is a multi-phase complex composite of mineral aggregates, asphalt binder, and air voids. Aggregate particles used in pavements typically have irregular shapes, random orientations, and varying sizes, ranging from less than 1μm up to 50 mm. Numerical simulation of such a material in a single scale requires a massive number of fine elements and degrees of freedom, thus greatly increasing the computational cost. The search for an accurate analysis method which can account for the microstructure of the mixture at a reasonable computational cost has led researchers in many disciplines to the multi-scale modeling approach.

Multi-scaling has gained considerable popularity in recent years (Lutif, et al., 2009; Aigner, et al., 2009; Valenta, et al., 2010; Yin, et al., 2011). Lutif et al. (2009) used an efficient scale-linking technique to simulate the mechanical response of asphalt mixtures. The global scale was homogenized from local scale responses and the model was implemented in a finite element platform. It was shown that this approach can significantly reduce the computational effort compared to single scale modeling. In another study, Aigner et al. (2009) proposed a multi-scale model to explain the complex behavior of asphalt mixture. The elastic-viscoelastic "correspondence-principle" was employed to determine the viscoelastic properties of the mastic, mortar, and asphalt concrete from rheological properties of the asphalt binder. They showed that good agreement can be achieved between results from multi-scale models and experimental results. Valenta et al. (2010) conducted a two-step homogenization to determine the material properties of the asphalt mixture. Their study showed the limitations of the classical micro-mechanical models. Yin et al. (2011) proposed a multi-scale model by combining macro-scale and meso-scale, showing that multi-scale modeling can reduce the cost of the computation by reducing the amount of elements in the numerical model.

Many researchers applied image analysis and micro-mechanical FE methods to study the effect of micro-structure of asphalt concrete on the complex behavior of asphalt mixture (Sepehr et al., 1994; Kose et al., 2000; Sadd et al., 2004; Dai et al., 2006; Dai et al., 2007; Abu Al-Rub et al., 2011; Sefidmazgi et al., 2012). Dai et al. (2006) used a mixed FEM to study the behavior of asphalt mixtures by using rigid elements and continuum elements for aggregates and asphalt mastics, respectively. The properties of the asphalt mastic elements were specified through a user material (UMAT) subroutine within the ABAQUS code to incorporate the linear and damage-coupled viscoelastic constitutive behavior of continuum phase. Dai et al. (2007) presented a micromechanical FE model to predict the viscoelastic creep stiffness of asphalt mixtures. The creep stiffness of the sand mastic of the asphalt mixture was used as the input viscoelastic properties of the model. They showed that the model gives reasonable prediction of the mixture creep stiffness across the reduced loading time. Abu Al-Rub et al. (2011) indicated that the effects of aggregate shape, aggregate distribution, and aggregate volume fraction should be considered in modeling the behavior of asphalt mixtures. Although these studies show the importance of micro-structure of asphalt mixture, lack of an appropriate approach for consideration of the contact between the fine aggregate particles in FE framework makes them dependent on costly and time consuming tests at the mortar scale level (mixture of the fine aggregates with binder).

A number of research works have been done on micro-mechanical modeling of asphalt mixtures using Discrete Element Modeling (DEM) (Dai et al., 2007; Chang et al., 1999; Luo et al., 2010). In DEM, aggregates are represented as discrete elastic particles and their interactions are controlled by the response of an asphalt thin film. Some of the most important benefits of the discrete element modeling are the ability to handle complex and changing contact geometries and large displacements. The disadvantage of this method is the huge computational cost and power which is needed for this type of modeling.

Most current asphalt mixture multi-scale models face challenges with regards to accurate implementation of aggregate geometry, by either utilizing artificially generated geometric aggregate or, if scanned aggregate geometries are used, it is only at the coarsest scale of analysis due to limitations in imaging precision. Furthermore, as mentioned above there are still some critical problems regarding particle interactions in FE micro-mechanical modeling of asphalt mixtures. In the present study the developed multi-scale aggregate geometries are significantly advanced for the finer scales through use of real particle shapes and gradations. Besides, a simple aggregates contact rule is defined and implemented through UMAT in the FE software to simulate the particle-to-particle contact behavior at all scales.

2.0 Multi-Scale Analysis Approach

Early in this study it was recognized that even the highest resolution 2D scanning and filtering techniques can only accurately detect a certain range of aggregate sizes. This is mainly due to obscuring of fine particles during the cutting process through smearing of the binder and loss of a portion of very fine imaged particles after application of filters during image analysis. Thus it was understood that such particles should be included in analysis of finer scales if a realistic simulation of mixture mechanical behavior is desired. Asphalt mixtures in this study were therefore considered at four interconnected scales: the binder scale, for which mechanical response is directly measured through use of a Dynamic Shear Rheometer: the "mastic-scale," which contains fillers ranging from 1μm up to 75μm in a binder matrix; the "mortar-scale" containing fine aggregates ranging from 75μm to 1.18mm in a mastic matrix, and the "asphalt mixture-scale," which contains all larger aggregates (those detected using simple 2D scanners and binary imaging filters) in a continuous mortar matrix.

In order to specify the asphalt mixture properties at each scale, the FE analysis software ABAQUS v6.12 (2012) was used to conduct the simulations. For the mastic and mortar scales the artificially generated images were used to model the scales. For the asphalt mixture scale a flatbed scanner was used to capture color images and the image processing software IPAS2 was used to convert the color images to binary images following a filtering procedure described by Coenen et al. (2012). In the multi-scale approach, each scale is modeled heterogeneously to determine the bulk material response properties at the analyzed scale. The properties calculated at each scale are then inputted into a homogeneous matrix at the next-higher level (Figure 1). The multi-scale procedure developed in this study as well as a summarization of the model calibration protocol and required input properties is shown in a flowchart in Figure 2. In the following sub-sections the homogenization and up-scaling process of the developed multi-scale model is described in detail.

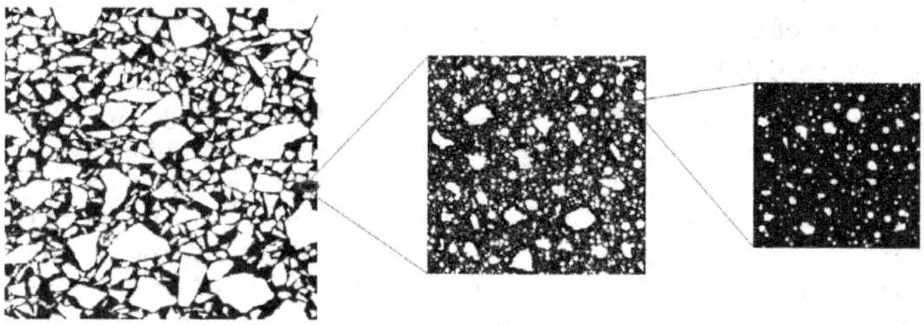

(a) (b) (c)

Figure 1. *Multi-scale analysis scheme: (a) Asphalt mixture-scale (large aggregates in mortar matrix) (b) Mortar-scale (fine aggregates in mastic matrix) (c) Mastic-scale (fillers in binder matrix).*

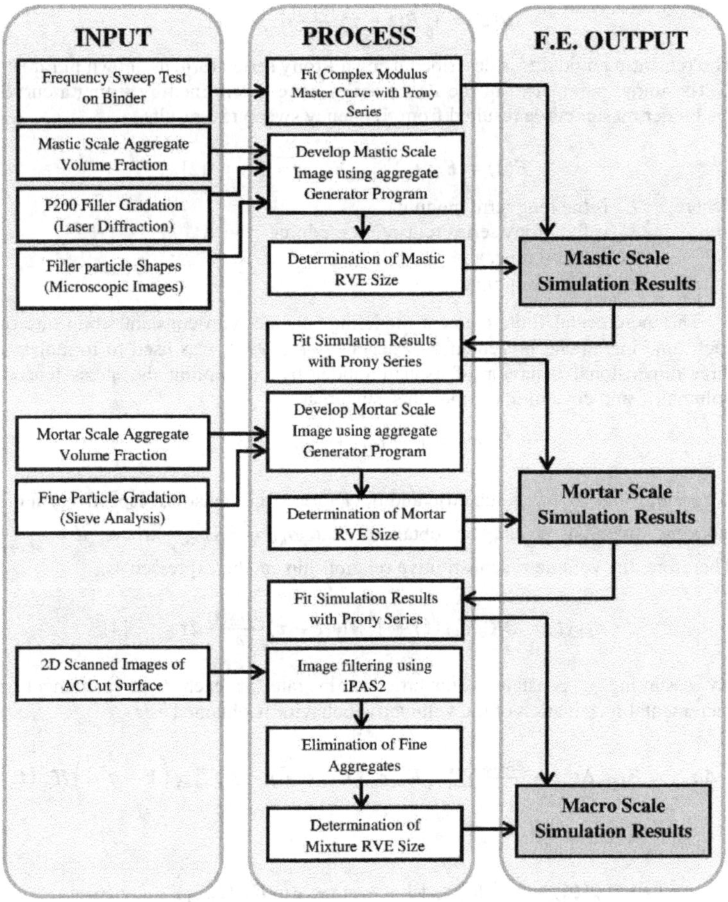

Figure 2. *Flowchart of calibration and utilization of the multi-scale model developed in this study.*

2.1. Numerical Implementation of Binder Constitutive Model

The one-dimensional linear constitutive behavior of the asphalt binder can be expressed by Boltzmann superposition integral as shown in Equation 1.

$$\sigma(t) = \int_0^t E(t - \tau) \frac{d\epsilon(\tau)}{d\tau} d\tau, \qquad [1]$$

The relaxation modulus is described using a Prony series form as shown in Equation 2. The initial parameters of the Prony series were determined by fitting a curve to the binder master curve resulted from frequency sweep test results.

$$E(t) = E_\infty + \sum_{m=1}^{M} E_m . e^{\frac{-t}{\tau_m}}. \qquad [2]$$

Where: E_∞ is the long term modulus
E_m is the Prony series relaxation modulus
t is loading time, and
τ_m is relaxation time

The incremental finite-element modeling scheme with constant strain rate over each time increment, presented by Zocher et al. (1997), was used to formulate the three-dimensional behavior of asphalt binder by uncoupling the stress tensor to volumetric and deviatoric components, such that:

$$\sigma_{ij} = (\frac{\sigma_{kk}}{3} \delta_{ij}) + \hat{\sigma}_{ij}, \qquad [3]$$

Where σ_{kk} and $\hat{\sigma}_{ij}$ are volumetric and deviatoric stress tensors, respectively. For an isotropic material it can be obtained that $\sigma_{kk} = 3K\epsilon_{kk}$, where $K = \frac{E}{3(1-2\upsilon)}$. Therefore, the volumetric constitutive relationship can be expressed as:

$$\sigma_{kk}(t) = 3K_\infty \epsilon_{kk}(t) + \int_0^t 3K(t - \tau) \frac{d\epsilon_{kk}(\tau)}{d\tau} d\tau. \qquad [4]$$

By assuming a constant volumetric strain rate at each time increment, the incremental formulation of the volumetric behavior is obtained as:

$$\Delta\sigma_{kk} = 3K_\infty \Delta\epsilon_{kk} + \frac{3\Delta\epsilon_{kk}}{\Delta t} \sum_{m=1}^{M} K_m \tau_m (1 - e^{\frac{-\Delta t}{\tau_m}}) - 3\sum_{m=1}^{M} \left(1 - e^{\frac{-\Delta t}{\tau_m}}\right) H_m(t_n), \qquad [5]$$

and

$$H_m(t_n) = \frac{\Delta\epsilon_{kk}}{\Delta t} K_m \tau_m (1 - e^{\frac{-\Delta t}{\tau_m}}) + e^{\frac{-\Delta t}{\tau_m}} H_m(t_{n-1}). \qquad [6]$$

For the deviatoric stress-strain relationship it can be obtained that $\hat{\sigma}_{ij} = \sigma_{ij} - \frac{\sigma_{kk}}{3}\delta_{ij} = 2G(\varepsilon_{ij} - \frac{\varepsilon_{kk}}{3}\delta_{ij}) = 2G\hat{\varepsilon}_{ij}$, where $G = \frac{E}{2(1+\upsilon)}$. The deviatoric constitutive behavior can be expressed as:

$$\hat{\sigma}_{ij}(t) = 2G_\infty \hat{\epsilon}_{ij}(t) + \int_0^t 2G(t - \tau) \frac{d\hat{\epsilon}_{ij}(\tau)}{d\tau} d\tau. \qquad [7]$$

The incremental formulation can be expressed as:

$$\Delta\hat{\sigma}_{ij} = 2G_{\infty}\Delta\hat{\epsilon}_{ij} + \frac{2\Delta\hat{\epsilon}_{ij}}{\Delta t}\sum_{m=1}^{M} G_m\tau_m\left(1 - e^{\frac{-\Delta t}{\tau_m}}\right) - 2\sum_{m=1}^{M}\left(1 - e^{\frac{-\Delta t}{\tau_m}}\right)R_m(t_n), \quad [8]$$

and

$$R_m(t_n) = \frac{\Delta\hat{\epsilon}_{ij}}{\Delta t}G_m\tau_m\left(1 - e^{\frac{-\Delta t}{\tau_m}}\right) + e^{\frac{-\Delta t}{\tau_m}}R_m(t_{n-1}). \quad [9]$$

The procedure described above is implemented in the well-known FE analysis software ABAQUS through a user-defined subroutine, UMAT, and is used to conduct multi-scale FE simulations.

2.2 Mastic Scale

Contrary to common simplifications in many models in which the filler particles are considered to be of a uniform size smaller than the 0.075 mm sieve size, particle sizes can vary significantly at this scale (as shown in Figure 3), thus affecting the accuracy of the results. The "Laser Diffraction" technique was used to experimentally determine the gradation of the filler particles in this study. In this technique a group of dispersed particles is irradiated by a laser beam, and the particle size distribution is calculated by measuring the intensity of light scattered from that group of particles. This method is widely used in many industries as a particle sizing technique for materials ranging in size from hundreds of nanometers up to several millimeters (Cyr and Tagnit-Hamou, 2001). In this study granite stone filler is used and its gradation is shown in Figure 3(a). In addition to the gradation, the realistic shape and geometry of the filler particles is captured through microscopic imaging and a database of 230 different filler particle shapes and geometries was developed. Figure 3(b) shows an example of the microscopic images used for development of the database.

A MATLAB (2011) based program was written to randomly generate the mastic binary images. The mastic generation code, based on the gradation and filler volume fraction of the mastic-scale, randomly chooses some of the filler shapes in the database and makes the mastic binary images. Based on the simulation results for an asphalt mastic with filler volume fraction of 0.4, the optimum representative volume element (RVE) size was determined to be a 225 μm × 225 μm area, which was used for all subsequent analysis at the mastic scale. As in the binder phase, a Prony series model with four generalized Maxwell models fitted to the complex modulus curve was used to provide the homogenized input parameters for the next higher scale.

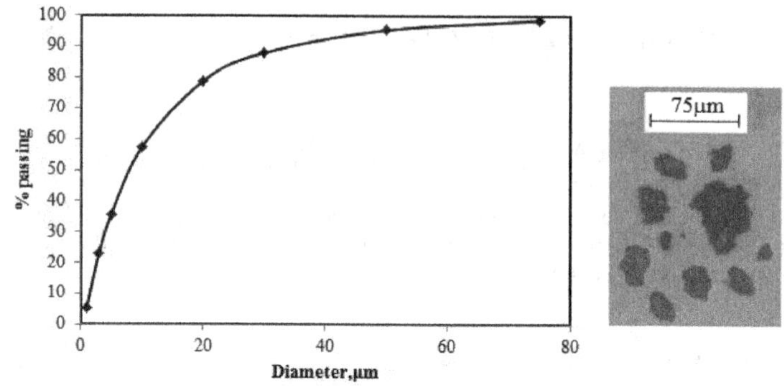

(a) (b)

Figure 3. *(a) Gradation of the granite stone filler measured by laser diffraction methodology (b) Scanned microscopic image of granite stone filler.*

2.3 Mortar Scale

The "mortar scale," the next higher level after the "mastic scale," consists of fine aggregates larger than 0.075 mm (#200 sieve opening) and smaller than 1.18 mm (No. 16 sieve opening). The gradation of the fine aggregates in the mortar scale is determined from the total aggregate gradation used for the mixture. The fine aggregate generation program developed as part of this study was used at the mortar scale to generate the mortar images based on the fine aggregate gradation and shape.

For determination of the input properties for the mortar scale of the multi-scale model, the homogenized viscoelastic properties of the mastic scale was assessed by fitting the Prony series to the simulated complex modulus versus frequency curve of the mastic. These properties were used as material input values for the continuous matrix of mortar scale of the multi-scale simulations. The homogenization process assumes mastic isotropy and homogeneity, when inputted as the continuous phase in the mortar scale.

The fine aggregate volume fraction at the mortar scale was determined based on the mixture gradation and design volumetric properties. It was calculated that for regular asphalt mixture designs the volume fraction of the fine aggregates in mortar scale varies from 0.40 to 0.55 depending on the coarseness of overall gradation. In comparison, the volume fraction of fillers in the mastic scale ranged from 0.10 to 0.40. Coarser overall mixture aggregate gradations result in lower volume fraction of fines in the mortar and mastic scales. Based on convergence of the relaxation modulus curves, an optimum RVE of 3.5 mm × 3.5 mm, approximately three times larger than the maximum size of fine aggregates in mortar scale, was determined. Homogenized mortar scale mechanical properties were identified by fitting the Prony series parameters to the complex modulus versus frequency curve of the mortar RVE. Results are inputted as material properties in the continuous matrix of the asphalt mixture scale, assuming an isotropic and homogeneous mortar matrix at this scale.

2.4 Asphalt Mixture Scale

The asphalt mixture scale is the highest level scale in the multi-scale analysis scheme used in the present study. A flatbed scanner was used to capture color images at a resolution of 20 μm/pixel of HMA cross sections. 2D digital image processing techniques were used to convert the color images to binary images. The continuous phase of the asphalt mixture scale images consists of the homogenized isotropic mortar scale. The imaging techniques used in this study cannot accurately capture aggregates smaller than 1.18 mm, mostly due to a portion of imaged particles smaller than 1.18 mm being lost during the application of filters when converting to binary. Therefore, a MATLAB code was used to eliminate the aggregates smaller than 1.18 mm from the images (Figure 4), and instead including this size range in the mortar and mastic scales.

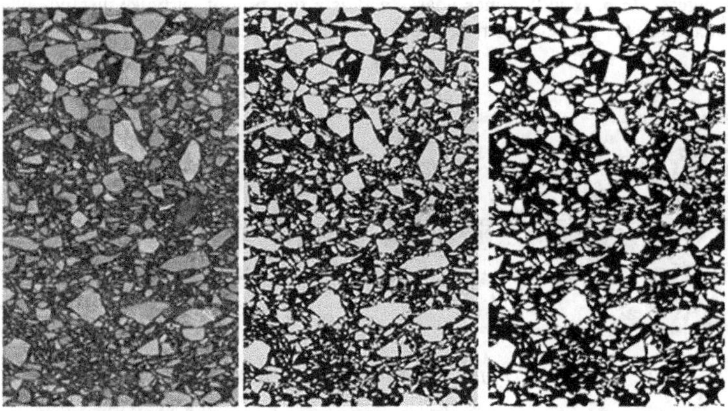

(a) (b) (c)

Figure 4. *Images of the mixture. (a) Scanned image of HMA section (b) Scanned image after image filtering processes (c) Scanned image after elimination of aggregates smaller than 1.18 mm.*

Due to the lack of color contrast between air voids and the background asphalt, coupled with the unavoidable filling of most air voids on the cut surface due to binder smearing during the cutting process, air voids cannot be reliably captured using current 2D imaging techniques. In the present study it was decided to overcome this deficiency by introducing artificially generated circular air voids in the macro scale image. Masad et al. (2002) studied the air void distribution in asphalt mixtures, concluding that the air void distribution of gyratory compacted samples followed a bathtub shape distribution, such that larger air voids were located near the top and bottom of the specimens, with the air voids in central portion having a relatively consistent smaller size. The same distribution was used in this study to distribute the air voids at the asphalt mixture scale (Masad et al., 2002). The air voids are considered as empty spaces in the FE simulation program.

2.5 Micro-Mechanical Modeling of Aggregate-Aggregate Contact

As mentioned earlier, one of the most important shortcomings of the micro-mechanical models presented by the researchers is the lack of an efficient contact law for the inter-particle behavior of two particles connected by a binder layer at lower scale levels. In order to simulate the binder behavior at the proximity zones, a simple contact law is defined based on the binder layer film thickness (Figure 5). In this approach the binder film elements are stiffened as stiff as an aggregate element when the binder film thickness tends to zero. In other words, the stiffening happens when the binder element compressive strain tends to unity. Figure 5 depicts the contact law used in the simulation. As shown, the stiffness of the binder element at the aggregate proximity zones is assumed as a function of particles distance which increases as the particles come into eventual contact.

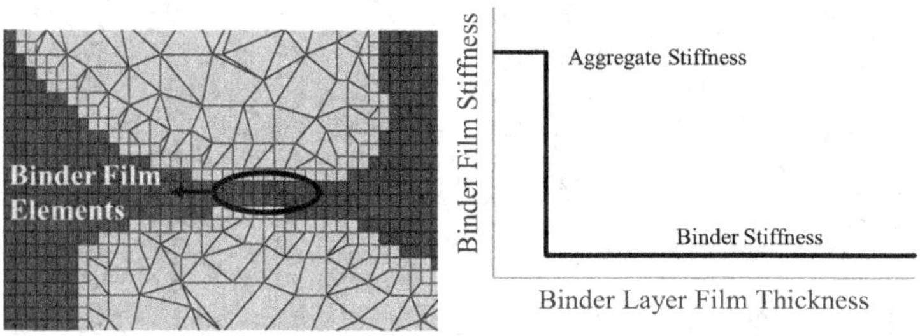

Figure 5. *Depiction of the contact law applied to the binder film elements.*

The above mentioned contact law was defined in the ABAQUS user material subroutine and used in all simulations. This novel approach of using proximity to represent contact is not a perfect assumption as there could still be relative slip (movement) between aggregates. Figure 6 shows that implementation of the particle-to-particle contact law into FE model can significantly affect the response of a mastic with 40% filler volume fraction under 10 kPa uniaxial creep loading. The results shown demonstrate that after a certain loading time, the results from the simulation considering contacts between filler particles begin to deviate significantly from the results of the simulation with no contact consideration. This behavior is due to the evolution of the particle contact structure as a result of mastic deformation. In other words as binder in the mastic deforms, filler particles get in contact (or proximity), which artificially stiffens the binder in proximity zones and reduces mastic scale strain.

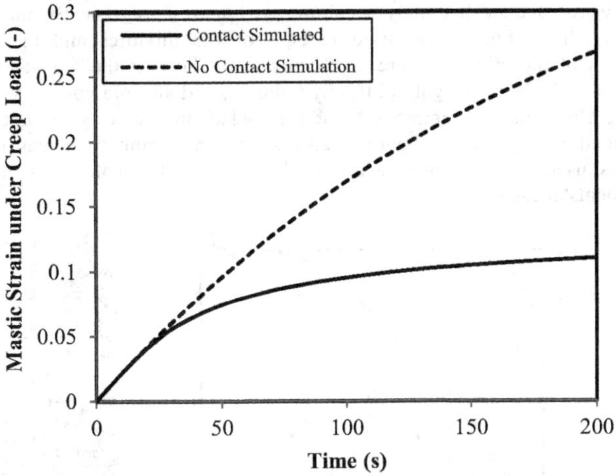

Figure 6. *Creep deformation of a 40% filler volume fraction mastic.*

To address the effect of mastic scale particle contacts in the input properties of the mortar scale continuous phase, the nonlinear superposition, which allows the dependency of relaxation modulus to strain level, is used:

$$\sigma(t) = \int_0^t E(t-\tau, \epsilon(\tau)) \frac{d\epsilon(\tau)}{d\tau} d\tau. \qquad [10]$$

A separable Kernel was considered for the relaxation modulus, such as $E(t,\epsilon)=E(t).\Omega(\epsilon)$, in which the relaxation modulus is a product of a time-dependent part and a strain-dependent part. This method is called quasi-linear viscoelasticity (QLV), which is originally proposed by Fung (1972). In order to determine the relaxation modulus components, two FE test simulations are performed on the desired scale: (a) determination of the scale complex modulus curve at a low strain amplitude level to make sure there is no effect of particle contact on the material response, and fitting a Prony series to the curve to quantify the time-dependent part of the mastic relaxation modulus; (b) creep loading test to determine the strain-dependent part of the relaxation modulus using Equation 10 while the time-dependent part is known from previous simulation.

The same FE test procedure is performed on the mortar scale to capture the particle contact effects on response of this scale. The obtained mortar scale relaxation modulus was used to provide the homogenized input parameters for the asphalt mixture scale.

3.0 Numerical Simulation and Model Validation

Two asphalt mixtures with significantly different gradations, categorized as fine and coarse were used in this study, as shown in Figure 7. Styrene Butadiene-Styrene (SBS) modified binder was used in both of the mixtures and the volumetric properties of the mixtures were kept constant for both of the gradations. Asphalt content was 5.1% by weight (12.0% by volume), and air void content was 6.5% by volume. Consequently almost 83% of the asphalt mixtures by volume would be constituted of aggregate. Figure 7 also shows the scanned images of the two mixtures used in this study, while Table 1 shows the volume fraction of the components at each scale.

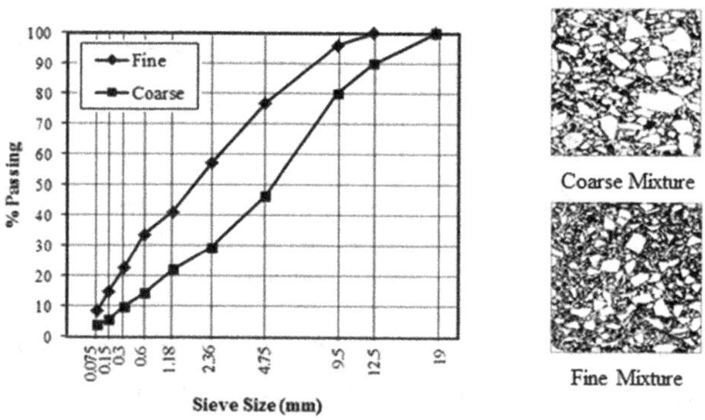

Figure 7. *Aggregate gradation and 2D scanned images of the mixtures.*

Table 1. *Volume fraction of components at each scale.*

Scale / Mixture	Asphalt Mixture Scale			Mortar Scale		Mastic Scale	
	Aggregates	Mortar	Air Void	Fine Aggregates	Mastic	Fillers	Binder
Coarse	0.645	0.290	0.065	0.49	0.51	0.20	0.80
Fine	0.484	0.451	0.065	0.59	0.41	0.40	0.60

In order to determine the prediction power of the multi-scale model, the total multi-scale response of the model from the binder through the macro scale was used to predict and simulate the mechanical response of the fine and coarse asphalt mixture samples, 100 mm in diameter and 150 mm high, placed under 345 kPa step stress and recovery periods. Each cycle consisted of 0.1 s of loading and 0.9 s recovery period. Results were compared to experimentally derived mixture testing results using a uniaxial testing machine at 46°C. The developed multi-scale approach is conducted at each scale and the results are presented in the following subsections.

3.1 Binder Experimentation and Mastic and Mortar Homogenization

The frequency sweep test was conducted to characterize the linear viscoelastic properties of the SBS modified binder. The shifted complex modulus and fitted master curve at the reference temperature 46°C are shown in Figure 8. Table 2 summarizes the model parameters of the binder obtained from frequency sweep test results.

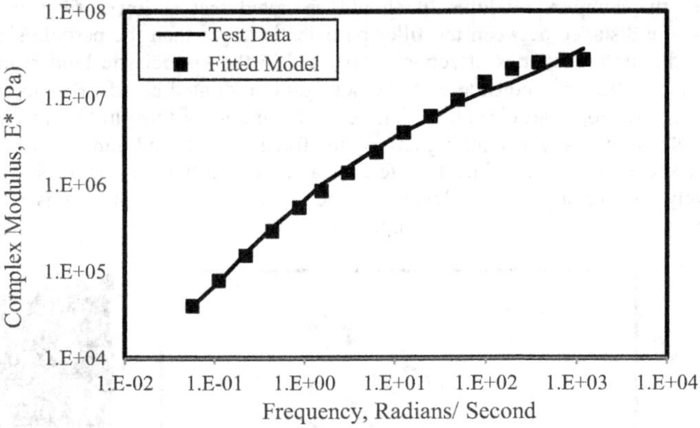

Figure 8. *Shifted complex modulus and fitted master curve for SBS modified binder.*

Table 2. *Linear viscoelastic model parameters.*

Prony Series Parameters	
E_∞= 600.0 Pa	
E_1= 6963.4 Pa	τ_1= 1s
E_2= 505505.7 Pa	τ_2= 0.1 s
E_3= 3273660.7 Pa	τ_3= 0.01 s
E_4= 23478594.0 Pa	τ_4= 0.001 s

To validate and confirm the accuracy of the FE model's mechanical response prediction, the experimentally derived complex modulus master curve of the SBS modified asphalt mastics with volume fractions of 0.2 and 0.4 were compared to those obtained using the FE modeling. A constant Poisson's ratio of 0.49 was assumed for the binder material while linear elastic properties were used to model the aggregate phase, setting the Young's modulus and Poisson's ratio to 25 GPa and 0.25, respectively (Dessouky et al., 2006). In the FE simulation, uniaxial sinusoidal strain-controlled loads with different loading frequencies, ranging from 0.01 to 100 Hz, were applied to the asphalt mastic samples and their complex moduli were calculated using the simulation results. The complex modulus diagrams of the mastics as well as the mastic models are shown in Figure 9.

The results indicate that there is a good agreement between experimental data and FE modeling results. The analysis shows that the mastic complex modulus does not increase proportionally to the increment of the filler content. It shows that adding 20% filler to the asphalt binder rather increases the complex modulus, however; adding an additional 20% filler to the asphalt mastic significantly increases the complex modulus. It should be noted that at lower filler volume fractions the distance between the filler particles is larger than the particle size; in such conditions the filler is effectively "diluted" in the viscoelastic binder matrix. Particle interactions are considered to be negligible in diluted blends. By increasing the volume fraction, particle interaction networks gradually form in the mastic. As the particle networks expand at higher volume fractions, the load transfer efficiency of the network increases, thus transferring a larger portion of the mastic load, effectively decreasing strain level in the binder matrix and consequently significantly increasing the mastic complex modulus.

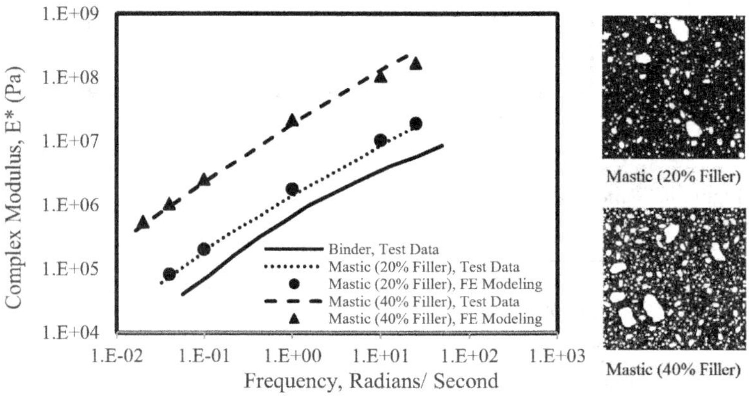

Figure 9. *Complex modulus diagrams of SBS modified binder and mastics at 46°C.*

A Prony series model fitted to the complex modulus curve was used to provide the homogenized input parameters for the mortar scale. The Poisson's ratio of the mastic phase was calculated through the model response. It was found by taking the negative transverse strain and dividing it by the axial strain. Similar to the mastic scale, a Prony series fitted to the mortar simulated complex modulus curve of the mixtures was used to provide the homogenized input parameters of the continuous phase of the asphalt mixture scale which is the last scale in the multi-scale approach.

Table 3 summarizes the homogenized Prony series parameters as well as the Poisson's ratio of the mastic and mortar scale of both of the fine and coarse graded mixtures.

Table 3. *Homogenized model parameters of lower scales of fine and coarse graded mixtures.*

Component	*SBS Modified Binder*	*Mastic of Coarse Mixture*	*Mastic of Fine Mixture*	*Mortar of Coarse Mixture*	*Mortar of Fine Mixture*
E_∞ *(MPa)*	0.0006	0.0017	0.0037	0.0077	0.0097
E_1 *(MPa)*	0.007	0.009	0.025	1.011	1.514
τ_1 *(s)*	1	1	1	1	1
E_2 *(MPa)*	0.51	1.43	23.93	29.14	167.84
τ_2 *(s)*	0.1	0.1	0.1	0.1	0.1
E_3 *(MPa)*	3.3	9.8	125.6	130.2	1080.2
τ_3 *(s)*	0.01	0.01	0.01	0.01	0.01
E_4 *(MPa)*	23.5	76.2	408.0	539.2	1572.4
τ_4 *(s)*	0.001	0.001	0.001	0.001	0.001
Poisson's Ratio	0.49	0.47	0.43	0.40	0.34

3.2 Mixture Scale Modeling

Aggregate volume fraction in asphalt mixture scale for coarse and fine graded mixtures was determined to be respectively 0.645 and 0.484 (Table 1). The RVE determination for the macro scale was carried out similar to that of the mastic and mortar scales, resulting in the selection of an optimum RVE of 60 mm × 60 mm for this scale.

Figure 10 and Figure 11 show the multi-scale simulation results of asphalt mixture permanent strain under the creep and recovery loading compared with test data for the fine and coarse graded mixtures. The comparisons indicate that the developed multi-scale model without consideration of particle contacts is only capable of predicting mixture mechanical properties with an acceptable accuracy at initial cycles of loading. The figures demonstrate that after the initial loading cycles, results begin to deviate from experimental measurements. This is believed to be due to evolving of the aggregate structure as a result of mixture consolidation and the increasing mechanical engagement of aggregate particles, thus increasing the importance of fully accounting for aggregate to aggregate contact mechanism. Figure 10 and Figure 11 also show that implementation of a contact law into the FE framework significantly improves the power prediction of the multi-scale model. As shown, the multi-scale model developed in this study predicts the deformation of asphalt mixture with an acceptable accuracy at primary and secondary zones of rutting. The analysis of the results showed that the FE model predicted permanent strain after 70 cycles of loading is 30% and 16% higher than the experimentally obtained one for fine and coarse graded mixtures, respectively. The results indicate that the over-prediction is greater in the simulation of the fine graded mixture with higher number of particle contacts at lower scales, compared to that of coarse mixture, which had a lower particle volume fraction at mastic and mortar scales.

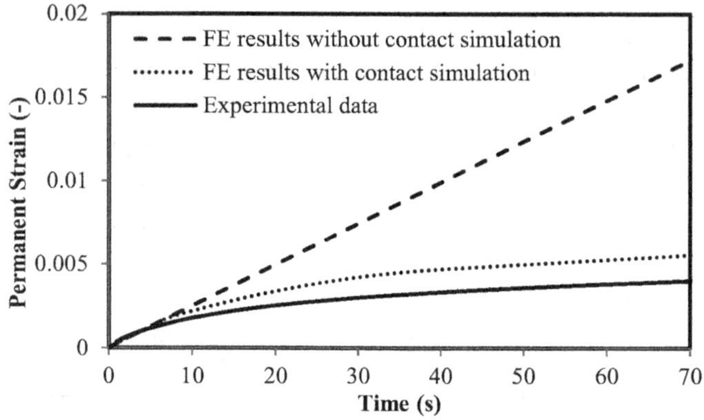

Figure 10. *Fine mixture permanent strain under continued cyclic loading.*

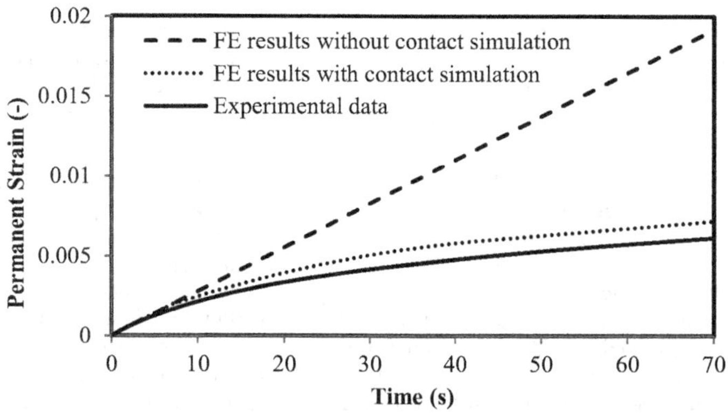

Figure 11. *Coarse mixture permanent strain under continued cyclic loading.*

Modeling aggregate contacts has proven to be a challenging process in a multi-phase finite element matrix, as aggregate pixels are always separated by binder pixels thus complicating modeling of contacts. Future work by the authors will focus on improving the multi-scale model to enable it to predict the asphalt mixture rutting behavior at tertiary zone when the aggregate structure gets unstable due to movement and rotation of the particles at the contact points.

4.0 Practical Applications of the Developed Model

This study developed a multi-scale model of asphalt mixture containing the three main phases of asphalt pavements, binder, aggregate, and air voids. To show the utility of this simulation technique, the two mixtures that were tested and discussed in the previous section were used to study effect of binder modification on performance of the asphalt mixtures. Two types of binder were used in this study: an unmodified PG 64-22 and an SBS modified PG 76-22 binder. Figure 12 shows the simulated total permanent strain of the asphalt mixtures after 70 cycles of creep and recovery loading. As shown, a two-grade bump in the grade of the asphalt binder reduces the permanent strain of both of the mixtures by approximately 20%. The results are in agreement with experiment based studies on SBS binder modification effect on asphalt mixture performance (Ping et al., 2008; Bahia, 2012).

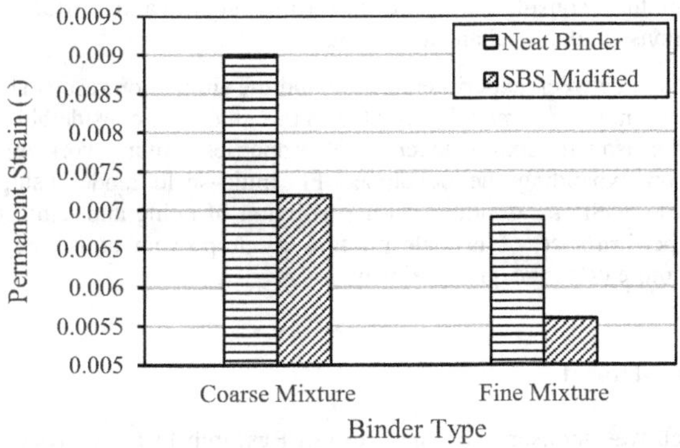

Figure 12. *Binder modification effect on permanent strain of fine and coarse graded mixtures as determined using the FE simulation model.*

Asphalt mixture designers need to make decisions on selecting binder, filler or even aggregate gradations. The abovementioned example shows that instead of the time consuming experimental trial and error process that designers have to use today, this virtual simulation tool can cut down on experimental testing and on time required to make decisions. The authors will focus on extending this tool for studying aging and damage effects and thus predict long term behavior of asphalt mixture in a more effective, practical way.

5.0 Conclusion

In this study a novel multi-scale model of asphalt mixture containing the three main phases of binder, aggregate, and air voids was developed to predict mechanical behavior of asphalt mixture under loading. The up-scaling approach used includes virtual generation of asphalt mastic and mortars using a library of images of actual fillers and fine aggregate particles mixed with the asphalt binder. It also included a virtual generation of air voids in the mortar phase during the analysis of the mixture scale. The study focused on relating the macro-scale behavior of asphalt mixture to rheological properties of asphalt binder, particle gradation, particle shape, and particle volume fraction at the mastic and mortar scale.

The model developed and presented in this study is unique in terms of overcoming the challenge faced in current asphalt mixture multi-scale models with regards to accurate implementation of fine aggregate geometry and shapes. Also the challenge of simulating particle contacts at mastic scales is addressed by implementation of a simple contact law into ABAQUS through the user material subroutine and transferred to the next higher scale by application of nonlinear superposition. The results indicate that evolution of the aggregate structure, as a result of mixture consolidation and distortion, significantly affects the asphalt mixture response under mechanical loading.

Comparison between experimental data and the simulation results shows that the finite element multi-scale model presented in this study has considerable promise for predicting the asphalt mixture mechanical properties. Future work by the authors will focus on expanding the developed FE multi-scale model using 3D X-ray tomography as well as extending it for prediction of aging and damage effects on long term performance of asphalt mixture by improvement of the constitutive models in a more effective, practical way.

6.0 Acknowledgment

This research was sponsored by the Western Research Institute "Asphalt Research Consortium". This support is gratefully acknowledged. Authors would also like to acknowledge the insightful input and contributions of Dr. Nima Roohi, Dr. Hassan Tabatabaee, and Dr. Raul Velasquez. The results and opinions presented are those of the authors and do not necessarily reflect those of the sponsoring agencies.

7.0 References

ABAQUS: Version 6.12. (2012). Providence, RI.

Abu Al-Rub, R.K., T. You, E. Masad, and D.N. Little. "Mesomechanical Modeling of the Thermo-Viscoelastic, Thermo-Viscoplastic, and Thermo-Viscodamage Response of Asphalt Concrete." *International Journal of Engineering Science* 28 (2011).

Aigner, E., R. Lackner, and C. Pichler. "Multiscale Prediction of Viscoelastic Properties of Asphalt Concrete." *Journal of Materials in Civil Engineering* 21 (2009).

Bahia, H., A. Golalipour and A. Coenen. "Comparing Effects of Crumb Rubber and Synthetic Polymers on Hot Mix Asphalt Performance." Asphalt Rubber Conference, Munich, Germany (2012).

Chang, G. K., and J. N. Meegoda. "Micromechanical Model for Temperature Effects of Hot-Mix Asphalt Concrete." *Transportation Research Record: Journal of the Transportation Research Board* 1687.1 (1999): pp. 95–103.

Coenen, A. R., M. E. Kutay, N. R. Sefidmazgi, and H. U. Bahia. "Aggregate Structure Characterisation of Asphalt Mixtures using Two-Dimensional Image Analysis." *Road Materials and Pavement Design* 13, no. 3 (2012): pp. 433–454.

Cyr, M., and A. Tagnit-Hamou. "Particle Size Distribution of Fine Powders by LASER Diffraction Spectrometry. Case of Cementitious Materials." *Materials and Structures* 34, no. 6 (2001): pp. 342–350.

Dai, Q., and Z. You. "Prediction of Creep Stiffness of Asphalt Mixture with Micromechanical Finite-Element and Discrete-Element Models." *Journal of Engineering Mechanics* 133, no. 2 (2007): pp. 163–173.

Dai, Q., M. H. Sadd, and Z. You. "A Micromechanical Finite Element Model for Linear and Damage-Coupled Viscoelastic Behaviour of Asphalt Mixture." *International Journal for Numerical and Analytical Methods in Geomechanics* 30, no. 11 (2006): pp. 1135–1158.

Dessouky, S., E. Masad, D. Little, and H. Zbib. "Finite-Element Analysis of Hot Mix Asphalt Microstructure using Effective Local Material Properties and Strain Gradient Elasticity." *Journal of Engineering Mechanics* 132, no. 2 (2006): pp. 158–171.

Fung, Y. C. "Stress-Strain-History Relations of Soft Tissues in Simple Elongation." *Biomechanics: Its Foundations and Objectives* 7 (1972): pp. 181–208.

Kose, S., M. Guler, H. U. Bahia, and E. Masad. "Distribution of Strains within Hot-Mix Asphalt Binders: Applying Imaging and Finite-Element Techniques." *Transportation Research Record: Journal of the Transportation Research Board* 1728, no. 1 (2000): pp. 21–27.

Luo, H., H. P. Zhu, Y. Miao, and C. Y. Chen. "Simulation of Top-Down Crack Propagation in Asphalt Pavements." *Journal of Zhejiang University Science A* 11, no. 3 (2010): pp. 223-230.

Lutif, J. E., F. V. Souza, Y. R. Kim, J. B. Soares, and D. H. Allen. "Multiscale Modeling to Predict Mechanical Behavior of Asphalt Mixtures." *Transportation Research Record: Journal of the Transportation Research Board* 2181, no. 1 (2010): pp. 28–35.

Masad, E., V. K. Jandhyala, N. Dasgupta, N. Somadevan, and N. Shashidhar. "Characterization of Air Void Distribution in Asphalt Mixes using X-Ray Computed Tomography." *Journal of Materials in Civil Engineering* 14, no. 2 (2002): pp. 122–129.

MATLAB: Version 7.12. (2011). The MathWorks Inc.

Ping, W. V., and Y. Xiao. "Evaluation of SBS Polymer Binder Effect on Resilient Modulus Properties of Florida HMA Mixtures." *24th OCTPA Annual Conference and NACGEAI International Symposium on Geo-Trans*, pp. 1–8. 2001.

Sefidmazgi, N. R., L. Tashman, and H. U. Bahia. "Internal Structure Characterization of Asphalt Mixtures for Rutting Performance using Imaging Analysis." *Road Materials and Pavement Design* 13, no. sup1 (2012): pp. 21–37.

Sadd, M. H., Q. Dai, V. Parameswaran, and A. Shukla. "Simulation of Asphalt Materials using Finite Element Micromechanical Model with Damage Mechanics." *Transportation Research Record: Journal of the Transportation Research Board* 1832, no. 1 (2003): pp. 86–95.

Sepehr, K., O. J. Harvey, Z. Q. Yue, and H. M. El Husswin. "Finite Element Modeling of Asphalt Concrete Microstructure." *Proceedings, 3rd International Conference on Computer-Aided Assessment and Control Localized Damage*. 1994.

Valenta, R., M. Sejnoha, and J. Zeman. "Macroscopic Constitutive Law for Mastic Asphalt Mixtures from Multiscale Modeling." *International Journal for Multiscale Computational Engineering* 8, no. 1 (2010).

Yin, A., X. Yang, S. Yang, and W. Jiang. "Multiscale Fracture Simulation of Three-Point Bending Asphalt Mixture Beam Considering Material Heterogeneity." *Engineering Fracture Mechanics* 78, no. 12 (2011): pp. 2414–2428.

Zocher, M. A., S. E. Groves, and D. H. Allen. "A Three-Dimensional Finite Element Formulation for Thermoviscoelastic Orthotropic Media." *International Journal for Numerical Methods in Engineering* 40, no. 12 (1997): pp. 2267–2288.

Exploring Low Temperature Performance in Black Space

David J. Mensching[a], Geoffrey M. Rowe[b], Jo Sias Daniel[a], and Thomas Bennert[c]

[a]*Department of Civil and Environmental Engineering, University of New Hampshire, 33 Academic Way, Durham, NH*

[b]*Abatech, Inc., P.O. Box 356, Blooming Glen, PA*

[c]*Center for Advanced Infrastructure and Transportation, Rutgers University, 93 Road 1, Piscataway, NJ*

ABSTRACT. *It is well known that cracking presents major challenges to the effective design and maintenance of asphalt concrete pavements. In colder climates, transverse cracking caused by thermal loading is a major distress mode. Recently, researchers have made effective use of Black Space diagrams to model stiffness and relaxation changes as they relate to non-load associated cracking. However, most of the research to this point has focused on the binder space. An encouraging Black Space-based approach to detecting cracking susceptibility is the Glover-Rowe (G-R) parameter, which requires the complex modulus and phase angle be known at a particular temperature-frequency combination. Recently, the parameter has been adopted to fit the low temperature Superpave binder criteria, as the 4 mm dynamic shear rheometer geometry becomes more practical for asphalt binder testing. In this study, the primary objectives are to 1) validate the G-R approach for binders with laboratory and field performance measurements; 2) develop an indicator parameter for low temperature cracking of mixtures using a combination of relevant material properties; and 3) develop an alternate approach to study mixture performance in Black Space. Binders and mixtures from Alberta, Indiana, New Hampshire, New Jersey, New York, Vermont, and the Strategic Highway Research Program are examined with various field and laboratory cracking metrics. The G-R intermediate temperature parameter correlates well with the overlay tester results, while low temperature laboratory tests show mixed results with respect to the Black Space location. The G-R low temperature parameter can be used to determine a critical cracking temperature that agrees with the Superpave binder specification. A preliminary evaluation of a mixture-based Black Space parameter is included. The results presented may serve as a guide to discussion of specification development and comparison of low temperature behavior among varying length scales.*

KEYWORDS: *hot mix asphalt, asphalt binders, reclaimed asphalt pavement, low temperature performance, thermal cracking*

The presentation was made by Dr. Mensching.

This is a reproduction of an article whose final and definitive form has been published in *Road Materials and Pavement Design*© 2015 Taylor & Francis. The article is available online at: http://dx.doi.org/10.1080/14680629.2015.1077015

1.0 Introduction

It is well known that cracking presents major challenges to the effective design and maintenance of asphalt concrete pavements. In colder climates, transverse cracking caused by thermal loading is a major distress mode. As temperatures decrease, thermal stress increases and can eventually exceed the tensile strength of the material. The relaxation modulus of asphalt binders controls the ability of the mixture to alleviate thermal stress build-up and resist thermal cracking. Although relaxation modulus is difficult to obtain experimentally, the principles of linear viscoelasticity can be applied to convert from other, more easily-measured material properties (i.e., complex modulus and creep compliance).

Researchers have sought to classify non-load associated cracking resistance through the use of an index parameter. Field studies have shown that ductility, measured at an intermediate temperature, correlates well with observed cracking (Clark, 1958; Doyle, 1958; Kandhal, 1977). Critical envelopes were designed to correlate with field measurements of non-load associated cracking (Kandhal, 1977). Using a mechanical-empirical relationship with observed cracking, Glover developed a parameter that relates storage shear modulus (G') and dynamic viscosity (η') of the binder to ductility at a common temperature-frequency combination. The rationale stems from the ease associated with a typical complex shear modulus (G*) test, as opposed to binder ductility testing (Glover et al., 2005).

A study by Anderson et al. (2011) identified parameters to relate ductility and binder properties to non-load associated cracking for airport pavements. The findings of the study identified three parameters to evaluate changes in relaxation properties with aging: 1) the Glover parameter (G'/(η'/G')); 2) the difference between the continuous low temperature binder grade measured using bending beam rheometer (BBR) creep stiffness (S(t)) and the log-log slope (m-value) of the creep curve ($\Delta T_c = T_{c,m(60)} - T_{c,S(60)}$); and 3) the rheological index (R). Three binders with different anticipated aging levels were tested using dynamic shear rheometer (DSR), BBR, and ductility testing, along with the standard procedures for determining continuous low temperature performance grade. The authors found that while the Glover parameter and ΔT_c were the best indicators of durability, there appeared to be value in using the rheological index, R, of which each binder had a unique relationship with the Glover parameter at a 0.005 rad/s frequency. The authors also observed a flattening complex shear modulus master curve with aging, which can be represented through the rheological index.

In a prepared discussion to the Anderson et al. (2011) study, Rowe offered a simplification of the Glover parameter. This Glover-Rowe (G-R) parameter (Equation 1) uses the G* and phase angle (δ) of the asphalt binder at 15°C and a frequency of 0.005 rad/s.

$$\frac{G^*(\cos\delta)^2}{\sin\delta} \qquad [1]$$

Since the parameter is calculated at a fixed frequency, Rowe expressed the parameter purely in terms of G* and δ, allowing users to plot the ductility-based failure planes in Black Space, which compares the change in complex modulus with a change in phase angle. Rowe asserts that the use of Black Space encourages clear illustrations of aging impacts. Figure 1 shows a sample dataset plotted with the G-R parameter (referred to as the G-R intermediate temperature measure) set to the converted ductility limits established in Glover et al. (2005). A value of 180 kPa corresponds to the onset of non-load associated cracking, while a value of 450 kPa or larger relates to significant cracking issues. The sample data in Figure 1 is from a study evaluating different binder aging levels (pressure aging vessel (PAV) and rolling thin-film oven (RTFO)). The diagram shows that as the materials are exposed to additional aging, their location in Black Space shifts to a stiffer, more elastic condition. There are several factors that influence cracking resistance, one of the most important being associated with irreversible oxidative aging. Oxidative aging is directly attributed to increases in stiffness and decreases in ductility, which decrease cracking resistance. The Superpave fatigue parameters are also plotted, showing that all binders pass the criteria, even after 60 hours of pressure aging. However, the G-R parameter predicts that the PG 76 binder will have cracking issues (Bennert, 2014).

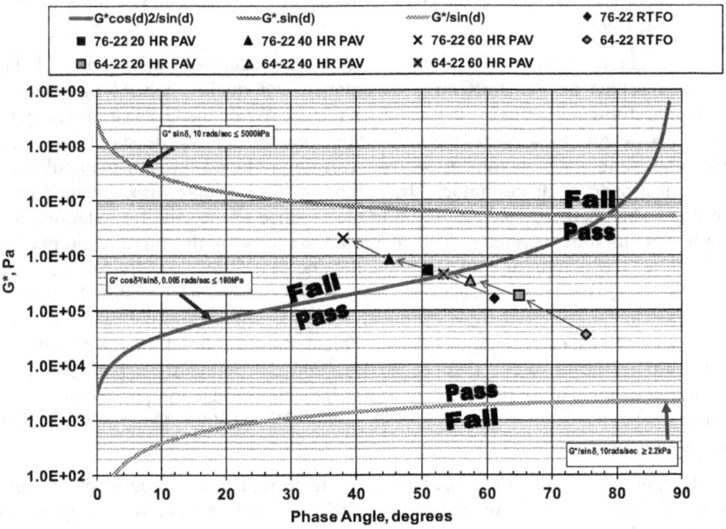

Figure 1. *Sample Glover-Rowe Black Space diagram (Bennert, 2014).*

King et al. (2012) further reinforced the importance of an asphalt binder's location in Black Space as a performance parameter for age-induced cracking. Using the G-R parameter, they argue that the standard Superpave parameter (G*sinδ) developed in the Strategic Highway Research Program (SHRP) is not an adequate damage indicator for cracking. The authors recommend development of a "damage region" in Black Space which combines the DSR outputs (G* and δ) with the creep stiffness and slope values from the bending beam rheometer (BBR). This is a concept currently being explored by Rowe (2014b).

But are these binder characterizations enough to fully explain the field behavior of asphalt mixtures? The research presents a need for a mixture-specific parameter due to impacts of aggregate-asphalt interactions on performance. Recently, Tabatabaee et al. (2012) found that the number of aggregate contact points influences the coefficient of thermal contraction for the mixture. This relationship causes a change in the accumulation of thermal strain, which impacts viscoelastic properties at low temperatures. Also, Alavi et al. (2013) found that aggregate source had a significant effect on the evolution of thermal stress and strain measured during low temperature evaluation of asphalt mixtures in the laboratory.

But are there other elements which factor into cracking resistance, specifically low temperature cracking? For instance, binder tests can capture aging effects to a degree through RTFO and the PAV, but do not encompass variability in performance due to construction. Do binder-mixture conversion techniques (i.e. Hirsch and Christensen-Anderson-Marasteanu (CAM) models) translate to the easy-to-understand Black Space diagram (Christensen et al., 2003; Marasteanu and Anderson, 1999)?

Throughout industry, contractors and owner agencies strive for a balance between robust and cost-effective acceptance and quality control procedures to predict performance, usually on a subjective level. While the G-R parameter appears to be a powerful tool for cracking indication, the approach was developed for asphalt binders. While some mixture properties can be predicted through manipulation of material properties of the binder, the authors believe a more direct method can be derived. This method would build upon the premise of the G-R formulation, but be applied to mixture variables. The increased scrutiny of the mixture properties is critical as the community moves towards increased levels of reclaimed asphalt pavement (RAP) in field sections. Typically, binders which include RAP are extracted and recovered using chemical solvents, which result in a fully-blended condition. The literature suggests that complete mobilization of the RAP binder does not normally occur at the time of mixing, and that in reality a partially-blended condition dependent on the individual materials and production parameters is likely (McDaniel et al., 2000; Soleymani et al., 2000; Oliver, 2001; Mogawer et al., 2012; Mensching et al., 2014).

To this point, little study has been done to formulate this type of parameter for mixture performance. Kim and Wen (2002) found that fracture energy at 20°C exhibited high correlations with field performance of WesTrack sections. The researchers used indirect tension (IDT) loading to evaluate several mixture parameters with respect to fatigue cracking. Researchers have also used disk-shaped compact tension test (DCT) data to recommend a threshold performance specification value based on fracture energy (Dave and Koktan, 2011; Marasteanu et al., 2012). Recently, Romero and Jones (2013) utilized BBR mixture testing to develop a surrogate Black Space using the creep stiffness and m-value obtained from the test as opposed to the traditional modulus and phase angle space. The authors postulated that a thermal stress failure envelope can be displayed in this space, but the envelopes have not yet been identified. These works aid in the rationalization of a mixture-based Black Space parameter which can serve as an adequate indicator of non-load associated cracking performance.

This paper focuses specifically on thermal cracking, as the structural considerations surrounding fatigue performance are vastly more complex and require extensive analysis to characterize. With this in mind, the objectives of this paper are to:

1. Validate the G-R approach for materials with laboratory and field performance measurements,
2. Develop an indicator parameter for low temperature cracking of mixtures using a combination of relevant material properties, and
3. Develop an alternate approach to study mixture performance in Black Space.

It is believed that the theory and application described herein will provide users with a simple cracking indicator robust enough to capture measured trends in conventional and RAP materials. Several mixtures from Alberta, Indiana, New Hampshire, New Jersey, New York, and Vermont are utilized, as well as the SHRP core binders. Laboratory and field cracking information is presented to aid in the correlation and evaluation of failure envelopes.

2.0 Theoretical Approach

In order to satisfy the objectives outlined above, knowledge of viscoelastic behavior and rheological modeling is needed. Furthermore, any developed mixture parameter must have a sound, fundamental backing. The premise circulates around the principle that a traditional Black Space diagram for mixture, comprised of dynamic modulus ($|E^*|$) and δ, is a valuable tool in representing two critical properties for determination of thermal cracking resistance: stiffness and stress relaxation. It is expected that as temperature decreases, an asphalt mixture becomes stiffer and more elastic, resulting in reduced relaxation capabilities (decreased slope of relaxation modulus master curve). The increasing prevalence and ease of using $|E^*|$ and δ for mixture performance is also a major reason for development of a parameter that requires nothing more than a cracking measure and a master curve in the linear viscoelastic region. In this section, the underlying theory and assumptions behind the binder parameter evaluations and the proposed mixture parameter are explained. Later, the theory described here will be applied to laboratory and field-measured cracking data to establish critical envelopes for low temperature performance indication.

2.1 Interconversions and Master Curve Construction

In developing the Black Space diagram, material master curves are required. It is well-known that when measurements are taken in the linear viscoelastic region, material properties (i.e., |E*| and creep compliance) can be converted among each other to gain a more complete understanding of the mechanical behavior. In this study, the RHEA™ software (Rowe and Sharrock, 2000) has been used extensively for the analysis of the rheology data, which is based upon the methods developed by Gordon and Shaw (1994). Shift factors move the position of an isothermal curve with reference to a specific temperature (T_R) to another location on the time or frequency axis. In the linear viscoelastic region, the same shift factor applies for all the viscoelastic parameters.

The amount of shift applied to two adjacent isotherms depends upon a numerical optimization that results in the two isotherms being represented by a single curve with the closest possible representation by a single function. Two types of shift are considered for asphalt materials; 1) adjustment for density considerations and 2) adjustment for loading time and frequency.

The adjustment for density results in a slight vertical shift to each isotherm. The rationale for this alteration is that to correctly describe the frequency response of a material, the behavior can be related in the manner described by Rouse (1953). When the stiffness at a given temperature is shifted to another temperature, the response at that new temperature will be associated with a material that would have a different density. To correctly account for this, the stiffness is adjusted according the expression in Equation 2:

$$S(T_R, t) = \frac{T_R}{T}\frac{\rho(T_R)}{\rho(T)} S\left(T, \frac{t}{a_T}\right) \qquad [2]$$

Where: $S(T_R, t)$ = stiffness at reference temperature (T_R) and loading time (t);
T = temperature;
ρ = density; and
t/a_T = shifted time.

If density shifts are not implemented, significant errors can result when interpreting master curve data over a large range of temperatures (Rowe et al., 2011). Equation 2 is shown in the format applied to BBR data. For a shift to DSR data, the parameters G*, G′ or shear loss modulus (G″) are substituted for the bending stiffness.

The adjustment for loading time results in a horizontal shift to the isotherm in relation to the reference isotherm. The method adopted follows the traditional approach of shifting the modulus values (either relaxation shear modulus (G(t)) or G' and G") along the horizontal axis to form smooth curves of modulus. The shift procedures for producing a master curve have been developed by various researchers. Gordon and Shaw (1994) defined various methods that can be applied for the characterization of viscoelastic materials. These methods have formed the basis of the techniques employed in the RHEA™ software. For the production of master curves, the following general steps are executed:

- Determine an initial estimate of the shift for each pair of isotherms using a linear fit and checking against a modified Williams-Landel-Ferry (WLF) calculation using Ferry's original "universal" constants (Williams et al., 1955).
- Iteratively refine the pairwise shifts using weighted least squares polynomial fits.
- Allow the order of the polynomial to be an empirical function of the number of data points and the decades of time/frequency covered by the pair of isotherms.
- Sum the optimized shift factors for each successive pair from zero at the lowest temperature to obtain a distribution of shifts with temperature above the lowest.
- The shift at T_R is interpolated and subtracted from every temperature's shift factor, causing T_R to become the origin of the shift factors.

In the software implementation of Gordon and Shaw's (1994) methods, a cubic spline fit has been used for the final shift to the reference temperature. Importantly, the implementation of the Gordon and Shaw method for producing shift factors makes no assumptions regarding the final shape of the master curve (relationship between stiffness and time or frequency) or the relationship between shift factor and temperature. Thus both loading rate dependency and temperature dependency are calculated independently. These methods, implemented in Rowe and Sharrock (2000) also allow for gaps in isothermal data to be captured through the use of a "wild" shift factor.

The relaxation (generalized Maxwell model) and retardation (generalized Kelvin-Voigt model) spectra analysis (Baumgaertel and Winter, 1989) can also be applied to data from the BBR enabling the interconversion to dynamic data. The procedure adopted to perform this interconversion is explained below:

- The BBR data is fitted with the Christensen-Anderson (CA), Christensen-Anderson-Sharrock (CAS) and Christensen-Anderson-Marasteanu (CAM) models (Rowe et al., 1999) to identify the fit with the lowest error. This master-curve is adopted.
- The Hopkins and Hamming (1957) method to convert to relaxation modulus (E(t)) is used.
- The E(t) data is then fitted with a CAM model using the glassy modulus determined from the initial fit. This gives a function which describes the E(t) data.
- The discrete spectra is calculated for the E(t) fitted function.
- The reciprocal of the observed times are then substituted into the function to estimate the E', E" data points.
- The data points are shifted using the original shift values obtained along with a reverse density correction to obtain dynamic isotherms corresponding to the original data.
- Extensional data is then output, converting to G′, G″ (or G*).

The use of BBR data converted to dynamic data assists with the generation of master curves to a) define the properties in a stiffness range were rheometers have problems with strain resolution and b) produce reliable data using the 8 mm geometry (typically used for the lower temperatures and higher stiffness range in DSR testing). Using converted BBR data generally assists in better defining the time-temperature properties over a greater range than would be otherwise possible. If data is being combined from the two measurement types, it is important to ensure that comparable data is being combined (e.g., aging condition). Researchers are aware that differences exist in the isothermal storage times between the BBR and DSR test methods, but these are anticipated to be minimal since the maximum times used in both tests for storage and conditioning are similar.

In the analysis of data it is imperative to consider the quality of the master curves produced and when and how to exclude data from the analysis. In previous work (Rowe and Sharrock, 2000), the root mean square (rms%) fit has been used as a statistic to assess the quality of fit to the data. The error at any point between the calculated modulus and a model fit to that frequency or time has been calculated on a log basis and expressed as a percentage error and averaged for the data set. For BBR data sets and master curve generation, the error produced with a three parameter model, such as with the CAM model, the rms% error should be 2.25% or less (Rowe et al., 1999). If only two isotherms are used the error should be less than 1.25% as indicated in the standard ASTM D6816 (2002). The goodness of fit relies upon the ability of a binder to exhibit those properties throughout the range considered. The CAM model, which lends itself to the BBR data, is shown in Equation 3:

$$S(T,t) = S_g\left[1 + \left(\frac{t}{\lambda}\right)^{\beta}\right]^{\frac{-\kappa}{\beta}} \qquad [3]$$

Where: $S(T, t)$ = bending stiffness at a temperature (T) and reduced time (t);
S_g = glassy bending stiffness modulus;
λ = reduced time at maximum curvature;
t = loading time of interest;
β, κ = fitting parameters.

In terms of complex shear modulus, typically obtained from the DSR, the expression for the CAM equation is presented in Equation 4:

$$G^*(T,\omega_r) = G_g\left[1 + \left(\frac{\omega_0}{\omega_r}\right)^{\beta}\right]^{\frac{-\kappa}{\beta}} \qquad [4]$$

Where: $G^*(T, \omega_r)$ = complex shear modulus at temperature (T) and reduced frequency (ω_r);
G_g = glassy modulus;
ω_0 = crossover frequency;
β, κ = fitting parameters.

In most master curve analyses, the value of κ is equal to unity. This parameter is related to the slope of the viscous asymptote. The relationship for phase angle can be obtained from the differential of the modulus with respect to the frequency (Christensen, 1992; Rowe, 2009) and is represented by Equation 5:

$$\delta = \frac{90}{1+(\frac{\omega}{\omega_0})^{\beta}} \quad [5]$$

Where: $\delta(T, \omega)$ = phase angle at temperature (T) and reduced frequency (ω_r).

In the original SHRP research, Anderson et al. (1994) proposed a parameter that is defined as the rheological index (R). Equation 6 gives the relationship between R and the master curve at the crossover frequency (ω_0):

$$R = \log G_g - \log G^*(\omega_0) \quad [6]$$

A modified Kaelble shift factor function is used in this method to develop expressions since it is more inclusive of data types (Rowe and Sharrock, 2011). The shift factor can then be multiplied by a particular isothermal frequency to find the corresponding reduced frequency. The modified Kaelble function is defined in Equation 7:

$$\log a_T = -C_1\left(\frac{T-T_d}{C_2+|T-T_d|} - \frac{T_R-T_d}{C_2+|T_R-T_d|}\right) \quad [7]$$

Where: aT = time-temperature shift factor;
T_d = defining temperature for inflection point;
T_r = reference temperature;
C_1 and C_2 = fitting constants.

For the mixture analysis, many of the same manipulations involving the $|E^*|$, δ, and a_T formulations are similar. In this study, the generalized logistic, or Richards curve, is used to fit the master curve (Richards, 1959). The differential method described by Rowe (2009) is utilized to determine the phase angle at a particular temperature-frequency combination. Equations 8 and 9 detail the Richards curve, with its appropriate differential form. Note that particular manipulations of the Richards shape parameters describe asymptotic and inflection point characteristics of the master curve (Rowe et al., 2011).

$$\log|E*| = \delta + \frac{\alpha}{[1+\lambda e^{\beta+\gamma \log(\omega)}]^{1/\lambda}} \quad [8]$$

$$\delta(\omega) = -90\alpha\gamma \frac{\alpha}{[1+\lambda e^{\beta+\gamma \log(\omega)}]^{(1+\frac{1}{\lambda})}} \quad [9]$$

Where: α, β, γ, λ = shape parameters;
δ = lower asymptote of modulus master curve;
δ+α = upper asymptote of modulus master curve;
-β/γ = inflection point frequency.

2.2 Considerations for Black Space Diagram

The Black Space diagram, a plot of complex stiffness modulus versus phase angle, is a valuable rheological tool because it enables users to assess how stiffness and elasticity of a material are related with or without using the time temperature superposition principle (TTSP) to convert to the reduced frequency or time domain (Airey, 2002). Since the phase angle describes the relaxation (King et al., 2012; Rowe, 2014b), the use of a Black Space diagram can lend itself to low temperature considerations in a similar manner to that used by Leahy et al. (1994) for results from the BBR. In Glover et al. (2005), the parameter formulated was based upon a relationship between G', η', and ductility to determine an appropriate temperature-frequency combination. As it relates to mixture Black Space and low temperature cracking indication, a particular temperature-frequency combination will allow for the most comprehensive assessment of the performance.

Since the G-R cracking parameter as defined by Anderson et al. (2011) is measured at a condition of 15°C and 0.005 rad/s to capture the performance associated with thermal or block cracking (King et al., 2012; Rowe et al., 2013), the authors first evaluated the same combination for the mixture. In the determination of the G-R low temperature parameter for a binder purchase specification for cold temperature cracking, several approaches exist. These include the testing scheme developed by Anderson et al. (2011) in which three isotherms are collected from the DSR and used to construct a partial master curve from which parameters may be determined or alternate methods as proposed by Rowe (2014b) could be considered including the use of a 4 mm plate in the DSR.

Recently, Rowe (2014a, b) converted the Superpave low temperature specification ($S \leq 300$ MPa, $m \geq 0.300$ at 60 s) to corresponding G* (111 MPa) and δ (26.2°) values for a Black Space-based alternative to the standard criteria. As shown in Figure 2, this G* and δ corresponds to the intersection of the S and m failure planes in Black Space. A G-R value of 184 MPa, which will be referred to as the G-R low temperature threshold, corresponds to this point shifted to a frequency of 0.005 rad/s. Rowe then suggested that the G-R parameter can be applied to the cold temperature cracking specification when the value is set at a G-R low temperature threshold and temperature is made to be the dependent variable.

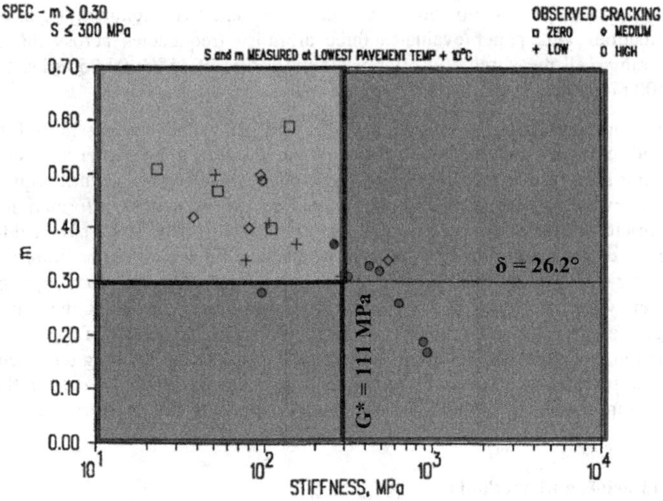

Figure 2. *Schematic of failure planes for Superpave low temperature criteria with equivalent complex modulus and phase angle values (from Rowe, 2014a, b).*

Due to the interaction of the asphalt binder with aggregate, the Black Space plot for a mixture shows a peak phase angle value at intermediate stiffness, as shown in Figure 3. The aggregate structure begins to dominate behavior at high temperatures, due to the low stiffness and viscous flow of the asphalt binder whereas at lower temperatures the mix volumetrics and binder stiffness control the behavior. The highest value of phase angle in this plot is associated with the inflection point in the master curve using the terms described by Equation 8 ($-\beta/\gamma$). It can be seen in this example that this occurs around a stiffness of 1,000 MPa.

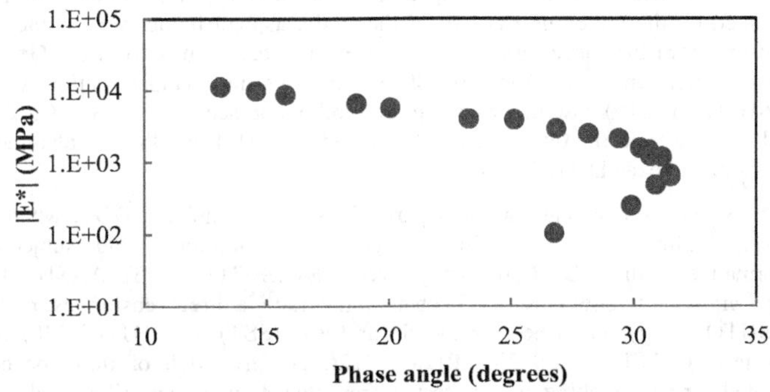

Figure 3. *Sample Black Space diagram for mixture.*

The challenge is to identify the representative temperature-frequency combination. This paper evaluates three arbitrary frequencies across the master curve range, all measured at a reference temperature of 15°C: 0.005 rad/s, 5 rad/s, and 500 rad/s.

In a previous paper (Mensching et al., 2014), typical cooling rates for New England sections were evaluated. A low temperature event, no matter how severe of a cooling rate, results in stiffening of the asphalt binder which in turn requires the need to evaluate properties at a faster loading rate to obtain information at an appropriate stiffness for assessment in the Black Space diagram. Additional evidence of the importance of frequency selection for a low temperature mixture parameter exists in the reality of varying relaxation response with loading rate/time. In other words, during a cooling event the relaxation modulus of the mixture is changing as the temperature changes. This complicates the issue if, for performance evaluation, a single temperature and frequency combination is being considered for use within the Black Space. Significant care must be taken to ensure that the temperature-frequency combination is actually relevant to field performance.

3.0 Materials and Methods

In order to establish a well-rounded exploration of low temperature performance, laboratory and field-measured cracking measurements are desired. With this in mind, materials from a variety of locations were used in the analysis. The first dataset comes from the New Jersey Long-Term Pavement Performance (LTPP) SPS-5 sections. As part of the LTPP SPS-5 sections, a 0% RAP mixture and a 30% RAP mixture were overlaid at different thicknesses (50 and 125 mm) using two different surface preparation conditions (milled and unmilled surface). Pavement distress was measured yearly during the service life of the sections. Prior to the end of service life, Bennert and Maher (2014) procured field cores from the 15-year-old pavement and loose mix that was stored at the LTPP Materials Reference Library in Sparks, Nevada. Intermediate and low temperature asphalt binder and mixture tests were conducted on the materials. As part of the study, asphalt binder was extracted and recovered from the loose mix and field cores. The test results from the G* master curves (American Association of State Highway and Transportation Officials (AASHTO) T 315) are described in the following section (AASHTO, 2012h). Cracking data from the overlay tester (OT) (Texas DOT Tex-248-F) is also included in this study (Texas DOT, 2014).

The second set of materials was part of a study for Indiana DOT, where three national highway sites at different ages were monitored and subjected to performance grading, DSR frequency sweeps (AASHTO T 315), AASHTO R 49 critical cracking temperature (CCT) characterization via direct tension testing (DTT) (AASHTO T 314), and the Superpave Shear Tester (SST) (AASHTO T 320), among other tests (AASHTO, 2012h; 2012a; 2012f; 2012g). Each of these pavements exhibited surface cracking and the results from this study are described in detail in a report prepared by Pellinen et al. (2004).

The third set of materials are from a report prepared for the Federal Highway Administration (FHWA) Binder Expert Task Group (ETG), based on results from the Lamont Test Road in Alberta (Bouldin et al., 1999; Bouldin et al., 2000). Seven binders were used in the study with varying material properties, crude sources, and binder grades. This analysis has been extended with results obtained from retesting the SHRP core asphalts (Anderson et al., 2013).The Lamont binders were graded according to the Superpave criteria (AASHTO M 320), which specifies the use of the DSR, BBR (AASHTO T 313), and DTT to identify high and low temperature performance grade, respectively (AASHTO, 2012i; 2012e). The G* master curve for the binder was developed from a frequency sweep (0.1 rad/s to 10 rad/s) at 6°C increments from 10°C to 64°C. However, to improve the CAM model fit, data with stiffness values lower than 100,000 Pa were generally excluded from the analysis in accordance with the recommendations developed by Rowe (2014b) which resulted in exclusion of temperatures generally above 30°C. The Lamont binder data available for reanalysis was largely obtained on RTFO samples with some minimal data collected over PAV conditions which was suitable for production of master curves and the evaluation of the G-R low temperature parameter. In the development of the critical cracking method with the Lamont data, no PAV aging was performed. The RTFO aging condition was used because the cracking in the field occurred within two years of construction (Bouldin et al., 1999).

In order to evaluate the relationship between laboratory cracking and the Black Space outputs, mixture data from Transportation Pooled Fund Study (TPF) 5(230) are also included. Material information from the eighteen Phase I mixtures are outlined elsewhere (Mensching et al., 2014). For all of the Pooled Fund mixtures, the master curves (|E*| and δ) (AASHTO T 342), thermal stress restrained specimen test (TSRST) (AASHTO TP10), OT, and the IDT low temperature strength (AASHTO T 322) results were utilized when available (AASHTO, 2012c; 2012b; 2012d). As done in Mensching et al. (2014), CCT of the mixtures were predicted according to a variation of TCModel (Hiltunen and Roque, 1994) by Christensen (1998). The materials from TPF-5(230) are all plant-produced, laboratory-compacted materials.

4.0 Results and Application

The details provided below are separated into binder and mixture sections for ease of reference. The G-R binder parameters are evaluated and compared between field and laboratory cracking data. The mixture section provides preliminary findings as it relates to the formation of a mixture-based Black Space parameter. Throughout the explanation of results and Black Space applications, various failure parameters will be used. Readers are asked to consult other works for listings of all field and laboratory test values used in this study (Bennert and Maher, 2014; Pellinen et al., 2004; Bouldin et al., 1999; Bouldin et al., 2000; Daniel et al., 2014; Mensching et al., 2014).

4.1 Binder Comparisons

The first subsection below presents the comparison of the G-R intermediate temperature based parameter with laboratory and field cracking measurements for each set of data. These comparisons are done in Black Space. The second subsection compares a low temperature critical cracking value determined using the G-R approach with other CCT values.

4.1.1 Glover-Rowe Intermediate Temperature-Based Parameter

In this section, the G-R intermediate parameter is evaluated against transverse/thermal cracking measurements from the New Jersey, Indiana, and Lamont field sections. Overlay test results are correlated with Black Space location for the New Jersey and Pooled Fund materials. It is likely that the intermediate temperature parameter will have mixed agreement with low temperature cracking data, as the stiffness-relaxation combination to best describe this type of distress is at high modulus, low phase angle locations in Black Space. The authors believe that in order to detect performance in the most accurate fashion possible, every distress type will correspond to a different zone in Black Space.

New Jersey SPS-5 Sections

Figure 4 shows the transverse cracking results for the New Jersey 50-mm overlay sections in terms of a transverse cracking deduct. Deduct values are a more-standardized way of converting distress measurements separated by severity levels into a composite value. The field performance results indicate that the 30% RAP mixture shows a larger degree of transverse cracking than the virgin mixture. As the service life of the pavement increases, the amount of transverse cracking in both the virgin and 30% RAP sections increases and a difference in performance is observed between the milled and unmilled sections. At 15 years, the total distress in the 30% RAP sections is greater than that observed in the virgin sections, but the distress has progressed to block cracking which is not captured in the transverse deduct value shown in Figure 4.

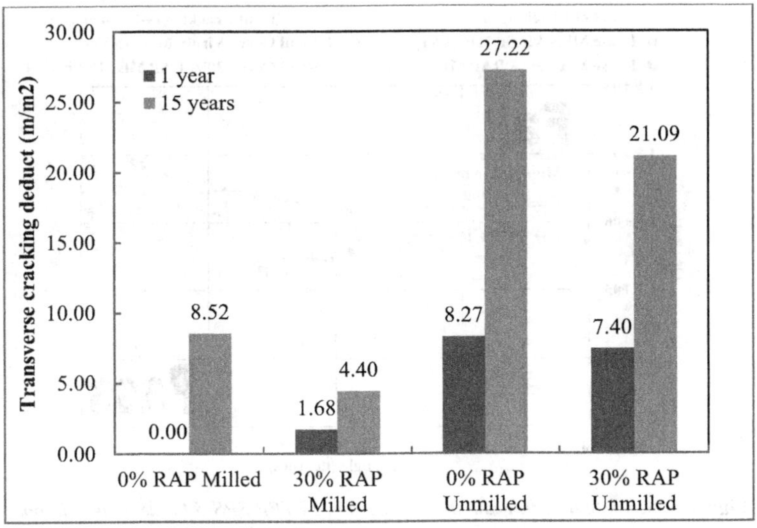

Figure 4. *Transverse cracking for New Jersey's SPS-5 50-mm overlay sections (after Bennert and Maher, 2014).*

The G* master curve data from the New Jersey LTPP SPS-5 sections plotted in Black Space are shown in Figure 5. The dotted line corresponds to the "onset of cracking" zone, where the G-R value equals 180 kPa, while the solid line represents the "significant cracking" limit (G-R value equals 450 kPa). For the analysis, it was assumed that the stored loose mix had aged properties of a one-year-old pavement, while the field cores were taken at the end of the service life of the pavement (about 15 years old). The progression of transverse field cracking can be tracked in the Black Space diagram by combining the transverse cracking noted in Figure 4 with the Black Space information shown in Figure 5 for the milled surface condition. The data label next to the Black Space data point is the transverse cracking deduct value measured and calculated from the visual distress survey conducted. As shown in Figure 5, the Black Space diagram was capable of tracking the transverse cracking distress, and matches quite well to the G-R intermediate temperature cracking thresholds at the end of the pavement life. It is noted that the 30% RAP field core transverse cracking deduct should theoretically be much higher as this section migrated into block cracking. Although the Black Space diagram is not able to differentiate the progression of transverse cracking to block cracking, the migration of the Black Space data into a more severe area of cracking clearly follows the pavement distress observed in the test sections. The same trend is found when evaluating the unmilled surface pavement section. However, the magnitude of the actual transverse cracking deduct is higher, which appears to be directly related to the surface treatment. Refer to Figure 4 for the cracking deduct values of the unmilled sections.

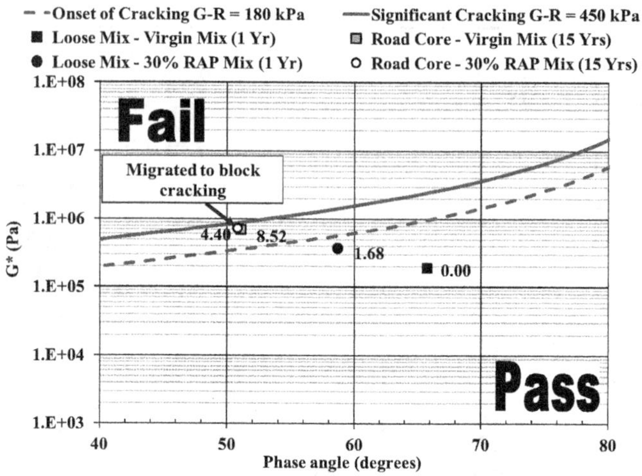

Figure 5. *Black Space diagram for New Jersey's LTPP SPS-5 binders superimposed with transverse cracking field measurements for milled pavement.*

As shown in Figure 5, the virgin binder is firmly in the pass region, indicating that at approximately one year, there should not be any non-load associated cracking. Meanwhile, the 30% RAP mixture (one year) is close to the "onset of cracking" zone, indicating that cracks could initiate shortly. This represents a slight discrepancy as thermal cracks were observed at this time, but the G-R intermediate temperature parameter does not detect it. The test data for the field cores show that both mixtures should be exhibiting cracking at 15 years.

Based on the data presented using the New Jersey LTPP SPS-5 test sections, it appears that the Black Space diagram analysis is able to track the accumulation of transverse cracking magnitudes in the milled and unmilled pavement sections. The actual magnitude of the transverse cracking deduct values appears to be influenced by the surface condition prior to paving (i.e., milled or unmilled). To some degree, the test data was also able to validate the proposed thresholds recommended by Rowe (2011). Recall the thresholds are best suited for intermediate temperature conditions, such as the conditions possible during age-induced cracking events.

The G-R intermediate parameter was also evaluated against OT results on the New Jersey binders (Figure 6). As expected, cycles to failure decrease with pavement age and increase in RAP content. The 0% RAP mixture experienced a larger change in field performance and cycles to failure with age than the 30% RAP material. The binder location in Black Space relates well to the OT results, indicating again that the Black Space concept is related to the cracking performance of asphalt mixtures.

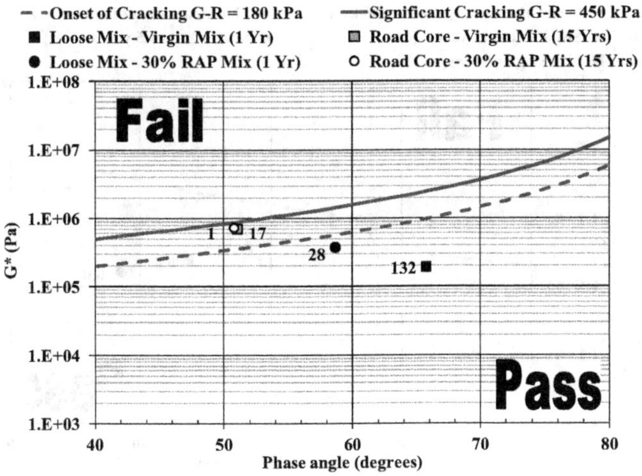

Figure 6. *Black Space diagram for New Jersey's LTPP SPS-5 binders superimposed with overlay tester fatigue life.*

Indiana Sections

The primary distress mode for the Indiana sites was surface cracking, and all sites showed some degree of preexisting surface cracking. Only one site had transverse cracking which may have been related to thermal effects. However, it is possible that this may also have been influenced by the larger overall degree of cracking on this site compared to the other locations. As shown in Figure 7, the data clearly shows that Site 1 is the poorer performing mixture. Although not shown for the sake of brevity, the AASHTO R 49 CCTs do not match with the Indiana sites as well. The AASHTO R 49 temperature ranks Site 1 as the poorest, but Sites 2 and 3 are ranked incorrectly. Overall the analysis of this study suggests that this approach may have some use in the assessment of load associated surface and/or transverse cracking on these types of sites.

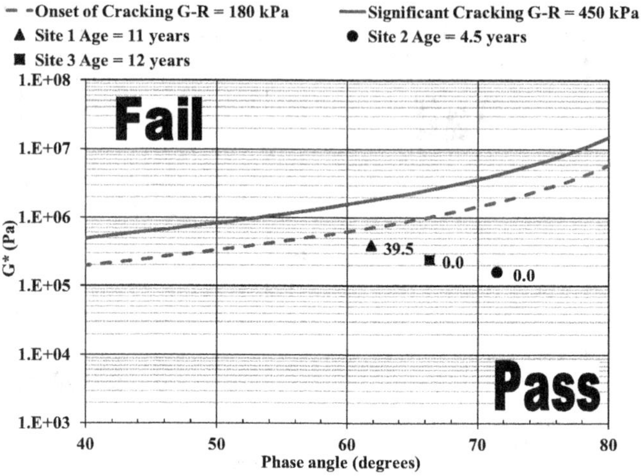

Figure 7. *Black Space diagram with transverse cracking deduct values for Indiana sections.*

Lamont Sections

In Figure 8, the Lamont binders are plotted at the intermediate G-R condition. The binders are named L1 through L7 to follow the scheme by Bouldin et al. (1999; 2000). Also shown are the cracks recorded for each section (number of cracks per kilometer.) The Black Space location at 15°C and 0.005 rad/s does not appear to track well with observed cracking for these binders, showing that the G-R parameter may not relate to low temperature performance when evaluated at a frequency of 0.005 rad/s at 15°C. One possible reason for the discrepancy may be the climate and traffic level, where the pavement section is exposed to more extreme cold temperatures and the dominant distress mechanism was presumed to be single event thermal cracking (Bouldin et al. 1999).

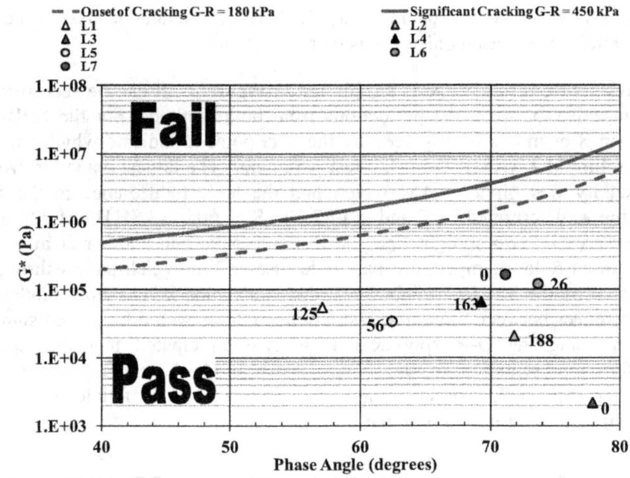

Figure 8. *Black Space diagram with transverse cracking frequency for Lamont sections.*

4.1.2 Glover-Rowe Low Temperature-based Parameter

With consideration to low temperature cracking it is not expected that the G-R parameter as developed by Anderson et al. (2011) and Rowe (2011) would be applicable. For a differing distress type, involving different temperatures and rates of loading some adjustments would be anticipated. The following section discusses the applicability of the G-R CCT, where the threshold value is held constant and the temperature is the dependent variable. The G-R low temperature binder grade refers to the CCT (the temperature at which G-R = 184 MPa) minus 10°C. In this section, only the Lamont and SHRP binders are presented due to a lack of available binder data for the other projects.

Lamont Sections and SHRP Asphalts

In this analysis the CCT was obtained for the seven Lamont sections (Bouldin et al., 2000) and the SHRP core asphalts which have been recently retested by the Asphalt Institute (Anderson et al., 2013).

In the determination of the CCT using the G-R low temperature parameter, master curves were developed using both BBR and DSR data. Data below the stiffness of 100,000 Pa was generally excluded from the data sets as recommended by Rowe (2014b). However, since several of these data sets were older, particularly for the Lamont data, this requirement was relaxed in a few cases. In addition, for data sets for Lamont sections 6 and 7 only BBR data was available and consequently the calculations were based only on this data.

The results from this analysis are plotted in Figure 9. It can be seen that good agreement exists between the predictions from both methods. For the SHRP core asphalts, an S or m has been placed by the data point to indicate which parameter was controlling the AASHTO M 320 test parameter. The largest deviation from the line of equity was binder AAM which had the largest difference in the failure temperature as determined by S and m (7.2°C). Since the AASHTO M 320 defines the failure area by two straight lines (at S = 300 MPa and another at m = 0.300) whereas the G-R low temperature parameter uses a line which passes through the apex created by the M 320 straight lines, the expectation is that those binders with large differences between S and m passing temperatures would be graded somewhat differently using the G-R approach. It is also anticipated that the G-R low temperature representation of the failure space might be more consistent with performance since a 90° corner is typically unrepresentative of field behavior and rather the transitions are more gradual.

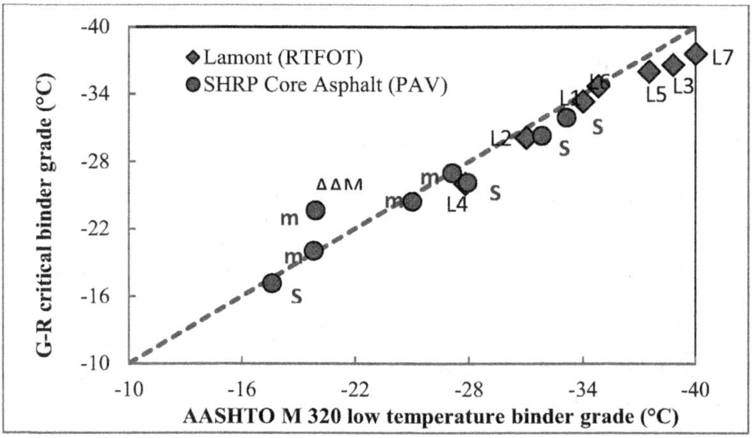

Figure 9. *Correlation between Glover-Rowe low temperature binder grade and AASHTO M 320 continuous binder grade.*

Table 1 compares the field cracking of the Lamont sections against the G-R critical temperature. Ideally, the warmest temperature would correspond with the poorest performing sections. While not an exact match, the G-R CCT agrees very well with the observed cracking at the Lamont Test Road and separates the relatively good performing sections from the relatively poor performing sections.

Table 1. *Comparison of Glover-Rowe low temperature criteria with Lamont binders.*

Binder	L1	L2	L3	L4	L5	L6	L7
AASHTO M320 CCT (°C)	-24.0	-21.0	-28.8	-17.8	-27.5	-24.8	-30.0
G-R CCT (°C)	-23.3	-20.1	-26.6	-16.0	-26.0	-24.7	-27.6
Number of transverse cracks per km	125	188	0	163	56	26	0
AASHTO M 320 rank/G-R rank/field rank	5/5/5	6/6/7	2/2/1	7/7/6	3/3/4	4/4/3	1/1/1

4.2 Mixture Comparisons

In comparing a mixture-based Black Space to field performance, the authors used field and laboratory parameters from several projects. In this section, comparisons between mixture Black Space, field cracking, and available laboratory tests are presented for the Indiana and New Jersey sections. Comparisons between mixture Black Space and laboratory testing are also presented for the Pooled Fund mixtures.

4.2.1 Indiana and New Jersey Mixture Black Space

In Figure 10, field performance (measured as the transverse cracking deduct value) is compared against the Black Space values for the New Jersey SPS-5 and Indiana materials, measured at a temperature of 15°C and frequencies of 500, 5, and 0.005 rad/s. The 30% RAP mixture has a lower maximum phase angle than the virgin mixture in the region of the inflection point frequency. When comparing the New Jersey Black Space points with the field cracking (Figure 10), DCT, and OT measurements (Figure 11), the modified G-R value for mixture ($E^*\cos^2(\delta)/\sin(\delta)$) detects the poorer performing 30% RAP mixture as frequency decreases (or moves closer to the inflection point) since there is more separation in Black Space. The OT results follow the trend seen in the DCT and field measurements, meaning that as frequency decreases there is additional scatter or separation in Black Space, allowing differences between the materials to become more pronounced.

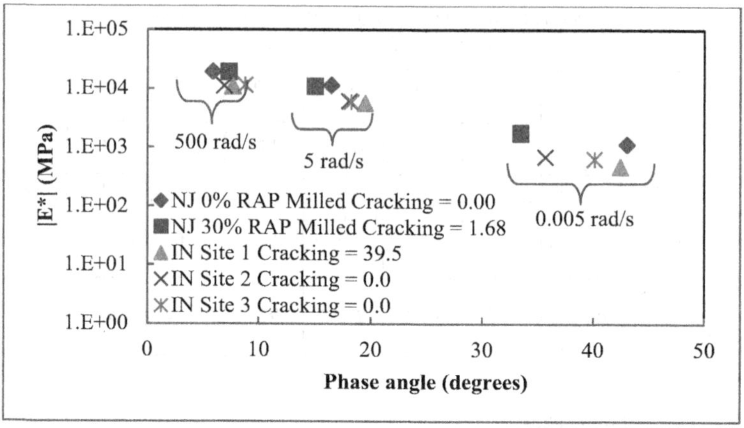

Figure 10. *Mixture Black Space diagram with transverse cracking deduct values for New Jersey 50 mm milled sections at reference temperature of 15°C and frequencies of 500 rad/s, 5 rad/s, and 0.005 rad/s.*

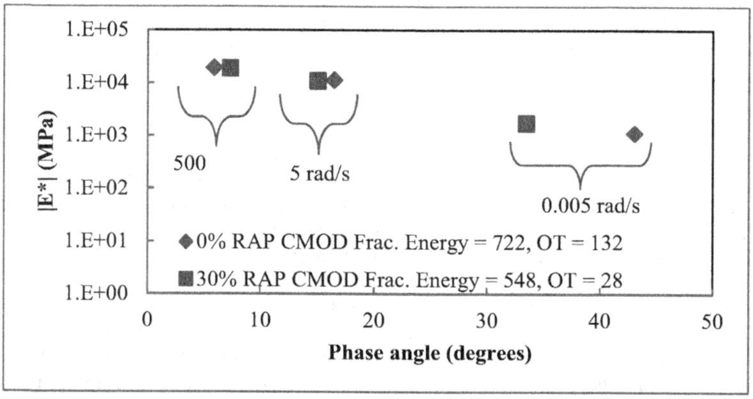

Figure 11. *Mixture Black Space diagram with crack mouth opening displacement fracture energy values (J/m^2) and overlay test cycles to failure for New Jersey mixtures at 15°C and frequencies of 500, 5, and 0.005 rad/s.*

The Indiana mixtures in Figure 10 also show the greatest differences (or scatter) in Black Space at 0.005 rad/s. The Black Space curves for these mixtures (not shown for brevity) confirm that as the measurements proceed down the reduced frequency axis (towards higher temperatures), the phase angle readings have greater scatter among the three mixtures, which should allow for better detection of differences in performance. The modified G-R mixture value would dictate that Site 2 is the poorest performing material due to a higher stiffness and lower phase angle, but the field measurements show Site 1 is the worst performer. However, recall that while Site 1 experienced transverse cracks, all three sections show surface cracking distress. It is also worth noting that the Indiana |E*| and δ master curves produced higher rms% error values than the New Jersey mixtures, reinforcing the need to be cautious when developing master curves.

4.2.2 Exploration of Mixture-Based Black Space Parameter

The Pooled Fund mixture data is used to explore a mixture-based Black Space parameter for low temperature cracking detection. No field performance data is available at this time, so laboratory tests with a critical temperature output were used. The first step in the process was to fit the |E*| and δ master curves. It is important to note that Figures 12–14 use the failure temperature of the TSRST and TCModel to shift to a |E*| and δ corresponding to a 15°C reference temperature at frequencies of 500, 5, and 0.005 rad/s for each of the Pooled Fund mixtures. The Black Space plots corresponding to the TSRST (Figure 12a) and the TCModel simulation (Figure 12b) at 10°C starting temperature and 5.6°C/h cooling rate are shown.

The R^2 value of the logarithmic trendline increases as the frequency decreases. This is likely due to the location on the |E*| and δ master curves. At the higher frequencies, the glassy modulus and plateau sections of the master curve are being used so the total range in stiffness being obtained when analyzed on a log scale is smaller. Based on previous discussion in this paper, it is possible that the optimum frequency for a mixture-based parameter may lie closer to the inflection frequency (or maximum phase angle).

Once an appropriate frequency is found, the development of the failure bands with respect to a cracking measure (laboratory or field) can be initiated. In this paper, the authors chose the shifted 0.005 rad/s frequency at 15°C and calculated the modified G-R value for each mixture for each laboratory test (Figure 13). This frequency was chosen due to the enhanced ability to detect Black Space differences as the curve progresses towards the maximum phase angle. The hypothesis is that particular laboratory (or field) tests may be captured by the mixture G-R parameter better than others. This would aid in the definition of the failure bands for a mixture-based G-R parameter. The figure below shows that performance groupings may exist, and there may be useful commonalities shown by the mixture G-R parameter (e.g., the PG 52-34 binder possibly performing better).

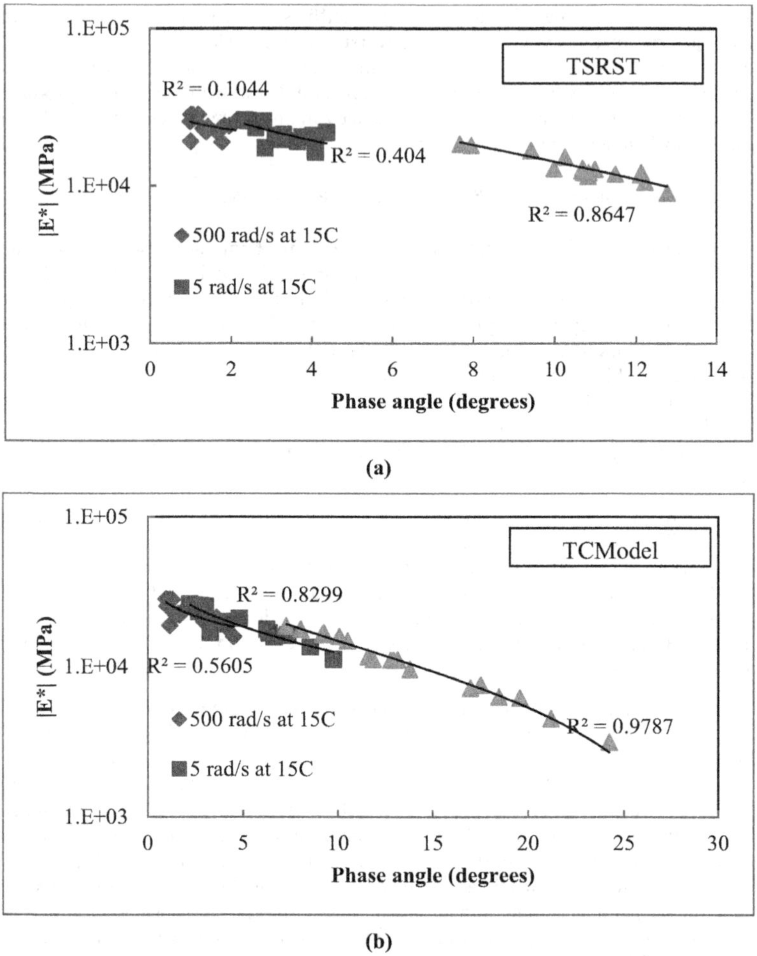

Figure 12. *Pooled Fund mixture Black Space diagrams, measured at failure temperature for a) TSRST and b) standard TCModel setup.*

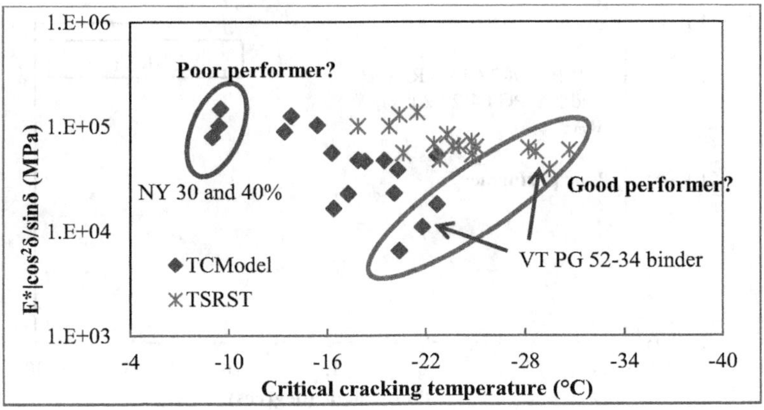

Figure 13. *Modified mixture Glover-Rowe values against critical cracking temperature for Pooled Fund mixtures.*

In Figure 14, the |E*| and δ at the CCT equivalent of the 15°C, 0.005 rad/s is shown for the Pooled Fund mixtures. The failure bands would be plotted in Black Space to identify the poorly performing mixtures based perhaps on a degree of separation in the diagram. However, some kind of field validation for any failure band definition is required regardless of whether or not a laboratory test is chosen for parameter development.

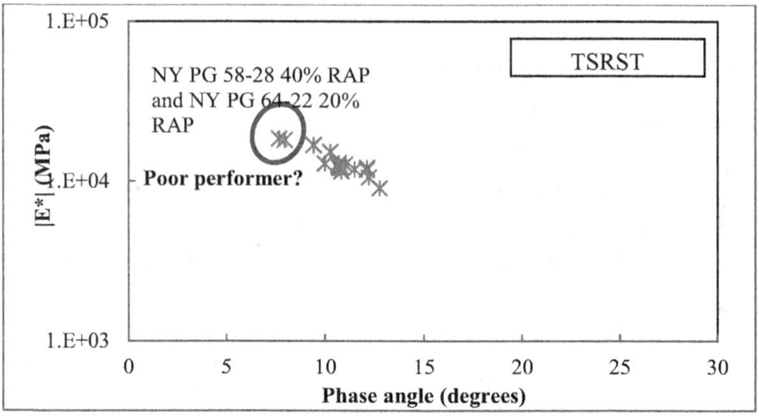

(a)

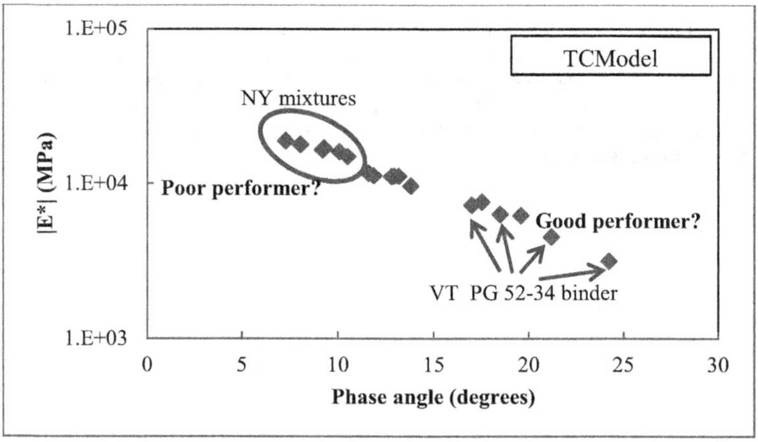

(b)

Figure 14. *Mixture Black Space diagrams for Pooled Fund materials, with hypothetical failure band definition scheme for a) TSRST and b) TCModel at the CCT and shifted 0.005 rad/s frequency.*

5.0 Conclusions and Recommendations

This study serves two purposes to the asphalt materials community: 1) to evaluate an emerging method to examine non-load associated cracking, and 2) to introduce the concept of a mixture-based parameter for low temperature performance that only requires a dynamic modulus test at a few temperatures and a measure of field performance, which introduces the ability to study the usage of the parameter for similar materials. The paper applies the G-R parameter to over 15 materials at low and intermediate temperature conditions to compare the Black Space to a cracking measure. Also, over 20 mixtures are used in a preliminary evaluation of a mixture-based Black Space parameter. Based on the results and discussion above, the following conclusions and recommendations have been drawn:

1. The G-R intermediate parameter, measured at 15°C and 0.005 rad/s, possesses the ability to match performance from the OT (an intermediate temperature test) quite well. The OT holds the potential to test non-load associated cracking in the form of reflective cracks. The results presented here may introduce a method to use the DSR and OT in tandem to assess thermal reflective cracking resistance.

2. The New Jersey sections seemed to track well in Black Space between start and end of life measurements when the binders were compared to the G-R intermediate parameter. The virgin and 30% RAP binders started at different locations in Black Space, but ended virtually at the same location. This is an interesting finding worth exploring for other RAP mixtures—perhaps a relationship exists regarding a predetermined level of additional damage RAP materials can absorb.

3. The Indiana mixtures show a relationship in Black Space. Possible reasons include higher variability in mixture testing, and the influence of preexisting pavement distress on the formation of thermal cracks.

4. All Lamont binders passed the intermediate temperature thresholds, but all but two of the seven sections cracked in the field. The intermediate temperature condition was able to distinguish one of the uncracked sections in Black Space, while the other uncracked section was not separated in the space from the cracked sections. This finding is likely tied to the notion that for every distress type, a particular Black Space region will best detect changes in performance. The Lamont binders exhibited thermal cracks in the field, which are better tied to the G-R low temperature thresholds. The G-R low temperature CCT provided a ranking very similar to that observed in the field performance, demonstrating its effectiveness with observed thermal cracking measures.

5. The G-R CCT calculation holds potential for virgin materials. Both the Lamont and SHRP core asphalts agree with the AASHTO M 320 continuous low binder grade very well.

6. An exploration into a mixture-based Black Space parameter is presented. The results show that measuring the $|E^*|$ and δ at frequencies closer to the inflection point (peak of mixture phase angle master curve) may hold more promise when it comes to performance prediction at low temperatures. The authors postulate that an exceedingly low temperature coupled with a very fast

loading rate may in fact result in a stiffness-elasticity condition that is essentially at the glassy asymptote. At this location, variability within the data may increase substantially with small changes in phase angle, causing a negative impact to the effectiveness of the analyzed portion of the Black Space diagram.

In closing, the G-R parameter at the intermediate condition shows an inconsistent ability to monitor low temperature performance in Black Space. However, the low temperature condition of the G-R parameter (which centers on a conversion of the Superpave low temperature criteria) holds promise for low temperature performance detection.

Future work is needed to assess the effectiveness of the parameter with varying amounts of recycled materials. This paper presents preliminary data in the development of a mixture-based Black Space parameter, which would utilize the increasingly prevalent $|E^*|$ and δ master curves. Once this analysis tool is in place, the influence of each length scale (e.g., mixture, mastic, fine aggregate matrix, and binder) on cracking performance can be explored in Black Space through additional research efforts. The authors believe the relationship between stiffness and relaxation can not only be applied to low temperature performance in Black Space, but also to structural distresses, such as classical fatigue cracking. Delving into a fatigue cracking parameter would require incorporation of structural elements, and could hold the potential to significantly reduce testing frameworks across the industry. The authors recommend future work be explored by a team of researchers, such as in an ETG setting, to identify the Black Space regions where the G-R parameter will be most effective per distress type.

6.0 Acknowledgements

The authors would like to acknowledge the participating state agencies in TPF 5(230) "Evaluation of Plant-Produced High-Percentage RAP Mixtures in the Northeast": New Hampshire, Maryland, New Jersey, New York, Pennsylvania, Rhode Island, and Virginia. Thanks are also extended to FHWA for acting as a project sponsor. Acknowledgements are given to Callanan and Pike Industries for supplying the mixtures. Gratitude is also expressed towards Dr. Walaa Mogawer for conducting the TSRST on the TPF 5(230) mixtures.

7.0 References

Airey, G.D., "Use of Black Diagrams to Identify Inconsistencies in Rheological Data," *Journal of Road Materials and Pavement Design*, Vol. 3, No. 4, 2002, pp. 403–424.

American Association of State Highway and Transportation Officials, "Determination of Low-Temperature Performance Grade (PG) of Asphalt Binders," AASHTO R 49, Washington, D.C., 2012a.

American Association of State Highway and Transportation Officials, "Method for Thermal Stress Restrained Specimen Tensile Strength," AASHTO TP 10, Washington, D.C., 2012b.

American Association of State Highway and Transportation Officials, "Standard Method of Test for Determining Dynamic Modulus of Hot-Mix Asphalt Concrete Mixtures," AASHTO T 342, Washington, D.C., 2012c.

AASHTO, "Standard Method of Test for Determining the Creep Compliance and Strength of Hot-Mix Asphalt (HMA) Using the Indirect Tensile Test Device," AASHTO T 322, Washington, D.C., 2012d.

AASHTO, "Standard Method of Test for Determining the Flexural Creep Stiffness of Asphalt Binder Using the Bending Beam Rheometer (BBR)," AASHTO T 313, Washington, D.C., 2012e.

AASHTO, "Standard Method of Test for Determining the Fracture Properties of Asphalt Binder in Direct Tension (DT)," AASHTO T 314, Washington, D.C., 2012f.

AASHTO, "Standard Method of Test for Determining the Permanent Shear Strain and Stiffness of Asphalt Mixtures Using the Superpave Shear Tester (SST)," AASHTO T 320, Washington, D.C., 2012g.

AASHTO, "Standard Method of Test for Determining the Rheological Properties of Asphalt Binder Using a Dynamic Shear Rheometer (DSR)," AASHTO T 315, Washington, D.C., 2012h.

AASHTO, "Standard Specification for Performance-Graded Asphalt Binder," AASHTO M320, Washington, D.C., 2012i.

Alavi, M.Z., Hajj, E.Y., Morian, N.E. and Sebaaly, P.E., "Low Temperature Characterization of Asphalt Mixtures by Measuring Viscoelastic Properties under Thermal Loading," Proceedings of ISCORD 2013: Planning for Sustainable Cold Regions, 2013, pp. 404–415.

American Society of Testing and Materials, "Standard Practice for Determining Low-Temperature Performance Grade (PG) of Asphalt Binders," ASTM D6816, West Conshohocken, PA, 2002.

Anderson, D.A., Christensen, D.W., Bahia, H.U., Dongre, R., Sharma, M.G., Antle, C.E. and Button, J., *Binder Characterization and Evaluation, Volume 3: Physical Characterization*, SHRP A-369 Report, Strategic Highway Research Program, National Research Council, Washington, D.C., 1994.

Anderson, M., Anderson, D., Bahia, H., Baumgardner, G., McGennis, B., Planche, J-P., Reinke, G. and Rowe, G., "Asphalt Binder ETG Intermediate Temperature Parameter Task Force," Status Report to Asphalt Binder ETG, 2013.

Anderson, R.M., King, G.N., Hanson, D.I. and Blankenship, P.B., "Evaluation of the Relationship between Asphalt Binder Properties and Non-Load Related

Cracking," *Journal of the Association of Asphalt Paving Technologists*, Vol. 80, 2011, pp. 615–663.

Baumgaertel, M. and Winter, H.H., "Determination of Discrete Relaxation and Retardation Time Spectra from Dynamic Mechanical Data," *Rheologica Acta*, Vol. 28, 1989, pp. 511-519.

Bennert, T., "Rejuvenating Agents with RAP in HMA—Quarterly Meeting Presentation," New Jersey Department of Transportation, 2014.

Bennert, T. and Maher, A., "Forensic Study on the Cracking of New Jersey's Long-Term Pavement Performance Specific Pavement Study Sections," *Transportation Research Record: Journal of the Transportation Research Board*, Vol. 2371, Transportation Research Board of the National Academies, Washington, D.C., 2014, pp. 74–86.

Bouldin, M.G., Dongre, R., Sharrock, M.J., Rowe, G.M., Anderson, D.A., Marasteanu, M., Dunn, L., Zanzotto, L. and Kluttz, R.Q., "A Comprehensive Evaluation of the Binders and Mixtures Placed on the Lamont Test Sections," Report for the Federal Highway Administration Binder ETG, 1999.

Bouldin, M.G., Dongre, R., Rowe, G.M., Sharrock, M.J. and Anderson, D.A., "Predicting Thermal Cracking of Pavements from Binder Properties: Theoretical Basis and Field Validation," *Journal of the Association of Asphalt Paving Technologists*, Vol. 69, 2000, pp. 455–496.

Christensen, D.W., *Mathematical Modeling of the Linear Viscoelastic Behavior of Asphalt Cements*, Ph.D. Dissertation, Pennsylvania State University, State College, PA, 1992.

Christensen, D. "Analysis of Creep Data from Indirect Tension Test on Asphalt Concrete," *Journal of the Association of Asphalt Paving Technologists*, Vol. 67, 1998, pp. 458–492.

Christensen, D.W., Pellinen, T. and Bonaquist, R., "Hirsch Model for Estimating the Modulus of Asphalt Concrete," *Journal of the Association of Asphalt Paving Technologists*, Vol. 72, 2003, pp. 97-121.

Clark, R.C., "Practical Results of Asphalt Hardening on Pavement Life," *Journal of the Association of Asphalt Paving Technologists*, Vol. 27, 1958, pp. 196–208.

Daniel, J.S., Bennert, T., Kim, Y.R., Mogawer, W., Congalton, A., Mensching, D., Sabouri, M. and Elwardany, M., *Evaluation of Plant Produced RAP Mixtures in the Northeast: Phase I Interim Report*, Transportation Pooled Fund Study 5(230), Federal Highway Administration, 2014.

Dave, E.V. and Koktan, P., *Synthesis of Performance Testing of Asphalt Concrete*, Report MN/RC 2011-22, Minnesota Department of Transportation, 2011.

Doyle, P.C., "Cracking Characteristic of Asphalt Cement," *Journal of the Association of Asphalt Paving Technologists*, Vol. 27, 1958, pp. 581–597.

Glover, C.J., Davison, R.R., Domke, C.H., Ruan, Y., Juristyarini, P., Knorr, D.B. and Jung, S.H., *Development of a New Method for Assessing Asphalt Binder Durability with Field Validation*, Report FHWA/TX-05/1872-2, Texas Department of Transportation, 2005.

Gordon, G.V. and Shaw, M.T., *Computer Programs for Rheologists*, Hanser/Gadner Publications, Cincinnati, OH, 1994.

Hiltunen, D.R. and Roque, R. "A Mechanics-Based Prediction Model for Thermal Cracking of Asphaltic Concrete Pavements," *Journal of the Association of Asphalt Paving Technologists*, Vol. 63, 1994, pp. 81–117.

Hopkins, I.L. and Hamming, R.W., "On Creep and Relaxation," *Journal of Applied Physics*, Vol. 28, No.8, 1957, pp. 906–909.

Kandhal, P.S., "Low-Temperature Ductility in Relation to Pavement Performance," *Low Temperature Properties of Bituminous Materials and Compacted Bituminous Paving Mixtures*, ASTM STP 628, American Society for Testing and Materials, 1977, pp. 95–106.

Kim, Y.R. and Wen, H., "Fracture Energy from Indirect Tension Testing," *Journal of the Association of Asphalt Paving Technologists*, Symposium, 2002, pp. 779–793.

King, G., Anderson, M., Hanson, D. and Blankenship, P., "Using Black Space Diagrams to Predict Age-Induced Cracking," *7th RILEM International Conference on Cracking in Pavements*, Delft, The Netherlands, 2012, pp. 453–464.

Leahy, R.B., Harrigan, E.T. and Von Quintus, H., *Validation of Relationships between Specification Properties and Performance*, SHRP A-409 Report, Strategic Highway Research Program, National Research Council, Washington, D.C., 1994.

Marasteanu, M.O. and Anderson, D.A., "Improved Model for Bitumen Rheological Characterization," *Eurobitume Workshop on Performance Related Properties for Bituminous Binders*, Brussels, Belgium, 1999.

Marasteanu, M., Teshale, E.Z., Moon, K.H., Turos, M., Buttlar, W., Dave, E. and Ahmed, S., *Investigation of Low Temperature Cracking in Asphalt Pavements National Pooled Fund Study—Phase II*, Task 3 Report, Minnesota Department of Transportation, St. Paul, MN, 2012.

McDaniel, R.S., Soleymani, H., Anderson, R.M., Turner, P. and Peterson, R., *Recommended Use of Reclaimed Asphalt Pavement in the Superpave Mix Design Method*, National Cooperative Highway Research Program Web Document 30, Project D9-12, National Research Council, Washington, D.C., 2000.

Mensching, D.J., Daniel, J.S., Bennert, T., Medeiros, Jr., M.S., Elwardany, M.D., Mogawer, W., Hajj, E.Y. and Alavi, M.Z., "Low Temperature Properties of Plant-Produced RAP Mixtures in the Northeast," *Journal of Road Materials and*

Pavement Design, Association of Asphalt Paving Technologists Special Issue, DOI: 10.1080/14680629.2014.926617, 2014.

Mogawer, W., Bennert, T., Daniel, J.S., Bonaquist, R., Austerman, A. and Booshehrian A., "Performance Characteristics of Plant Produced High RAP Mixtures," *Journal of Road Materials and Pavement Design*, Vol. 13, Supplement 1, Special Issue: Papers from the 87th Association of Asphalt Paving Technologists Annual Meeting April 1-4, 2012, 2012, pp.183–208.

Oliver, J.W.H., "The Influence of the Binder in RAP on Recycled Asphalt Properties," *International Journal of Road Materials and Pavement Design*, Vol. 2, No. 3, 2001, pp. 311–325.

Pellinen, T., Rowe, G. and Biswas, K., *Evaluation of Surface (Top Down) Longitudinal Wheel Path Cracking*, Report FHWA/IN/JTRP-2004/6, Indiana Department of Transportation, Indianapolis, IN, 2004.

Richards, F.J., "A Flexible Growth Function for Empirical Use," *Journal of Experimental Botany*, 10(29), 1959, pp. 290–300.

Romero, P. and Jones, Z.L., *Implementation of Low Temperature Tests for Asphalt Mixtures to Improve the Longevity of Road Surfaces*, Report MPC-13-260, North Dakota State University Upper Great Plains Transportation Institute, Fargo, ND, 2013.

Rouse, Jr., P.E., "A Theory of the Linear Viscoelastic Properties of Dilute Solutions of Coiling Polymers," *Journal of Chemical Physics*, Vol. 21, No. 7, 1953, pp. 1272–1280.

Rowe, G., "Phase Angle Determination and Interrelationships within Bituminous Materials," *7th International RILEM Symposium on Advanced Testing and Characterization of Bituminous Materials*, Book 1 (Edited by Loizos, Partl, Scarpas and Al-Qadi), Rhodes, Greece, 2009, pp. 43–52.

Rowe, G., "Evaluation of the Relationship between Asphalt Binder Properties and Non-Load Related Cracking," Prepared Discussion, *Journal of the Association of Asphalt Paving Technologists*, Vol. 80, 2011, pp. 649–663.

Rowe, G., "Analysis of SHRP Core Asphalts New 2013/14 Test Results," presented at Federal Highway Administration Binder Expert Task Group, San Antonio, TX, 2014a.

Rowe, G., "Interrelationships in Rheology for Asphalt Binder Specifications," submitted to Canadian Technical Asphalt Association, 2014b.

Rowe G.M. and Sharrock, M.J., "Development of Standard Techniques for the Calculation of Master Curves for Linear Viscoelastic Materials," *1st International Symposium on Binder Rheology and Pavement Performance*, University of Calgary, Alberta, Canada, 2000.

Rowe, G.M. and Sharrock, M.J., "Alternate Shift Factor Relationship for Describing the Temperature Dependency of the Viscoelastic Behavior of Asphalt Materials,"

Transportation Research Record, Vol. 2207, Transportation Research Board of the National Academies, Washington D.C., 2011, pp. 125–135.

Rowe, G.M., Baumgardner, G. and Sharrock, M.J., "Application of Rheological Models to Modified Binders," Presentation to the 48th Petersen Asphalt Research Conference, Western Research Institute, Laramie, WY, 2011.

Rowe, G.M., Sharrock, M.J., Bouldin, M.G., & Dongre, R.N., "Advanced Techniques to Develop Asphalt Master Curves from the Bending Beam Rheometer," *Journal of Petroleum and Coal*, 43(1), 1999, pp. 54–59.

Rowe, G.M., King, G. and Anderson, M., "The Influence of Binder Rheology on the Cracking of Asphalt Mixes on Airport and Highway Projects," *4th International Conference on Asphalt Materials*, Guangzhou, China, 2013.

Soleymani, H.R., Anderson, M., McDaniel, R. and Abdelrahman, M., "Investigation of the Black Rock Issue for Recycled Asphalt Mixtures," *Journal of the Association of Asphalt Paving Technologists*, Vol. 69, 2000, pp. 366–390.

Tabatabaee, H., Velasquez, R., Arshadi, A. and Bahia, H., Investigation of Low Temperature Cracking in Asphalt Pavements National Pooled Fund Study—Phase II, Task 5 Report, University of Wisconsin-Madison, 2012.

Texas Department of Transportation, "Test Procedure for Overlay Test," Tex-248-F, Austin, TX, 2014.

Williams, M.L., Landel, R.F. and Ferry, J.D., "The Temperature Dependence of Relaxation Mechanisms in Amorphous Polymers and Other Glass-Forming Liquids," *Journal of the American Chemical Society*, Vol. 77, 1955, pp. 3701–3706.

8.0 Discussion

GAYLE KING: David, very valuable paper. It's going to take several PhD theses to sort through these Black Space related cracking issues. I do have some comments. Could you bring up the slide showing the intermediate temperature Glover-Rowe results, where none of the binders in the Lamont Test Road would have been predicted to fail?

DAVID MENSCHING: Yes.

GAYLE KING: The problem here is that Lamont is in a very cold Canadian climate, whereas the Glover-Rowe parameters defining the damage zone were derived for a typical Pennsylvania climate, assumed to be PG 58-28. As for any performance-graded binder parameter, DSR test conditions or failure limits must be adjusted for pavement temperature. Since the low pavement temperature for Lamont is so much colder than Pennsylvania, much softer grades of asphalt would be used for paving. Without adjusting DSR test temperature or frequency, the aged binder cracks in Lamont long before reaching the failure limits defined for Pennsylvania. Your successful approach here ties a Glover-Rowe failure approach to the low pavement temperature, which necessarily adjusts the test temperature to the low pavement temperature for any climate.

DAVID MENSCHING: Yes.

GAYLE KING: Mike Anderson's 2011 AAPT paper suggests that unmodified binders have similar Black Space relationships at intermediate and low test temperatures, so either Glover-Rowe or the BBR ΔT can rank materials for aged-induced surface cracking (block cracking). You show the Glover-Rowe approach works at low pavement temperatures as automatically tied to climate, so G-R should rank materials similarly at intermediate temperatures so long as a comparable climate adjustment is made. Geoff and I have been discussing ways to grade adjust the intermediate temperature version of G-R, including an off-set from the low pavement temperature (T(low)+43°C) or the midpoint of the climate temperature range. Another option is to hold the test temperature at 15°C and adjust frequency with climate. It doesn't bother me that the Lamont performance was not ranked correctly when applying the intermediate temperature G-R test conditions and failure limits reported in the literature. It still should work if test conditions are correctly adjusted for other climates.

DAVID MENSCHING: Thanks.

GAYLE KING: By choosing to use the low temperature of the PG grade, you eliminate the climate problem, but using DSRs would require instrument upgrades such as 4 mm plate geometry and compliance correction. I really like your approach for adapting BBR data to create Glover-Rowe like failure zones rather than ΔT failure criteria. Conceptually, your idea for using low temperature failure criteria to predict age-induced surface cracking also fits our observations for damage in unconfined mixes after long-term aging. (King et al., Rilem Cracking Conference, 2012, Delft)

DAVID MENSCHING: Right. I think that's a good point. When you're using this type of analysis, you do have to be aware of the limitations of it. It was really calibrated, from what I understand, for sections in Pennsylvania. So when you're looking at an area like Alberta, where it's much colder than Pennsylvania, you may need to make adjustments to calibrate accordingly, as you're saying. And that effect is captured here. Thank you.

GAYLE KING: The other big problem here is there is a real difference in results and ranking sometimes between the low and the intermediate test temperatures. But as long as we're meeting time-temperature superposition and Christensen-Anderson model conditions where the model works, it doesn't matter where we run the test, as long as all the conditions for a good R-value time-temperature position are met. But waxes in asphalt melt on heating, so if you change the temperature, you change the liquid. When you change the liquid, you change your location in black space, and you change the rankings.

DAVID MENSCHING: I agree.

GAYLE KING: Another concern has to do with boundary conditions for equating crack prediction at low vs. intermediate test temperatures. When rheological measurements are made in the linear range for binders that fully satisfy time-temperature superposition and Christensen-Anderson model assumptions, relative material rankings should hold over a broad range of test conditions. That is, curves for R-value in Black Space have uniform space if a common glassy modulus is assumed. Waxy asphalts complicate the problem, because waxes gradually melt on heating, and they precipitate/crystallize on cooling. Temperature alters the structure of the liquid, such that time-temperature superposition is no longer valid. The unfortunate consequence of wax on asphalt rheology is that there is no constant R-value. The relative ranking of different binders now changes with test temperature because the relative location in Black Space has changed.

DAVID MENSCHING: Okay. Thank you.

GAYLE KING: A second comment regarding the frequencies you used: 0.005, 5, and 500 rad/sec.

DAVID MENSCHING: Yes. Right around here. Which slide do you want to look at?

GAYLE KING: It doesn't really matter. Go back to a slide that had more data points, and the individual frequency sweeps are not separated due to phase changes caused by wax.

DAVID MENSCHING: Okay.

GAYLE KING: If you plot multiple frequency sweeps for a single binder in Black Space, it's easy to see which frequencies fall within the linear range, because those data points will fall exactly on the Black Space mastercurve (R-value). For his cracking parameter, Glover chose a very low DSR test frequency of 0.005 rad/sec, which is well within the linear test regime. At a much higher 5 rad/sec, the data still appears to be within assumptions for linearity. At 500 rad/sec, the individual frequency sweeps clearly diverge from the full mastercurve, so assumptions for linearity and a uniform R-value are no longer valid. Glover's test frequency of 0.005 rad/sec requires a long time for a single test, perhaps 20 minutes, and is overly conservative. 500 rad/sec is clearly beyond the linear range of the material, and therefore too high. It would be helpful to have research that would adapt the Glover-Rowe approach to a friendlier test frequency that remains within the linear region but speeds up the test, perhaps 0.1 or 1 rad/sec. Dr. Rowe showed how to adapt the G-R failure limits for frequency in his prepared discussion to Dr. Mike Anderson's 2011 AAPT paper.

DAVID MENSCHING: Right.

GAYLE KING: A third comment. You mentioned TSRST as validation for low temperature crack prediction. TSRST is certainly appropriate for prediction of transverse thermal cracking, as well as for validation of BBR specifications. The models for transverse thermal cracking are in place, and they're good, but they're all derived from the assumption that tensile stresses develop as cooling mixes shrink while confined in one dimension. As a reminder, the Glover-Rowe approach was developed to explain block cracking (age-induced surface cracking) in pavements, not transverse thermal cracking. There is no comparable pavement confinement in highly cracked pavement surfaces, so Glover-Rowe performance rankings do not necessarily agree with TSRST measurements. A new cracking model is needed to explain why aged asphalt mixes crack without the confinement imposed by TSRST. Binder phase angle should be much more important to healing processes that occur between multiple thermal cycles than to a single-event tensile load to failure. That's another PhD thesis I'll leave to Dr. Daniel.

DAVID MENSCHING: Yes.

GAYLE KING: So I would say 500 is absolutely unacceptable, 0.005 that's in the recommendations now we know is linear but is so slow it takes you 20 minutes to a couple hours to run the test, so that's not acceptable.

DAVID MENSCHING: Right.

GAYLE KING: We know we can go higher. Five's probably a little high, 0.1 we've got legacy data. It's fast. We know we're still linear. So that's still to be worked out and anything you could do to help sort that out would be good.

DAVID MENSCHING: What we've seen in additional work is, at least for what I'm defining as a typical New England event, our E* point is corresponding closely to 20 °C 1 Hz, so again in the linear range. But, yes, 500 rad/s, you know, the phase angles are so low, you are right; the assumptions of linearity needed are probably invalid at that point.

GAYLE KING: Third comment and that is you mentioned TSRST and obviously we use a BBR for transverse thermal cracking, but this parameter was originally developed to be more for block cracking, that is, where we don't have confinement in a pavement, and therefore it's not anticipated that this will necessarily match TSRST rankings.

DAVID MENSCHING: Okay, right. Thank you for your comments.

GEOFFREY ROWE: David, nice paper.

DAVID MENSCHING: Nice paper to you too!

GEOFFREY ROWE: I guess I'm allowed to discuss our paper. Just a couple of comments about some of these mixtures just to build on what Dave said. What we wanted to do was go back and look at some of the old sites, so the Indiana work was some work that Tehri Pellinen and I had been involved with Gerry Huber and various other people from Indiana DOT who helped us select some of those test sites. We were sort of curious to go back and see how these roads worked with that Glover-Rowe framework. In addition, we also thought we should look at the Lamont Test Road. The Lamont site was a classic thermal cracking problem. The Glover-Rowe concept was initially developed around block cracking and durability/aging cracking, but we can apply the same concept to thermal cracking. So what we did is that we took data from the Bending Beam Rheometer and performed an interconversion to G* and phase angle. We then know that there is a certain value of G* to which the S(t) corresponds. In addition, there is a value of the phase angle to which that the m-value corresponds. These appear to be pretty good interconversions and with those and we end up with the point on the "Black space," like David showed, which is a small upper left hand corner of that "Black space" where if you are in that box on the upper left hand corner, you actually have the same calibration of your thermal cracking parameter that Rita Leahy and her coworkers had in the SHRP report referenced. The data points in that plot were the validation points, where from Rita's report that she developed is part of the validation effort of the SHRP work. So in this work there are two different concepts. The biggest thing that we're trying to understand is the effect of stress and buildup, which is E* or G* dominated, and then the relaxation of that stress, which is phase angle dominated (or as in the Bending Beam Rheometer, S and m dominated). One of the references we made for this work was some work done by Pedro Romero in Utah. He's at Salt Lake City and had done a similar sort of exercise like this, and he had presented a report that was published by the DOT. It was about a year ago now so that's available. He had looked at a similar sort of concept but taking data from a bending beam where he used Mihai's technique (the technique that's been developed by the University of Minnesota) and looked at S and m. That's a very useful report and that's online, and it sort of builds on this concept. This concept is building upon this stiffness and relaxation concept. We know we can do more with it but it's not everything. I know Frank Fee and various other folks have talked about the importance of mixture and aggregate and binder and the interactions that we get with all of these together. I think what we're trying to do is to explore this concept. This is why the title of the paper was called Exploring. We're not saying it's a perfect world; we're just trying to understand and see how these things might fit together.

AMIR GOLALIPOUR: Very nice presentation, David and Geoff. There is a lot of information in a very short time.

DAVID MENSCHING: Thanks.

AMIR GOLALIPOUR: One question that I have is in follow up of Gayle's question. What was the idea behind choosing these testing frequencies? Because as you can see, it's going from an extreme to another extreme, from very, very low frequency to very high frequency, and there is only one point in the middle range. I was just wondering why you chose these two extremes. Why not have some more points in the middle range to probably capture the material behavior more clearly?

DAVID MENSCHING: There wasn't a whole lot of science behind the selection, if I'm being honest. We chose three different frequencies that we thought would capture different response regions along the black space curve. So the 0.005 rad/s, that happens to be the frequency selection of the Glover-Rowe, but it's also an area where you're approaching inflection. Five hundred, that's very stiff. That's where you going to be approaching that glassy region. So, yes, I agree that it would be better to have more points, and in our present work we're doing more of that.

AMIR GOLALIPOUR: Especially at 500 rad/s frequency, like impact kind of behavior, so I'm not sure if this is really simulating material's behavior in the field. Also, a quick comment for your slide 29. I'm seeing that for that green line, 0.005 radians/second, it looks you used kind of exponential fitting. However, for all other graphs, you had linear type of relationship. Was there was any specific reason you changed the fitting method and treating the data differently here than other graphs?

DAVID MENSCHING: They're the same type of fit, logarithmic fits in this case.

AMIR GOLALIPOUR: For example if you go to slide 28, it looks like linear fitting to me. It didn't look to me exponential fitting here, but you changed the fitting. I was just wondering if maybe there is some reason that I don't know about it.

DAVID MENSCHING: Yes. Well, I think in this case the difference between TCModel and the TSRST was we were starting to see more points toward the inflection point of the Black Space curve. Because you're talking about a situation where TCModel critical cracking temperature predictions generally were warmer. So when we're plugging these frequencies in, we're actually shifting ourselves further towards the inflection point frequency on the black space curve. So I think that's part of it. The real purpose of these slides was to assess the variability, you know, how much jumping around you have in your points, and then ultimately what that separation is. And then bring that further to a comparison with the field and field and lab performance. In this case, it was lab performance because the Pooled Fund materials aren't in the surface course.

AMIR GOLALIPOUR: And just one more quick comment. When you're looking at the variability, one of the things that I'm sure you considered is the sensitivity of the instrument because at 500 rad/s, you can see, as you mentioned, your phase angle is very low, so a little bit change would be like a 100% difference; but when you're going to a larger phase angle, you may not see that type of behavior.

DAVID MENSCHING: Thank you for the comment.

AMIR GOLALIPOUR: Thank you for the good study.

GEOFFREY ROWE: Just a quick comment on these frequencies. Dave, I think you can probably confirm that these are reduced frequencies from master curves.

DAVID MENSCHING: Yes, they are.

GEOFFREY ROWE: So it's not as if we're actually running the test at that frequency, that these are reduced frequencies from master curve.

DAVID MENSCHING: Correct. Sorry for the confusion.

GEOFFREY ROWE: So all the measurements are sort of within that linear region because we used the normal procedures. Some of this work was done ten years ago now so maybe the procedures now are a little more accurate and better refined, but basically, they were all reduced.

DAVID MENSCHING: Right.

Evaluation of Dynamic Modulus in Asphalt Paving Mixtures Utilizing Small-Scale Specimen Geometries

Benjamin F. Bowers[a], **Brian K. Diefenderfer**[a], **and Stacey D. Diefenderfer**[a]

[a] *Virginia Center for Transportation Innovation and Research, Charlottesville, VA 22903*

ABSTRACT. *The results of dynamic modulus testing have become one of the primarily used performance criteria to evaluate the laboratory properties of asphalt mixtures. This test is commonly conducted to characterize asphalt mixtures mechanistically using an asphalt mixture performance tester (AMPT) as developed in NCHRP Project 9-29. The test specimen geometry consists of a cylinder with a 100 mm diameter and a 150 mm height. This geometry is practical for laboratory-prepared specimens produced using a gyratory compactor. However, the specimen scale is problematic when the test specimen is prepared from field cores and the investigator wants to isolate the testing to a single asphalt mixture material/layer. This is because most asphalt mixture layers, especially surface and intermediate layers, are placed with a thickness less than 150 mm. This study investigated the use of small-scale specimens as an alternative means to conduct dynamic modulus testing of asphalt mixture materials. To validate the small-scale approach, the dynamic modulus measured on small-scale cylindrical specimens was compared to the dynamic modulus measured on full-size cylindrical specimens (100 x 150 mm) using asphalt mixtures with a nominal maximum aggregate size (NMAS) of 9.5, 12.5, 19.0, and 25.0 mm. Small-scale specimens with a diameter and height of 38 x 135 mm, 50 x 135 mm, 38 x 110 mm, and 50 x 110 mm were evaluated. From the findings of this study, for 9.5 and 12.5 mm NMAS mixtures, any of the four small-scale geometry dimensions tested appears to be suitable alternatives to the full-size specimen. For 19.0 and 25.0 mm NMAS mixtures, the two small-scale geometries with a diameter of 50 mm appear to be suitable alternatives to the full-size specimen.*

KEYWORDS: *dynamic modulus, asphalt mixture, small scale, asphalt mixture performance tester, AMPT*

The oral presentation was made by Dr. Bowers.

1.0 Introduction

Determination of dynamic modulus has become one of the most common methods for evaluating asphalt mixture properties, in part because of the ability to use the resultant modulus values for performance prediction. This determination can be readily achieved using the asphalt mixture performance tester (AMPT) as recommended by NCHRP Project 9-19 (Witczak et al., 2002) and developed in NCHRP Project 9-29 (Bonaquist, 2008). Testing often is performed in accordance with AASHTO TP 79 (Standard Method of Test for Determining the Dynamic Modulus and Flow Number for Asphalt Mixtures Using the Asphalt Mixture Performance Tester (AMPT)) (American Association of State Highway Transportation Officials [AASHTO], 2013), on cylindrical specimens with a 100 mm diameter and a 150 mm height, which are typically prepared from laboratory-compacted gyratory specimens. However, conducting dynamic modulus tests on field-placed asphalt mixtures in accordance with the AASHTO TP 79 test procedure is problematic, as most asphalt mixture layers are placed in the field at a thickness less than 150 mm. In this case, the fabrication of a 150 mm tall test specimen requires that either the test specimen be created from a single core containing multiple layers/materials, or from multiple cores wherein the desired layer is stacked until a sufficient height is achieved or that an alternate test geometry be employed. Limited research has been conducted to investigate the use of alternate test geometries for performance tests (Kim et al. 2004, Kutay et al., 2009; Li and Gibson, 2013; Pellinen et al., 2006; Witczak et al., 2000). These alternative geometries may be beneficial because of the typical as-constructed layer thickness for asphalt pavements.

Kim et al. (2004) developed a method to conduct dynamic modulus testing in the indirect tensile mode by loading a disk-shaped specimen with a diameter of 150 mm and a thickness as thin as 38 mm. The method was compared with that specified in AASHTO TP 79 using a sample of 12 North Carolina asphalt mixtures with a nominal maximum aggregate size (NMAS) ranging from 9.5 to 25 mm and asphalt binder performance grades (PG) 64-22 and 70-22. Visual analysis showed good agreement between the dynamic modulus mastercurves for the two specimen geometries. Statistical analyses were conducted by performing *t*-tests (with unequal variances) for each mixture at five reduced frequencies and showed that of the 60 tests conducted, 90% showed no statistical difference between the two test specimen geometries. In addition, the authors computed a percentage difference between the moduli from the two geometries; of the 288 combinations evaluated, 70% of the tests had a difference less than 10%.

Pellinen et al. (2006) examined another approach to the testing of road cores by using rectangular prism specimens, sized 80 x 75 x 130 mm that could be fabricated from existing road cores for forensic and/or post-construction analysis and comparison. When compared to full-size cylindrical specimens, the rectangular specimens were slightly stiffer, although an analysis of variance (ANOVA) at a level of significance of $\alpha = 0.05$ found no significant differences between the two geometries. The authors noted the discrepancy between a lack of significance in statistical analysis and the observed differentiation between sample types, attributing it to the small sample size and high degree of variability in the dynamic modulus test. The authors cited a coefficient of variation (COV) as high as 47% between samples as being the cause of the acceptance of the null hypothesis despite the general trends. Based on this finding, the authors stated that a 10% to 15% modulus difference can be considered acceptable when different specimen geometries are examined. They concluded that more testing was needed to validate these results but that the initial findings indicated that using rectangular specimens is a feasible way to test pavement cores.

Witczak et al. (2000) conducted a study to determine the minimum test specimen dimensions for uniaxial dynamic modulus testing to provide accurate material characterization independent of the NMAS and specimen size. Cylindrical specimens with diameters of 70, 100, and 150 mm and height to diameter (H/D) ratios of 1.0, 1.5, 2.0, and 3.0 were evaluated using the dynamic modulus and repeated load permanent deformation tests. Dynamic modulus testing was conducted at a high stiffness condition (4°C, 16 Hz, 1 MPa nominal stress) and a low stiffness condition (40°C, 0.1 Hz, 34 kPa nominal stress) in accordance with ASTM D3497. Tests were conducted on surface and base mixtures with a NMAS of 12.5 and 19 mm and 37.5 mm, respectively, with air void content tolerances of ±0.5% and ±1.0%, respectively. A two-way ANOVA considering specimen diameter and H/D ratio was performed. The study showed that there were no significant differences between means for all specimens. However, upon further study, the authors decided that the ANOVA could not determine the difference between means because of the spread of the data, particularly below a H/D ratio of 1.5. They concluded that dynamic modulus tests should not be conducted with an H/D ratio less than 1.5. When examining the phase angle, the authors determined that the specimen diameter did have a significant effect but ultimately concluded that more research needed to be conducted to study this effect.

Li and Gibson (2013) compared the results of dynamic modulus testing using small-scale specimens with a 38 mm diameter and both a 110 and 140 mm height with the dynamic modulus from full-size specimens. Small-scale cylindrical specimens were prepared by coring vertically or horizontally through 150 mm diameter field cores or gyratory-prepared samples. All sets of specimens were tested in accordance with AASHTO TP 79. Testing was conducted at temperatures of 4.4, 21.1, 37.8, and 54.4°C and frequencies of 25, 10, 5, 1, 0.5, and 0.1 Hz on three replicates per mixture. Ten mixtures with a NMAS from 4.75 to 19.0 mm were tested. The binder PG ranged from 58-34 to 82-22. The study showed that, in general, the dynamic modulus of the small-scale specimens was less than that of the full-sized specimens. The difference between the dynamic modulus of the 110 mm height small-scale and full-size specimens was generally less than 20%. The difference between the dynamic modulus of the 140 mm height small-scale and full-size specimens was generally greater than 20%.

2.0 Objective

The objectives of this research project were to (1) investigate the feasibility of conducting dynamic modulus tests using the AMPT on small-scale specimens produced from asphalt mixtures using different small-scale specimen sizes than used previously; (2) compare the results of dynamic modulus testing using small-scale specimens versus full-size specimens; and (3) determine an optimum small-scale specimen geometry from the four small-scale geometries tested. By meeting these objectives, a determination can be made as to whether small-scale specimens can be used as an alternate geometry for dynamic modulus testing when full-size specimens are not possible.

3.0 Materials and Methods

3.1 Custom Hardware Fabrication

Prior to testing, some test fixture components had to be custom machined to accommodate the reduced specimen size. Specifically, custom arms for the linear variable-displacement transducer (LVDT) stud gluing jig and testing platens for the AMPT were manufactured from aluminum; photographs of each are shown in Figures 1 and 2, respectively. A unique set of gluing jig arms was manufactured for the small-scale specimens because of their decreased diameter. The upper portion of each gluing jig arm had to be extended so that the arm could apply pressure to the LVDT stud as it was being glued to the specimen. For the 38 and 50 mm diameter specimens, the upper portions of the arms were extended by approximately 32 and 25 mm, respectively. All other dimensions for the custom arms were the same as the stock gluing arms. Custom testing platens for the AMPT were also machined to facilitate centering the specimen during testing. A unique set of platens was fabricated to match the diameter of the small-scale specimens. Removable spacer blocks, used as the lower support for the small-scale specimens, were also manufactured for the different specimen heights. Testing was conducted using stock LVDT gauges and studs; the LVDT stud spacing was 70 mm.

Figure 1. *Photograph of LVDT stud gluing jig showing custom arms.*

Figure 2. *Photograph of AMPT testing platens with removable spacer blocks for 38 and 50 mm diameter small-scale specimens.*

3.2 Specimen Preparation

From each mixture, eight to ten 150 mm diameter by 200 mm height samples were prepared in a gyratory compactor. Three were used to fabricate full-size (100 mm diameter, 150 mm height) specimens in accordance with AASHTO PP 60 (Preparation of Cylindrical Performance Test Specimens Using the Superpave Gyratory Compactor (SGC)) (AASHTO, 2013); the remaining samples were used to produce the small-scale specimens. The small-scale specimens were fabricated by coring perpendicular to the long axis of the gyratory-prepared sample as shown in Figure 3; generally three small-scale specimens were cored from each gyratory sample. In some cases, the height of the gyratory sample was less than 200 mm, and from these samples, only two small-scale specimens were produced. The small-scale specimens were fabricated such that, at a minimum, the top and bottom approximately 1 inch of the gyratory sample was avoided. This procedure was followed to maintain uniform air void contents between specimens (following findings of Chen et al. [2013]). The small-scale specimens were then trimmed to height using a conventional wet saw to produce three samples each at 38 mm and 50 mm diameter by 110 mm height and three each at 38 mm and 50 mm diameter by 135 mm height. Figures 4 and 5 show photographs of the gyratory sample after the small-scale specimens were cored and specimens prepared for testing, respectively.

Figure 3. *Core drill sample holder used to extract small-scale cylindrical specimens from a gyratory sample.*

Figure 4. *Gyratory samples after small-scale cylindrical specimens were cored.*

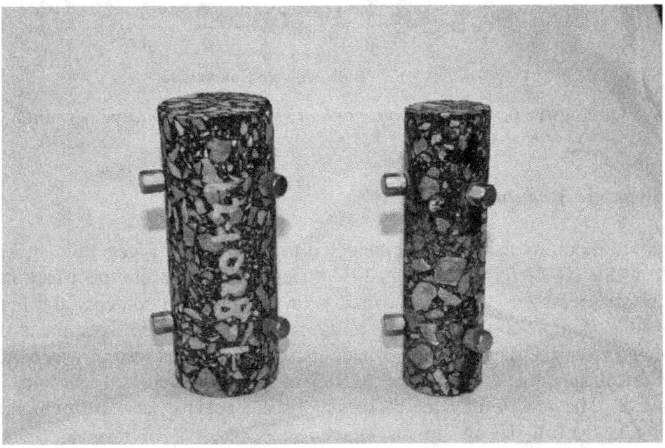

Figure 5. *Examples of small-scale cylindrical specimens prepared for testing.*

All specimens were prepared from plant-produced mixtures. Gyratory samples were compacted either immediately after production while still hot or at a later time after reheating. The intent was to produce gyratory samples with air void contents such that the cored full-size AMPT specimen (100 mm diameter by 150 mm height) contained 7% to 8% air voids. However, some specimens (small scale and full size) were produced with air void contents as high as 12%. Some specimen replicate groups exceeded the ±0.5% of the average design air void as specified in AASHTO PP 60. Figure 6 shows the cumulative distribution of the bulk-specimen air void content from all specimens with respect to specimen geometry. From Figure 6 it can be seen that the 50th percentile bulk air void content ranged from approximately 6.5% to 7.0% with a range of approximately 2% to 12%.

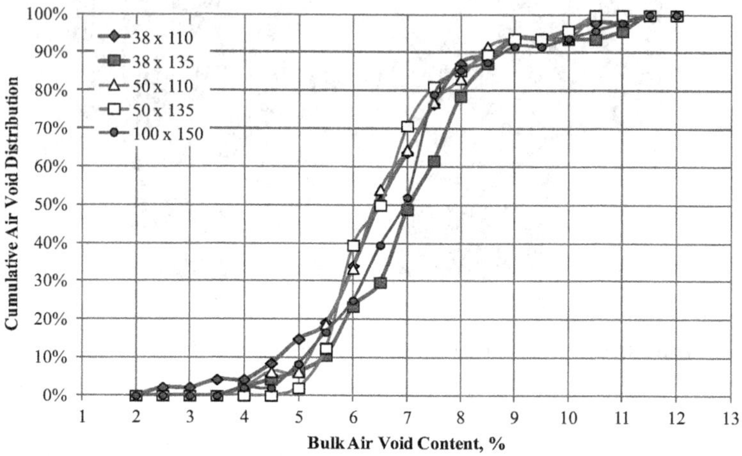

Figure 6. *Bulk air void content shown as cumulative percentage.*

3.3 Dynamic Modulus Testing

Dynamic modulus testing was conducted using an AMPT generally in accordance with AASHTO TP 79 (AASHTO, 2013). Deviations from the specification included using test specimens with higher-than-allowable air void contents and higher-than-allowable sample-to-sample variability. Data from these specimens were not removed from the data set. Small-scale dynamic modulus testing was conducted on cylindrical specimens with 110 and 135 mm heights and both 38 and 50 mm in diameter. The test results for these small-scale specimens were compared to the results for the full-size specimens.

Originally, the study sought to investigate further the work of Li and Gibson (2013) by testing specimens with a height of 110 and 140 mm. Although Li and Gibson used only one specimen diameter (38 mm), an additional diameter (50 mm) was investigated to determine if the larger diameter might better correlate with the full-size specimen geometry for larger NMAS mixtures. It was found that a small-scale specimen height greater than 135 mm could not be achieved for a 50 mm diameter specimen because of the curvature of the specimen ends from the parent 150 mm diameter gyratory sample. Since the two ends must be trimmed to be flat and parallel, the maximum height that could be obtained was approximately 135 mm for the 50 mm diameter specimens and slightly more than 140 mm for the 38 mm diameter specimens. To maintain consistency in the height of the 38 and 50 mm diameter specimens, 135 mm was chosen as the maximum. Figure 7 shows a photograph of a 38 mm diameter by 110 mm height specimen with LVDT gauges attached in the AMPT.

Figure 7. *Close-up of a 38 mm diameter by 110 mm height specimen within the AMPT test frame.*

As shown in Table 1, sixteen asphalt mixtures were used to fabricate small-scale and full-size specimens for dynamic modulus testing. The Lab Mix ID is given as xx-yyyy where xx is the last two digits of the calendar year in which the mixture was sampled and yyyy is the consecutive number of the mixture as recorded in the laboratory (starting with 1000). These asphalt mixtures represent typical mixtures used in Virginia. The NMAS ranged from 9.5 to 25 mm. The binder performance grades included PG 64-22, 70-22. and 76-22. The asphalt mixtures included 15 dense-graded mixtures and one stone matrix asphalt (SMA) mixture. For the first five mixtures tested (12-1028, 13-1070, 13-1088, 12-1041, and 12-1052), testing was first completed on 135 mm height specimens, which were then trimmed down to 110 mm in height. The testing at 135 mm height was conducted at temperatures of 4.4, 21.1, and 37.8°C. Since there was a concern that the specimens could be damaged during testing, the 54.4°C temperature was not used. The testing was then repeated after trimming with the addition of the 54.4°C test temperature. For the remainder of the study, unique 135 and 110 mm height specimens were prepared and testing was conducted at all four temperatures for each diameter/height combination.

Table 1. *Asphalt mixtures tested.*

Nominal Maximum Aggregate Size, mm	Lab Mix ID	Binder Performance Grade
9.5	12-1028[a]	70-22
	13-1070[a,b]	64-22
	13-1088[a]	64-22
	13-1107[c]	64-22
	13-1119	64-22
	13-1121	64-22
	13-1125	64-22
12.5	12-1041[a]	76-22
	12-1052[a]	70-22
	13-1118	64-22
19.0	13-1104	64-22
	13-1106	64-22
	13-1128	64-22
	14-1006	70-22
25.0	13-1108[c]	64-22
	13-1120	64-22

[a]*Testing was not conducted at 54.4°C for 135 mm height specimens (110 mm height specimens were trimmed from 135 mm height specimens).*
[b]*Stone matrix asphalt mixture.*
[c]*Testing was not conducted at 54.4°C for any specimen.*

4.0 Analysis and Discussion of Results

Figure 8 shows an example of the dynamic modulus testing results for a single asphalt mixture; the results are included for both 38 and 50 mm diameter specimens having a height of 110 mm versus the results of a full-size specimen. For each of the four test temperatures there are six data points corresponding to the six test frequencies. Each data point represents the average dynamic modulus of three replicates. The data are plotted with respect to the line of equality. Surrounding the line of equality are two sets of dashed lines representing ±10% and 20% difference from the line of equality. Figure 8 shows that the data at 4.4, 21.1, and 37.8°C for both specimen diameters for this particular mixture were nearly on the line of equality, indicating that there is little difference between the dynamic modulus from full-size or small-scale specimens for this mixture at either specimen diameter at the three lowest test temperatures. The data from the 54.4°C test temperature showed that the small-scale specimens were approximately 10% to 20% stiffer than the full-size specimens at the 50 and 38 mm diameters, respectively.

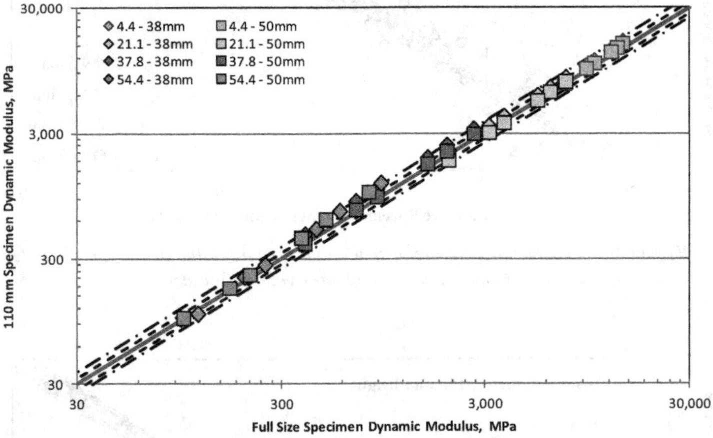

Figure 8. *Example showing the average dynamic modulus from one mixture using 110 mm height small-scale specimens versus full-size specimens.*

4.1 Results from 135 mm Height Specimens

Figures 9 and 10 present the results of dynamic modulus tests for 135 mm height small-scale specimens at 38 and 50 mm diameters, respectively, versus the results for full-size specimens. From each figure it can be seen that the agreement of dynamic modulus between small-scale and full-size specimens generally decreases with decreasing stiffness. By visual inspection, the agreement of dynamic modulus between small-scale and full-size specimens generally improved with increasing small-scale specimen diameter when comparing the 38 mm diameter small-scale specimen results (Figure 9) to the 50 mm diameter small-scale specimen results (Figure 10). Of note in Figure 9 is the deviation from the line of equality for the samples at low stiffness, which may be attributed to the higher variability encountered when testing at 54.4°C.

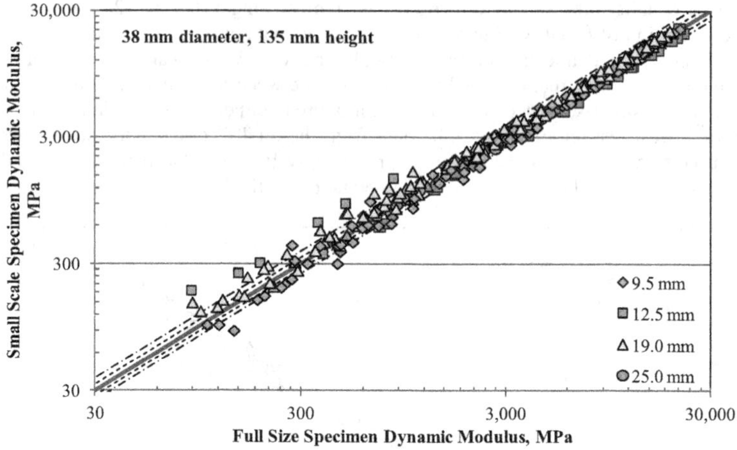

Figure 9. *Results showing the average dynamic modulus of 38 mm diameter by 135 mm height small-scale specimens versus full-size specimens.*

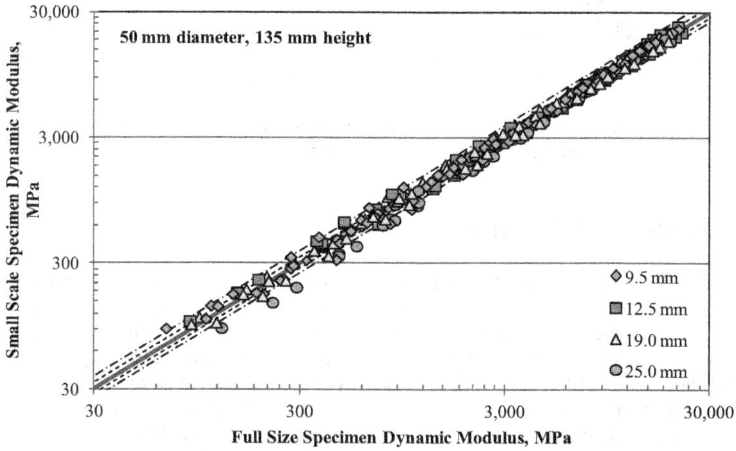

Figure 10. *Results showing the average dynamic modulus of 50 mm diameter by 135 mm height small-scale specimens versus full-size specimens.*

4.2 Results from 110 mm Height Specimens

Figures 11 and 12 present the results of dynamic modulus tests for 110 mm height small-scale specimens at 38 and 50 mm diameters, respectively, versus full-size specimens. From each figure it can be seen that the agreement of dynamic modulus between small-scale and full-size specimens generally decreases with decreasing stiffness, as was found for the 135 mm height small-scale specimens. By visual inspection it was less clear whether the agreement of dynamic modulus between small-scale and full-size specimens increased with increasing small-scale specimen diameter when moving from the 38 mm diameter small-scale specimen results (Figure 11) to the 50 mm diameter small-scale specimen results (Figure 12).

In addition, Figures 11 and 12 show the influence of repeated testing on small-scale specimens. The first five mixtures evaluated (12-1028, 13-1070, 13-1088, 12-1041, and 12-1052 as noted in Table 1) were first tested as 135 mm height specimens at 4.4, 21.1, and 37.8°C. Following this, they were trimmed to 110 mm height specimens by removing approximately 12.5 mm from each end of the specimen using a wet saw. The specimens were then retested at 4.4, 21.1, 37.8, and 54.4°C as 110 mm height specimens. During analysis it was found that the outlying data, designated with markers having no fill in Figures 11 and 12, came from these trimmed specimens. The higher variability is likely due to the small-scale specimens experiencing damage during testing at the 135 mm height or during the trimming procedure. After the data from the five retested mixtures were reviewed, unique specimens were fabricated at each height for the remainder of the study.

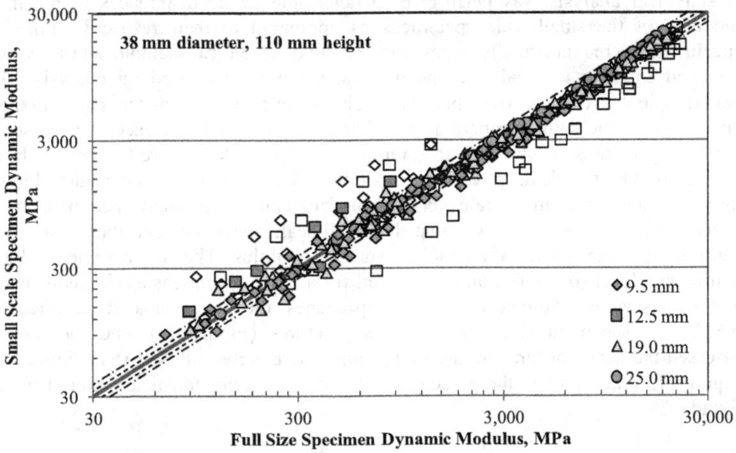

Figure 11. *Results showing the average dynamic modulus of 38 mm diameter by 110 mm height small-scale specimens versus full-size specimens.*

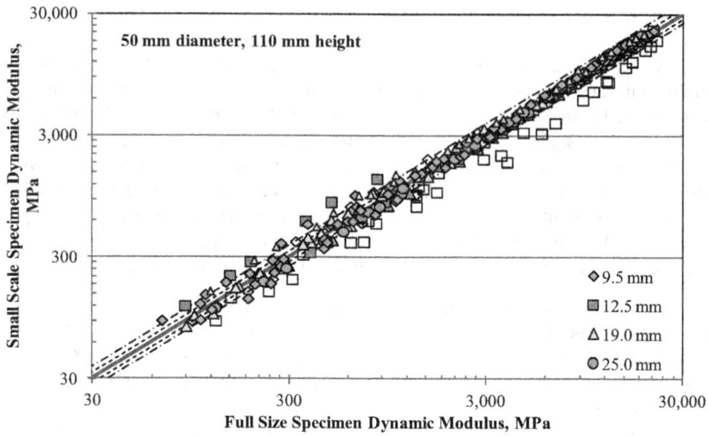

Figure 12. *Results showing the average dynamic modulus of 50 mm diameter by 110 mm height small-scale specimens versus full-size specimens.*

4.3 Assessing Agreement

A statistical analysis was performed to determine if the differences in dynamic modulus of the small-scale specimens as compared to their respective full-size specimens were statistically significant. A two tailed two-sample *t*-test, which assumed equal or unequal size and unequal variance, was used for analysis. The two-sample *t*-test was chosen because each specimen was from the same mixture and tested at the same temperatures and frequencies, with the exception of some omitted specimens at 54°C due to sampling problems which were believed to have been caused by the elevated testing temperature. The treatment was considered to be the specimen size (small-scale versus full-size) and independent specimens were tested. The null hypothesis was that the small-scale specimens and their respective full-size specimens had equal mean dynamic modulus. The test assumed a 95% confidence level ($\alpha = 0.05$) and accounted for all 237 specimens at all frequencies. The results were examined using two approaches: the first evaluated the effect of NMAS on specimen size across all temperatures (Figure 13), and the second assessed the effect of temperature on specimen size across all NMASs (Figure 14). Figures 13 and 14 show the percentage of tests where no statistical difference was found.

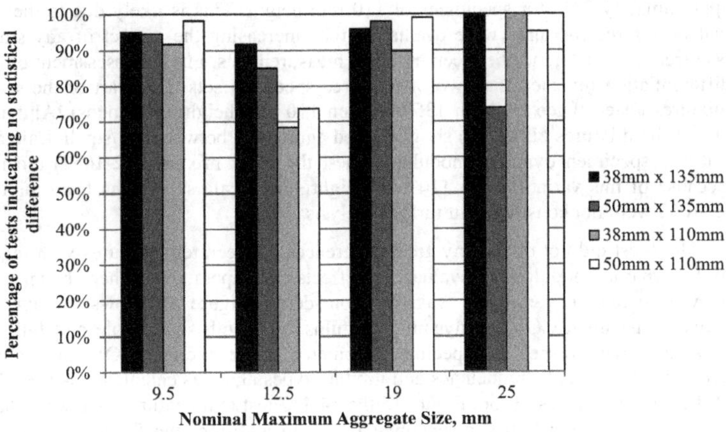

Figure 13. *Percentage of tests indicating no statistical difference with respect to nominal maximum aggregate size for each specimen geometry.*

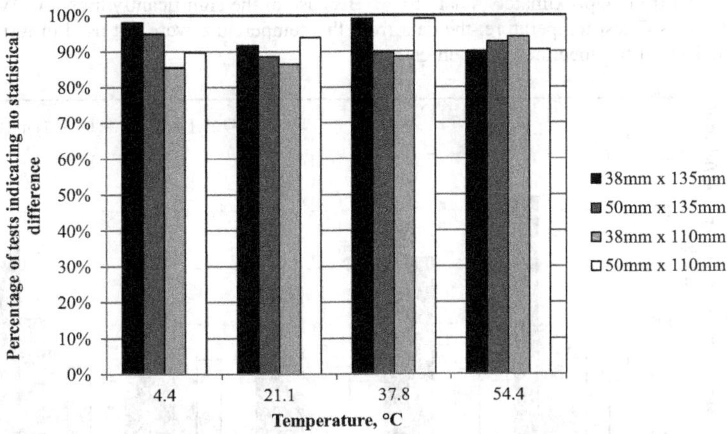

Figure 14. *Percentage of tests indicating no statistical difference with respect to temperature for each specimen geometry.*

The *t*-test showed that the NMAS and temperature did not significantly affect the similarity of dynamic modulus between the small-scale specimens and their respective full-size specimens. The 12.5 mm NMAS specimens did have a lower occurrence of tests indicating no significant differences in dynamic modulus values between small-scale and full-size specimens than most other mixtures at approximately 70% for specimens of 110 mm height. This is likely due to the fact that only three mixtures were evaluated, thus increasing the impact of any single occurrence of difference between modulus measurements. Further assessment of this differentiation revealed that two of the three specimen sets were part of the set of mixtures tested twice (as both 135 and then 110 mm height specimens). Although one of the mixtures (12-1041) showed good agreement between the small-scale and full-size specimen dynamic modulus values, the other mixture (12-1052) did not. Because of this variation, the 110 mm height small-scale specimens from mixture 12-1052 were not considered in further analysis.

The *t*-test did not reveal any great differences between temperatures with regard to the dynamic modulus. To evaluate the effects of temperature further, the inherent between-specimen variability was also considered. Figure 15 shows the average between-specimen COV for dynamic modulus from both small-scale and full-size specimens with respect to specimen geometry. The average COV shown was determined across all frequencies and mixture types and was calculated as the COV of the three replicates. From Figure 15 the 54.4°C test temperature is shown to have the highest average between-specimen COV; in addition, the COV is shown to decrease with decreasing test temperature. The individual COVs calculated for three replicates at the 4.4, 21.1, and 37.8°C test temperatures ranged from approximately 1% to 38%. The COVs calculated for three replicates at the 54.4°C test temperature ranged from approximately 4% to 51%. Because of the significantly higher COV at the 54.4°C test temperature, the data from this temperature were not used in further analysis of the specimen geometries.

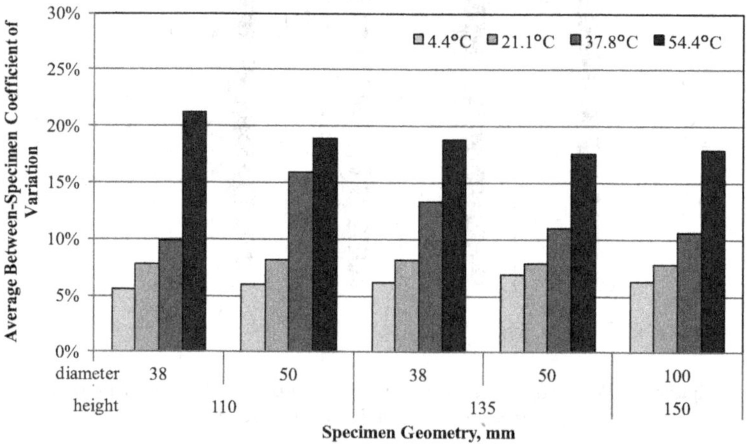

Figure 15. *Average between-specimen coefficient of variation for dynamic modulus with respect to specimen geometry.*

4.4 Determining an Optimum Small-Scale Geometry

Following the indication from the t-tests that the small-scale specimens were a viable alternative to full-size specimens for dynamic modulus testing, further analysis was conducted to determine if using one of the four small-scale geometries would be preferred to using the others. These efforts were carried out for each of the four NMAS evaluated based on the assumption that NMAS will impact the quality and accuracy of material characterization based on the proportion of cross-sectional area of a test specimen that a single large aggregate particle may occupy.

The approach to determine an optimum small-scale specimen geometry compared the ratio of the small-scale specimen dynamic modulus to the full-size specimen dynamic modulus (henceforth referred to as the *SS/FS ratio*) for each mixture. The SS/FS ratio was calculated as the average dynamic modulus from the three small-scale specimen replicates divided by the average dynamic modulus from the three full-size specimen replicates. An SS/FS ratio of 1.0 indicates that the average dynamic modulus from small-scale and full-size specimens is equal; an SS/FS ratio of 1.1 or 0.9 indicates that the average dynamic modulus from the small-scale specimens is 10% stiffer or less stiff than the full-size specimens, respectively. The SS/FS ratio was calculated for each test temperature and frequency for each of the 16 mixtures. The only exception included the data from one 12.5 NMAS mixture that was trimmed down to a 110 mm height, as discussed previously. The data for those specific 110 mm height specimens were removed from the SS/FS analysis.

Figures 16 through 19 present the average SS/FS ratio of all mixtures for each of the respective small-scale geometries at a test frequency of 10 Hz. The data for the other frequencies followed a similar trend (with respect to specimen geometry) with the exception of the 12.5 mm NMAS mixture at the 0.1 Hz frequency, which had a ratio of 0.82 for both of the 110 mm specimen heights.

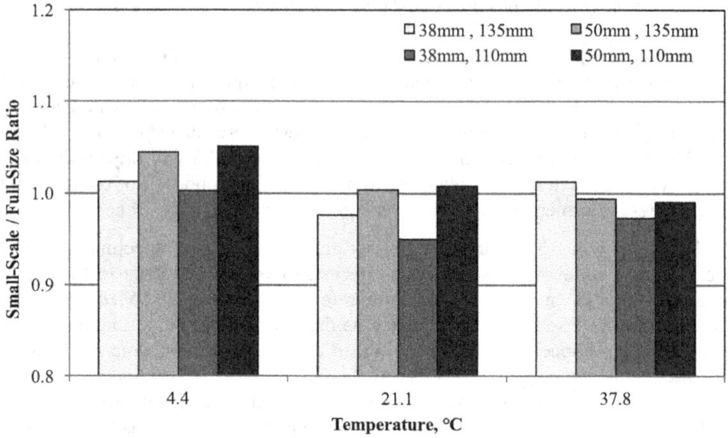

Figure 16. *Small-scale/full-size dynamic modulus ratio at 10 Hz and three temperatures for 9.5 mm NMAS mixtures.*

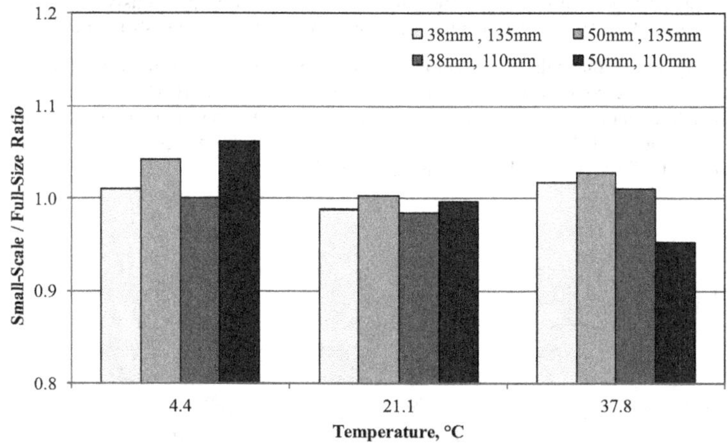

Figure 17. *Small-scale/full-size dynamic modulus ratio at 10 Hz and three temperatures for 12.5 mm NMAS mixtures.*

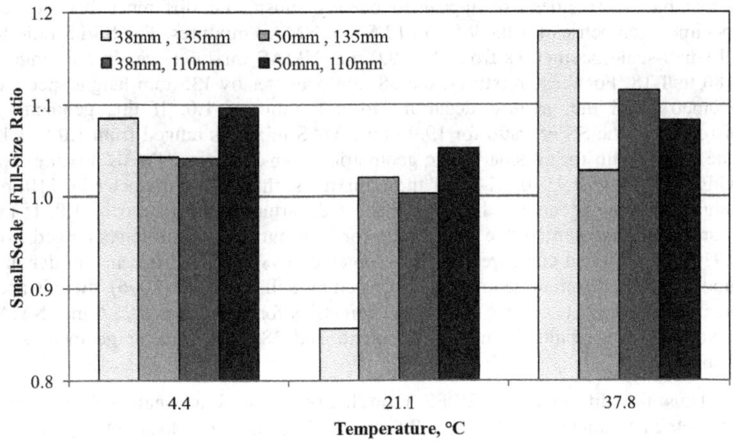

Figure 18. *Small-scale/full-size dynamic modulus ratio at 10 Hz and three temperatures for 19.0 mm NMAS mixtures.*

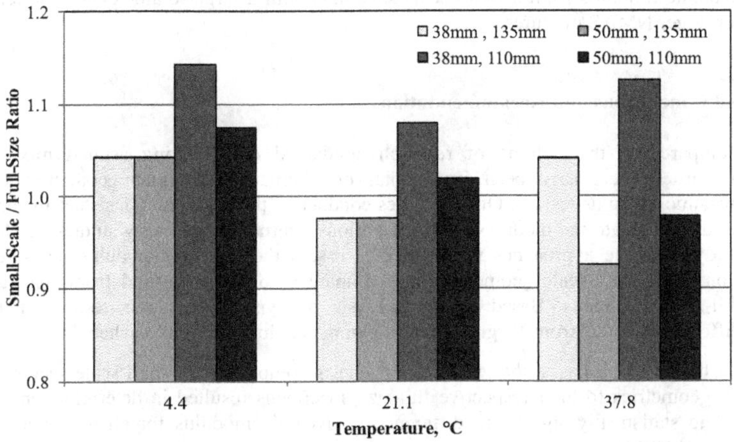

Figure 19. *Small-scale/full-size dynamic modulus ratio at 10 Hz and three temperatures for 25.0 mm NMAS mixtures.*

At the 10 Hz test frequency, the SS/FS ratio for all small-scale specimen geometries from the 9.5 mm and 12.5 mm NMAS mixtures was within the range of 0.95 to 1.06 (i.e., about a 6% difference). In light of the observation by Pellinen et al. (2006) that a 10% to 15% difference in dynamic modulus between geometries is acceptable, there does not appear to be any substantial difference between the specimen geometries for the 9.5 and 12.5 mm NMAS mixtures. The SS/FS ratio for all small-scale geometries from the 19.0 mm NMAS mixtures was in the range of 0.86 to 1.18. For these mixtures, the 38 mm diameter by 135 mm height specimen geometry had the greatest departure from a ratio of 1.0. If this geometry is disregarded, the SS/FS ratio for 19.0 mm NMAS mixtures ranged from 1.0 to 1.11. The SS/FS ratio for all small-scale geometries from 25.0 mm NMAS mixtures was within the range 0.93 to 1.14. For these mixtures, the 38 mm diameter by 110 mm height specimen geometry had the greatest departure from a ratio of 1.0. If this geometry is disregarded, the SS/FS ratio for 25.0 mm NMAS mixtures ranged from 0.93 to 1.07. When compared to the acceptable 10% to 15% difference in dynamic modulus from different geometries suggested by Pellinen et al. (2006), there is not a great difference between the specimen geometries for the 19.0 and 25.0 mm NMAS mixtures if the results from the aforementioned 38 mm diameter geometries are ignored.

Thus the difference in SS/FS ratio between the four small-scale specimen geometries evaluated for 9.5 and 12.5 mm NMAS mixtures does not appear to be significant. However, for 19.0 and 25.0 mm NMAS mixtures, the 38 mm diameter had the largest departure from a SS/FS ratio of 1.0, although the effect of height appeared to be mixed. This supports the initial inclination that measuring the dynamic modulus from a small-scale specimen with a larger diameter is beneficial for larger NMAS mixtures.

5.0 Conclusions and Recommendations

Compared to the volume of research conducted on assessing asphalt mixture properties, there have been few studies on alternative specimen geometries for dynamic modulus testing. Of the studies conducted, there was no consistent method used to evaluate the modulus from the various alternative geometry studies. In this study, multiple approaches were taken to assess the dynamic modulus measured from four small-scale geometries and compare it to that measured from standard full-size specimens. Based on the analysis of dynamic modulus testing of 16 different mixtures from Virginia, the following conclusions were reached:

- In almost all cases, the results of the *t*-tests comparing all small-scale specimen geometries to their respective full-size specimens resulted in determinations of no statistically significant difference in dynamic modulus for all temperatures and NMAS.
- Between-specimen variability was highest for dynamic modulus data obtained at the 54.4°C test temperature. Considering the average magnitude of this variability, further evaluation is recommended to investigate the suitability of testing small-scale specimens at 54.4°C.
- The departure from an SS/FS ratio of 1.0 was similar for all small-scale specimen geometries for 9.5 and 12.5 mm NMAS mixtures.

- The departure from an SS/FS ratio of 1.0 was greatest for small-scale specimens with a 38 mm diameter for 19.0 and 25.0 mm NMAS mixtures.
- Based on the SS/FS ratio, there is no difference between small-scale specimens having a height of 110 or 135 mm for 9.5, 12.5, 19.0, or 25.0 NMAS mixtures.
- The small-scale specimen geometries studied herein are viable alternatives to full-size specimens for dynamic modulus testing and can be used where production of full-size specimens is problematic.

Based on the analyses conducted in this study, it is recommended that 50 mm diameter small-scale specimens be used for 19.0 and 25.0 mm NMAS mixtures and that either 38 or 50 mm diameter small-scale specimens be used for 9.5 and 12.5 mm NMAS mixtures. The use of small-scale geometry specimens offers potential benefits such as the ability to use dynamic modulus evaluation in the context of forensic analyses or other investigations requiring the use of in situ pavement layers. Although the testing conducted herein shows that there is a good relationship between small-scale and full-size specimens with respect to dynamic modulus, the authors recommend that further investigation into the effects of air voids, NMAS, and aggregate anisotropy be conducted.

6.0 Acknowledgments

The authors acknowledge the assistance of Sara Link, University of Virginia; Charles Schwartz and Sadaf Khosravifar, University of Maryland; Mostafa Elseifi, Louisiana State University; Troy Deeds, Donnie Dodds, Benjamin Earl, Ken Elliton, Kim Snead, and Linda Evans, Virginia Center for Transportation Innovation and Research (VCTIR); Trenton Clark, Virginia Asphalt Association; and the personnel at S. L. Williamson, Superior Paving, Lee Hy Paving, Independence Construction Materials, Colony Construction, and Virginia Paving.

7.0 References

American Association of State Highway and Transportation Officials. "Standard Specifications for Transportation Materials and Methods of Sampling and Testing." 33rd Edition and AASHTO Provisional Standards. Washington, DC, 2013.

Bonaquist, R. *Refining the Simple Performance Tester for Use in Routine Practice*. NCHRP Report 614. Transportation Research Board of the National Academies, Washington, DC, 2008.

Chen, J., Huang, B., and Shu, X. "Air-Void Distribution Analysis of Asphalt Mixture Using Discrete Element Method." *Journal of Materials in Civil Engineering*, Vol. 25, No. 10, 2013, pp. 1375–1385.

Kim, Y. R., Seo, Y., King, M., and Momen, M. "Dynamic Modulus Testing of Asphalt Concrete in Indirect Tension Mode." *Transportation Research Record: Journal of the Transportation Research Board*, No. 1891. Transportation Research Board of the National Academies, Washington, DC, 2004, pp. 163–173.

Kutay, M. E., Gibson, N., Youtcheff, J., and Dongre, R. "Use of Small Samples to Predict Fatigue Lives of Field Cores; Newly Developed Formulation Based on Viscoelastic Continuum Damage Theory." *Transportation Research Record: Journal of the Transportation Research Board*, No. 2127. Transportation Research Board of the National Academies, Washington, DC, 2009, pp. 90–97.

Li, X., and Gibson, N. "Using Small Scale Specimens for AMPT Dynamic Modulus and Fatigue Tests." *Journal of the Association of Asphalt Paving Technologists*, Vol. 82, 2013, pp. 579–615.

Pellinen, T. K., Xiao, S, and Raval, S. Y. "Dynamic Modulus Testing of Thin Pavement Cores." *Journal of ASTM International*, Vol. 3, No. 4, 2006, pp. 144–157.

Witczak, M. W., Bonaquist, R., Von Quintus, H., and Kaloush, K. "Specimen Geometry and Aggregate Size Effects in Uniaxial Compression and Constant Height Shear Tests." *Journal of the Association of Asphalt Paving Technologists*. Vol. 69, 2000, pp. 733–793.

Witczak, M. W., Kaloush, K., Pellinen, T., El-Basyouny, M., and Von Quintus, H. *Simple Performance Test for Superpave Mix Design*. NCHRP Report 465. Transportation Research Board of the National Academies, Washington, DC, 2002.

8.0 Discussion

IMAD AL-QADI: Ben, congratulations. This is a very interesting presentation and indeed good work. I think this brings us back to the idea of using height to diameter ration of 2 or a little greater. I really like that in the work that you presented today. In the field, usually we do compaction at 2 and greater for 19 mm and it can be 1 inch for 9.5 mm. So, this will allow proper specimens.

BEN BOWERS: Correct.

IMAD AL-QADI: So why not consider a larger diameter for the 19 and 25 mm rather than the 25? I think this can be addressed easily.

BEN BOWERS: Right, and I think that that's a great suggestion. One thing I didn't talk about was the study done by Li and Gibson and presented at the 2013 AAPT meeting. They looked at a 38 mm diameter and 140 mm height. One problem we ran into is that when we core small-scale specimens out of a pavement core, if you cut to 140 mm and 50 mm diameter, you have rounding of the specimen ends because you're at the edges of the core. So I think that looking at a 110 mm height and maybe trying a larger diameter would be a great approach.

IMAD AL-QADI: And you can just keep the ratio of the diameter to the height constant.

BEN BOWERS: Absolutely.

IMAD AL-QADI: Congratulations again. Good work.

BEN BOWERS: Thank you. I appreciate it.

ALI HEKMATFAR: Thank you for your great presentation. I have several questions. So the first one is how many samples you ran for each dynamic modulus, for each mixture?

BEN BOWERS: For each mixture for each sample size, we had triplicate specimens. So let's just take one size, let's say the 50 mm x 110 mm height. For all 16 mixtures, we took that sample size and did it three times.

ALI HEKMATFAR: Triplicates. All right.

BEN BOWERS: And we did that for all four sample sizes and for the full size, so there was like 237 total samples tested in this work.

ALI HEKMATFAR: And how did you delete your outliers from your statistical analysis?

BEN BOWERS: That was the thing. And that was one of the things I wanted to consider. In the t-test, there was some variation, but I didn't cut anything out. I wanted to let everything be fair game and then start to whittle out samples as I went further. The t-test passed everything. There was one sample when I started looking… Let's see. Here at the t-test, you remember this where I told you about the 12.5 mm NMAS and how those were cut from 135 mm to 110 mm? Out of those two, one of them was still in line with the one that was full size and with the one that wasn't trimmed down. However, one of them was way out, and so I did remove that sample. And I meant to mention that. It is in the paper that we removed that sample, just because we knew exactly why it was wrong, due to the fact that this sample was trimmed and retested and the rest were not.

ALI HEKMATFAR: All right. As far as I remember, based on the specification, coefficient of variation within replicates, depends on how many specimens you're running. If I'm wrong, just correct me. For example, if you run three specimens for each mixture, the coefficient of variation should be around 9%.

BEN BOWERS: Right. One of the deviations that I mentioned from the spec was the fact that the specimen-specimen variation was outside of that. So you are correct, there were some that were outside of that.

ALI HEKMATFAR: All right. And the last question, how did you attach the bolts to the small specimens?

BEN BOWERS: So if we run back to this slide, if you look here at the top, we mentioned we used a unique set of gluing jig arms. Essentially what we did was we extended the top of the jig arm, and by the top I mean if it's an L, the bottom of the L, was extended so that it could actually touch the specimen. We kept the same gauge length that we'd been running on everything, which was 70 mm.

ALI HEKMATFAR: Thank you. Great work.

BEN BOWERS: Absolutely. Thank you for your questions.

SHADI SAADEH: Ben, great work, great presentation. I have two questions for you.

BEN BOWERS: Yes, sir.

SHADI SAADEH: First thing is you know that air-void distribution within the sample is dependent on where it is compacted and would be different throughout the specimen. So given the way you extracted your samples, I'm assuming that there will be a huge difference between, say, sample that came from the top of the specimen compared to that in the middle or that at the bottom.

BEN BOWERS: Right.

SHADI SAADEH: However, I didn't see that in your results of the coefficient of variation, I was expecting way higher than that. I'd like to hear your comment about that, but I have another comment about it too.

BEN BOWERS: Okay.

SHADI SAADEH: The high coefficient of variation at 54 degrees, is similar to the full scale specimens, which is related to the fact that the aggregate skeleton plays a significant or more significant role when we go to a higher temperature as compared to that at lower temperature.

BEN BOWERS: Absolutely.

SHADI SAADEH: And that skeleton or the way it is aligned is directly related to the air-void distribution and where we get our sample from.

BEN BOWERS: Yes, sir. Do I have a comment? Yes. I didn't know if you were finished.

SHADI SAADEH: Yeah, I'm done.

BEN BOWERS: So we did consider air-void distribution. Some of the work we looked at was by [Jingsong] Chen back in 2013. The work came out of the University of Tennessee, where they looked at how air-void distribution comes together with discrete modeling. We made sure to steer clear of the tops and bottoms of the gyratory sample. We also tried to test specimens throughout the gyratory sample and take that into consideration. I did mention air-void distribution. Air voids is something that should be further investigated. It's not something we built into our original test matrix. We started thinking about this as we were testing. But I do think that's a great point. You made the comment about the 54.4°C specimens – we did include the full size and that's where the highest COV was, as well. I think my concern is that with the smaller geometry, that aggregate skeleton, there's less there, and so it would make sense to consider a lower "high" temperature. It's just a suggestion or something that I think someone should or we may look further into.

SHADI SAADEH: Just one follow-up.

BEN BOWERS: Absolutely.

SHADI SAADEH: Did you look for a representative volume element, some work that was done before, on the size of the sample? I mean, because you went to a 25 mm on a sample that is 100 mm. So it depends how we cut it and how we core it, and that relates also to the very little air-void distribution, very little air voids in our small sample. My concern is if we really try to prove this concept, that we can derive properties of a material from a small scale and apply it to a bigger scale.

BEN BOWERS: Right.

SHADI SAADEH: Your data shows that the air voids doesn't matter.

BEN BOWERS: Right, it was interesting.

SHADI SAADEH: That's scary.

BEN BOWERS: I was surprised, as well.

SHADI SAADEH: Thank you, thank you.

BEN BOWERS: All of these specimens were fabricated from plant mix. We had the same folks making them. We tried to make everything as repeatable as humanly possible, as well. But the data... That was the curious thing. And I agree with you. It's a curious thing that air void contents didn't control. That or the effect of air voids was not high enough... It's a factor, but the effect isn't high enough to make enough of a difference to statistically reject the specimens. And that's another thing to consider. I know that it may be a little counterintuitive, but perhaps within the range that these specimens were to each other, it may not have caused an effect. But, again, we're talking in hypotheticals now, so I don't want to... And I think I'm on record.

SHADI SAADEH: Thank you very much.

BEN BOWERS: Absolutely. Thank you for your comments.

LOUAY MOHAMMAD: Nice work, Ben.

BEN BOWERS: Thank you.

LOUAY MOHAMMAD: Could you go to the slide where you showed the variation, I believe, F-test, t-test, whatever you did? Okay, I think it's this one.

BEN BOWERS: This one or the temperature?

LOUAY MOHAMMAD: Either one. Now if I understand correctly, you had axial dynamic modulus measurements.

BEN BOWERS: Yes, sir.

LOUAY MOHAMMAD: And then you had small scale dynamic modulus measurements. Thus, you have companion measurements of axial E* and small scale E* measurements, right?

BEN BOWERS: Yes, sir.

LOUAY MOHAMMAD: Thus, did you look at a paired t-test rather than lumping other effects, i.e., nominal maximum aggregate size, temperature, etc., in performing your analysis to come up with these percentages (axial E* vs small scale E*)? Then one can count the number of frequent events that occurs when they (axial E* vs small scale E*) are equal, they are less, or they are greater. Then you are not masking other contributing factors.

BEN BOWERS: Right. So for the sake of clarity, what I did was I paired the small scale of a given size of a given mix with the full size of a given size of a given mix.

LOUAY MOHAMMAD: Okay.

BEN BOWERS: And I did that for each individually. So I ran the statistics that way, and then I took a cumulative of all of those tests, and that's what's presented here. Does that make sense?

LOUAY MOHAMMAD: Yes, but then you are averaging.

BEN BOWERS: I'm taking averages of the different mixes.

LOUAY MOHAMMAD: Yes, but you're still averaging, right? When you do cumulative, you're averaging.

BEN BOWERS: Right. It is considering…

LOUAY MOHAMMAD: And then you have high and low, and I don't know what's in this high and low.

BEN BOWERS: Right.

LOUAY MOHAMMAD: What I would like to see is that when you have pairwise t-test, it's just a single event. Then, you are comparing only two things going on at the same time. When you have a lot of frequencies and a lot of temperature, I would recommend that you would consider looking at those individually and then that will give you a better picture of the differences.

BEN BOWERS: Okay.

LOUAY MOHAMMAD: And second is that the data presented is what it is.

BEN BOWERS: Right.

LOUAY MOHAMMAD: And that defies what you learn in a textbook about isotropy.

BEN BOWERS: Right.

LOUAY MOHAMMAD: So you're telling me that properties measured in the vertical direction will be the same as the ones measured in the horizontal direction. This indicates that either the test is not sensitive to capture anisotropy or there's something else going on. Because we do know that this material is anisotropic, yet you're showing that the properties are similar. As I said, data is what it is as data.

BEN BOWERS: Right.

LOUAY MOHAMMAD: I would be interested to see 1) your comment, and 2) are you doing future work to investigate this isotropy-anisotropy effect?

BEN BOWERS: What am I doing?

LOUAY MOHAMMAD: Yes.

BEN BOWERS: So one place I'm wondering if it's going to show up is, I told you currently when we go out to a job, we make standard samples using the Superpave gyratory compactor of the mix on site. And then what we do is we go out and we collect a core, and we core sideways. Because the difference here is the compactive effort and the way it was compacted. And so I think that that's where we may see some differences. Now you're going to have to take into account whether or not your air voids and your densities and everything are equivalent. But I'm wondering if that's one way we're going to be able to approach this and see that difference. I'd love to know if you have any other ideas.

LOUAY MOHAMMAD: Well, we can discuss afterwards.

BEN BOWERS: I agree, yes.

LOUAY MOHAMMAD: Another thing since the motivation of testing small scale samples is because your pavement is thin.

BEN BOWERS: Right.

LOUAY MOHAMMAD: So, of course, alternatively, there is the indirect tensile dynamic modulus (IDT E*) test method, which many individuals have performed.

BEN BOWERS: Yes.

LOUAY MOHAMMAD: As a part of NCHRP Project 9-48, we compared IDT E* versus axial E* and found that there are certainly differences for the many mixtures we evaluated. Also, we developed shift factors to estimate one from the other and so on. I suggest you may consider doing similar activities.

BEN BOWERS: Absolutely. Thank you.

RANDY WEST: Nice job, Ben. I congratulate you and your co-authors on a very nice study.

BEN BOWERS: Thank you.

RANDY WEST: Also to follow along with kind of the question about anisotropy that we have in samples whether it's cores or gyratory samples. I was trying to see if you had in your paper a description of the aggregate shapes, any kind of flat and elongated information. I didn't see that, so I wondered if you would be able to comment on that or maybe to add some information about the variety of mixes that you had and whether or not you had aggregates that had high flat and elongated ratios or at least within the specification range that we deal with? My concern would be that if you only dealt with very cubical aggregates, it may not represent all the types of aggregates that we see across the country.

BEN BOWERS: That's a great point. And, unfortunately, I can't comment on that. I'm not sure. I haven't looked at the aggregate types; we took what we were given. But Stacey might have a comment.

STACEY DIEFENDERFER: All of the mixes tested meet VDOT specifications on F&E; these require a maximum of 10% at the 5:1 ratio on the coarse aggregate fraction (retained on #4 sieve) of the mix. The SMA mixes require F&E to be measured on material retained on the #4 sieve for the mix as well; it must meet a maximum of 20% 3:1 ratio and 5% 5:1 ratio. We don't have the detailed mix information at this point, because the contractors shipped us a box of mix and a G_{mm} for compaction, told us the mix type, and left it at that. However, everything was collected from production mix on VDOT contracts, so there is basic aggregate type information available from the mix designs – it's mostly a matter of tracking backwards to that point. In hindsight, that probably should have been compiled during the study, although time was a factor. In essence, we wanted real world mixtures to get real world results and these mixtures are representative of what was produced across a span of several months.

RANDY WEST: Okay.

STACEY DIEFENDERFER: So they'd be representative of standard mixes that would meet acceptance.

RANDY WEST: Right, within Virginia.

STACEY DIEFENDERFER: Yes.

RANDY WEST: I don't know if that represents all…

STACEY DIEFENDERFER: Our aggregate requirements are not particularly unusual. These would have all been diabase, granite, limestone, in standard surface mixes.

RANDY WEST: Thank you.

BEN BOWERS: Thank you, Randy.

EMIN KUTAY: Certainly excellent work, very comprehensive. I have two comments. One of them is about representative volume element comment that we just had from Shadi. Your study is really a study of representative volume element. You're really redefining the concept of representative volume element for dynamic modulus. Typically, rule of thumb is three times the NMAS.

BEN BOWERS: Yes.

EMIN KUTAY: You're showing here that that's not necessarily true. So I think that is something that we need to emphasize, so that rule of thumb doesn't necessarily apply to dynamic modulus. My second comment is actually related to what Tom said 'under linear conditions.' The 2009 paper that we had where we investigated cracking performance of 38 mm diameter, 110 mm diameter. That was a very limited study, only included five mixes, ALF mixes, and one nominal maximum aggregate size. But what that study showed was we ran push-pull tests, fatigue tests and looked at the C versus S curves, damage corrected – six curves. They were identical. Small sample diameter, they were identical to large scale diameter samples. Even then, when you are talking about cracking, that limited study showed that the representative volume element rule of thumb, you don't have to follow that rule. You can violate that rule fine. And unlike Louay, I'm going to broadcast an idea and say that go ahead and probably run push-pull or fatigue tests and see if that is true or not.

BEN BOWERS: Thank you. Thank you for your comments.

MARTIN VAN DE VEN: Delft University. I'm also very surprised that there is not a difference in vertical direction or horizontal direction. As you probably know, Texas A&M has even a factor in their mechanistic design method for the anisotropy. In addition to that you often also hear stories that if you take a core in the driving track and if you core that horizontally in the driving direction it is different compared to perpendicular on the driving direction… I assume you take cores in the tracks of the road or not?

BEN BOWERS: We take them randomly.

MARTIN VAN DE VEN: Yes, but also in tracks?

BEN BOWERS: Yes.

MARTIN VAN DE VEN: And did you check if in the driving direction or perpendicular to that if there is a difference?

BEN BOWERS: That's an interesting idea. That would be something to look at, definitely.

MARTIN VAN DE VEN: I heard that there is a difference, but I've never really seen it.

BEN BOWERS: I could see where that would have an effect.

MARTIN VAN DE VEN: And then adding to that, the small diameter from the roads, we all know the variability at several spots on the road. So if one is unlucky with taking a core and that small horizontal core in that core hits, for example, the 25 size hits a segregation spot or whatever, then you get very bad results, right?

BEN BOWERS: Right, I agree. And so we actually when we're doing this testing with cores, we pull – much to the dismay of some – we pull a couple of extras because you do, you run into that problem. And then we go back through with our technicians and identify the ones that… Because we don't want to get into testing two materials when we're supposed to be, you know.. And that definitely would play a factor. So thank you for that comment. That's definitely something worth noting.

MATTHEW CORRIGAN: Ben, I was curious about your temperature selection. I think I saw the temperatures you selected were basically the direct inputs to the Pavement ME software?

BEN BOWERS: Yes.

MATTHEW CORRIGAN: That's probably why you moved forward in that direction. The data showed the most difference between the full scale and the small scale specimens at the high temperature and low loading rate. Based on some recent work I've been doing on a separate issue, we've identified limitations at that high test temperature and with the instrumentation that's used to measure it using the on specimen glued gauge points.

BEN BOWERS: Okay.

MATTHEW CORRIGAN: I was curious if you had separated out that high temperature data at the 54.4°C and how the rest of your results and statistics would play out? The current AASHTO provisional standards for the AMPT evaluation have eliminated that high temperature recommendation so you don't have that error within your calculations.

BEN BOWERS: It's not something we did in our statistical evaluation, but it would be interesting to see, especially if that's the direction things are going. I mean, it definitely was where we had our most variability, and it sounds like everybody's seeing that.

MATTHEW CORRIGAN: Right. Not knowing the details about your instrumentation specifically, it could be the issue with the gauge points moving and the instrumentation at the high test temperature.

BEN BOWERS: Okay, thank you. I appreciate the comments.

AUTHOR CLOSURE

After the discussion at the 2015 AAPT Annual Meeting outlined in the above commentary, the authors performed a paired t-test to analyze the data. The results of the paired t-test were similar to those of the unpaired t-test and did not change the final conclusions. The authors see merit in both approaches and understand why some may find the paired versus the unpaired t-test more indicative of the statistical significance of the comparative specimens. For more information, please contact the authors.

Improved Hirsch Model for Estimating the Modulus of Hot Mix Asphalt

Donald W. Christensen[a], Ramon Bonaquist[b]

[a]*Advanced Asphalt Technologies, LLC, 834 Webster Drive, State College, PA.*

[b]*Advanced Asphalt Technologies, LLC, 40 Commerce Circle, Kearneysville, WV.*

ABSTRACT: This paper presents an improved version of the Hirsch model for estimating the modulus of hot-mix asphalt (HMA) from asphalt binder modulus and mixture composition. The original Hirsch model was developed in 2002 and has been shown by several independent researchers to be reasonably accurate. However, the authors believed that the model could be improved by addressing several issues: (1) simplifying the Hirsch equation mathematically; (2) including aggregate specific gravity as a predictor, which indirectly accounts for variations in aggregate modulus; (3) including strain level as a factor; (4) recalibrating the model using a data set gathered using the asphalt mixture performance tester (AMPT) following current standard protocols; and (5) addressing steric hardening during testing. These goals were achieved, although it appears that under normal testing conditions strain sensitivity and steric hardening are not significant factors affecting HMA dynamic modulus. Verification of the model using a limited amount of independently collected data suggests that the improved model eliminates or reduces a tendency of the original model to underestimate HMA dynamic modulus values.

KEYWORDS: Hirsch model, modulus, dynamic modulus, aggregate specific gravity, aggregate modulus, non-linearity, strain sensitivity, steric hardening

1.0 Introduction

The purpose of this paper is to present an improved version of the Hirsch model for estimating the modulus of hot-mix asphalt (HMA) based upon binder properties and mixture composition. The Hirsch model for HMA is theoretically based on a series element and a parallel component in combination, as shown in Figure 1. In the original research on the application of the Hirsch model to HMA, various different approaches to the specific construction of the model and calibration were evaluated, including, for instance, treating the phases present as mastic, aggregate and air or as asphalt binder, aggregate and air. In the final model, the three phases were simply the asphalt binder, the aggregate and air voids (Christensen et al., 2003).

The oral presentation was made by Dr. Christensen.

This is a reproduction of an article whose final and definitive form has been published in *Road Materials and Pavement Design*© 2015 Taylor & Francis. The article is available online at: http://dx.doi.org/10.1080/14680629.2015.1077635

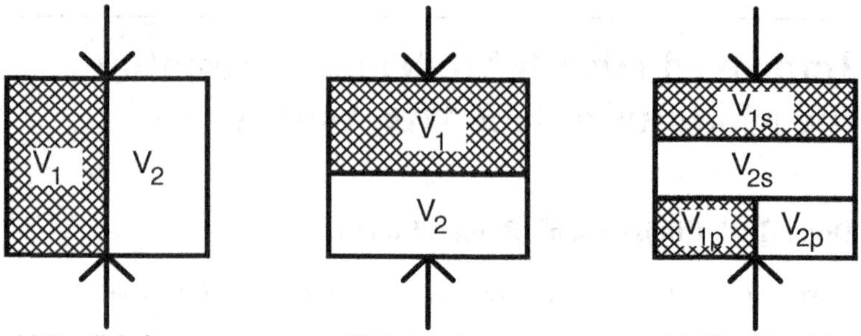

(a) Parallel phases (b) Series phases (c) Hirsch model

Figure 1. *Schematic representation of composite models for parallel, series, and Hirsch (combination) arrangement of phases (Christensen et al., 2003).*

The most widely used version of the Hirsch model expresses HMA dynamic modulus, $|E^*|mix$, as a function of binder dynamic shear modulus, mix voids filled with asphalt, mix voids in the mineral aggregate and the contact factor Pc:

$$|E^*|_{mix} = Pc\left[29{,}400\,(1 - VMA/100) + 3\,|G^*|_{binder}\left(\frac{VFA \times VMA}{10{,}000}\right)\right] + (1 - Pc)\left[\frac{1 - VMA/100}{29{,}400} + \frac{VMA}{3\,VFA\,|G^*|_{binder}}\right]^{-1} \qquad [1]$$

$$Pc = \frac{\left(0.138 + \dfrac{VFA \times 3|G^*|_{binder}}{VMA}\right)^{0.58}}{36.2 + \left(\dfrac{VFA \times 3|G^*|_{binder}}{VMA}\right)^{0.58}} \qquad [2]$$

Where:

$|E^*|_{mix}$ = dynamic complex modulus of an HMA mixture, in extension, MPa
$|G^*|_{binder}$ = dynamic complex modulus of the asphalt binder in the mixture, in shear, Pa
VMA = voids in the mineral aggregate for the HMA mixture, volume %
VFA = voids filled with asphalt for the HMA mixture, volume %

The model has been independently evaluated by a number of researchers with generally good results, but the authors felt that there was potential for improving upon the original model (Dongre, 2005; Pellinen et al., 2007; Mohammad et al., 2007). This feeling is in part the result of some independent evaluations that seemed to suggest that the Hirsch model sometimes exhibits poor accuracy (Mohammad et al., 2007; Sakhei Far, 2011). The potential improvements addressed in this research were in four areas: (1) simplification of the Hirsch model; (2) possible changes in modulus with time (steric hardening); (3) variations in the aggregate modulus; and (4) changes in modulus with stress (or strain) level. It should be noted that the overall objective in refining the Hirsch model is not necessarily to develop a model capable of estimating the modulus of HMA under any possible combination of conditions (temperature, frequency, stress, age at test) but to provide a highly accurate model for estimating the dynamic modulus of HMA as measured using the proposed standard procedure in conjunction with the Asphalt Mixture Performance Tester (AMPT), that is, to provide a good estimate of the modulus as would be measured using AASHTO TP 79. This is important in limiting the scope of the research, since the temperature, strains and stresses are fairly well defined by this test method.

The use of recycled asphalt pavement (RAP) and recycled asphalt shingles (RAS) in asphalt concrete and the use of warm mix asphalt (WMA) have become widespread in the U.S. over the past decade, and this warrants a few comments concerning the application of the Hirsch model to mixes containing RAP and/or RAS or produced using WMA technology. There is no reason—theoretical or practical—that the Hirsch model cannot be applied to mixes containing RAP and/or RAS, as long as the modulus of the effective binder is known. Effective binder in this case means the binder in the mix that represents a blending of the binder from the RAP/RAS and any binder added to the mix at the plant; binder that remains adsorbed in the RAP/RAS and does not blend with newly added binder is not included as effective binder. No mixes containing RAP/RAS were included in developing the model, in part because in the original data was largely gathered prior to the widespread use of RAP/RAS, but also because including such materials would interject significant additional uncertainty in calibration of the model since the only method of determining binder modulus in RAP or RAS is by performing rheological measurements on extracted binders, and the extracted binder may not precisely represent the properties of the binder dissolved from the RAP/RAS in the actual mix. Application of the Hirsch model to mixes containing significant amounts of RAP/RAS would normally involve either measuring or estimating the properties of the RAP/RAS binder that leach into the mix and blend with newly added binder. This unfortunately will both decrease the utility of the model and also increase the error involved. The potential problems in applying the Hirsh model to WMA are similar but probably not as complicated. There are two potential problems in applying the Hirsh model to WMA: (1) WMA additives might affect the modulus of the binder; and (2) WMA might tend to age differently (probably more slowly) than comparable mixes produced using hot-mix technology. Some further comments on application of the Hirsh model to mixes containing RAP and/or RAS and to WMA are given towards the end of the paper.

Table 1. *Properties for mix designs used in developing revised Hirsch model.*

Aggregate	**NMAS** *mm*	**Gsb**	**Asphalt Content** *Wt. %*	**Compaction** *Gyrations*	**VMA at Gyration**	**Voids at Gyrations**
Pennsylvania Limestone	9.5	2.758	5.9	75	18.7	7.3
New York Gravel	19	2.603	6.2	75	15.3	3.1
Wisconsin Granite	12.5	2.647	4.9	75	15.2	5.3
Wisconsin Gravel	12.5	2.717	5.2	75	16.6	6.2
Kentucky Limestone	12.5	2.670	5.7	75	15.6	4.2
Mississippi Grave/Chert	12.5	2.438	5.4	75	14.3	7.0
Virginia Diabase	9.5	2.862	5.0	75	17.2	6.6
Virginia Limestone	9.5	2.656	6.8	75	19.9	8.6
Florida Limestone	12.5	2.347	7.0	100	17.6	6.9

Table 2. *Aggregate gradations for mixes used in developing revised Hirsch model.*

Aggregate	*Percent Passing, %, for Sieve Size (mm):*										
	25	**19**	**12.5**	**9.5**	**4.74**	**2.36**	**1.18**	**0.60**	**0.30**	**0.15**	**0.075**
PA Limestone	100	100	100	97	67	41	25	15	10	7	5.5
NY Gravel	100	95	79	73	49	33	20	13	8	5	3.1
WI Granite	100	100	96	85	63	47	36	26	13	6	4.1
WI Gravel	100	100	96	86	64	48	36	25	12	5	3.5
KY Limestone	100	100	95	87	64	40	25	17	10	6	4.8
MI Chert	100	100	95	90	65	44	31	24	11	6	5.3
VA Diabase	100	100	100	95	52	38	28	20	12	7	4.9
VA Limestone	100	100	100	93	49	32	21	14	9	7	5.2
FL Limestone	100	100	96	90	58	35	21	13	8	5	4.1

2.0 Materials and Methods

The volumetric properties and related data for the mixes used in developing the revised Hirsch model are shown in Table 1; the aggregate gradations are shown in Table 2. The binder properties (parameters for the Christensen-Anderson model for calculating $|G^*|$) are summarized in Table 3 (Christensen and Anderson, 1992). The materials used were selected specifically to provide a robust data set for developing an improved version of the Hirsch model, and included aggregates with a wide range in specific gravity values and asphalt binders with a wide range in consistency.

3.0 Modification of the Hirsch Model for HMA

As discussed above, the Hirsch model is a mechanical analogue for estimating the modulus of HMA. It consists of a parallel element of asphalt, aggregate and air, and a series element with the same composition. In applying the Hirsch model to HMA it is assumed that the contact factor varies with temperature, so that in effect, the relative proportions of material in the HMA that behaves as if they are in a series versus parallel arrangement also changes with temperature.

Table 3. *Christensen-Anderson model parameters for the asphalt binders used in developing the revised Hirsch model.*

Binder Grade	*Christensen-Anderson Model Parameters (RTFOT Residue)*			
	Glassy Modulus, Gg *Pa*	**Crossover Frequency, ωc** *rad/s*	**Rheological Index, R**	**Transition Temperature, Tc** *°C*
PG 58-28	3.44E+08	2.17E+02	1.50	-11.2
PG 64-22	2.05E+09	2.56E+02	2.02	-8.6
PG 70-10	2.74E+08	1.03E+01	1.09	-0.1
PG 70-28	1.00E+09	1.68E+01	2.76	-11.1
PG 76-22	4.19E+09	2.14E+01	2.81	-3.2

However, in analyzing the data generated during this research it was discovered that the effect of the parallel element on modulus values estimated with the Hirsch model are always negligible. This doesn't mean that the Hirsch model does not apply; again, it means only that the effects on modulus of the HMA material behaving as a parallel element are negligible when compared to the effects of the material behaving as a series element. This leads directly to the following, modified version of the Hirsch model:

$$|E^*|_{mix} = Pc\left[E_{agg}\left(1 - VMA/100\right) + 3\,|G^*|_{binder}\left(\frac{VFA \times VMA}{10{,}000}\right)\right] \qquad [3]$$

$$Pc = \frac{\left(H_1 + \dfrac{VFA \times 3|G^*|_{binder}}{VMA}\right)^{H_3}}{H_2 + \left(\dfrac{VFA \times 3|G^*|_{binder}}{VMA}\right)^{H_3}} \quad [4]$$

Where H_1, H_2 and H_3 are constants determined through calibration and the other variables are as defined previously.

The standard error of the logarithms of $|E^*|$ for the modified model was, after rounding, exactly the same as the original model (0.0905). Furthermore, the standard error of the difference between the predicted $|E^*|$ values for the two models was extremely small—less than 0.2%—indicating that these forms of the Hirsch model are for practical purposes interchangeable. Again, it should be emphasized that this modified version of the Hirsch model does not mean that the original formulation is not valid, only that the second component of the model—the part representing the series element in the mechanical analog—is negligible compared to the series component.

Some additional comments are warranted at this point concerning the contact factor (Equations 2 and 4). Experience has suggested that the form of the contact factor is reasonable and that when used in conjunction with the basic Hirsch model (Equations 1 and 3) it has produced good estimates for HMA modulus. However, it is after all an empirical function and it is possible that better approaches exist for estimating the contact factor. Specifically, the authors felt that the logistic function (or some modified version of it) might prove to be more useful that the exponential function currently used in the Hirsch model. In fact, the current form of the contact factor is very similar to the logistic function. However, the logistic function in a generalized form is more flexible than the currently used equation for calculating the contact factor and for that reason is used in the recalibrated version of the Hirsch model as discussed later in this document:

$$P_c = H_1 + (1 - H_1)\frac{\exp[H_2 + f(|G^*|, VTM, VMA, VFA, \varepsilon)]}{1 + \exp[H_2 + f(|G^*|VTM, VMA, VFA, \varepsilon)]} \quad [5]$$

Where H_1 and H_2 are constants to be determined in calibration of the Hirsch model (not identical to H_1 and H_2 for Equation 4), and $f(|G^*|, VTM, VMA, VFA, \varepsilon)$ is some function of the binder dynamic shear modulus $|G^*|$, one or more volumetric factors such as VMA and VFA and strain level. The coefficient H_1 represents an absolute minimum value for P_c, independent of other predictor variables. Because the Hirsch model is intended to be based on HMA volumetrics, aggregate gradation and surface area are not included in the model. To date, this has not proven to be a significant obstacle to the accuracy of the model.

4.0 Aggregate Modulus in the Hirsch Equation

In the original Hirsch model, the constant 29,400 represents an average value for aggregate modulus, in MPa. As mentioned above, it was felt that there might be a relationship between aggregate specific gravity and the modulus, as used in the Hirsch model. To evaluate this hypothesis, the dynamic modulus of eight mixtures was measured at -10°C. A binder was selected with exceptionally high modulus values at low temperatures (a PG 70-10) so that the mixture would be as stiff as possible; data under these conditions would provide the best estimate of the aggregate modulus. It should be noted that an attempt was made to evaluate this hypothesis using the original data set used to develop the Hirsch model, but this was unsuccessful; this data set had a relatively narrow range of aggregate modulus values. It also appeared that the variability in the modulus data was somewhat higher than is now commonly observed, not because of errors in experimental methods but because the HMA modulus testing equipment and protocols have been improved.

Equation 3 can be rearranged to give an equation expressing the aggregate modulus, E_{agg}, as a function of binder modulus,

$$E_{agg} = \frac{|E^*|_{mix}}{P_C} - \frac{3|G^*|_{binder} \times VMA \times VFA}{10{,}000\left(1 - VMA/100\right)} \qquad [6]$$

Where the variables are as defined above. Equation 6 was used to estimate aggregate modulus for the eight mixes/aggregates used in this study. The data used in this calculation are summarized in Table 4, which in the interest of brevity includes data for 1 Hz only; the complete data set can be found in the final report for the research project (Christensen and Bonaquist, 2014). Figure 2 is a plot of estimated aggregate modulus as a function of aggregate specific gravity. Note that for each mixture there are three data points, representing modulus data at 0.1, 1 and 10 Hz. The figure includes a power law fit through the data, which was used because it fit the data well and because it behaves properly at the lower bound, where a specific gravity of zero (if possible) should result in an aggregate modulus of zero. The power law exponent suggests that aggregate modulus increases approximately with the square of the specific gravity. Although the r^2 value is not exceptionally high at 37%, it certainly supports the hypothesis that aggregate modulus is related to specific gravity. It should also be noted that the calculation of aggregate modulus using the equation for P_C from the original Hirsch model is only an estimate and inexact. Providing a better estimate of aggregate modulus in the Hirsch model should significantly improve the accuracy of the model, especially at low temperatures. In the final recalibration of the Hirsch model, rather than using the relationship from Figure 2 to estimate aggregate modulus, a sub-model will be included allowing the power law relationship between aggregate modulus and specific gravity to be determined from the full set of data.

Table 4. *Data for estimating aggregate modulus.*
Frequency = 1 Hz, binder |G| = 237 MPa*

Aggregate	Mix |*E**| *MPa*	Pc	VMA	VFA	Aggregate Gsb	Aggregate Modulus *MPa*
Wisconsin Granite	21,990	0.721	16.6	58.5	2.647	36,510
NY Gravel	15,360	0.711	18.9	61.6	2.603	26,540
Mississippi Chert	24,520	0.760	12.2	61.3	2.438	36,700
Wisconsin Gravel	19,060	0.720	17.2	60.1	2.717	31,890
Pennsylvania Limestone	25,090	0.721	18.0	63.7	2.758	42,330
Florida Limestone	16,920	0.709	16.9	53.9	2.347	28,620
Virginia Limestone	22,410	0.714	19.8	65.7	2.656	39,050
Virginia Diabase	31,520	0.749	15.4	70.2	2.862	49,650

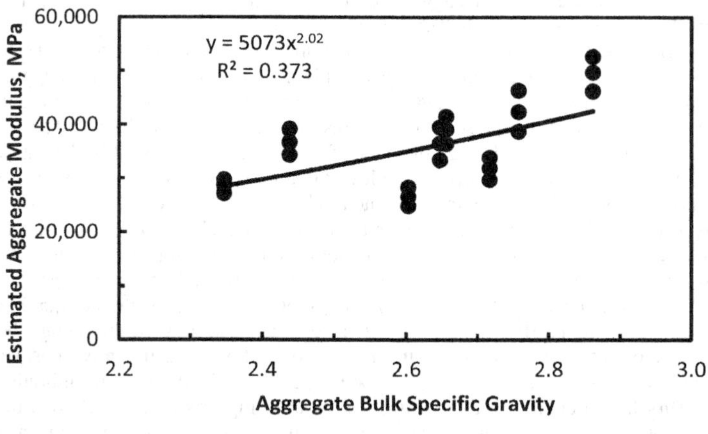

Figure 2. *Plot of estimated aggregate modulus as a function of aggregate bulk specific gravity.*

5.0 Strain Sensitivity of HMA Dynamic Modulus

The purpose of the strain sensitivity experiment was to determine if moderate deviations from the standard strain level used in HMA dynamic modulus testing would significantly affect the resulting modulus value, and if so, to develop a sub-model for predicting this effect that can be incorporated into the Hirsch model.

The Hirsch model is intended to be a surrogate for standard dynamic modulus testing of HMA mixtures, as performed following AASHTO TP 79. Therefore, a large variation in strain values for this experiment was not needed, only a relatively modest one to account for relatively small variations from the standard target strain of 100×10^{-6}.

In the strain sensitivity experiment, eight different aggregates were combined with three different binders to produce a total of 15 different mixtures. In general, each aggregate was combined with two different binders; one aggregate was combined with a single binder. The combinations of aggregates and binders used in this experiment are listed in Table 5. Each of these mixtures was tested at a temperature of 40°C at frequencies of 0.01 and 1 Hz. Only high temperature testing was performed because it was believed that strain sensitivity would be most apparent under these conditions. This is based on a very extensive set of dynamic modulus master curves developed by T. K. Pellinen using multiple stress levels; this study clearly demonstrated that the degree of non-linearity in the modulus of asphalt concrete increases with decreasing modulus (Pellinen, 2001). Two strain levels were used for each combination of aggregate, binder and frequency: a standard strain level of about 100×10^{-6}, and a higher level typically about double this value (a few tests were conducted at strains of 300 to 400×10^{-6}).

The final analysis of strain sensitivity is addressed in the recalibration of the Hirsch model, discussed below, using the data described here along with data from the other experiments and several other sources. However some discussion of the general approach to strain sensitivity is warranted. As mentioned above, strain sensitivity manifests itself more as the modulus of HMA decreases; that is it should be expected that strain sensitivity in asphalt concrete will increase at higher temperatures and slower loading rates (Pellinen, 2001). Although intuitively one might expect that increased strain should cause a decrease in the modulus of HMA (as it generally does in asphalt binder), this is not necessarily the case. At the levels typically used to measure the dynamic modulus of HMA, it is possible that an increase in strain might cause an increase in modulus; a decrease in modulus with strain might eventually be observed at relatively high strain levels. Furthermore, it should be considered that the strain levels used in the standard dynamic modulus procedure are relatively low, having been intentionally selected to limit strain sensitivity and to provide maximum repeatability and reproducibility in the resulting test data. A further complication in identifying and characterizing strain sensitivity in HMA modulus measurements is the modest accuracy of these measurements; the coefficient of variation associated with inter-laboratory precision (repeatability) has been reported to be about 15 to 20 percent. This means that in order to be detectable, strain sensitivity in HMA must produce changes in modulus significantly greater than this. It should be considered that variation of the applied strain during dynamic modulus measurement of HMA in the vicinity of 100 to 400 × 10^{-6} might in fact have a statistically negligible effect on the resulting data. This should not necessarily be considered a "negative" conclusion, as it would suggest that the strain level used in the current protocol is in fact appropriate.

Table 5. *Materials used in the strain sensitivity experiment.*

Aggregate	**NMAS** *mm*	**Binder Grade**
Wisconsin Granite	12.5	PG 70-10
		PG 70-28
NY Gravel	12.5	PG 70-10
		PG 70-28
Mississippi Chert	12.5	PG 70-10
		PG 64-22
Wisconsin Gravel	12.5	PG 70-10
		PG 70-28
Pennsylvania Limestone	9.5	PG 70-10
		PG 64-22
Kentucky Limestone	12.5	PG 64-22
Virginia Limestone	9.5	PG 70-10
		PG 70-28
Virginia Diabase	9.5	PG 70-10
		PG 64-22

Because strain sensitivity appears to be more prominent at low modulus levels, where the aggregate structure dominates the behavior of HMA, it can be concluded that strain sensitivity is largely the result of the effect of strain on aggregate structure. For this reason, the most logical approach to addressing strain sensitivity in the Hirsch model is to incorporate a term for strain sensitivity into the P_C function (Equations 4 and 5 above). Probably the most likely such expression would be a power law function:

$$\Delta_{\varepsilon} = K_{\varepsilon 1}\varepsilon^{K_{\varepsilon 2}} \qquad [7]$$

Where $K_{\varepsilon 1}$ and $K_{\varepsilon 2}$ are power law constants characterizing the strain sensitivity effect, Δ_{ε}. As discussed in the section below, there are various possible alternatives for the P_C function and the way in which strain sensitivity is characterized within those functions.

Prior to the rigorous analysis of strain sensitivity that will be included in the recalibration of the Hirsch model, a preliminary analysis of the data is in order. Figure 3 is a plot of relative dynamic modulus as a function of average peak-to-peak strain for the data generated in the strain sensitivity experiment. Relative modulus was calculated as the modulus for a given mix (at a given temperature and frequency) measured at a given strain divided by the average modulus for all strain levels at the selected temperature and frequency. Relative modulus was used in order to remove variability due to differences among mixes, which might otherwise overwhelm any non-linearity and make it difficult to identify. The only sources of variability in Figure 3 are experimental error (noise) and non-linearity. Therefore significant non-linearity should be observable as some trend in the data: increasing with strain, decreasing with strain, showing a maximum or minimum with respect to strain, and so forth. Such trends do not appear to exist (the r^2 value of 6% is very low), suggesting that for testing conditions used in this study (and representing current standard protocols for measuring $|E^*|$) that there is not significant strain sensitivity relative to typical measurement errors, and that it is in fact not necessary to address strain sensitivity within the Hirsch model given that the objective of the model is to estimate HMA $|E^*|$ values as they would be measured using the AMPT following standard procedures. It should be emphasized that the conclusion here is not that there is no non-linearity in asphalt concrete modulus, but that under standard test conditions modest changes in applied strain will not normally produce significant changes in modulus in comparison with typical experimental error. Improved precision in modulus measurements might make any non-linearity easier to observe and render them statistically significant.

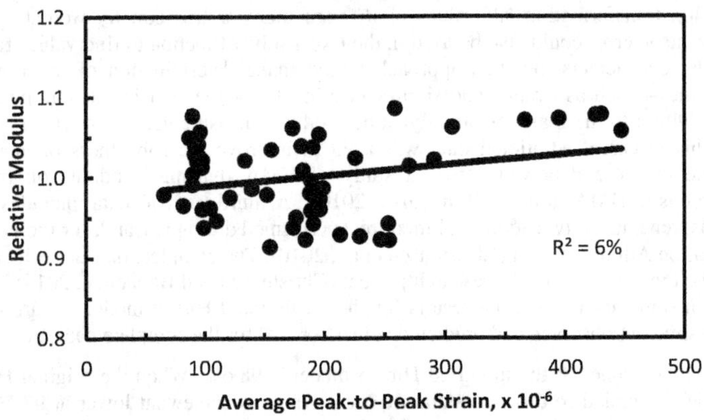

Figure 3. *Plot of relative modulus as a function of average peak-to-peak strain for data from strain sensitivity experiment.*

6.0 Recalibration of the Hirsch Model

The solver utility in Microsoft Excel ("Solver") was used to determine the coefficients in a variety of models based upon Equations 3, 5, 7 and an expression for aggregate modulus as a function of aggregate specific gravity. Because Solver does not provide detailed statistics on model coefficients, such as standard deviation, t-ratio and level of significance p, these parameters were calculated separately using standard statistical methods. Of the various models evaluated, the overall best model was as follows:

$$|E^*|_{mix} = Pc\left[H_6 G_{agg}^{H_7}\left(1 - VMA/100\right) + 3|G^*|_{binder}\left(\frac{VFA \times VMA}{10{,}000}\right)\right] \quad [8]$$

Where:

$$P_c = H_1 + \left(1 - H_1\right)\frac{\exp\left[H_2 + H_3 \ln\left(\frac{VFA}{100} \times |G^*|\right) + H_4 VMA + H_5 \ln(\varepsilon)\right]}{1 + \exp\left[H_2 + H_3 \ln\left(\frac{VFA}{100} \times |G^*|\right) + H_4 VMA + H_5 \ln(\varepsilon)\right]} \quad [9]$$

And H_1, H_2, H_3...H_7 are coefficients fitted to the model, and the other variables are as defined previously. In fitting the model, a log transformation was used, that is, the square of the errors between the predicted value of log $|E^*|$ and the measured value of log $|E^*|$ was minimized. This is necessary because of the wide range in values of $|E^*|$; if the error in the absolute value of $|E^*|$ is minimized, rather than the log of $|E^*|$, then the model fitting is heavily weighted towards high modulus values. By using the log transformation, all values of $|E^*|$ are more or less equally weighted. The percentage error could also be used in the Excel solver function to find values for the model coefficients, but this approach would make determination of other model statistics—such as t-ratios and significance levels for the coefficients—impossible. In addition to the materials and dynamic modulus measurements performed as part of this research, additional data was used to improve the robustness of analysis. Some of these data were from a study of HMA dynamic modulus on twelve Wisconsin HMA mixes (Bonaquist, 2010). A third set of data includes $|E^*|$ measurements on two additional mixtures, performed during research for the Federal Aviation Administration (Christensen et al., 2010). The complete data set is included in the Final Report for the research project (Christensen and Bonaquist, 2014). Table 6 is a summary of the final results for the recalibrated Hirsch model. Figure 4 is a plot showing observed and predicted values for $|E^*|$ for the complete data set.

The r^2 value for the modified Hirsch model is 98.6%. When the original Hirsch model is applied to this new data set, the r^2 value is somewhat lower at 97.7% (in both cases, adjusted for degrees of freedom). This represents about a 20% reduction in the mean square error for the new model compared to the original. There is also a slight difference in bias for the two models, although in both cases the bias is small: -1.4% for the new model, vs. +2.9% for the original. However, in comparing the two models it must be emphasized that the revised Hirsch model includes terms for both applied strain and aggregate specific gravity (a surrogate for aggregate modulus).

Therefore, differences in the predictions for the model will likely depend on the applied strain and the aggregate modulus. Large differences in predicted values should only be expected when the applied strain deviates from the current standard of 1×10^{-4} and/or the aggregate specific gravity deviates significantly from a value of about 2.6.

Table 6. *Summary of statistics for recalibrated Hirsch model.*

Coefficient		Estimated Value	Variance	Std. Dev.	t-value	p-value
H_1	Logistic minimum	0.0060	1.21×10^{-06}	0.00110	5.44	0.0000
H_2	Logistic constant	0.6628	6.47×10^{-04}	0.02544	26.06	0.0000
H_3	Ln (binder |G*|)	0.5861	2.80×10^{-04}	0.01672	35.05	0.0000
H_4	VMA	-0.1287	1.24×10^{-04}	0.01114	11.55	0.0000
H_5	ln (Strain)	-0.1706	3.02×10^{-07}	0.00055	310.50	0.0000
H_6	Aggregate G_s	7,960	6.75×10^{6}	2,600	3.06	0.0025
H_7	Aggregate exponent	1.5874	0.0810	0.2846	5.58	0.0000

Analysis of Variance (Dependent Variable: log of |E|)*

Source	Sum of Squares	D.F.	Mean Square
Regression	90.548	6	---
Error	1.2785	232	0.00554
Total	91.826	237	---
r^2, adjusted for degrees of freedom	98.6 %		

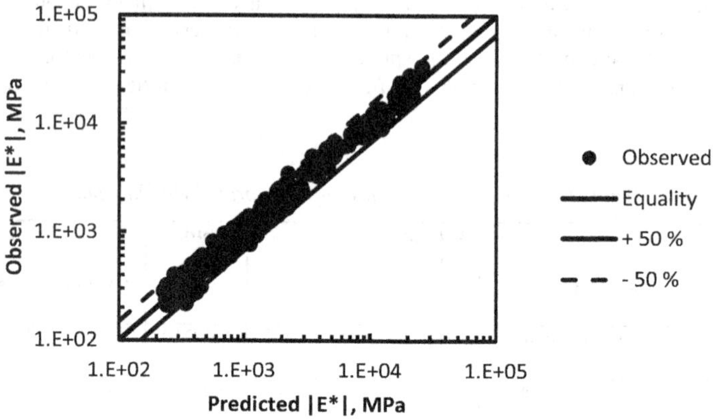

Figure 4. *Plot showing observed and predicted values for |E*|.*

In order to fully evaluate the recalibrated model, it is necessary to examine residual plots, which are shown below in Figures 5 through 10. The first of this series, Figure 5, shows model residuals (observed - predicted log |*E**|, %) as a function of predicted |*E**| (log scale). The residuals appear to be independent of log |*E**|. Also included in this plot are estimated maximum expected values for single-operator measurements and for between laboratory measurements of |*E*|*, as reported by Bonaquist (2011). Because these are model predictions, rather than actual measurements, it is not clear which precision statement should apply. Intuitively, it might be expected that the single-operator precision should apply, but the Hirsch model is based on volumetric composition—and the differences between measured and actual volumetric composition could be quite large. Single-operator precision, after all, is normally based on replicate specimens made from the same batch of hot mix, with a single set of volumetric compositional factors. The same problem exists with aggregate batching and binder handling, where differences in the precise storage and heating protocol could lead to differences in the binder modulus within the HMA specimen. It appears that about half the observations are within the expected single-operator precision, while most of the remaining observations are within the between-laboratory precision. All considered, the accuracy of the recalibrated Hirsch model is probably approaching the limits of what is possible for a semi-empirical approach to estimating HMA modulus.

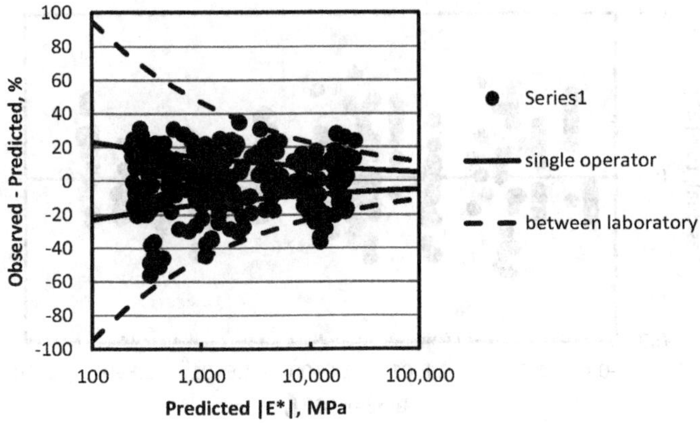

Figure 5. *Residuals as a function of predicted |E*| value, with maximum expected single-operator and between laboratory confidence intervals.*

It appears in Figure 5 that somewhat more residual terms are negative than are positive, but this is because the model was fitted using a log transformation. In Figure 6, which shows residuals for log |*E**| as a function of binder modulus (log scale), the residuals appear well balanced and independent of modulus. Figures 7 through 10 are related plots showing residuals as a function of VMA, strain, aggregate specific gravity and VFA, respectively. For the most part these plots suggest that the residuals are independent of the various predictor variables. Figure 10 does seem to show some curvature in the relationship between residuals and VFA, but there is significant scatter in this plot and it is not clear that the relationship is real. The r^2 value for a quadratic fit through this plot is quite low (2 %), so it was concluded that the residuals are in fact independent of VFA.

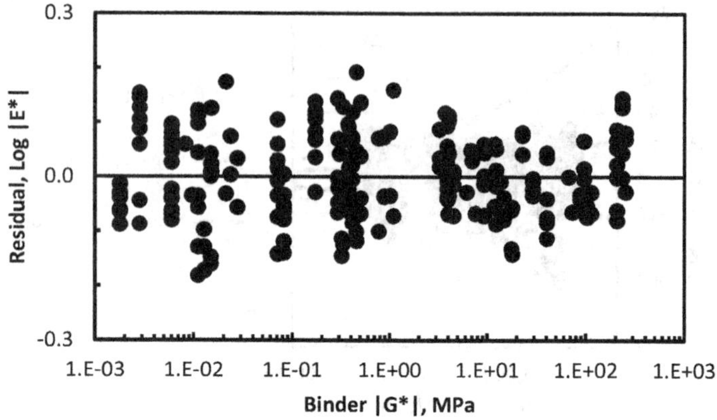

Figure 6. *Residuals as a function of binder dynamic shear modulus, |G*|.*

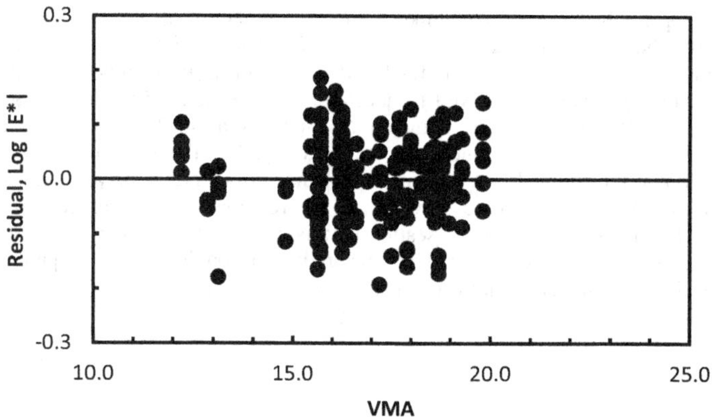

Figure 7. *Residuals as a function of VMA.*

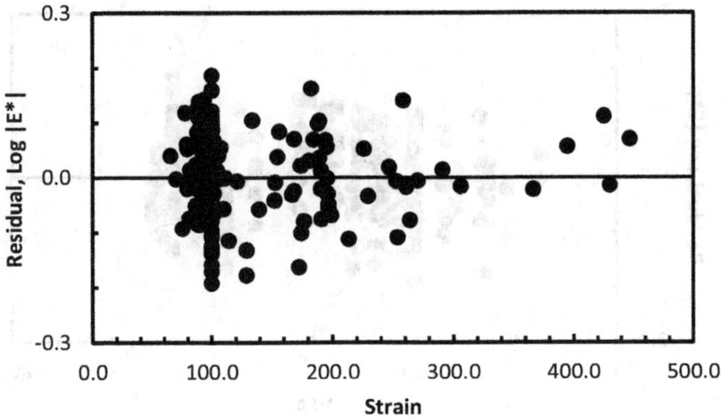

Figure 8. *Residuals as a function of strain.*

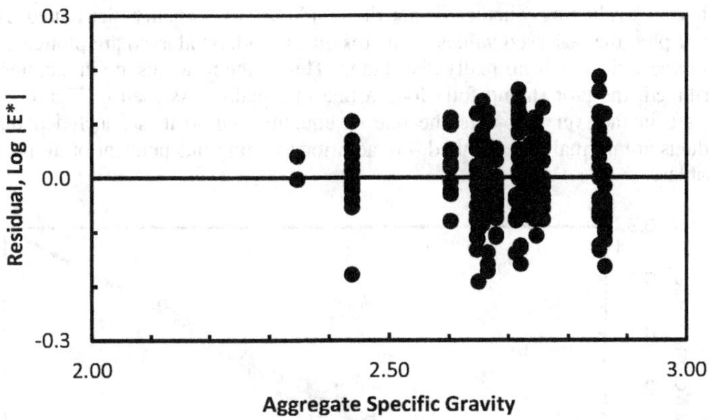

Figure 9. *Residuals as a function of aggregate specific gravity.*

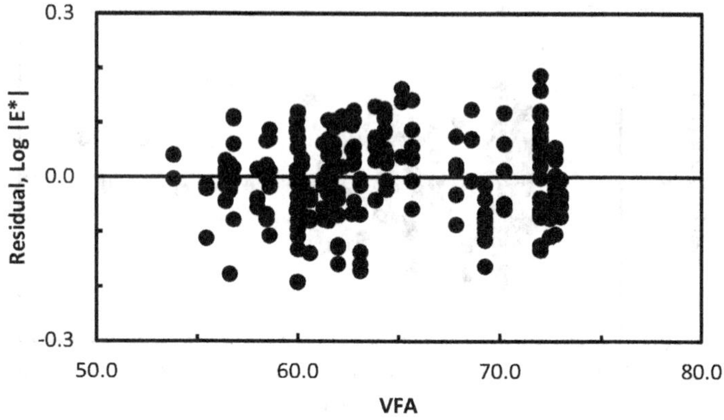

Figure 10. *Residuals as a function of VFA.*

Figure 11 is a normal probability plot of the model residuals. This is a standard method of evaluating whether or not the residuals are normally distributed. In this type of plot, the observed values of the residuals (horizontal axis) are plotted against their expected value if normally distributed. Thus if the residuals are in fact normally distributed, the plot should fall along a line of equality. As seen in Figure 11, the data are in fact very close to the line of equality and so it is concluded that the residuals are normally distributed – in addition to being independent of all predictor variables.

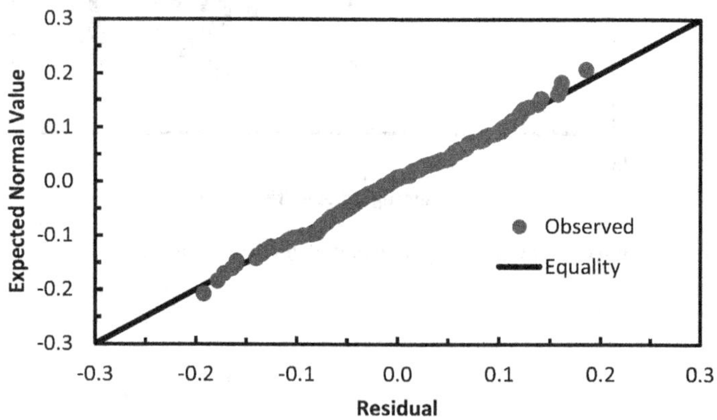

Figure 11. *Normal probability plot of model residuals.*

7.0 Evaluation of the Recalibrated Model Using the Original Data Set

As explained above, the data set used in the development of the original Hirsch model had several shortcomings. These included stain values that deviated significantly from those currently used in standard HMA |*E**| measurements and the use of a somewhat different testing protocol than is now used. However, the quality of the data is reasonable and served well in the original model development. Therefore a reasonable check on the recalibrated model would be to evaluate its accuracy using the original data set. Figures 12 and 13 are plots of residuals as a function of predicted |*E**| for the recalibrated and original Hirsch models, respectively, using the data set that the original Hirsch model was based upon. The pattern of residuals is very similar, but the residuals for the recalibrated model are shifted downward somewhat—that is, the new version appears to produce slightly higher estimates for |*E**| compared to the original model. The standard error (log sense) was 0.109 for the recalibrated Hirsch model and 0.088 for the original; the new model gives a slightly higher standard error, as should be expected given the bias in the predictions compared to the original model.

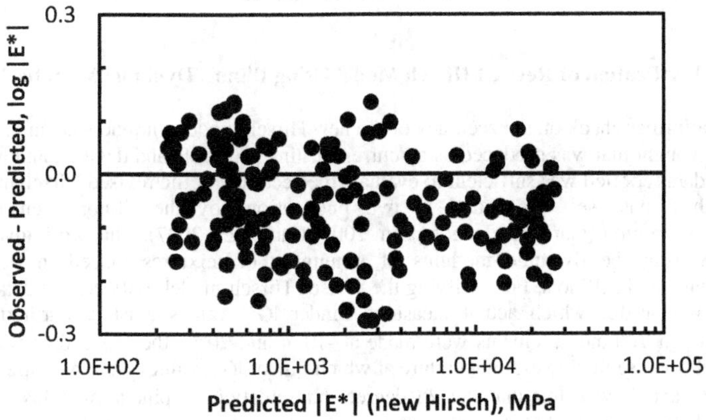

Figure 12. *Residuals for recalibrated Hirsch model plotted as a function of* |*E**|, *using original data set.*

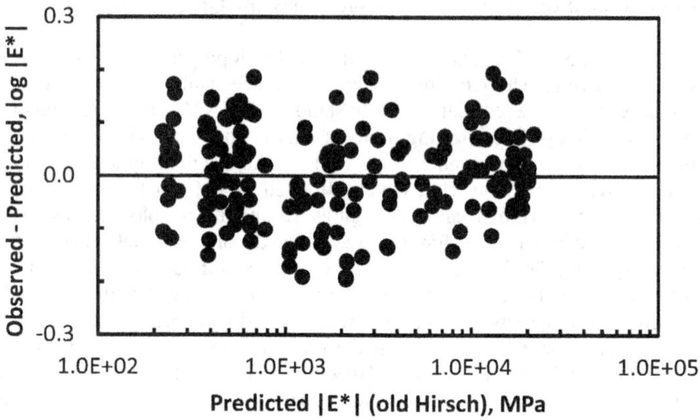

Figure 13. *Residuals for original Hirsch model plotted as a function of |E*|, using original data set.*

8.0 Verification of Revised Hirsch Model Using Illinois Dynamic Modulus Data

As a further check on the accuracy of the new Hirsch model, an independent data set was sought that was produced using current testing protocols and devices, and where the data reported was sufficient to evaluate the accuracy of the revised Hirsch model. Such a data set was found in two publications by the Illinois Center for Transportation (Garcia and Thompson, 2007; Carpenter, 2007). This work involved measuring the dynamic modulus of a number of mixtures placed in flexible pavements in Illinois. For verifying the revised Hirsch model, only certain data was used—that for which actual measured binder |*G**| values existed. Furthermore, although |*E**| measurements were made at -10, 4 and 20°C, the -10°C data was not used because the lowest temperature at which binder |*G**| values were measured was 16°C and it was felt that the 26 degree extrapolation of binder modulus values needed to estimate |*E**| at -10°C was excessive and not a good test of the new model. Data at 10 and 0.1 Hz was used for verification. One of the Illinois mixes contained RAP and it was not clear if the reported binder properties for this mix were for the actual blended binder for this mix, and so these data were deleted. Also, one of the mixes did not have volumetrics on the tested core—only the compacted cylinder—the differences can be substantial. Therefore, data for this specimen was not included in the verification. Almost all data required for applying the new Hirsch model was reported for the final set of five mixtures; the only data not available was the strain level during measurement; it was assumed to be 100×10^{-6}, but could have deviated substantially from this value. Figure 14 is a plot showing errors in predicted |E*| values (observed – predicted) for the final set of data, for the revised and original Hirsch models. This figure also includes the D2S confidence interval calculated using formula reported by Bonaquist (2011) with a sample size of n = 3, as was used in the |E*| measurements in this case. The limits are similar to those shown in Figure 5 for inter-laboratory confidence limits, but Figure 5 was based on n = 2.5, since the sample size included both n=2 and n=3. This figure shows that predictions for the revised Hirsch model appear to be unbiased, whereas the predictions made using the original Hirsch model are approximately 20% too low. Based upon this comparison, and Figures 12 and 13 presented earlier, it can be concluded that the revised Hirsch model tends to produce |*E**| estimates somewhat higher than the original Hirsch model. At least one previous research effort suggested that the original Hirsch model tended to under-predict dynamic modulus (Dongre, 2005), so it is possible that the revised version has eliminated or reduced this bias. The improvement would probably be most notable for mixes made using dense, high-modulus aggregates, since the new version adjusts the aggregate modulus term according to aggregate specific gravity. The overall accuracy of the revised Hirsch model appears to be acceptable, approaching the limits of D2S inter-laboratory precision. It is possible that the accuracy of the Hirsch model (and other models that rely largely on mixture composition) might be limited by the precision of HMA volumetric measurements.

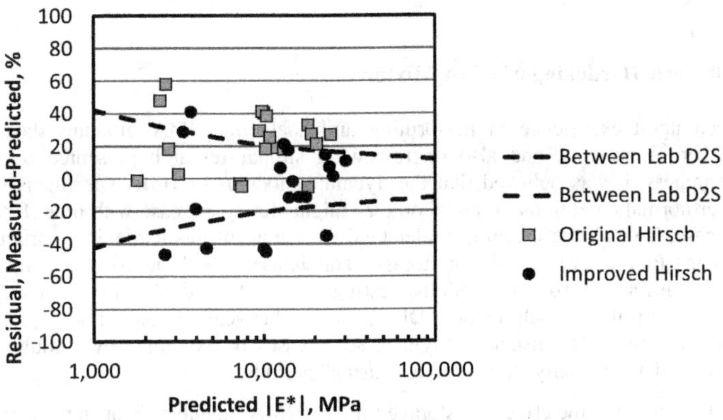

Figure 14. *Residuals for revised and original Hirsch, using data reported by the Illinois Center for Transportation (Garcia and Thompson, 2007; Carpenter, 2007).*

9.0 Estimating Phase Angle

The simplest and best model for predicting phase angle was found to be a quadratic equation using log of |*E**| as the predictor variable. This relationship is shown graphically in Figure 15. The standard error of the phase angle prediction was found to be 3.2%; the overall accuracy of the model should therefore be considered to be about ± 6%.

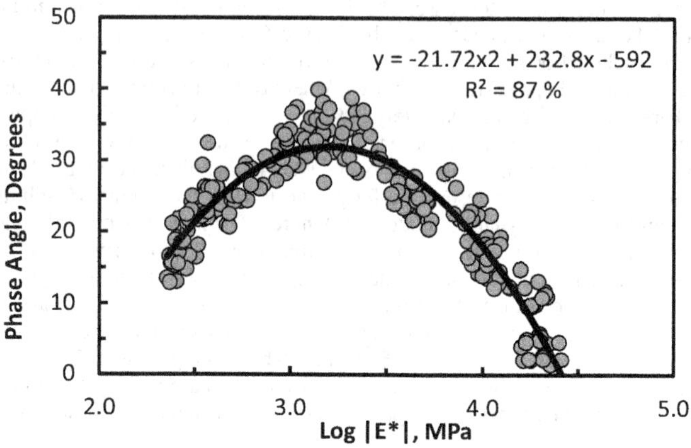

Figure 15. *Phase angle as a function of log |E*|.*

10.0 Steric Hardening of HMA Mixtures

Based upon experience in performing and analyzing HMA modulus data in a number of projects, and also on reviewing similar research presented by other researchers, it was believed that the dynamic modulus of HMA specimens stored under normal conditions in the laboratory might slowly increase with time. If this is the case, it would be useful to understand the nature of this hardening, particularly over the first several months of storage. This is useful both in terms of controlling the conditions of routine modulus testing of HMA, and also in evaluating the accuracy of the Hirsch model. Discrepancies between measured and predicted modulus values, for instance, might in some cases be explained by modulus tests performed at unusually short or long “curing” periods.

To determine the effects of storage time on HMA modulus, frequency sweeps on the HMA specimens were performed at curing times of 3, 7, 28 and 90 days. Engineers familiar with testing of portland cement concrete will recognize these times as standard ages for compressive strength tests. The aggregates and asphalt binders used are listed in Table 7. Details on the mix designs, aggregate gradations and binder properties are presented above.

Table 7. *Materials used in the steric hardening experiment.*

Aggregate	NMAS, *mm*	Binder Grade
Wisconsin Granite	12.5	PG 64-22
NY Gravel	12.5	PG 70-28
Mississippi Chert	12.5	PG 64-22
Wisconsin Gravel	12.5	PG 70-28
Pennsylvania Limestone	9.5	PG 64-22
Kentucky Limestone	12.5	PG 70-28
Virginia Limestone	9.5	PG 70-28
Virginia Diabase	9.5	PG 64-22

In analyzing the data, it was decided to make use of both the Christensen-Anderson (CA) model for binder modulus and the Hirsch model for aggregate modulus. Specifically, any change in modulus observed with curing time was modeled as a change in the crossover frequency (ω_c) of the binder. The crossover frequency is defined as the point in the binder master curve at which the phase angle is 45°; it also coincides with the point at which the viscous asymptote intersects the glassy modulus. It can and should be thought of as the location parameter for the CA model and is indicative of the overall hardness of an asphalt binder – as the crossover frequency increases the overall hardness of a binder decreases. Put another way, it was assumed that all changes in mix modulus were the result of changes in the binder modulus, and that the changes in binder modulus occurred only through changes in the location of the master curve, and not through changes in the shape of the master curve or in the transition temperature. Any steric hardening observed could then be modeled as a change in binder crossover frequency with curing time. Alternately, changes in crossover frequency due to hardening can also be modeled using a shift factor in which the applied frequency is multiplied by a factor that depends in this case on the elapsed time from specimen fabrication to testing.

Three replicate specimens were prepared for each combination of aggregate and binder. Dynamic modulus measurements were made on each specimen at frequencies of ranging from 0.01 through 10 Hz at temperatures of 4, 20 and 40°C. The CA model parameters and the Hirsch model were then used to fit the modulus data for each mix at each curing age. Because this analysis was done prior to the development of the revised Hirsch model, the original version was used in the analysis. The log glassy modulus was set at 9.00 for all cases and held constant; this produced more consistent results and better convergence in the analyses. The values of the other CA model parameters—*R*, *Tg* and crossover frequency—were found through non-linear least squares, using the Microsoft Excel solver function. Furthermore, it was found that significantly better fits were obtained if the aggregate modulus term in the Hirsch model was allowed to vary (but remaining constant for all conditions for a given mix). The error term minimized was the difference between log measured $|E^*|$ and log predicted $|E^*|$.

The results of this analysis are summarized in Figures 16 through 18 below. Figure 16 shows binder crossover frequency as a function of curing time (log scale). There appears to be significant variability in these data, but there does appear to be a slight downward trend in the crossover frequency, corresponding to a slight increase in binder and mixture modulus. Figure 17 is a plot of the log of the hardening shift factor—that is, the change in crossover frequency relative to seven days—as a function of curing time (log scale). This makes the downward trend in crossover frequency more apparent, although there is still considerable noise in the data. Figure 18 is a plot of the average log hardening shift factor as a function of curing time. It is proposed that the variability in shift factors seen in Figure 17 is primarily due to experimental noise, and that the average hardening function shown in Figure 18 can probably be applied to a wide range of mixes. It should be pointed out that the amount of hardening observed here is quite small. For example, the Wisconsin Granite mixture, which exhibited one of the strongest hardening trends, had a hardening shift factor of -0.16 at 90 days relative to 7 days. The corresponding increase in $|E^*|$ was 8.6%. Applying the same ratio of shift factor to % increase in modulus, the trend shown in Figure 17 would mean that the average increase in modulus due to steric hardening would be about 5% at 90 days. Applying the same logic, limiting curing times for dynamic modulus testing to between 3 and 28 days should result in no more than about a 3% error due to steric hardening effect. These relatively small levels of hardening can be compared to a typical coefficient of variation in the steric hardening data of about 10%. For three replicates, the corresponding D2S precision for these experimental data would be about 16%. Given the modest amount of hardening observed in this experiment, the variability observed in the hardening shift factor among different mixes and at different times is totally consistent with experimental error, and probably not due to inherent differences in the behavior of different mixes and/or asphalt binders. It can be concluded that under normal laboratory storage conditions, and at curing times of up to three months (or somewhat longer), steric hardening is a relatively minor factor in the dynamic modulus of HMA mixtures and need not be specifically addressed within the Hirsch model or other methods of estimating $|E^*|$ values.

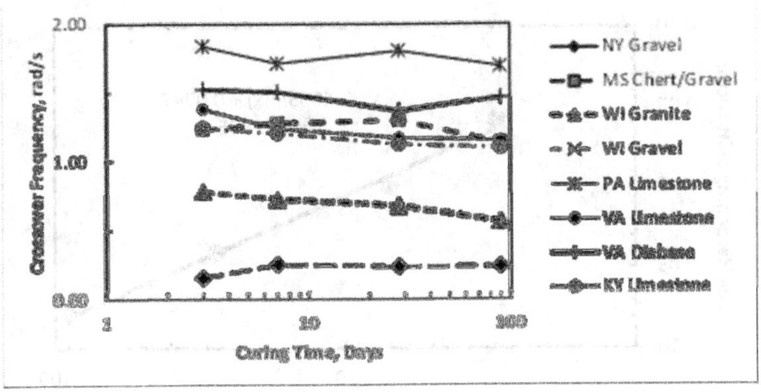

Figure 16. *Plot of binder crossover frequency as a function of curing time for eight HMA mixes.*

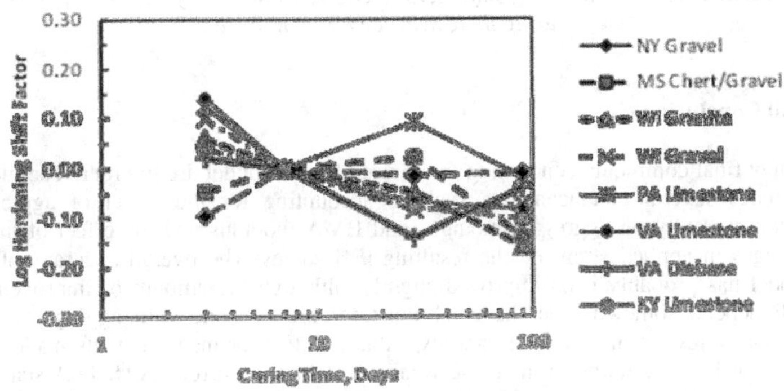

Figure 17. *Log hardening shift factor (relative to seven days) as a function of curing time for eight HMA mixes.*

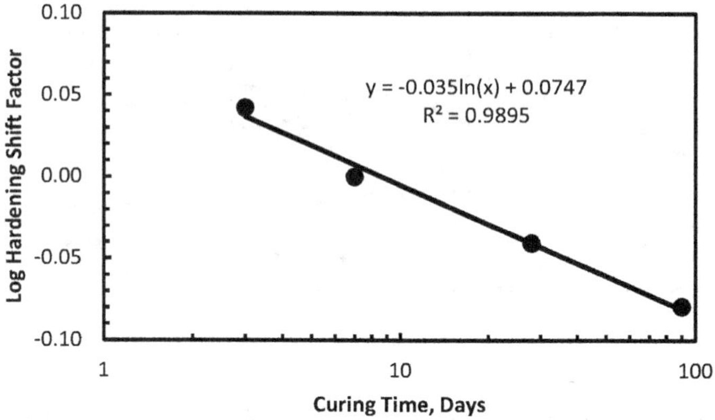

Figure 18. *Log hardening shift factor (relative to seven days) as a function of curing time, average for eight mixes.*

11.0 Conclusion

A few final comments concerning the revised Hirsch model are in order. The model includes several enhancements, such as accounting for the effect of aggregate specific gravity on aggregate modulus and HMA modulus and the effect of minor changes in applied strain on the resulting $|E^*|$ values. The overall accuracy of the model has probably been improved slightly, although the amount of improvement will depend somewhat on the conditions; for mixes made with aggregates with unusually low or high specific gravity values, and/or for measurements made with strain values deviating from those established in the current AASHTO standard procedure, the improvement could be significant. Otherwise, the recalibrated and original Hirsch model probably will produce essentially equivalent predictions. That being said, the comparison of current precision values for $|E^*|$ measurements with errors in the recalibrated Hirsch model would suggest that it would be very difficult to produce a model for predicting HMA modulus that was significantly better. This is not necessarily because of the superiority of the current version of the Hirsch model, but because of limitations in the accuracy of measurements of HMA modulus and volumetric factors such as VMA, VFA and air voids.

It must also be emphasized at this point that the recalibrated Hirsch model as presented in this report is intended to be a surrogate for $|E^*|$ values determined using procedures reasonably close to the current AASHTO standard. This means, for instance, that the model cannot be used to predict modulus values determined using different geometries (flexure or shear, for instance) or cylindrical specimens with a substantially different aspect ratio than set forth in AASHTO TP 79. The applied strain should probably be between 80 and 300×10^{-6} in order to apply the Hirsch model with confidence. Furthermore, the model is only intended for use in predicting modulus values in the unconfined state. Finally, the binder $|G^*|$ values used as input for the Hirsch model should either be values measured at the temperature and frequency of interest, or predicted using the Christensen-Anderson model (or the Christensen-Anderson-Marasteanu model, or some other model providing comparable predictions of $|G^*|$), again, at the temperature and frequency of interest. Binder modulus values after rolling thin film oven aging should normally be used, but values determined using binder modulus values after other aging regimes will be accurate for the given binder properties. Using other binder modulus values—such as those predicted from penetration and viscosity values, or using loading frequencies other than those of interest, will often produce extremely large errors.

As discussed in the introduction, the Hirsch model, both in its original form and in its revised form, is fully applicable to asphalt concrete mixes containing RAP and/or RAS, and or made using WMA technology. However, applying the Hirsch model in these cases is somewhat more complicated than in standard HMA mixes containing no recycled materials. When using the Hirsch model to predict modulus values of mixes containing RAP and/or RAS, the modulus of the effective blended binder must be used. This can be done using several methods. The binder can be extracted from the RAP and then tested in the DSR to obtain modulus values at the desired temperatures and frequencies. These modulus values could then be used along with values for the new binder and appropriate blending charts (or other similar technique) to estimate the modulus of the blended binder. Alternately, the extracted RAP binder and the new binder could be blended in the appropriate proportions and then tested with the DSR to determine $|G^*|$ values.

Application to WMA is in some ways simpler. There are two questions that must be addressed. The first is have the WMA additives changed the modulus of the binder, and if so, how much? The second question is does the WMA exhibit the same degree of hardening with time as a typical HMA mix? The first question could potentially be addressed by blending the binder and additive and then testing for dynamic shear modulus. Unfortunately, there is no guarantee that the additive will behave the same in this situation as it would in the actual mix. The second question, unfortunately, can only be answered through testing. In practice, in applying the Hirsch model to WMA at this time, the best approach would be to calibrate the Hirsch model using limited modulus testing, using the following procedure as an example:

1. Determine the modulus values of the binder at the desired temperature and frequencies.
2. Fit the binder modulus values to the CA or other equivalent model.
3. Perform a modulus test on the mix at a single temperature and several frequencies.
4. Using modulus values estimated using the CA or other model, estimate mixture modulus using the Hirsch model and compare these to the

measured values. Vary the crossover frequency (or model location parameter) until the error between the predicted and measured mixture modulus values is minimized.

5. Using modulus values estimated with the CA or other model, and the adjusted crossover or location parameter, predict mixture modulus values at desired temperatures and frequencies using the Hirsch model.

Although this procedure requires some mixture testing, it reduces it by using a calibration of the Hirsch model to account for the potential unusual behavior of WMA. This same approach could also be applied to mixes containing RAP and/or RAS, or to WMA containing these materials. Additional research is needed to identify and refine the best approaches to using the Hirsch and other predictive models for asphalt concrete modulus to materials containing RAP and/or RAS, to WMA and to mixes containing unusual additives or admixtures.

Some mention should be made of the usefulness of the Hirsch model (and other similar predictive models) in evaluating mixes containing RAP and/or RAS and WMA. For example, by testing the modulus of the new binder in a mix, the mix made with only the new binder, and the mix containing different types and amounts of RAP and/or RAS, the modulus of the RAP binder and of the effective blended binder can be estimated. These values can then be compared with established binder rheological specifications to determine if the blended binder meets these requirements. The properties of the RAP/RAS binder determined in this way could also be compared to the properties determined on the extracted RAP/RAS binder to evaluate how well the properties of such extracted binders represent the properties of the RAP binder in situ.

12.0 Summary

The recalibrated Hirsch model presented in this report is an improvement over the original model, but not necessarily a great improvement. It will provide improved estimates of HMA $|E^*|$ values, especially when a mix contains aggregates with unusually high or low specific gravity values, or when the strain values used in the measurement deviate somewhat from those specified in AASHTO TP 79. In most cases, the estimates made using this improved model should be close to those of the original. Steric hardening of HMA mixes does occur, but using normal specimen fabrication and handling protocols—where specimens are tested within a few weeks of compaction—the effects of steric hardening should be minimal. The analysis presented in this report suggests that dynamic modulus measurements of HMA mixtures should be done at between 3 and 28 days after mixing and compaction. Even at curing times of 90 days, the increase in modulus caused by steric hardening will typically be about 5%.

The accuracy of the revised Hirsch model is probably approaching what is currently possible given the limitations of modulus and volumetric composition measurements for HMA. The Hirsch model will provide modulus estimates with accuracy close to that of measured values, if typical inter-laboratory precision is considered. Therefore, laboratories with limited experience in HMA modulus measurements will probably obtain more reliable values using the Hirsch model than with actual measured values and at much lower cost. The Hirsch model is also useful in providing preliminary estimates of modulus for pavement design and also for estimating modulus values of HMA mixtures for research projects.

It is not clear if further significant improvements are possible in the Hirsch model for HMA modulus as presented in this report. It is possible that consideration of several other factors might improvement the accuracy of the model slightly. Such factors might include compaction level, aggregate gradation and aggregate angularity/internal friction.

13.0 Conclusions and Recommendations

The proposed revised Hirsch model is an improvement over the original model in several ways. It is based on a database of recent $|E^*|$ measurements, made with current equipment and protocols. It includes terms for applied strain level and aggregate specific gravity, which is linked to the aggregate modulus term in the Hirsch equation. The newer model should therefore provide better modulus estimates for mixes using aggregates with unusually high or low specific gravity values, or when strains applied during measurement of the modulus differ significantly from the current standard of about 100×10^{-6}. The revised model appears to provide somewhat higher estimates for HMA dynamic modulus, suggesting that the original model tends to under-predict $|E^*|$ values. As with the original model, dynamic modulus values predicted with the revised Hirsch model can be used for research purposes, for pavement design and analysis, and in forensic studies. Agencies not experienced in performing $|E^*|$ measurements on HMA mixtures will probably obtain more accurate estimates using the improved Hirsch model than by actually measuring modulus values, although the use of RAP and/or RAS in many mixes makes such an application problematic in many cases. Additional research is needed on the best approach to applying the Hirsch model and other similar predictive methods to mixes containing RAP and/or RAS and to WMA.

Effects of steric hardening on HMA $|E^*|$ measurements under normal conditions are relatively small. Such effects should typically be no more than about 3% for measurements made on specimens between 3 and 28 days after mixing and compaction. Because of the small magnitude of this effect, it has not been directly addressed in the revised Hirsch model.

14.0 Acknowledgement

This paper is based upon work supported by the Federal Highway Administration (FHWA) under Agreement No. DTFH61-07-00009. Although the National

Cooperative Highway Research Program (NCHRP) did not provide direct support for the research presented in this paper, the work that lead to this research was originally funded by the NCHRP as part of NCHRP Projects 9-25 and 9-31. Any opinions, findings, and conclusions or recommendations expressed in this publication are those of the authors and do not necessarily reflect the views of the FHWA or the NCHRP. The authors acknowledge the significant research concerning HMA permeability performed by previous researchers, and which produced much of the data used in the analysis, as cited within the body of the paper.

15.0 References

Bonaquist, R. F. *Wisconsin Mixture Characterization Using the Asphalt Mixture Performance Tester (AMPT) on Historical Aggregate Structures*, Report WHRP 09-03, Springfield, VA: National Technical Information Service, January 2010, 111 pp.

Bonaquist, R. F. *Precision of the Dynamic Modulus and Flow Number Tests Conducted with the Asphalt Mixture Performance Tester*, NCHRP Report 702, Washington, D. C.: Transportation Research Board, 2011, 200 pp.

Carpenter, S. H., *Dynamic Modulus Performance of IDOT Mixtures*, Report FHWA-ICT-07-008, Springfield, VA: National Technical Information Service, December 2007, 91 pp.

Christensen, D. W. and Bonaquist, R. F., *Improved Models of Performance Related Properties for Hot-Mix Asphalt*, Draft Final Report to the Federal Highway Administration under Agreement No. DTFH61-07-00009, Sterling, VA: Advanced Asphalt Technologies, LLC, 2014, 127 pp.

Christensen, D. W., and Anderson, D. A., "Interpretation of Dynamic Mechanical Test Data for Paving Grade Asphalt Cements." *Journal of the Association of Asphalt Paving Technologists*, Vol. 61, 1992.

Christensen, D. W., Bennert, T., McQueen, R. D. and Brar, H., "Superpave Gyratory Compaction Requirements for FAA's Hot Mix Asphalt Specification," *2010 FAA Worldwide Airport Technology Transfer Conference*, April 2010, 10 pp.

Christensen, D. W., Pellinen, T., and Bonaquist, R. F., "Hirsch Model for Estimating the Modulus of Asphalt Concrete," *Journal of the Association of Asphalt Paving Technologists*, Vol. 72, 2003, pp. 97–121.

Dongre, R., L. Meyers, D'Angelo, J., Paugh, C and Gudimettla, J., "Field Evaluation of Witczak and Hirsch Models for Predicting Dynamic Modulus of Hot-Mix Asphalt," *Journal of the Association of Asphalt Paving Technologists*, Vol. 74, 2005, pp. 381–434.

Garcia, G. and Thompson, M., *HMA Dynamic Modulus-Temperature Relations*, Report FHWA-ICT-07-006, Springfield, VA: National Technical Information Service, April 2007, 97 pp.

Mohammad, L. N., Saadeh, S., Obulareddy, S. and Cooper, S., "Characterization of Asphalt Mixtures Using Simple Performance Tests," *ASTM Journal of Testing and Evaluation* (online), Vol. 36, Issue 1, January 2008, 12 pp.

Pellinen, T. K., *Investigation of the Use of Dynamic Modulus as an Indicator of Hot-Mix Asphalt Performance*, A Dissertation Presented in Partial Fulfillment of the Requirements for the Degree Doctor of Philosophy, Arizona State University, May 2001, 803 pp.

Pellinen, T., Zofka, A., Marasteanu, M. and Funk, N., "Asphalt Mixture Stiffness Predictive Models," *Asphalt Paving Technology 2007, Journal of the Association of Asphalt Paving Technologists*, Vol. 76, 2007, pp. 575–622.

Sakhei Far, M. S. *Development of New Dynamic Modulus Predictive Models for Hot Mix Asphalt Mixtures*. A dissertation submitted to the Graduate Faculty of North Carolina State University in partial fulfillment of the requirements for the degree Doctor of Philosophy. Raleigh, NC: North Carolina State University Department of Civil Engineering, 2011, 276 pp.

16.0 Discussion

EMIN KUTAY: How many different unique HMAs have you used in calibrating your model?

DON CHRISTENSEN: I'm sorry. Can you say that again?

EMIN KUTAY: How many different hot-mix asphalts have you used while calibrating this new model?

DON CHRISTENSEN: I'm still having a little problem. How many different asphalt binders?

EMIN KUTAY: No, mixtures. Different mixtures.

DON CHRISTENSEN: Oh. Well, we had nine different aggregates and five different binders, and I'm not sure of the exact number of mixes, to be honest with you. It was probably on the order of 30 maybe. I'd have to go back and look at the paper, look at our data.

EMIN KUTAY: Just as a follow-up, what's your opinion of the applicability of this new model to gap-graded SMAs and mixtures with crumb rubber modified asphalt? Have you included any of those?

DON CHRISTENSEN: We didn't have any SMAs in here. My feeling is it would probably work fine for SMA because the volumetrics would be similar and that's what this is all based upon. Crumb rubber is a different matter. I'm not sure what the effect there would be, but my guess would be that that may not work as well for something containing crumb rubber or any other kind of, say, particulate modification.

GEOFFREY ROWE: Very nice paper, Don, as usual. I've got a couple of questions because I'm trying to understand the equations fully. If I look back at one of your slides, you show, I think, the variability reducing—that one here—so you have a better prediction at the stiffer parts of the moduli. I think that would be correct.

DON CHRISTENSEN: This is the slide that you're referring to?

GEOFFREY ROWE: Yes, that's correct.

DON CHRISTENSEN: Are you asking does it do a better job at higher modulus?

GEOFFREY ROWE: Better job at higher moduli.

DON CHRISTENSEN: You know, if I look at this I would say maybe, but I wouldn't bet the house on it.

GEOFFREY ROWE: Okay.

DON CHRISTENSEN: It's not clear to me. I think it's possible and I think in general there is more variability in modulus measurements at higher temperature, so that way it wouldn't surprise me.

GEOFFREY ROWE: I wanted to ask you, how does your functional form compare to the sigmoids that we're using in the Witczak model? Your equations are arranged in a different format, so you don't have the frequency term in the equation as such. You have the binder stiffness term in the equation. So are you seeing a similar sort of sigmoidal shape?

DON CHRISTENSEN: Yeah, that's a good question. In this model and even in the original, in both of them, there was a sigmoidal function, and that's what this P_C, that's where the sigmoid comes in. In fact, this is what's called a logistic function—it's a well-known statistical function, I'm sure you're familiar with it—and it does produce a very nice sigmoid going from zero to one. I think any model that you're looking at for hot mix modulus is probably going to have some kind of sigmoidal function in there.

GEOFFREY ROWE: Yes. Would I be correct in saying that your lower asymptote of that type of sigmoid that you would have in that function also the binder stiffness term built into that?

DON CHRISTENSEN: The lower asymptote here—and I forgot to mention. I'm glad you brought that up—but if you look at that function for P_C, that's essentially what that 0.006 is. That's a lower kind of limiting value, although you also have to take into account that aggregate modulus in the upper equation, but you can calculate what that lower limit is, and theoretically it's the modulus you would have for the mix if the modulus of the binder were to be zero.

GEOFFREY ROWE: Yes, which would occur as G* approaches zero in that top equation.

DON CHRISTENSEN: Yes, correct. And I did want to point out. I'm glad you brought it up. I tried to make that value, that 0.006, a function, say, of VMA. I thought it should be or maybe strain, and I tried a couple different variations there, but none of them worked, and the best results were simply to make it a constant.

GEOFFREY ROWE: The other question and my final sort of discussion was the phase angle relationship that you use in this paper is the sort of polynomial fit to the data that you have, whereas in your previous work on binder, you took the differential approach of the stiffness with respect to frequency in development of your phase angle for the binder modulus. We know that same approach works for the mixes; you know we looked at that in our work on mixes. I think it is more encompassing, if you know what I'm saying. It's less empirical. I don't know if you've got any thoughts or comments on that.

DON CHRISTENSEN: That's another interesting point, a very good question, and I did look at that. And I'd have to go back through some of spreadsheets and notes. I thought that would give better predictions for phase angle. I'm not sure if everyone followed that. The idea here is if you take a log-log plot of E* versus frequency – and correct me if I'm wrong – that should be proportional to the phase angle. It is for binders, and it should be for mixes.

GEOFFREY ROWE: Yes. The differential of the stiffness with respect to frequency.

DON CHRISTENSEN: Correct.

GEOFFREY ROWE: Yes.

DON CHRISTENSEN: But I looked at that, and, again, I'm not 100% sure because I did this a while ago, but I think I did look at this and the predictions weren't that good. But perhaps I did something wrong, and I'll probably go back and look at this again. Because I think I had hoped that that would provide better phase angle predictions, and it's a very good question. But it didn't seem to, but perhaps I made an error, and I think it's worthwhile going back and looking at it again. I'll let you know what I find out.

GEOFFREY ROWE: We looked at it with the Witczak form of the sigmoid and then my version of the sigmoid by adding an additional parameter to the sigmoid to change the inflection point. Both of those seemed to give a reasonable assessment of that when we looked at a wide range of different materials. There are several different ways of trying to get that differential.

DON CHRISTENSEN: Yeah, and now that I think about it, it's possible that I was actually looking at the actual measurements and they may have been too noisy for this to work, whereas if we take the predicted values and take the differential of those, that may provide better estimates.

GEOFFREY ROWE: Sure. I'd be happy to share my data with you, as well, if you want that.

DON CHRISTENSEN: Yeah, that would be helpful because I think that's a good question, and maybe we'll be able to work this in to the final version of the paper if it works out.

GEOFFREY ROWE: Thanks, Don. Nice paper.

PHILLIP BLANKENSHIP: On the data that you were showing on the error where you're storing the samples long term, again, a minute point on that. Was that one mixture or was that on all your mixtures you were looking at?

DON CHRISTENSEN: This one.

PHILLIP BLANKENSHIP: That's the one.

DON CHRISTENSEN: Yeah, and I know you've done some work on this same topic, and I think your results were somewhat different.

PHILLIP BLANKENSHIP: Right. We discussed it, but I've never really seen the data on that.

DON CHRISTENSEN: Well, I think at that time I didn't have all my results yet.

PHILLIP BLANKENSHIP: Yup.

DON CHRISTENSEN: So these at the upper right, those are all the mixes, I think eight different mixes, and then at the bottom, it's the average. And I attribute this kind of up and down to noise. Maybe that's not correct, although it wouldn't make sense to me. You look at these and some of them actually show it hardening and then softening, and I don't believe that that happens. So, to me, when I see something like that, I think it's noise. We did eight mixes, and we didn't see a lot of hardening here and a lot of difference between these mixes, although partly this is because of the noise in the data. And I mentioned, probably thinking of the work you did, there may be mixes that harden more than this. There may be mixes maybe of highly absorptive aggregates where there is substantially more stearic hardening. And I'm not even sure you should call it stearic hardening in that case because potentially I think what happens is you have preferential absorption of binder components by the aggregate, say something like a low-density chert, for instance, although I think we included one of those in our aggregates thinking that that would be one that would harden a lot and it didn't.

PHILLIP BLANKENSHIP: I applaud your work. It's a very confusing topic for us. This is one of those things when Mike Anderson – Mike, I'll be sure to strike this from the comments when I get to edit – when Mike goes, Phil, try this. And so we looked at it. We did the same thing. You know, we looked at it at one day, three days, 10, 21, 28, 84 days. And we only looked at it with one aggregate, one asphalt. And we went back and repeated it thinking that because we didn't have the noise and it was really a tight data set and it was statistically significant, and the data did what you're showing up there, and it went down and it came back up. I don't know what's going on, but what you're showing is exactly what we saw in some of that data.

DON CHRISTENSEN: And you repeated it.

PHILLIP BLANKENSHIP: And we repeated it thinking it was an error on our end, and I was wrong, it did it again. And then what we saw, though, was on an aggregate absorption, water absorption of about 1 to 1.5%. We saw a stiffening at the dynamic modulus of about 20 to 25% over three to four weeks. And so, yeah, I follow where you're going on the average and I'm glad you showed the top right graph. I don't really know what's going on either, other than some absorption, but it definitely makes me think on when do we cut our samples and when we test them and so on.

DON CHRISTENSEN: I think what we conclude here is there is still some work to be done on this topic.

PHILLIP BLANKENSHIP: Absolutely.

DON CHRISTENSEN: Maybe what we're seeing here is two different effects. Maybe there's an aggregate absorption effect and there's a binder stearic hardening effect. Maybe these go in different directions in some cases.

PHILLIP BLANKENSHIP: So in the studies that we did going forward, as you know, on the ruggedness study, we labeled everything to where we said we're just going to lock everything in and test it after three weeks to try to cut out some of that noise. But there's definitely a lot of noise that we saw between day one and ten, and so I appreciate your work.

DON CHRISTENSEN: Thank you. And I'll have to take another look at the work you did and maybe we can reconcile this in some way.

MARTIN VAN DE VEN: Sorry I don't know maybe but the databases that you use, is that always the same test from which you get values?

DON CHRISTENSEN: I'm sorry?

MARTIN VAN DE VEN: You said you use databases, big databases, and the test to determine that stiffness, is that always the same test?

DON CHRISTENSEN: These were all the same, they were done, I believe, on the same piece of equipment by the same, I don't know if it was the same technician but if it wasn't one guy it was maybe only two. So it was all very carefully controlled.

MARTIN VAN DE VEN: Yeah, but was it also four-point bending tests or tension-compression tests?

DON CHRISTENSEN: Oh, I'm sorry. I referred to AASHTO TP 79 thinking everyone would know what that is.

MARTIN VAN DE VEN: Okay, I am also sorry for not knowing.

DON CHRISTENSEN: I was using the uniaxial dynamic modulus. We're applying a sinusoidal strain to the specimen and then measuring the load. They're taken out of gyratory specimens, cored out, and then sawed at the ends and so forth. Does that answer your question?

MARTIN VAN DE VEN: Yes, thanks. I'm just asking because at Delft University we do a lot of tensile and compression tests and we determine the stiffness in tension and compression. You also know that probably. It can differ quite a lot.

DON CHRISTENSEN: And that's why I pointed out at the beginning of the paper that this is meant to only predict modulus values that you would get from this particular test because, yes, we've seen big differences. If you go to a shear test, a bending test, these are all going to give you—actually, depending on the temperature and so forth— the differences can be greater than an order of magnitude.

MARTIN VAN DE VEN: Absolutely, I fully agree.

DON CHRISTENSEN: I mean not minor, these are big differences. And then someone else brought up anisotropy, but we don't have to open that can of worms. So, yeah, that's a good point, yeah.

Binder Composition and Intermediate Temperature Cracking Performance of Asphalt Mixtures Containing Recycled Asphalt Shingles

Samuel B. Cooper, Jr.[a], Ioan Negulescu[b], Sreelatha S. Balamurugan[c], Louay Mohammad[d], and William H. Daly[c]

[a]*Technology Transfer and Training, Louisiana Transportation Research Center and Dept. of Civil and Environmental Engineering, Louisiana State University, Baton Rouge, LA 70808.*

[b]*Department of Textiles, Apparel and Merchandising, Louisiana State University and LSU AgCenter, Baton Rouge, LA 70803*

[c]*Dept. of Chemistry, Louisiana State University, Baton Rouge, LA 70803*

[d]*Louisiana Transportation Research Center and Dept. of Civil and Environmental Engineering, Louisiana State University, Baton Rouge, LA 70803*

ABSTRACT. *The use of recycled asphalt shingles (RAS) as a partial replacement for petroleum-based virgin asphalt binder has received considerable attention in recent years. The objective of this study is to correlate the molecular structure of asphalt binders of conventional asphalt mixtures as well as of mixtures containing recycled asphalt shingles (RAS) with their cracking potential at intermediate temperature. Laboratory testing evaluated the molecular composition of asphalt binders obtained from asphalt mixtures evaluated in this study using gel permeation chromatography (GPC), extent of aging using Fourier Transform Infrared (FTIR) spectroscopy, and fracture resistance of laboratory produced mixtures using the Semi-Circular Bending (SCB) test at intermediate temperature. Molecular fractionation through GPC of RAS samples confirmed the presence of associated asphaltenes in greater concentrations than recycled asphalt pavement (RAP) samples. High concentrations of high molecular weight asphaltenes decrease the fracture resistance of the asphalt mixtures. The use of rejuvenating agents, Cyclogen-L and Hydrogreen, did not reduce the concentration of the highly associated asphaltenes, and thus they failed to improve the cracking resistance of the asphalt mixtures evaluated in this study.*

KEYWORDS: *hot-mix asphalt, stone mastic asphalt, recycled asphalt shingles, gel permeation chromatography, FTIR spectroscopy, semi-circular bend test.*

The oral presentation was made by Mr. Daly.

This is a reproduction of an article whose final and definitive form has been published in *Road Materials and Pavement Design*© 2015 Taylor & Francis. The article is available online at: http://dx.doi.org/10.1080/14680629.2015.1077013

1.0 Introduction

1.1 Recycled Asphalt Shingles

Asphalt shingles are the most popular roofing material and are utilized in roughly 67% of the United States residential roofing market. There are approximately 11 million tons of asphalt shingles disposed of each year in the United States. The shingle waste is comprised of ten million tons of installation scraps (tear-offs/post-consumer) from re-roofing and one million tons of manufactured shingle waste (post-industrial waste) (Northeast Recycling Council, Inc., 2011). Recycled Asphalt Shingles (RAS) are comprised of asphalt binder (19–36%); fiberglass or cellulose backing (2–15%); sand-sized aggregate and ceramic-coated natural rock (20–38%); and mineral filler or stabilizer that includes dolomite, limestone, and silica aggregates (8–40%) (Northeast Recycling Council, Inc., 2011). Recent estimates indicate that RAS contain 15–35% of asphalt binder, which may provide an annual savings of $1.1 billion and reduce nonrenewable energy consumption in the United States (Northeast Recycling Council, Inc., 2011; Gevrenov, 2008). The use of RAS in asphalt mixtures reduces the negative impacts on the environment associated with the extraction, transportation, and processing of virgin materials. It also preserves valuable landfill space while allowing state agencies/owners to offset some of the extra costs caused by the increase in aggregate and petroleum prices.

Since the early 1990s, a number of research studies evaluated the use of RAS in hot-mix asphalt (HMA) and its influence on the mix mechanical behavior. The use of RAS in HMA is expected to provide significant benefits to the asphalt industry and highway agencies by reducing the amount of virgin asphalt binder that is added to the mixture. Air-blown asphalt is typically used in the manufacturing of asphalt shingles; this type of asphalt binder has a greater viscosity than regular asphalt binder used in HMA (Foo et al., 1999). Button et al. (1996) evaluated the influence of adding 5–10% of asphalt shingles on the mechanical properties of asphalt mixtures as compared with untreated mixes. The use of RAS resulted in decreased tensile strength and creep stiffness of the mixture, but it improved the mix resistance to moisture damage.

In spite of the potential benefits of the use of RAS in asphalt mixtures, many states are still cautious in the use of high RAS content in asphalt mixtures because of construction and durability concerns. To ensure the successful use of RAS, confidence in the mixture design procedure requires addressing many concerns related to the interaction between virgin and recycled materials, which has an effect on the durability of the produced mixture.

1.2 Asphalt Characterization

The composition of asphalt differs depending on the crude source and the refining process. The components of asphalt have been characterized as maltenes, asphaltenes, naphthene aromatics and polar aromatics by extraction techniques to separate the components. The maltenes are low molecular weight (MW) hydrocarbons, which act as dispersing agents for the higher molecular weight components. The nature of the asphaltenes (Mullens, 2011), the most aromatic of the heaviest components of asphalt, is the major factor controlling the properties of asphalt binders. Asphaltenes, as the higher MW component, by virtue of their molecular size are the bodying agent for the maltenes and have a significant influence on asphalt performances (Rostler and White, 1970). The largest "molecules" are assemblies of smaller molecules held together by one or more intermolecular forces. Through changes in the polarity of the solvent used in the analysis, the ability of the samples to undergo self-assembly by different interactive mechanisms has been probed (Jennings et al., 1993). Therefore, by analyzing the asphaltenes of asphalt paving combinations, such as asphalt cements and polymer-modified binders, with or without reclaimed asphalt materials (RAP, RAS), one might correlate physical performance of mixtures containing these materials with the content and MW magnitude of asphaltenes species.

The differences in the molecular weights of maltenes and asphaltenes have prompted efforts to separate these components using Gel Permeation Chromatography (GPC). GPC provides a simple separation of molecules in a sample according to their sizes or, more specifically, their hydrodynamic volumes. This molecular size excluding technique can be likened to a sieving process in which the largest materials elute first, followed by successively smaller molecules. The GPC ability to separate by molecular size rather than by some complex property such as solubility or absorptivity is one of the great advantages of the technique. This feature made GPC especially suited for fractionating complicated mixtures like crude oil residua, asphalts, and asphaltenes for almost 50 years now (Altgelt, 1965; Dickie & Yu, 1967; Snyder, 1969). GPC very uniquely mirrors the quantitative distribution of all species present in a binder, such as maltenes, asphaltenes, and polymers. The instrument signal, viz., the difference between the refractive indices of the eluting solution containing the asphalt and that of the solvent (ΔRI) is plotted versus the eluting volume (mL). The molecules of larger size are excluded first, allowing the differentiation of asphalt species on the scale of MW = 10^6-10^2 Daltons. A correlation of the eluting volume with the molecular weight of the eluting fraction is achieved using narrow molecular weight standards (Federal Highway Administration, 2013).

Efforts to predict the properties of asphalts using GPC have been reported. Rather than estimate the actual molecular weight of the eluting fractions, the GPC chromatograms have been divided into three regions: large molecular size (LMS), medium molecular size (MMS), and small molecular size (SMS). Researchers stated that the LMS and SMS regions are significant with respect to predicting pavement performance (Rostler & White, 1970; Jennings et al., 1993; Wahhab et al., 1999; Elseifi et al., 2012; Hassan et al., 2014). Although the division of the chromatograms into arbitrary regions can be done, we prefer to calibrate the GPC chromatograms and identify the maltenes, asphaltenes and polymer components on the basis of their molecular weight ranges (Lamontagne et al., 2001; Mouillet et al., 2008). Using

MW regions, it is possible to divide the LMS fraction into ranges which change when the asphalt ages or is modified.

As presented in Figure 1 for a PG 64-22 binder, quantification of asphalt components is readily made by determination of area of the respective eluted fraction calculated based on the fact that the area under the curve represents 100% of the sample molecules injected into the GPC system. The total GPC curve can be de-convoluted to show the contributions of the asphalt components using commercially available software Origin 7.

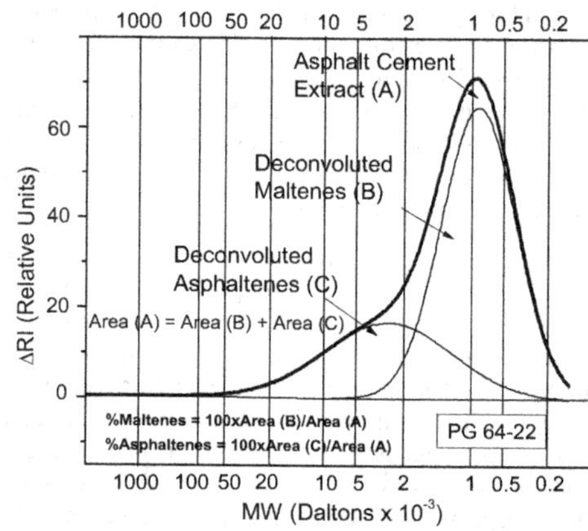

Figure 1. *Determination of maltenes and asphaltenes content of PG 64-22 binder by de-convolution of the GPC curve.*

Earlier determinations by osmometry indicated that the average MW of maltenes (as heptane soluble binder fraction) is 700-900 Daltons and that of asphaltenes (as heptane insoluble binder fraction) ranges between 2,000 and 10,000 Daltons (Zhou et al., 2013a). These MW data have been confirmed by the GPC method, which became a routine technique in Louisiana for analysis of asphalt binders (Daly et al., 2013). Since the MW of polymers used in asphalt industry is higher than 10,000 Daltons, the polymer and asphalt components of polymer modified asphalt cements could be separated completely with accurate determination of the molecular weight of species achieved by calibration with standard polystyrenes of narrow MW (Figure 2).

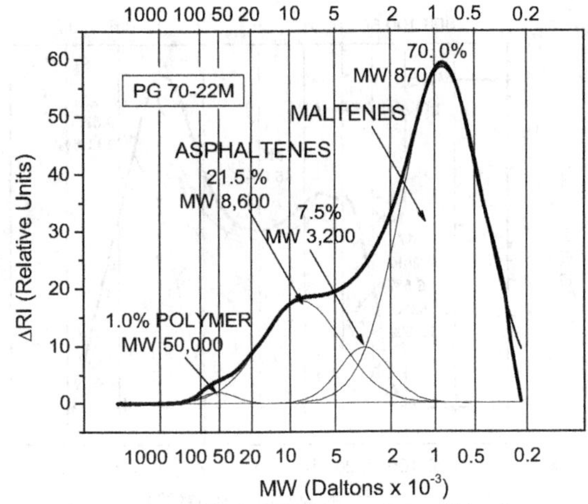

Figure 2. *GPC elution curve of PG 70-22M containing 1% polymer extracted from a mixture aged for 5 days at 85°C.*

RAS material can be generated from either post-consumer (shingles that have been in service on roofs and have been removed), RAS-P, or manufactured waste (produced during manufacture of new shingles), RAS-M. The material used to manufacture roofing shingles is highly oxidized blown asphalt as confirmed by the GPC chromatograms; the assemblies of asphaltenes species from blown asphalt and RAS-M are practically identical (Figure 3). In addition to oxidation, the blowing process increases the asphalt aromaticity (conjugation) and average molecular size, which improves opportunities for self-assembly. A bi- or tri-modal peak shape showing the presence of two or three distinct populations of molecular sizes is regarded as evidence of intermolecular association in the LMS region on the left of the chromatogram (Jennings et al., 1993). Over 25% of associated asphaltenes in blown asphalt and MWS have an apparent average MW of 10K-50K Daltons. Asphalt binders extracted from MWS and post-consumer waste shingles (PCWS) typically have different properties because of the aging of the latter that occurs on a roof. A further major concern with using RAS relates to the variability in the properties of the RAS materials originating from different sources (Hassan et al., 2014; Lamontagne et al., 2001).

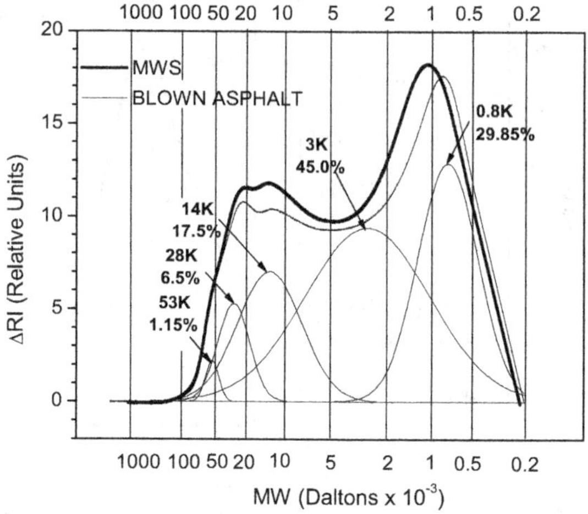

Figure 3. *GPC data of blown asphalt and RAS M extracted binders.*

1.3 Fourier Transform Infrared Spectroscopy (FTIR)

Characterization of oxidative asphalt aging with FTIR has been studied extensively in the past few years. The formation of carbonyl (C=O) containing molecules, which can be identified in the FTIR spectrum, has been correlated with standard asphalt binder aging techniques, RTFO and PAV (Lamontagne et al., 2001). It is well established that the main process occurring during this period is the oxidation of asphalt molecules which then leads aggregation due to the strongly interacting oxygen-containing molecule (Mouillet et al., 2008; Bowers et al., 2013). Since the main process is oxidation, the oxidized species can be used to quantify the amount of aging. FTIR spectra of the aged samples show a peak around 1700 cm^{-1} which is the characteristic of C=O species. A typical spectrum of an aged asphalt sample is shown in Figure 4, showing the key absorption bands.

In our previous investigations related to aging of styrene-butadiene-styrene (SBS) copolymer modified asphalt cements, we have used FTIR in order to gain a relative understanding of oxidation, which is directly related to asphalt binder aging (Negulescu et al., 2006). It was observed that the area of the carbonyl absorbance occurring at 1695 cm^{-1} increased as compared to that of the C-C absorbance occurring at 1455 cm^{-1}. The ratio of the C=O and C-C vibrations gave a relative comparison of how much oxidation was occurring. It was called carbonyl index. As the C=O index increased, there was a higher level of oxidation in the asphalt binder, and a stiffening of the binder was observed (Negulescu et al., 2006). Since both

MWS and PCWS are highly oxidized materials, it is expected that the carbonyl indices of paving asphalts incorporating RAS-M and RAS-P would be large.

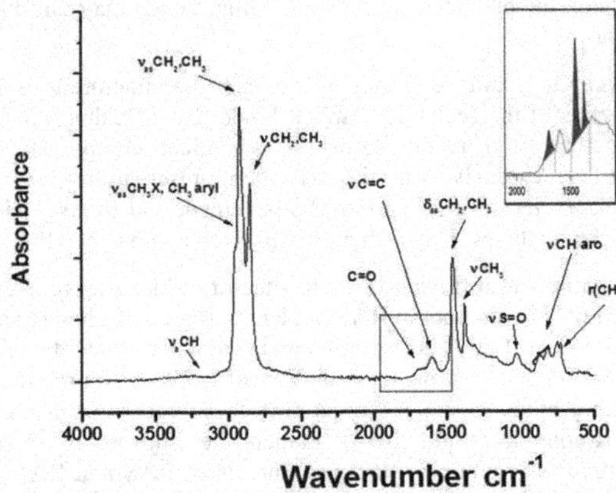

Figure 4. *Typical FTIR spectrum of an aged asphalt.*

A correlation of C=O index and the size and distribution of asphaltenes given by the maltenes/high end asphaltenes (MW>10K Daltons) ratio might be attempted, to a limited extent, to predict the field performance at intermediate temperatures as reflected by the value of critical value of the J-integral (J_c). The carbonyl index for each sample discussed in this paper is reported in Tables 2 and 3 (vide infra).

1.4 Performance of RAS in Asphalt Pavements

The industry has been addressing the stiffness and blending concerns by using softer binders when using higher RAP/RAS contents. Since fatigue cracking is influenced more by the intermediate temperature binder properties, using soft (modified) binders is an effective method to improve cracking resistance of RAS mixes. Zhou et al. conducted a comprehensive investigation of HMA mixes containing RAS (2013b). This study characterized the RAS asphalt binder including the evaluation of blending charts for virgin/RAS binders. In addition, the impact of RAS content on the optimum asphalt binder content and respective engineering properties on mixtures containing RAS was evaluated. It was concluded that the use of RAS had no significant influence on dynamic modulus, but it improved mixture resistance to rutting and moisture damage. However, mixtures containing RAS had very poor cracking resistance as compared to mixtures containing no RAS. The authors explored two approaches to improve cracking resistance of mixtures containing RAS. It was stated that using soft binder and increasing the design density can improve cracking resistance. In terms of rutting and moisture damage, the use of soft binders was superior to increasing the design density. When using the softer binder and low air void approaches, one should be aware that, if the RAS is not well blended into the mixture or if segregation occurs during mixing and/or placement, there will be spots on the pavement with "softer" mix, which may fail due to rutting (Zhou et al., 2013a).

Asphalt rejuvenating and softening agents are manufactured to restore the rheological properties of the reclaimed asphalt binder by diffusing into it, restoring its colloidal structure, and reconstituting its chemical components. Therefore rejuvenators have been extensively used in pavement preservation to revive the hard and oxidized top layer by penetrating into the pavement and fluxing with the aged binder to balance the maltenes to asphaltenes ratio (Newcomb et al., 1984).

It is unquestionable that diffusion is the key factor in blending rejuvenators with asphalts. Diffusion of rejuvenators in RAP binders has been of interest since several decades ago. Although there is experimental evidence that the rejuvenators penetrated the RAP binders (Carpenter and Wolosick, 1987; Noureldin and Wood, 1987) corresponding evidence on the impact of these agents on RAS is limited. Soohyok et al. (Soohyok et al., 2014) studied the impacts of three different rejuvenators on mixtures containing various contents of RAS and RAP. They used loaded wheel tests (LWT), overlay test, repeated load test, and dynamic modulus to characterize the mixtures. It was concluded that the rejuvenators improved the cracking resistance, moisture susceptibility, and rutting resistance compared to the control mixture. However, the ranking of the three rejuvenators used in the study depended on mixture types and properties evaluated.

With the increased interest in using RAS, the use of recycling agents is considered essential in order to soften and/or to rejuvenate the aged and stiff binders in RAS. Recycling agents are classified as two types: rejuvenating agents and softening agents. Softening agents lower the viscosity of the aged binder while rejuvenating agents are intended to restore the rheological and chemical properties of the aged binder (Soohyok et al., 2014). Examples of softening agents include asphalt flux oil, lube stock, and slurry oil. Examples of rejuvenating agents include lubricating and extender oils, which contain a high proportion of maltenes constituents and low saturate contents that do not react with asphaltenes (Soohyok et al., 2014). The design and production of asphalt mixtures containing RAS requires provisions to assure that the final product will meet and/or exceed the expected pavement life as required by construction and performance specifications.

2.0 Objectives and Scope

The objective of this study is to correlate the molecular structure of asphalt binders of conventional asphalt mixtures, as well as of mixtures containing RAS, with their cracking potential.

A compilation of various asphalt mixtures with a 12.5-mm nominal maximum aggregate size (NMAS) was evaluated in this study. In this study, the mixtures 64CO, 70CO, 76CO, and SMACO were the control mixtures containing no RAS utilizing PG 64-22, PG 70-22M, and PG 76-22M respectively. The mixtures 70PG5P, SMA5PHL and SMA5P contained 5% RAS-P, no recycling agents, PG 70-22M and PG 76-22M for SMA mixtures respectively. Mixture 70PG5M contained 5% RAS-M, no recycling agents, and PG 70-22M. The mixture 70PG5PHG contained 5% RAS-P, 5% Hydrogreen, and PG 70-22M. The mixture 70PG5PCY contained 5% RAS-P, 12% Cyclogen-L, and PG 70-22M. The mixture 64PG3P20RAP contained 3% RAS-P, 20%RAP, no recycling agents, and PG 64-22. Mixture 52PG5P contained 5% RAS-P, no recycling agents, and PG 52-28.

The paper will discuss the use of GPC to analyze the differences between asphaltenes found in virgin asphalt binders and RAS samples, RAS-M samples, and RAS-P samples. Cracking potential of the corresponding asphalt cements was evaluated using the SCB test at intermediate temperature. The percent content of polymers, asphaltenes and maltenes in aged polymer modified binders, as well as that of asphalts containing RAS-M and RAS-P, was reported. The potential impact of two rejuvenating agents, naphthalene oil (Cyclogen-L) and an abietic acid derived oil called Hydrogreen was analyzed by GPC.

3.0 Methodology

3.1 Materials

Limestone, gravel, granite, and natural sand aggregates commonly used in Louisiana were utilized in this study. RAS was incorporated into the asphalt mixtures at 5% by total weight of the mixture. Two types of rejuvenators (recycling agents) were used in this study, naphthalene oil (Cyclogen-L) and an abietic acid derived oil (Hydrogreen). The recycling agents were incorporated into asphalt mixtures at various percentages by total weight of RAS material in the mixture. The asphalt mixtures were prepared using an SBS polymer modified asphalt cement binder meeting Louisiana specifications for PG 70-22M and PG 76-22M (Louisiana Department of Transportation and Development, 2006). In addition, an unmodified PG 64-22 asphalt cement binder and a soft PG 52-28 binder was used.

3.2 Mixture Design

The details of the designed mixtures can be found elsewhere (Cooper et al., 2014).

During the asphalt mixture design process, the optimum asphalt cement content is determined by varying the virgin %AC content while utilizing the same composite aggregate blend until volumetric and densification criteria are met. With the optimum %AC of the mix and the percentage of virgin AC utilized being known, the actual %RAS AC contribution in the total binder can be calculated for each mixture. In addition, the actual shingle asphalt availability binder factor, the %RAS in the binder, and the recycled binder ratio (RBR), can be determined. The RBR is defined as the actual %RAS contribution divided by the design optimum asphalt cement content. For economic benefits it is important to increase the RBR. Recycling agents were employed to increase the RBR for mixtures containing PCWS. Table 1 is a recapitulation of the %RAS binder availability and the RBR for the mixtures evaluated in this study.

Table 1. *%RAS AC binder availability.*

Mixture ID	Optimum Design AC [%]	RAS AC Contribution [%]		AC (Virgin) [%]	Virgin Binder in Design AC [%]	Actual %RAS AC Availability Factor	Recycled Binder Ratio, [%]
		Available	Actual				
64CO	5.3	0.0	0.0	5.3	100.0	0.0	0.0
70CO	5.3	0.0	0.0	5.3	100.0	0.0	0.0
76CO	5.3	0.0	0.0	5.3	100.0	0.0	0.0
SMACO	6.2	0.0	0.0	6.2	100.0	0.0	0.0
70PG5P	5.3	1.4	0.5	4.8	90.6	35.7	9.4
SMA5P	6.2	1.4	0.6	5.6	90.3	42.9	9.7
SMA5PHL	6.2	1.4	0.6	5.6	90.3	42.9	9.7
70PG5M	5.3	1.3	0.6	4.7	88.7	46.2	11.3
70PG5PHG	5.3	1.4	1.4	3.9	73.6	100.0	26.4
70PG5PCY	5.3	1.4	1.2	3.9	77.4	85.7	22.6
64PG3P20RAP	5.0	1.5	1.5	3.5	70.0	100.0	30.0
52PG5P	5.3	1.4	0.7	4.6	86.8	50.0	13.2

3.3 Extraction of Binder from RAS Samples

The RAS samples and/or mixtures containing RAS were extracted with refluxing toluene under nitrogen using a Soxhlet extractor. The solution of asphalt cement in toluene was cooled to room temperature and then filtered to remove most of the fine particles of sand present. The filtered solution was allowed to stand overnight, was decanted, and was concentrated under a vacuum using a rotary evaporator. The concentrated asphalt cement solution in toluene was then dried for 36 to 48 hours in a vacuum oven first at room temperature (ca. 24 hrs), then at 50°C for 12 hrs.

3.4 Fourier Transform Infrared Spectroscopy (FTIR)

The FTIR spectra for the samples were obtained using a Bruker Alpha FT-IR spectrometer (Alpha), which uses a diamond single reflection attenuated total reflectance (ATR). An OPUS 7.2 data collection program was used for the data analysis. The following settings were used for data collection: 16 scans per sample, spectral resolution 4 cm^{-1}, and wave number range 4000-500 cm^{-1}. Approximately 1% solution of mix samples was made in carbon disulfide (CS_2) solvent and filtered using a 0.2 μ filter. A few drops of the solution were kept on the diamond crystal and the solvent was allowed to evaporate. Spectrum was collected after the complete evaporation of the solvent. The carbonyl index was calculated from the band areas measured from valley to valley as shown in Equation 1 (Marsac et al., 2014). This was done using the OPUS spectroscopy software provided with the Bruker FTIR instrument.

$$Carbonyl\ Index\ \% = \frac{Area\ of\ the\ carbonyl\ centered\ around\ 1700\ cm^{-1}}{\sum Area\ of\ the\ spectral\ bands\ between\ 1490\ and\ 1320\ cm^{-1}} \qquad [1]$$

3.5 Gel Permeation Chromatography (GPC)

GPC was performed using an EcoSEC high performance GPC system (HLC-8320GPC) of Tosoh Corporation, equipped with a differential refractive index detector (RI) and UV detector. A set of four microstyragel columns of pore sizes 200 Å, 75 Å (2 columns) and 30 Å from Tosoh Bioscience was used for the analysis. Tetrahydrofuran (THF), at a flow rate of 0.35 mL/ min, was used as the solvent. Columns were calibrated using polystyrene standard mixtures PStQuick B (MW= 5480000, 706000, 96400, 10200, 1000), PStQuick E (MW= 355000, 37900, 5970, 1000), and PStQuick F (MW= 190000, 18100, 2500, 500) from Tosoh Bioscience. The asphalt in the mix samples was extracted using THF solvent and the solutions were filtered using 0.45 micron Teflon filters. The concentration of asphalt solution was 0.5%.

3.6 Semi-Circular Bend (SCB) Test

Cracking potential was evaluated using the SCB test procedure based on fracture mechanics (FM) principles (Anderson, 2005) and suggested by Wu et al. (2005), Figure 5. The critical strain energy release rate, also called the critical value of J-integral (J_c), was used to describe the mixture's resistance to fracture:

$$J_c = -\left(\frac{1}{b}\right)\frac{dU}{da} \qquad [2]$$

Where: J_c= critical strain energy release rate (kJ/m^2);

b = sample thickness (m);

a = notch depth (m);

U = strain energy to failure (kJ); and

dU/da = change of strain energy with notch depth (kJ/m).

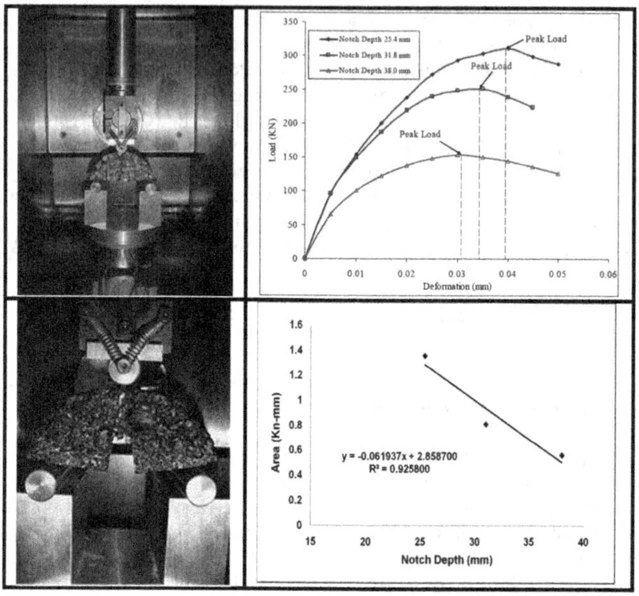

Figure 5. *Semi-circular bending test set-up.*

To determine the critical value of J-integral (J_c) using Equation (2), semi-circular specimens with at least two different notch depths should be tested to determine the change of strain energy with notch depth (dU/da). In this study, three notch depths of 25.4 mm, 31.8 mm, and 38 mm were tested to increase the accuracy of slope calculation (dU/da) by fitting a regression line to the change of strain energy with notch depth.

The semi-circular specimen was loaded monotonically until fracture failure under a constant cross-head deformation rate of 0.5 mm/min in a three-point bending load configuration. The load and deformation were continuously recorded. This test was performed at a temperature of 25°C. The area under the loading portion of the load deflection curves up to the maximum load was measured for each notch depth; this represents the strain energy to failure, *U*. The average values of *U* were then plotted versus the different notch depths to compute a regression line slope, which gives the value of (*dU/da*). The J_c was computed by dividing the *dU/da* value by the specimen thickness. High J_c values are desirable for fracture-resistant mixtures.

Specimens were compacted in the Superpave gyratory compactor (SGC) to an air void level of 7 ± 0.5%. Triplicate specimens were utilized for this test. In general, the coefficient of variation was within 15% for the samples tested.

4.0 Results and Discussion

4.1 Binder Characterization

The GPC traces of extracted binders from the Texas RAS used in this study are shown in Figure 6. Discernible differences between RAS samples are evident when one compares the maltenes component to the high end asphaltenes (MW>10K Daltons) component of the extracts. The ratio of the areas shown in the de-convoluted chromagrams identifies possible problems with compatibility of component species when blended with virgin asphalts (vide infra). For example, this ratio for the binder extracted from RAP originating from the Texan source is ~55/33. In contrast, the corresponding ratio for a Minnesota RAS-P extract is quite different, viz., ~80/20. The high molecular weight asphaltene content in the Texas RAS-P suggests that this material will be less compatible with virgin asphalt binders.

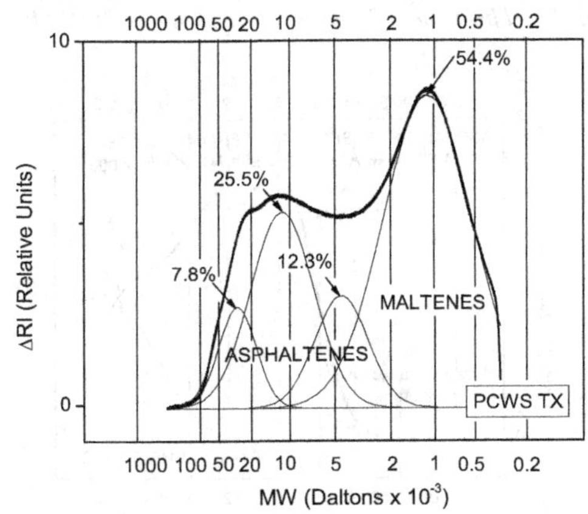

Figure 6. *GPC traces of RAS binders extracted from RAS-P of Texas origin.*

The asphaltene fraction isolated by heptane precipitation of the asphalt binder isolated from a Texas RAS-M sample was examined using GPC. The MW of molecular components of MWS (RAS-M) asphaltenes shown in Figure 7 surpasses greatly that of a similar precipitation of a PG 64-22 binder in Figure 8. Asphaltenes from PG 64-22 can be separated into two fractions, molecules with an average MW of 2,000 Daltons (60%) and associated asphaltenes with a peak average MW of 6,700 Daltons (22%). The asphaltenes from RAS-M could be separated into three fractions with average MWs >3,000 Daltons (52%), 12,000 Daltons (32%) and 24,500 Daltons (15%). Thus, associated asphaltenes with molecular weights higher than the associated asphaltenes in PG 64-22 comprise 47% of the RAS-M sample.

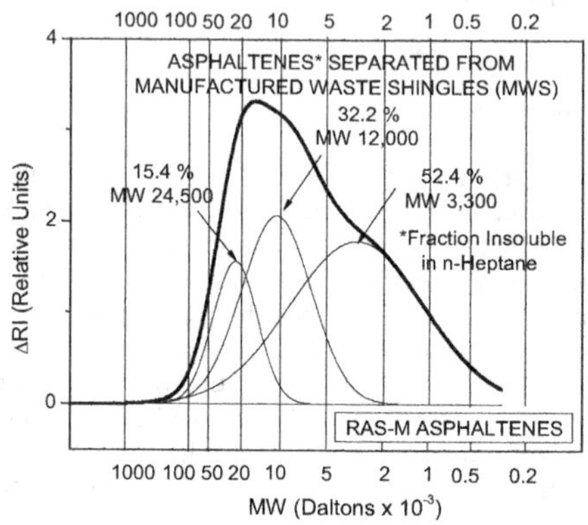

Figure 7. *Average MW distributions of n-heptane insoluble asphaltenes species isolated from manufactured waste shingles (RAS-M).*

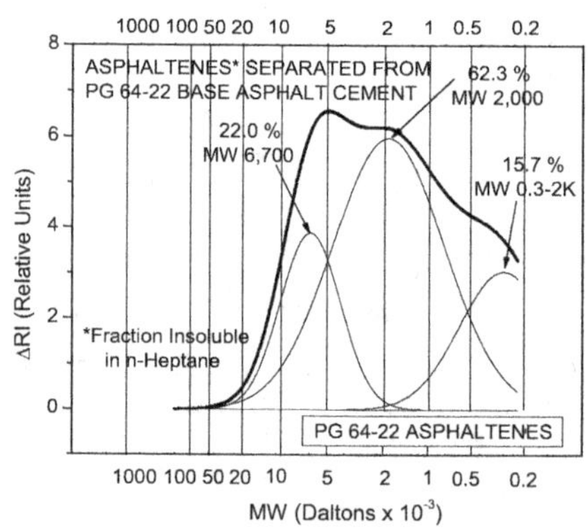

Figure 8. *Average MW distributions of n-heptane insoluble asphaltenes species isolated from the PG64-22 binder.*

The GPC chromatograms in asphalt binder extracted from PG 70-22M and a binder mixture containing 5% RAS-P (70PG5P) are presented in Figures 9 and 10, respectively. The contribution of extremely oxidized components in RAS-P is apparent in Figure 9. The 70PG5P mixture contains over 18% of species with an average MW>10K Daltons, out of which ~ 6% are of MWs averaging 33,000 Daltons. In contrast, a mixture prepared with 5% RAS-M, mixture 70PG5M, contained only 3% of species with average MWs of 24,000 Daltons. The maltenes/high end asphaltenes (MW>10K Daltons) ratio of 70PG5M (66/3) is significantly different from that of mixture 70PG5P (61/18), suggesting a higher potential compatibility of mixtures containing RAS-M.

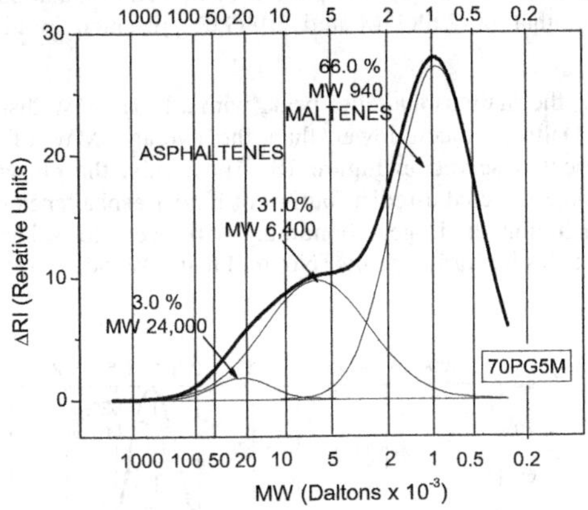

Figure 9. *MW distribution of molecular species of 70PG5M binder extracted from PG 70-22M containing 5% RAS-M.*

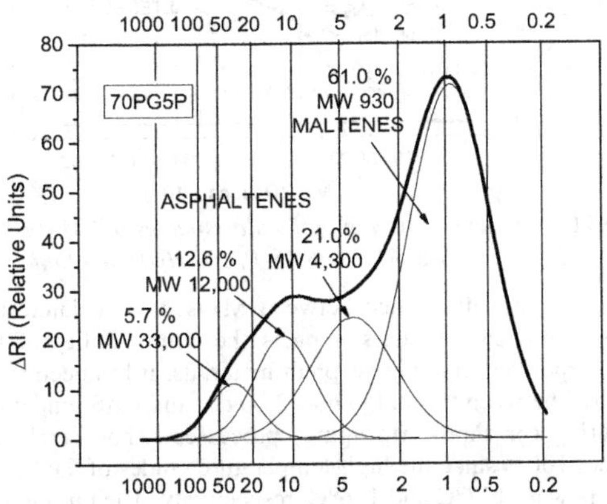

Figure 10. *MW distribution of molecular species of 70PG5P binder extracted from PG70-22P containing 5% RAS-P.*

The high MW end of associated RAS asphaltenes can exceed even 100,000 Daltons, as the de-convoluted GPC chromatogram of mixture SMA5PHL depicts in Figure 11. The binder of this mixture is obtained from PG76-22M asphalt and 5% RAS-P. Both the residual polymer species (SBR) from the binder (0.5%) and the highest average MW asphaltenes species of RAS-P (1.5%) can be identified. The presence of the SBR polymer in the mix enhances the compatibility of the highly associated asphaltenes. However, the pattern of this RAS-containing binder differs completely from that of 70PG5M and 70PG5P presented in Figures 9 and 10, respectively.

In Figure 11, the main asphaltenes peak, with a broad MW distribution, appears at around 5K Daltons, much lower than the average MW of associated RAS asphaltenes present observed in Figure 10. Apparently, the presence of Ca^{2+} ions from hydrated lime effected a redistribution of binder asphaltenes because the polar interactions, including hydrogen bonding, disrupted the self-assembly among asphaltene molecules by forming salts (Noureldin and Wood, 1987).

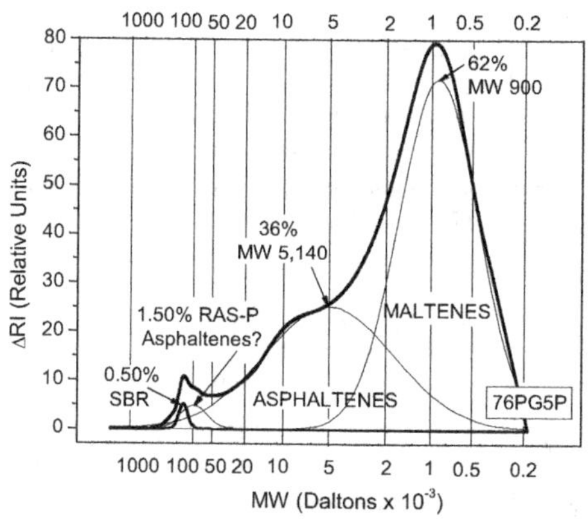

Figure 11. *MW distribution of molecular species of SMA5PHL binder (5.6% PG76-22M, 5% RAS-P, 3% Hydrated Lime).*

The extremely large difference between MWs of associated asphaltenes from RAS and those from base binders impacts the compatibility of the mixes when shingles are incorporated in paving asphalt materials. It has been reported that some blending occurred between virgin PG 64-22 binder and RAS binder during the HMA mixing and curing (or short-term aging) processes (Zhou et al., 2013a). But the blending was not 100% since the high-temperature grades of the RAS-M and RAS-P extracted binders were 122°C and 166°C, respectively. The mixing temperature for a PG 64-22 binder is around 143°C (290°F). Zhou et al. (2013b) state that "extremely, impractically high temperature is required in order to make the RAS binder flow and comingle with virgin binder" whether using tear-off asphalt shingles or manufactured waste asphalt shingles. Moreover, earlier investigations on cross-blending of asphaltenes and maltenes fractions among several asphalts (White et al., 1970) indicated that the asphaltene fractions are not as equally interchangeable as the maltenes components, and that the effect of both molecular weight and the chemical nature of the asphaltenes must be taken into account to predict properties of asphalts from chemical composition. Two types of segregation causing phase separations can occur when blending asphalts containing dissimilar asphaltenes and maltenes fractions: one is simply separation as occurs in supersaturated solutions when components are ejected because of insufficient solvent (e.g., flocculation in blends of high asphaltenes content), and the other is rejection of a component when its amount exceeds the mutual compatibility limit in the form of physical ejection of a liquid from a gel (Rostler and White, 1970; White et al., 1970). In view of these observations, the large difference between the MW of asphaltenes fractions of the base PG 64-22 (ca. 20% of maximum MW ≈ 7-8,000 Daltons) and the asphaltenes present in waste shingles (RAS-M and RAS-P), ~ 40% of MW >12,000 Daltons, with 15% MW ≈ 25-30,000 Daltons) makes difficult the dispersion of large RAS asphaltenes associations by the maltenes of the base asphalt with which the shingles will be blended. To this aim, one has to consider also the maltenes/high end asphaltenes (MW>10K Daltons) ratio mentioned above; the higher the ratio, the better. It has been shown earlier that an increase in the binder content of LMW (i.e., MW<3K), or in other words, of the content ratio of maltenes/asphaltenes, resulted in an increase in its elongation properties at intermediate and low temperatures (Shen et al., 2006).

4.1.1 Intermediate Temperature Cracking Performance of Asphalt Mixtures Containing Recycled Asphalt Shingles (RAS)

Since the asphaltenes content in asphalts is related to the stiffness, the question arises: when blending virgin asphalt cement with RAS, will the asphaltenes content of the resulting binder follow the additive rule and will the J_c will change correspondingly? The answer is no in most cases because the non-polar maltenes of the virgin asphalt are not compatible with the very highly oxidized RAS asphaltenes species. It has been shown that the virgin/RAS binder blending was nonlinear, unlike the well-known virgin–RAP binder linear blending (Zhou et al., 2013b). RAS-P binders were much stiffer than RAS-M binders (Zhou et al., 2013a; Soohyok et al., 2014). Compared with RAS-P, RAS-M binders had much less impact on properties of blended virgin/RAS binders (Zhou et al., 2013b, Cooper et al., 2014). Cooper reported that the addition of 5% RAS-P decreased the intermediate temperature cracking resistance (expressed by critical strain energy J_c) of a PG 70-22M binder when compared to that of a similar mixture in which RAS-P has been substituted with RAS-M (Cooper et al., 2014).

Analyzing the GPC data obtained for the same materials investigated by Cooper et al., i.e., 70PG5M and 70PG5P presented in Figures 7 and 8 respectively, we consider that the main reason for higher stiffness of the 70PG5P RAS-P binder reported by the authors is the degree of association of its large MW end asphaltenes (~ 6% of MW 33,000 Daltons), which is higher than that found in the 70PG5M binder containing 5% RAS-M (3% MW 24,000 Daltons). It has been pointed out before that the ability of asphalts to form an intermolecular network by associations could lead to cracking with time and under cold conditions (Soohyok et al., 2014).

Table 2 contains the data for a series of asphalt binders extracted from RAS containing mixtures investigated in this work. The total percent content of asphaltenes with MW larger than 20K Daltons is listed together with the values of carbonyl index and of J_c integral in order to find a correlation between these data to predict, to a limited extent, the field performance of considered mixtures.

Table 2. *Stiffness of mixtures and carbonyl index of related RAS binders.*

Designation	Binder	RAS-P %	RAS-M %	RAP %	Total MW >20K %	C=O Index X10^3	Jc kJ/m^2
64CO	PG64-22	0	0	0	4.7	5.9	0.37
70CO	PG70-22M	0	0	0	1.0	6.8	0.50
76CO	PG76-22M	0	0	0	1.3	10.2	0.60
SMACO	PG76-22M	0	0	0	6.2	8.1	0.72
70PG5P	PG70-22M	5	0	0	5.7	21.2	0.40
SMA5P	PG76-22M	5	0	0	8.4	8.8	0.63
SMA5PHL	PG76-22M	5	0	0	2.0	3.7*	0.75
70PG5M	PG70-22M	0	5	0	3.1	12.0	0.50
70PG5PHG	PG70-22M	5	0	0	11.5	17.1	0.23
70PG5PCY	PG70-22M	5	0	0	8.7	13.0	0.36
64PG3P20RAP	PG64-22	3	0	20	3.0	12.4	0.33
52PG5P	PG52-28	5	0	0	5.7	11.7	0.23

**Presence of lime interferes with analysis*

A comparison of the critical strain energy (J_c) data for mixtures of Table 2 evaluated in this study versus the content of asphalt species with MW larger than 20K Daltons, the MW threshold of asphaltenes related to the stiffness of asphalt binders (Daly et al., 2011), is presented in Figure 12. Higher J_c values are desirable for fracture-resistant mixtures. A minimum J_c value of 0.50 to 0.65 kJ/m^2 is typically used as a failure criterion (Wu et al., 2005). However, a minimum threshold J_c value of 0.5 kJ/m^2 is being considered in Louisiana as an acceptance criterion for mixture design. Data listed in Table 2 show that rather large carbonyl indices and low J_c values, less than the minimum Louisiana acceptance of 0.5 kJ/m^2, were registered for mixtures containing PG 64-22 and PG 70-22M binders in which 3-6% asphalt species had MW > 20K Daltons (asphaltenes). However, the PG 76-22M mixtures containing sufficient polymers to accommodate RAS retained satisfactory J_c values of at least 0.6 kJ/m^2. The presence of just 0.5% high MW polymer in SMA5PHL can successfully prevent the stiffening effect of 1.5% high end large asphaltenes shown in Figure 11.

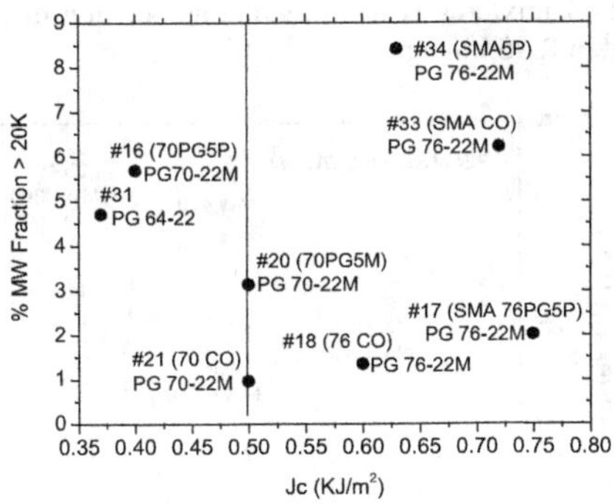

Figure 12. *Plot of J_c versus the content of asphalt fractions with MW>20K Daltons.*

A correlation might be found between the GPC distribution of molecular weight of binders and of their polarity as expressed by the oxygen content (carbonyl index, C=O) and the intermediate temperature performance (expressed by J_c). High MW polymers can be used to mitigate the stiffness of the resultant binder in RAS-P mixtures. A similar conclusion has been drawn for mixtures prepared with polymer-modified binders which were the best performers against fracture versus mixtures with high RAP content (40%) and the one prepared with straight unmodified binder (Elseifi et al., 2012).

4.1.2 Characterization and Intermediate Temperature Cracking Performance of Asphalt Mixtures Containing RAS and Rejuvenating Agents

As shown above, RAS asphaltenes are highly associated with apparent molecular weights approaching 100K Daltons. The rejection of high MW RAS components (MW>20K Daltons, see Figures 9 and 10) when its amount exceeds the mutual compatibility limit with added virgin asphalt, may be the main cause for the drop in stiffness for the RAS mixtures investigated by Mogawer et al. (2013) and low cracking resistance of RAS-P containing binders when polymers are not present as shown in Figure 10. Association might be reduced by adding rejuvenators in order to improve the intermediate temperature performance (expressed by critical strain energy, J_c).

One of the recycling agents chosen for the present study was Hydrogreen, which is an esterified derivative obtained from rosin, a by-product of the pulp and paper industry. This environmentally green rejuvenator is a low molecular product with the MW distribution shown in Figure 13. Its oxygen content is reflected by a significant carbonyl index (C=O = 0.04). Only 25% of its species matches the molecular weight of maltenes from an asphalt binder (MW ≈ 800-1500 Daltons). The other recycling agents considered were a naphthenic oil (Cyclogen-L), also a rejuvenator, and an asphalt cement meeting a PG 52-28, a softening agent. The anticipated role of the rejuvenators for RAS mixtures is to lower the association of high end large MW asphaltenes present in RAS binders.

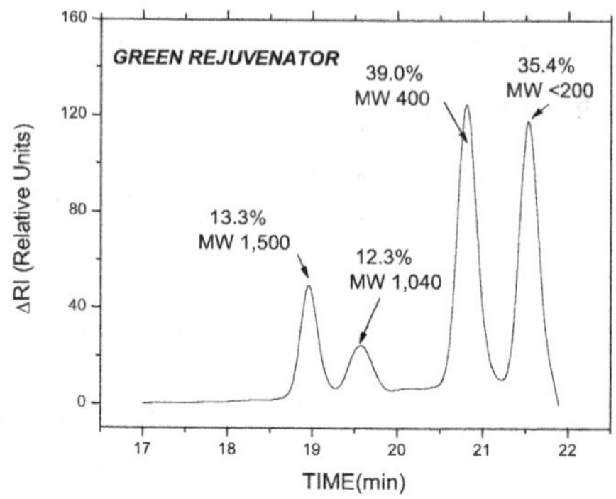

Figure 13. *Molecular mass distribution of Hydrogreen green rejuvenator.*

However, addition of 5% Hydrogreen to a PG 70-22 binder containing 5% RAS-P (70PG5PHG) does not seem to affect the distribution of high MW fractions derived from RAS-P (Figure 14).

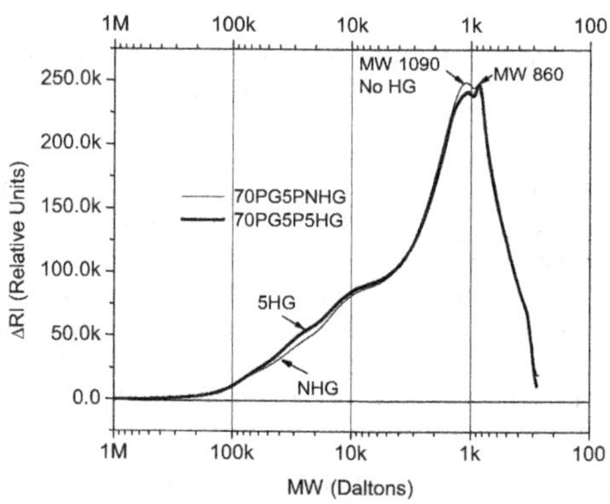

Figure 14. *GPC traces of PG 70-22M containing 5% RAS-P with (70PG5PHG) or without (70PG5PNHG) 5% HG by total weight of mix.*

At the same time, data listed in Table 3 and Figure 15 show that the addition of 5% Hydrogreen to a PG 70-22 binder containing 5% RAS-P (70PG5PHG) did not eliminate the high MW asphaltenes and did not preclude the increase of this fraction by aging of the mixture for five days at 85°C (SCB aging). It even seems that this rejuvenator promotes aging as shown by the increase of asphaltenes from 10.0% to

11.5%, the corresponding reduction of the content of maltenes, and the increase in carbonyl index. These factors may have precluded the decrease of J_c from 0.37 to 0.26 kJ/m^2.

Table 3. *Stiffness of mixtures and carbonyl index of related RAS binders containing recycling agents.*

Sample ID	Total VHMW % Polymer &Highly Associated Asphaltenes 1000K-19K	Total HMW % (Asphaltenes) 19K-3K	Maltene % < 3K	C=O Index x10^3	Jc KJ/m^2
#50 70PG5PHG	9.59	28.06	62.34	13.9	0.37
#43 aged 5days @85C:	11.50	29.06	59.43	17.1	0.26
#51 70PG5P12CY	8.30	25.98	65.38	9.9	0.13
#44 aged 5days @85C	8.77	27.17	64.06	13.0	0.36
#52 52PG5P	4.62	24.78	70.59	9.1	0.17
#45 AGED 5days @85C	5.68	26.10	68.22	11.7	0.24

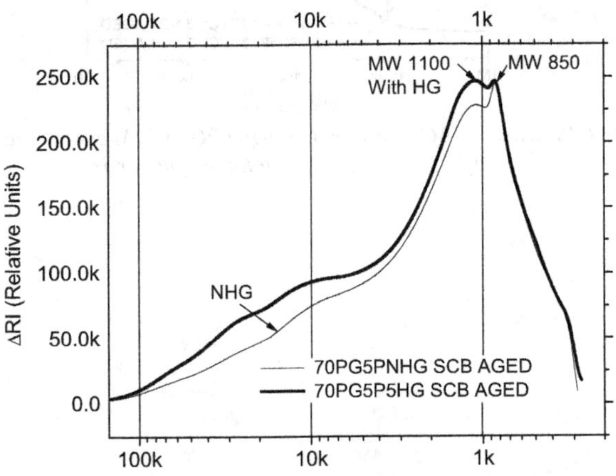

Figure 15. *GPC traces of SCB aged PG 70-22M containing 5% RAS-P with (70PG5PHG) or without (70PG5PNHG) 5% HG by total weight of mix.*

Blending and aging of 70PG/RAS-P binders with other rejuvenating agents thought to improve the low temperature performance of the mixtures provided similar results. Adding softening agents instead of the Hydrogreen rejuvenator did not seem to alter the MW distribution of asphalt components (Table 3). GPC traces and MW distribution remained practically the same after SCB aging for both Cyclogen-L containing PG 70-22M binder and the PG 58-22 binder (Figures 16 and 17). While the C=O index increased accordingly after aging, the cracking resistance expressed by critical strain energy, J_c, remained below the accepted limit (J_c<0.5 kJ/m^2) with decreased values for SCB aged mixtures, except for 70PG5P mixture containing 5% PCWS with 4.6% PG 70-22M asphalt cement by total weight of mix (Figure 18).

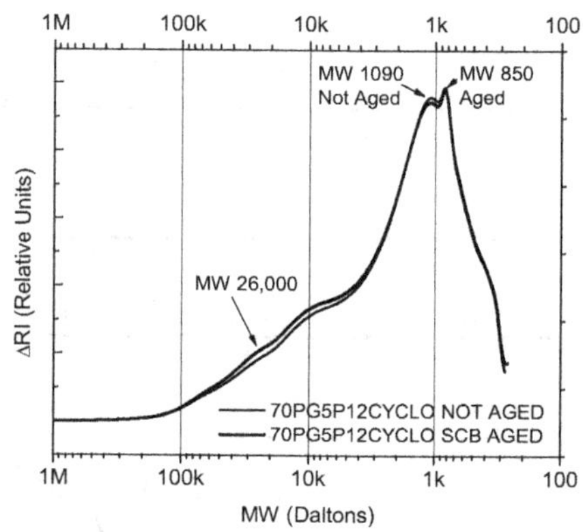

Figure 16. *GPC traces of SCB aged and un-aged PG70-22M containing 5% RAS-P with 12% Cyclogen-L by total weight of mix.*

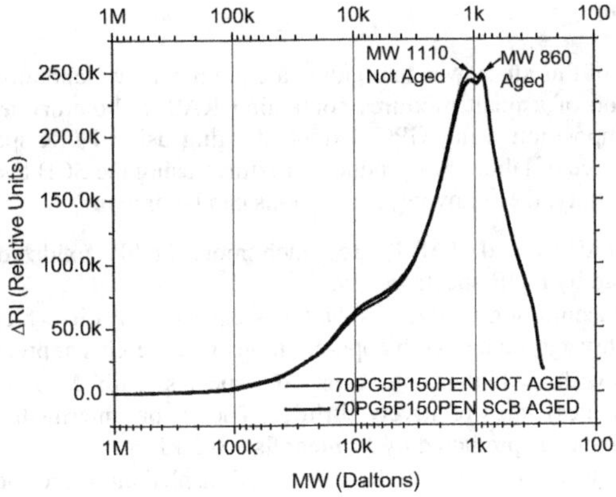

Figure 17. *GPC traces of SCB aged and un-aged PG52-28 containing 5% RAS-P with 4.6% PG 52-28 by total weight of mix.*

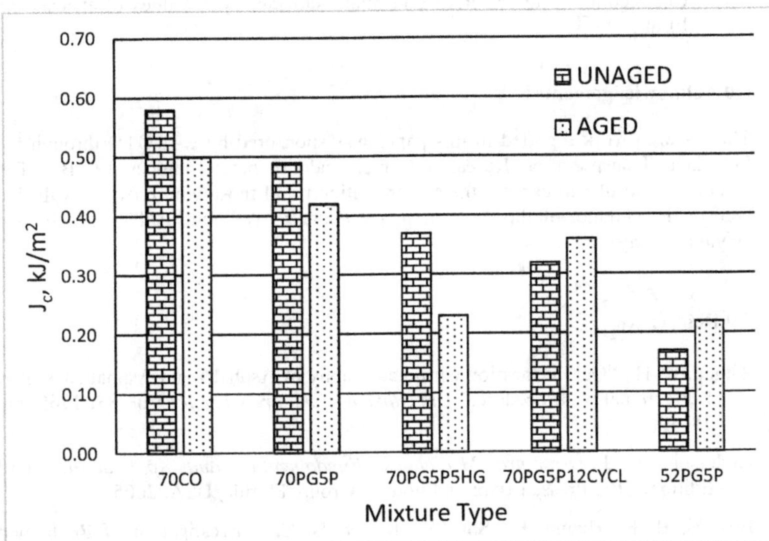

Figure 18. *J_c vs asphalt mixture types.*

5.0 Conclusions

The objective of this study was to conduct a comprehensive laboratory evaluation of the composition of asphalt mixtures containing RAS. Laboratory testing evaluated molecular composition using GPC, extent of aging using FTIR spectroscopy, and fracture resistance of laboratory produced mixtures using the SCB test. Based on the results of this study, the following conclusions can be drawn:

- Both RAS-M and RAS-P are much more highly oxidized than RAP as indicated by FTIR spectroscopy.
- The concentration of RAS asphaltenes exceeds 40% in which 25% of these are highly aggregated with apparent molecular weights approaching 100K.
- Mixes with a high content of AC species with MW > 20K Daltons (asphaltenes) are relatively brittle. The poor intermediate temperature performance is predicted by J_c integrals < 0.5 kJ/m^2.
- The high molecular weight associated asphaltenes are not significantly dissociated by adding rejuvenators. The use of rejuvenators negatively impacted intermediate temperature performance for the mixtures evaluated in this study.
- AC mixtures containing block copolymeric species with MW>70K Daltons can accommodate 5% RAS and retain satisfactory J_c values of at least 0.5 kJ/m^2.

6.0 Acknowledgements

The research work reported in this paper was sponsored by LADOTD through the Louisiana Transportation Research Center under contract number 12-3B. The authors would like to express their appreciation to all those who provided valuable help in the conduct of this project, in particular to William "Bill" King, Jr, and Gaylon Baumgardner.

7.0 References

Altgelt, K. H., "Gel Permeation Chromatography of Asphalts and Asphaltenes. Part I. Fractionation Procedure," *Die Makromolekulare Chemie*, Vol. 88, 1965, pp. 75–89.

Anderson, T. J., *Fracture Mechanics—Fundamentals and Application*, Third Edition, CRC Press, Taylor and Francis Group, Florida, USA, 2005.

Bowers, B. F., Huang, B., Shu, X., Miller, B. C., "Investigation of Reclaimed Asphalt Pavement Blending Efficiency Through GPC and FTIR," *Construction and Building Materials*, Vol. 50, 2013, pp. 517–523.

Button, J. W., Williams, D., and Scherocman, J. A., *Shingles and Toner in Asphalt Pavements*, Report No. FHWA/TX-96/1344-2F, Texas Transportation Institute, College Station, TX, 1996.

Carpenter, S. H., and Wolosick, J. R., "Modifier Influence in the Characterization of Hot-Mix Recycled Material," *Transportation Research Record*, No. 777, Transportation Research Board, Washington, D.C., 1987, pp. 15–22.

Cooper, Jr., S. B., Mohammad, L. N., and Elseifi, M. A., "Laboratory Performance of Recycled Asphalt Shingles," Transportation Research Board, National Research Council, TRB Paper No. 14-4383, Washington, D.C., 2014.

Daly, W. H., Negulescu, I. I., Mohammad, L. N., and Chiparus, I., "A Comparative Analysis of Modified Binders: Original Asphalts and Materials Extracted from Existing Pavements," Louisiana Transportation Research Center Report No. FHWA/LA11/462, *Technology Today* (A publication of the Louisiana Transportation Research Center), Vol. 25 (4), 2011.

Daly, W. H., Negulescu, I. I., and Balamurugan, S., "Implementation of GPC Characterization of Asphalt Binders at Louisiana Materials Laboratory," Project Capsule 10-6B, 2013.

Dickie, J. P., and Yu, T. F., "Macrostructure of the Asphaltic Fractions by Various Instrumental Techniques by Various Instrumental Methods," *Analytical Chemistry*, Vol.39, 1967, pp. 1847–1852.

Elseifi, M. A., Salari, S., Mohammad, L. N., Hassan, M., Daly, W., and Dessouky, S., "A New Approach to Recycle Asphalt Shingles in Hot Mix Asphalts," *Journal of Materials in Civil Engineering*, Vol. 24(11), 2012, pp. 1403–1411.

Federal Highway Administration, *Implementation of GPC Characterization of Asphalt Binders at Louisiana Materials Laboratory*, Report Number FHWA/LA.13/505, LTRC Project No. 10-6B, October 2013.

Foo, K. Y., Hanson, D. I., and Lynn, T. A., "Evaluation of Roofing Shingles in Hot Mix Asphalt," *Journal of Materials in Civil Engineering*, 1999, Vol. 11(1), pp. 15–20.

Gevrenov, J., "Recycling Shingles into Roads, US EPA Perspective," presentation made at Missouri Showcase, Utilization of Recycled Asphalt Shingles in Hot-Mix Asphalt, 2008.

Hassan, M. M., Asce, M., Lodge, A., Mohammad, L. N., and King, Jr., B. W., "Variability and Characteristics of Recycled Asphalt Shingles Sampled from Different Sources," *Journal of Materials in Civil Engineering*, Vol. 26(4), 2014, pp. 748–754.

Jennings, P. W., Pribanic, J. A., Raub, M. F., Smith, J. A., and Mendes, T. M., *Advanced High Performance Gel Permeation Chromatography Methodology*, SHRP-A-630, Strategic Highway Research Program, National Research Council, Washington, DC., 1993.

Lamontagne, J., Dumas, P., Mouillet, V., Kister, J., "Comparison by Fourier Transform Infrared (FTIR) Spectroscopy of Different Ageing Techniques: Application to Road Bitumens," *Fuel*, Vol. 80(4), 2001, pp. 483–488.

Louisiana Department of Transportation and Development, *Louisiana Standard Specifications for Roads and Bridges*, Baton Rouge, Louisiana, 2006.

Marsac, P., Piérard, N., Porot, L., Van den bergh, W., Grenfell, J., Mouillet, V., Pouget, S., Besamusca, J., Farcas, F., Gabet, T., and Hugener, M., "Potential and Limits of FTIR Methods for Reclaimed Asphalt Characterization," *Materials and Structures*, 47, 2014, pp. 1273–1286.

Mogawer, W. S., Booshehriana, A., Vahidia, S., Austerman, A., "Evaluating the Effect of Rejuvenators on the Degree of Blending and Performance of High RAP, RAS, RAP/RAS Mixtures," *Road Materials and Pavement Design,* Vol. 14, Issue Supplement 2, 2013, pp. 193–213.

Mouillet, V., Lamontagne, J., Durriew, F., Planche, J-P., Lapalu, L., "Infrared Microscopy Investigation of Oxidized and Phase Evolution in Bitumen Modified with Polymers," *Fuel*, Vol. 87, 2008, pp. 1270–1280.

Mullens, O. C., "The Asphaltenes," Annual Rev. *Analytical Chemistry*, Vol. 4, 2011, pp. 393–418.

Negulescu, I., Mohammad, L., Daly, W., Abadie, C., Cueto, R., and Daranga, C., "Chemical and Rheological Characterization of Wet and Dry Aging of SBS Copolymer Modified Asphalt Cements: Laboratory and Field Evaluation," *Journal of the Association of Asphalt Paving Technologists*, Vol. 75, 2006, pp. 267–318.

Newcomb, D. E., Nusser, B. J., Kiggundu, B. M., and Zallen, D. M., "Laboratory Study of the Effects of Recycling Modifiers on Aged Asphalt Cement," *Transportation Research Record*, No. 968, Transportation Research Board, Washington, D.C., 1984, pp. 66–77.

Northeast Recycling Council, Inc., "Asphalt Shingles Waste Management in the Northeast Fact Sheet," December 2011, http://www.nerc.org/documents/asphalt.pdf.

Noureldin, A. S., and Wood, L. E., "Rejuvenator Diffusion in Binder Film for Hot-Mix Recycled Asphalt Pavement," *Transportation Research Record*, No. 1115, Transportation Research Board, Washington, D.C., 1987, pp. 51–61.

Rostler, F. S., and White, R. M., "Fractional Components of Asphalts-Modification of the Asphaltenes Fractions," *Proceedings of the Association of Asphalt Paving Technologists*, Vol. 39, 1970, pp. 532–574.

Shen, J., Amirkhanian, S., and Xiao, F., "High Pressure Gel Permeation Chromatography of Aging of Recycled Crumb Rubber-Modified Binders with Rejuvenating Agents," *Transportation Research Record: Journal of the Transportation Research Board,* No. 1962, Transportation Research Board of the National Academies, Washington, D.C., 2006.

Snyder, J. R., "Determination of Asphalt Molecular Weight Distributions by Gel Permeation Chromatography," *Analytical Chemistry*, Vol. 39, 1969, pp. 1223–1227.

Soohyok, I., Zhou, F., Lee, R., and Scullion, T., "Impacts of Rejuvenators on Performance and Engineering Properties of Asphalt Mixtures Containing Recycled Materials," *Construction and Building Materials*, Vol. 53, 2014, pp. 596–603.

Wahhab, H. I. Al-Abdul, Asi, I. M., Ali, F. M., and Al-Dubabi, I. A., "Predictions of Asphalt Rheological Properties using HP-GPC," Journal of Materials in Civil Engineering, Vol. 11, 1999, pp. 6–14.

White, R. M., Mitten W. R., and Skog, J. B., "Fractional-Components of Asphalt Compatibility and Interchangeability of Fractions Produced from Different Asphalts," *Proceedings of the Association of Asphalt Paving Technologists*, Vol. 39, 1970, pp. 492–531.

Wu, Z., Mohammad, L., Wang, L. B., and Mull, M. A. "Fracture Resistance Characterization of Superpave Mixture using the Semi-Circular Bending Test," *Journal of ASTM International (JAI)*, Vol. 2, No. 3, 2005.

Zhou, F., Li, H., Hu, S., Lee, R., Scullion, T., Claros, G., Epps, J., and J. Button, "Evaluation of Use of Recycled Asphalt Shingles in HMA," *Journal of the Association of Asphalt Paving Technologists*, Vol. 82, 2013a, pp. 367–402.

Zhou, F., Li, H., Lee, R., Scullion, T., and Claros, G. "Recycled Asphalt Shingle Binder. Characterization and Blending with Virgin Binders," Transportation Research Record: Journal of the Transportation Research Board, No. 2370, Transportation Research Board of the National Academies, Washington, D.C., 2013b, pp. 33–43.

8.0 Discussion

IMAD AL-QADI: Thanks for an excellent presentation. Now if we want to add the recycling agent to the binder and assume that we are dealing with a PG 70-22, as you showed, or 64-22, and we add 5% RAS to that one and we know that the 5% RAS is at least a PG 112, it can be a PG 114. If we're going to use the simple equation by NCAT, that will bring your PG to maybe 80-86, God knows. And then after that, we're adding polymer to it. So here I'm having a stiff binder, isn't it?

BILL DALY: That's correct. It will be very stiff. The polymer actually should reduce the stiffness somewhat, but more importantly, it's compatibility. What I'm more concerned about is these things are going to phase separate, and phase separation is going to be a disaster.

IMAD AL-QADI: I understood this. What my concern is is the workability now of the mix because I'm dealing here with what may be flexible but it's a very, very stiff binder too. Would you agree? I'm not a chemist, by the way, so you can say whatever you want.

BILL DALY: No, no, and I'm not an engineer. But I agree, it's going to be a stiff mix, no question about it. And a small amount of polymer may be enough. I mean, we've been adding 1 or 2% and maybe 0.1% is enough, just to give you some compatibility.

IMAD AL-QADI: No, no, no. I love the presentation, but I'm just thinking about this out loud. But it met the J_c values too, right? The threshold of the J_c values?

BILL DALY: Yes.

IMAD AL-QADI: Okay.

BILL DALY: We haven't solved that problem in terms of trying to get up to that threshold with a recycling agent.

IMAD AL-QADI: Thank you again.

LOUAY MOHAMMAD: Based on what we've done, we used a PG 70-22 binder. I think you've seen that. If you look at the percent contribution that comes from the RAS binder and the scenario that you described, we have not seen in the laboratory any problem with mixing/workability. However, we followed a process of the addition of RAS that will facilitate workability.

GRANT WOLLENHAUPT: Thank you for the presentation. Very interesting to hear a chemist's perspective on things. One question, well two comments, but the first question is how did you derive the dosing of the rejuvenating agent that you guys used?

LOUAY MOHAMMAD: By the recommendation of the manufacturer for most of the recycling agents evaluated. Further, we continued to increase dosage until there was no substantial contribution coming out of the RAS binder.

GRANT WOLLENHAUPT: Because looking at what Fujie and Walaa had done with optimizing the dosage range, it looked, especially on what Fujie showed, that the aging characteristics were worse at lower dosage rates but when you increased on some of the rejuvenators, looking at that Glover-Rowe parameter, it actually increased the performance and decreased the aging susceptibility, so to maybe try to optimize the rejuvenator dosage. Because looking at what Walaa did too, obviously you saw his TSRST results and when they optimized their dosage rate, you saw improved characteristics with the TSRST as compared to what you guys were showing with a decreased performance of that. So maybe there's a key component here. With some of the chemistry, there may be some detrimental effects of freeing up that additional binder and you see that in the performance test, but with a proper dosage rate, maybe there's an offset characteristic that's happening that wasn't quite measured yet.

LOUAY MOHAMMAD: Thank you for the comment. As stated, the dosage level followed manufacturers' recommendation.

GRANT WOLLENHAUPT: Right, as was Dr. Zhou's when he started, and then when he optimized it to get performance, it was different than every manufacturer's recommendation.

LOUAY MOHAMMAD: Yes. Reminder our approach is a mix-based one where we use similar VFA level as explained in the presentation. We do not optimize based on binder extraction/blending because this is not reality.

GRANT WOLLENHAUPT: Yeah, and that then goes back to the mixture performance test with you yesterday. Maybe some of those have to be moved, then you can optimize performance and you can optimize cost without trying to hold the volumetric constant.

BILL DALY: Thank you.

GALE PAGE: I've listened to presentations on this work at least three times, and I've tried to understand it. And I'm not a chemist, I'm an engineer. And you out in the audience, as well as me, probably don't quite understand the implications of this work which I believe are important. Let me give you my two cents on the bottom line of this work having listened to it three times and that is not all recycling agents are created equal.

BILL DALY: Correct.

GALE PAGE: Not all recycling agents rejuvenate the old binder to an original condition. Some recycling agents will, you use the term, mobilize or free up the binder so that it's floating around with the new binder but it doesn't change the chemistry so you have more asphalt available but it's still the same old asphalt….

BILL DALY: It's not good asphalt.

GALE PAGE: Right, it's the same brittle stuff it was.

BILL DALY: Right.

GALE PAGE: So I think that's important to understand because, as I said, not all recycling agents can be considered equal. Some of them have different effects, and this presentation was the first that I've seen to show those different effects. And this work also shows the effect of polymer on some of these very hard brittle asphalts that may be freed up now to blend with the new asphalt. The polymer, I like to use the term, covers a multitude of sins, so even though you still have that very hard aged asphalt binder, the polymer can overcome some of the resultant brittle properties of the binder, particularly when you're looking at the mix tests. So that's my two cents.

BILL DALY: And you got it. I think you got it right. Thank you.

Cyclic Loading Behavior of Asphalt Concrete Mixture Using Disk-Shaped Compact Tension (DC(T)) Test and Released Energy Approach

Chaiwat Na Chiangmai[a], and William G. Buttlar[a]

[a]Department of Civil Engineering, University of Illinois, Urbana-Champaign, IL

ABSTRACT. This study involves the evaluation of fracture behavior under cyclic loading using the disk-shaped compact tension (DC(T)) test and a released energy based analysis approach. The cyclic DC(T) test was developed based on the monotonic DC(T) test (Wagoner and Buttlar, 2005); however, some modifications to the geometry and testing mode were necessary to facilitate the cyclic fracture test. The research was motivated to explore possible extensions of the DC(T) test device to consider cyclic fracture phenomena such as cyclic thermal cracking, block cracking and reflective cracking. Five different asphalt concrete mixes were tested under cyclic loading at four test temperatures (-12, 0, 10, and 20°C). After an extensive exploratory stage, the load-controlled testing mode utilizing a sine waveform and a frequency of 0.5 Hz with no rest period were selected as the standard testing parameters for this study. In addition, a peak load obtained from the monotonic DC(T) test was used as a reference value for determining loading magnitudes of the cyclic DC(T) test for a given mixture and test temperature. For data analysis, a released energy approach was introduced as a key concept for characterization of the cyclic fracture data generated in this study. Stemming from this approach, a released energy rate parameter, R2, was identified with the characteristic of mixture and temperature independence. By correlating a fracture energy based parameter to released energy rate (R2), it was shown that cyclic loading behavior could be predicted based upon three different data sets deriving from the DC(T) test: one involving a comprehensive cyclic loading testing suite; a slightly simpler method involving a limited number of required cyclic tests; and a highly simplified approach where cyclic fracture behavior was predicted from monotonic fracture test results alone (standard DC(T) fracture energy). All three prediction methods were shown to be plausible, but as expected, the more rigorous the testing suite, the more accurate the prediction. Finally, practical extensions of the work for design and analysis are provided.

KEYWORDS: cyclic, fracture, DC(T), asphalt, fracture energy, released energy, hysteresis loop, reflective cracking, block cracking, fatigue cracking.

This paper is for publication only.

1.0 Introduction

Pavement cracking continues to be a critical challenge in pavement engineering, particularly as higher recycling rates of materials containing aged/oxidized asphalt products are being motivated by the desire increase pavement sustainability. Concepts involving released energy (i.e., energy release rate) have been extensively employed to study cracking behavior in asphalt concrete materials, as it provides an energy-based approach to examine how material properties change or degrade under cyclic loading (Van Dijk et al., 1972). This method is generally based on cyclic loading data expressed in terms of load and displacement or stress and strain curves, which exhibit changing hysteresis loops as load cycles progress (Figure 1). Currently, approaches of initial dissipated energy (IDE) by Rowe (1993) and SHRP (1994), total dissipated energy (TDE) by Chomton and Valayer (1972) and Tayabali et al. (1992), and ratio of dissipated energy change (RDEC) by Ghuzlan and Carpenter (2000) and Shen and Carpenter (2005), have utilized the released energy concept to evaluate the behavior of asphalt concrete under cyclic loading. Many of the early approaches were predicated on arbitrary failure limits (Hveem, 1955), such as a 50-percent reduction in initial stiffness by Pell (1967), Monismith and Deacon (1964), Ramsamooj (1991) and Ghuzlan and Carpenter (2000). Other approaches have been introduced to study fatigue and fracture behavior in asphalt mixtures by Majidzadeh et al. (1971), Lytton et al. (1993), and Birgisson et al. (2002), some based on complex, continuum-damage-mechanics based approaches by Kim et al. (1997 and 2004) and Kutay (2011). However, the aforementioned approaches were not directed at studying cyclic fracture behaviors such as those present in reflective cracking or cyclic thermal cracking phenomena, which is the subject of this investigation.

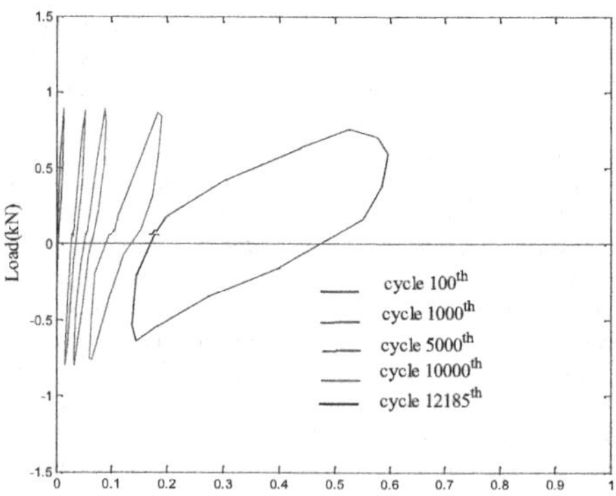

Figure 1. *Loading and unloading paths or hysteresis loops in cyclic loading.*

Pavements are subjected to repeated loading from both vehicular and thermal loading. As a result, pavement performance is strongly linked to the material's ability to withstand these repeated loading forms. In addition to traditional fatigue cracking, thermal, block, and reflective cracking involve large, discrete crack patterns in pavements induced by repeated loading which are costly to repair and maintain, as described by Nunn (1989) and Kim and Buttlar (2002). Unlike traditional fatigue studies, discrete cracking modes generally require a traditional fracture testing geometry to elicit realistic pavement fracture behavior. Discrete fracture tests such as the disk-shaped compact tension test (Wagoner et al., 2005) are now standardized and being implemented in the United States to control low temperature cracking. However, dealing with cyclic, discrete cracking is a more complex endeavor, yet to be completely understood let alone standardized. In this study, a modified DC(T) test geometry and released-energy based approach for data analysis were developed as a step towards a better understanding of this important phenomenon, following on preliminary work reported in the thesis of Na Chiangmai (2010).

2.0 Test Setup

This section presents a detailed summary of the adjustments made to the standard DC(T) test (ASTM D7313-07) in order to facilitate reliable cyclic loading tests. In early attempts, the standard DC(T) configuration was employed to perform a cyclic loading test; however, undesirable breakage at the loading holes occurred occasionally, especially for 19-mm or larger nominal maximum aggregate size (NMAS) mixtures. As a result, the standard DC(T) geometry was modified to reliably facilitate the cyclic DC(T) test. Figure 2 presents a comparison of the standard and modified cyclic DC(T) geometries. In addition, tight fitting, Teflon-coated loading pins were used to facilitate load reversal.

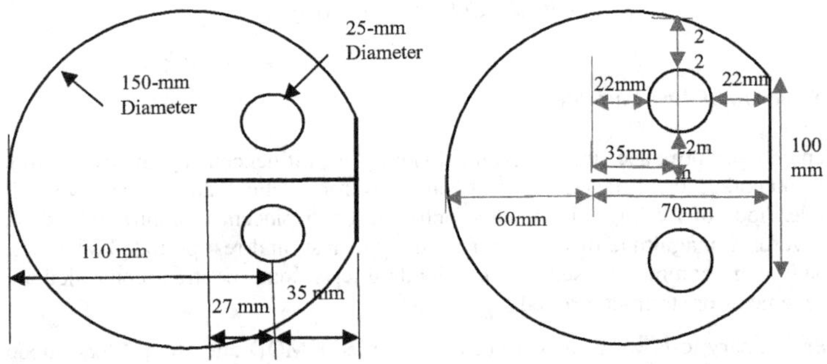

Figure 2. *Standard (left) versus modified, cyclic (right) DC(T) geometries.*

According to the standard DC(T) test procedure, a crack mouth opening displacement (CMOD) gauge was used to control and measure separation of the specimen during fracture. Since the study proposed to evaluate cyclic loading behaviors at low and intermediate test temperatures as high as 20°C, significant effects of material compliance were anticipated in the CMOD measurement. It was believed that separation of the material as characterized at the crack tip opening displacement (CTOD) location was a more suitable approach to studying fracture mechanisms in the proposed application. However, unstable test results occurred when CTOD test control was attempted.

As a compromise, a new test control gauge location was considered; a so-called 1-cm offset CTOD location. This location was found to strike a reasonable balance between material compliance issues and unstable test control. A modified DC(T) geometry and 1-cm offset CTOD location were proposed to perform both monotonic and cyclic loading tests for the remainder of this study, as illustrated in Figure 3.

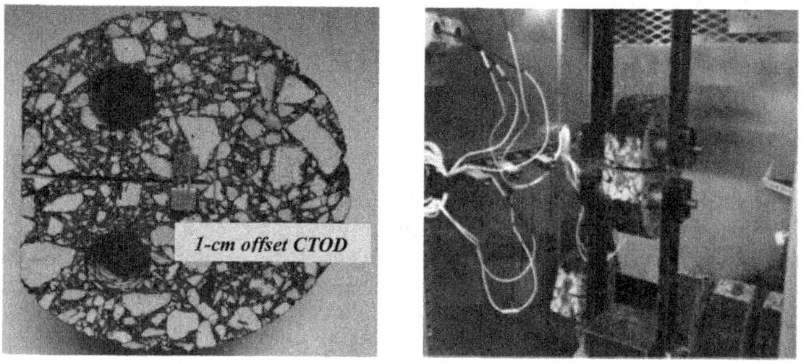

Figure 3. *Modified DC(T) geometry with 1-cm offset CTOD location (left) and modified DC(T) test setup (right).*

2.1 Selection of Test Variables

Pavement performance under cyclic loading is influenced by many factors. Experimentally, these factors or test variables for cyclic loading tests typically include mode of loading (either load-controlled or displacement-controlled), shape of waveform, magnitude of loading, frequency of test, and rest period. Two modes of loading are commonly used in cyclic loading tests: load- or stress-controlled and displacement- or strain-controlled.

Preliminary tests were carried out in controlled CMOD and control load mode. Reversing (from positive to negative set points) and non-reversing CMOD and load control waveforms were considered. For this study, a load-controlled, reversing, sinusoidal waveform was selected for the cyclic DC(T) test, using the modified DC(T) geometry and 1-cm offset CTOD location previously described. Ultimately, this testing mode was selected based on practical considerations and robustness of data produced. The load control mode produced stable crack growth across a broad range of load levels, whereas CMOD control did not lead to significant crack growth except for high initial CMOD levels, which sometimes produced rapid failure. In addition, load control is a much simpler mode for cyclic test control and more desirable for a practitioner-friendly test. Reversing loads were selected in order to avoid negative creep behavior, which would otherwise be present if only tensile loads were applied. This is a distinct advantage of the DC(T) over other proposed cyclic tests, such as the cyclic indirect tensile test (IDT) or semi-circular bend test (SCB), which cannot be practically subjected to reversing load forms. The subjects of fatigue endurance limit, rest periods, and healing were not considered in this stage of the study, but are certainly important directions to consider in future research.

2.2 Magnitude of Loading

Cyclic fracture behavior is obviously strongly related to the magnitude of repeated loading, where repeated loads of higher magnitude are associated with relatively lower numbers of cycles to failure (N_f), and vice-versa. In this study, magnitudes of loading for the cyclic DC(T) test were determined based upon peak loads measured in the monotonic DC(T) test, using the modified DC(T) geometry presented herein. The peak monotonic load value was used to establish the upper limit or 100% loading (tension and compression) amplitude levels for the sinusoidal waveform in the cyclic tests. For example, Figure 4 presents the loading waveform for the 60% loading case.

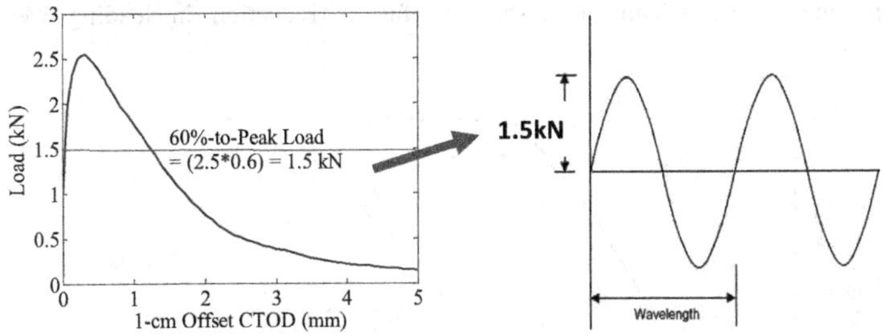

Figure 4: *Example of loading magnitude definition for cyclic DC(T) test.*

2.3 Materials and Experimental Design

Five asphalt concrete mixes were used in the study, including a 4.75-mm NMAS stress absorbing membrane interlayer (SAMI) mix, a 9.5-mm NMAS PG 64-22 mix, a 19-mm NMAS PG 58-28 mix, a 9.5-mm NMAS recycled asphalt shingle (RAS) mix, and a 19-mm mix made with foamed asphalt. The SAMI, PG 64-22, and foamed mixes were loose mixes received from the plant during other research investigations. The PG 58-28 mix was mixed and compacted in the laboratory, while the RAS mix was characterized using core samples obtained from the field. A plot of the five mixture gradations are shown in Figure 5. Although limited in scope, a fairly broad range of mixtures were considered in this initial study, with a relatively large number of test variables and test replicates to enhance veracity of results.

In terms of the experimental program for the cyclic DC(T) test, the following test parameters were used:

- Test temperatures: -12, 0, 10, and 20°C
- Mode of loading: load-controlled
- Wave shape: sinusoidal, reversing
- Frequency: 0.5 Hz
- Rest period: none
- Number of replicates: 10-13 reps per mixture for each test temperature
- Magnitude of loading: different percentages of peak load as a percentage of monotonic peak load for a given mixture and test temperature.

The highest repeated loading magnitude used was in the range of 85-90% of the peak monotonic fracture load, based on experience. Lower load levels at intervals in the range of 5-10% were used, until more than 50,000 cycles were required to fail the specimen (after which, no tests with further reduction in loading were attempted).

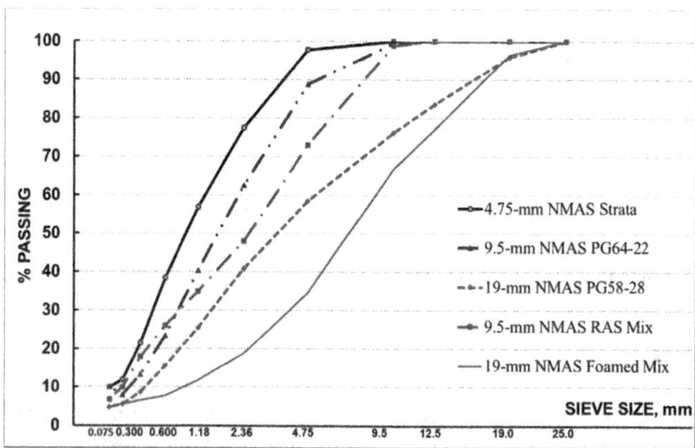

Figure 5. *Gradations of five study mixtures represented on a 0.45-power chart.*

3.0 Test Results

Applied load and measured CTOD were obtained from the modified DC(T) test, producing hysteresis loops when plotted, as expected. Figure 6 illustrates typical load-displacement behavior, comparing monotonic and cyclic DC(T) test responses. Note that the monotonic DC(T) test was controlled in the normal fashion with constant CMOD displacement opening rate, whereas a load-controlled mode of loading was used for the cyclic DC(T) test. Both loading approaches created the fracture processes of crack initiation, crack propagation, and complete failure of the test specimen, as shown in Figure 6. Interestingly, when average monotonic test results were plotted against a typical cyclic test result, the monotonic test result appears as a failure envelope for the cyclic test on the tension side of response. This suggests a unique relationship in the load-separation behavior of asphalt mixtures tested in a similar geometry regardless of monotonic vs. cyclic loading. This in turn suggests that pavement models involving progressive fracture under varied load levels may possibly be simplified based on this observed fracture behavior. An interesting observation in this behavior is that the machine control loop was unable to 'close the loop' in the final load cycles, as the material behaved according to the failure envelope, which eventually dropped below the prescribed repeating load level.

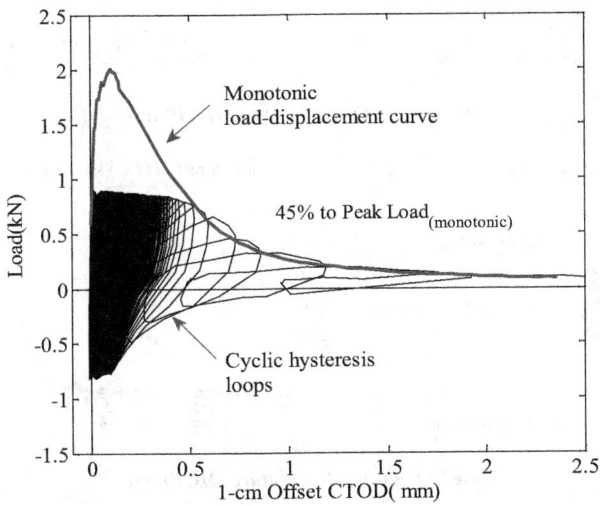

Figure 6. *Example of comparison of load and displacement relation between monotonic and cyclic DC(T) test results.*

3.1 Released Energy Development (RED) Model

A key aspect of data analysis was to determine cyclic fracture parameters and to attempt to develop prediction models to describe fracture behavior under cyclic loading for eventual use in simplified design approaches considering cyclic fracture. The released energy occurring during each loading cycle, which was defined as the area within the hysteresis loop in a full loading cycle, was plotted against number cycles. Figure 7 presents plots of released energy versus number of cycles (right), associated with a given the load-displacement plot (left) for two different loading magnitudes: high cycle fracture (HCF) and low cycle fracture (LCF) tests for the 9.5-mm NMAS PG 64-22 mix tested at 20°C. Similar procedures were employed for the other study mixtures and test temperatures.

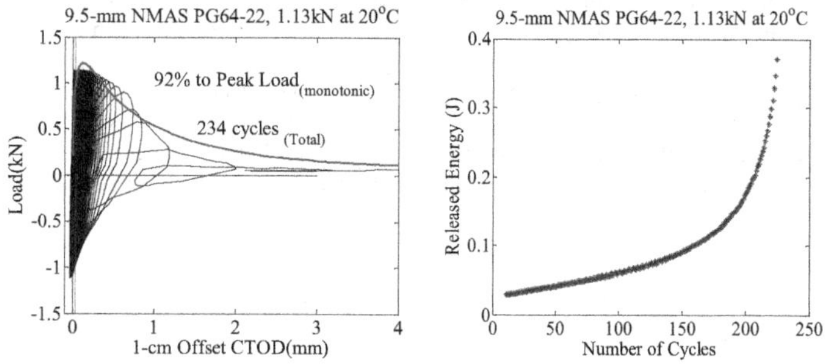

Case 1: Low-Cyclic Fracture (LCF) test

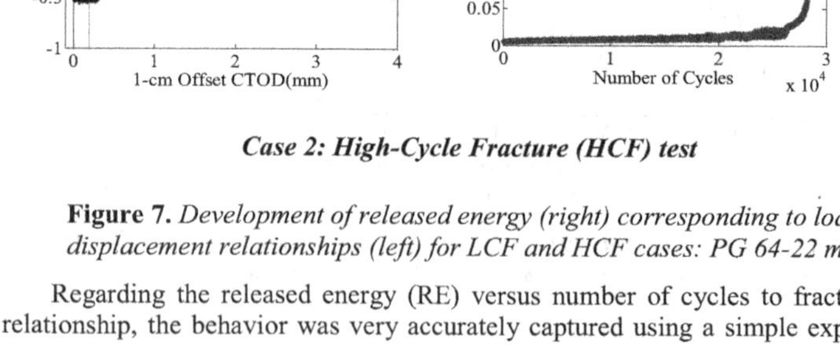

Case 2: High-Cycle Fracture (HCF) test

Figure 7. *Development of released energy (right) corresponding to load-displacement relationships (left) for LCF and HCF cases: PG 64-22 mix.*

Regarding the released energy (RE) versus number of cycles to fracture (N_c) relationship, the behavior was very accurately captured using a simple exponential function (Figure 8). A two-term exponential function, termed herein as the released energy development (RED) model is given as:

$$Released\ Energy\ (RE) = R1e^{(R2xN_c)} + R3e^{(R4xN_c)}$$

where N_C represents cycle number, and R1, R2, R3, and R4 are material coefficients defined as released energy factors or parameters.

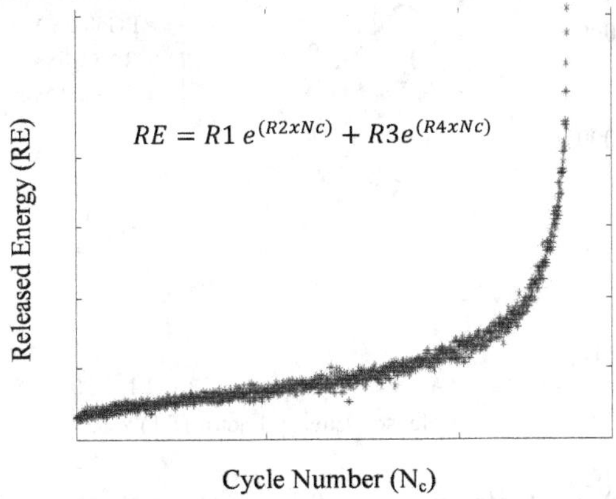

Figure 8. *Typical released energy development (RED) curve describing the cyclic fracture behavior of an asphalt concrete mixture.*

3.2 Characteristic N_f-R1 Relationship

The released energy factor, R1 was examined for different loading magnitudes for a given mixture. The R1 coefficient describes the initial released energy during testing. For the same mix and test temperature, by plotting the R1 versus N_f on the semi-logarithm scale, the initial released energy (R1) values for different loading magnitudes were found to be correlated to the total number of cycles to failure N_f. As a result, Figure 9 and Figure 10 present the N_f-R1 relationship for all study mixes performed at -12°C and 20°C, respectively.

A relationship between N_f and R1 was then captured using a power-law function, as follows:

$$N_f = C1\left(\frac{1}{R1}\right)^{C2}$$

Where N_f is total number of cycles to fracture, R1 is the released energy factor obtained from the RED model, and C1and C2 are material coefficients. A clear ranking of study mixtures was observed, with the SAMI mixture requiring significant energy to produce complete fracture, and the RAS mixture requiring significantly less energy to failure.

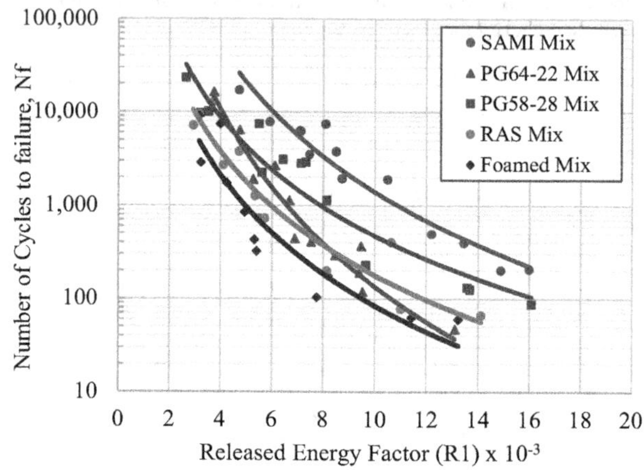

Figure 9. *N_f-R1 relationship for all mixes at test temperature of -12°C.*

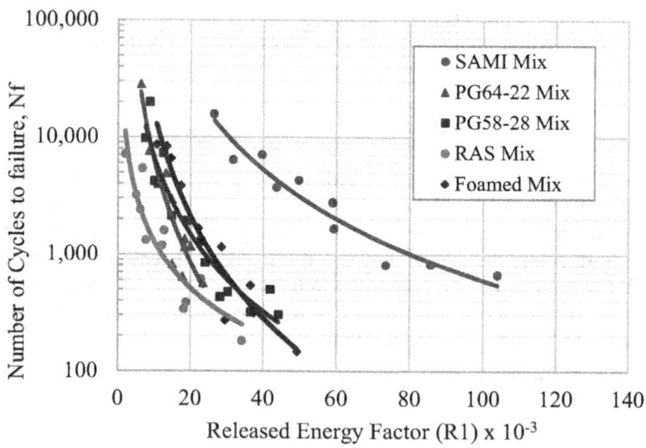

Figure 10. *N_f-R1 relationship of all mixes at test temperature of 20°C.*

3.2.1 Effect of Mixtures on N_f-R1 Relationship

Figures 9 and 10 demonstrate that the N_f-R1 relationship is mix-dependent. A comparison of the material coefficients, C1 and C2, for each study mixture is presented in Figure 11.

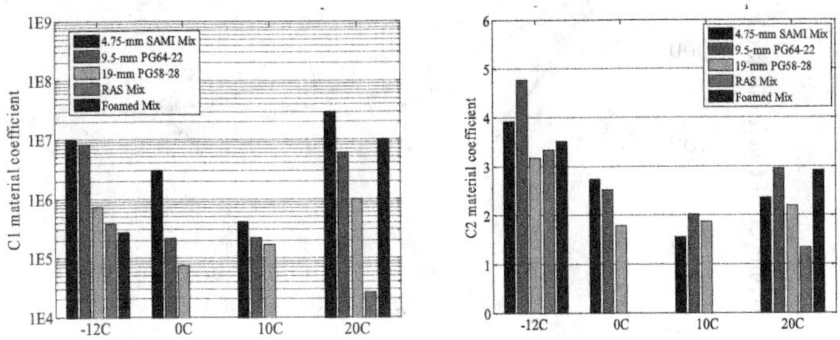

Figure 11. Comparison of C1 and C2 parameters: all study mixtures.

The material coefficient C1 appears to be related to the relative fracture resistance of the study mixes. By considering the same level of the initial released energy factor (R1), the C1 values were highly correlated to the number of cycles to failure (N_f), where the higher C1, the longer the service life. As a result, Table 1 presents a rank of the mixtures in terms of service life, associated with the C1 values. On the other hand, material coefficient C2, which was in the range of 1 to 5 as presented in Figure 11 (right), did not appear to be related to expected relative mixture performance.

Table 1. *Ranking of mixture performance associated with C1 coefficient.*

Mixture	Temperature			
	-12°C	0°C	10°C	20°C
4.75-mm NMAS SAMI	1	1	1	1
9.5-mm NMAS PG 64-22	2	2	2	2
19-mm NMAS PG 58-28	3	3	3	3
9.5-mm RAS with Eco-binder	4	n/a	n/a	5
19-mm NMAS Foamed mix	5	n/a	n/a	4

Note: lower numbers represent better performance (longer service life).

3.2.2 Effect of Temperatures on N_f-R1 Relationship

The N_f-R1 relationship was also influenced by test temperatures. By comparing the C1 and C2 coefficients for the same mixtures across different test temperatures, as presented in Figure 12, it was found that C1 and C2 values decreased with decreasing temperature.

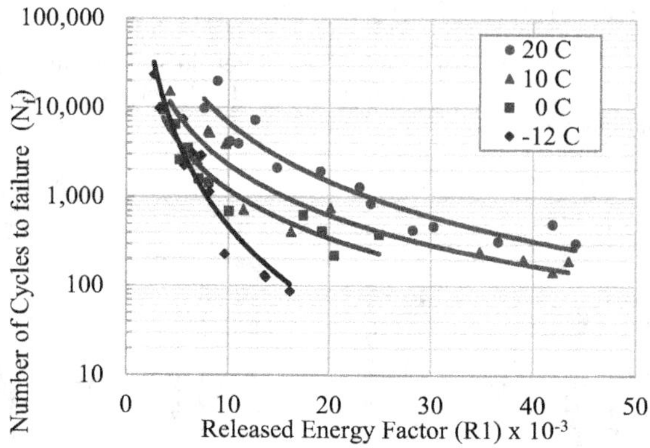

Figure 12. *Comparison of N_f-R1 curves of PG 58-28 mix for all temperatures.*

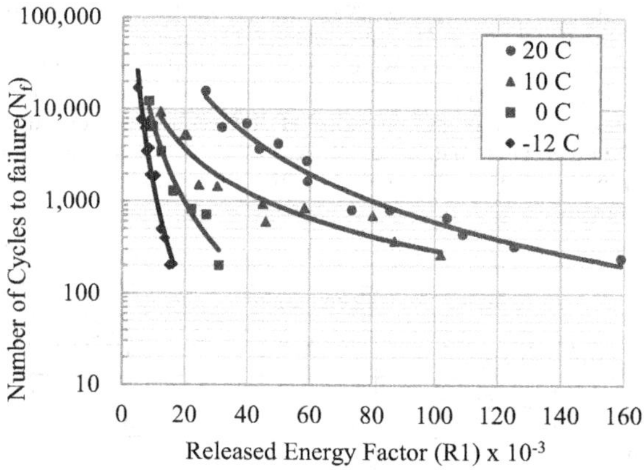

Figure 13. *Comparison of N_f-R1 curves of SAMI mix for all temperatures.*

Figures 12 and 13, for example, present the Nf-R1 curves for the PG 58-28 and SAMI mixes for all test temperatures, which display similar trends. A 'disadvantage' of the N_f-R1 relationship, from the standpoint of test simplification, is that the model depends on mix type and temperature. As a final note, selection of testing loads such that a practical minimum N_f of 100 cycles is recommended to maintain a stable N_f-R1 relationship.

3.2.3 Unique Characteristic of N_f–R2 Relationship

Similar to the other released energy factors of the RED model, a relationship between the R2 coefficient and the total number of cycles to failure (N_f) was investigated to evaluate cyclic fracture behavior and a possible means towards test simplification. By definition, the R2 exponential coefficient indicates the growth rate of the released energy. Figure 14 presents a plot of N_f versus the R2 parameter on a log-log scale. Interestingly, a unique N_f -R2 relationship was found for all study mixtures and test temperatures. As a result, the R2 parameter has been identified as a candidate cyclic fracture criterion, and was used to develop simplified cyclic fracture prediction models. The apparently unique N_f -R2 relationship was mathematically described using the following power-law function:

$$N_f = 1.18\left(\frac{1}{R2}\right)^{0.977}$$

Where N_f and R2 are as defined earlier.

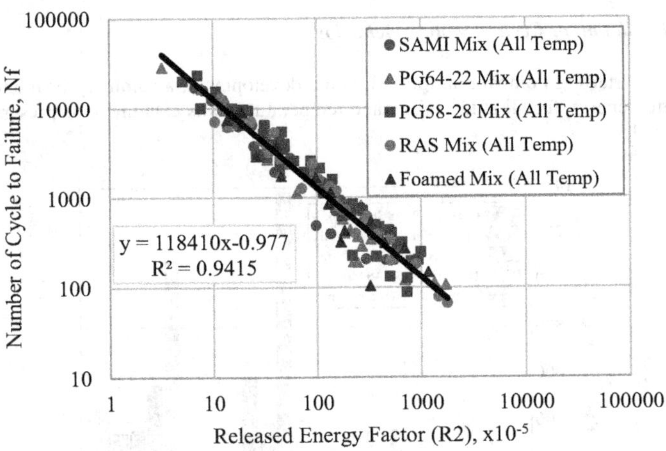

Figure 14. *Unique N_f -R2 relationship for all study mixtures and temperatures.*

Using the proposed model, number of cycles to failure (N_f) can be predicted from the R2 parameter. This particular model was constructed based on a total number of 188 cyclic fracture tests, conducted on the five study mixes at four test temperatures and several loading levels. Since both power-law coefficients were found to be close to 1, the N_f-R2 relationship can be simplified as:

$$N_f \approx \frac{1}{R2}$$

In this approximation, N_f is slightly under-predicted, which leads to a conservative result. Furthermore, it is believed that the N_f-R2 model could be used to predict a cyclic-fracture-based 'endurance limit.' Note that tests taking less than 100 cycles were not included in the formation of the N_f-R2 relationship.

4.0 Prediction of Cyclic Loading Behavior

The main objectives of the study were not only to evaluate the cyclic fracture behavior of asphalt concrete mixtures, but to also attempt to develop a prediction model for cyclic fracture behavior from reduced cyclic testing data sets, or ideally, from monotonic fracture results alone. Consequently, prediction models for the cyclic fracture behavior were attempted based on three potential scenarios: (1) a full suite of cyclic DC(T) tests; (2) monotonic plus several cyclic DC(T) tests, or; (3) monotonic DC(T) test results alone.

4.1 Prediction of Cyclic Loading Behavior

As a starting point for simplified model development, a summary of monotonic fracture energy for all study mixes and temperatures was examined, as presented in Figure 15.

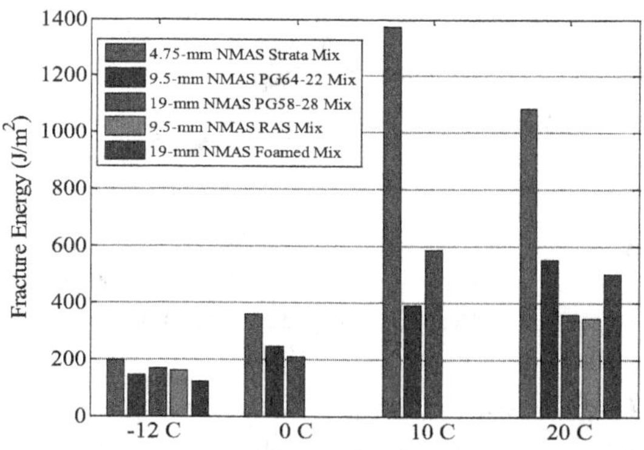

Figure 15. *Summary of fracture energy as monotonic fracture parameter.*

4.2 Prediction with Full Suite of Cyclic DC(T) Tests

By plotting N_f versus G_f -R2 ratio as the released energy ratio, a cyclic fracture prediction model was developed. For example, Figure 16 presents a plot of the prediction model of the mixes, conducted at test temperatures of -12°C and 20°C. Using all available study data, a correlation between N_f and released energy ratio was developed and can be mathematically described using a power-law equation given as:

$$\text{Model A:} \quad N_f = A\left(\frac{G_f}{R2}\right)^B$$

Where N_f represents number of loading cycles, G_f and R2 are the fracture energy and released energy rate for the same mixture and temperature, respectively, and A and B are material coefficients.

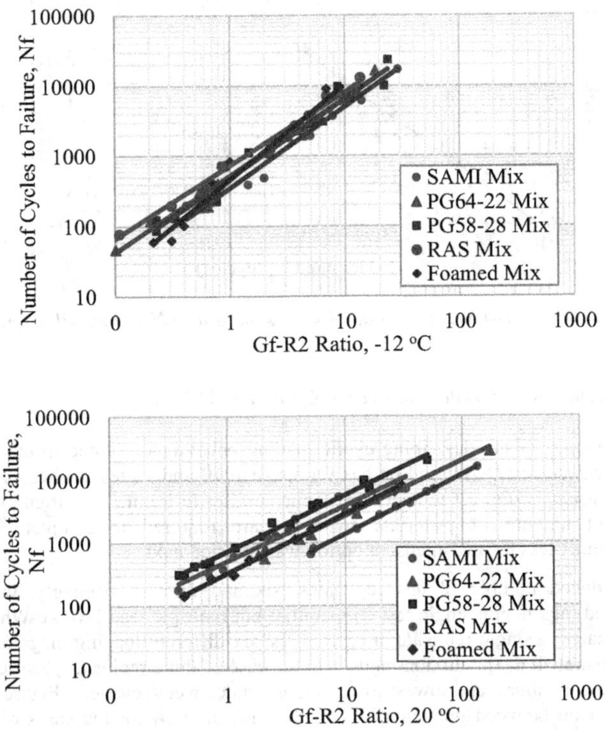

Figure 16. *Plot of N_f versus G_f-R2 ratio for mixes tested at -12°C and 20°C.*

Regarding the cyclic DC(T) test, values of R2 can be derived from the RED model. Then, the predicted N_f can be computed using the prediction model of the N_f and released energy ratio relationship, as presented by Model A. Based on the analysis of study data, Figure 17 presents a plot of the predicted N_f from model A versus the measured N_f directly obtained from the experimentation. The results showed that the predicted N_f matches actual data with a root-mean-square error (RMSE) of 15%.

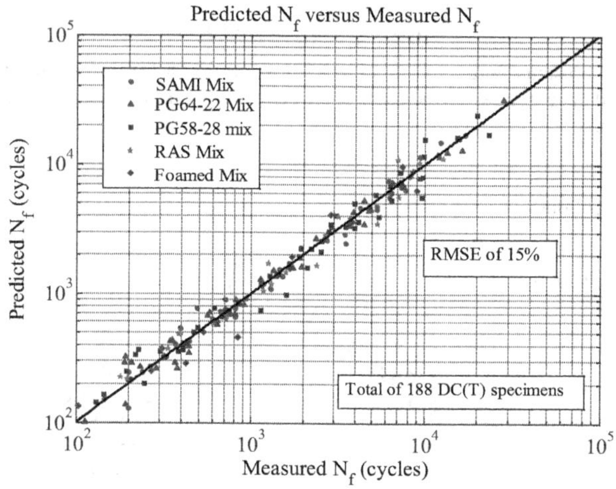

Figure 17. *Predicted N_f using model A versus measured N_f across all temperatures.*

4.3 Prediction with Limited Number of Cyclic DC(T) Tests

The prediction model describing cyclic loading behavior presented in the previous section was constructed based on having between 10 and 13 test replicates for each mixture at a specific test temperature. Thus, similar to traditional fatigue testing, a significant amount of materials and time are required to complete such an evaluation. A less testing-intensive option was pursued next.

A reduced number of test replicates, six and three respectively, were next considered (again, at a single test temperature but multiple load levels). When using six replicates, a single replicate at each of the six different loading magnitudes was used to establish the prediction model. In the case of three replicates, test results for the highest, medium, and lowest loading magnitudes were selected. Figure 18 plots a comparison between predicted N_f and the measured N_f for the cases of six and three replicates, respectively. A difference in the root-mean-square error (RMSE) for the case with six and three replicates was quite small (1% difference), and only slightly higher than the prediction made using the full data set. Therefore, it appears that cyclic fracture behavior can be predicted with reasonable accuracy by testing at a single temperature, with one replicate at each of three different load levels.

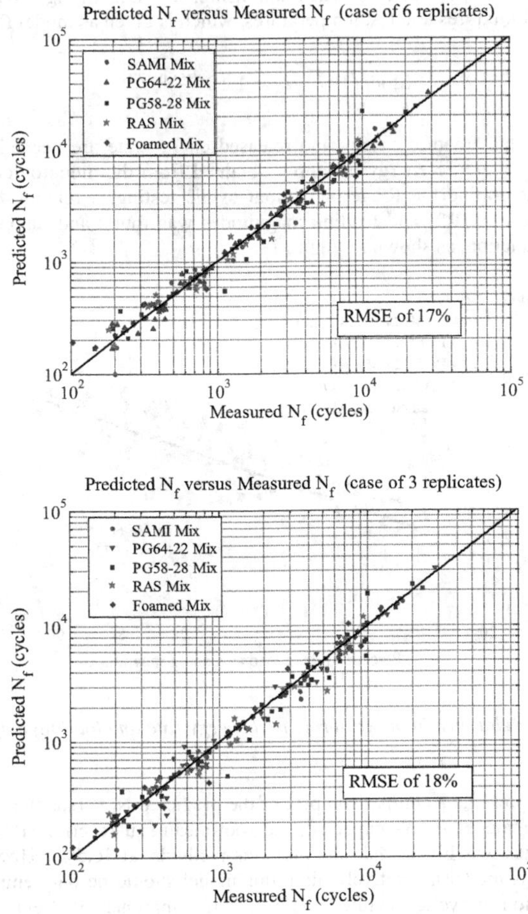

Figure 18. *Predicted N_f versus measured N_f for 6 and 3 replicates.*

4.4 Prediction without Cyclic DC(T) Test (Monotonic Fracture Energy Only)

Referring back to the unique N_f-R2 relationship, a single fitting model could be derived for all mixtures and test temperatures, which is given as model B:

$$\text{Model B:} \quad N_f = 1.18\left(\frac{Gf}{R2}\right)^{0.977}$$

In this model, fracture energy and released energy rate are needed in order to determine N_f. Fracture energy is simply obtained from the monotonic DC(T) test. In order to determine the R2 term without cyclic testing, a relationship between energy release rate (R2) and loading magnitudes was found and mapped using an exponential function, as shown in Figure 19.

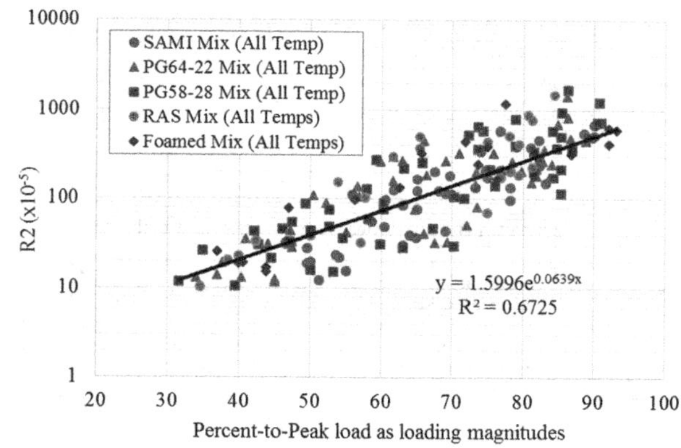

Figure 19. *Relationship between released energy rate and loading magnitudes.*

Figure 20 presents a comparison plot of the predicted N_f versus the measured N_f obtained from model B. As expected, the root-mean-square error (RMSE) of the predicted values was higher than the previous methods, at 36.5%. However, since the error in this model is relatively high, this model should be only employed for a rough evaluation of cyclic fracture behavior, or if comprehensive field validation of the simplified method is conducted and found to produce acceptable accuracy.

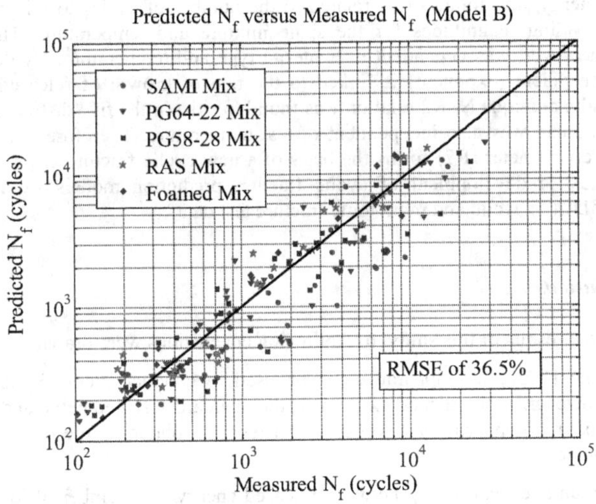

Figure 20. *Predicted N_f versus measured N_f of model B (no cyclic testing required) for all test temperatures.*

5.0 Summary, Conclusions and Recommendations

This section presents the main study findings, the study conclusions, and recommendations for research implementation and further research.

5.1 Summary

Pavements are subjected to repeated loading from both vehicular and thermal loading; therefore, pavement performance is related to the material's resistance to cyclic loads. In addition to traditional fatigue cracking, thermal, block, and reflective cracking involve large, discrete crack patterns in pavements induced by repeated loading. These cracking forms permit ingress of moisture, lead to significant roughness, and are costly to repair and maintain. Unlike traditional fatigue studies, discrete cracking modes generally require a traditional fracture testing geometry to elicit realistic pavement fracture behavior. In this study, a modified DC(T) test geometry and released-energy based approach were developed to study discrete, cyclic cracking phenomena. A unique relationship between released energy and number of load cycles appears to exist, which forms the basis for a released energy development (RED) model to characterize cyclic loading behavior in the proposed test. The R1 coefficient in this model was linked to initial released energy, and found to be related to the number of cycles to failure among different loading magnitudes for the same mixture and temperature. The N_f-R1 relation, however, was dependent on material type and temperature. By definition, the R2 coefficient (exponent) is linked to the rate of growth of released energy. Interestingly, a unique N_f-R2 relation was found to exist in the five distinct mixtures investigated and was also independent of test temperature. Therefore, the released energy rate parameter (R2) forms the basis of a new cyclic fracture criterion. This led the way to development of cyclic fracture prediction models from various subsets of DC(T) cyclic and monotonic fracture test data.

5.2 Conclusions

Based on the results of this study, the following conclusions were drawn:

1. The DC(T) test apparatus can be used to conduct cyclic fracture tests; however, a slight modification to the standard test geometry and loading fixtures is necessary to obtain consistency in obtaining the desired failure plane.
2. A unique relationship between released energy and number of load cycles appears to exist, which forms the basis for a released energy development (RED) model to characterize cyclic loading behavior in the proposed test.
   a. The N_f-R1 relation, was found to be dependent on material type and temperature.
   b. A unique N_f-R2 relationship was found to exist in the five distinct mixtures investigated, and was also independent of test temperature.
3. Cyclic fracture prediction models can be developed using various subsets of DC(T) cyclic and monotonic fracture test data: one involving a comprehensive cyclic loading testing suite; a slightly simpler method involving a limited number of required cyclic tests; and a highly simplified approach where cyclic fracture behavior was predicted from monotonic fracture test results alone (standard DC(T) fracture energy). All three prediction methods were shown to be plausible, but as expected, the more rigorous the testing suite, the more accurate the prediction, as characterized by the root-mean-square error statistic.

5.3 Recommendations

Of the three predictive models presented, the intermediate level (involving three replicate cyclic fracture test replicates), appears to be the best compromise between test expediency and accuracy. However, the monotonic fracture-energy-based prediction model might prove to be a useful first-order estimator of cyclic fracture behavior in forensic investigations and other engineering studies.

Modification of commercially available DC(T) apparatus to run the proposed test is recommended as a next step in research implementation. Additionally, application of the proposed test and models to a broader set of asphalt mixtures, and in comprehensive field validation studies are recommended. For instance, forensic investigation of good- and poor-performing asphalt overlays could form the basis for a practitioner-friendly approach to the design and control of asphalt mixtures to withstand reflective cracking. Similar approaches could be employed in studies of block cracking and thermal fatigue cracking. Finally, studies to investigate effect of rest periods and healing on cyclic fracture behavior should be conducted.

6.0 References

Birgisson, B., Roque, R., Sangpetngam, B., and Zhang, Z., "Hot Mix Asphalt Fracture Mechanics: A Fundamental Crack Growth Law for Asphalt Mixtures," *Journal of the Association of Asphalt Paving Technologists*, Vol. 71, 2002, pp. 816–827.

Chomton, G., and Valayer, P.J., "Applied Rheology of Asphalt Mixes Practical Application," *Proceedings Third International Conference on the Structural Design of Asphalt Pavements*, London, 1972.

Ghuzlan, K., and Carpenter, S.H., "An Energy-Derived/Damage-Based Failure Criteria for Fatigue Testing," *Transportation Research Record No. 1723*, Washington, D.C., 2000, pp. 131–141.

Hveem, F.N., "Pavement Deflections and Fatigue Failures," *Highway Research Bulletin 114*, Highway Research Board, Washington, D.C., 1955, pp. 43–87.

Kim, J., and Buttlar, W.G., "Analysis of Reflective Crack Control System Involving Reinforcing Grid over Base-Isolating Interlayer Mixture," *Journal of Transportation Engineering,* American Society of Civil Engineers, Vol. 28, No. 4, 2002, pp. 375–384.

Kim, Y.R., Lee, H.J., Little, D.N., "Fatigue Characterization of Asphalt Concrete Using Viscoelasticity and Continuum Damage Theory," *Proceedings of Asphalt Paving Technologists*, Vol. 66, 1997, pp. 520–569.

Kim, Y.R., Daniel, J.S., Bisirri, W., "Fatigue Characterization of Asphalt Concrete Using Viscoelasticity and Continuum Damage Theory," *Association of Asphalt Paving Technologists*, 2004, pp. 557–583.

Kutay, E., "Practical Fatigue Characterization of Asphalt Mixtures Using PP_VECD (Push-Pull Test Based Viscoelastic Continuum Damage Analysis) Software." In Advanced Models to Characterize and Design Asphalt Pavements: Implementation and Application Examples, webinar, Transportation Research Board, the National Academies, October 2011.

Lytton, R.L., Uzan, H., Fernado, E.G., Roque, R., Hiltuen, D., and Stoffels, S.M., *Development and Validation of Performance Prediction Models and*

Specifications for Asphalt Binders and Paving Mixes, SHRP A-357, Strategic Highway Research Program, Washington, D.C., 1993.

Majidzadeh, K., Kauffmann, E.M., and Rasmsamooj, D.V., "Application of Fracture Mechanics in the Analysis of Pavement Fatigue," *Proceedings of the Association of the Asphalt Paving Technologists*, Vol. 40, 1971, pp. 227–246.

Monismith, C.L., and Deacon, J.A., "Fatigue of Asphalt Paving Mixtures." *Transportation Engineering Journal, Proceedings of the American Society of Civil Engineers*, Vol. 95, No. TE2, 1964.

Na Chiangmai, C., *Fatigue-Fracture Relation on Asphalt Concrete Mixtures*, M.S. Thesis, University of Illinois at Urbana-Champaign, IL, 2010.

Nunn, M.E., Investigation of Reflective Cracking in Composite Pavements in the United Kingdom, *Proceedings of the RILEM Conference on Reflective Cracking (Liege)*. London: E & FN Spoon, 1989, pp. 143–153.

Pell, P.S., "Fatigue of Asphalt Pavement Mixes," *Proceedings Second International Conference on the Structural Design of Asphalt Pavements*, Ann Arbor, Michigan, 1967, pp. 456–483.

Ramsamooj, D.V., "Prediction of Fatigue Life of Asphalt Concrete Beams from Fracture Tests," *ASTM Journal of Testing and Evaluation*, Vol. 19, 1991, pp. 231–239.

Rowe, G.M., "Performance of Asphalt Mixtures in the Trapezoidal Fatigue Test," *Proceedings Association of Asphalt Paving Technologists*, Vol. 62, 1993, pp. 344–384.

Shen, S., and Carpenter, S. H., "Application of Dissipated Energy Concept in Fatigue Endurance Limit Testing," *Journal of Transportation Research Record: Transportation Research Board*, No. 1929, 2005, pp. 165–173.

SHRP, *Fatigue Response of Asphalt-Aggregate Mixes*, A-404, Strategic Highway Research Program, National Research Council, 1994.

Tayebali, A.A., Rowe, G.M., and Sousa, J.B.. "Fatigue Response of Asphalt-Aggregate Mixtures." *Proceedings of Asphalt Paving Technologists*, Vol. 61, 1992, pp. 333–360.

Van Dijk, W., Moreaud, H., Quedeville, A, and Uge P., "The Fatigue of Bitumen and Bituminous Mixes," *Proceedings Third International Conference of the Structural Design of Asphalt Pavements*, London, 1972, pp. 354–366.

Wagoner, M.P., Buttlar, W.G., and Paulino G.H., "Disk-Shaped Compact Tension Test for Asphalt Concrete Fracture," *Experimental Mechanics*, Vol. 45, 2005, pp. 270–277.

AAPT Leading Edge Forum: Implementing Specifications for New Technologies

Gale Page
King of Asphalt Consulting, representing the Florida Department of Transportation

Steve Davis
Washington State Department of Transportation

Ross A. Bentsen
Quigg Engineering Inc., representing the Illinois Tollway

Nathan Morian
Nevada Department of Transportation

Chris Abadie
Louisiana Department of Transportation & Development

1.0 Discussion—Gale Page

GRANT WOLLENHAUPT: Thanks for the presentation, Gale. I enjoyed it. Going back to the prequalification requirements of contractors who messed up on the warranty instead of bonding, what's the contractors' feedback when that's been in place?

GALE PAGE: I'm going to tell you, we were surprised that they accepted the 3 year Value Added (warrantee) spec, and there really hasn't been a lot of negative feedback. We've had contractors go out and fix areas that needed to be fixed. And, again, it's only in three years, so there hasn't been a lot of negative reaction after it's been implemented. We implemented it, I believe, in 2002.

GRANT WOLLENHAUPT: Have there been any contractors who then weren't qualified to bid because of work? Has that issue come up and then been resolved and that worked out well you haven't gotten there yet.

GALE PAGE: We haven't had any issues.

GRANT WOLLENHAUPT: All right. Thank you.

GALE PAGE: It's worked. So if you can try it, you may like it.

REBECCA McDANIEL: When do you use that value-added specification? Is it on all projects or interstates?

GALE PAGE: On which ones?

REBECCA McDANIEL: The ones that Grant was just asking about.

GALE PAGE: Oh, the three-year Value Added spec? It's on everything. It's in our standard spec book, so if you go online, you can actually read the wording, and it's pretty simple. And it lists the threshold values and everything in the spec, as I recall.

REBECCA McDANIEL: Okay. Thank you.

GALE PAGE: As I remember writing it anyway.

CHRIS ABADIE: Thank you, Gale. Great presentation and review. I do recognize Florida is a leader in a lot of things in asphalt and I do attribute that at least partly to your leadership. We want to thank you for that. All right, now I'm going to quit talking. I want to ask you, is there something that you would have liked to have seen implemented that for some reason or other did not go that way?

GALE PAGE: Something implemented? No, I think the only thing that I've seen that I had some questions about, which I think is going to be resolved in the future, is Florida had a pretty tight binder specification with a 0.5% mass loss and that kept out a lot of strange binder from the Northeast coming into Florida. Anyway, that has been changed to 1% mass loss in order to make sure that Florida is drawing from a wider market, that supply wouldn't be limited. So I think eventually that people are going to find out that, again, from a cracking standpoint what is it in the binder that you need to have, and I think we're going to get that resolved with a better cracking test for the binder. Hi, Brother.

GAYLE KING: Gale, you made a tongue and cheek comment, "innovative agency engineers don't get recognition or financial reward for the risks they take." In fact, I sat on the other side of the table from you and Chris (Abadie) many times in the past when hoping to introduce PMA and other new technologies. It was clear that there were only a few agency engineers like yourselves that were willing to take those risks, perhaps because you felt secure enough in your positions to be able to do so.

GALE PAGE: Yeah, what could they do? They could only fire me.

GAYLE KING: Well, that's the problem, though. Many engineers recognize that failure while taking risks can get them fired. The problem with the public sector incentive system is that taking successful risks that truly advance technology yield little if any reward. In private industry, there are bonuses and incentives for those that take risk by adopting new technology that so positively impacts the bottom line. My question to you is, "Where does the impetus to create change of that magnitude come from in the public sector. And is there anything we could do as an organization to make that happen?

GALE PAGE: I tell you, there's a great satisfaction in what I as an agency person is able to do, because you're kind of like a big fish in a small pond. And if you can make something happen, I think that's its own reward, and you can see it. And if you don't see it and feel it in yourself, then I guess you're in the wrong business. Right, Chris.

GAYLE KING: Shouldn't there be something more?

GALE PAGE: Oh, sure. But I think that the people that like their work and like what they're doing, just like you do, will continue to do that and be involved, just like yourself in the emulsion industry whether you get paid for it or not, right?

GAYLE KING: That's a great answer, and we need more like you. In private industry, I suspect people respond more to bonuses than the common good will.

CHUCK HUGHES: Gale, I'm sure you appreciate the cooperation that you've gotten from the industry because without that cooperation, you wouldn't be able to achieve very much, and that's a very important factor and you're very lucky.

GALE PAGE: Absolutely. And we've been very fortunate in Florida to have hot mix industry people working with us, and also from a liquid standpoint the asphalt suppliers, we've been fortunate to have outstanding relations with them, even though our asphalt specs at the time I was there were a little more restrictive than the other surrounding states.

2.0 Discussion—Steve Davis

LARRY ILG: How many grade increases did you have to face as the state now that you've upped your RAP contents?

STEVE DAVIS: When we're getting that data from the contractor and they're running the binder recovery process, we're seeing stuff up in the 90s. And so as we're seeing those virgin grades come in, and, obviously that's all we're going to see when it goes to production, but we're see about 3 different grades. They're down 52s, 58s, and 58-34 is the biggest one we see on the east side of our state, which is funny because we took that out as a specification grade binder about 10 years ago. We just thought it was too soft. I think I have a slide in here. So it's pretty faint, but you can see that's Washington State and basically where it goes from the orange to green is the Cascade mountain range, so probably 75% of the State of Washington lives in the orange and 25 lives in green, with Seattle, Tacoma, Olympia, Everett, Vancouver all on the west side.

GALE PAGE: You know what my biggest problem was when I was with the DOT was personnel issues. All this other stuff is kind of fun. That's why I was smiling. But I have a question regarding the IDT requirement. You said you went from 150 to 175, and you said it was too strict so you raised the IDT requirement.

STEVE DAVIS: Yes.

GALE PAGE: Am I missing something? Usually, you have a problem meeting a minimum IDT requirement, so was that a max IDT requirement?

STEVE DAVIS: Yeah, it was a max.

GALE PAGE: Oh, a max. Okay.

STEVE DAVIS: I'm sorry if I misspoke or confused anyone.

GALE PAGE: I'm familiar with having a minimum IDT, rather than a max.

STEVE DAVIS: Oh, no. No, we're max.

CHRIS ABADIE: Thank you for a nice presentation. The general contracting community has approached us when we talk about the need for maybe a little more asphalt in the mixture, and they said if you want more asphalt, just pay for asphalt separately, and our neighboring state of Arkansas does that. I was curious as to how you pay for your hot mix. Is it by the ton or do you pay for asphalt separately?

STEVE DAVIS: It's by the ton, yeah. Everything is completed mix and I see one of our suppliers here. We have very strict requirements of our suppliers and quality control plans and onsite assessments and all that, but when it comes down to it, we actually don't even have a contract with our asphalt suppliers. So if your PAV DSR is out on 10 samples in a row, we're going after the price of the mix (price adjustment) and we're going after the paving contractor, and what the paving contractor does from there is between them and the asphalt supplier. They have the contract with the supplier.

BAOSHAN HUANG: University of Tennessee. I can understand you set a upper limit for IDT because you put in RAP. You tend to have higher IDT but it will be more brittle.

STEVE DAVIS: Yes.

BAOSHAN HUANG: Your IDT values are between 150 to 175 psi, right?

STEVE DAVIS: Yes, psi.

BAOSHAN HUANG: It's not a very high value and it depends on the size of your specimen. Could you clarify the size of your specimen for your IDT test?

STEVE DAVIS: Yeah, and I didn't want to get too crazy on this, but we use ASTM D6931, but we use 62 mm specimen.

BAOSHAN HUANG: A 62 mm specimen?

STEVE DAVIS: Yes, 62, the, same as our Hamburg.

BAOSHAN HUANG: For the height?

STEVE DAVIS: Yes. Oh, 6 inch, 150 mm is the diameter of the specimen.

BAOSHAN HUANG: What's the thickness?

STEVE DAVIS: It's 62 mm.

BAOSHAN HUANG: Okay. Thank you.

STEVE DAVIS: Yeah. Our specification does have a change unit for that. And that's another topic, but we've been through that quite a bit and it's much easier for us production wise and we've done a lot of comparison testing.

BAOSHAN HUANG: One more thing.

STEVE DAVIS: Yes, go ahead.

BAOSHAN HUANG: You cut the specimen or just compact the specimen to the height?

STEVE DAVIS: We do not cut the specimen; we compact it.

BAOSHAN HUANG: Thank you.

AUTHOR'S CLOSURE

There seemed to be some confusion regarding the IDT specification on the WSDOT High RAP/RAS mixes. The IDT specification is 175 psi maximum on a 150 mm diameter specimen, compacted to 62 mm. WSDOT has implemented this as a maximum specification as a counter to our Hamburg specification. WSDOT allows a maximum of 40% binder replacement on their High RAP/RAS mixtures and we realize that these mixtures are going to be much stiffer than our conventional mixtures. The IDT specification was implemented to avoid cracking susceptibility.

3.0 Discussion—Ross Bentsen

JIM SCHEROCMAN: Ross, I haven't even seen you in a few years, but you haven't changed.

ROSS BENTSEN: It has been a while, Jim.

JIM SCHEROCMAN: Question for you on fractionization of RAP or breaking it up. Do you use all of it? If a contractor's got a pile of RAP and you crush it into two different sizes, do you still use all of it?

ROSS BENTSEN: Generally, that's the balance that the contractors are fighting is being to use it. They're not able to put as much fine. Generally, you're getting with the mixes we have, depending on how they fractionate it, you're getting about 40 to 50% fine FRAP and they're not able to quite get all of that into the mixes. So they aren't using everything that they can as far as when they have a stockpile when they fraction it; they're generally having some fine FRAP left over.

JIM SCHEROCMAN: Because if you use all of it, you're going to get the same binder content. You're going to get smaller pieces perhaps, but the binder content going into the mix from the RAP isn't going to change if you use all of it, right?

ROSS BENTSEN: Right, sure.

JIM SCHEROCMAN: So my question is, what's the real advantage to fractionization?

ROSS BENTSEN: We looked at it from a quality control standpoint, as well as changing their mindset on how to process the RAP that comes from millings and putting it into stockpiles. Instead of the previous process, which was just crushing everything down to minus 3/8 inch—that crushing operation was creating a lot of dust—and so we changed their mindset to be able to screen it first. And some contractors are actually screening it and anything they use that doesn't get into the basically minus ½ inch to minus ¼ inch, they actually have a third stockpile, so they're not crushing anything. And that is helping as far as the dust control. And then, to me, having more bins for, say, 40% of a particular material, is providing more quality control, just like we do with the virgin aggregates.

JIM SCHEROCMAN: I guess that most contractors that I'm familiar with use it all, so if you use it all then whether you put in in one pile or two piles doesn't make a lot of difference in the long run.

ROSS BENTSEN: Okay, sure.

MARTIN VAN DE VEN: The warm mix, did you use waxes or what was the system that you used for the warm mix?

ROSS BENTSEN: The evaluation we did actually had one wax, one chemical, and one foaming, and so we were able to look at all three.

MARTIN VAN DE VEN: All three. Okay, thanks.

ROSS BENTSEN: And as far as where are the contractors in warm mix, much like nationally, most of them are using the foaming process, some are still using the chemical, and I'm guessing nationally that's probably in the same realm. Yes, Randy.

RANDY WEST: Ross, I may have missed it in your presentation, but I was curious about whether or not for the RAS binder or for the RAP binder, you consider all of that binder to be effective in the calculations of mix design or do you use a reduced factor for that?

ROSS BENTSEN: We are currently, assuming all the recycled binder goes into that, we are using the standard VMA calculations for that and are not saying it's 85% effective or those approaches that some agencies are taking.

4.0 Discussion—Nathan Morian

MICHAEL DUNNING: Very good presentation. We've had 15% recycle now for about 20 years, Your presentation was interesting. I appreciate what you're showing us here for me to think about. I don't know how much cracking might be going on because we put NovaChip right on top of it and it just seals everything up. That's the whole purpose on new asphalt and for maintenance. But I really appreciate what you're doing here. We also use 93% minimum compaction. On your board, it appeared that when you did your design, you did the control with your zero RAP. Then you performed the design with RAP in the laboratory. Then you reviewed the field compaction requirements. For the control area you had 92% compacted for one and then 93% for another. If I hear you right, that means I have 93% minimum requirement for RAP materials, then if I had a control strip of no RAP it may compact to about 94%.. So you're saying that with or without RAP there is a difference between compaction in the field. Is that what you're saying here?

NATHAN MORIAN: So actually you're talking about Average Production Monitoring data?

MICHAEL DUNNING: Yes.

NATHAN MORIAN: Effectively this is the relative density from the Gmm for each respective mixture. Those are taken on a lot basis and that's how we measure the compaction in the field with a nuclear density gauge. And so, yes, on the mat itself the average of that kind of shotgun pattern on the previous graph was almost 94%, and we had just under 93% with the RAP mixture, for the two different sections. So those were field measures, not requirements or specifications limits.

MICHAEL DUNNING: Right, okay. Appreciate that, thank you.

FRANK FEE: Your fatigue testing that you're doing on the RAP mixes and all of the mixes, I presume that's a beam fatigue?

NATHAN MORIAN: Yes, four-point bending.

FRANK FEE: Have you considered running some kind of a fracture promulgation test, such as an OT test or an SCB or something like that? We have seen that there's really a difference between the characteristics you'll get from that kind of test, than you will with a traditional beam fatigue for all of the reasons you know about.

NATHAN MORIAN: The simple answer is yes, we have considered it. Have we done it? No, but we are planning to evaluate the brittleness using the UTSST on the field core samples. Thanks, Frank.

WALAA MOGAWER: Good presentation, Nathan. You might have touched on it, but why is there a 0.7% difference in the binder content? You're only using 15% RAP and you're assuming 100% blending.

NATHAN MORIAN: In the general case, yes. Essentially, this is what came from the design, and I didn't bring the mix designs with me. I would have to get a little more detailed background information. So I don't have a good answer for you. But we run the Hveem design. All of our mix designs are done in-house, so this is the mix design process that we as NDOT typically follow. The contractors propose the aggregate bin percentages and then we effectively give them the asphalt content based on the Hveem and Lottmans. We call them Lottmans, but it's a TSR requirement. So as to the actual difference, I couldn't tell you for sure without specifically looking at the design input information.

WALAA MOGAWER: Thank you.

HONGBIN XIE: Very good presentation. I've got a question. Can you move the slide back to 25? Just a clarification, at the bottom of the Table, the fatigue performance, when you say control less than RAP, what does that mean?

NATHAN MORIAN: So that means that the expected fatigue life or the fatigue resistance is lower from the mechanistic standpoint. So if you look at the fatigue plot itself...

HONGBIN XIE: Which one has shorter life, RAP or control?

NATHAN MORIAN: So if you look at the fatigue relationship, strain vs the number of cycles to failure, the bottom line is the RAP and the top is the control mix. The virgin mix, we called it control.

HONGBIN XIE: So the RAP mix had a longer life?

NATHAN MORIAN: According to this figure, yes. But I mention it in this form because the RAP is so much stiffer that when we plug it into a mechanistic evaluation we see a difference. We are essentially doing a simplified version of the MEPDG calculations. And so when you include the stiffness of the RAP mix, you get a much lower strain. I should have put that figure in the presentation, but the strain is lower for the RAP mix because it's stiffer. So, you get less deflection because the mix is stiffer and so you enter this table at a lower strain value under isothermal conditions, resulting in a longer fatigue life with the RAP mix. That's how we simplify it. We don't vary the temperature.

HONGBIN XIE: So you're really not doing the strain control, but more like by the stress control.

NATHAN MORIAN: Well, sort of. These were tested under strain control, but yes there is an interaction between the strain level from the mechanistic evaluation and stiffness of the mixture, it's a balance. So when you input stiffness into the calculation, the strain you get is just based upon linear elastic analysis.

HONGBIN XIE: All right. Thank you.

NATHAN MORIAN: I can't remember exactly if we used ELSYM5 or it was probably 3D-Move. More than likely it was a linear elastic calculation done with 3D-Move, not the dynamic loading case.

5.0 Discussion—Chris Abadie

MICHAEL DUNNING: We have open grade. Well, actually, we have stopped using it, but in the past we used open grade. I know NDOT still uses it. The point is that at about five to six years out, it stops working for reducing of the spray because the voids fill with fines. Of course, we don't have much water there.

CHRIS ABADIE: Yes, but did you see this line I showed on macrotexture?

MICHAEL DUNNING: No.

CHRIS ABADIE: The reason I showed that slide is because what's important in safety is the macrotexture and OGFC holds its macrotexture for life.

MICHAEL DUNNING: Okay, very good. Thank you.

CHRIS ABADIE: Go ahead.

BOB SAUBER: Yeah, my comment on the open graded, I loved it when I was with the New Jersey DOT. I'm with myself now, RWS Consulting.

CHRIS ABADIE: Congratulations.

BOB SAUBER: But anyway, yeah, the maintenance guys would complain, complain about the open graded for winter maintenance snow removal. And the problem with open graded in a freeze environment is when they spread rock salt, the rock salt sits on the surface and the water drains through so it doesn't form the brine solution that forms on a dense graded mix. But when I was there, I always convinced the commissioner, the couple of snow days were offset by the many rain days and that reduced spray and the safety and the skid resistance with the open graded. And I asked those guys, the maintenance guys, "Well, why did you stop complaining?", and they said the same thing. After a few years, the surface kind of clogged up and the snow removal wasn't a big issue. But I agree with the macrotexture issue that it really is a much safer mix, and I don't know why people in non-freeze climates aren't using more open graded. The best success we saw was with the crumb rubber mix, you know, the asphalt rubber. Not the terminal blend but where you bring a blending unit in. We were specifying a minimum of 8.5% asphalt in the mix. I have a picture that Trap Rock gave me where they took up a night joint and it rolled up like a carpet. It had tremendous flexibility. We started out using a Georgia porous European type mix. We had done it for years, but the asphalt rubber mix is really an improvement. Sixteen years we got out of some of the old McDonald process that Joe Smith had done, and it only got milled up because an adjacent section was taken up.

CHRIS ABADIE: We have our own version of the crumb rubber that we allow for OGFC, and probably 30% of our OGFC has crumb rubber.

BOB SAUBER: Yeah, it gives you a good smooth and quiet ride. … And it's easy mix to put down. There's no air void requirement.

CHRIS ABADIE: I'd like more asphalt in it.

BOB SAUBER: You need a good, heavy tack coat application. That's one of the important things, right?

CHRIS ABADIE: Yes, that is why we require the spray paver, to enable sufficient quantities of tack coat. Okay. Thank you.

AAPT/ISAP International Forum: Cold Recycling of RAP

Agency Perspective on Cold In-Place Recycling
Benjamin Bowers, *Virginia Center for Transportation Innovation and Research*

RILEM Proposal for RAP Characterization
Eshan Dave, *RILEM - International Union of Laboratories and Experts in Construction Materials, Systems and Structures*

Cold Recycling of RAP with Bituminous Emulsions - Characterization and Experiences
Todd Thomas, *Colas Inc.*

Bituminous Emulsion and Recycling: a French Overview Vincent Gaudefroy, *IFSTTAR - Institut Français des Sciences et Technologies des Transports, de l'Aménagement et des Réseaux*

Sustainability of Cold Recycling
Everett Crews, *MWV Asphalt Innovations*

See page 647 for "An Agency Perspective on In-Plane Pavement Recycling" paper and discussion.

2.0 Discussion—Eshan Dave

MIKE HEITZMAN: I do agree we need to have a better understanding of the asphalt binders when we're doing our in-place recycling. My question to you, though, is how are you going to get a sample ahead of the project to actually represent the actual gradation?

ESHAN DAVE: I'm not sure if I understand your question. Actual gradation in the sense of… Oh, because the milling is going to generate…

MIKE HEITZMAN: That's right.

ESHAN DAVE: And that definitely is a challenge. So the idea here was to… I mean it definitely suits in an easier manner to a central plant type of approach where you already have a stockpile. But, yeah, it's a challenge. I don't think I have a solution for that.

MIKE HEITZMAN: Because most of the cases, you're going to take cores and try to crush them down but nothing you do will represent as milling.

ESHAN DAVE: Exactly. That's not the same as milling.

MIKE HEITZMAN: Thank you.

ESHAN DAVE: Thanks.

CHRIS ABADIE: Thank you for a good session and good presentation. I see that you have probably various volumetrics when you have various temperatures. Would you have from your mix design some recommended in-place densities associated?

ESHAN DAVE: Yeah. That's the idea. I think that's the next step, but the idea was to collect enough information here so that when you have your final spec, you can make recommendation regarding the density.

CHRIS ABADIE: Thanks.

NIMA ROOHI: Very nice presentation. I have a question about the cohesion test. I think I didn't clearly understand the necessity of the test or the correlation with any sort of material characteristics in the design stage. You say you compact the material at three temperatures and then test it like you do TSR, right?

ESHAN DAVE: Correct.

NIMA ROOHI: And is that representing activation of asphalt that is in the RAP due to cold-in-place recycling or hot recycling?

ESHAN DAVE: Yeah. So essentially it does not. The goal here is not necessarily trying to get, say, a material performance parameter. The goal here is to essentially be able to make sure if you have one sample you get this set of numbers, you get another set of numbers for another sample whether you can say they are similar or not. And get some information that you can then use in finding out things such as your optimal emulsion amount or foaming amount, things of that nature.

NIMA ROOHI: I was wondering if you could get that information from the binder testing or not. Because you have binder testing in your asphalt as well.

ESHAN DAVE: Yes, but, again, you have to keep in mind this is cold recycling, so not all your binder that you may… The way we did the bitumen binder content or binder test is either chemical or ignition, and that's going to give you the whole spectrum, everything that's in there, but everything that is in there may not be readily available to the mix.

NIMA ROOHI: Thank you.

GABRIELE TEBALDI: One of the points behind this work is to have a test that in few hours allows you, in the plant or in the field, to check if the RAP that is coming, mainly in the plant, is something similar with the RAP that you used in your laboratory to do your test for the mix design.. Something simple that allows you to say, "Okay, the RAP that I have is similar with the RAP that I used for the mix design so I can go ahead" or "this is something different from what I used for mix design: I have to be careful. This is the key point.

NIMA ROOHI: Yeah, but the question is in what aspect, in what aspect the RAP is similar to the RAP in the field work.

GABRIELE TEBALDI: The only thing that you can do is to check the material that you have. If your material is different from the one that you had in the laboratory, your mix design doesn't work. This is the purpose of the test. It is not the test to investigate. It is not a fundamental test. It is just a test for quality control, to check if the RAP in the field it's different from the material that you tested in your mix design.

ESHAN DAVE: Thank you.

GALE PAGE: Just a comment regarding the breakdown during milling. We did a rather extensive study in Florida and we did establish factors for the breakdown of the RAP during the milling, but it's still is an issue to get a representative sample or how to break down the RAP from a core in order to represent milling. It's still a difficult situation. Most contractors are able to do that and get representative samples from cores to do in their mix design, but we don't specify how they do it.

ESHAN DAVE: Thanks.

MARTIN VAN DE VEN: I understand if you do cold-in-place recycling and you generate RAP from low quality or it is a thin layer, that you use it. But if you have milled RAP well controlled from the road, then my experience is that in most cases you reuse it in hot mix. That's the highest possible reuse or recycling.

ESHAN DAVE: Yeah.

MARTIN VAN DE VEN: What's your opinion on that?

ESHAN DAVE: As we saw in previous presentation, I think obviously as you mentioned, depending on quality of RAP, usually the way we had been doing things in hot recycling, you are seeing better return on your value, but I think we are on a transition if you think of it that way. I think next 10 years, next 15 years, we are going to see if not same, probably even more with the amount of cold recycling.

MARTIN VAN DE VEN: Also, when you mill it, you bring it to a plant and you collect it? That was my question.

ESHAN DAVE: I'm not sure I understood it.

MARTIN VAN DE VEN: Well, if you mill it and you bring it to a certain cold mix plant that cost also money.

ESHAN DAVE: Then obviously with in-place you are definitely saving on that cost material.

MARTIN VAN DE VEN: Yah, I understand that, or you could better use it in a hot mix plant for recycling in a high quality asphalt mixture. .

ESHAN DAVE: Yes. Thanks.

ELIE HAJJ: I have a very quick question. What does the red mean on the slide?

ESHAN DAVE: Thanks for asking. Those are results from FHWA from their survey and this I got from their website. It seems to have been updated last year. Red means states that have done multiple cold recycling jobs, green means they have a few of them, and blue means they haven't done any recently. Obviously, it's not my data, so if you have any questions about it. And white means no response. Thank you.

3.0 Discussion—Todd Thomas

MANSOUR SOLAIMANIAN: Excellent presentation, Todd. One question, when you were talking about design, I guess you mentioned 30 gyrations is typical to use?

TODD THOMAS: Yes.

MANSOUR SOLAIMANIAN: We've been doing designs at 50 gyrations. So my first question is what's the base for choosing 30 gyrations?. And number two, based on 50 gyrations for indirect tension, we have established the threshold at around 45 to 50 psi on the IDT strength. So if you go to 30 gyrations, would you still keep that level of strength or do you have any number in mind that that we should go by in terms of pass-fail criteria on IDT strength?

TODD THOMAS: Steve Cross has done some research in the past on relating the gyratory compaction level with the in-place density and had come up with a value of somewhere between 30 and 35 gyrations. My experience has said that's pretty close. You might get that density even further into its service life even within just a few months, which is when the cores were taken to verify that. But from an ITS criteria, at 30 gyrations about 40 or 45 psi, I believe is the number that's typically used, and for Marshall stability at 40 C, it's 1250 pounds.

MANSOUR SOLAIMANIAN: The comparison of densities from the laboratory gyration versus field that your mention maybe has been the base to come up with design gyrations as 30 to 35. In that case, has it been mostly looking at density itself or looking at the void content?

TODD THOMAS: That was looking at void content.

MANSOUR SOLAIMANIAN: What kind of void levels?

TODD THOMAS: Typically, air voids are like 9 to 14%. There are some exceptions below and above that value, but that would be fairly standard.

MANSOUR SOLAIMANIAN: Great. Well, thank you.

TODD THOMAS: Thank you.

CHRIS ABADIE: Thank you, Todd, for a nice presentation. I didn't see a lot in the last three presentations, but really in general for cold-in-place recycling I find that you have a high potential to have a crack resistance or a crack mitigating mixture. I wanted to know your thoughts on that. Is that something that you would agree with or it needs more work?

TODD THOMAS: Yes, I would generally agree with that. I mean, first of all, we're removing the cracks through the process, doing it 3 to 5 inches thick. But if you look at some of the thermal cracking data with the creep compliance, you'll see measured thermal cracking, at least the crack initiation temperatures of easily, easily meeting what the climate would require. So I don't have a real scientific explanation for that, but we're not running this through a plant, we're not aging it, we're adding a new material to it and that is helping that out. I wish I could say if those air voids at 9 to 14% are good or bad. I've heard both. I've heard some people say that they have some kind of cushioning effect when it comes to resisting that cracking. Again, that's not a very scientific way of saying it, but I've heard it explained that way.

CHRIS ABADIE: We have experienced resistance in reflective cracking when using cold-in-place recycling, but only in the field without the laboratory numbers to support it. Thank you.

TODD THOMAS: Thank you, Chris. Randy?

RANDY WEST: Todd, thanks very much for bringing this presentation. We're still learning a lot about these processes, and at NCAT, we're trying to catch up to speed to a lot of stuff that you guys have already done and folks like Steve Cross have already done. But that question that Mansour asked about kind of the compaction level in the laboratory is one of the basic things you have to answer at the very beginning because that drives your optimum asphalt content and your optimum moisture content and then impacts other additive contents as well.

TODD THOMAS: Yes, absolutely.

RANDY WEST: And then also obviously drops the strength criteria that you would set. We're seeing some numbers lower than 30 gyrations for the Virginia material that Ben Bowers presented on earlier, but that's really as far as we've gone at this point. So that's just a comment. Another comment that I have is the comment that you made about considering that the asphalt in the RAP is not active; you consider it black rock. That's a little bit disappointing, I guess, that we're taking a material that has value, the asphalt binder that has a tremendous value, and we're completely neglecting that value in these processes. I really don't think that it's all black rock. So I wonder if we might be missing the opportunity to rejuvenate that binder with some emulsified products or with some of the new rejuvenators that are out there, so we are starting to explore that option.

TODD THOMAS: And I appreciate you bringing that up. I probably used the wrong term because certainly my experience has shown, and I've seen and I know others have seen, that if you look at the hardness of the asphalt or even the asphalt content, that's going to have a big effect on what the final properties are. I've seen some of the western US designs that have extremely, extremely hard binder in the RAP. It's going to take more new bituminous material to help activate it. But I've seen other projects where the binder was so active that it needed very little new bituminous for it. So, yes, I apologize for the way I stated that, but I agree, there's still value in the asphalt that's in that RAP.

RANDY WEST: Just to follow up on the comment of trying to identify how much of that binder in RAP might be active. A group of us toured Japan a couple months ago. They were looking at RAP for hot mix recycling, but there were a couple things they were doing that I thought were good for us to kind of learn from. One is, they basically take the RAP and compact the RAP into Marshall size specimens and look at, in essence, a modulus. And if that modulus is certainly below a certain value, they feel like it is active. And if it's really stiff, then they consider it not to be active. Initially they recovered binder from the RAP material, and when it has less than 20 penetration, they consider that not to be a binder that would be active.

TODD THOMAS: Okay. That's good to know.

MARTIN VAN DE VEN: Nice presentation. You mentioned the much lower stiffness than for hot mix asphalt; that is in general the conclusion?

TODD THOMAS: Yes.

MARTIN VAN DE VEN: Quite some very low stiffness values reported. If you have to use this low stiffness to determine the contribution of the cold-mix layer to the pavement thickness design, this could be a challenge. Do you think that fatigue is an issue in this type of mixture?

TODD THOMAS: That's a great question. Fatigue is hard to measure in the laboratory. I think that the research that's being done right now in the test track and other places, at least for the foamed cold central plant, will tell us a lot about that.

MARTIN VAN DE VEN: Okay, let us wait the field results.

TODD THOMAS: Thank you.

PHILLIP BLANKENSHIP: Great body of work. I have a comment and then a question. I see CIR often come through our laboratory. I really believe that CIR is more engineered today than many of our hot mix designs. We have put many tests around its performance. If you could go back and pick your top two or three tests – and you showed the Hamburg already – what would you say that we need to be looking at to promote CIR performance? Is it a different set of tests to reduce raveling? What are your top two or three performance tests that you would recommend to this audience?

TODD THOMAS: I would say it would be some of what I showed here, I mean dynamic modulus from a characterization for pavement design. I think an early cure test like raveling is a very useful test to at least understand that the emulsions have early breaking characteristics and are gluing things together early on in the life. And I think that moisture resistance testing with vacuum saturation, soaking, and doing essentially a TSR, is a very valuable test too. So it's not feasible to do dynamic modulus for mix design, but you can do the other tests mentioned pretty easily for mix design.

PHILLIP BLANKENSHIP: Thanks, Todd.

TODD THOMAS: Thanks, Phil.

4.0 Discussion—Vincent Gaudefroy

MICHAEL DUNNING: We like cold mix as they work very well for us. One question I have here is it takes time to cure. On steep grades, is there a problem with that?

VINCENT GAUDEFROY: I'm sorry, I don't understand your question.

MICHAEL DUNNING: Before it cures.

VINCENT GAUDEFROY: What was your question?

MICHAEL DUNNING: Well, the question is it takes time to cure cold mix, okay on the steep road as well.

VINCENT GAUDEFROY: We recommend to not cover the mixture at least later three to six weeks. This time has to allow the layer to dry and the bitumen to develop these properties of cohesion. According to specific regional practices and for low traffic, the time could be extended up to several months (winter period).

MICHAEL DUNNING: As far as what? Three to six weeks before it cures, before getting truck traffic on that?

VINCENT GAUDEFROY: No. The traffic is opened just before the implementation step, but we recommend covering the new CMA layer after a short curing period of about three to six weeks to protect the layer from the traffic and set a surface course made of CMA or HMA.

MICHAEL DUNNING: So I'm looking at still having a cure time evolving there for what, about a month or so, or two months for full cure?

VINCENT GAUDEFROY: It depends on the traffic and also the localization on the sections. In few cases a full cure is quickly obtained after one year but the general behavior observed in France is an increase of the mixture stiffness during the next two or three summers after the implementation step and a stabilization of the complex modulus between 2500 and 4000 MPa measured at 15°C and 10 Hz.

MICHAEL DUNNING: Okay.

GABRIELE TEBALDI: Can I add a comment? Curing is depending not only for the traffic condition, what in a sense is the so-called post compaction phase, that helps to remove the water but is important to consider how much active filler you may use in your mixture. Okay, be careful not to go too much up with too much cement mix or it can become brittle. And the temperature. So the different temperature is a different condition.

MICHAEL DUNNING: On a slope, with a curing process going on, you've got steep truck traffic in a mountain area. Is there a problem with that?

GABRIELE TEBALDI: This could be a problem. I know a technique. It was not my idea but it was an idea of one of my colleagues that could be used. He used the idea to put on the top of the mixture OGFC in order to protect the mixture and allow the water to go out. So it was like a reverse use of cold mixture to allow the water to go inside. The technique was used by Kim Jenkins in San Paulo I27 Highway to allow to open the road immediately after the work.

AAPT/ISAP International Forum: An Agency Perspective on In-Place Pavement Recycling

Brian K. Diefenderfer[a] **and Benjamin F. Bowers**[a]

[a]*Virginia Center for Transportation Innovation and Research, Virginia Department of Transportation, Charlottesville, VA 22903*

ABSTRACT: *In-place pavement recycling, a series of techniques used for rehabilitating and constructing pavement structures, has seen increased use in recent years. However, many highway agencies have not yet fully adopted the processes despite the growing literature showing examples of successful projects, cost savings, and environmental benefits. This paper discusses the processes by which the Virginia Department of Transportation used the results of internal research and efforts of other agencies to develop specifications and usage guidelines for in-place pavement recycling.*

KEYWORDS: *cold in-place recycling, cold central plant recycling, full-depth reclamation, in-place recycling*

1.0 Introduction

This paper is a summary of a presentation given at the AAPT/ISAP International Forum on Cold Recycling of RAP during the Association of Asphalt Paving Technologists Annual Meeting on March 9, 2015, by Dr. Bowers. The implementation of in-place pavement recycling practices by the Virginia Department of Transportation (VDOT) through specification and research activities are summarized herein; the paper is organized in a manner similar to that of the presentation.

The oral presentation was made by Dr. Bowers.

1.1 Glossary

The following definitions or explanations of common terms related to in-place recycling are provided for clarity:

- CIR—Cold In-Place Recycling (CIR) is a process used to rehabilitate the upper portions of the bound layers of an asphalt pavement and is typically performed at depths of 2 to 6 inches.
- CCPR—Cold Central-Plant Recycling (CCPR) is a process in which the recycled material is milled from a roadway and brought to a centrally located recycling plant that incorporates the recycling agents into the material.
- FDR—Full-Depth Reclamation (FDR) is used to correct severe structural deterioration that is deep within the pavement structure and is typically performed at depths of 4 to 12 inches. This includes both the bound layer(s) and a portion of the underlying unbound materials.
- Cold Recycler—The machine that is used to perform the CIR process. The cold recycler consists of a down-cutting drum that pulverizes the bound layers to the desired depth and pumping systems to deliver the recycling agent and water to the pulverized materials.
- Recycling agent—Foamed or emulsified asphalt is used as the recycling agent in the CIR and CCPR processes to bind the pulverized material particles together.
- Reclaimer—The machine that is used to perform the FDR process. The reclaimer consists of an up-cutting drum that pulverizes the bound layers and unbound materials to the desired depth and pumping systems to deliver the stabilizing agent and water to the pulverized materials.
- Stabilizing agent—Used to bind the pulverized material particles together for the FDR process. The stabilizing agent may consist of foamed or emulsified asphalt, hydraulic cement, lime, fly ash, etc., or combinations of these depending on the properties of the materials being stabilized.
- Fillers—Fillers are used to increase the resistance to moisture damage, promote the adhesion of asphalt binder to the RAP particles, and improve the dispersion of foam droplets within foamed asphalt recycled mixtures. Fillers may consist of *natural* fillers (e.g., stone dust) or *active* fillers (e.g., hydraulic cement, lime, fly ash, etc.). Active fillers can provide further stiffness to the mixture.
- Expansion ratio—The ratio of the maximum volume of foamed binder relative to the original volume of binder.
- Half-life—The time taken for the foamed binder to collapse to half of its maximum volume.

- Breaking time—The time taken for the binder in an emulsified asphalt to separate from the water phase and produce films of binder on the particles of the recycled material.
- Curing time—The time taken to displace water from an asphalt recycled/stabilized layer and the resultant increase in stiffness and strength properties.

1.2 Background

Previous studies have shown in-place pavement recycling to be an effective means for rehabilitating asphalt pavements in a manner that can reduce costs, decrease environmental impacts, and increase the speed of construction (Nataatmadja, 2001; Thenoux et al., 2007; and Stroup-Gardiner, 2011). However, despite these advantages, many departments of transportation (DOTs) have not implemented these processes for many reasons. From an agency perspective, the reasons include unfamiliarity with the processes, difficulty in determining appropriate engineering properties for quality assurance, limited performance data for high-volume roadways, and inconsistencies in specifications across different agencies. VDOT has worked to overcome these obstacles to implementing widespread use of in-place pavement recycling through its research and specification development.

In-place pavement recycling methods include the following processes: cold planing (CP), hot in-place recycling (HIR), cold recycling (CR), and full-depth reclamation (FDR). CP is a familiar process to many pavement practitioners in that it removes deteriorated layers from a pavement and allows for placement of new material without significantly changing the final surface elevation. CP truly becomes a recycling process when the milled materials, known as reclaimed asphalt pavement (RAP), are re-used in another pavement. HIR is a process used to address pavement deterioration within the upper 1 to 2 inches of the pavement by adding rejuvenating agents to the heated and scarified pavement.

CR includes the techniques of cold in-place recycling (CIR) and cold central-plant recycling (CCPR). CIR recycles the upper 3 to 6 inches of an existing asphalt pavement by pulverizing, adding recycling agents, and repaving the recycled mixture. The CIR process is commonly performed by a train of equipment that includes a tanker (containing hot asphalt for foaming or emulsified asphalt), a cold recycler, a paver (if the cold recycler does not have an on-board paving screed), and rollers. An example of a cold recycler performing CIR is shown in Figure 1. In addition, mobile sizing and crushing equipment may be included in some applications. CCPR is a process in which the recycled material is milled from a roadway (or provided by an existing stockpile) and brought to a centrally located mobile recycling plant that incorporates the recycling agent. An example of the CCPR mobile plant is shown in Figure 2. The recycling agents used for CIR and CCPR are a foamed or emulsified asphalt binder. Active fillers such as hydraulic cement, hydrated lime, fly ash, or cement or lime kiln dust are often added to increase the resistance to moisture damage, promote the adhesion of asphalt binder to the RAP particles, and improve the dispersion of foam droplets within a foamed asphalt CIR/CCPR material among other reasons (Asphalt Academy, 2009). CCPR is typically performed at depths of 3 to 6 inches and can be placed in multiple lifts.

Figure 1. *A cold recycler performing the cold in-place recycling technique.*

Figure 2. *A cold central-plant recycling mobile plant.*

FDR corrects deeper structural deficiencies within an existing pavement by stabilizing the bound layers and a portion of the underlying unbound materials. Figure 3 shows an example of a reclaimer performing the FDR process. The depth of pulverization depends on the thickness of the bound layers but typically ranges from 6 to 12 inches. As with CR techniques, the stabilizing agents used in FDR techniques are a foamed or emulsified asphalt binder. In addition, active fillers such as hydraulic cement, hydrated lime, fly ash, or cement or lime kiln dust are often added to increase the resistance to moisture damage, reduce the plasticity index, promote the adhesion of asphalt binder to the granular particles, and improve the dispersion of foam droplets within a foamed asphalt FDR material among other reasons (Asphalt Academy, 2009). Generally, the materials used as recycling agents for CIR/CCPR or as stabilizing agents for FDR may be the same; however, they are termed differently.

Figure 3. *A reclaimer performing full-depth reclamation.*

CIR can be beneficial to an agency in that materials can be recycled in-place without the need to haul them in and out of a work zone. Since the recycled materials are not removed from the site, but instead are left in place, safety may be enhanced by not subjecting the traveling public to large drop-offs at the edges of milled pavement sections. CCPR can also be an advantageous recycling method for DOTs in several unique ways. First, the milled material can be removed from the roadway and processed in the mobile plant while the underlying layers are stabilized or replaced as needed. Second, existing stockpiles of RAP can be recycled into structural layers for newly constructed pavements or for the rehabilitation of existing ones. As an example, a recently completed study showed that Virginia paving contractors have approximately 4.7 million tons of RAP (Hoppe et al., 2015). From this, the CCPR process could be used to produce a one-lane-width base layer 6 inches thick for approximately 2,100 miles. It is unknown how much stockpiled RAP exists in the United States. Third, when thicker applications are required, CCPR materials can be paved in multiple layers. FDR is a beneficial method when structural issues within the existing pavement reach beyond the bound layers and into the unbound layers such as the aggregate layer or subgrade. FDR is often used independently to reconstruct a pavement section but the recycled content of the entire pavement structure may be increased when FDR is used in conjunction with CCPR to rebuild the pavement.

2.0 Virginia's In-Place Pavement Recycling Research Experiences

Prior to 2008, VDOT employed in-place pavement recycling techniques (primarily FDR) infrequently and without formal documentation of the processes or outcomes. Beginning in 2008, VDOT began a research program which sought to establish a materials properties catalog using mechanistic laboratory tests. Three FDR sections, with foamed and emulsified asphalt and hydraulic cement as stabilizing agents, were constructed on three low-volume primary routes. Deflection testing, by use of a falling weight deflectometer, was conducted over a two-year period to assess the potential for continued curing. These studies found that FDR continued to gain strength over two years after construction; the strength gain was attributed to curing of the recycled material (Diefenderfer and Apeagyei, 2011). This was the first of many VDOT lead research efforts focused on in-place pavement recycling. The major notable efforts tested these techniques on a high-volume interstate corridor and an accelerated testing facility.

2.1 Interstate 81—Staunton, Virginia

In 2011, VDOT completed the first pavement recycling project in the United States that used three in-place recycling techniques on the interstate system. The two-lane section of I-81 that was rehabilitated carries approximately 6,400 trucks per day (approximately 28% of the total volume of daily traffic). Because the location and severity of deterioration were different between the two lanes, the lanes were rehabilitated differently. Since the left lane carries less traffic, it had less deterioration. For the left lane, the upper 2.75 inches of the existing pavement was milled and CIR was used to recycle the next 4.25 inches using a combination of foamed asphalt and hydraulic cement. It is important to account for the bulk-volume expansion of the material; in this case, recycling 4.25 inches resulted in the desired 5-inch recycled layer. This lane was capped with 2 inches of a new asphalt mixture approximately two days after construction of the CIR layer.

The right lane was in much worse condition and had distresses that originated within the clayey subgrade that had been used as a capping material during construction in the early 1960s. The sequence of construction in the right lane began with milling approximately 10 inches of the existing 11 to 12 inches of asphalt mixture. The RAP generated from this process was moved to one end of the project and stockpiled. Following milling, FDR was completed on the next 12 inches, consisting of the bottom 1 to 2 inches of asphalt pavement, approximately 6 to 8 inches of aggregate base, and the top approximately 2 to 4 inches of subgrade. After about 4 to 6 hours of curing, the stockpiled RAP was processed in a CCPR plant using foamed asphalt and hydraulic cement and repaved over the FDR layer. The depth of the CCPR layer was 8 inches in one section and 6 inches in the other. After the CCPR layer was allowed to cure for approximately 8 to 16 hours, an asphalt mixture layer was placed on top at a thickness of 2 inches in the 8-inch CCPR section and at a thickness of 4 inches in the 6-inch CCPR section. After 6 to 8 weeks while all ancillary work was completed (including edge drain installation, guard rail adjustments, etc.), a final 2-inch asphalt mixture overlay was placed over the entire project. A schematic showing the cross section of the I-81 recycling project is shown in Figure 4.

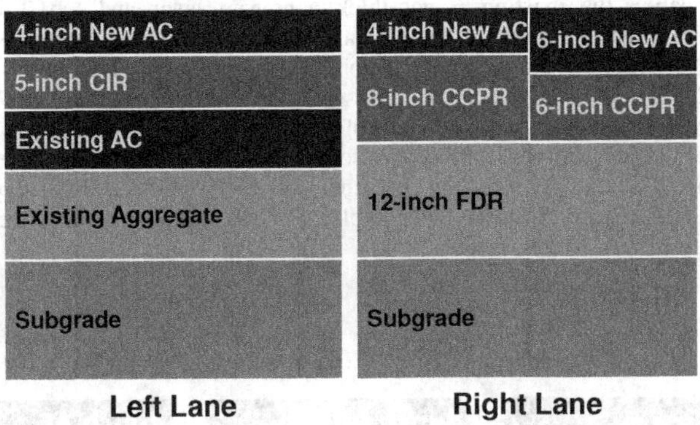

Figure 4. *Completed cross section of I-81 recycling project. AC = asphalt mixture; CIR = cold in-place recycling; CCPR = cold central-plant recycling; FDR = full-depth reclamation.*

The I-81 recycling project was significant for VDOT in that it represented one of its first major in-place recycling projects. The project was also significant nationally in that it was one of the highest truck-traffic volume routes where in-place pavement recycling was used. More details about this project can be found in Diefenderfer et al. (2012). A following study by VDOT showed that after nearly 3 years and approximately 8 million equivalent single-axle loads (ESALs), the pavement was performing well with no visually observable distresses. Rut depths were less than 0.25 inch, and the ride quality in terms of the International Roughness Index ranged from 45 to 55 inches per mile and is considered *excellent* according to VDOT standards (Diefenderfer et al., 2015 and VDOT, 2012).

2.2 National Center for Asphalt Technology (NCAT) Test Track Sections

Following construction of the I-81 project, VDOT partnered with the National Center for Asphalt Technology (NCAT) to place three sections using in-place recycling as part of the 2012 Test Track. All three sections contained a 5-inch-thick layer of CCPR that was produced from RAP left over from the I-81 recycling project. The RAP was transported to the NCAT track and crushed to break up the large clumps that had conglomerated. The CCPR that was produced had a maximum particle size of about 9.5 mm, which is considerably finer that the grading envelope recommended by Wirtgen (2010). In two of the sections, the CCPR was placed on top of the existing aggregate base and then capped with 2 inches of a dense-graded asphalt mixture followed by 2 inches of a stone matrix asphalt (SMA) mixture in one section and 4 inches of a dense-graded asphalt mixture followed by 2 inches of SMA mixture in the other. The reason the two sections were constructed was to observe any potential difference in performance associated with a difference in overlay thickness. In the third section, a reclaimer was used to simulate FDR by cement-treating the existing 6-inch-thick aggregate layer and top 2 inches of subgrade (by the definition of the Asphalt Recycling and Reclaiming Association [ARRA]), FDR also includes some portion of the overlying asphalt layer). Again, a 5-inch CCPR layer was placed in this section. The cement-stabilized layer and CCPR were also capped with 2 inches of a dense-graded asphalt mixture followed by 2 inches of SMA mixture. Since two sections had a 4-inch overlay, differences in performance resulting from the presence of the stabilized base layer were investigated. Figure 5 shows the completed cross section for the three sections placed at the NCAT Test Track.

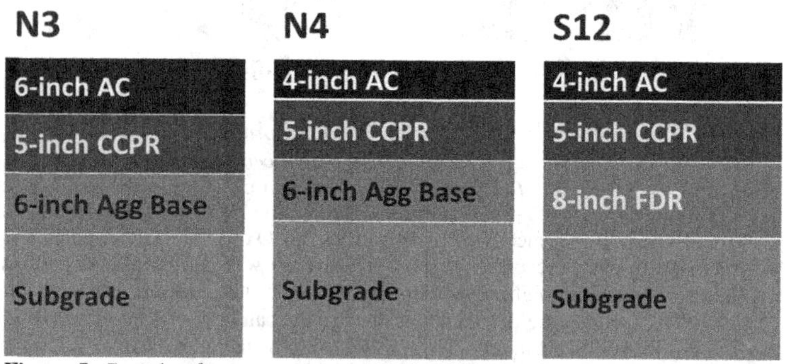

Figure 5. *Completed cross sections of 2012 NCAT Test Track in-place recycling sections N3, N4 and S12. AC = asphalt mixture; CCPR = cold central-plant recycling; FDR = full-depth reclamation.*

At the end of 2014, the 2012 NCAT Test Track sections had experienced 10 million ESALs of truck traffic and the three recycled sections showed no signs of deterioration based on the surface-observable condition. Figure 6 shows the results of strain measurements taken at the bottom of the CCPR layer. The figure shows that the thicker asphalt overlay reduced the measured strain (Section N3 vs. Section N4) but an even greater reduction in strain was produced by the presence of the stabilized base (Section S12 vs. Section N4). VDOT will continue testing of these three recycled sections as part of the 2015 NCAT Test Track.

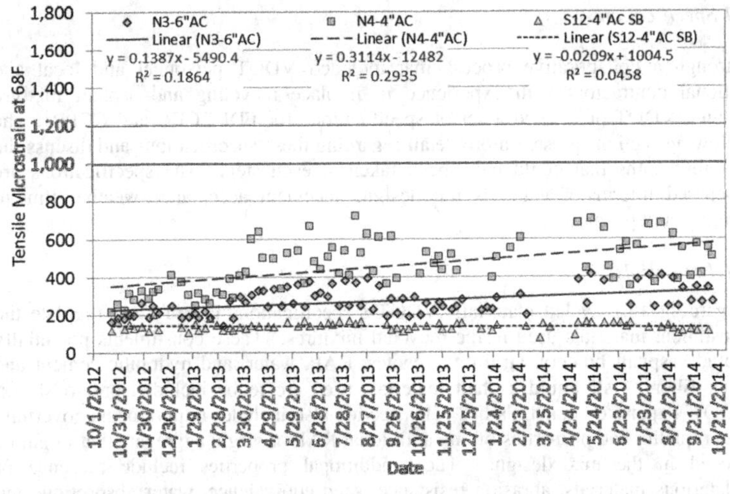

Figure 6. *Tensile strain measurements from NCAT Test Track. Section labels can be compared with schematic from Figure 5. Figure courtesy of David Timm, Auburn University.*

3.0 Specifications and Usage Guidelines

The I-81 pavement recycling project was completed using project-specific specifications. These documents tended to be prescriptive and were used for the first few recycling projects completed by VDOT. To increase the use of pavement recycling techniques for future projects, VDOT thought a set of universal specifications that were performance-based and internal guidance documents were needed. Starting around 2012 and continuing into 2015, VDOT established a task force to produce these documents; the task force consisted of VDOT representatives from the districts and central materials and research divisions. The task force made much use of the experiences gained from the I-81, NCAT, and other recycling projects completed within Virginia. In addition, the experiences and knowledge from other agencies and the recycling and asphalt paving industries were sought and proved to be very valuable. The most difficult parts in completing the specifications and internal guidance documents were determining which parts of the specifications from other agencies were practical and which tended to call on proprietary methods. In addition, the task force worked to ensure that the newly developed documents protected VDOT's interests in achieving quality future recycling projects without being overly burdensome to the contracting community.

3.1 Specifications

Through a collaborative process that included VDOT personnel and local and national contractors with experience in in-place recycling and asphalt mixture paving, VDOT developed a set of specifications for FDR, CIR, and CCPR. The following sections present more detail regarding these specifications and discuss the different paths that could have been taken at each step. The specifications are discussed in terms of materials, mix design, placement, acceptance, weather, time to overlay, and payment.

3.1.1 Materials

Regarding all recycled materials, VDOT's specifications identify and regulate the constituent materials used in the recycled mixtures. These constituents potentially include asphalt binders, virgin aggregates, RAP, water, and hydraulic cement and other fillers. All liquid asphalt materials used are to be included previously on VDOT's approved materials list. This section also includes requirements governing water quality and properties of any additional RAP or virgin aggregate that might be needed in the mix design. These additional properties include presence of deleterious materials, abrasion resistance, sand equivalence, water absorption, and maximum particle size. Of particular interest, VDOT specifies the maximum particle size of any additional RAP or virgin aggregate as 50 mm for FDR and 37.5 mm for CIR and CCPR. Currently, VDOT allows the use of up to 50% virgin aggregate in a recycled mixture following that the above requirements and the mix design properties are met. The use of virgin aggregate was allowed with the rationale that contractors located in rural areas of the state (where less existing RAP is found) could also produce CCPR mixtures. However, the agency assumes that a recycled mixture containing 100% RAP would be less expensive than one in which some portion of virgin aggregate is used.

3.1.2 Mix Design

The specifications for each recycled material project require the contractor to develop a mix design from samples obtained from the project site. The mix design includes the quantity of any recycling/stabilizing agent, fillers (if used), and water, as well as the target gradation and target field density. In addition, the expansion ratio and half-life for materials using foamed asphalt and the minimum breaking time for materials using emulsified asphalt are required.

VDOT requires the mix design for FDR to report the liquid limit, plastic limit, plasticity index, soil classification, and moisture density relationships for all of the materials being reclaimed. These properties can be used by an agency to confirm which stabilizing agent should be used, as indicated by the *Basic Asphalt Recycling Manual* (ARRA, 2015). Since the primary stabilizing agent may be hydraulic cement, foamed asphalt, or emulsified asphalt, there are different strength requirements for each stabilizing agent. When hydraulic cement is used, VDOT requires a 7-day compressive strength of at least 250 psi but not more than 450 psi. Although these values may seem low by comparison to many other agency specifications, they are intentionally set to minimize the potential for reflection cracking. In addition, they follow the recommendation by the Portland Cement Association (PCA) for FDR using cement (Luhr et al., 2005). The requirements for FDR using foamed or emulsified asphalt also require a minimum indirect tensile strength or Marshall stability value, respectively, as described below for CIR and CCPR mixtures.

Since the two recycling agents for CIR and CCPR are foamed asphalt and emulsified asphalt, the performance indicators from the mix design phase were separated into two components depending on the recycling agent employed. Mixtures using either recycling agent type require the contractor to report the moisture density relations (in accordance with AASHTO T 180, Method D) and the gradation (in accordance with AASHTO T 27). Those mixtures using emulsified asphalt are governed by their Marshall stability (ASTM D5581/AASHTO T 245), retained stability (ASTM D5581/AASHTO T 245), raveling (ASTM D7196), and thermal cracking properties (AASHTO T 322). For those mixtures using emulsified asphalt, VDOT requires a minimum Marshall stability of 2,500 lb for 6-inch-diameter specimens and 1,250 lb for 4-inch-diameter specimens. Mixtures using foamed asphalt are governed by their indirect tensile strength (ITS) (AASHTO T 283, Section 11), retained indirect tensile strength (AASHTO T 283, Section 11), and expansion ratio and half-life properties (in accordance with the procedures outlined in Wirtgen Cold Recycling Manual, 2010). VDOT chose to follow recommendations found in Wirtgen (2010) by selecting a dry ITS value of 45 psi (compacted at 75 blows per face in a Marshall hammer or 30 gyrations in a gyratory compactor) and a retained ITS of 70% for mixtures using foamed asphalt.

During construction, Wirtgen (2010) suggests that a washed gradation in accordance with AASHTO T 27 be conducted on materials that have been pulverized but not yet stabilized. This process is sensible for construction but may be problematic in the laboratory. If 4-inch-diameter specimens are to be fabricated, often particles that do not pass a 19-mm sieve are removed and so there is some question on whether this accurately reflects the properties in the field. In addition, many times if a sample of RAP is collected from an existing stockpile or a series of cores are collected from an existing project, they must be broken down into a representative gradation that reflects the process achieved by the recycler/reclaimer. This is often controlled by the experience of the testing lab and may be difficult for an agency to specify.

3.1.3 Placement

For all three recycling processes, VDOT requires the contractor to submit a quality control (QC) plan that includes the responsible personnel for each step of the QC process, a QC sampling plan, a description of the testing laboratory responsible for the QC testing, procedures in the event any corrective action is needed, and a contingency plan outlining the procedures to follow in the event problems arise during construction.

During placement activities, VDOT thought it was not practical to require specific equipment manufacturers/types, so the specifications state that the equipment must be capable of processing the materials to the required depths, automatically metering the dosage of stabilizing or recycling agent, and producing a material that can meet the requirements of the mix design. For CIR, VDOT requires a cold recycler with a down-cutting milling head that can process a full lane width in no more than two passes. The machine must be able to adjust the recycling agent application rate relative to the forward speed of the machine and have the ability to turn off nozzles so that the recycling agent is not applied twice when multiple passes overlap. If any fillers are applied in a dry condition, they must be spread by use of a material distributor truck. For CCPR plants, the specifications call for certain tolerances in the application of recycling agents and any fillers. The CCPR plant must also have the ability to self-report the recycling agent(s) and water quantities used during production. Pavers for CIR and CCPR projects are required to be typical asphalt materials pavers that meet VDOTs traditional asphalt mixture placement requirements.

Since one of the driving factors in achieving a quality project is the density achieved during construction, VDOT decided to require specific roller weights for the recycling projects. For CIR and CCPR projects, two rollers are required at a minimum; at least one being a double steel-wheeled vibratory roller weighing not less than 24,000 lb and at least one being a pneumatic tire roller having a minimum weight of 50,000 lb. At this time, VDOT does not specify roller sizes for FDR projects.

For all three recycling processes, VDOT also requires that a test strip be constructed ranging from 1,000 to 2,500 ft in length. The test strip is to be placed at the design width and thickness, and the design stabilizing/recycling agent content from the mix design is to be used. The quality of the test strip materials is to be evaluated by VDOT before proceeding with the actual project. The location of the test strip may be off-site or on the actual project. The primary drawback to requiring a test strip is that the quality tests required (as discussed in the next section) are often multiday tests. Thus, VDOT decided on this requirement knowing that the contractor would likely factor in multiple mobilizations if equipment is required to construct a test strip and then not begin actual construction for up to another week.

3.1.4 Acceptance

VDOT uses the following attributes to accept the recycled materials during construction: density, depth, gradation, stabilizing/recycling agent dosage rate, stability/strength testing, and binder properties. The density of a constructed layer is required to be measured with a nuclear density gauge operating in direct transmission mode to the full depth of the layer. Diefenderfer and Apeagyei (2014) showed that differences in the bulk density, ITS, and resilient modulus (at certain temperatures) between the top and bottom of cores retrieved from thicker layers could be statistically significant. For relatively thinner layers (e.g., 4 to 5 inches or less), an agency could likely allow the indirect mode as the direct transmission method can be labor intensive when the source-rod pin is driven into compacted material. Currently, VDOT requires the average density of the compacted materials within a 1,000-ft sublot to be 97.0% or greater of the approved mix design density (i.e., the maximum density reported for a given stabilizing/recycling agent and water combination) for 100% payment. The value to which the density measurement is compared can be a point of contention. An agency may choose to require that the comparisons be made to a value derived at mix design to add a level of conservatism to the acceptance process. However, the mix design value represents an idealized condition that may not well represent localized changes in material composition and underlying support, which can influence achieved density. On the other hand, a contractor may wish to be subject to a density requirement based on the maximum density achievable during either a test strip or the beginning of a day's production. A contractor might argue that this field-derived value better represents the conditions available at construction.

Measuring the compacted depth of the recycled layer is often performed by locally trenching a small area and measuring the layer. This process cannot be performed by retrieving a core immediately after construction as the recycled material will likely disintegrate during the coring process because of the limited curing time.

The gradation test can be conducted at many steps during the process; immediately after pulverization, immediately after stabilization, or immediately after paving/compaction. Wirtgen (2010) suggests that a washed gradation in accordance with AASHTO T 27 be conducted on materials that have been pulverized but not yet stabilized. This is easily done for CCPR mixtures in that material may be collected after mixing in the pug mill but without any stabilizer added. However, the process is a bit more difficult for CIR mixtures. To collect a pulverized but not yet stabilized sample, the cold recycler must first pulverize the material in-place for a short distance, and then a sample can be collected. To continue production, the recycler must then back up and re-pulverize the section but add the stabilizing agent. This work can often be performed during changes in the binder tanker, likely once or twice per day, to minimize work interruptions.

The stabilizing/recycling agent dosage rate can be estimated by measuring the quantity of binder used on the project and compared to an approximate amount of materials recycled within the road. Alternatively, an agency may wish to measure the dosage rate directly by testing within an asphalt ignition oven in accordance with AASHTO T 308. While the direct measurement is likely preferred by an agency, the ignition oven will report a binder content that includes both the added binder for the recycling process and the binder from the parent RAP in addition to any RAP aggregate breakdown. This can be overcome by performing two burns and subtracting the binder content of an unstabilized sample from the binder content of a sample that has been stabilized.

VDOT currently requires the strength or stability to be at least 99% of the mix design value for 100% payment. Details of the strength/stability testing were discussed earlier; however, the construction process adds another degree of complexity. Test specimens are often fabricated from loose recycled materials obtained during construction. These may be fabricated by a standard number of blow counts in the field Proctor or Marshall hammer or a standard number of gyrations in a gyratory compactor. If these standard numbers were used during the mix design process, repeating their use here will allow for a check of the strength/stability versus those values from the mix design process. However, using a standard number of blows/gyrations for all projects may not adequately reflect the actual conditions from construction. One alternative would be to fabricate specimens to a desired density (such as the as-placed density), rather than a standard number of blows/gyrations. There is still much debate on which process is the most desirable to an agency.

The properties of the asphalt binder used in recycling projects may be assessed in many ways. Typically, for a project using foamed asphalt, the half-life and expansion ratio are used as the controlling factors. For a project using emulsified asphalt, the binder is usually only required to meet the 98% reliability for the cold temperature at the top of the recycled layer using LTPPBind software. In addition, some agencies may require that a full performance grade (PG) analysis be conducted. One benefit of conducting a PG analysis is that it may allow the agency to determine if a modified binder was used; binder modification is often detrimental to the required foaming properties.

3.1.5 Weather

Although not always addressed by agency specifications, the weather during construction can play a significant role in the construction quality and curing rate of the recycled materials. VDOT currently allows recycling operations when the atmospheric and materials (i.e., the RAP source materials) temperatures are a minimum of 50°F. It is also recommended that material not be placed when the weather forecast calls for freezing temperatures within 48 hours of placement.

3.1.6 Time to Overlay

The time that a contractor must wait until a recycling project can be overlaid with an asphalt mixture course or a surface treatment is often defined qualitatively. Typical specifications may state that the recycled layers may not be covered until the moisture content has reached a maximum of 2% to 3%, 50% of the optimum moisture content, following proof rolling, approval by the project engineer, or some timeframe that may range from two to as many as 21 days. In a conference presentation describing a survey on current practices relating to curing CIR and FDR mixtures that was sent to local and state highway agencies for the Transportation Engineering Road and Research Alliance (TERRA), it was found that while 78% of respondents had some criteria for evaluating curing, 61% stated that determining the time to place an overlay was a concern or an issue within their agency (Cross, 2015).

3.1.7 Payment

In the specifications, VDOT elected to pay for the recycling processes and any stabilizing or recycling agent separately. For design purposes, VDOT requires that a foamed or emulsified asphalt content of 2.5% be used for bidding purposes with any differences from the actual amount used to be adjusted after the project is awarded and the mix design is determined. For FDR, any hydraulic cement, lime, or other cementitious materials are required to be bid at a quantity of 5.0%. Using a predetermined stabilizing/recycling agent content helps in not giving an advantage to a contractor who bids using a lower content. FDR and CIR are paid by the area recycled at the planned depth, and CCPR is paid by ton of recycled material.

3.2 Agency Usage Guidelines

At the same time the specifications were developed, VDOT prepared an internal guidance document to help VDOT pavement designers select those projects where recycling should be considered that would have the highest chance of success. The developed usage guidelines recommend that VDOT pavement designers consider the current distress rating (and its recent history), project length, maintenance history, and existing pavement structure when selecting recycling projects. Specifically, the guidelines suggest that projects identified as requiring reconstruction or rehabilitative maintenance (often denoted as requiring at least 4 inches of pavement manipulation) be initially screened as candidate sites for recycling. For CIR projects, a minimum length of five lanes-miles is suggested; for FDR a minimum length of three lane miles is suggested; there is no minimum length for a CCPR project. The maintenance history should be viewed to determine if previous maintenance efforts did not reach their desired service life or if the amount of patching required exceeds approximately 15% of the project surface area. A combination of the project length, maintenance history, and distress types help to indicate if recycling is a viable alternative.

Since VDOT currently operates within the 1993 AASHTO *Pavement Design Guide* (AASHTO, 1993), VDOT's usage guidelines document currently specifies the structural number for FDR as 0.25 and for CIR and CCPR as 0.35. These values were based on the experiences of other agencies and the results of FWD testing on VDOT recycling projects. No mechanistic values are yet specified.

The guidelines document states that recycling projects be surfaced by one or more layers of an asphalt mixture whose thickness is determined by the required structural capacity from design. For low-volume routes, surface treatments are acceptable. For all non-interstate routes, if the design traffic volume exceeds 10 million ESALs, the recycled layer is to be surfaced with a multilayer asphalt mixture overlay having a minimum thickness of 3.5 inches. For all interstate routes, the recycled layer is to be surfaced with a multilayer asphalt mixture overlay having a minimum thickness of 3.5 inches regardless of traffic. Determining the required overlay thickness is an area of current research study.

4.0 Resources

The documents listed here were found to be helpful in the specification and internal guidance document development process. This should not be considered a complete list of relevant literature; however, much of the relevant literature is cited within these documents.

4.1 Background Information

These documents provide background and insight into the design, construction, and testing associated with in-place recycling technologies, as well as a context for in-place recycling and the state of the practice.

- *Basic Asphalt Recycling Manual (BARM)* by ARRA and the Federal Highway Administration
- *Cold Recycling: Wirtgen Cold Recycling Technology* by the Wirtgen Group.
- *NCHRP Synthesis 421: Recycling and Reclamation of Asphalt Pavements Using In-Place Methods* by Mary Stroup-Gardiner.

4.2 Cost Savings

The following document provides information about potential savings associated with in-place recycling techniques. Nevada DOT cites savings of approximately $600 million over 20 years using CIR and FDR in lieu of traditional reconstruction.

- "Cold In-Place Recycling and Full-Depth Reclamation Projects by Nevada Department of Transportation: State of the Practice" by Sohila Bemanian, Patty Polish, and Gayle Maurer, in *Transportation Research Record: Journal of the Transportation Research Board*, No. 1949, pp. 54–71.

4.3 Longer-Term Performance

- "Ten-Year Performance Review of In Situ Hot-Mix Recycling in Ontario" by Tom Kazmierowski, Pamela Marks, and Stephen Lee, in *Transportation Research Record: Journal of the Transportation Research Board,* No. 1684 pp. 194-202
- "Asphalt Pavement Rehabilitation Treatment: Analysis of 20 Years of Performance" by Dennis A. Morian, Yanqing Zhao, Janice Arellano, and Donald E. Hall, in *Transportation Research Record: Journal of the Transportation Research Board,* No. 1905 pp. 36-43.
- "Initial Performance of Virginia's Interstate 81 In-Place Pavement Recycling Project" by Brian K. Diefenderfer, Benjamin F. Bowers, and Alex K. Apeagyei, in *Transportation Research Record: Journal of the Transportation Research Board,* In Press.

4.4 Life-Cycle Analysis and Energy Savings

- "Life-Cycle Environmental Analysis for Evaluation of Pavement Rehabilitation Options" by Stephen A. Cross, Warren H. Chesner, Henry G. Justus, and Edward R. Kearney, in *Transportation Research Record: Journal of the Transportation Research Board,* No. 2227, pp.43-52.
- "Sustainable Pavements: Environmental, Economic, and Social Benefits of In Situ Pavement Recycling" by Andrew E. Alkins, Becca Lane, and Tom Kazierowski, in *Transportation Research Record: Journal of the Transportation Research Board,* No. 2084, pp. 100-103.
- "Life-cycle Cost Analysis of Base Course Using Cold In-place Recycling: Case Study" by Weiguang Zhang, Jun Yang, Xiaoxiang Fan, Ruochong Yang, and Bin Yu, in *Geo-Frontiers, 2011* pp. 4428-4437.

4.5. Material Properties

The NCHRP Project 9-51 report outlines material properties for CCIR, CCPR, and FDR that can be used in the pavement design process. Cores were collected and characterized from projects around the United States and Canada for analysis.

- NCHRP 9-51 Project Report: *Material Properties for Cold In-Place Recycled and Full-Depth Reclaimed Asphalt Concrete for Pavement Design.* Expected publication date early 2016.

5.0 Summary

Virginia's experience with in-place recycling techniques has been largely positive. There are great potential environmental, economic, and technical benefits to using these approaches in lieu of traditional paving materials. Further, prior to the I-81 project in Virginia, there had been limited literature supporting the use of these recycling techniques in the United States on high truck-volume interstate facilities. The research findings from the I-81 study in conjunction with the ongoing NCAT study support the use of these techniques on roadways with high truck volumes and have actually shown that the material properties of these mixtures may outperform traditional expectations (e.g., structural coefficient values). This paper summarizes the perspectives from the Virginia Department of Transportation in the development of specifications and usage guidelines for in-place recycling projects.

6.0 Acknowledgments

The authors acknowledge the assistance of Angela Beyke, Rob Crandol, Affan Habib, David Kaulfers, Tom Tate, Chaz Weaver, and Mike Wells, VDOT; Sara Link, University of Virginia; Charles Schwartz and Sadaf Khosravifar, University of Maryland; Mike Marshall, Wirtgen GmbH; Todd Thomas, Colas Solutions; Steve Cross, Oklahoma State University; David Stowell, Slurry Pavers, Inc.; Richard Ferron; Buzz Powell, National Center for Asphalt Technology; David Timm, Auburn University; Trenton Clark, Virginia Asphalt Association; Troy Deeds, Donnie Dodds, Benjamin Earl, Ken Elliton, Kim Snead, and Linda Evans, Virginia Center for Transportation Innovation and Research (VCTIR).

7.0 References

Alkins, A. E., Lane, B., and Kazmierowski, T., "Sustainable Pavements: Environmental, Economic, and Social Benefits of In Situ Pavement Recycling," *Transportation Research Record: Journal of the Transportation Research Board, No. 2084*, Transportation Research Board of the National Academies, Washington, DC, 2008, pp. 100–103.

American Association of State Highway and Transportation Officials, *Guide for Design of Pavement Structures*, Washington, DC, 1993.

American Association of State Highway and Transportation Officials, *Standard Specifications for Transportation Materials and Methods of Sampling and Testing*, 33rd Edition, and AASHTO Provisional Standards, Washington, DC, 2013.

Asphalt Academy, *Technical Guideline: Bitumen Stabilized Materials*, 2nd Edition, Pretoria, South Africa, 2009.

ARRA, *Basic Asphalt Recycling Manual*, 2nd Edition, Asphalt Recycling and Reclaiming Association, Annapolis, MD, 2015.

ASTM International, *Annual Book of ASTM Standards, Volume 04.03: Road and Paving Materials; Vehicle-Pavement Systems*, West Conshohocken, PA, 2013a.

ASTM International, *Annual Book of ASTM Standards, Volume 04.08: Soil and Rock*, West Conshohocken, PA, 2013b.

Cross, S., "Survey on Current Practice of Cure Time for CR and FDR." Presentation to 19th Annual Transportation Engineering and Road Research Alliance (TERRA) Pavement Conference, University of Minnesota, accessed: http://www.terraroadalliance.org/events/conference/2015/documents/design_cross.pdf on May 18, 2015.

Cross, S. A., Chesner, W. H., Justus, H. G., and Kearney, E. R., "Life-Cycle Environmental Analysis for Evaluation of Pavement Rehabilitation Options," *Transportation Research Record: Journal of the Transportation Research Board, No. 2227*, Transportation Research Board of the National Academies, Washington, DC, 2011, pp.43–52.

Diefenderfer, B. K., and Apeagyei, A. K., *Analysis of Full-Depth Reclamation Trial Sections in Virginia*, Report No. VCTIR 11-R23, Virginia Center for Transportation Innovation and Research, Charlottesville, 2011.

Diefenderfer, B. K., Apeagyei, A. K., Gallo, A. A., Dougald, L. E., and Weaver, C. B., "In-Place Pavement Recycling on I-81 in Virginia," *Transportation Research Record: Journal of the Transportation Research Board, No. 2306*, Transportation Research Board of the National Academies, Washington, D.C., 2012, pp. 21–27.

Diefenderfer, B. K., and Apeagyei, A. K., I-*81 In-Place Pavement Recycling Project*, Report No. VCTIR 15-R1, Virginia Center for Transportation Innovation and Research, Charlottesville, 2014.

Diefenderfer, B. K., Bowers, B. F., and Apeagyei, A. K., "Initial Performance of Virginia's Interstate 81 In-Place Pavement Recycling Project," Accepted for publication in *Transportation Research Record: Journal of the Transportation Research Board*, Transportation Research Board of the National Academies, Washington, DC, 2015.

Hoppe, E. J., Lane, D. S., Fitch, G. M., and Shetty, S., *Feasibility of Reclaimed Asphalt Pavement (RAP) Use as Road Base and Subbase Material*, Report No. VCTIR 15-R6, Virginia Center for Transportation Innovation and Research, Charlottesville, 2015.

Kazmierowski, T., Marks, P., and Lee, S., "Ten-Year Performance Review of In-Situ Hot-Mix Recycling in Ontario," *Transportation Research Record: Journal of the Transportation Research Board, No. 1684*, Transportation Research Board of the National Academies, Washington, DC, 1999, pp. 194–202.

Luhr, D. R., Adaska, W. S., and Halsted, G. E., *Guide to Full-Depth Reclamation (FDR) with Cement*, Portland Cement Association, Skokie, IL, 2005.

Morian, D.A., Oswalt, J., and Deodhar, A., "Experience with Cold In-Place Recycling as a Reflective Crack Control Technique: Twenty Years Later," *Transportation Research Record: Journal of the Transportation Research Board, No. 1869*, Transportation Research Board of the National Academies, Washington, DC, 2004, pp. 47–55.

Morian, D. A., Zhao, Y., Arellano, J., Hall, D. E., "Asphalt Pavement Rehabilitation Treatment: Analysis of 20 Years of Performance," *Transportation Research Record: Journal of the Transportation Research Board, No. 1905*, Transportation Research Board of the National Academies, Washington, DC, 2005, pp. 36–43.

Nataatmadja, A., "Some Characteristics of Foamed Bitumen Mixes," *Transportation Research Record: Journal of the Transportation Research Board, No. 1767*, Transportation Research Board of the National Academies, Washington, DC, 2001, pp. 120-125.

Stroup-Gardiner, M., *Recycling and Reclamation of Asphalt Pavements Using In-Place Methods*, NCHRP Synthesis 421, Transportation Research Board of the National Academies, Washington, DC, 2011.

Thenoux, G., Gonzalez, A., and Dowling, R., "Energy Consumption Comparison for Different Asphalt Pavements Rehabilitation Techniques Used in Chile," *Resources, Conservation and Recycling*, Vol. 49, 2007, pp. 325–339.

Virginia Department of Transportation, Maintenance Division, *State of the Pavement Report 2012*, Richmond, 2012

Wirtgen GmbH, *Cold Recycling Technology*, 3rd Edition, Windhagen, Germany, 2010.

Zhang, W., Yang, J., Fan, X., Yang, R., and Yu, B., "Life-Cycle Cost Analysis of Base Course Using Cold In-Place Recycling: Case Study," Proceedings of Geo-Frontiers Congress, American Society of Civil Engineers, 2011, pp. 4428–4437.

8.0 Discussion

MICHAEL DUNNING: Okay, you had a central mix plant. Are you able to use also, in lieu of that, an existing plant for asphalt concrete?

BENJAMIN BOWERS: This has been attempted in a traditional plant but to my knowledge it has not been successfully performed. In the central plant you're using a pugmill, so I think it probably depends on your setup. You're also not applying any heat whatsoever, so that's the other thing to consider.

MICHAEL DUNNING: Right. That sounds good. We have drum mix plants, not pugmills.

BENJAMIN BOWERS: Yes.

MICHAEL DUNNING: The reason why I bring it up is that most of our large contractors are actually suppliers. To get their project, it would be better for them to go back to their plant than make a central plant. So it sounds like they need a pugmill versus a drum plant?

BENJAMIN BOWERS: Yes, it's pugmill versus a drum.

MICHAEL DUNNING: I mean, you could not use a drum?

BENJAMIN BOWERS: I don't think so. That is not to say that it couldn't be successful at some point, but likely with significant plant modifications.

MICHAEL DUNNING: Okay.

GABRIELE TEBALDI: You have to use a specific plant, because otherwise the risk to do a disaster is right behind the corner. Because with cold recycling, you have to deal with water and you have to deal with filler. So you need a plant that allows you to manage those two elements, and generally the old plant does not. So it's better to have a specific design and equipment for that.

BENJAMIN BOWERS: Thanks, Gabriele.

MIKE HEITZMAN: Ben, good presentation. I think we as the asphalt community need to start looking at those mechanical properties of these recycling and rehabilitation techniques. My comment to you on your list of acceptance criteria, why was smoothness not included?

BENJAMIN BOWERS: I definitely think that's something that does need to be considered, and so retroactively I may add that. Thank you for pointing that out.

MIKE HEITZMAN: Right. And what we're doing in Iowa, I told my cold-in-place recycler, "I don't want a roller coaster when I have to put my asphalt down, so I'm going to set some criteria that you're going to have to live to." It won't be the same as the asphalt, but we have to have a fairly good template when we get done with that.

BENJAMIN BOWERS: Absolutely. That's a great point. Thank you.

JOHN HARVEY: Great summary. A question, and it may go to NCAT as well. When you're looking at your FDR sections, are you tracking damage through your back-calculated stiffnesses? On our test sections in California, we've been pretty amazed that the asphalt may crack off the top because we have very thin overlays, but in our FDR with foamed asphalt and cement we're not picking up any damage after up to eight years.

BENJAMIN BOWERS: Right. We are not seeing damage at this point. We've followed the Interstate-81 test sections for three years and only found an increase in stiffness when considering the entire pavement structure, i.e. we did not back-calculate the stiffness for the FDR layer only. However, these do have thicker overlays of 3-plus inches.

JOHN HARVEY: That ties into those mechanical performance equations.

BENJAMIN BOWERS: Right.

JOHN HARVEY: Okay, thanks.

BENJAMIN BOWERS: Do you have a comment, Randy?

RANDY WEST: Just to answer John's question about what we're monitoring at the test track sections using the CCPR sections. We are monitoring the back-calculated properties, as well as measuring strain response of the sections under traffic. And same as you have observed in your sections, there's been no deterioration so far, so that's why we're getting this great performance that Ben has mentioned. And we're going to continue to monitor the structural responses and do the back calculations to see if we see that deterioration as you're describing.

JOHN HARVEY: That's great. If we can get these into the mechanistic design procedures and get them out into the mainstream as fast as possible, I think that's pretty important.

BENJAMIN BOWERS: Absolutely. Thank you. I would also like to note that the NCHRP 9-51 *Material Properties of Cold In-Place Recycled and Full Depth Reclamation Asphalt Concrete for Pavement Design* study will provide inputs for stiffness and rutting resistance. But there are no deterioration models for these pavements.

ELIE HAJJ: For the cold-in place recycling, do you have any target in-place air voids or density, and if you do, how are in-place air voids being measured?

BENJAMIN BOWERS: The density we've been targeting is 132 pcf-ish. The average from the NCHRP 9-51 study was approximately 129 pcf.

ELIE HAJJ: Right, but in terms of air voids, any target or specified value?

BENJAMIN BOWERS: Oh, in terms of air voids, I don't remember off the top of my head. And, again, I wasn't at the project.

ELIE HAJJ: The key is how to measure the in-place air voids of CIR because you cannot use a nuclear gauge density. In some cases, it was hard to reach the appropriate air voids level. We have seen some CIR mixtures with in-place air voids in the range of 20 to 25%.

BENJAMIN BOWERS: Okay.

ELIE HAJJ: Some contractors do a very good job and they get in-place air voids down in the range of 13%.

BENJAMIN BOWERS: Right.

ELIE HAJJ: Additionally for a performance-based mix design you will need to target the appropriate air void levels of the CIR.

BENJAMIN BOWERS: Right.

ELIE HAJJ: From our experience in Nevada where CIR has been used for some time, we start seeing field failure with the CIR the minute you get to the high in-place air void levels.

BENJAMIN BOWERS: I believe that, yeah. And my understanding is we did use the nuclear gauge to monitor that.

ELIE HAJJ: On the CIR?

BENJAMIN BOWERS: Yes.

ELIE HAJJ: I believe you will have some issues to get the appropriate measurements for in-place air voids. That's something to look into, especially in the QC/QA process, what would be the appropriate method to get an early indication of the in-place air voids.

BENJAMIN BOWERS: As opposed to asphalt mixtures, industry practice is that the air voids are not the controlling property, density is. When using the nuclear density gauge the operator must keep in mind that the density will likely change with depth, so direct transmission is preferred. This has worked effectively for us. Thank you.

MARTIN VAN DE VEN: Yes, I would like to ask a question, I did not hear about curing at all. These are mixes you are talking about?

BENJAMIN BOWERS: So in our experience with curing, with the foaming process we haven't had to wait very long. With the emulsion, of course, you want to make sure it breaks, so you'd have to wait for that to occur and monitor that. However, some emulsions can be set to also break very quickly. But each recycling agent cures over time.

MARTIN VAN DE VEN: Exactly, yeah. But my question was about curing and how long you need for that before placing the next layer.

BENJAMIN BOWERS: So we cored, and if you core too quickly, you pull it out and it falls apart.

MARTIN VAN DE VEN: Yah, yah.

BENJAMIN BOWERS: But if you wait a couple of weeks the cores will stay together… But at that point you've already paved the surface over and they're driving on it.

MARTIN VAN DE VEN: That's what I wanted to know. So you pave very fast after production? This can be very dangerous.

BENJAMIN BOWERS: Yeah, pretty quickly. You don't hang out for a long time before you pave.

MARTIN VAN DE VEN: Okay, thanks. But be careful with bringing the next layer too fast in time.

BENJAMIN BOWERS: But, again, we're kind of getting to the….. I see you over here swaying. I mean, you know, there's definitely differing opinions on that. From the Interstate-81 project, the first overlay went on about 12-15 hours after the CCPR and CIR were placed, which is very fast. Many agencies require a "curing" period of anywhere from three to seven days which can significantly lengthen a project and may not always be needed, depending on the recycling agent.

GABRIELE TEBALDI: Generally, it is better to wait at least to the optimum moisture content go down to 50%.

BENJAMIN BOWERS: Right. For lack of anything else to specify, the general requirements for moisture are usually to wait until the moisture content is 50% of optimum or 2-2.5%.

GABRIELE TEBALDI: And one quick comment, be careful with the compaction because the compaction you have in the field is not exactly what you have in the lab. The compaction in the lab with this kind mix is quite a black hole right now because how to manage the water in the compaction in the lab is something that must be investigated.

BENJAMIN BOWERS: Right.

MARTIN VAN DE VEN: So I think curing is an important issue. Thank you.

BENJAMIN BOWERS: Jim, did you have a question?

JIM SCHEROCMAN: Yes, sir. Most pavements that need rehabilitation are in pretty miserable shape and have significant differences in their performance or their condition from one location to another. Do you have any recommendations as to how often you should sample the existing pavement to check the thickness of the layers or the condition of the layers or how much mud has pumped up into the aggregate base for example? Do we just take one test per mile or what do we do?

BENJAMIN BOWERS: I don't have a recommendation for that. I would do it with relative frequency. I mean, you want to make sure you're not getting two miles down the road and all of a sudden you're only getting 3 inches not 5, but in terms of that frequency, I think it's kind of up the person, the inspector on the job, people of that nature. Does that answer your question? The new *Basic Asphalt Recycling Manual* from the Asphalt Recycling and Reclaiming Association has recommendations for this.

JIM SCHEROCMAN: Doesn't it have to be done before the project is let to contract rather than letting an inspector decide at the time of construction?

BENJAMIN BOWERS: Oh, I see. So you're asking, depending on how large your job is, how often you need to core and check what you have?

JIM SCHEROCMAN How do you determine what's already there so you know what to do?

BENJAMIN BOWERS: We've been coring. So we identify what the sections are and we have cores frequently throughout those sections. And for the most part, unless, of course, you're dealing with like a continually reinforced concrete, which is going to be its own bear.

JIM SCHEROCMAN Please don't swear.

BENJAMIN BOWERS: I know, bad words in this room. But what I've seen in my limited experience is that we've been coring to see what we've had on these sections, and, sure, there are areas that may not be fit for these processes. We've seen projects where CIR was proposed that were really better suited for FDR because the existing structure wasn't very thick. But in terms of the frequency, that has been left up to the district, who goes out and does the coring as they would any time they're going to go out and do any kind of rehabilitative maintenance.

JIM SCHEROCMAN: Many times there's a lot more variability in the in-place roadway than we want to have and it obviously affects what we can do.

BENJAMIN BOWERS: I agree with that.

STACEY DIEFENDERFER: I was just going to respond to Jim's comment because I was around when the I-81 project and a number of others were constructed. What VDOT does, generally, in choosing the project is to follow a set of recommendations developed by the agency which includes looking at maintenance records. Then, a lot of times we run GPR and core to get a good indication of what the pavement structure is prior to actually setting up the contract. VDOT has been going out and doing pretty extensive investigation for these types of projects. Once the contractor gets the project, they are responsible for developing the mix design but only after the agency determines that it has the potential to be successfully completed. So there's work done throughout the process and a good bit of it is done up front so that you can get a reasonable bid and have reasonable expectations from the get-go and then that's expanded during the beginnings of the project working with the contractor to make sure that it is successful. So that sort of information is available.

AAPT Symposium: Rejuvenators

Gaylon Baumgardner and Jo Daniel, Presiding

Asphalt Rejuvenation - A Historical Perspective
Steve Escobar, *Asphalt Pavement & Recycling Technologies*

Asphalt Rejuvenation - A Contractor Perspective
Colin Durante, *Pavement Technology, Inc.*

Asphalt Rejuvenation - An Agency Perspective
Louay Mohammad, *Louisiana State University*

Dose Optimization - Balancing Stiffness and Flexibility
Bob Frank, *RAP Technologies LLC & Compliance Monitoring*

Rejuvenator Characterization, Blend Characteristics, and Proposed Mix Design Method
Fujie Zhou, Soohyok Im, *Texas A&M University*; David Morton, *APAC-Texas*; Robert Lee, *Texas Department of Transportation*; Sheng Hu, Tom Scullion, *Texas A&M University*

Investigating the Aging Mitigation Capabilities of Rejuvenators in High RAP Mixtures Using Black Space Diagrams, Binder Rheology and Mixture Tests
Walaa Mogawer, *University of Massachusetts Highway Sustainability Research Center*; Thomas Bennert, *Rutgers University Center for Advanced Infrastructure and Transportation*; Alexander Austerman, *University of Massachusetts Highway Sustainability Research Center*; Christopher Ericson, *Rutgers University Center for Advanced Infrastructure and Transportation*

Discussion—Louay Mohammad

GAYLE KING: Louay, nice presentation. One slide identified 'slurry oil' as one of the rejuvenators.

LOUAY MOHAMMAD: Okay.

GAYLE KING: To a refiner like Bob McGennis, "slurry oil" would be the decant oil bottoms from a catalytic cracker. Such an oil would contain many un-branched polycyclic aromatic molecules, and would be accompanied by a Material Safety Data Sheet detailing serious potential health risks. Most refiners no longer make "slurry oil" available to the market. Do you know if this product really comes from cat cracker bottoms?

LOUAY MOHAMMAD: The slurry oil shown in the slide was a type of softening agents. However, as the next slide showed this agent was not used in the results presented

GAYLE KING: I hope you said no.

LOUAY MOHAMMAD: Well, I said that this type of softening agent was not used. Further, recycling agents used in the laboratory follows required environmental handling procedures.

GAYLE KING: And use is with very good ventilation, like fume hoods. Fumes are the biggest problem.

LOUAY MOHAMMAD: Yes, following proper environmental and occupational requirements such as proper attire, ventilation and mask, etc.

GAYLE KING: Seriously, if your MSDS has those warnings, it shouldn't be used in the field anyway. I would recommend removing that product from the laboratory and take the term "slurry oil" off the slide.

LOUAY MOHAMMAD: As stated earlier, this agent was not used in the study.

GAYLE KING: Let me add a second health caution for the audience. I recently reviewed a paper in which coal derived oils (coming from the destructive distillation of coal) were being used as asphalt rejuvenators. These coal tar oils typically contain Benzo-a-pyrenes, which are much more hazardous yet than PNAs in cracked petroleum streams like "slurry oil". Worker health studies in Europe found high health risks when coal tar was added to paving grade asphalt. MSDS warnings are important.

LOUAY MOHAMMAD: No materials will enter our laboratory without their MSDS sheets.

GAYLE KING: Very good.

LOUAY MOHAMMAD: We pay a great deal of attention to handling of materials.

GAYLE KING: I know you do. That's an important comment for everyone here.

LOUAY MOHAMMAD: This was meant to be a literature review of candidate recycling agents.

GAYLE KING: Okay. Understood.

LOUAY MOHAMMAD: But thanks for the comment.

Discussion—Bob Frank

AMIR GOLALIPOUR: Thanks for the interesting presentation. I had just a quick comment. You mentioned that basically using the bio-based oils or rejuvenators are more effective than the petroleum ones. Typically, that trend is correct, but as you showed the trend is linear so it's basically a viscosity driven trend so it depends on the viscosity of your oil. So we did a study in UW MARC under ARC project two years ago. And if your petroleum oil has a lower viscosity than bio-based oil, that trend in effect is changing basically. So the linear trend that you showed is solely based on the viscosity of your oil. So if your viscosity of oil is lower, basically, you will have more efficient oil.

BOB FRANK: Thank you very much for your comment. Some of Martin Zaumanis' work was viscosity based and spoke to that question. His research indicated that there might be more going on, that when we're looking at different recycling agent chemistries, it's not just a simple matter of viscosity blending. We can't just use a viscosity blending chart to determine optimum dose. A blending chart might be sufficient to product anticipated viscosity of a particular combination.

AMIR GOLALIPOUR: When I mentioned the viscosity, even the rheological properties like complex modulus, it is basically a relation with the viscosity. When I mentioned viscosity, I meant basically the stiffness of that oil has a different chemical interaction, but usually when you're just looking at the original basically blending, that effect should be linear. We did also a study with aging, and we basically developed some of the models that can take into effect the aging effect of that.

BOB FRANK: Okay. Thank you for that.

AMIR GOLALIPOUR: Thank you.

Panel Discussion

GEOFFREY ROWE:... in the SHRP spec. By default, the very fact that we have an intermediate temperature at a PAV condition and bending beam data, we actually have by default built the R value into the specification. It was in the Strawman specs and the concept of having a temperature susceptibility parameter or a rheological index was in that Strawman spec that came out. They didn't like that for implementation, but it did actually get in there by the backdoor, so we had it in there. And then I look at other things going on in Europe, and round about 2000 there were some papers published looking at the phase angle in roads in France and how that related to cracking. Some of my former colleagues in the UK picked that up and they came up with this concept of a viscoelastic transition temperature, and they said that's related to cracking. So I looked at this and I said, "Well, guess what. The VET temperature that they have is no more than the R value." So R value, believe it or not, is a very important parameter for trying to characterize the values of the performance of the binder and its propensity to crack. So when we rejuvenate, we want to basically look at trying to restore our rheological properties, trying to restore the R value. And this was the concept that we had. And Charles Glover was working on and other folks like Mike Anderson and his team, trying to understand how these things change and age in a black space type environment where you consider both the rheological index, the R value, the phase angle, the stiffness modulus. So this is where this all comes from. There's a big amount of work behind it, so it's not just something which has been a developed in a vacuum. It includes papers that go back to the 1980s with work done in France, work done in Europe, even in the SHRP context.

So if I looked at the discussion that Frank provided us with the binders that you were looking at, if you were to look at those instead of looking at them as individual properties, I'm certain you would see those ones that you had no delta there. It would basically be the fact that you haven't restored the R value and the crossover frequency in the correct way. So there's a lot of information here. What we have to do as a community is put this all together, and I really think it's a very positive symposium and I think we can do a lot with the information presented. So that's my discussion, and I'd welcome any comments back.

NIMA ROOHI: Thank you, Geoff. Any comments from the panel? Thank you Geoff.

JOHN D'ANGELO: Geoff, you're right there has been significant work on the rheological parameters, but I'll tell you the one thing, I think the mistake is being made when we look at the R value and the crossover frequency, at a set temperature all the time. And that's where the big mistake is. The crossover frequency and the R value, it's a geometry issue, and if the master curve is shifted at a different temperature, it's going to change some. So I think if we're really going to look at these parameters, it has to be done at equal stiffness so the temperatures are not always going to be the same. So when looking at that equal stiffness value, then the crossover frequency and the R value start to really make much better comparisons than just saying everything's going to be at 20 degrees or 25. It's really got to be tied to the stiffness value.

GEOFFREY ROWE: John, you make a very good point. The important thing about the R value – and I think Don said this many times in his work and we've done this – to look at this again is that if you're fitting a rheological index to a binder property, you effectively only need one or two points and you can get the rheological index. But the important thing is you have to be in a certain stiffness range. So I wrote this into a CTAA paper just recently. If we keep within a stiffness range of about 10^5 to 10^9 Pascals, then we can fit a Christensen-Anderson model very well and it really doesn't matter how you determine that R value. That's the key range. We've set up several procedures for doing that, one with a bending beam and the intermediate temperature data. Mike Anderson used a different procedure in the AAPT paper he presented. We can it several different ways. The important thing is you have to be within that stiffness region.

GAYLE KING: Referring to Dr. Rowe's earlier comments, I would add a caution. The existing Glover-Rowe failure envelope was developed for a PG 58-28 climate.

The PG specifications for rutting and thermal cracking adjust the DSR/BBR test temperatures to the high and low pavement temperature for the local climate. Likewise, test parameters for the Glover-Rowe parameter must be tied to climate if G-R is to be used as a predictive parameter for age-induced surface cracking. Because G-R captures cracking damage in an aged surface, PAV aging time needs to be extended to more accurately reflect the physical state of the binder when cracking occurs. Much age-induced cracking research on binders is now done after 40 hour PAV, or longer.

This recycled binder formulation concept which targets cracking with an aging-time-to-failure is novel. Previous blending protocols emphasized high temperature rutting modulus, but failed to capture the problematic lower binder phase angles that come with oxidation. Nor were long-term aging tests used to predict cracking damage in either binders or mixtures. It's convenient to see critical blending data in one column where it can be easily assessed. Well done.

Fujie, you answered Frank Fee's question correctly, in that one step in the design process did blend the extracted binder from tear-off shingles with the virgin binder before testing. I think everyone in the room believes that the highly oxidized asphalt in TOAS will not completely mix with a rejuvenator or virgin binder in an asphalt recycling process, maybe not at all. I sense the audience may not understand the purpose of this test within the context of NCHRP 9-58's broader binder formulation strategy, so I would like to extend the answer Frank with a YES, but.....

The project uses a fundamental premise for performance. A recycled binder blend satisfy minimum rutting requirements, yet maximize the pavement service life before age-induced surface cracking becomes a problem. A second premise is that the virgin binder will act as the mobile or continuous phase that dominates mixture performance. But that mobile phase will stiffen as it blends with the aged asphalt, either during HMA mixing or more gradually after the mixture cures. Oxidation will cause the mobile phase of the binder to stiffen even more. Rutting is most likely to occur very early in the pavement life, before aging stiffens the binder, and before the softer virgin binder or rejuvenator become fully blend with the aged asphalt. To inhibit cracking, softer binders with higher phase angles are preferred. But complete blending yields the hardest initial binder, and a phase angle that has been substantially reduced by the presence of highly oxidized molecules. Ongoing oxidation in the pavement further compounds the problem. The Glover-Rowe parameter appears to rank the path to damage quite well if the binder is fully blended. Hence, long-term aging for a completely blending binder represents a worst-case-scenario for cracking. This test provides a target for the minimum amount of virgin binder that might be needed to satisfy pavement life cycle needs for surface cracking. A mixture rutting test, such as Hamburg WTD, sets the maximum amount of soft binder that can be tolerated for rutting purposes. With max and min limits set on binder ratios, long-term aging studies can be run on various mixture blends to evaluate cracking performance. Hopefully, laboratory mixture preparation and testing protocols will simulate field performance regardless of the amount of binder blending that occurs. Frank, does this answer your question, and are we on the right track?

FRANK FEE: The point that I think you need to deal with, in this lab evaluation stuff, is you have to try and understand – does it match reality? Can a plant produce that? Or the other way around – what does a plant produce? What kind of time and temperature is a plant going to provide? Where is that material? How long is it going to be in the silo? How are these materials going to react under those conditions? You're working in a vacuum in a lab. You set your own parameters. Do they meet reality? My concern here is to work backwards a little bit. Find out what's happening in the plant. Figure out what kind of conditioning happens from the start of the mixing bowl to the thing that came out of the plant and come to an understanding of that. And that's my point. You're doing great work, but is it in the realm of reality? That's my question.

FUJIE ZHOU: I would like to have Dr. Amy Epps Martin to address this issue because this is an ongoing work under NCHRP 9-58. Amy, go ahead.

AMY EPPS MARTIN: No, Frank. We are looking at reality. We're trying to use some of what we got from 9-52 in 9-58, so we are tying the conditioning and the aging in different climates and for different construction dates to the testing that we'll be doing on binders, mortars—I want to mention that we're doing the mortar testing that was developed in ARC at UNR—and mixtures. So we're trying to be as comprehensive as possible, but as you all know making an experiment design that you can complete with all of these variables is difficult. So we're going to try. But we are considering that, Frank, in the project that is using the materials from the Texas field project that Fujie talked about.

FUJIE ZHOU: One thing I would also add is on the NCHRP 9-58, we did compact the specimens at the plant with the mix just out of the silo. We tried as best as possible to simulate what's happening in the field. We also brought plant mix back to the lab, reheated and compacted the mix, and then characterized all the mixes. also in the third step, we took the cores just after the field compaction and we're taking cores again after one year of service, maybe two year, so we try our best to see how we match the lab to the field under NCHRP9-58.

DALE DECKER: And I actually have a question, not a dissertation, so you'll be happy to know that. The question is for Bob. You assume some linearity, some lack of variability, if you will, in some of the products. Contractors are regularly abused about not knowing the variability of RAP—questioned maybe is a better way to say it—and you assumed that there was linearity in the waste engine oil, as well as the waste vegetable oil. And my experience has been that those products are very variable—not very uniform is what I'm trying to say—and I just would like for you, if you would, please, to comment on what you've seen in terms of variability for the waste engine oil and the waste vegetable oil.

BOB FRANK: To be clear, the linearity was in the rheological properties for one project at different doses.

DALE DECKER: I understand that, but what you have seen overall?

BOB FRANK: In terms of variability between products, it's been my experience with waste oil, for example, because we have a lot more experience as an industry with the waste oil, primarily as a fuel rather than as a recycling agent.

DALE DECKER: Exactly.

BOB FRANK: And there we found that the variability was a function of the supplier and the markets that he served in his collection. When you identify a supplier that just does the Jiffy Lubes and doesn't do metal working fluids, his product is very uniform and of a high quality. It really goes to your sourcing being the best way to control the variability in these waste-derived products.

DALE DECKER: And that's my concern is because I have seen quite a bit of variability chemically, residue, and a lot of different things in terms of the waste oils that depending on whether it's an industrial waste oil or, as you say, Jiffy Lube. So what about the vegetable oil, is it the same issue? Is that McDonald's oil?

BOB FRANK: Less so. And, again, it depends on the supplier. But on the vegetable oil market, you have a very sophisticated market that has been put in place to feed the biodiesel industry that has very high quality standards. You have a supply chain on the vegetable oil side that generates a very consistent and uniform product because it's going into the biodiesel industry.

DALE DECKER: Okay. Thank you very much.

RANDY WEST: My question is to Mr. Calvert. I think that in your presentation you talked about the spray application of rejuvenators onto roadways to restore, basically to rejuvenate, the surfaces. So my question that I have is there is an issue with skid resistance for the pavements after they've had this application and how is that handled?

JOHN CALVERT: Good question. And, yes, there is an impact on skid values. The degree of impact, I guess, as far as the seriousness of it, basically, it goes to the posted speed limit of the roads that you've treated and how much did you drop the skid numbers and how long do they stay down? Our experience has been it's a very temporary drop in the skid values. Most of the application of rejuvenators across the nation have been on residential streets, which is the bulk of asphalt pavements across the nation. And we drop the skid values, say, 10 or 15 points for a day or two, but when you've got vehicles moving 20–25 miles per hour, it's not really a problem. Now when we get on higher speed roadways, what we do ahead of time, is we will run skid numbers on those pavements to see what are we dealing with. If we have initial skid numbers, say, in the 40–45 range, we'll address those before we put the rejuvenator down by surface abrading, doing shot blasting to bring the skid values up to the 60-70 ranges. What we've seen with some projects that we've done we started out with low skid numbers in the mid-30s. We took those numbers with the surface abrasion up to high 60s, and after the rejuvenator application, we were still in the high 40s, so it wasn't an issue. But almost any product that you put on a pavement, you're going to lower the skid numbers. It's a matter of how long are they going to stay down and do you have to do something to address them?

RANDY WEST: Okay. Thank you.

PAVEL KRIZ: I would like to take it back to fundamentals. I really don't have a question, rather a comment. I mean, all afternoon you were talking about rheology. You're trying to adjust the consistency with various percentages of various oils, but we really didn't talk about chemistry and I think that's very important. We talk about rejuvenation, but, in fact, this is no rejuvenation. I think one of you had explanation of what rejuvenation is from the dictionary. And when asphalt ages on the road or on the roof, those chemical changes are irreversible and permanent. We changed the molecules and they will never be new molecules again. Those molecules will be different. And what do we do? We put oil in it to dilute the system so we more or less same rheology as we had 20 years before. But the system is very different, and at least to a lot of challenges with phase separation, compatibility, and aggregate absorption and so on. So I just want to really make sure this forum understands that it's not about finding an oil and the percentage to hit the right rheology. We have a lot of issues in Canada with roads which use some rejuvenators and they crack in cold climate because there were some incompatibilities. Of course, the sources of those incompatibilities or challenges are not known or not that well understood. But this is something we really should look into before we go to roads where we use 10% of RAS with 15% of some oil we got God knows from where. Okay? Thank you.

LOUAY MOHAMMAD: I want to basically comment back to you. I recommend that you attend the Wednesday afternoon session where my colleague, Bill Daly, will present chemistry analysis of various binder compositions and their relationship, or lack off, to mixture intermediate temperature cracking performance.

WALAA MOGAWER: I just want to say I'm a civil engineer, so I'm not going to claim I know chemistry very well. But in our paper and in the slides I added a comment that we need to look at chemical analysis. I really don't know just from the testing if it's restoring what it lost. So we're using whatever tools are available, then when I get the products, I don't know what's in it. So someone will give me a product and tell me it's putting maltenes back in. I have no way of knowing that. But I agree with you that chemical analysis should supplement what we're doing.

MICHAEL DUNNING: I appreciate the comment before with Randy. And also to talk about your mentioning why we don't have as many areas that have rejuvenator going down as far as agencies. And as a comment, being an agency, we tried rejuvenator before. Maybe in those 25 mph zones, unfortunately, our residents go faster than that and they'll slide right through intersections and we've had accidents. And what happens is the commissioner gets notified, the director gets notified. It goes down the road to division manager and then I get notified. And so, of course, we say we're not going to use that. I appreciate… I would love to be able to use it in the future. I see in your demonstration here what it can do, and I'm sold on that. So now if I go up there again and we figure this all out. We don't have any skid tester machine. I guess we could go on our foot and kind of measure it that way and figure it out. But if we could figure it out, then I would like to use it again. So my question is then, at what point in time would you place it on a road? A fairly new road or what?

JOHN CALVERT: The point of using the rejuvenator is to try to catch that pavement while it's still in excellent condition, typically one to five years old, before you start seeing distress or raveling or signs of cracking. And a lot of agencies will take the roads that they had paved five years previously and that becomes their rejuvenator list for that year. And we've seen excellent results with that in the 43 years we've been doing this. The skid problems, a lot of it depends on the contractor doing the work. Did they do any testing ahead of time to see what was the actual dosage needed for that pavement? You can't treat every street at the same rate.

MICHAEL DUNNING: Yes, right. Yes.

JOHN CALVERT: And did they overshoot it a little bit and left too much material on the surface? But there are ways to mitigate that. And very often what we'll do when we do residential streets just to be sure there's not a skid issue, we will lightly dust the pavement with some sand.

MICHAEL DUNNING: Oh, yes. We did the same thing.

JOHN CALVERT: About 1-2 pounds per square yard and then we come back and remove it all. But, normally, the only times we've ever gotten into an issue where we had to work harder to get skid numbers back were when we got into a pavement that had a real low air-void content and it just would not go into the pavement, and then we actually had to put a blotter material out there to kind of wick up the material and get it off the surface, but that's very odd when that that happens. But your comments are right on base. Thank you.

MICHAEL DUNNING: I appreciate the ring test, as well, because another problem we have there is we have half-street sections for developments. They may change every 200 feet or half mile or quarter mile, so we have to do a ring test for every section and you can't just leave your truck and go all the way down for a half mile, which costs time and money and frustration, you might say. Well, thank you. I have another question. As far as the 92% RAP was used, I think Steve mentioned that. You mentioned replacing the Los Angeles at 92% RAP, how often would you have to do that? I guess I got the wrong person. Okay, yeah. How often would have to do that as far as replacing the road, every couple years, five years, 10 years with the 92% RAP?

BOB FRANK: The program goes very close to the standard, which is about a 10-15 year. Then they start the rehabilitation process. Basically, slurry seals is the big one in this area, if it's in residential. If it's in a major artery, they will look at overlays or whatever is appropriate at that time.

MICHAEL DUNNING: Okay. Thank you.

JOHN D'ANGELO: I've got a question I'm going to throw out there and hopefully I will get a free-for-all from all of you on this. Yesterday, the first paper right after lunch was the effects of mineral filler on changes in molecular size distribution of asphalt during oxidative aging, and part of that discussion was – and it's something we've all looked at for a long time—is that these very highly polar molecules bond onto the aggregates and guess what? They're extremely hard to get off. I mean even to the point when we try to extract them, it's hard to get them off. And I know a lot of people know a lot more about chemistry than me, so I'll open it up to them and you guys and whatever to discuss this. But then, so when we look at trying to rejuvenate, it's not just a matter of do I put a rejuvenator in an extracted asphalt and check the binder properties, I have to look at how that's going to react, how that material is bonded to aggregate, what's going on with that. I think Pavel has done some great work on diffusion of those things and maybe we'll get him to discuss it more. But, again, he did it based on just the binders, not one of those materials bonded to the aggregate. And I think it has a big effect, and I think it's key why some of the mix testing is really more important to figure this out, and then it has to be aged. But now I've just done that and I'll let you guys go for it.

WALAA MOGAWER: Thanks for the comment, John.

JOHN D'ANGELO: It wasn't a comment it was a question.

GALE PAGE: John, I didn't understand your comment as question either. Regardless, we've been doing hot mix recycling in Florida since the late 1970s, and I don't think we've ever had a failure due to the recycling process, and I think that's because we did it right from the very beginning. But I would like to address the whole issue of spray application of rejuvenators and its lack of use by agencies where you're trying to rejuvenate the old surface. I can tell you that Florida has asked people to come down and demonstrate their products. I know at least since the late 1970s and as recently, I think, as 10 years ago Florida has evaluated products. It's not that we don't have a market. We've got a big market. The shoulders on our Interstates usually are not milled out during a rehabilitation project. But we like our shoulders to look nice afterwards, and we have asked people to come down and try their products out to give us a rejuvenated shoulder, make it look it black, and assure we don't have cars slipping off. We've even had products where they've put a bauxite treatment on top of the rejuvenator to assure skid resistance. But yet, as I recall, at least three of studies that we did on the Interstates there was not much rejuvenation of the surface and there were problems with skid resistance and problems in the construction process. So, again, it's not that the agencies don't have a market, we need a product that works.

NIMA ROOHI: All right, thank you for the comment. You want to respond to that? Okay.

JOHN CALVERT: Thanks, Gale. We participated in some of those studies with FDOT, and you're correct. Depending on the type of product that's referred to as rejuvenator, you can drastically and permanently lower your skid numbers. And what our experience has been is the maltene-based rejuvenator had a minimal impact on the skid numbers or a shorter impact on the skid numbers in the time that it took to get back to the original than other products. The shoulders is a place that we're always leery of because that's a place when the pavements are wet and people get on the shoulder, they're trying to brake. So you have to watch your skid numbers. Thank you.

NIMA ROOHI: All right. I have just one question. On the aging study that you did, you guys are proposing that aging effect to be included in selection of rejuvenating agent. Is that right?

FUJIE ZHOU: We think aging the blended binder (including extracted RAP binder or RAS binder, possibly rejuvenator, and virgin binder). The goal is to try to get a similar aging resistance to the virgin binder. That's the whole goal. And we do think that it's a very important component when you're evaluating the blended binder.

NIMA ROOHI: All right, but practically speaking, are you proposing like—I'm destroying the words here—but are you proposing including … like taking a look at the mass loss as a representation, taking a look at the RTFO aging process? Because you were talking about different hours of aging, which I know how much people are going to like them to wait that long plus running lots of tests. And I thought maybe… I don't know if you've done this or not, but you have plotted your G star or phase angle versus number of hours and using that trend that you get, you may just need like two tests. And so if it's like linear you have two tests, two data points. And then from that you can figure out if your aging susceptibility is sufficiently good for you or not. But I was wondering if you have any comments on the practical aspect. If you're going to take into account this as part of binder selection in the mix design process, how you're going to consider that order.

FUJIE ZHOU: At this stage, we're talking about research. We're trying to do whatever we can to best characterize the blended binder and also select a dosage rate of rejuvenator. But after you do more and more research and finally you may pick up one simplified procedure, as you just mentioned. But for the time being, in the research stage we need a deeper study and then you can finalize it with more and more data accumulated. That's my thinking.

NIMA ROOHI: Okay, thank you.

GAYLE KING: Fujie, if you don't mind, I'd like to add to that. The PAV aging test was set at 20 hours primarily as a compromise so binder certifications could be delivered one day after sampling. At 20 hours, the rate of oxidation is relatively constant, so binder properties at longer aging times can be extrapolated with moderate success if the rate is known. Unfortunately, two oxidation mechanisms of differing rates control early aging, so the entire 0-20 hour PAV aging period should not be extrapolated to longer aging times. However, typical binders do not reach the physical damage that caused age-induced surface cracking until the asphalt has aged in the PAV oven for 40 to 60 hours. I think the binder ETG now recommends that researchers use at least 40 hour PAV for long-term binder aging research studies. RAP taken from cracked pavements is often harder than the 40-hour or even 60-hour PAV residues. The phase angles of these highly oxidized binders are very low, and most softening agents do not restore the phase angle to that of the original virgin binder, even when the binder moduli match. Phase angle restoration is crucial to the success of recycling. Appropriate long-term laboratory aging tests for binders and mixtures which capture phase angle loss and recovery are critical.

NIMA ROOHI: Thank you.

AMIR GOLALIPOUR: To add to the comment that Gayle had, as you know, the aging of the binder and the rheology of the binder is not something that happens linearly for different aging. So looking at a linear trend for basically chemical aging of the binder, I'm not sure if it would be the best solution to kind of predict that behavior because you can look at all the publications about chemical aging and the reaction that happens for the binder during the aging, usually it's a nonlinear trend and it's not easy to predict it. And that's the reason that you have RTFO PAV. Otherwise, you could just have one of those and then have a straight line off it. Thank you.

NIMA ROOHI: Thank you.

MICHAEL DUNNING: I see a lot of research going on here for the RAP and the RAS. But as far as an agency, for us to make a determination of what to do, what kind of equipment are you looking for us to buy or even use our existing equipment that we can then determine what to do with our pavements? Research is great, but are we headed towards a practical method for us to take a look at our pavements?

RICHARD STEGER: I could probably address that.

BOB FRANK: There are several procedures that agencies use. For the City of Los Angeles, if you look at their Brown book, it covers it in detail exactly how they do their evaluations. And if you can't get the Brown book, give me a call and I'll make sure you get a copy of it. But he's outlined his whole procedure. Like I said, it's been used for over 30 years. It's in finite detail what he did originally to get it started and what they constantly look at. Now, it's a little less, but basically they have a very good program to design it.

MICHAEL DUNNING: Great. I'm seeing that you're using different kind of rejuvenators, etc., for a product to see how it affects your low temperature and high temperature application in the field. How do we determine this? Is it the same process that I can use that you're talking about in Los Angeles by which to try different types of rejuvenators and see if it works okay with respect to your low temperature PG grades and high temperature PG grades?

BOB FRANK: He does include that, but also he goes further into his own mix designs. And basically at the end of the day, the recycled mix is going to be an equivalent to virgin mix in every aspect of the mix going out to the field – VMA, voids filled with asphalt. He has everything covered, but he does a very strong job and the QA side is that he holds the contractor responsible to what he doesn't design. That is the key to their success.

MICHAEL DUNNING: Yes, yes it is.

BOB FRANK: And basically it's been like over 30 years, but there are several contractors that do that, but it's been phenomenal, but he does do an excellent job in QA. And if you don't do QA… well.

MICHAEL DUNNING: I understand you there. I'll see you after.

BOB FRANK: Sure.

MICHAEL DUNNING: Thank you.

NIMA ROOHI: Well, I think Louisiana has done a decent job on this as well.

LOUAY MOHAMMAD: I am not familiar with the conversation with the design manual of the city of Los Angeles. However, it is important to recognize that volumetric design requirements are not sufficient to ensure acceptable performance. Therefore, any design process used to determine the combination of asphalt binder content and aggregate structure to meet a criteria needs to be complemented by a mechanical test. For example, if an agency is concerned with durability or high temperature, then proper performance test(s) should be conducted to evaluate their performance. That is the message that I would like to leave.

RICHARD STEGER: InVia Pavement Technologies has a material that can be used as a material—it was mentioned by Fujie in his presentation—as there are other materials that can do this. And we have also done enough of these Glover-Rowe analyses, as Fujie and Walaa demonstrated in both of their presentations, and we have developed a model that can be used to estimate a dosage rate. And as Louay Mohammad just said that this estimation of your initial dosage rate should then be followed up with a mechanical test to determine whether or not the mixture is actually going to meet the criteria for the climate that you're placing it in. So there is a method and there is a way to get to that, and it does require some evaluation. You have to extract and recover the RAP, the RAS, and virgin binders. So there is a relatively simple way to get to the answer that all of these people have been asking about. So it's not out of the question to be able to do it. You can do it relatively quickly. And then the piece right now that's really missing is the mechanical test to confirm that this dosage rate is correct. We've done some follow-on testing to some of the mixtures that Fujie discussed, and we found that some of those materials fell short of what we were expecting, so we may have to adjust the model some.

FUJIE ZHOU: One thing to add after Richard's comments. And the thing I would like to mention is all the data I presented in the paper and today are from the lab study. It is not the field plant mix. In the field we only had a few short test sections, and ideally we can get what we want, but in most cases, we couldn't get what we want. We all know, when you have very short test sections, what the plant produced is all you got. So that's why Richard is saying his test result is a little bit below what we hoped. That's the reality. That's the plant mix he tested and we are also in the process of testing all the field sections, the plant mixes from the five test sections. We're almost done with that part. But everything I presented today is in the lab design phases. Thank you.

GEOFFREY ROWE: I just want to follow up a little bit on what Richard said. I mean, the concept that we had with that criteria was sort of looking at a particular set of binders, so it's not held in stone that we have to be at those values. The real concept is the same as we use in the bending beam rheometer – it's about stress and relaxation. We're going to have to validate things. And I think the point that various authors have made here today, that we have to do some work with the mix as well, is a very well taken point. We can't ignore the mix. And John D'Angelo made the comment about the paper with the fillers and understanding a lot more than just the binder. But this is a way that we can almost make a marketplace in understanding our rejuvenators. I think that it's very important to understand what's going with rejuvenation, and we see the same sort of effects working with the surface seals. We did some work in Denmark looking at the penetrating seals, and you can see the same sort of concepts. But we can't neglect the mix, and that's a very good concept. And there isn't a firm value that we've set that, some sort of magic number, so this number could change. It's really no more the concept of stress and relaxation that Dave Anderson and Don Christensen and others looked at in SHRP.

WALAA MOGAWER: This is a very good comment, Geoff, and I agree with you completely. The mixes always have to be part of it, but sometimes we don't know which test to use and if they're going to give us the same conclusion. But in addition to this, I go back to chemistry. I think that should be supplemented with some chemical evaluation. What that is, I don't know.

GEOFFREY ROWE: I agree completely. There was a very useful ISAP symposium at TRB I think two years ago, and we can restore things rheologically, but we do see different structures in chemistry. I don't understand what that means, but I take the point and absolutely agree with you.

WALAA MOGAWER: Thanks.

NATHAN MORIAN: I think that's a good point that we've all been making. I was initially going to respond to what Michael Dunning was saying that we do a lot of this research that takes a very long time. So, what tests do we do? How are we really evaluating these materials? I think the simple answer we are really looking for is, will all of this research go to something we can use as an agency? I've had to field this question before. We're really trying to understand the fundamentals and the background information, and how all of this goes into play. If we add, for example, a rejuvenator, we can soften the viscosity, we can reduce the binder grade, but when you start to incorporate aging, then the chemical structure starts to play a role. Most of the time, we don't really know what that effect is going to be. We have some understanding from early chemistry work that Claine Petersen has done decades ago. We understand a bit about what that does, but when you start adding RAP, now we add shingles, and now we add a rejuvenator, what if we put a polymer in there? These systems start to get very, very complicated. So eventually we've completely changed the system from what we thought we understood. All of the research is eventually trying to understand what's going on in these systems. It's a very complicated material. When we try to summarize it down to 1 time PAV, 2 times PAV, a lot of times we end up getting an oversimplification of the real fundamental changes that are happening in our material. So I guess the short answer is, yes, we are working towards a purchase spec. To agree with Geoff, Gayle, and John it's not as easy as picking two points on a given rheology plot, looking at an R value, and moving from there. So it's an effort that we're all working on. But I just want to bring out the point that there's a lot of background information feeding into the solution and it gets tremendously more complicated the more we blend these dissimilar materials, when both the physical and chemical structures are potentially affected.

MICHAEL DUNNING: Thank you for clarifying.

WALAA MOGAWER: That is very well said. And that's why I go back to the rejuvenator itself and the analysis of it. There are so many products, so it's complicated. You're adding it because you have RAP in your mix, which is another component you're adding. So unless we know that the product is doing the job for us, we're going to go the wrong way. Now, as far as the PAV and 2 PAV, 3 PAV, it's the tools we have that we're using. But I agree with you, it's more than that.

GRANT WOLLENHAUPT: From a contractor perspective, our agencies in Missouri and Kansas are losing money all the time, and they can't afford to keep our infrastructure going. So as contractors, we're always trying to look for a way to make our money go farther and still put down good roads. For us, that meant we've been dabbling in rejuvenators for six years now. We've put down two million tons of asphalt with rejuvenators, and while we don't have the ability to quantify the performance, we do know that those roads are holding together better than the roads that we put down at the same time with RAP and RAS. And what that means is what you guys are doing up there with performance testing, I think is great. Fujie, what you're doing, testing your control. Walaa, what you said you're going to do in your future research. You know, test the performance of the mix that contractors are putting down now. We have to start somewhere. I know Wisconsin's made big jumps in moving forward with performance-based specifications. And I know these two contractors would've gotten up and said something about it, but they're from Wisconsin and they want to go have some beer and extend things. But we appreciate as contractors what you guys are doing, but we've got to move forward sometime. Our roads, our safety, our infrastructure just can't wait for the answers. So let's baseline what we're doing right now and make it a little better tomorrow.

NIMA ROOHI: Yeah, I can say Wisconsin beer is a good rejuvenator as well.

GAYLE KING: There have been a number of comments about chemical compatibility and its importance to recycling. We are really struggling with this issue in the high RAP/RAS NCHRP study 9-58. It's not just a problem of selecting the right compatibility test. The most incompatible molecules are the highly oxidized asphaltenes that cluster into large agglomerates. If these molecules fully dissolved into the continuous asphalt phase, they could indeed create compatibility problems. However, these molecules are the least likely candidates for complete blending in field recycling conditions. Asphaltenes from tear-off shingle waste could be highly incompatible with a rejuvenator, but only if the TOAS asphalt becomes fully integrated into the binder phase. There is no way to make or obtain a partially-blended asphalt for a compatibility test.

The choice of compatibility test method is also not obvious. I look forward to speaking further with Dr.'s Daly and Mohammad about their Gel-Permeation Chromatography method. And we have reviewed the many compatibility papers from WRI and others. There are classic tests like SARA or Heithaus, and automated research tools like AFT, AFM, SAR-AD. The problem is to find an inexpensive, practical mix design tool that can spot problem blends. The good news: Gerry Reinke gave presentations at the two most recent binder ETG meetings in which he compared SARA results from an Iatroscan to the BBR ΔTc parameter. His findings from a limited data set indicate that less compatible asphalts will have poor ΔTc (i.e. lower phase angles at a given modulus). These results are encouraging, in that a physical cracking parameter such as ΔTc or Glover-Rowe might even pick up chemical incompatibility as a partial cause for early surface cracking. To those running chemical tests for compatibility, please consider how your results relate to physical binder properties in Black Space.

JEAN-PASCAL PLANCHE: Since Gayle was talking about WRI, I have to say something about WRI, right? There are lots of tools now which are available from a chemical standpoint, and I'm not going to talk about them in details, because it isn't the right audience, I think, and everybody's tired at the end of the meeting. But there are lots of tools available today. We've made enormous progress, so that we can get results in just a couple of hours nowadays. Talking about SARA, we have developed the SARA-AD that gives an awful lot of information on asphalt chemical composition. Then we have developed software that does chemometrics. And then from a few data points, we can get very strong, powerful and very reliable models. So the tools are there. It's just a matter of using them with enough data so that it leads to robust correlations and models. But don't worry about that, we'll sort this out – this is just a matter of time.

AAPT Symposium: Rejuvenator Characterization, Blend Characteristics, and Proposed Mix Design Method

Fujie Zhou[a], Soohyok Im[a], David Morton[b], Robert Lee[c], Sheng Hu[a], and Tom Scullion[a]

[a]Texas A&M Transportation Institute, College Station, Tx 77843

[b]APAC-Texas, Dallas, Tx 75207

[c]Texas Department of Transportation, Austin, Tx 78701

ABSTRACT. *The use of reclaimed asphalt pavement (RAP) and recycled asphalt shingles (RAS) can significantly reduce the increasing cost of hot-mix asphalt paving, conserve energy, and protect the environment. In most cases, they also improve rutting resistance of asphalt mixes. However, in Texas premature cracking has become a serious concern for mixes containing RAP/RAS. This paper presents the latest effort on improving field performance of RAP/RAS mixes through adding rejuvenators to the mixes. This paper first describes the use of a dynamic shear rheometer with large parallel plates to characterize five representative rejuvenators with widely varying viscosity properties. The blending characteristics of rejuvenator/RAP/RAS/virgin binders were then investigated. It was found that three of the rejuvenators investigated in this study are very effective at rejuvenating aged binders. It was also found that 10% (or less) of rejuvenator is enough to make the aged RAP/RAS/virgin binder blend meet the binder specification of a PG 70-22 virgin binder. Additionally, a regional linear-blending concept was proposed and validated in this study. Specifically, aging characteristics of the rejuvenated binders were assessed using the Glover-Rowe parameter and associated damage curves. A relative aging time approach related to the Glover-Rowe damage onset and significant cracking curves was explored in this study, which is important to ensure that the rejuvenated binders have similar (or even better) long term performance to virgin binders. A rejuvenator/RAP/RAS/virgin binder mix design method was proposed for project-specific service conditions in this paper. The proposed mix design method includes 1) the selection of rejuvenator type, 2) the determination of the range of rejuvenator through binder rheological and aging characterization, and 3) the selection of rejuvenator content through a laboratory balanced mix design approach and performance evaluation system. The Hamburg wheel tracking test and associated criteria are used to control rutting/moisture damage and the Overlay test (OT) and the required OT cycles are employed for controlling cracking. Additionally, five field test sections were constructed on SH31 in north Texas to demonstrate and verify the proposed mix design method approach.*

KEYWORDS: *rejuvenator, RAP, RAS, mix design.*

The oral presentation was made by Professor Zhou.

1.0 Introduction

In the asphalt paving industry, the cost of building asphalt pavements continues to rise. The costs of raw materials, equipment and the fuel to run that equipment have significantly increased over the last few years. As a result, the asphalt industry continues to find ways to stretch road building dollars while attempting to maintain the quality of asphalt mixes. One of the ways is to use recycled materials. Both reclaimed asphalt pavements (RAP) and recycled asphalt shingles (RAS) have been increasingly used in asphalt mixes to replace virgin asphalt and aggregate materials in order to reduce mix costs. Last year in Texas, over 1 million tons of RAP and 150,000 tons of RAS were used in asphalt mixes. The trend is to use more and more recycled materials in asphalt mixes, because it not only reduces the construction cost and improves rutting resistance of asphalt mixes, but it is also improves sustainability by aiding in the reduction of greenhouse gas emissions; and using RAS in hot-mix asphalt (HMA) also conserves valuable landfill space. However, recycled binders from both RAP and RAS are often severely aged and substantially stiffer than the virgin binders. As the percentage of recycled binder increases in asphalt mixes, the ratio of the aged binder to the total binder increases, resulting in a stiffer mix that often has a lower resistance to cracking, which is one of the major concerns with RAP/RAS mixes. It is critical to address the premature cracking problem in the mix design phase in order to most effectively use these recycled materials.

Generally, four approaches have been utilized to address the potential premature cracking issue of RAP/RAS mixes: 1) limiting RAP/RAS usage; many state Departments of Transportation (DOTs) in the United States set an upper limit for RAP/RAS usage, such as maximum allowable binder replacement of, for example, 20% for surface mixes; 2) Using soft virgin binders, especially on the low temperature grade (i.e. PG XX-28, PG XX-34); 3) Increasing design density (lowering design air voids) or reducing N_{design}; and 4) Rejuvenating RAP/RAS binder with recycling agents or rejuvenators in the mix design process.

The Texas Department of Transportation (TxDOT) has implemented the first three approaches in its new asphalt mix specification. Recently, TxDOT initiated a research study at the Texas A&M Transportation Institute (TTI) with an overall objective of best using rejuvenators to improve RAP/RAS mix performance. Rejuvenators have been used to restore the performance properties of oxidized pavements for hot-in-place recycling and as surface treatment emulsions to preserve aged asphalt pavements. However, they have not been widely used in HMA or warm-mix asphalt (WMA) mixtures in recent years. Most national research on rejuvenators dates back to late 1970's and early 1980's (Davidson et al., 1977; Davidson et al., 1978; Dunning and Mendenhall, 1978; Escobar and Davidson, 1979; Epps et al., 1980; Kari et al., 1980). The method of identifying the required grade and amount of rejuvenators was also established on the basis of the same viscosity tests (Epps et al., 1980). Clearly, the use of rejuvenators needs to be updated to reflect today's performance grade (PG) binder specification, advanced asphalt mix performance tests, and the use of warm-mix asphalt (WMA) technology.

To achieve this main objective the following steps were undertaken by the researchers;

- Characterize rheological, chemical, and aging characteristics of rejuvenators,
- Investigate characteristics of rejuvenator/RAP/RAS/ virgin binder blends,
- Propose a balanced mix design method for asphalt mixes containing RAP/RAS/rejuvenators, and
- Demonstrate the rejuvenator mix design method through field test sections.

Details of each of these are presented below. This is followed by a summary and conclusions at the end of the paper.

2.0 Rejuvenator Characterization: Rheological, Chemical, and Aging Characteristics of Rejuvenators

Before discussing laboratory tests and associated results, this section presents some background information about rejuvenators.

2.1 RAP Test Sections on IH40 and Observed Field Performance

The structure of asphalt binder is often treated as a complex colloid in which insoluble asphaltenes are dispersed in the soluble maltenes. With sufficient maltene components, the asphaltene micelles under applied stress can move smoothly within asphalt. As asphalt ages, the maltenes are transformed to asphaltenes through oxidation, which results in an unbalanced ratio of maltenes to asphaltenes, and consequently a hardened asphalt mix and brittle pavement (Boyer, 2000; Karlsson, 2002). The purpose of using a rejuvenator is to 1) restore the aged asphalt characteristics to a consistency level appropriate for construction purposes and for the end use of the mixture; 2) restore the aged asphalt to its optimal chemical characteristics for durability, and 3) provide sufficient additional binder to coat new aggregate and to satisfy mix design requirements (Epps et al., 1980). According to Carpenter and Wolosick (1980), the working mechanism (or diffusion process) of a rejuvenator consists of the following four steps:

1. The rejuvenator forms a very low viscosity layer that surrounds the asphalt-coated aggregate where the binder layer is highly aged.

2. The rejuvenator begins to penetrate into the aged binder layer, decreasing the amount of raw rejuvenator that coats the particles and softening the aged binder.

3. No raw rejuvenator remains, and the penetration continues, decreasing the viscosity of the inner layer and gradually increasing the viscosity of the outer layer.

4. After a certain time, equilibrium is approached over the majority of the recycled binder film.

Over the years many materials have been suggested as rejuvenators or recycling agents, such as aromatic oils, paraffinic oils, napthenic oils, waste engine oils, waxes plus fatty acids, and tall oils. (Note that some of these rejuvenators listed here may not be used in recycled asphalt mixes due to environmental concerns or potential health concerns.) To better classify available materials as different groups, the Pacific Coast User-Producer Group evaluated a variety of rejuvenators in late 1970s and early 1980s (Kari et al., 1980), and its research results led to ASTM D4552, "Standard Practice for Classifying Hot-Mix Recycling Agents." The rejuvenators are classified into six grades (or groups) mainly through viscosity measured at 60°C (140°F), as shown in Table 1.

Table 1. *Physical properties of hot-mix recycling agents (ASTM D4552-10).*

Test	ASTM test method	RA1	RA5	RA25	RA75	RA250	RA500
Viscosity, 60°C, mm^2/s (cSt)	D2170 or D2171	50-175	176-900	901-4500	4501-12500	12501-37500	37501-60000
Flash point, COC, °C [°F]	D92	219 [425]	219 [425]	219 [425]	219 [425]	219 [425]	219 [425]
Saturates, wt, %	D2007	Max. 30	Max. 30	Max. 30	Max. 30	Max. 30	Max. 30
Tests on residue from RTFO or TFO 163°C [325 °F]: 1)Viscosity ratio*	D2872 or D1754	Max. 3	Max. 3	Max. 3	Max. 3	Max. 3	Max. 3
2)Wt change, %		Max. 4	Max. 4	Max. 3	Max. 3	Max. 3	Max. 3
Specific gravity	D70 or D1298	Report	Report	Report	Report	Report	Report

*Note: *Viscosity ratio=viscosity of residue from RTFO or TFO Oven Test/Original Viscosity.*

2.2 Rejuvenator Selection and Associated Laboratory Rheological and Chemical Tests

The five rejuvenators selected in this study are Hydrogreen, Evoflex, EAR, AC0.6, and AC5. These five rejuvenators include two traditional asphalt cement based rejuvenators (AC0.6, and AC5) and three new rejuvenators recently formulated (Hydrogreen, Evoflex, EAR). As it can be seen later, these five rejuvenators cover all six RA grades and beyond.

As discussed previously, neither ASTM D2170 nor D2171 is often used today in the field of asphalt binder testing. Instead, the dynamic shear rheometer (DSR) is widely employed to characterize asphalt binder properties. Therefore, in addition to the viscosity tests required by ASTM D4552, TTI researchers explored using the DSR for characterizing rheological properties of five rejuvenators ranging from RA1 to RA500. Both AASHTO T 315 PG grade test and frequency sweep test at various temperatures and frequencies were performed on the five rejuvenators. Specifically, a set of 50-mm diameter parallel plates are used here for those liquid rejuvenators. Additionally, a Fourier transform infrared-attenuated total reflectance spectroscopy (FTIR-ATR) was utilized to measure aging characteristics of the five rejuvenators. Note that the aging characteristics are evaluated through the integrated carbonyl area under the carbonyl absorbance band wavenumber from 1820 to 1650 cm^{-1} widely used for oxidative aging characteristics (Jemison et al., 1992; Glover et al., 2005; Yut et al., 2012).

2.3 Rejuvenator Test Results and Analyses

2.3.1 Viscosity and RA Grade

Table 2 shows viscosity test results of the five rejuvenators. Compared to the RA grade specification (Table 1), the five rejuvenators cover five RA grades and one of them, Evoflex, is even beyond RA1 named RA1-minus.

Table 2. *Laboratory test results of the five rejuvenators.*

Parameter	Aging condition	Rejuvenators				
		Evoflex	**Hydrogree n**	**ERA**	**AC0.6**	**AC5**
Viscosity, 60°C, mm^2/s (or cSt)	Original	47.6	58.7	227	7300	53700
	RTFO aged	125.0	106.0	401	13300	14230 0
Viscosity ratio	RTFO/original	2.6	1.8	1.8	1.8	2.6
RA grade (or group)		RA1 Minus	RA1	RA5	RA75	RA500

2.3.2 PG Grade and Complex Shear Modulus Master Curves

Both Evoflex and Hydrogreen are liquid, so it may not be appropriate to grade them using the Superpave PG system. However, the researchers graded them following the PG high temperature grading criterion so that an overall comparative ranking can be drawn for rejuvenators, virgin binders, RAP binder, and RAS binders. Figure 1 shows the overall rankings of PG high temperature grades of the five rejuvenators, virgin binders, RAP binders, manufacture waste asphalt shingle (MWAS) binders, and tear-off asphalt shingle (TOAS) binders. It is obvious that a lower RA grade of rejuvenators (RA1 or even RA1 minus) should be selected for balancing stiff RAS and RAP binders so that a workable asphalt binder blend could be reached through blending. The stiffer the RAP/RAS binder, the lower RA grade rejuvenators required. Note that compatibility between rejuvenators and RAP/RAS binders should not be ignored, although it is beyond the scope of this paper.

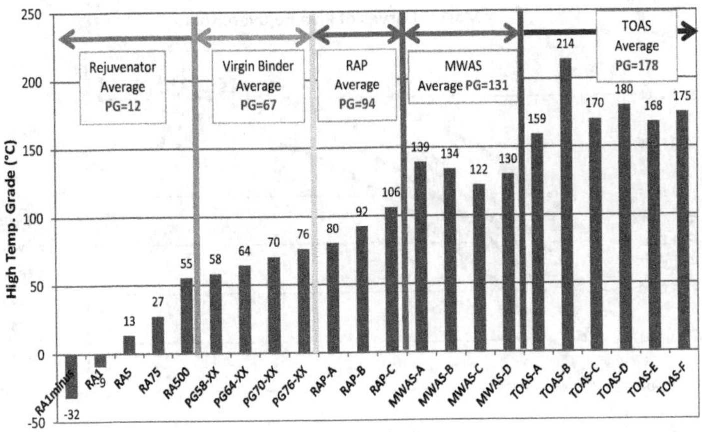

Figure 1. *PG high temperature grades of rejuvenators, virgin, RAP, RAS binders.*

The complex shear moduli (G*) of each rejuvenator at five temperatures (-10, 5, 20, 40, 60°C) and seven frequencies (varying from 1 to 100 rad/s) were measured using a DSR. A set of 50-mm parallel plates were utilized for measuring G* values for both Evoflex and Hydrogreen. The measured G* values were shifted to a reference temperature of 20°C using the WLF equation and modeled with the CAM model (Marasteanu and Anderson, 1999) as shown below. The shifted master curves for the five rejuvenators are shown in Figure 2. Table 3 lists the CAM model parameters and the WLF parameters C_1 and C_2.

$$\text{CAM model: } G^*(\omega) = G_g\left[1 + \left(\frac{\omega_c}{\omega}\right)^v\right]^{-\frac{w}{v}}$$

$$\text{WLF shift factor: } log\ a(T) = \frac{-C_1(T-T_{ref})}{C_2+T-T_{ref}}$$

Where G^* is complex shear modulus (Pa) at frequency ω (rad/s), G_g is glassy modulus (=1GPa), ω_c is the crossover frequency (rad/s), v and w are shape parameters, and R ($=log2/v$) is the rheological index; $a(T)$ is the shift factor at temperature T, T_{ref} is the reference temperature, and C_1 and C_2 are the WLF parameters.

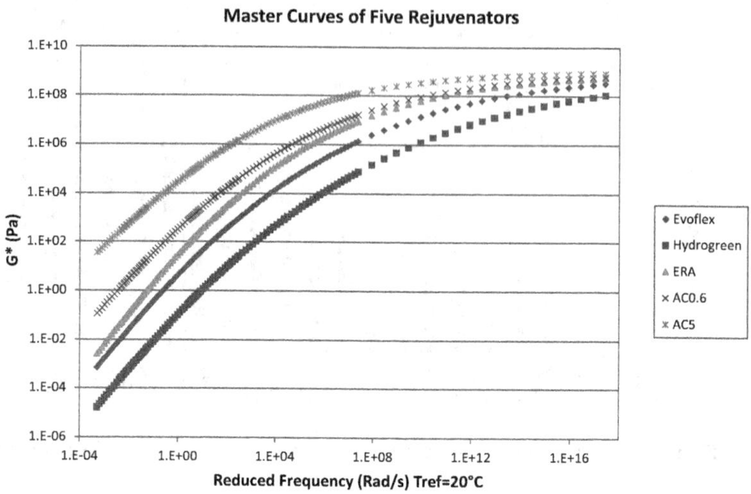

Figure 2. *Frequency sweep test results: master curves of five rejuvenators.*

Table 3. *CAM model parameters and WLF shifting parameters for each rejuvenator.*

Rejuvenator	**CAM Model Parameters**				**WLF Shifting Parameter**	
	w	v	R (*log2/v*)	ω_c	C_1	C_2
Evoflex	1.76	0.082	3.67	100.0	6.65	109.24
Hydrogreen	1.76	0.070	4.32	307.0	0.99	38.94
ERA	1.47	0.115	2.62	10000.0	5.64	85.69
AC0.6	1.46	0.100	3.00	304.6	1.30	41.12
AC5	1.17	0.124	2.42	303.9	15.52	163.67

The data presented in Figures 1 and 2 indicated that it is feasible to use the DSR to characterize rejuvenators. Since the DSR has been the main instrument to characterize asphalt binders for the last two decades, the authors highly recommend that the viscosity-based rejuvenator specification be updated with the complex shear modulus measured by the DSR. Clearly, more studies are needed in this area.

2.3.3 FTIR Aging Characteristics

One of the concerns with rejuvenators is aging resistance. This study employed FTIR-ATR to characterize aging characteristics of the five rejuvenators at the following four different aging conditions: 1) original, 2) RTFO aging, pressure aging vessel (PAV) 20hr aging, 3) PAV 40hr aging, and 4) PAV 80hr aging. Meanwhile, one virgin asphalt binder, PG 70-22, was added for comparison purposes. The measured carbonyl areas for all rejuvenators and the virgin binder are shown in Figure 3. Generally, a larger carbonyl area corresponds to severe aging. Several observations can be made from Figure 3. (1) Both Evoflex and Hydrogreen are completely different from the other three rejuvenators, ERA, AC0.6, and AC5. Such high carbonyl areas for both Evoflex and Hydrogreen may result from their main component, tall oils. The other three rejuvenators are closer to the virgin PG 70-22 binder. (2) When comparing to the original conditions (before aging), all five rejuvenators except AC0.6 have 50% higher carbonyl area after 80 hr aging, which is comparable to the virgin PG 70-22 binder. Thus, the five rejuvenators have similar aging resistance to the virgin PG 70-22 binder, although some of them may start with a very high carbonyl area in the beginning.

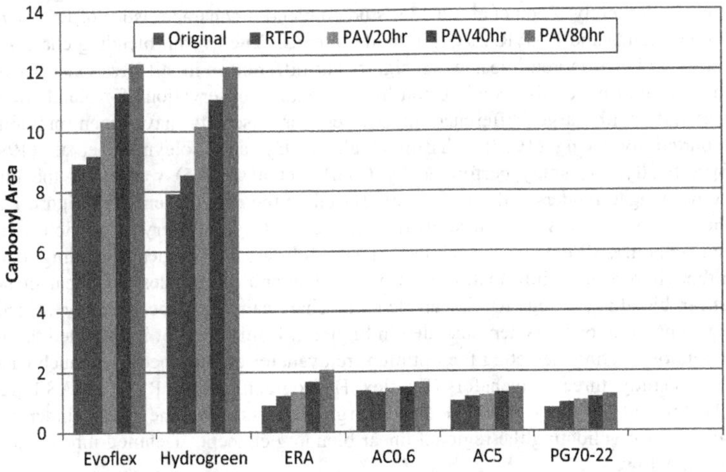

Figure 3. *Aging characteristics of five rejuvenators and one asphalt binder.*

In summary, this section describes the rheological and chemical properties of rejuvenators and their aging characteristics, which are important when selecting rejuvenator type suitable for certain RAP/RAS binders. But the more important question is how to determine the amount of a rejuvenator required for a binder blend to meet specification requirements. The next section discusses characteristics (the rheological and aging properties) of rejuvenator/RAP/RAS/ virgin binder blends with an ultimate goal of determining the amount of rejuvenator required to meet the specification.

3.0 Characteristics of Rejuvenator/RAP/RAS/Virgin Binder Blends

The goal of determining rejuvenator content is to ensure that the rejuvenator/RAP/RAS/virgin binder blend meets the specified PG high and low temperature grade requirements, and has similar (or even better) aging resistance than the originally specified virgin binder. Therefore, the rheological PG grade (blending chart) and aging characteristics of rejuvenator/RAP/RAS/virgin binder blend are discussed below.

3.1 Regional Linear Blending Chart for Rejuvenator/RAP/RAS/Virgin Binder Blends

Blending characteristics between virgin and RAP binders have been well studied, and the linear blending chart is valid for determining either RAP or virgin binder content. Recently, Zhou et al. (2013a) studied blending characteristics of RAS/virgin binder blends and RAP/RAS/virgin binder blends. The overall blending chart is not linear, rather it is non-linear when significant differences in stiffness exist between the blending materials. Similar non-linear blending observations for blends of two materials with large difference in stiffness or viscosity have been previously reported by Irving (1977), Chaffin et al. (1995), and Soleymani et al. (1999). Specifically, the study performed by Chaffin et al. (1995) was about the blend between aged binders and rejuvenators. Based on the data presented in Figure 1 and the previous work described above, it is very possible that rejuvenator/RAP/RAS/virgin binder blends follow a non-linear blending curve, although very little information is currently available. Regardless of linear or non-linear blending, a regional linear-blending chart can likely be used for selecting rejuvenator amount, as demonstrated in Figure 4. If this concept is validated through laboratory testing, selecting the optimum rejuvenator amount becomes much easier. In this study, three rejuvenators (Evoflex, Hydrogreen, and EAR), two RAS binders (TOAS and MWAS), two PG 64-22 virgin binders, and one RAP binder were selected for validating the regional linear blending concept. Detailed information is given below.

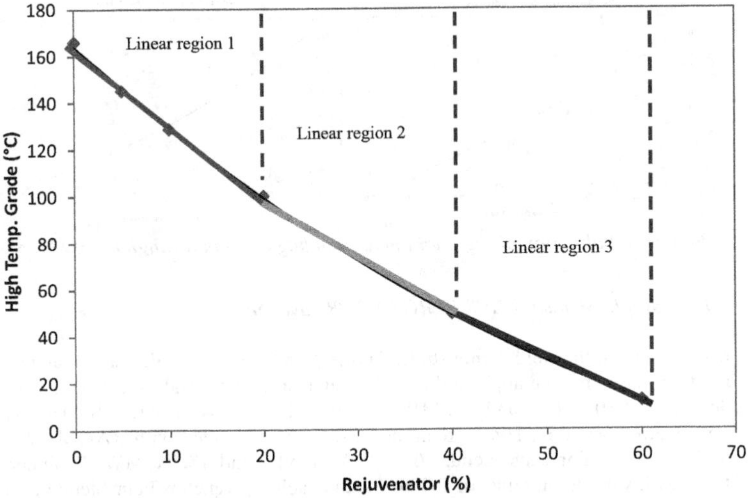

Figure 4. *Regional linear blending concept.*

3.1.1 Binary Blends: PG 64-22/Hydrogreen and TOAS/Hydrogreen

As the first step of the validation process, the researchers chose one rejuvenator, Hydrogreen (having the smallest shear modulus [see Figure 2]); one virgin binder, PG 64-22 (the most frequently used binder grade for RAP/RAS mixes in Texas); and one TOAS binder (having average stiffness of TOAS binders). Six and seven Hydrogreen blending ratios were selected for the validation, respectively. For simplicity, only the high temperature PG grade of each blend was evaluated. Figure 5 presents the high temperature PG grade measured for blends of PG 64-22/Hydrogreen and TOAS binder/Hydrogreen. It can be clearly seen that the regional linear blending concept is valid for any region for both cases as long as it is within the 20% range. For practical applications, a region of 20% rejuvenator is large enough for asphalt mixes. Normally, less than 10% rejuvenator is needed to make the RAP/RAS mixes meet the binder specification (i.e., PG 70-22). Therefore, this concept is validated for binary blends. Next validation focuses on multiple blends and both high and low PG grades.

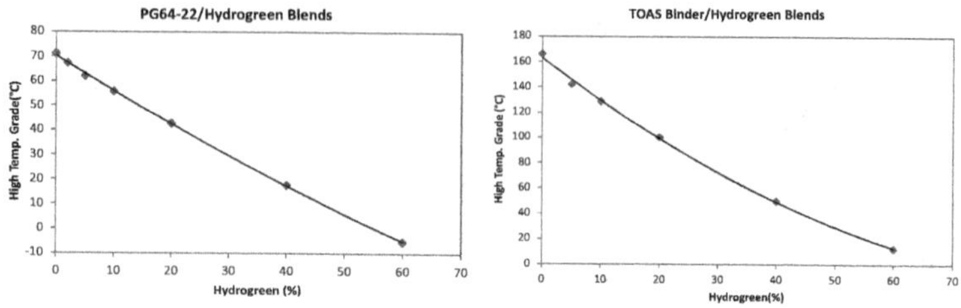

Figure 5. *Validation of regional linear blending concept through binary blends.*

3.1.2 Multiple Blends: RAP/RAS/PG 64-22/Rejuvenators

Materials from the field demonstration project on SH31 (to be discussed later) were used for validating the applicability of the regional linear blending concept. The mix design used 10% RAP and 5% MWAS and a Lion PG 64-22 virgin binder. Three rejuvenators, Evoflex, Hydrogreen, and ERA, were blended with RAP/MWAS/PG 64-22 binders. For each blend, 10.8% RAP binder and 18.4% MWAS binder are fixed with varying amounts of PG 64-22 and each rejuvenator. Four blending ratios of each rejuvenator to the total binder (by weight) were used in this study, namely 0, 2, 5, and 10%. A total of ten blends, as shown in Table 4, were graded through the Superpave PG system. Figure 6 shows PG high and low grades for each blend. It is clear that the linear blending concept is valid for all ten blends. Meanwhile, it can be seen that there is no need for the rejuvenator content to go beyond 10%, because 10% rejuvenator is high enough to make the final blend meet the specification requirements for both high and low PG grades (say PG 70-22) for Texas conditions on SH31.

Table 4. *Multiple blends for RAP/RAS/PG 64-22/rejuvenators.*

RAP binder/RAS binder/virgin binder	**Hydrogreen**	**Evoflex**	**ERA**
10.8%/18.4%/70.8%	0%	0%	0%
10.8%/18.4%/68.8%	2%	2%	2%
10.8%/18.4%/65.8%	5%	5%	5%
10.8%/18.4%/60.8%	10%	10%	10%

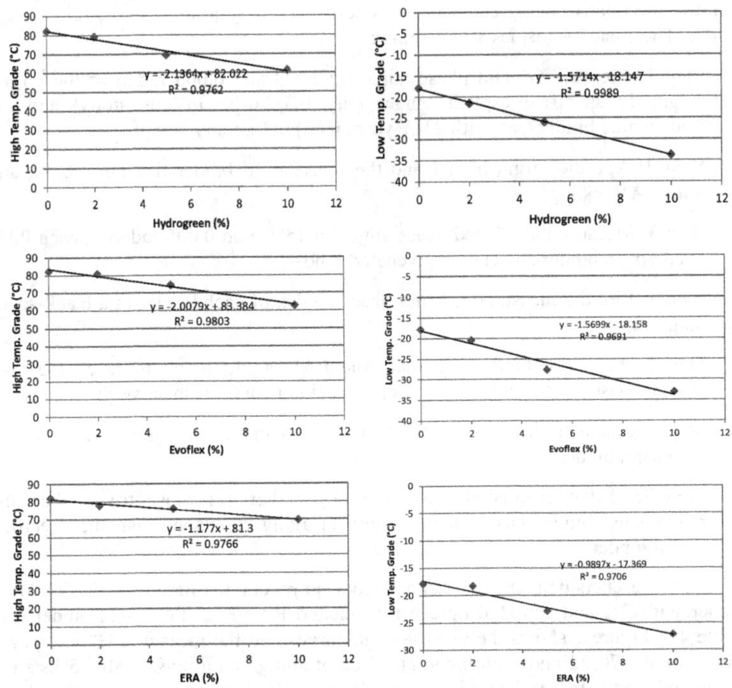

Figure 6. *Validation of regional linear blending concept for multiple blends.*

3.2 Aging Characteristics of Rejuvenator/RAP/RAS/Virgin Binder Blend

As clearly shown in Figure 6, all three rejuvenators can effectively rejuvenate aged binders. However, the long-term effectiveness of rejuvenators has been questioned. Therefore, this issue must be addressed when selecting rejuvenator type and amount. In the last several years the combination of Glover-Rowe (G-R) damage parameter ($\frac{G^*(cos\delta)^2}{sin\delta} = 180kPa - damage\ onset;\ \frac{G^*(cos\delta)^2}{sin\delta} = 450kPa - significant\ cracking$) and black space diagram has been widely used for evaluate asphalt binder cracking resistance (Rowe, 2011; King et al., 2012). Anderson et al. (2011) proposed to use 20, 40, and 80hr PAV aging conditions for analyzing long-term aging characteristics of asphalt binders and potential for block cracking. Although it is not clearly known what the 20, 40, and 80hr PAV aging truly represent in terms of aging in the field, different aging characteristics of asphalt binders can be clearly identified in the black space diagram. Muncy and Steger (2014) recently extended the G-R parameter to quantify the relative aging characteristics between a virgin binder and a rejuvenated binder. The basic idea is described below:

- Step 1: Measure G* and phase angle (at 15°C and 0.005 rad/s) of the virgin, originally specified asphalt binder (say PG 70-22 in this study) and the rejuvenated binder (say with 2% Hydrogreen) before any aging;
- Step 2: Age the virgin binder and the rejuvenated binder through 20, 40, and 80hr PAV testing;
- Step 3: Measure the G* and phase angle (at 15°C and 0.005 rad/s) of each PAV aged virgin binder and each rejuvenated binder.
- Step 4: Plot the measured G* and phase angles of each binder in a black-space diagram;
- Step 5: For each binder, calculate the PAV aging hours to reach the G-R damage onset curve and the significant cracking curve, respectively;
- Step 6: Calculate the ratios of the PAV aging hours of the rejuvenated binder to the virgin binder;
- Step 7: If the calculated ratios are larger than (or close to) 1, then the rejuvenated binder has better (or similar) aging resistance than the specified virgin binder.

The researchers demonstrate this new concept by comparing the virgin PG 70-22 binder with 2% and 5% Hydrogreen rejuvenated RAP/RAS/PG 64-22 binder (see Table 4). Figure 7 shows the G-R damage curves and the measured G* and phase angles of PG 70-22 and each blend at different aging conditions. Table 5 lists the calculated hours for each binder to reach the G-R damage onset curve and the significant cracking curve, respectively. Apparently, 5% Hydrogreen rejuvenated binder has similar (if not better) aging resistance than the virgin PG 70-22 binder. Although the 2% Hydrogreen rejuvenated binder meets Superpave PG high and low grades requirement, caution should be exercised due to its poor aging characteristics.

In summary, this section evaluated the blend characteristics of rejuvenator/RAP/RAS/virgin binders. The results presented above clearly indicate that the regional linear-blending concept is accurate enough for practical application to determine the required rejuvenator content for meeting specification requirements. Additionally, aging characteristics represented by the G-R parameter should be evaluated to ensure long-term performance of rejuvenated binders. Ideally, the rejuvenated binder should have similar or even better aging resistance than the original virgin binder. These findings constitute part of the proposed mix design method for asphalt mixes containing RAP/RAS/rejuvenators, which is described in the next section.

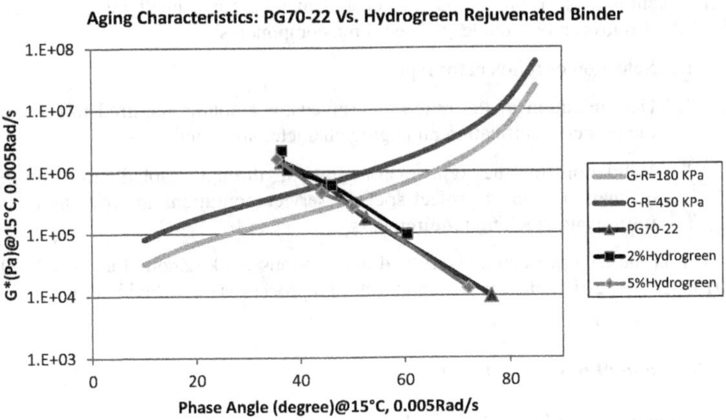

Figure 7. *Aging characteristics of virgin PG 70-22 and Hydrogreen rejuvenated binder.*

Table 5. *PAV aging hours to reach the G-R damage curves.*

Binder	G-R Damage Onset Curve		G-R Significant Cracking Curve	
	Required PAV hours	Rejuvenated /Virgin	Required PAV hours	Rejuvenated /Virgin
Virgin PG 70-22	26	N/A	40	N/A
2% Hydrogreen rejuvenated binder	14	0.54	20	0.50
5% Hydrogreen rejuvenated binder	23	0.88	44	1.10

4.0 Proposed Mix Design Method for Mixes Containing RAP/RAS/Rejuvenator

It is a well-known fact that properties of the rejuvenated asphalt binder are critical for obtaining a successful mix design. However, more importantly, it is the properties of the entire asphalt mixture that ultimately determine field performance.

Consequently it is proposed that the design method for asphalt mixes containing RAP/RAS/rejuvenator includes at least three components:

1. Selection of rejuvenator type,
2. Determination of the range of rejuvenator amounts required to meet both the binder specification and aging characteristics, and
3. Validation of the rejuvenator amount through asphalt mix property evaluation to meet project specific service conditions in terms of mixture rutting and cracking requirements.

Based on the data presented above and the previous work (Zhou et al., 2009, 2012, 2013a, 2013b, 2014), the three-step mix design process is described below.

4.1 Selection of Rejuvenator Type

The amount and hardness of the asphalt in an aged asphalt mix are considered when selecting the RA grade. The general rule for selecting rejuvenator type is to use a lower RA grade for stiff RAP and RAS binders. Meanwhile, the compatibility between the selected rejuvenator and RAP/RAS/virgin binder should be considered, although it is beyond the scope of this paper.

4.2 Determination of Rejuvenator Range through Binder Tests

As discussed in Section 3, the following three steps are required to determine the range of rejuvenators:

- Step 1: Determine the range of rejuvenators based on the specific project PG binder requirement. As clearly demonstrated in Section 3.1, the regional linear blending rule for both Superpave PG high and low grades is valid for rejuvenator/RAP/RAS/virgin binder blending, which significantly reduces the blending work. Only two blending ratios of the rejuvenator to RAP/RAS/virgin binder are needed. The high PG grade of blended binders controls the maximum rejuvenator amount, and the low PG grade of the blended binders determines the minimum rejuvenator amount (Shen and Ohne, 2002). In case of rejuvenator Hydrogreen (see Figure 6), it will range from 2% to 5%.

- Step 2: Verify aging characteristics of rejuvenated binder through comparing to virgin binder. The proposed detailed process was described in Section 3.2. At the end of the process, a range of rejuvenator amounts will be determined so that the rejuvenated binder has similar or even better aging resistance to the original virgin binder. In case of rejuvenator Hydrogreen (see Figure 7), the range will be from 2% to 5%.

- Step 3: Select the final range of rejuvenator amount for mix property evaluation based on Steps 1 and 2. In case of rejuvenator Hydrogreen, the final range for mixture testing will be larger than 2% but less than 5%.

4.3 Selection of Rejuvenator Amount for Project-Specific Service Conditions through Mixture Tests

In last ten years, TTI researchers have focused on developing the balanced mix design method for project-specific service conditions (Zhou et al. 2007, 2009, 2010, 2012, 2013b, 2014). A successful mix needs to have balanced rutting and cracking resistance, and the rutting and specifically cracking requirements depend on site specific conditions. These include traffic, climate, pavement structure, layer thickness, and existing pavement conditions, particularly for asphalt overlays. The authors believe that the proposed balanced mix design system can also be used for designing mixes containing a rejuvenator. The only difference here is to determine the optimum rejuvenator content rather than virgin binder content. The revised design flowchart for rejuvenator/RAP/RAS/virgin binder mix design is shown in Figure 8. Note that TGC and SGC stand for Texas Gyratory Compactor and Superpave Gyratory Compactor, respectively.

To demonstrate and verify the proposed mix design method, five field test sections were constructed on SH31 of Tyler District, Texas. Detailed information is described in the next section.

5.0 Demonstration and Verification of the Balanced Mix Design Method Proposed for Rejuvenator/RAP/RAS/Virgin Binder Mix

Five field test sections were constructed by APAC-Texas on the outside eastbound lane of highway SH31, in TxDOT's Tyler District in June 2014. The purpose of these five test sections was to demonstrate and verify the proposed mix design method shown in Figure 8 for asphalt mixes containing rejuvenator/RAP/RAS. SH31 is a two-way divided highway with AADT of 9800 and 10% truck traffic. The estimated 20-year traffic load in 18 kips is 3.5 million ESAL (Equivalent Single Axle Load). These test sections are located between the east side of the city of Murchison and the west side of the city of Brownsboro, Texas (see Figure 9). For each test section 350 tons of asphalt mix were produced. On June 3, 2014, Sections 1, 2, and 3 were constructed. Sections 4 and 5 were paved on June 4, 2014. Section 4 is followed by Section 5.

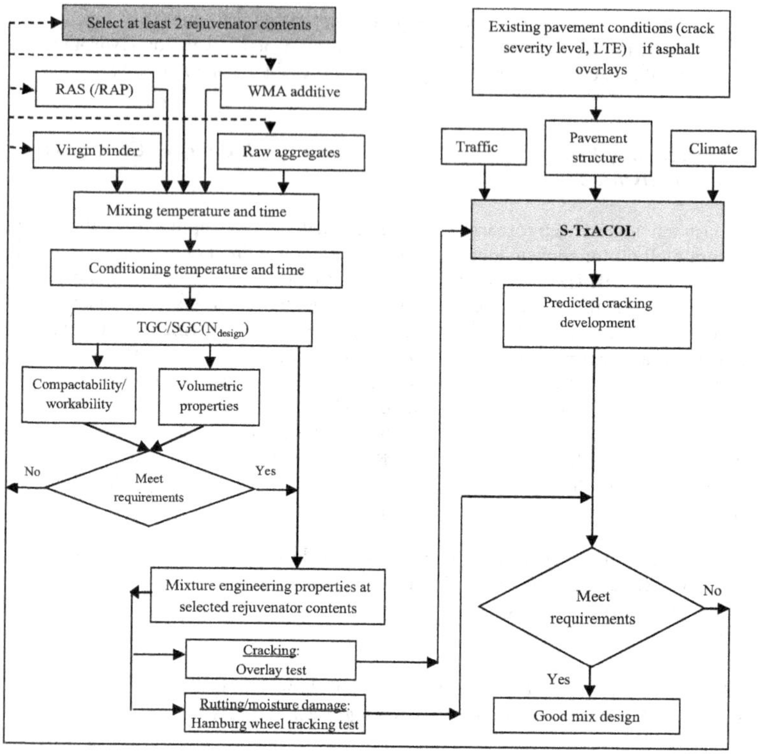

Figure 8. *Balanced rejuvenator/RAP/RAS/virgin binder mix design for project-specific service conditions.*

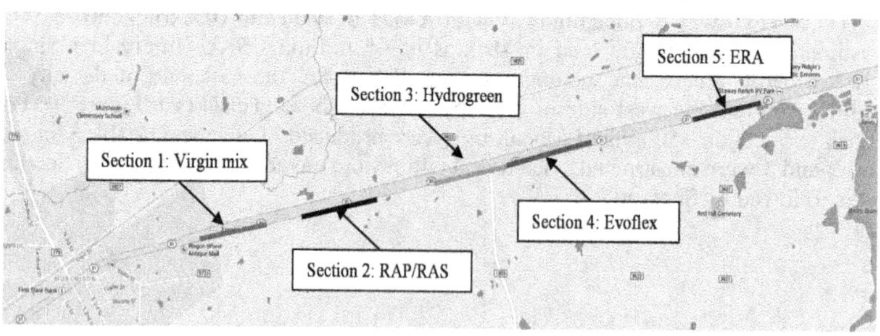

Figure 9. *Layout of five test sections on SH31.*

All five sections have the same pavement overlay structure: this includes, 1 in. of Crack Attenuating Mix (CAM) and 2 in. of dense-graded Type-C surface mix. The Type-C mix was modified as described below;

- Section 1: A virgin mix with PG 70-22 binder (no RAP/RAS/rejuvenator)
- Section 2: The control mix with PG 64-22 virgin binder, 10% RAP and 5% Manufacturer Waste Recycled Shingles (MWAS)
- Section 3: The same control mix with rejuvenator Hydrogreen
- Section 4: The same control mix with rejuvenator Evoflex (3.7% by weight of total asphalt binder) and Evotherm (0.3% of weight of total of asphalt binder)
- Section 5: The same control mix with rejuvenator ERA (1.3% of weight of total asphalt binder content)

APAC-Texas designed both the virgin PG 70-22 mix and the control section with RAP and MWAS following the TxDOT standard mix design procedure (Tex-204-F). The optimum asphalt binder contents for the virgin PG 70-22 mix and the control mix are 4.6% and 4.9%, respectively. For the control mix with RAP and MWAS, the asphalt binder replacement ratio is 29.2% – 10.8% from 10% RAP and 18.4% from 5% MWAS.

Both TTI and APAC-Texas worked together to complete the mix designs for the three rejuvenator test sections. The three-step mix design process described in Section 4 was followed, and detailed information is provided below:

- Step 1: Select rejuvenator types.

  First of all, TTI researchers characterized the extracted RAP and MWAS binders. The measured high temperature PG grades for the RAP and MWAS binders were 109 and 134°C, respectively. Both binders were very stiff. The low temperature PG grades for both binders could not be measured with the bending beam rheometer. Based on the results presented in Section 2, three out of five rejuvenators were selected for the test sections, and they are Hydrogreeen, Evoflex, and ERA.

- Step 2: Determine rejuvenator range through binder tests

  The goal was to have the rejuvenated binder meet the PG 70-22 binder specification. Based on the work described in Sections 3.1 and 3.2, it seems that 2-5% of rejuvenator can make the binders meet the PG 70-22 requirements in terms of PG high and low grades. The aging characteristics comparison between the PG 70-22 virgin binder and the rejuvenator binders are shown in Figures 7 and 10. Based on the measured data, the recommended range for mixture evaluation is 2-5% for both Hydrogreen and Evoflex, within which they can meet the PG binder specification requirements and have acceptable aging characteristics. However, this is not the case for ERA. Apparently, more ERA is needed to have similar aging resistance to the virgin PG 70-22. It was decided for the ERA section to add extra ERA to the original design binder content of 4.9 % rather than the binder replacement concept.

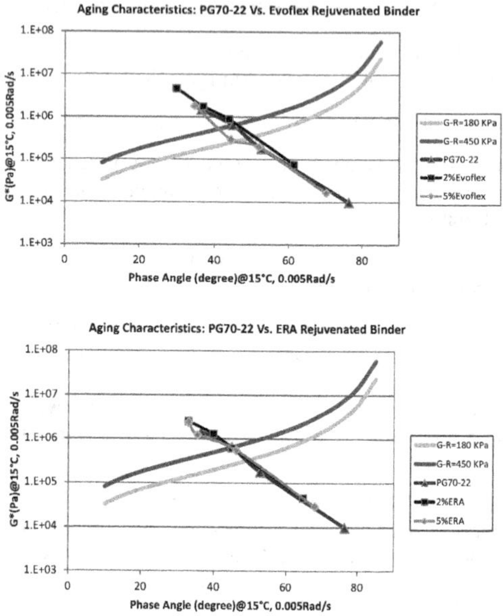

Figure 10. *Aging characteristics of virgin PG 70-22, Evoflex- and ERA rejuvenated binders.*

- Step 3: Select rejuvenator content optimized for project-specific service conditions through mixture tests

  Based on the results of Step 2, two trial rejuvenator contents for each rejuvenator were chosen for mixture testing, as listed in Table 6. Following the proposed mix design flowchart shown in Figure 8, this study focused on the engineering properties of asphalt mixes with rejuvenator only, although both compactability and volumetric properties are important as well. The Hamburg wheel tracking test was performed to evaluate rutting resistance of all five mixes, following Tex-242-F, "Test Procedure for Hamburg Wheel-Tracking Test (HWTT)". The cracking resistances of the same mixes were evaluated using the Texas Overlay test according to Tex-248-F, "Test Procedure for Overlay Test (OT)". Both the Hamburg and OT test results are shown in Figure 11. Note that the rut depth shown in Figure 11 is measured at 10000 passes. It can be observed from Figure 11 that the use of rejuvenators can significantly improve the cracking resistance of RAP/RAS mixes, and the measured Hamburg rut depths are also acceptable. Another interesting observation in Figure 11 is that the rejuvenated RAP/RAS mix can have better cracking resistance than the virgin mix.

Table 6. *Selected rejuvenator contents for mixture performance evaluation.*

Rejuvenator content	Hydrogreen (weight of total asphalt binder)	Evoflex (weight of total asphalt binder)	ERA (Extra addition, weight of total asphalt binder)
Low end	2.6%	3.7%	1.0%
High end	5.0%	5.0%	2.0%

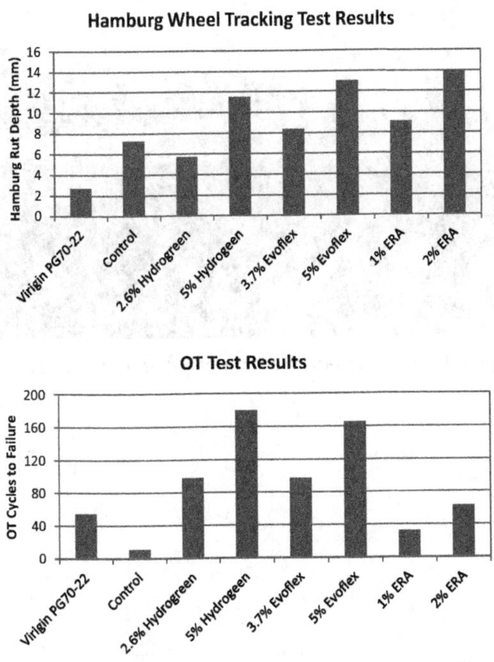

Figure 11. *Hamburg and OT test results of SH1 mixes.*

Regarding the cracking requirement, TTI researchers used the S-TxACOL overlay design program to calculate the required OT cycle for the 2-in. surface mix to last 15 years without significant cracking problems. The existing pavement structure after milling is 2-in. asphalt over an old jointed concrete pavement. Prior to milling there was lots of severe transverse and longitudinal cracking (see Figure 12). The loading transfer efficiency at cracks was assumed to be 70%. Based on the traffic, weather, pavement structure and FWD backcalculated layer modulus, the calculated minimum OT cycles for the surface mix to meet the 15-year design life is predicted to be 50 cycles. All mixes, except the control mix and the mix with 1% ERA, met this cracking requirement. Based on the laboratory test results, the final rejuvenator contents recommended for Hydrogreen, Evoflex, and ERA are 2.6%, 3.7%, and 1.3%, respectively. All five test sections with the five mixes designed under this study have been successfully constructed on SH31, as shown in Figure 13. TTI researchers are closely monitoring the performance of these five sections, and will report the observed field performance at later time.

Figure 12. *Existing pavement condition of SH31 before milling.*

Figure 13. *Construction of five test sections on SH31 (June 2014).*

In summary, this section demonstrates the proposed mix design method through actual field project, and the performance of these five sections on SH31 being observed will be used to verify both rutting and cracking requirements of asphalt mixes.

6.0 Summary and Conclusions

This paper evaluated five rejuvenators and blending characteristics of rejuvenator/RAP/RAS/virgin binders in terms of their rheological, chemical, and aging properties. A balanced mix design method was proposed for asphalt mixes containing RAP/RAS/rejuvenators. Additionally, five field test sections were constructed with the purpose of demonstrating and validating the proposed mix design method. The following conclusions and recommendations are made based on the research presented in this paper.

- It was feasible to use the DSR with large parallel plates to characterize liquid rejuvenators. Since the DSR has been the main instrument to characterize asphalt binders for the last two decades, the authors highly recommend that the viscosity-based rejuvenator specification be updated with complex shear modulus measured by the DSR.

- With respect to aging resistance of rejuvenators, although some of them may start with a very high carbonyl area in the beginning, the five rejuvenators had similar aging resistance when compared to the virgin binder investigated in this paper.
- The regional linear blending concept provides a means of selecting the amount of rejuvenator required to meet the specified binder specification. It was found that the three rejuvenators investigated in this study are very effective to rejuvenate the aged binders. An amount of 10% or less rejuvenator was needed to meet the binder grade requirement.
- Aging characteristics of virgin and rejuvenated binders were assessed using the Glover-Rowe damage parameter curves. A relative aging time approach related to Glover-Rowe damage onset and significant cracking curves was explored in this study, which is important to ensure that the rejuvenated binders have similar (or even better) long-term performance to virgin binders.
- A rejuvenator/RAP/RAS/virgin binder mix design method was proposed for project-specific service conditions. The proposed mix design includes 1) selection of rejuvenator type, 2) determination of the range of rejuvenator through binder rheological and aging characterization, and 3) selection of rejuvenator content through a balanced mix design and a performance evaluation system. The Hamburg wheel tracking test and associated criteria are used to control rutting/moisture damage and the Overlay Test and the required OT cycles determined from S-TxACOL cracking prediction with consideration of climate, traffic, pavement structure, and existing pavement conditions are employed for controlling cracking. Additionally, five field test sections were constructed on SH31 near Tyler, Texas, to demonstrate and verify the proposed method. Initial performance has been excellent and multi-year monitoring is underway.

7.0 Acknowledgements

The authors would like to acknowledge the financial support from TxDOT. Many other people who helped in different ways are highly appreciated.

8.0 Disclaimer

The contents and opinions of this paper reflect the views of the authors, who are solely responsible for the facts and the accuracy of the data presented herein. The contents of this paper do not necessarily reflect the official views or the policies of any agencies.

9. 0 References

Anderson, R.M., G.N. King, D.I. Hanson, and P.B. Blankenship, "Evaluation of the Relationship Between Asphalt Binder Properties and Non-Load Related Cracking,' *Journal of the Association of Asphalt Paving Technologists*, Vol. 80, 2011, pp. 615–649.

Boyer, E., *Asphalt Rejuvenators: Fact, or Fable*, Asphalt Institute, prepared for presentation at the Transportation System 2000 (TS2K) Workshop, San Antonio, Texas, Feb. 28-Mar. 3, 2000.

Carpenter, S., and J. Wolosick, "Modifier Influence in the Characterization of Hot-Mix Recycled Material," *Transportation Research Record, No. 777*, Washington, DC, 1980, pp. 15–22.

Chaffin, J.M., R.R. Davison, C.J. Glover, and J.A. Bullin, "Viscosity Mixing Rules for Asphalt Recycling," *Transportation Research Record No. 1507*, Washington, D.C., 1995, pp. 78–85.

Davidson, D., W. Canessa, and S. Escobar, "Practical Aspects of Reconstituting Deteriorated Bituminous Pavements," *Recycling of Bituminous Pavements*, STP 662, American Society for Testing and Materials, Edited by L. E. Wood, 1977, pp. 16–34.

Davidson, D., W. Canessa, and S. Escobar, "Recycling of Substandard or Deteriorated Asphalt Pavements–A Guideline for Design Procedures," *Proceedings of the Association of Asphalt Paving Technologists*, Vol. 46, 1977, pp. 496–525.

Dunning, R. and R. Mendenhall, "Design of Recycling Asphalt Pavement and Selection of Modifiers," *Recycling of Bituminous Pavements*, STP 662, American Society for Testing and Materials, Edited by L. E. Wood, 1997, pp. 35–46.

Epps, J.A., D.N. Little, R.J. Holmgreen, and R.L. Terrel, *Guidelines for Recycling Pavement Materials*, NCHRP Report 224, National Cooperative Highway Research Program, Washington, D.C., 1980.

Escobar, S. and D. Davidson, "Role of Recycling Agents in the Restoration of Aged Asphalt Cements," *Proceedings of the Association of Asphalt Paving Technologists*, Vol. 48, 1979, pp. 375–402.

Glover, C., R. Davison, C. Domke, Y. Ruan, P. Juristyarini, D. Knorr, and S. Jung, *Development of a New Method for Assessing Asphalt Binder Durability with Field Validation*, Report FHWA/TX-05/1872-2, Texas Transportation Institute, College Station, TX, 2005.

Irving, J., *Viscosities of Binary Liquid Mixtures: The Effectiveness of Mixture Equations*, Report 631, National Engineering Laboratory, Department of Industry of U.K., 1977.

Jemison, H., B. Burr, R. Davison, J. Bullin, and C. Glover, "Application and Use of the ATR-FTIR Method to Asphalt Aging Studies," *Petroleum Science and Technology*, Vol. 10, No. 4, 1992, pp. 795–808.

Kari, W.J., N.E. Andersen, D.D. Davidson, H.L. Davis, R.N. Doty, S.J. Escobar, D.L. Kline, and T.K. Stone, "Prototype Specifications for Recycling Agents Used in Hot-Mix Recycling," *Proceedings of the Association of Asphalt Paving Technologists*, Vol. 49, 1980, pp. 177–192.

Karlsson, R., *Investigation of Binder Rejuvenation Related to Asphalt Recycling*, Royal Stockholm, Sweden, Institute of Technology, PhD Dissertation, 2002.

King, G., M. Anderson, D. Hanson, and P. Blankenship, "Using Black Space Diagrams to Predict Age-Induced Cracking," *Proceeding of the 7th RILEM International Conference on Cracking in Pavements*, Vol. 4, 2012, pp. 453–463.

Marasteanu, M., and D. Anderson, "Improved Model for Bitumen Rheological Characterization," *Eurobitume Workshop on Performance Related Properties for Bituminous Binders*, Brussels, Belgium, 1999.

Mogawer, W.S., A. Booshehrian, S. Vahidi, and A.J. Austerman, "Evaluating the Effect of Rejuvenators on the Degree of Blending and Performance of High RAP, RAS, and RAP/RAS Mixtures," *Road Materials and Pavement Design*, 14(S2), 2013, pp. 193–213.

Muncy, D., and R. Steger, Personal Communication, April 2014.

Rowe, G.M., Prepared Discussion for the AAPT paper by Anderson et al., "Evaluation of the Relationship between Asphalt Binder Properties and Non-Load Related Cracking," *Journal of the Association of Asphalt Paving Technologists*, Vol. 80, 2011, pp. 649–662.

Shen, J. and Y. Ohne, "Determining Rejuvenator Content for Recycling Reclaimed Asphalt Pavement by SHRP Binder Specification," *International Journal of Pavement Engineering*, Vol. 3, 2002, pp. 261–268.

Soleymani, H.R., H.U. Bahia, and A.T. Bergan, "Time-Temperature Dependency of Blended and Rejuvenated Asphalt Binders," *Journal of the Association of Asphalt Paving Technologists*, Vol. 68, 1999, pp.129–152.

Yut, I., A. Bernier, and A. Zofka, "Development of a Compact Laboratory Aging Procedure for Asphalt Binders," *Journal of the Association of Asphalt Paving Technologists*, Vol. 81, 2012, pp.693–715.

Zhou, F., S. Hu, T. Scullion, M. Mikhail and L.F. Walubita, "A Balanced HMA Mix Design Procedure for Overlays," *Journal of the Association of Asphalt Paving Technologists*, Vol. 76, 2007, pp. 823–850.

Zhou, F., S. Hu, X. Hu and T. Scullion, *Mechanistic-Empirical Asphalt Overlay Thickness Design and Analysis System*, Research Report FHWA/TX-09/0-5123-3, Texas Transportation Institute, College Station, TX, 2009.

Zhou, F., S. Hu, and T. Scullion, "Advanced Asphalt Overlay Thickness Design and Analysis System," *Journal of the Association of Asphalt Paving Technologists*, Vol. 79, 2010, pp. 597–634.

Zhou, F., S. Hu, G. Das, R. Lee, T. Scullion, G. Claros, "Successful High RAP Mixes Designed with Balanced Rutting and Cracking Requirements," *Journal of Association of Asphalt Paving Technologists*, Vol. 81, 2012, pp. 477–505.

Zhou, F., H. Li, S. Hu, J.W. Button, and J.A. Epps, *Characterization and Best Use of Recycled Asphalt Shingles in Hot Mix Asphalt*, FHWA/TX-12/0-6614-2, Texas A&M Transportation Institute, College Station, TX, 2013a.

Zhou, F., H. Li, S. Hu, R. Lee, T. Scullion, G. Claros, J. Epps, and J. Button, "Evaluation of Use of Recycled Asphalt Shingles in HMA," *Journal of the Association of Asphalt Paving Technologists*, Vol. 82, 2013b, pp. 367–402.

Zhou, F., S. Hu, T. Scullion, and R. Lee, "Balanced RAP/RAS Mix Design for Project-Specific Conditions," *Journal of the Association of Asphalt Paving Technologists*, Vol. 83, 2014, pp.171–196.

10.0 Discussion

JOHN D'ANGELO: Interesting work, but I got a serious question for you. Your overlay test, you're looking at cracking, right?

FUJIE ZHOU: Yes.

JOHN D'ANGELO: Did you do any aging of the mix before you tested it or did you just do the typical two hours at mixing temperature?

FUJIE ZHOU: We followed the TxDOT standard procedure. We didn't do long-term aging. We just used what we call short-term aging in Texas.

JOHN D'ANGELO: With all the different additions to the mix you've dramatically changed the chemical structure of all these binders. You've added aged materials like RAS and RAP, plus the addition of rejuvenators. In the past when testing mixes with neat asphalts you could expect similar aging to some extent. Now you've got mixes with significantly different molecular structures with the combination of the aged binder material and rejuvenators so you can't assume they're going to age the same. Without actually using some simulated aging process, you really don't have a comparison between the properties of the different mixes for a real comparison to estimate real performance in the field with time.

FUJIE ZHOU: Thank you for your comment, John. I agree with you and that is something we are investigating. We also continue monitoring the field test sections and take cores at different stages, trying to see how the aging changes versus time.

GRANT WOLLENHAUPT: Fujie, good presentation. When you were looking at the Glover-Rowe parameter and the aging characteristics, did you look at an extracted binder of what the control mix would have been to see what the hours for failure criteria would have been for that?

FUJIE ZHOU: We didn't do that.

GRANT WOLLENHAUPT: Okay. Because you're comparing it to…

FUJIE ZHOU: Yeah, that's a very interesting point.

GRANT WOLLENHAUPT: Yeah, to the 70-22 and how often are you putting that mix down.

FUJIE ZHOU: Yeah. I think we will consider that for next test section. We're working on that now. Thank you.

AMY EPPS MARTIN: I just want to mention, John, that these same materials were gathered for NCHRP 9-58, and so we're utilizing these same materials. We're doing some of what you said in that project, both of us together.

FUJIE ZHOU: Thank you, Amy. Amy is leading the NCHRP 9-58, and is looking for long-term aging and all kind of other things. John, this is just the Texas part, but we're working on the NCHRP 9-58, and we're looking for long-term aging as well. Thank you.

GEOFFREY ROWE: Fujie, very nice work.

FUJIE ZHOU: Thank you.

GEOFFREY ROWE: I think what you're looking at with the concept of a time to certain aging condition is quite nice. I was talking with some colleagues in the United Kingdom a week or so ago, and they're looking at what they call an aging profile test for trying to look at how materials age. Mike Anderson and his co-workers had that sort of concept in their paper four or five years ago in AAPT. So I think that's a very nice concept and I congratulate you on that. I think what we need to is to understand how these products age over time and I think you're doing that so congratulations.

FUJIE ZHOU: Thank you. I fully agree with you and we need to look at more in terms of the black space and look at the parameters and see how things are marching to the left. That's very important. Thank you.

GERRY HUBER: Fujie, I would just like to confirm something that I thought I heard you say. This is a surface mix and did it have an asphalt binder content of 4.9%?

FUJIE ZHOU: Yes.

GERRY HUBER: So does the asphalt content strike you as being particularly low for a surface mix in Texas?

FUJIE ZHOU: That for the Type C is pretty normal.

GERRY HUBER: Yes, I realize it is pretty normal for Texas.

FUJIE ZHOU: For Texas. I'm not talking about all other states.

GERRY HUBER: Correct. Here's where I'm going with my questioning, and I realize this is an aside to what you're studying in terms of the rejuvenators. But one of the things that is a very hot topic in the industry today, of course, is cracking of asphalt pavements. And with the increased use of RAP and RAS in the last, let's say, five to ten years, there's been a lot of emphasis on cracking. What I'd like to do is ask your opinion about if the hot mix had been designed using VMA calculated with G_{sb} instead of a G_{se}, and if it had another 0.8 or 0.9% asphalt binder in it as a result, how much of an impact do you believe that might have on the cracking behavior and mitigation of the effect of the RAP and the RAS on it?

FUJIE ZHOU: Thank you, Gerry. Well, one thing, I'm a true believer that the amount of asphalt content is really critical for cracking, and we have to make sure that we get enough binder into our mix. That's why more than ten years ago in Texas we started to look at the cracking issue and our overlay test was upgraded to make sure that we have enough binder in the mix. In terms that you consider the VMA and true V_{be}, I think that's a good approach to go. But if you look at the RAP/RAS mixes, it is sometimes difficult to accurately determine all the specific gravities. That's another concern, I guess.

GERRY HUBER: Yes. What your study does is help to spotlight some of the issues that are happening nationwide. In my experience asphalt binder contents have been decreasing over the last decade, and I suspect that plays a significant role in the amount of cracking that we have in this country.

FUJIE ZHOU: Yes. Thank you.

FRANK FEE: Two questions. What value did you give for the binder in the RAS and what kind of temperatures did you use to prepare the mixes in the lab?

FUJIE ZHOU: The shingle, if I remember correctly, is the PG 134. Extracted RAS binder. This is a manufacturer waste shingles.

FRANK FEE: No, the percentage of the value of the binder that was in the RAS, what did you assign it, the full value that you extracted from it?

FUJIE ZHOU: Yes, we assigned the full value.

FRANK FEE: Okay. And what temperatures did you use in the laboratory to prepare the specimens?

FUJIE ZHOU: Are you talking about compaction temperature or are you talking about the mixing temperature?

FRANK FEE: Mixing. What mixing temperature did you use?

FUJIE ZHOU: The mixing temperature, I think, is the PG 70 corresponding to 300F.

FRANK FEE: Okay. That's probably fairly below the value of the melt point of the RAS material.

FUJIE ZHOU: But when they do this kind of mix, it is corresponding to the working binder.

AAPT Symposium: Investigating the Aging Mitigation Capabilities of Rejuvenators in High RAP Mixtures Using Black Space Diagrams, Binder Rheology and Mixture Tests

Walaa S. Mogawer[a], Thomas Bennert[b], Alexander Austerman[a] and Christopher Ericson[b]

[a]*University of Massachusetts Highway Sustainability Research Center (HSRC), Fall River, MA, 02723*

[b]*Rutgers University Center for Advanced Infrastructure and Transportation (CAIT), Piscataway, NJ 08854*

ABSTRACT. *This study was undertaken to better understand the effects of rejuvenators on high Reclaimed Asphalt Pavement (RAP) mixtures and to examine these effects after both short- and long-term aging to determine if rejuvenators can assist in mitigating the mixture aging due to the RAP. A two tier evaluation was undertaken: (1) examine the rheology of extracted and recovered binders and (2) mixture performance tests.*

Tests performed on extracted and recovered binders included rheological plots of shear modulus (G) versus phase angle (δ) (commonly known as a Black Space Diagram), rheological parameters derived from master curves using the Christensen-Anderson model (ω_o – R-value Space diagrams), Superpave Performance Grading (PG) results, critical cracking temperatures, and Multiple Stress Creep Recovery (MSRC) results. The results from these tests were compared to mixture test results for rutting and moisture damage susceptibility, fatigue cracking and low temperature cracking to determine if they correlate.*

Testing was conducted on a Superpave 9.5mm mixture incorporating 50% RAP by weight of aggregates and a base binder of PG 58-28. A control mixture using all virgin materials was also developed and tested for comparison purposes. A total of four different rejuvenators were evaluated including an aromatic oil, a paraffinic oil, an organic blend developed using green chemistry, and an organic blend based on organic oils.

Overall, the data indicated that Black Space diagram and the ω_o – R-value Space diagram were able to capture the aging mitigation effects of each asphalt rejuvenator. These effects were varied based on rejuvenator type. The magnitude of the aging was less severe for certain rejuvenators, which indicated that the type of rejuvenator can play a role in the degree of mitigation of aged binder. Generally the mixture experienced less aging with the use of a

The oral presentation was made by Dr. Mogawer.

rejuvenator regardless of aging period and moved toward the behavior of the control mixture and away from the 50% RAP mixture without rejuvenator. The Black Space diagram and the ω_o – R-value Space diagram agreed with the mixture tests for rutting and moisture damage, fatigue cracking and low temperature cracking. The binder grading results exhibited subtle variances based on the rejuvenator used and aging period. These variances were not always consistent. This indicated that PG tests may not have been as sensitive to aging; however Black Space analysis did show sensitivity to aging. Thus, because the Black Space diagram and the ω_o – R-value Space diagram are simple to develop, these diagrams have the potential to be used as a tool to evaluate the effect of rejuvenators on the mitigation of aging of high RAP mixtures.

KEYWORDS: reclaimed asphalt pavement, asphalt rejuvenator, aging, black space diagram, performance

1.0 Introduction

The use of reclaimed asphalt pavement (RAP) in paving mixtures has been a common practice in the United States. State transportation agencies and producers of hot-mix asphalt (HMA) have the experience to produce mixtures at low RAP contents that exhibit comparable performances to virgin mixtures during their life span. However, state transportation agencies are often reluctant to use higher RAP contents, generally greater than 20% by weight of total mixture (Copeland, 2011). Using higher amounts of RAP in new paving mixtures presents a concern that the resultant mixtures may be prone to more load and non-load associated cracking and adhesion/cohesion failures during the service life of the pavement. This is due to the asphalt binder contained in the RAP being significantly oxidized/hardened due to aging. Oxidation is a chemical reaction between the asphalt binder and oxygen that occurs at all stages of a paving mixture's life (production, construction, and in-service). Oxidation changes the molecular structure of the asphalt binder resulting in a stiffer binder that has a reduced capacity to relieve stresses. Increased aging has been shown to contribute to the reduction of the stress relaxation capacity of the binder which is part of what leads to decreased cracking resistance of the mixture (Molenaar et al., 2010; Daniel, 2013). Using larger amounts of RAP in new paving mixtures will compound these issues as increased amounts of aged RAP binder will be subsequently exposed to additional aging during production, construction, and in-service. A common strategy to mitigate the effect of the aged binder on the performance of mixtures with RAP is to use a softer virgin binder when higher than typical RAP contents of 15-20% are utilized.

Although conventional soft binders have been used in an attempt to compensate for the stiffness of aged RAP binders, several studies have shown that asphalt rejuvenators can allow more RAP to be incorporated in asphalt mixtures than a softer binders alone (Shen et al., 2007a and b; Shen et al., 2005; O'Sullivan, 2011; Al-Qadi et al. 2007; Brownidge and Grady, 2010). Asphalt rejuvenators are manufactured to restore the rheological properties of RAP binders by diffusing into them and restoring their colloidal structure and chemical components (Shen et al., 2005) (Brownidge and Grady, 2010). Rejuvenators have been extensively used in pavement surface preservation to revive hard and oxidized surface layers by penetrating into the pavement and fluxing with the aged binder to balance the maltenes to asphaltenes ratio (Brownidge and Grady, 2010). Even with these benefits, some state agencies do not use rejuvenators in new paving mixtures because of potential rutting-related concerns (O'Sullivan, 2011).

This study was undertaken in an effort to better understand the effects of rejuvenators on high RAP mixtures (50% RAP by weight of mixture), and it examined these effects after both short- and long-term aging to determine if rejuvenators can assist in mitigating aging in high RAP mixtures. A two tier evaluation was undertaken: (1) examine the rheology of extracted and recovered binders and (2) mixture performance tests. Tests performed on extracted and recovered binders included rheological plots of shear modulus (G*) versus phase angle (δ) (commonly known as a Black Space Diagram), rheological parameters derived from master curves using the Christensen-Anderson model, Superpave Performance Grading (PG) results, critical cracking temperatures, and Multiple Stress Creep Recovery (MSRC) results. The results from these tests were compared to mixture test results for rutting and moisture damage susceptibility, fatigue cracking and low temperature cracking to determine if they correlate. Overall, the results of the study provide a better understanding of the interrelationship between rejuvenators and aging in high RAP mixtures.

2.0 Background on the Rheological Methods Used to Assess Binder Aging

The performance grade (PG) system uses the Dynamic Shear Rheometer (DSR) to measure high and intermediate binder performances. It provides a shear modulus (G*) and a phase angle (δ) which is used to measure the elastic (recoverable) and viscous (non-recoverable) behaviors of binders at high temperatures, and elastic (recoverable) and cracking (non-recoverable) behaviors of binders at intermediate temperatures. Creep stiffness (S) and creep slope (m-value) from a Bending Beam Rheometer are used to measure low-temperature performance along with the critical cracking temperature provided by a Direct Tension test. The PGs provided by these tests and also the Multiple Stress Creep Recovery (MSCR) test were used in this study to evaluate changes in binder rheology due to aging and the effect of the rejuvenators.

Recently, G* and δ have also been used to generate a rheological plot commonly referred to as Black Space Diagram. Researchers have illustrated the use of Black Space Diagram to evaluate the changes in binder rheology due to aging (King et al., 2012; Anderson et al., 2011). Another analysis that can be used to evaluate these changes is the Christensen-Anderson Model (CAM). These two methodologies were the focus of this study.

2.1 Black Space Diagram and the Glover –Rowe Damage Parameter

Figure 1 illustrates a Black Space Diagram that shows the current performance grade (PG) parameter for fatigue cracking (G*sinδ) in addition to a new Black Space function defined by a new parameter, named the Glover-Rowe parameter, in the form of $G^*(\cos\delta)^2/(\sin\delta)$ (Anderson et al., 2011). This parameter was developed based on the Glover fatigue cracking parameter, G'/(η'/G'), which was found to have a high correlation to the ductility of the asphalt binder (Glover et al., 2005). It is determined from intermediate temperature DSR testing. The advantage of this Glover-Rowe parameter is that as long as the test frequency (ω) is known, variables G* and δ can be plotted to create a damage curve in black space. Based on the work of Anderson et al., (2011) and Rowe (2014), preliminary thresholds have been proposed to determine when non-load associated cracking, specifically block cracking, may begin (Damage Onset) and when there will be definite cracking problems (Significant Cracking). A typical cause of block cracking is the inability of asphalt binder to expand and contract with temperature cycles because of the aging of the asphalt binder. These thresholds have $G^*(\cos\delta)^2/(\sin\delta)$ values of 180 kPa and 450 kPa, respectively, when tested at 15°C (59°F) and a loading frequency of 0.005 rad/sec.

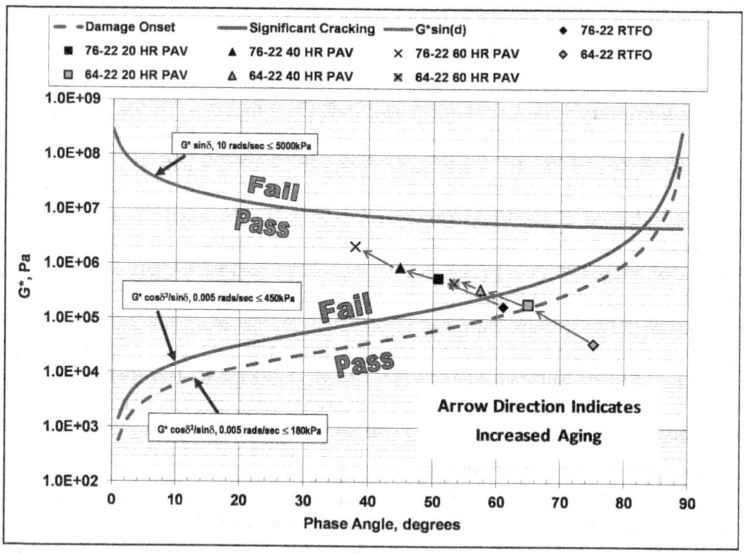

Figure 1. *PAV-aged binders passing through the Glover-Rowe damage zone.*

Similar to the work described by Anderson et al. (2011), Figure 1 illustrates data for two different binders (PG 64-22 and PG 76-22) that were aged in a pressure aging vessel (PAV) for 0, 20, 40, and 60 hours. The PAV aging was done after all binders were short-term aged in a rolling thin film oven (RTFO). The purpose of the longer PAV aging times was to create a more highly-aged sample in the laboratory. Using the new Glover-Rowe parameter and presenting the data in Figure 1, the RTFO aging for each binder started at the lower right location in the Black Space diagram; each additional aging period caused the rheological properties to move to the upper left of the diagram (increase in the stiffness and reduced phase angle for each binder). It should be noted that even after 60 hours of PAV aging, the asphalt binders still "Pass" the current G*sinδ Superpave specification. For extracted and recovered binders from high RAP content mixtures with consequently higher amounts of aged binder in a mixture, binder testing data is expected to follow the same trend illustrated in Figure 1. This trend is a rheological response to move towards the upper left of the Black Space diagram. Daniel (2013) confirmed this trend on extracted and recovered asphalt binder from plant produced mixtures in Vermont. The trend in the Black Space Diagram indicated that as RAP percentage increases for the same mixture, the G* and δ data migrates from the lower right to the upper left of the Black Space. Therefore, if a rejuvenator is assisting in restoring the rheological properties of aged RAP binder, the trend should be reversed on the Black Space Diagram (movement toward lower right of the diagram). This movement in Black Space, combined with the Glover-Rowe damage parameters, will be utilized in this study to understand if rejuvenators can assist in mitigating the aging experienced by binders extracted and recovered from high RAP mixtures. Additionally, a quantification of the binders' resistance to non-load associated cracking can be made utilizing the Glover-Rowe damage parameter.

2.2 Christensen-Anderson Model (CAM) Master Curve Parameters

Another tool that can be used to investigate the effect of rejuvenators on the rheological properties of aged binders is to use traditional rheological master curves of G* versus loading frequency. The Christensen-Anderson Model (CAM; Christensen and Anderson, 1992) is a very useful tool because the master curve parameters (ω_o, R, and T_d) have specific physical significance. The Cross-over Frequency, ω_o, is a measure of the overall hardness of the binder. As this frequency increases, the hardness of the binder decreases which is desirable for rejuvenated binders. The Rheological Index, R-value, is an indicator of the rheological type. It is defined as the difference between the log of the glassy modulus and the log of the dynamic modulus at the cross-over frequency. As R-value increases, the master curve becomes flatter indicating a more gradual transition from elastic behavior to steady-state flow. Normally, R-value is higher for oxidized asphalt. Accordingly, the R-value is expected to increase with oxidization. Therefore, for rejuvenators to be effective, the R-value for the overall aged plus virgin binder should decrease. Figure 2 shows the parameters used to define the shape of the binder master curve.

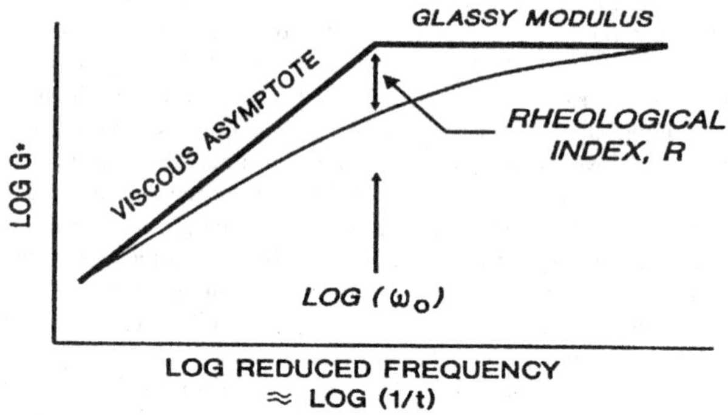

Figure 2. *Functional form of Christensen-Anderson asphalt binder master curve model (Christensen and Anderson, 1992).*

Figure 3 shows the ω_o and R-value for the same test data presented earlier in Figure 1. In this case, ω_o and R-value are plotted in their own space (ω_o – R-value Space). The PG64-22 and PG76-22 asphalt binders migrate from the upper left to the lower right of the ω_o – R-value Space as the magnitude of aging increases. The same trend can be expected as the RAP content of the asphalt mixture increases.

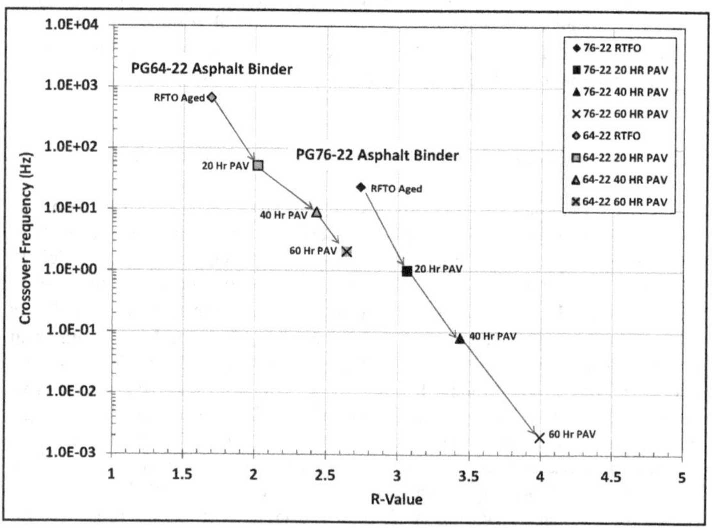

Figure 3. *Crossover frequency – R-value space: PG 64-22 and PG 76-22 asphalt binders after different aging levels.*

The master curve test results and analysis indicate that G* and δ, as well as the functional form of the master curve itself (ω_o and R-value), can be utilized to evaluate aging in an asphalt binder. Since aging can be clearly identified using this method, it is hypothesized that the same testing and analysis procedures can be utilized to evaluate the effectiveness of rejuvenators for mitigating the effects of aged asphalt binder in high RAP mixtures.

3.0 Objective

The primary objective of this research was to investigate if rheological plots of G* versus δ, commonly known as a Black Space Diagram, and binder rheological parameters taken from master curves using the Christensen-Anderson Model (CAM) can be used to evaluate the effects of rejuvenators on aging mitigation in high RAP mixtures. The results from these evaluations were compared to mixture test results for rutting and moisture damage susceptibility, fatigue cracking and low temperature cracking to determine if they correlate.

4.0 Experimental Plan

In order to address the objectives of this project, an experimental plan was developed as shown in Figure 4.

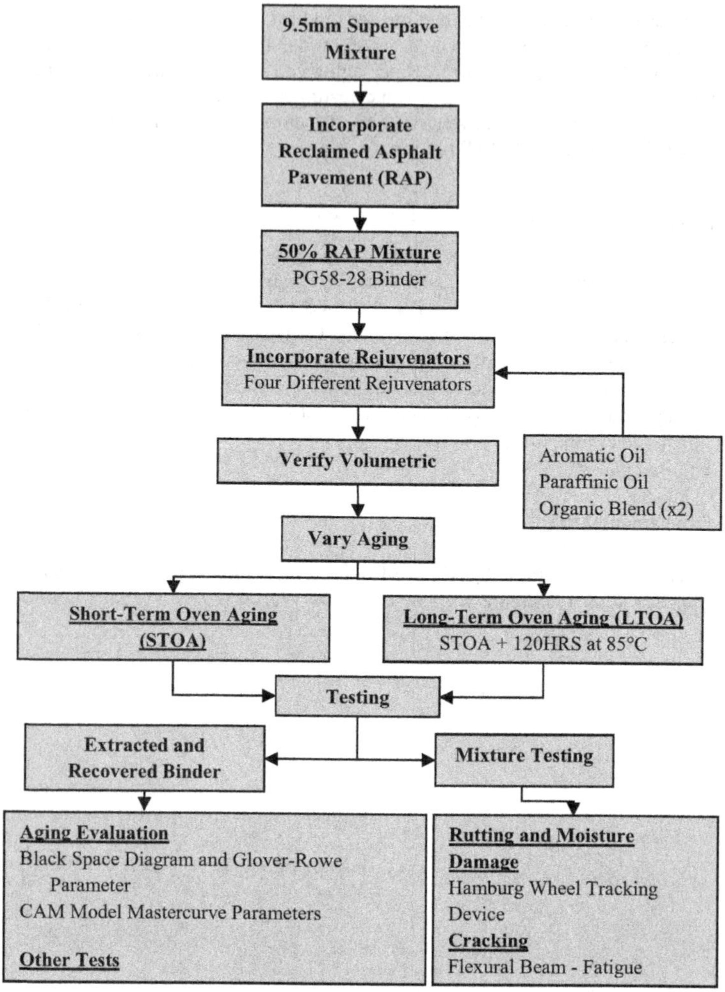

Figure 4. *Experimental plan.*

5.0 Materials

All materials for this study (binders, aggregates, and RAP) were obtained from sources in the state of Massachusetts.

5.1 Rejuvenators

Four asphalt rejuvenators were selected so that different types were represented including those already commonly used and those based on emerging green technologies. Details about each rejuvenator are outlined in Table 1.

Table 1. *Rejuvenator descriptions and details.*

Rejuvenator ID	Description	Details
AO	Aromatic Oil	Commonly used in pavement preservation activities
PO	Paraffinic Oil	Commonly used in pavement preservation activities
OB1	Organic Blend	Green chemistry product
OB2	Organic Blend	Organic oils based

The proposed dosage of each asphalt rejuvenator was based on a previous study (Mogawer et al., 2013) and also to achieve a resultant binder in the high RAP mixtures with similar binder grade to the control mixture used in the study. Based on the design binder content and percent asphalt binder in the RAP, this dosage equaled 9.0% rejuvenator by weight of recycled binder. This dosage was utilized throughout the study. Each rejuvenator was added directly to the pool of heated binder immediately prior to mixing for each specimen fabricated.

5.2 Asphalt Binder

A PG 58-28 binder was utilized for designing and evaluating the high RAP mixtures. As outlined previously, the conventional method for introducing higher RAP contents into asphalt mixtures is to use a softer PG grade binder. The PG 58-28 binder was used in an attempt to offset the potential mixture stiffening due to the use of high percentage of RAP in the mixtures. The PG 58-28 was selected because it was the softest grade available. Based on the viscosity of the binder, the mixture mixing temperature was 150°C (300°F) and the compaction temperature was 138°C (280°F).

5.3 Aggregates

The aggregates utilized were from a crushed stone source in Wrentham, Massachusetts. Two aggregate stockpiles were obtained: 9.5 mm crushed stone and stone dust. Each aggregate stockpile was tested to determine their gradations, which are shown in Table 2. Sieve analysis was completed in accordance with American Association of State Highway and Transportation Officials (AASHTO) test method T 11, "Standard Method of Test for Materials Finer Than 75-µm (No. 200) Sieve in Mineral Aggregates by Washing," and T 27, "Standard Method of Test for Sieve Analysis of Fine and Coarse Aggregates" (AASHTO, 2014).

Table 2. *Virgin aggregate stockpile and post ignition RAP aggregates properties.*

Sieve Size	9.5 mm	Stone Dust	RAP Aggregates Post Ignition
19.0 mm	100	100	100
12.5 mm	99.4	100	100
9.5 mm	93.8	100	100
4.75 mm	29.7	99.7	76.8
2.36 mm	5.2	83.7	57.6
1.18 mm	2.8	57.1	43.3
0.600 mm	2.3	38.6	31.1
0.300 mm	2.1	24.9	19.8
0.150 mm	1.8	15.9	12.1
0.075 mm	1.5	10.9	8.3
Binder Content, % (AASHTO T308) =			5.6%

5.4 Reclaimed Asphalt Pavement (RAP)

RAP was obtained from the same contractor as the aggregates. The RAP stockpile was fractioned in the laboratory in order to meet the gradation requirements for this study. The binder content of the RAP was determined using the ignition oven in accordance with AASHTO T 308, "Determining the Asphalt Binder Content of Hot Mix Asphalt (HMA) by the Ignition Method" (AASHTO, 2014). The aggregates in the RAP remaining post ignition were tested to determine their properties which are shown in Table 2.

6.0 Description of Mixture Aging Procedures

For this study, the major variable under investigation was the impact of rejuvenators on the aging properties of high RAP mixtures and their binders after different periods of aging. The three aging schemes utilized are described below. For the binder and mixture performance evaluations, testing and analysis was conducted after the short-term and long-term aging.

6.1 Mixture Design and Volumetric Verification Aging

For mixture design and volumetric property verification, specimens of each mixture were batched, mixed and aged at the respective compaction temperature previously noted for 2 hours ± 5 minutes in a loose state in accordance with AASHTO R 30, "Standard Practice for Mixture Conditioning of Hot Mix Asphalt (HMA)," Section 7.1, Mixture Conditioning for Volumetric Mixture Design (AASHTO, 2014). Specimens were stirred after 60 ± 5 minutes to maintain uniform conditioning. After aging, specimens were immediately compacted.

6.2 Short-Term Oven Aging (STOA)

STOA was conducted in accordance with AASHTO R 30, "Standard Practice for Mixture Conditioning of Hot Mix Asphalt (HMA)," Section 7.2 Short-Term Conditioning for Mixture Mechanical Property Testing (AASHTO, 2014). For STOA aging, specimens of each mixture were batched, mixed and aged at 135 ± 3°C (275± 5°F) for 4 hours ± 5 minutes in a loose state. Specimens were stirred after 60 ± 5 minutes to maintain uniform conditioning. After aging, specimens were returned to the compaction temperature and immediately compacted.

6.3 Long-Term Oven Aging (LTOA)

LTOA was conducted in accordance with AASHTO R 30, "Standard Practice for Mixture Conditioning of Hot Mix Asphalt (HMA)," Section 7.3 Long-Term Conditioning for Mixture Mechanical Property Testing (AASHTO, 2014). For LTOA aging, specimens were first subjected to STOA. Next, the compacted STOA specimens were cooled for 16 ± 1 hour. In accordance with AASHTO R 30, Subsection 7.3.4 Long-term Conditioning of Prepared Test Specimens, the compacted specimens were conditioned in an 85 ± 3°C (185 ± 5°F) oven for 120 ± 0.5 hours (5 days). After aging, the specimens were allowed to cool at room temperature which took approximately 16 hours. The specimens were not handled or disturbed until after completely cooled to room temperature.

7.0 Mixture Design

The target gradation for the mixtures utilized in this study is shown in Table 3 and Figure 5. This target gradation met the requirements for a 9.5 mm Superpave mixture in accordance with AASHTO M 323, "Superpave Volumetric Mix Design," and AASHTO R 35, "Superpave Volumetric Design for Hot Mix Asphalt" (AASHTO, 2014).

Table 3. *Target mixture gradation and specification.*

Sieve Size	Sieve Size (mm)	Target Gradation for All Mixtures	Superpave 9.5mm Specification
3/4"	19.0	100	-
1/2"	12.5	100	100 min
3/8"	9.5	98	90-100
No. 4	4.75	85	90 max
No. 8	2.36	58	32-67
No. 16	1.18	42	-
No. 30	0.600	27	-
No. 50	0.300	15	-
No. 100	0.150	9	-
No. 200	0.075	6.0	2-10
Binder Content		6.5%	-

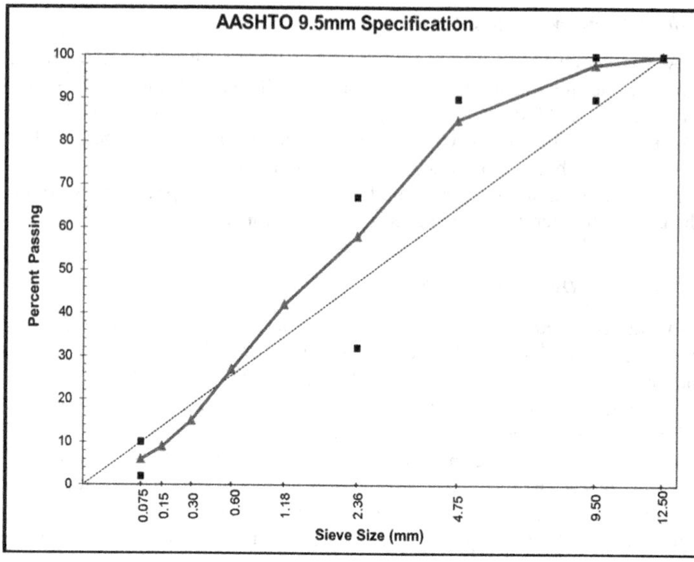

Figure 5. *Mixture design gradation.*

A control mixture utilizing all virgin materials was designed using the PG 58-28 binder. RAP was then used to replace 50% of the virgin aggregates with RAP aggregates. The aggregate gradations for the control mixture and the 50% RAP mixture were identical. Because they were identical, mixture design verifications for the 50% RAP mixtures incorporating each rejuvenator were performed at the design binder content determined for the control mixture. Verifications were completed assuming 100% contribution of the RAP binder; hence the virgin binder content utilized for the verification was lowered to account for the recycled binder content.

The design Equivalent Single Axle Loads (ESALs) for this project was selected as 0.3 to <3 million which is consistent with surface course mixtures in New England. The design Superpave gyratory compactive effort for this ESAL level was N_{design} = 75 gyrations.

Dried virgin aggregates were utilized for the mixture design and subsequent mixture verifications at the mixing and compaction temperatures previously noted. To incorporate the RAP into the mixtures, a procedure that was used in a prior study utilizing similar materials was followed (Mogawer et al., 2011). This procedure was utilized to eliminate moisture in the RAP stockpile material and to optimize the blending between the aged and virgin binders in the mixture. The procedure steps were:

1. The RAP was air dried until a constant mass was achieved.

2. The RAP was further dried for two days at 60°C (140°F).
3. The RAP was added to the heated dry aggregates two hours prior to adding the binder during the mixing process.

The average results of the mixture design and volumetric verifications are shown in Table 4. Relative to the control mixture, the 50% RAP mixture with no rejuvenator had higher air voids (+1.1%) and a lower Voids Filled with Asphalt (VFA). This may be an indication that good blending is occurring between the virgin and RAP binders with the RAP binder dominating the resultant binder and subsequently leading to the mixture being more difficult to compact. Conversely, the discrepancies in volumetric properties could also be an indicator that incomplete blending or no blending is occurring between the virgin and RAP binders leading to lower effective asphalt content and thus higher air voids. The addition of the rejuvenators did improve the volumetric properties by decreasing the difference in the air voids between the control and the 50% RAP mixture from +1.1% to a maximum difference of +0.48%. This maximum difference was considered acceptable (i.e., the mixtures were equivalent) since an air voids tolerance of ±0.5% is common during production. The effect of these rejuvenators on the rheological properties of the binders described in the next few sections might provide insights into the reasons for this improvement. Also note that the Voids in Mineral Aggregate (VMA) percentage for the 50% RAP mixtures with rejuvenators were slightly below the minimum of 15.0% but deemed acceptable for further study because they were within typical allowable production tolerances.

Table 4. *Mixture design and volumetric verification results.*

Mixture	Air Voids, %	Voids in Mineral Aggregate VMA, %	Voids Filled with Asphalt VFA, %
Specification	*4.0%*	*15% min.*	*65-78*
Control – PG58-28	3.63	15.9	77.2
50% RAP + No Rejuvenator	4.77	15.3	68.7
50% RAP + Aromatic Oil	3.90	14.9	73.8
50% RAP + Paraffinic Oil	4.11	14.7	72.0
50% RAP + Organic Blend 1	3.95	14.3	72.3
50% RAP + Organic Blend 2	4.06	14.5	72.0

8.0 Binder Testing and Results

The asphalt binders of the various mixtures were extracted and recovered in accordance with AASHTO T 164, "Standard Method of Test for Quantitative Extraction of Asphalt Binder from Hot Mix Asphalt (HMA)," (AASHTO, 2014) and ASTM D5404, "Standard Practice for Recovery of Asphalt from Solution Using the Rotary Evaporator" (ASTM, 2012). After the recovery process, each asphalt binder was tested for its performance grade (PG) in accordance with AASHTO M 320, "Standard Specification for Performance-Graded Asphalt Binder," (AASHTO, 2014) and by the Multiple Stress Creep Recovery (MSCR) in accordance with AASHTO T 350-14, "Standard Method of Test for Multiple Stress Creep Recovery (MSCR) Test of Asphalt Binder Using a Dynamic Shear Rheometer (DSR)" (AASHTO, 2014). Critical cracking temperatures were determined in accordance with AASHTO R 49, "Determination of Low-Temperature Performance Grade (PG) of Asphalt Binders" (AASHTO, 2014). G* master curves were also measured and utilized to evaluate the overall stiffness properties of the asphalt binders as well as their relative aging characteristics. The G* master curve data was generated using the 4.0 mm geometry for the DSR as per the recommendations by Sui et al. (2010 and 2011) and Farrar et al. (2013).

Prior to actual testing of each recovered asphalt binder, the DSR was run in oscillatory mode within strain ranges of 0.005 to 0.2% to ensure the asphalt binder sample would be tested in the linear region at the respective test temperature. The following test temperatures were used in the following order: -15, -5, 5, 15, 25, 35, 45, 60, and 75°C. The following range of loading frequencies were utilized to cover two decades of loading times and conducted in the following order: 157, 100, 62.8, 31.4, 10, 6.28, 3.14, 1.0, 0.63, 0.1, and 0.063 rad/sec. The master curves were generated using the RHEA™ software. The CAM model parameters were determined using the entire dataset of test frequencies and temperatures. The Glover-Rowe parameter was determined by curve shifting and not measured directly at 0.005 rad/sec at 15°C.

8.1 Aging Mitigation Evaluation

As discussed earlier, the Glover-Rowe parameter, defined as $G^*(\cos\delta)^2/(\sin\delta)$, has been found to be sensitive to the aging of asphalt binders. Based on work conducted by Anderson et al. (2011), as well as additional analysis by Rowe (2014), thresholds have been found that correlate well to non-load associated cracking (i.e., Block Cracking). One of the causes for block cracking is the aging of the asphalt binder. Figure 6 shows the Glover-Rowe parameter plotted in Black Space at a test temperature of 15°C (59 °F) and loading frequency of 0.005 rad/sec for the control mixtures. As indicated in the figure, as the Control mixtures were aged from STOA to LTOA, the test data migrates from the lower right of the Black Space towards the upper left. The figure also shows that the virgin PG 58-28 Control mixture aged more than the 50% RAP mixture based on the magnitude of the movement in Black Space. This would be expected as the 50% RAP mixture already has a significant amount of aged/oxidized asphalt binder that would be minimally aged due to the laboratory aging procedure. It is also interesting to note that the 50% RAP mixture at LTOA condition passes the initial "Onset of Cracking" threshold shown in the figure, while the virgin PG 58-28 mixture stays below the threshold in the "PASS" zone. A summary of all of the Black Space and Master Curve parameters can be found in Table 5.

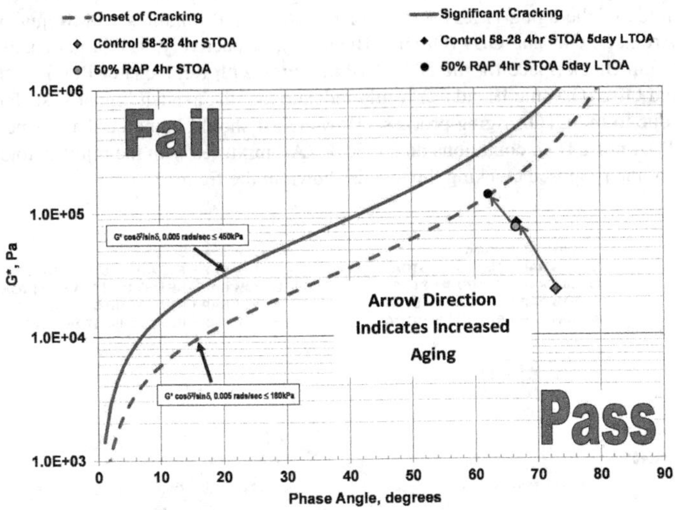

Figure 6. *Black Space diagram for control and 50% RAP mixtures – short-term and long-term aging.*

Table 5. *Black Space and master curve parameters.*

Mixture	Aging	G* @ 15°C and 0.005 rad/sec (Pa)	δ @ 15°C and 0.005 rad/sec (degrees)	Crossover Frequency (ω_o)	Rheological Index (R-value)
Control PG58-28	STOA	23,320	72.9	732.2	1.92
	LTOA	80,780	66.7	103.0	2.04
50% RAP – No Rejuvenator	STOA	74,840	66.6	113.4	2.15
	LTOA	136,830	62.2	31.6	2.27
50% RAP + Aromatic Oil	STOA	41,689	68.8	234.7	2.09
	LTOA	69,730	66.2	98.5	2.15
50% RAP + Paraffinic Oil	STOA	22,320	70.4	405.5	2.08
	LTOA	79,550	63.2	44.0	2.20
50% RAP + Organic Blend 1	STOA	59,070	67.8	136.6	2.06
	LTOA	65,260	65.8	79.8	2.15
50% RAP + Organic Blend 2	STOA	25,890	68.8	205.4	2.04
	LTOA	75,220	63.7	56.5	2.27

The Black Space diagram for the STOA and LTOA rejuvenator mixtures is shown in Figure 7. The results again show a migration of the test data. However, the magnitude of the aging is less severe for the 50% RAP mixtures with Rejuvenator AO (Aromatic Oil) and OB1 (Organic Blend #1), respectively. The larger change in location in Black Space for the 50% RAP mixtures with Rejuvenator PO (Paraffinic Oil) and OB2 (Organic Blend #2) would indicate that the asphalt binder is stiffening more due to the LTOA aging process. However, it should be noted that for neither the STOA nor LTOA conditions do the 50% RAP mixtures with the rejuvenators fail either of the proposed cracking thresholds shown in the figure.

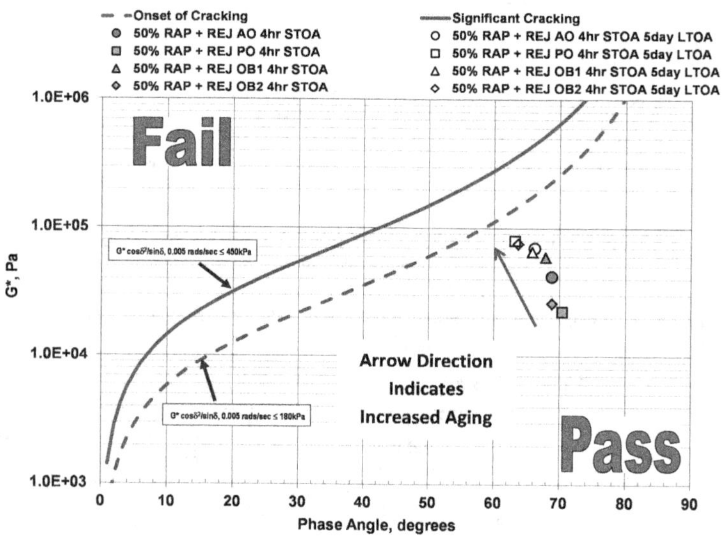

Figure 7. *Black Space diagram for rejuvenator mixtures – short-term and long-term aging.*

Figures 8 and 9 show the Black Space diagrams for the STOA and LTOA conditions, respectively. Along with the rejuvenator mixtures, the figures also contain the virgin PG 58-28 Control mixture and the 50% RAP with no rejuvenator mixture. When comparing the relative aging in the figures, the more effective rejuvenator would result in Black Space parameters closer to virgin PG 58-28 Control mixture. Meanwhile, rejuvenators that are less effective would result in Black Space parameters closer to the 50% RAP mixture. As indicated by the "green" arrow in Figure 8, each of the rejuvenators used provides some aging mitigation of the aged binder associated with the 50% RAP mixture and moves the respective samples away from the cracking thresholds. Based on the STOA condition, Figure 8 indicates that Rejuvenator PO (Paraffinic Oil) provides the largest mitigation of aging as its Black Space parameters are very similar to the virgin PG 58-28 Control sample and moved furthest away from the 50% RAP sample.

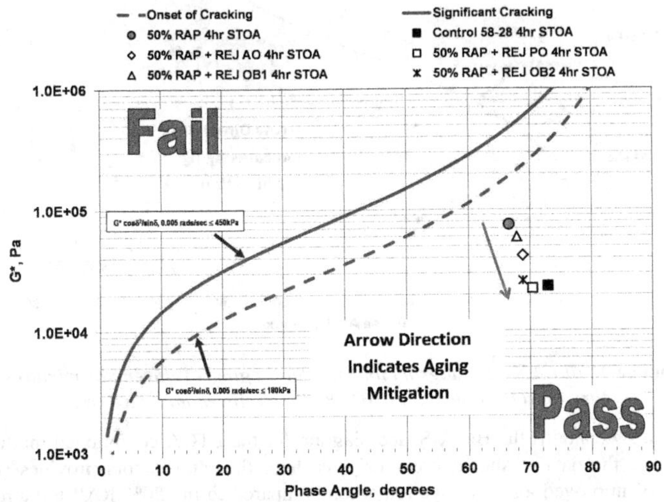

Figure 8. *Black Space diagram for short-term aging – rejuvenator mixtures compared to control and 50% RAP no rejuvenators mixture.*

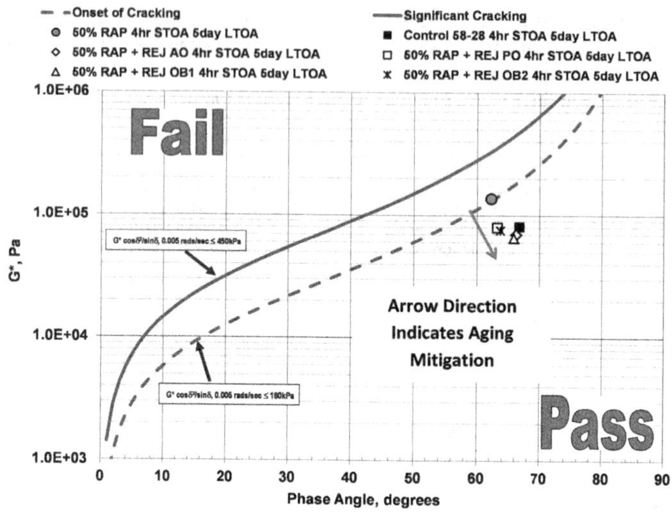

Figure 9. *Black Space diagram for long-term aging – rejuvenator mixtures compared to control and 50% RAP no rejuvenators mixture.*

Figure 9 contains the Black Space diagram for the LTOA conditioned mixtures. Similar to the dataset shown previously, each of the rejuvenators provides some degree of improved aging mitigation when compared to the 50% RAP mixture. In the case of the LTOA conditioned mixtures, Rejuvenators AO (Aromatic Oil) and OB1 (Organic Blend #1) resulted in the best performance as both of these mixtures were closer to, if not slightly better than, the virgin PG 58-28 Control. This would indicate that Rejuvenators AO (Aromatic Oil) and OB1(Organic Blend #1) are still providing a level of rejuvenating even after long-term aging conditions – something that Rejuvenators PO (Paraffinic Oil) and OB2 (Organic Blend #2) appear to be limited at accomplishing.

8.2.1 CAM Model – Crossover Frequency – R-value Space as an Aging Index

Another proposed method for assessing aging and aging mitigation of asphalt binders is through the evaluation of the CAM model parameters Cross-over Frequency (ω_o) and Rheological Index (R-value). As presented earlier, as the parameters plotted in ω_o – R-value Space migrate from the upper left to the lower right of the space, it clearly tracks the aging of the asphalt binder. Therefore, asphalt binders that have undergone a rejuvenation, or aging mitigation, should migrate back towards the upper left.

Figure 10 shows the ω_o – R-value Space for the control and 50% RAP no rejuvenator mixtures. As one would expect, as aging increased from STOA to

LTOA, the test data moves downward and towards the right, indicating that aging of the asphalt binder has taken place. The magnitude of the change in the virgin PG 58-28 Control mixture is greater than the 50% RAP mixture as approximately half of the asphalt binder in the 50% RAP mixture has already been highly oxidized.

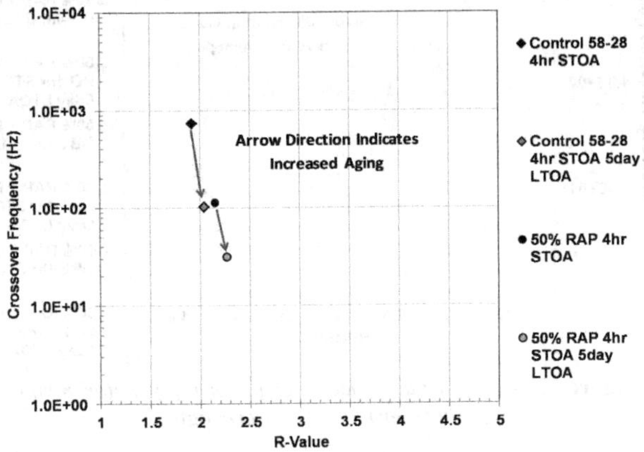

Figure 10. *ω_o – R-value Space diagram for control and 50% RAP no rejuvenators mixtures – short-term and long-term aging.*

The rejuvenator test samples are shown in Figure 11. Again, a downward right movement is observed in the test data, indicating that binder aging has taken place. However, different magnitudes in the downward movement are observed between the different rejuvenated samples. The greatest change was found when using Rejuvenator PO (Paraffinic Oil), which would indicate that this particular rejuvenator may not maintain the same rejuvenating potency over the long-term performance of the asphalt mixture. Meanwhile, the mixtures using Rejuvenator AO (Aromatic Oil) and OB1 (Organic Blend #1) show the least change in aging between the STOA and LTOA conditions.

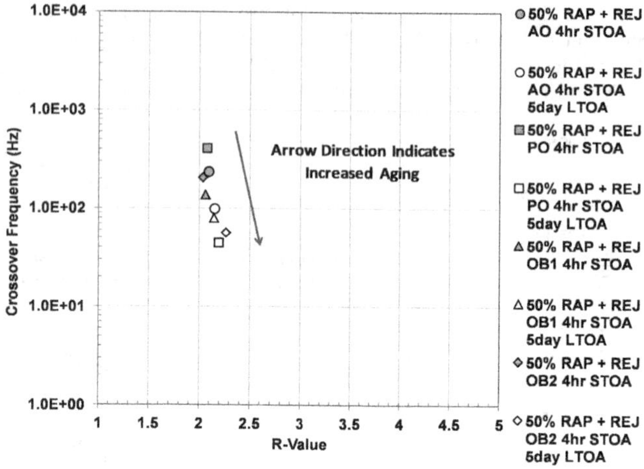

Figure 11. *ω_o – R-value Space diagram for rejuvenator mixtures – short-term and long-term aging.*

Figures 12 and 13 show the mitigation of aging when using the different rejuvenators and comparing the rejuvenators in the ω_o – R-value Space with the virgin PG 58-28 Control and the 50% RAP mixture. In Figure 12, the STOA samples show an upward left movement away from the 50% RAP sample (no rejuvenator) and towards the virgin PG 58-28 Control sample. The same type of change in the ω_o – R-value Space is also observed in Figure 13 for the LTOA samples; however, the magnitude of the movement away from the 50% RAP (no rejuvenator) sample is not as great. The results shown in Figures 12 and 13 indicate that all four of the rejuvenators evaluated in this study provide some type of aging mitigation to the 50% RAP mixture. However, the amount of aging mitigation witnessed varies among the different rejuvenators, as well as the degree of mixture aging.

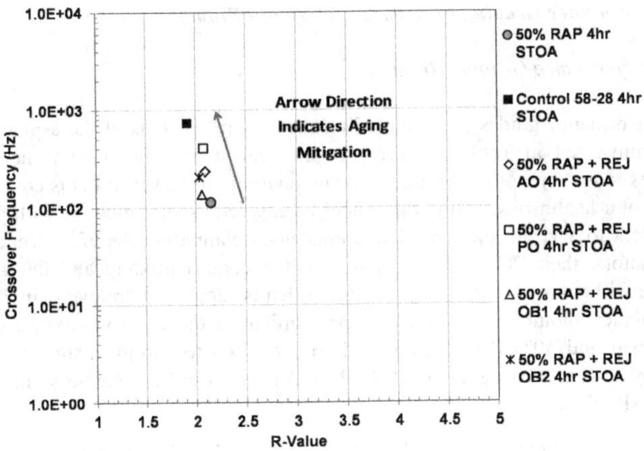

Figure 12. *ω_o – R-value Space diagram for short-term aging – rejuvenator mixtures compared to the control and 50% RAP no rejuvenators mixtures.*

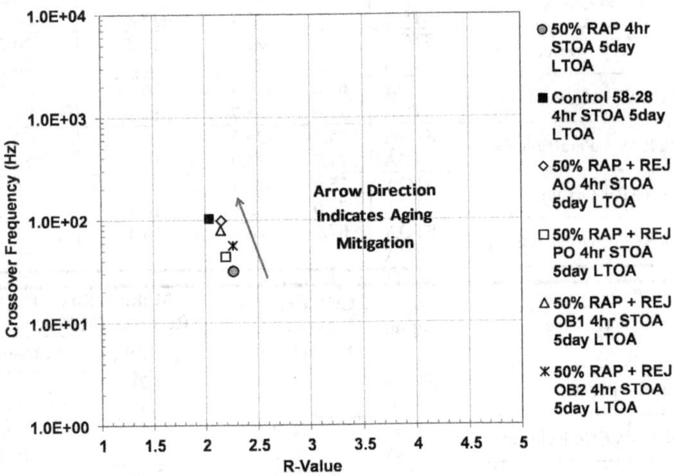

Figure 13. *ω_o – R-value Space diagram for long-term aging – rejuvenator mixtures compared to the control and 50% RAP no rejuvenators mixtures.*

8.2 Performance Grading of Recovered Asphalt Binders

8.2.1 Performance Grading Results

The performance grades (PG), including the continuous PGs of the asphalt binders, are summarized in Table 6. From a practical standpoint, most if not all state agencies would be willing to utilize a rejuvenating product if it: (1) is cost effective, (2) is not a health risk, and (3) produces a recycled asphalt mixture with properties equivalent to a virgin mixture. Therefore, when comparing the effectiveness of the rejuvenators, their PGs were compared to the control mixture and the 50% RAP mixture without rejuvenator. A rejuvenator that is capable of "rejuvenating" the aged RAP binder should result in PGs very similar to the control mixture or fall in between it and 50% RAP mixture without rejuvenator. Rejuvenators that are not effective in rejuvenating the aged RAP binder will result in PGs very similar to the 50% RAP mixture.

Table 6. *Recovered asphalt binder PG grade results.*

Mixture	Aging	Continuous Grade (°C)			PG Using AASHTO R29 & M320
		High	Low	Intermediate	
Control – PG58-28	STOA	64.5	-29.4	16.5	64-28
	LTOA	69.7	-27.7	17.8	64-22
50% RAP + No Rejuvenator	STOA	71.4	-27.8	18.8	70-22
	LTOA	75.5	-25.5	21.3	70-22
50% RAP + Aromatic Oil	STOA	68.0	-28.4	16.7	64-28
	LTOA	71.6	-28.7	17.8	70-28
50% RAP + Paraffinic Oil	STOA	65.2	-29.5	15.3	64-28
	LTOA	71.7	-25.0	17.7	70-22
50% RAP + Organic Blend 1	STOA	75.5	-27.7	19.8	70-22
	LTOA	73.0	-25.4	18.4	70-22
50% RAP + Organic Blend 2	STOA	67.5	-28.4	16.3	64-28
	LTOA	71.9	-27.1	17.5	70-22

Mixture	Aging	Crit. Cracking Temp (°C) AASHTO R 49	Multiple Stress Creep Recovery (MSCR) @64°C	
			J_{nr} (1/kPa)	% Recovery
Control – PG58-28	STOA	-28.1	4.21	0.7
	LTOA	-26.8	1.80	3.0
50% RAP + No Rejuvenator	STOA	-26.9	1.47	4.6
	LTOA	-25.1	0.71	10.1
50% RAP + Aromatic Oil	STOA	-28.6	2.38	2.3
	LTOA	-28.3	1.37	4.8
50% RAP + Paraffinic Oil	STOA	-30.1	3.77	0.6
	LTOA	-28.8	1.37	5.1
50% RAP + Organic Blend 1	STOA	-26.5	1.76	2.6
	LTOA	-27.9	1.23	5.8
50% RAP + Organic Blend 2	STOA	-30.6	2.69	2.0
	LTOA	-27.9	1.30	5.2

8.2.2 Recovered Asphalt Binder High Temperature Performance

As would be expected, Table 6 shows that the control mixture had the lowest high temperature PG, as well as the largest non-recoverable creep compliance (J_{nr}) at both STOA and LTOA. Rejuvenator PO (Paraffinic Oil) provided high temperature properties very similar to the control mixture including the J_{nr} values.

8.2.3 Recovered Asphalt Binder Low Temperature Performance

It is often the fatigue and low temperature properties that are of concern with high RAP content mixtures. As noted previously, a rejuvenator utilized in the 50% RAP mixture should provide low temperature properties similar to or better than the Control mixture. Based on Table 6, the low temperature PG from AASHTO R 29 indicates that Rejuvenator AO (Aromatic Oil) restored the low temperature properties of the 50% RAP mixture to those of the control mixture. The low temperature PG, determined using AASHTO R 49, indicates that three 50% RAP mixtures with rejuvenators provided critical cracking temperatures similar to or better than the control mixture. The rejuvenators were AO (Aromatic Oil), PO (Paraffinic Oil), and OB2 (Organic Blend #2).

9.0 Mixture Tests

The results from the binder evaluation were compared to mixture test results for rutting and moisture damage susceptibility, fatigue cracking and low temperature cracking to determine if they correlate.

9.1 Rutting and Moisture Damage

Rutting and moisture damage testing was conducted in accordance with AASHTO T 324, "Hamburg Wheel-Track Testing of Compacted Hot-Mix Asphalt (HMA)" (AASHTO, 2014). This test is utilized to determine the failure susceptibility of a mixture due to weakness in the aggregate structure, inadequate binder stiffness, or moisture damage (AASHTO, 2014). In this test, a mixture is submerged in heated water (typically 40-50°C) and subjected to repeated loading provided by a 705N (158lb) steel wheel. As the steel wheel loads the specimen, the corresponding rut depth of the specimen is recorded. The rut depth versus numbers of passes of the wheel is plotted to determine the Stripping Inflection Point (SIP) as shown in Figure 14. The SIP gives an indication of when the test specimen begins to exhibit moisture damage (stripping).

Gyratory specimens for this test were fabricated using the Superpave gyratory compactor (SGC) to an air void level of 7.0±1.0% as required by AASHTO T 324. Mixtures were tested after undergoing both STOA and LTOA aging. Testing was conducted at a test temperature of 50°C (122°F). The specimens were tested at a rate of 52 passes per minute after a soak time of 30 minutes at the test temperature. Testing terminated at 20,000 wheel passes or when visible stripping was noted. Table 7 presents the HWTD results.

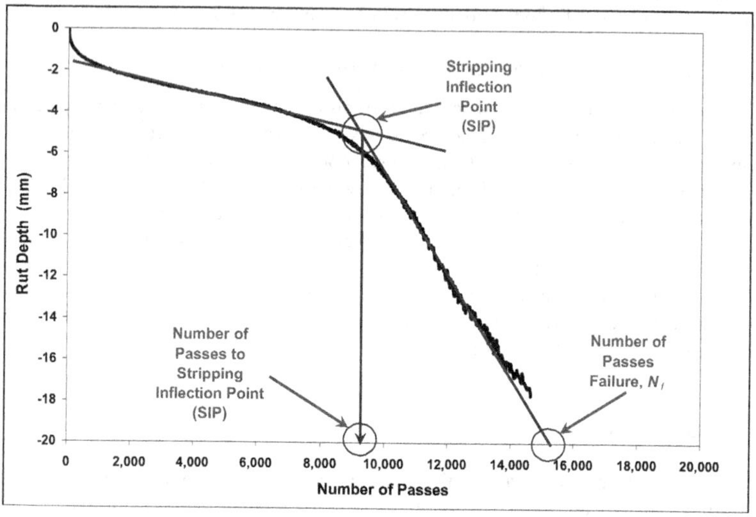

Figure 14. *Determination of HWTD stripping inflection point (SIP).*

The data in Table 7 illustrates, for mixtures with STOA aging, that the control mixture failed the HWTD test fairly quickly while the 50% RAP mixture with no rejuvenator passed the test with no inflection point and minimal rutting (< 2 mm). If the rejuvenators used in this study are effective in mitigating the aging of the aged RAP binder, it would be expected that the performances of the 50% RAP mixtures with rejuvenators fall between the performance of the control mixture and 50% RAP mixture with no rejuvenator. This is expected because the overall binder in 50% RAP mixtures with rejuvenators should be stiffer than the control binder (PG 58-28) and softer than the overall binder in the 50% RAP mixture without rejuvenator. After STOA and LTOA the data indicates, as presented in Table 7, that stripping inflection points and the rut depths of the 50% RAP mixtures with the rejuvenators did fall between them, and it agrees with the Black Space and ω_o – R-value Space diagrams that suggested these rejuvenators are mitigating the aging of the RAP binder. Also a relationship between MSCR and the HWTD rutting was not found. This was most likely due to the fact that when the HWTD test is conducted wet and there is a combined effect of permanent deformation and stripping that occurs.

Table 7. *Rutting and moisture susceptibility HWTD results.*

Mixture	Aging	Stripping Inflection Point	Rut Depth at 10,000 Passes (mm)	Rut Depth at 20,000 Passes (mm)
Control PG58-28	STOA	5,900	>20	>20
	LTOA	12,000	3.00	>20
50% RAP - No Rejuvenator	STOA	NONE	1.09	1.80
	LTOA	NONE	0.91	1.45
50% RAP + Aromatic Oil	STOA	9,800	5.06	>20
	LTOA	14,700	1.65	12.21
50% RAP + Paraffinic Oil	STOA	14,500	1.46	12.76
	LTOA	15,200	3.13	14.15
50% RAP + Organic Blend 1	STOA	8,500	7.61	>20
	LTOA	13,600	1.92	5.20
50% RAP + Organic Blend 2	STOA	12,900	1.92	16.70
	LTOA	12,500	3.10	17.73

9.2 Cracking

9.2.1 Fatigue Cracking - Four-Point Flexural Bending Beam AASHTO T321

One of the most common and historically used laboratory test procedures to evaluate the fatigue cracking performances of asphalt mixtures is the four-point flexural beam fatigue test, and it is the only standard procedure. The AASHTO test protocol is AASHTO T 321, "Determining the Fatigue Life of Compacted Hot Mix Asphalt (HMA) Subjected to Repeated Flexural Bending" (AASHTO, 2014). In order to investigate the relative fatigue cracking performances of the mixtures used in this study, the four point bending beam fatigue tests were conducted following this procedure.

Slabs with dimensions of 150 mm wide, 180 mm tall, and 450 mm long (6 in. wide, 6 in. tall, and 17.5 in. long) were fabricated for each mixture using the IPC Global Pressbox slab compactor. From each slab, beams with dimensions of 63 mm wide, 50 mm tall, and 380 mm long (2.5 in. wide, 2 in. tall, and 15 in. long) were cut such that the sides had smooth faces. The air voids of the final cut specimens were 7±1%. Beam specimens were conditioned at the test temperature of 15°C (59°F) for at least two hours prior to testing. A 15°C (59°F) test temperature was selected as it represents the intermediate temperature for the Northeast.

Each beam fatigue test was conducted in strain control at a loading frequency of 10Hz applied using a sinusoidal waveform. Specimens were tested at strain levels of 250 με, 500 με, and 750 με. Initially, all mixtures were tested at the 250 με. However, at this strain level, the 50% RAP mixture with the rejuvenators reached over six million cycles with less than 20% loss in the initial stiffness measured at 50 cycles. Therefore, the two higher strain levels were selected based on trial and error with the aim of achieving more than 10,000 cycles at failure. At 500 $\mu\varepsilon$ and 750 $\mu\varepsilon$, all mixtures lost 50% of their initial stiffness closely after at least 10,000 cycles. The number of cycles to failure was determined by fitting an exponential function to the flexural stiffness versus number of cycles relationships and then evaluating the number cycles that it took to decrease the initial stiffness by 50%. The beam fatigue testing results are shown in Figures 15 and 16.

The data in Figures 15 and 16 indicates, for the majority, that the number of cycles to failure for the 50% RAP mixtures with rejuvenators improved as compared to the 50% RAP mixture without rejuvenator. This occurred for both aging conditions. In one case (50% RAP + Aromatic Oil LTOA) at 750 $\mu\varepsilon$, the average cycles to failure marginally decreased. Overall, the data suggested that the rejuvenators are helping in "rejuvenating" the aged RAP asphalt binder, leading to an overall binder softer than the overall binder in the 50% RAP mixture without rejuvenator. This agrees with the Black Space and the ω_o – R-value Space diagrams that suggested these rejuvenators are softening the aged RAP binder. It should be noted that these observations are made based on the fact that the fatigue characteristics of mixtures are a function of the overall binder stiffness.

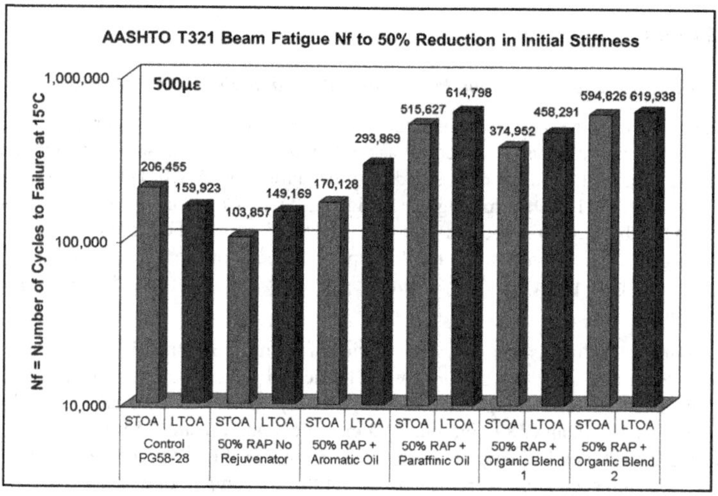

Figure 15. *Fatigue cracking results from the flexural beam fatigue - 500$\mu\varepsilon$.*

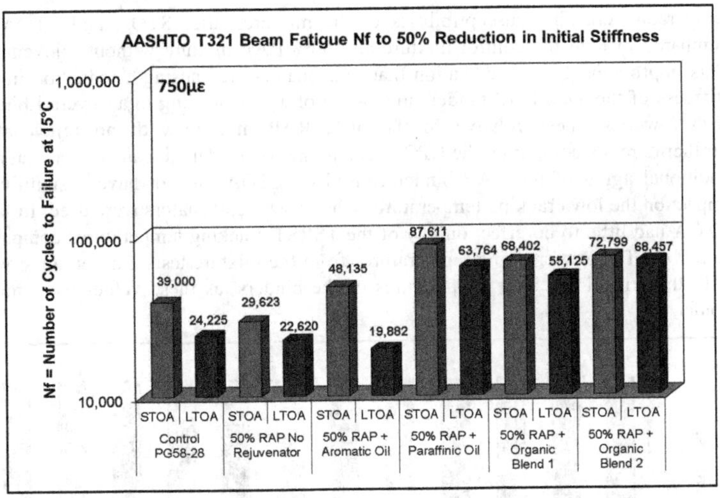

Figure 16. *Fatigue cracking results from the flexural beam fatigue - 750με.*

9.2.2 Low Temperature Cracking - Thermal Stress Restrained Specimen Test (TSRST)

To assess the effect of aging and rejuvenators on the low temperature cracking susceptibilities of the mixtures, each mixture was tested in the Thermal Stress Restrained Specimen Test (TSRST) device in accordance with AASHTO TP 10-93 (AASHTO, 1993) with the exception that Superpave gyratory compacted specimens were utilized.

In the TSRST test, an asphalt mixture specimen is cooled at a constant rate (-10°C/hour) while its original length is held constant by the TSRST device. As the specimen gets colder, it tries to contract but cannot which results in the accumulation of thermal stresses. Eventually the thermal stresses exceed the tensile strength capacity of the specimen resulting in specimen fracture (crack). The temperature at which this fracture occurs is recorded and noted as the low cracking temperature of the mixture.

SGC specimens 185 mm (7.3 in) tall by 150 mm (5.9 in) in diameter were fabricated for each mixture. TSRST specimens were then cored and cut to a final height of 160 mm tall (6.3 in.) by 54 mm (2.1 in.) in diameter. The air voids of the final cut specimens were 7±1%. The TSRST results are presented in Figure 17.

The incorporation of the 50% RAP into the mixture made the low cracking temperature warmer relative to the control mixture by 4°C and 3°C after STOA and LTOA, respectively. The incorporation of the rejuvenators helped improve the low temperature cracking susceptibilities of the mixtures after STOA and LTOA as compared to both the control mixture and 50% RAP mixture without rejuvenator. This improvement is an indication that rejuvenators are causing a reduction in the stiffness of the aged RAP binder (mitigation of aging) leading to an overall binder with lower stiffness relative to the 50% RAP mixture with no rejuvenator. Furthermore, by comparing the LTOA data to the STOA data, it can be seen that any additional aging of the RAP binder caused by LTOA did not have a significant impact on the low cracking temperatures when these rejuvenators were used. In fact, LTOA had little to no effect on any of the TSRST cracking temperatures compared to STOA. The low cracking temperatures from the mixture tests did not agree well with the critical cracking temperatures of the binders, as these values were much colder.

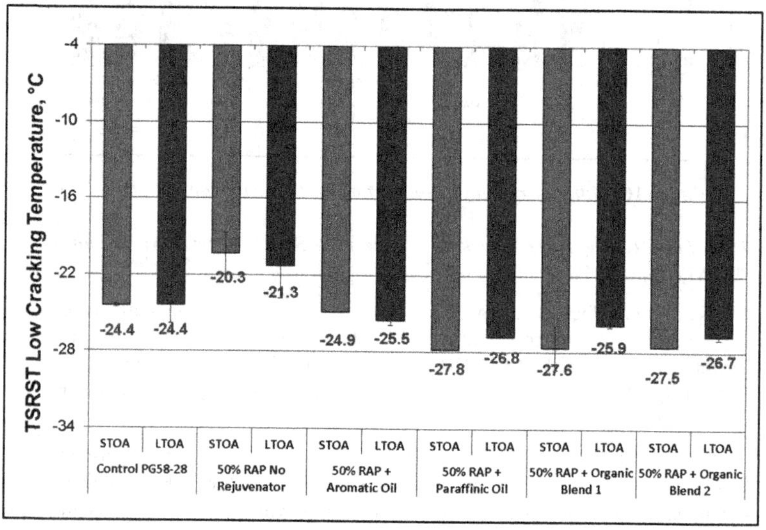

Figure 17. *Low temperature cracking results from the TSRST.*

10.0 Conclusions

This study was undertaken to better understand the effects of rejuvenators on high RAP mixtures (50% RAP), and it examined these effects after both short- and long-term aging to determine if rejuvenators can assist in mitigating aging in high RAP mixtures. A two tier evaluation was undertaken: (1) examine the rheology of extracted and recovered binders and (2) mixture performance tests. Based on the testing and analysis the following conclusions were made:

- The Black Space parameters for the virgin PG 58-28 control mixture and the 50% RAP mixture with no rejuvenator showed that when these mixtures were aged from STOA to LTOA, the test data migrated from the lower right of the Black Space towards the upper left, indicating that aging of the asphalt binder had taken place. The same trend was observed when rejuvenators were used. However, the magnitude of the aging was less severe for certain rejuvenators. This indicates that the type of rejuvenator can play a role in the degree of mitigation of aged binder.
- For STOA conditioned mixtures, Rejuvenator PO (Paraffinic Oil) provided the largest mitigation of aging as its Black Space parameters were very similar to the virgin PG 58-28 control mixture and was furthest away from the 50% RAP mixture without rejuvenator.
- For LTOA conditioned mixtures, Rejuvenators AO (Aromatic Oil) and OB1(Organic Blend #1) provided the largest mitigation of aging as their Black Space parameters were closer to, if not slightly better than, the virgin PG 58-28 control mixture. This indicates that these two rejuvenators are still providing an effective level of rejuvenation even after long-term aging – something that Rejuvenators PO (Paraffinic Oil) and OB2 (Organic Blend #2) appear to be limited at accomplishing.
- The ω_o – R-value Space parameters for the virgin PG 58-28 control mixture and the 50% RAP mixture with no rejuvenator showed that when these mixtures were aged from STOA to LTOA, the test data migrated downward and towards the right, indicating that aging of the asphalt binder had taken place. The magnitude of the change in the virgin PG 58-28 control mixture was greater than for the 50% RAP mixture because approximately half of the asphalt binder in the 50% RAP mixture had already been highly oxidized.
- The ω_o – R-value Space indicated that all four rejuvenators provided some type of aging mitigation. However, similar to the Black Space diagram, the amount of aging mitigation varied among the different types of rejuvenators.
- The binder grading results exhibited subtle variances based on the rejuvenator used and aging period. These variances were not always consistent and may be due in part to variability in the test procedures. This indicates that the PG tests may not have been sensitive to aging. Since the Black Space analysis did show sensitivity to aging, it may provide more beneficial information about aging mitigation for these types of mixtures. Additionally, the smaller sample size (about 1 gram of binder) needed for a Black Space diagram requires less effort and time to obtain compared to the sample size needed for the standardized PG tests. This could allow for easier field aging evaluation as a single core would be required to obtain the test sample, whereas PG tests would require multiple cores.
- After both STOA and LTOA, the HWTD stripping inflection points and the rut depths for the 50% RAP mixtures with the rejuvenators fell between the

values for the 50% RAP mixture with no rejuvenator and the virgin PG58-28 control mixture. This shows that the rejuvenators are helping in mitigating the aging of the RAP binder which agreed with the Black Space and the ω_o – R-value Space diagrams.

- After both STOA and LTOA, the higher fatigue lives provided in the beam fatigue test for the 50% RAP mixtures with the rejuvenators showed that the rejuvenators are helping in mitigating the aging of the RAP binder, which agreed with the Black Space and ω_o – R-value Space diagrams.
- After both STOA and LTOA, the lower TSRST cracking temperatures for the 50% RAP mixtures with the rejuvenators compared to both the 50% RAP mixture without rejuvenator and virgin PG58-28 control mixture showed that the rejuvenators are helping in mitigating the aging of the RAP binder. Furthermore, by comparing the LTOA data to the STOA data, it was found the any additional aging of the RAP binder caused by LTOA did not have a significant impact on low temperature cracking when these rejuvenators were used. In fact, LTOA had little to no effect on any of the TSRST cracking temperatures compared to STOA. The low cracking temperatures from the mixture tests did not agree well with the critical cracking temperatures of the binders, as the binder values were much colder.
- Overall, the data indicated that the Black Space diagram and the ω_o – R-value Space diagram have the potential to be used as tools to evaluate the effects of rejuvenators on the mitigation of aging of high RAP mixtures. These two diagrams agreed with the mixture tests for rutting and moisture damage, fatigue cracking and low temperature cracking. Further work using chemical analysis of the binders is needed to reinforce the work presented here and add to the understanding of the ability of each rejuvenator to restore the properties of an aged asphalt binder after different periods of aging.

11.0 References

Al-Qadi, L.I., Elseifi, M., and Carpenter, H.S., *Reclaimed Asphalt Pavement – A Literature Review*, Publication FHWA-ICT-07-001, Illinois Center of Transportation, March 2007.

American Association of State Highway and Transportation Officials, *AASHTO Provisional Standards*, American Association of State Highway and Transportation Officials (AASHTO), Washington, D.C., May 1993.

American Association of State Highway and Transportation Officials (AASHTO), *Standard Specifications for Transportation Materials and Methods of Sampling and Testing*, American Association of State Highway and Transportation Officials, Washington, D.C. 34th, Edition, 2014.

American Society for Testing and Materials (ASTM), *Annual Book of ASTM Standards*, ASTM International, Section 4, Volume 04.03, 2012.

Anderson, M., King, G., Hanson, D., and Blankenship, P., "Evaluation of the Relationship between Asphalt Binder Properties and Non-Load Related Cracking," *Journal of the Association of Asphalt Paving Technologist*, Vol. 80, 2011, pp. 615-661.

Brownidge, J. and Grady, C., "The Role of an Asphalt Rejuvenator in Pavement Preservation Use and Need for Asphalt Rejuvenation," *Compendium of Papers from First International Conference on Pavement Preservation,* Chapter 5, 2010, pp. 351–364.

Christensen D. and Anderson D., "Interpretation of Dynamic Mechanical Test Data for Paving Grade Asphalt," *Journal of the Association of Asphalt Paving Technologists*, Vol. 61, 1992, pp. 67–116.

Copeland, A., *Reclaimed Asphalt Pavement in Asphalt Mixtures: State of the Practice*", Publication FHWA-HRT-11-021, Turner-Fairbank Highway Research Center, Federal Highway Administration, McLean, VA, April 2011.

Daniel, J.S."AAPT Symposium: Recycling with RAP, RAS, and Secondary Aggregates *Journal of the Association of Asphalt Paving Technologist*, Vol. 82, 2013, pp. 661–682.

Farrar, M. J., Salmans, S. and Planche, J-P."Recovery and Laboratory Testing of Asphalt Emulsion Residue – Application of Simple Aging Test and 4-mm Dynamic Shear Rheometer Test," *Transportation Research Record: Journal of the Transportation Research Board*, No. 2370, Transportation Research Board of the National Academies, Washington, D.C., 2013, pp. 69–75.

Glover, C.J., Davison, R.R., Domke, C.H., Ruan, Y., Juristyarini P., Knorr, D.B., Jung, S.H., *Development of a New Method for Assessing Asphalt Binder Durability with Field Evaluation*, Federal Highway Administration and Texas Department of Transportation, Report # FHWA/TX-05/1872-2, August 2005.

Kandel, P.S., "Low-Temperature Ductility in Relation to Pavement Performance," *ASTM STP 268: Low-Temperature Properties of Bituminous Materials and Compacted Bituminous Paving Mixtures*, C.R. Marek (Ed), American Society for Testing and Materials, Philadelphia, PA, June 1977, pp. 73–114.

King, G., Anderson, M., Hanson, D., and Blankenship, P., "Using Black Space Diagrams to Predict Age-Induced Cracking," *7th RILEM International Conference on Cracking in Pavements*, RILEM Book series Volume 4, 2012, pp. 453–463.

Mogawer, W.S., Austerman, A.J., Bonaquist, R., and Roussel, M., "Performance Characteristics of Thin Lift Overlay Mixtures Containing High RAP Content, RAS, and Warm Mix Asphalt Technology," *Transportation Research Record: Journal of the Transportation Research Board, No. 2208*, Transportation Research Board, Washington, D.C., 2011, pp. 17–25.

Mogawer, W.S., Booshehrian, A., Vahidi, S. and Austerman, A.J., "Evaluating the Effect of Rejuvenators on the Degree of Blending and Performance of High RAP, RAS, RAP/RAS Mixtures," *Journal of the Association of Asphalt Paving Technologists*, Vol. 82, 2013, pp. 403–436.

Molenaar, A., Hagos, E., and Van de Ven, M., "Effects of Aging on the Mechanical Characteristics of Bituminous Binders in PAC", *Journal of Materials in Civil Engineering*, Vol. 22, No. 8, American Society of Civil Engineers, 2010, pp. 779–787.

O'Sullivan, K., *Rejuvenation of Reclaimed Asphalt Pavement (RAP) in Hot Mix Asphalt Recycling with High RAP Content,* Master's Thesis, Worcester Polytechnic Institute (WPI), April 2011.

Rowe, G., "Analysis of SHRP Core Asphalts – New 2013/14 Test Results," presentation from the Binder Expert Task Group Meeting, San Antonio, TX, April 2nd, 2014.

Shen, J.A., Amirkhanian, S.J., and Lee, S.J., "The Effects of Rejuvenating Agents on Recycled Aged CRM Binders," *International Journal of Pavement Engineering*, Vol. 6, No. 4, December 2005, pp. 273–279.

Shen, J., Amirkhanian, S., and Miller, J., "Effects of Rejuvenating Agents on Superpave Mixtures Containing Reclaimed Asphalt Pavement," *Journal of Materials in Civil Engineering,* Vol. 19, No. 5, 2007a, pp. 376–384.

Shen, J.A., Amirkhanian, S.J., and Tang, B.M., "Effect of Rejuvenator on Performance-Based Properties of Rejuvenated Bitumen Binder and Mixtures," *Construction and Building Materials*, Vol. 21, No. 5, 2007b, pp. 958–964.

Sui, C., Farrar, M.J., Tuminello, W., and Turner, T., "New Technique for Measuring Low-Temperature Properties of Asphalt Binders with Small Amounts of Material," *Transportation Research Record: Journal of the Transportation Research Board,* No. 2179, Transportation Research Board, Washington, D.C., 2010, pg. 23–28.

Sui, C., Farrar, M.J., Harnsberger, P.M., Tuminello, W., and Turner, T., "New Low-Temperature Performance-Grading Method – Using 4-mm Parallel Plates on a Dynamic Shear Rheometer," *Transportation Research Record: Journal of the Transportation Research Board,* No. 2207, Transportation Research Board, Washington, D.C., 2011, pg. 43–48.

12.0 Discussion

JOHN D'ANGELO: Walaa, very nice presentation.

WALAA MOGAWER: Thank you.

JOHN D'ANGELO: Could you go to that first slide you have, though, of the Glover-Rowe parameter?

WALAA MOGAWER: The very first one.

JOHN D'ANGELO: Yes. The slide with the PG 64 and 76 on it. Yes, that one. First I wanted to ask you, these are just virgin binders, correct?

WALAA MOGAWER: Right. These were studies, data from Rutgers, that they've done before. It doesn't have any rejuvenators.

JOHN D'ANGELO: As I look this, then, the graph is saying that the PG 76 binder is just going to crack, correct? Starting off, it plots beyond the crack initiation line. So I'm wondering about your thoughts on that. I actually like the parameter myself, and I think possibly the direction of changes with aging is okay, but the criteria numbers they've got there aren't, especially comparing a neat 64-22 with the polymer-modified 76-22 we would typically get improved crack resistance with the PG 76. So I'm wondering if what we're really seeing here more than anything else is just the effect of a soft material against a hard material, not necessarily real crack resistance – it's just softening. Which is a good indication of a lot of things, but you've got to be careful about exactly what we think is going to crack and not crack. Now directionally, it might be very, very informative, but we better be careful with the criteria.

WALAA MOGAWER: I agree and I'll have Geoff maybe follow up. But as far as softening, I agree with you, John. That's why the second bullet where we need to look more into chemical analysis, the components of the resultant binder. This diagram is mainly for block cracking, which relies on aging, so I don't know if you want to add to this, Geoff.

GEOFFREY ROWE: I can see Gayle King back there, as well, and Gayle King and I wrote a paper about an airport in Newark, and Frank Fee was also complicit in this work, as well as several other folks in the room. But it was sort of an interesting situation because they had three binders used in their airport, a 64-22, I can't remember – I can look it up and I'll add it correctly to the discussion – but I think it was a 76-22 and a 70-22, even though one of them might have been a 28. The curious thing about that study – and this wasn't going to be my discussion, by the way – but the curious thing about that study is that both those modified ones had very poor performance and sort of raveled and tired and sort of were undesirable. So exactly the opposite to what John was sort of saying in terms of the performance they had onsite. And this was a Newark airport laid with a P401 mix, you know, very well compacted and everything else. And at the time we looked at this study, we did a whole bunch for work right about 2002, 2003, we couldn't figure out why we saw the cracking. And then Mike came up with this sort of concept of looking at data like this, and we looked at it again, and, yeah, it just explained the information. So, you know, that was quite interesting. Gayle, have I got it correctly?

GAYLE KING: That's right. It was 76-22. It should have been a good binder. But the extracted binder was in the damage zone, and the pavement cracked.

WALAA MOGAWER: Right.

GAYLE KING: And it was cracked.

WALAA MOGAWER: Actually, that paper got me interested in the black space and that's why we used it, yeah.

JOHN D'ANGELO: I can't deny your data, but I also can say that's this is likely an anomaly, because we've got too much other data showing the exact opposite and, actually, with a lot of mix data and everything else. It's just interesting. I think we've got to be very, very careful because there's a lot of things that go on in a mix that gets placed like overheating, you know, and a lot of other things, especially with polymer-modified, especially when they're first used. But anyway, it's something I think we need to probably look at a lot more.

WALAA MOGAWER: Thanks, John.

GEOFFREY ROWE: Let Frank make a comment because I can see Frank coming up.

FRANK FEE: Yeah, Geoff. We're pretty intimate with that project. You do have to understand that the mix plays a part here, and that the results that Geoff's looking at are on recovered materials from the field years afterwards. Now what happened with that mix from time zero to four years is a function also of the mixture and how it was used in that application. So you cannot remove that from the equation here. So there's a gap there that we don't know enough about, all right?

WALAA MOGAWER: Right.

FRANK FEE: So I assure you we know a lot about the binder that went into the mix, and it wasn't anything like the results indicated. So something happened in the aging characteristics of that mix in the field. We did a lot of evaluations of these mixes through other airports, and all that data with those same binders came out okay. So I just want to put that in there.

WALAA MOGAWER: Thanks, Frank. Thank you.

Recipients of W.J. Emmons Annual Awards

1949	J.A. Lettier, D.F. Fink, N. B. Wilson and F.F. Farley
1950	Alvyn Day
1951	E.C. Hughes and H.F. Hardman
1952	N.W. McLeod
1953	James E. Rice
1954	F.M. Williams and F.W. Kimble
1955	R.L. Griffin, T.K. Miles, and C.J. Penther
1956	O.B. Andersland and W.H. Goetz
1957	Jean Lefebvre
1958	Luke Corbett and R.E. Swarbrick
1959	Robert F. Baker and H.S. Papazian
1960	L.E. Corbett and R.E. Swarbrick
1961	Carl L. Monismith, K.E. Secor and E.W. Blackman
1962	M. Livneh and E. Shklarsky
1963	F.N. Hveem, E. Zube and J. Skog
1964	James F. Shook
1965	C.L. Monismith, G.A. Secor and K.E. Secor
1966	R.J. Schmidt and L.E. Santucci
1967	John Skog
1968	A.W. Sisko and L.C. Brunstrum
1969	P.S. Pell and I.F. Taylor
1970	R.G. Hicks and F.N. Finn
1971	John A. Deacon
1972	James L. Duthie
1973	Willem Heukelom
1974	T.R.J. Fabb
1975	M.W. Witczak
1976	A.I.M. Claessen, R. Ditmarsch and C.P. Valkering
1977	Frank R. Meyer, Edward T. Hignell, G.J.A. Kennepohl, Ralph C.G. Haas
1978	J.A.N. Scott
1979	V.P. Puzinauskas
1980	Harry S. Pink, Richard E. Merz and David S. Bosniack
1981	Dallas N. Little, Jr., R.J. Holmgreen and J.A. Epps
1982	G.R. Morris, J.A. DiVito and N.J. Chen
1983	Teh-Chang Lee, Ronald L. Terrel and Joe P. Mahoney
1984	John H. Tenison, Jr.
1985	C.L. Monismith, J.A. Epps and F.N. Finn
1986	Harold W. Muncy, Gayle N. King and J.B. Prudhomme
1987	G.A. Huber and G.H. Heiman
1988	Joseph L. Goodrich
1989	P.S. Kandhal, C. Lubold, Jr. and F.L. Roberts
1990	C.B. Dawley, B.L. Hogweide and K.O. Anderson
1991	John D'Angelo and Ted Ferragut
1992	Donald W. Christensen, Jr., and David A. Anderson

1993 Ted Vinson and D. Jung
1994 Ronald Reese
1995 Timothy Aschenbrener, Robert McGennis and Ronald Terrel
1996 Manual G. Rodríquez, Geoffrey Morrison, Joshua vanLoon, Simon Hesp
1997 Y. Richard Kim, Hyun-Jong Lee and Dallas Little
1998 Leslie Myers, Reynaldo Roque and Byron Ruth
1999 Reynaldo Roque, Zhiwang Zhang and Bhavani Sankar
2000 Mary Stroup-Gardiner and M. Law
2001 L. Tashman, E. Masad, B. Peterson and H. Saleh
2002 K.J. Jenkins, A.A.A. Molenaar, J.L.A. de Groot, M.F.C. van de Ven
2003 D. Christensen, T. Pellinen and R. Bonaquist
2004 R. Roque, B. Birgisson, C. Drakos and B. Dietrich
2005 R. Dongre, L. Myers, J. D'Angelo, C. Paugh
2006 Y.R. Kim, H. Lee, D. Little, Y. Kim
2007 B. Delaporte, H. DiBenedetto, P. Chavert, G. Gauthier
2008 E. Kutay, N. Gibson, J. Youtcheff
2009 A. LaCroix, Y.R. Kim, M. Sakhaeifar
2010 F. Zhou, S. Hu, T. Scullion
2011 A. Montepara, E. Romeo, M. Isola, G. Tebaldi
2012 Y. Chen, G. Tebaldi, R. Roque, G. Lopp, Y. Su
2013 N. Roohi, P. Teymourpour, H. Bahia
2014 F. Zhou, S. Hu, T. Scullion, R. Lee

Runners-up for W.J. Emmons Annual Award

1949 J.M. Rice and W.H. Goetz
1950 N.W. McLeod
1953 E.G. Williams and L.E. Gregg
1954 Charles Mack
1955 Ward K. Parr, Paul J. Serafin and Tom Humphries
1956 J.O. Izatt, C.C. Evans and C.T. Metcalf
1957 J.A. Dillard and R.L. Alwood
1958 Carl L. Monismith
1960 Thomas L. Speer
1961 Austin Douglas and Egons Tons, R.N. Traxler
1963 J. Skog and E. Zube, W.J. Kari and L. E. Santucci
1964 Bonner S. Coffman, David C. Kraft and Jorge Tamayo
1965 Kamran Majidzadeh and Herbert E. Schweyer
1966 Willem Heukelom
1967 V.P. Puzinauskas
1968 R.J. Schmidt, L.J. Painter, J.B. Skog and V.P. Puzinauskas
1969 Jon A. Epps and Carl L. Monismith
1970 P.F. Dickson and J.S. Corlew
1971 H.J. Fromm and J.T. Corkill
1972 H.J. Fromm and W.A. Phang
1973 Matthew W. Witczak

1974	P.J. van de Loo
1975	H. van Dijk
1976	F.N. Finn, K. Nair and J. Hilliar
1977	W. van Dijk and W. Visser
1978	Richard D. Barksdale
1979	H. Plancher, A.J. Hoiberg, S.C. Suhaka and J.C. Petersen
1980	R.C. Deen, H.F. Southgate and J.G. Mayes
1981	William Whitcomb, R.G. Hicks and Steven J. Escobar
1982	J.W.H. Oliver
1983	Ervin L. Dukatz, David A. Anderson and James L. Rosenberger
1984	George A. Way, John Eisenberg, James Delton and John Lawson
1985	W.R. Meier, Jr., and R.A. Jimenez
1986	J.L. Goodrich, L.H. Dimpfl and P.E. Graf
1987	J.P.J. van der Heide
1988	Bernard Brule, Yvonnick Brion, and Anne Tanquy
1989	E.R. Brown and S. Cross
1990	R. Roque and B.E. Ruth
1991	R.M. Anderson, D. Walker, J.A. Scherocman and L.E. Epley
	T.W. Kennedy and W.V. Ping , S.F. Brown, J.N. Preston and K.E. Cooper
1992	H.U. Bahia, D.W. Christensen and D.A. Anderson
1993	Gayle King, H. King, O. Harders, W. Arand, P. Chaverot and J. Planche
1994	Chris A. Bell, M.J. Felling and A. Wieder
1995	Basil Harris and Kevin Stuart
1996	William Buttlar and Reynaldo Roque
1997	E. Ray Brown, John Haddock, Rajib Mallick and Todd Lynn
1998	Hussain Bahia, Timothy Friemel, Pehr Peterson and Jeffrey Russell
1999	H. Bahia, H. Zhai, K. Bonnetti and S. Kose
2000	R.M. Anderson, G.A. Huber, D.E. Walker and X. Zhang
2001	H. Bahia, H. Zhai, M. Zeng, Y. Hu and P. Turner
2002	Jo Sias Daniel and Y. Richard Kim
2003	D. Christensen and R. Bonaquist
2004	K. Nam and H. Bahia
2005	W. Buttlar and E. Dave
2006	R.M. Anderson, M. Huner
2007	E. Masad, E. Arambula, R. Ketcham, A. Abbas and A. Epps Martin
2008	N. Kringos, A. Scarpas and A. De Bondt
	P.J. Yoo and I. Al-Qadi
2009	L. Mohammad, A. Bae, M. Elseifi, J. Button, J. Scherocman
2010	S. Caro, E. Masad, A. Bhasin, D. Little and M.S. Sanchez-Silva
2012	G. Baumgardner, J. Hemsley, W. Jordan III, I. Howard
2013	N. Sefidmazgi, P. Teymourpour, H. Bahia
2014	T. Niu, R. Roque, G. Lopp

Honorary Members

Honorary Members in the Association are those people recognized by the membership as having outstanding eminence and of long experience in the practice of some phase of asphalt paving technology.

Clifford Richardson	1927
Prevost Hubbard	1948
W.J. Emmons	1948
Ward K. Parr	1984
Francis N. Hveem	1985
Norman W. McLeod	1985
Carl L. Monismith	1988
J. York Welborn	1989
Fred N. Finn	1990
Woodrow J. Halstead	1991
Bernard A. Vallerga	1993
William H. Goetz	1994
Vaughn Marker	1994
Jon A. Epps	1996
Bob M. Gallaway	1996
William Kari	1999
Thomas Kennedy	1999
Moreland Herrin	2000
Charles Hughes	2000
R.A. Jimenez	2001
Eugene L. Skok	2001
Peter Pell	2002
Byron Ruth	2003
David A. Anderson	2004
Stephen F. Brown	2004
J. Claine Petersen	2004
Charles F. Potts	2004
V.P. Puzinauskas	2005
Lawrence Santucci	2005
Matthew Witczak	2007
Elton R. Brown	2008
R. Gary Hicks	2010
Prithvi Kandhal	2012
Frank Fee	2014
James A. Scherocman	2015

Award of Recognition

Presented to those members recognized by the Board of Directors as having contributed to the field of Asphalt Technology and the Association for many years.

William H. Campen	1972
Charles F. Parker	1973
E. T. Hignell	1974
Robert J. Schmidt	1975
John M. Griffith	1976
Charles R. Foster	1977
Richard H. Lewis	1980
Herbert E. Schweyer	1981
Francis N. Hveem	1981
Bernard E. Gray	1982
Harold J. Fromm	1984
Richard L. Davis	1985
Robert M. Nady	1986
James F. Shook	1987
J. Claine Petersen	1988
Bernard A. Vallerga	1989
David G. Tunnicliff	1990
Kenneth O. Anderson	1990
Ronald L. Terrel	1991
Jack N. Dybalski	1992
Jack E. Stephens	1992
Robert L. Dunning	1993
William A. Phang	1993
Eugene L. Skok	1994
Leonard E. Wood	1994
Roy S. Hodgson	1995
James A. Burton	1996
Charles W. Smoot	1996
Elizabeth (Betty) Skok	1996
Gerhard J.A. Kennepohl	1997
Stephen Brown	1998
Bernard McCarthy	1998
Ilan Ishai	1999
James Moulthrop	1999
Frank Fee	2000
Roger Olson	2000
John Weigel	2001
Thomas White	2001
Imants Deme	2002
Andreas Kemp	2002

Frederick Hugo	2003
Louay Mohammad	2003
Harold "Skip" Paul	2004
David E. Newcomb	2005
John Bukowski	2005
Manfred Partl	2007
Douglas Coleman	2008
Dallas Little	2008
Michael Heitzman	2010
Gerald Reinke	2010
Jo Sias Daniel	2012
Amy Epps Martin	2012
Hussain U. Bahia	2013
Rebecca McDaniel	2014
Gerald Huber	2015

AAPT Scholarship Fund

In 1992 the AAPT scholarship fund was established. There are now three levels of sponsorship.

Permanent Sponsor: Cumulative Contributions of $20,000 or more. These are:

Chicago Testing Laboratory, in honor of Ward K Parr
Southeast Asphalt User Producer Group
Douglas Coleman, Retired Michigan DOT

Annual Sponsor: Contribution of $500 or more in a calendar year. The 2013 Annual Sponsors are:

Louay Mohammad, Louisiana State University
Audrey Copeland, National Asphalt Pavement Association

Annual Contributor: Contributions of $100 or more in a calendar year. The 2015 Annual Contributors are:

David Anderson
Gaylon Baumgardner
John Baxter
Elton Brown
Wayne Byard
Dale Decker
Bob Frank
R. Gary Hicks
Stanley Herrin
Gerald Huber
Charles Hughes
Dennis Krivohlavek
Pascal Mascarenhas
Robert McGennis
James Moulthrop
David Newcomb
Joe Proctor
Todd Thomas
George Way

The scholarships that have been awarded are:

Randy Clark West	University of Florida	1992
Todd Scholz	University of Nottingham	1993
John Haddock	Auburn University	1994
Jo Ellen Sias	North Carolina State University	1995
Cynthia Lynn	Auburn University	1996
Robert L. Schmitt	University of Wisconsin	1996
Brian Birdsall	North Carolina State University	1997
Weng on Tam	University of Texas	1997
Leslie Myers	University of Florida	1998
Angela Waldhoff	University of Illinois	1998
Glen A. Malpass	North Carolina State University	1999
Stacy G. Williams	University of Arkansas	1999
Fenella Long	University of California-Berkeley	2000
Zhiming Si	Texas A&M University	2000
David Timm	University of Minnesota	2000
Ghassan Chehab	North Carolina State University	2001

Xue Li	University of Minnesota	2001
Daniel Sherman	University of Illinois	2001
Michael Waggoner	University of Florida	2001
Aslam Al-Omari	Washington State University	2002
Il-Seok Oh	Iowa State University	2002
Laith Tashman	Washington State University	2002
Young Guk Seo	North Carolina State University	2002
Aaron LeChance	University of New Hampshire	2003
Nelson Gibson	University of Maryland	2003
Mark King	North Carolina State University	2003
Xinjun Li	University of Minnesota	2003
Amit Bhasin	Texas A&M University	2004
Angela Priest	Auburn University	2004
Christopher Robinette	Michigan Tech University	2004
Benjamin Underwood	North Carolina State University	2004
Edith Arambula	Texas A&M University	2005
Audrey Copeland	Vanderbilt University	2005
Lucas-Jan Ebels	University of Stellenbosch	2005
Rodrigo Delgadillo	University of Wisconsin	2005
Andrew Braham	University of Illinois	2006
Eshan Dave	University of Illinois	2006
Niki Kringos	Delft University of Technology	2006
Adam Zofka	University of Minnesota	2006
Eli Fini	University of Illinois	2007
Pengcheng Fu	University of California- Davis	2007
Enad Mahmoud	Texas A&M University	2007
Elena Romeo	University of Florida	2007
Kamilla Vasconcelos	Texas A&M University	2007
Silvia Caro	Texas A&M University	2008
Raul Velasquez	University of Minnesota	2008
James Willis	Auburn University	2008
Mary Robbins	Auburn University	2009
Adriana Vargas-Nordcbeck	Auburn University	2010
Maryam Sakhaeifar	North Carolina State University	2010
Behzad Behnia	University of Illinois	2011
Arash Motamed	University of Texas	2011
Salman Hakimzadeh	University of Illinois	2012
Brian Hill	University of Illinois	2012
Matthew Gersch	Auburn University	2013
Mirella Villani	Delft University of Technology	2013
David Mensching	University of New Hampshire	2014
Fan Yin	Texas A&M University	2014
Rostyslav Shamboroskyy	Rutgers University	2015
Michael Vrtis	Auburn University	2015

Past Presidents

Francis P. Smith
Hugh W. Skidmore
Leroy M. Law
Walter J. Emmons
Prevost Hubbard
Charles A. Mullen
Henry L. Howe
Roger L. Morrison
George W. Craig
Alan W. Dow
Victor Nicholson
J. Strother Miller
Walt Hempelmann
Earl F. Kelley
Gene Abson
Claude L. McKesson
Thomas E. Stanton
Lloyd F. Rader
Charles M. Baskin
Victor A. Endersby
Roland Vokac
Walter H. Flood
Jewell R. Benson
Norman W. McLeod
John Goshorn
Frank R. Olmstead
Arnold J. Hoiberg
George H. Dent
Carl A. Carpenter
Henry G. Nevitt
William H. Goetz
M.F. MacNaughton
James E. Ward
James M. Rice
Frank M. Williams
Roy E. Bollen
John O. Izatt
J. York Welborn
W.L. Hindermann
Fred N. Finn
Carl L. Monismith
L.C. Krchma
C. Merritt Hewett
Woodrow J. Halstead
Ward K. Parr
Vaughn Marker
Charles W. Beagle
Bob M. Gallaway
Paul J. Serafin
Larry L. Kole
Moreland Herrin
R.A. Jimenez
Bernard F. Kallas
Jon A. Epps
William J. Kari
Charles S. Hughes
V.P. Puzinauskas
Richard E. Root
Byron E. Ruth
Richard W. Smith
Charles F. Potts
Thomas W. Kennedy
Lawrence E. Santucci
James A. Scherocman
Gale C. Page
David A. Anderson
H. Fred Waller, Jr.
R. Gary Hicks
Dale S. Decker
Robert B. McGennis
Gayle N. King
Prithvi S. Kandhal
E. Ray Brown
Gerald A. Huber
Larry L. Michael
Ervin L. Dukatz
John D' Angelo
Rita B. Leahy
Frank Fee
Stephen F. Brown
R. Michael Anderson
James S. Moulthrop
Reynaldo Roque
David Newcomb
Rebecca McDaniel
Ramon Bonaquist
Brian Prowell

Life Members

Michael Acott	1978	Jon A. Epps	1968
Waddah Akili	1978	Steven Escobar	1976
David A. Anderson	1972	Fred M. Fehsenfeld	1966
Kenneth O. Anderson	1970	Phillip D. Feliz	1983
Earl G. Angell	1962	Juan Fernandez del Campo	1972
Christian Angst	1980	Fred N. Finn	1955
Guneri Artunkal	1974	Milton Fletcher	1981
John Baxter	1974	C. Royce Fugate	1979
Gary Behnke	1981	Arthur M. Furber	1982
Chris A. Bell	1983	Thurmond J. Gates	1964
Wade B. Betenson	1970	Mike Geller	1971
Rod Birdsall	1979	Pedro Diaz Gonzalez	1952
Donald P. Blasland	1982	Geoffrey Griffiths	1979
Frank J. Bowery	1959	Michael N. Guerin	1977
Christopher W. Bowker	1980	Barry L. Gundersen	1982
Lennart S. Brenner	1964	Ralph C.G. Haas	1970
Pinckney J. Brewer	1958	James E. Hampton	1960
J. Donald Brock	1968	Douglas Hanson	1976
Stephen F. Brown	1978	Chris Hardwick	1984
Elton R. Brown	1980	William J. Harper	1973
John Bukowski	1979	James H. Havens	1959
Joe W. Button	1978	M.J. Hensley	1970
William T. Camm	1959	Moreland Herrin	1955
Robert E. Campbell	1970	Stanley M. Herrin	1982
Paul F. Cecchini	1967	R. Gary Hicks	1970
James Chehovits	1984	Roy S. Hodgson	1972
Mervin F. Clark	1970	Richard J. Holmgreen, Jr.	1982
Andrew C. Clayton	1983	Howard B. Hopps, Jr.	1951
William J. Coffman	1962	Eugene Y. Huang	1964
Michael I. Darter	1968	John E. Huffman	1969
Donald Cornelison	1985	Charles S. Hughes	1964
Stephen A. Cross	1985	Frederick Hugo	1969
J. Keith Davidson	1983	Matti S. Huhtala	1969
John F. Deatherage	1953	Robert Humer	1979
Dale S. Decker	1978	Ilan Ishai	1972
Imants J. Deme	1971	Rudolf A. Jimenez	1961
J. Henk Dijkink	1979	Edwin H. Jones	1966
Ervin Dukatz	1980	Robert H. Joubert	1972
Robert L. Dunning	1966	Nabil Kamel	1981
Jack N. Dybalski	1958	Prithvi S. Kandhal	1972
Robert A. Eaton	1973	James E. Kelly	1966
Frank A. Enloe	1967	Andreas Kemps	1981
Larry E. Epley	1970	Thomas W. Kennedy	1979

Life Members (cont.)

Gerhard J.A. Kennepohl	1975	V.P. Puzinauskas	1967
Irving Kett	1968	John T. Reinhard	1953
N. Paul Khosla	1980	Gerald Reinke	1982
Takashi Kobayashi	1972	J.F.(Jack) Renner	1956
Larry L. Kole	1961	M. Felix A. Rodriguez	1975
Dennis D. Krivohlavek	1982	Z. Luis Rodriguez	1974
Dah-Yinn Lee	1968	Richard E. Root	1971
Kang-Won Wayne Lee	1982	Reynaldo Roque	1983
Giorgio Legnani, Jr.	1963	Jerry G. Rose	1973
William M. Lehmkuhl	1942	Byron E. Ruth	1963
Rafael Limon	1980	Lawrence E. Santucci	1966
Robert A. Lindquist	1951	James A. Scherocman	1967
Dallas Little	1980	Gary H. Schmidt	1971
Robert P. Lottman	1960	Russell Schnormeier	1965
Robert L. Lytton	1983	Richard J. Schreck	1983
Donaldson MacLeod	1979	Kenneth E. Secor	1967
M.R. Malya	1957	James F. Shook	1962
J. Paul Martin	1952	Eugene L. Skok, Jr.	1965
Pascal Mascarenhas	1985	Charles W. Smoot	1974
Steven R. Marvin	1977	Anthony Stock	1981
Joseph Massucco	1984	Kai Kong Tam	1979
Jack R. Mathews	1970	Clifford A. Taylor	1966
Richard W. May	1983	Tom K.G. Teo	1973
Bernie M. McCarthy	1968	Marshall Thompson	1985
Robert B. McGennis	1984	Mang Tia	1982
Jacques de Medina	1969	Walter Tommasini	1973
Juan E. Ramos Medina	1963	Gerald S. Triplett	1965
John Miner	1981	Per Ullidtz	1979
Carl L. Monismith	1956	Cecil J. Van Til	1965
Gene R. Morris	1969	Harold L. Von Quintus	1982
James S. Moulthrop	1985	H. Fred Waller, Jr.	1975
David Newcomb	1980	George Way	1979
Kerry A. Nothnagel	1977	Harold H. Weber, Jr.	1984
Donald O'Connor	1981	Steven Weidenhammer	1979
Shoji Ogino	1977	John J. Weigel, Jr.	1984
John W.H. Oliver	1982	Alden L. West	1962
Roger C. Olson	1984	Thomas D. White	1973
Gale C. Page	1985	Robert A. Wilkinson	1966
Harold R. Paul	1984	Matthew W. Witczak	1970
Peter S. Pell	1968	Ken Wonson	1975
J. Claine Petersen	1966	Masaru Yamada	1984
Alfredo Pinilla	1972	John Yeaman	1972
Charles F. Potts	1970	Jia Yu	1977
Joe Proctor	1985	Ludo Zanzotto	1982

Index
Contributors and Discussors

Underlined page numbers refer to formal papers and presentations. Other page numbers refer to discussion presentations.